47/86/
743

Please check for disk when discharging
This book is to be returned on or before
the last date stamped below

06 MAY 19

Digital Signal Processing
Applications
with
the TMS320 Family

Volume 1

**Prentice-Hall and Texas Instruments
Digital Signal Processing Series**

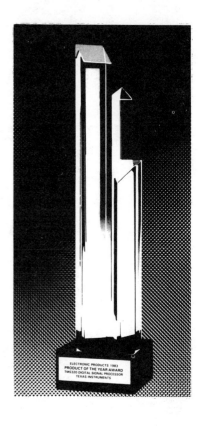

Digital Signal Processing Applications with the TMS320 Family

Volume 1

Kun-Shan Lin, Editor

Digital Signal Processing
Semiconductor Group
Texas Instruments

TEXAS
INSTRUMENTS

Prentice-Hall, Inc., Englewood Cliffs, New Jersey 07632

Published by Prentice-Hall, Inc.
A division of Simon & Schuster
Englewood Cliffs, New Jersey 07632

IMPORTANT NOTICE

Texas Instruments (TI) reserves the right to make changes in the devices or the device specifications identified in this publication without notice. TI advises its customers to obtain the latest version of device specifications to verify, before placing orders, that the information being relied upon by the customer is current.

In the absence of written agreement to the contrary, TI assumes no liability for TI applications assistance, customer's product design, or infringement of patents or copyrights of third parties by or arising from use of semiconductor devices described herein. Nor does TI warrant or represent that any license, either express or implied, is granted under any patent right, copyright, or other intellectual property right of TI covering or relating to any combination, machine, or process in which such semiconductor devices might be or are used.

The software code contained in this book is copyrighted and all rights are reserved by Texas Instruments, Inc. This code is intended for use on a Texas Instruments digital signal processor (TMS32010, TMS32020 . . .). No other use is authorized.

TRADEMARKS

The trademarks that have been mentioned in this book are credited to the respective corporations in the listing below.

TRADEMARK	CORPORATION	TRADEMARK	CORPORATION
Apple	Apple Computers, Inc.	Microstuf	Microstuf, Inc.
CP/M	Digital Research, Inc.	MS-DOS	Microsoft, Inc.
CROSSTALK	Microstuf, Inc.	PAL	Monolithic Memories, Inc.
DEC, DECtalk	Digital Equipment Corp.	PC-DOS	IBM Corporation
DFDP	Atlanta Signal Processors, Inc.	PDP-11,Q-Bus	Digital Equipment Corp.
Eclipse	Data General Corp.	TEKTRONIX	Tektronix Corporation
EZ-PRO	American Automation	UNIX	Bell Laboratories
IBM	IBM Corporation	VAX, VMS	Digital Equipment Corp.
ILS	Signal Technology, Inc.	VMEBUS	Motorola, Inc.
Intel	Intel Corporation		

ISBN 0-13-212466-1   025

Prentice-Hall International (UK) Limited, *London*
Prentice-Hall of Australia Pty. Limited, *Sydney*
Prentice-Hall Canada Inc., *Toronto*
Prentice-Hall Hispanoamericana, S.A., *Mexico*
Prentice-Hall of India Private Limited, *New Delhi*
Prentice-Hall of Japan, Inc., *Tokyo*
Prentice-Hall of Southeast Asia Pte. Ltd., *Singapore*
Editora Prentice-Hall do Brasil, Ltda., *Rio de Janeiro*

Contents

PART I   DIGITAL SIGNAL PROCESSING AND THE TMS320 FAMILY

PART II   FUNDAMENTAL DIGITAL SIGNAL PROCESSING OPERATIONS

DIGITAL SIGNAL PROCESSING ROUTINES

DSP INTERFACE TECHNIQUES

PART III   DIGITAL SIGNAL PROCESSING APPLICATIONS

TELECOMMUNICATIONS

COMPUTERS AND PERIPHERALS

Speech Coding/Recognition

Image/Graphics

Digital Control

FOREWORD

This book discusses the representation of signals in numeric or symbolic form and the manipulation of such representations using special high-speed microcomputers, i.e., the performance of digital signal processing. It is a valuable contribution that teaches us about where we are, and it suggests where we might be going in this rapidly changing field.

Like most technical activities, digital signal processing seems to have advanced in uneven steps over a long period of time. Scientists and engineers have been doing numerical computations since they first began to represent physical reality by mathematical models and theories. Newton used finite difference methods in the 17th century, and recently we have learned that Gauss had the fundamental idea of the FFT in the 19th century.

It was a long time from Newton and Gauss until the development of the digital computer, but in the late 1950's and early 1960's, computers reached a level of capability and accessibility that permitted extensive experimental research to test ideas and theories. A few researchers therefore began to explore the possibilities of digital signal processing (DSP). At first, their computational tools were mainframe computer systems and later dedicated minicomputers provided hands-on interactive computing. These researchers developed algorithms for signal processing, studied and developed techniques for design and implementation of discrete-time systems, and began to explore applications of digital signal processing in such areas as seismic exploration and speech communication. For awhile, the prevailing point of view was that digital computers provided a convenient way of simulating a system, which could ultimately be constructed with analog components. However, this soon gave way to the realization that completely digital systems could be constructed to implement almost any signal processing function with far greater sophistication than was ever possible with analog components.

As a result, we have just experienced an exciting 20-year period in which researchers in university and industrial research laboratories advanced the fundamental theory of digital signal processing to a rather high level while rationalizing their efforts with the claim, "Someday we will have cheap computers or dedicated digital systems to implement these ideas." Concurrently, another group of technologists was busy packing more and more circuitry onto smaller and smaller pieces of silicon, and they were probably saying something like, "Someday there will be a need for vast computational power in a small package at a low price." The TMS32010 and its rapidly growing family of descendents are such devices, and they owe their existence to the dual forces of the promise of the digital approach to signal processing and the integrated circuit and microprocessor revolution.

The DSP microcomputer has changed the perspective again, and we have already begun to take the next big step. The steps are coming more quickly and the potential impact is greater and greater. Now we want to know how to take advantage of this happy confluence of DSP technique and microelectronic technology. Because things change so rapidly in this area, there is a pressing need for ways to quickly learn how to utilize the new technology. This book is a response to the need. The book's purpose is to teach us about the issues and techniques in implementing digital signal processing algorithms on TMS320s and other chips of this class. It does its teaching through examples. Many of the chapters are really extended case studies, which we can apply directly in our own problem setting or which we can generalize in new directions.

The book is arranged in three parts. It begins with a brief introduction to the capabilities of the TMS320 family. The second part focuses upon the programming of DSP algorithms on TMS320 processors. The chapters of this part address such issues as fixed-point arithmetic, arithmetic quantization, computation speed and sampling rate, memory limitations, and interfacing to the outside world. These issues are not specific to the TMS320 – they arise when any fixed-point microprocessor is used to implement DSP algorithms. Indeed, by studying the problems posed by the limitations of any architecture, we surely can learn something about how to design a better DSP microcomputer. The third and final section of the book is concerned with applications of the TMS320 in telecommunications, speech processing, computer graphics, and control systems. The chapters in this section are interesting in their own right, but perhaps more important will be the new applications that they suggest. This section should be viewed as the first rather than the last word on applications of DSP microcomputers.

The present book is timely. It gives us insight into some of the important issues in the implementation and application of DSP systems using single-chip microcomputers like the TMS320. But what about the future? Certainly we will see a progression of more and more capable chips: faster chips, chips with larger memories, floating-point chips, and perhaps most importantly, chips with new architectures that facilitate multiprocessor implementations of DSP algorithms.

The development of engineering tools for the design of complex DSP systems utilizing DSP microcomputers has lagged behind the development of chips. While our knowledge of the theory and technique of digital signal processing has advanced to where we are now actively applying it in many areas, the computational tools available for the study and design of DSP systems have evolved from mainframe batch processors to the personal computer, another product of the microelectronic revolution. Presently available personal computer workstations are more powerful than the mainframe processors that supported the early research in digital signal processing. We are now beginning to use personal computers in innovative ways to facilitate the design, development, and debugging of DSP systems utilizing the new DSP microcomputers, but so far the potential of these systems has been vastly under-utilized in DSP design. Much more remains to be done.

Improvements in the chips and rapid advances in the development design tools will speed the application of DSP microcomputers. Some of the potential applications areas are obvious: military defense, voice and data communications, digital audio and television, robot vision and control, medical intrumentation, and aids for the handicapped. These are just a few areas where DSP techniques can lead to solutions that are more economical and far more innovative than ever before. Many of the new applications will involve combinations of ''traditional'' numerical signal processing and symbolic information processing. Such applications will no doubt call for new theories and techniques as well as new hardware architectures. Perhaps applications that we have not yet thought of will turn out to be the most interesting and important.

I like to think that the DSP microcomputer has saved DSP theory from the fate of passive network synthesis theory. Many brilliant engineers toiled mightily to learn how to go from a desired system function to a passive RLC network having that system function. Along came the cheap integrated circuit op-amp, and now most electrical engineers are unaware of the existence of passive network synthesis techniques; those who are have no need for them. On the other hand, DSP theory and technique and DSP microcomputers were made for each other. That is not to say that something will not come along to make them both obsolete, but until then, I am looking forward to seeing what we can do with what we have in our hands right now.

Ronald W. Schafer
Atlanta, Georgia
January 22, 1987

PREFACE

Digital Signal Processing (DSP) involves the representation, transmission, and manipulation of signals using numerical techniques and digital processors. It has been an exciting and growing technology during the past few years. Its applications have also been expanded vigorously to encompass not only the traditional radar signal processing but also today's digital audio processing (for consumer laser disc players).

In designing a DSP application, the designer faces the following immediate technical challenges:

- Selecting digital signal processors powerful enough to perform the task,
- Obtaining technical support as well as development tools from the semiconductor vendor,
- Creating DSP algorithms to execute the application, and
- Implementing these algorithms on the processors.

These difficult but not impossible tasks challenge design engineers. Each of these problems must be resolved before a successful DSP product can be introduced. The purpose of this applications book is to serve as a reference guide for engineers who need solutions to the above design problems in DSP applications. Readers will benefit from the DSP devices/tools introduced and the fundamental DSP operations and application examples presented in the book for either instant solutions or ideas for solutions to the above challenges.

This book consists of three major parts. The first part briefly introduces the device architectures, characteristics, support, and development tools for the first two generations of the Texas Instruments TMS320 digital signal processors. Readers who are not familiar with these processors should begin their reading with this part of the book. The second part of the book covers some of the common DSP routines, such as Finite Impulse Response (FIR) and Infinite Impulse Response (IIR) filters and Fast Fourier Transforms (FFT), implemented using the TMS320 devices. Hardware interfacing and multiprocessing with these devices are also included. The last part of the book is applications specific. Some typical DSP applications are selected and thoroughly discussed. These applications are divided into two categories: telecommunications, and computers and peripherals (including speech coding/recognition, image/graphics, and digital control).

The materials included in this book are primarily application reports, which have been generated by the digital signal processing engineering staff of the Texas Instruments Semiconductor Group. Some published articles and a technical report from MIT Lincoln Laboratory have also been reprinted herein to supplement the application reports in order to provide completeness of the subject matter. The application reports contain more complete theory and implementations (consisting of algorithms, TMS320 code, and/or schematics) than the reprinted articles.

Readers who desire to obtain more information regarding the device characteristics and programming of the TMS230 digital signal processors should refer to other publications, such as the appropriate devide User's Guide or Data Sheet. References are usually included at the end of each report. A comprehensive list of literature published since 1982 on the TMS320 digital signal processors is provided in the appendix for interested readers to gain further knowledge on various aspects of the TI processors.

The editor would like to credit the authors who contributed the application reports and articles included in the book. A special thanks and note of appreciation go to Maridene Lemmon for a thorough review of the entire manuscript. Many comments from members of the TI Semiconductor Digital Signal Processing staff, especially Maridene Lemmon and Mike Hames, have also greatly improved the structure of the book.

Kun-Shan Lin, Ph.D.

PART I
DIGITAL SIGNAL PROCESSING AND THE TMS320 FAMILY

1. Introduction

A brief overview of digital signal processing and its major application areas is presented in this section. A guide to how the book is organized and can be used is also provided.

Overview of Digital Signal Processing

In the last decade, Digital Signal Processing (DSP) has made tremendous progress in both the theoretical and practical aspects of the field.[1-5] While more DSP algorithms are being discovered, better tools are also being developed to implement these algorithms. One of the most important breakthroughs in electronic technology is the high-speed digital signal processors. These single-chip processors are now commercially available in Very Large-Scale Integrated (VLSI) circuits from semiconductor vendors. Digital signal processors are essentially high-speed microprocessors/microcomputers, designed specifically to perform computation-intensive digital signal processing algorithms. By taking advantage of the advanced architecture, parallel processing, and dedicated DSP instruction sets, these devices can execute millions of DSP operations per second. This capability allows complicated DSP algorithms to be implemented in a tiny silicon chip, which previously required the use of a minicomputer and an array processor.

With this VLSI advancement, innovative engineers in industry are discovering more and more applications where digital signal processors can provide a better solution than their analog counterparts for reasons of reliability, reproducibility, compactness, and efficiency. These digital signal processors are also highly programmable, which makes them very attractive for (1) system upgrades, in the case of advancements in DSP algorithms, and (2) multitasking where different tasks can be performed with the same device by simply changing its program. Because of these and many other advantages, digital signal processors are becoming more prevalent in areas of general-purpose digital signal processing, telecommunications, voice/speech, graphics/imaging, control, instrumentation, and the military. Table 1 lists some applications in these areas.

Table 1. Typical Applications of the TMS320 Family

GENERAL-PURPOSE DSP	GRAPHICS/IMAGING	INSTRUMENTATION
Digital Filtering	3-D Rotation	Spectrum Analysis
Convolution	Robot Vision	Function Generation
Correlation	Image Transmission/	Pattern Matching
Hilbert Transforms	Compression	Seismic Processing
Fast Fourier Transforms	Pattern Recognition	Transient Analysis
Adaptive Filtering	Image Enhancement	Digital Filtering
Windowing	Homomorphic Processing	Phase-Locked Loops
Waveform Generation	Workstations	
	Animation/Digital Map	

VOICE/SPEECH	CONTROL	MILITARY
Voice Mail	Disk Control	Secure Communications
Speech Vocoding	Servo Control	Radar Processing
Speech Recognition	Robot Control	Sonar Processing
Speaker Verification	Laser Printer Control	Image Processing
Speech Enhancement	Engine Control	Navigation
Speech Synthesis	Motor Control	Missile Guidance
Text to Speech		Radio Frequency Modems

TELECOMMUNICATIONS		AUTOMOTIVE
Echo Cancellation	FAX	Engine Control
ADPCM Transcoders	Cellular Telephones	Vibration Analysis
Digital PBXs	Speaker Phones	Antiskid Brakes
Line Repeaters	Digital Speech	Adaptive Ride Control
Channel Multiplexing	Interpolation (DSI)	Global Positioning
1200 to 19200-bps Modems	X.25 Packet Switching	Navigation
Adaptive Equalizers	Video Conferencing	Voice Commands
DTMF Encoding/Decoding	Spread Spectrum	Digital Radio
Data Encryption	Communications	Cellular Telephones

CONSUMER	INDUSTRIAL	MEDICAL
Radar Detectors	Robotics	Hearing Aids
Power Tools	Numeric Control	Patient Monitoring
Digital Audio/TV	Security Access	Ultrasound Equipment
Music Synthesizer	Power Line Monitors	Diagnostic Tools
Educational Toys		Prosthetics
		Fetal Monitors

Organization of the Book

This book is organized into three major parts. Part I contains a brief introduction to digital signal processing and a cursory review of the first two generations of the TMS320 digital signal processors, their characteristics, and the support available from Texas Instruments and third parties.

Part II consists of a collection of common DSP routines and interfaces for implementing DSP algorithms using the TMS320 processors. These DSP routines are: Finite Impulse Response (FIR)/Infinite Impulse Response (IIR) filters, Fast Fourier Transforms (FFT),

Pulse Code Modulation (PCM) companding, floating-point arithmetic, precision digital sine-wave generation, and matrix operations. Interfacing the TMS320 to external memory devices and microprocessors, and using the TMS320 in a multiprocessing environment are also included. These DSP routines and interfaces serve as the basis for developing DSP applications.

Part III is applications specific. Some typical DSP applications are encompassed and catagorized into two areas: telecommunications and computers and peripherals. In the telecommunications area, six applications are selected: telecommunications interfacing using the TMS32010, a single-chip TMS32020 echo canceller, implementation of the Data Encryption Standard (DES) algorithm using the TMS32010, 32-kbit/s Adaptive Differential Pulse Code Modulation (ADPCM) transcoders using the TMS32010, TMS32010 16-kbit/s subband coders, and a Dual-Tone MultiFrequency (DTMF) encoder/decoder using the TMS32010. For computers/peripherals, which has become a fast expanding area for numerous DSP applications, reprints of a series of articles on speech vocoding at 2.4-kbit/s and 9.6-kbit/s and speech recognition with the TMS32010 are included. An application report on the TMS32020 for image/graphics applications is also presented. A detailed report on the design of a digital control system with the TMS32010 digital signal processor concludes Part III.

Readers are encouraged to adapt these DSP routines, algorithms, and implementations to their applications using the TMS320. It is important to bear in mind that the materials given in each report and article only serve as examples. Further optimization of either the code or the circuit to meet specific performance/cost goals is possible. These goals are usually application-dependent. Readers need to make the appropriate tradeoffs to fit their specific design criteria.

A Digital Signal Processing Software Library is available, which includes code for the major DSP routines and applications in this book. This library can be ordered through TI using the following part numbers:

VAX/VMS (1600 BPI magnetic tape)	TMDC3240212-18
TI/IBM MS/PC-DOS (multiple 5¼" floppies)	TMDC3240812-12

In summary, even with a book of this size, it is impossible to cover all the DSP routines and applications. Because of the TMS320 family's excellent computation power, high programmability, and complete development tools, more applications will be created in the future. These applications will extend beyond the two areas selected for this volume. The TMS320 family of processors will continue to provide cost-/performance-effective solutions in the dynamic field of digital signal processing.

References

The following references in digital signal processing may be helpful to the reader of this book:

1. A. Oppenheim and R. Schafer, *Digital Signal Processing*, Prentice-Hall (1975).

2. L. Rabiner and B. Gold, *Theory and Application of Digital Signal Processing*, Prentice-Hall (1975).

3. A. Oppenheim (editor), *Applications of Digital Signal Processing*, Prentice-Hall, (1978).

4. *Issues of IEEE Transactions on Acoustics, Speech, and Signal Processing.*

5. *Proceedings of IEEE International Conference on Acoustics, Speech, and Signal Processing.*

2. The TMS320 Family

This section provides a brief description of the TMS320 digital signal processing family. The family's device architectures, characteristics, and features suitable for digital signal processing applications are discussed. To succeed in designing a DSP application, comprehensive development and support tools are required. Texas Instruments provides a whole family of tools to help DSP engineers. These tools are summarized in this section. Extensive publications produced by Texas Instruments include device specifications, device user's guides, and development tools reference guides. A comprehensive list of these publications is also included. Readers can obtain copies of them by contacting the TI Customer Response Center (CRC) hotline number, 1-800-232-3200 or by returning the literature request card included in the back of this book.

TMS320 Digital Signal Processing Family

The Texas Instruments TMS320 product line contains a family of digital signal processors, designed to support a wide range of high-speed or numeric-intensive DSP applications. These 16/32-bit single-chip microprocessors/microcomputers combine the flexibility of a high-speed controller with the numerical capability of an array processor, thereby offering an inexpensive alternative to a multichip bit-slice processor or an expensive commercial array processor.

The TMS320 family contains the first MOS microprocessor capable of executing five million instructions per second. This high throughput is the result of the comprehensive, efficient, and easily programmed instruction set and of the highly pipelined architecture. Special instructions, such as multiply/accumulate with fast data move, have been incorporated to speed the execution of DSP algorithms. A comprehensive set of general-purpose microprocessor instructions is also included. For example, the branch instructions encompass all the various conditions of the accumulator. Three different addressing modes are provided: direct, indirect, and immediate. A full set of Boolean instructions is included for testing bits. Bit extractions and interrupt capabilities are also part of the features of the TMS320 processors.

Architecturally, the TMS320 utilizes a modified Harvard architecture for speed and flexibility. In a strict Harvard architecture, the program and data memories lie in two separate spaces, permitting a full overlap of the instruction fetch and execution. The TMS320 family's modification of the Harvard architecture allows transfer between program and data spaces, thereby increasing the flexibility of the device. This architectural modification eliminates the need for a separate coefficient ROM and also maximizes processing power by maintaining two separate bus structures (program and data) for full-speed execution.

The Texas Instruments TMS320 family consists of two generations of digital signal processors. The first generation contains the TMS32010 and its offspring: TMS320C10, TMS32011, and TMS32010-25. The TMS32020 and TMS320C25 are the second-generation processors. The features described thus far are common among the processors in the family. Some specific features are added in each processor to provide different cost/performance tradeoffs. These features will be covered in the remaining part of this section. The flexibility of the TMS320 family gives a DSP designer alternatives in selecting a device in the family that can best serve his applications.

The two generations of digital signal processors can be plotted on a hypothetical performance and time scale, as shown in Figure 1. Several trends in digital signal processors can be observed from the figure. Offspring of the first two generations of processors are becoming available to provide better DSP system integration and cost/performance tradeoffs. Specifically, peripheral circuits are being integrated into the DSP device to reduce chip counts, board space, power consumption, and system cost. Because of the low-power and high-speed advantages of CMOS circuits, DSP devices are also being introduced in CMOS in addition to their NMOS counterparts. In the future, newer generations of DSP processors will also be needed to meet higher performance requirements for some applications and further expand the DSP horizon.

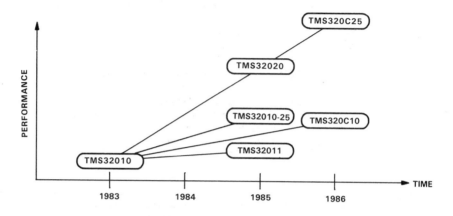

Figure 1. The Generations of Digital Signal Processors

TMS32010 Digital Signal Processor

The TMS32010[1] is the first-generation digital signal processor. Its hardware implements many functions that other processors typically perform in software. For example, this device contains a hardware multiplier that performs a 16 x 16-bit multiplication with a 32-bit result in a single 200-ns cycle. A hardware barrel shifter is used to shift data on its way into the ALU. Extra hardware has been included so that auxiliary registers, which provide indirect data RAM addresses, can be configured in an auto-increment/decrement mode for single-cycle manipulation of data tables. This hardware-intensive approach gives the design engineer the type of power previously unavailable on a single chip.

The TMS32010 has both microprocessor and microcomputer modes selectable via the device MC/$\overline{\text{MP}}$ pin. When used in the microcomputer mode, the TMS32010 is equipped with a 1536-word ROM, which is mask-programmed at the factory with a customer's program. It can also execute from an additional 2560 words of off-chip program memory at full speed. This memory expansion capability is especially useful for applications sharing the same subroutines. In this case, the common subroutines can be stored on-chip while the application-specific code is stored off-chip. When used in the microprocessor mode, the TMS32010 can execute full-speed 4096-word off-chip instructions. Figure 2 shows the functional block diagram of the TMS32010 processor. The device is fabricated in a 2.4μ NMOS technology and has a chip area of 51K square mil. It is produced in a 40-pin dual-in-line package and a 44-pin plastic-leadless-chip-carrier (PLCC) dissipating 950 mW (typically). The maximum clock frequency is 20.5 MHz for an instruction rate of five million instructions per second.

Some of the key features of the TMS32010 are:

- 200-ns instruction cycle
- 1.5K words (3K bytes) of program ROM
- 144 words (288 bytes) of data RAM
- 16 x 16-bit parallel multiplier
- External memory expansion to 4K words (8K bytes) at full speed
- Interrupt with context save
- Barrel shifter
- On-chip clock
- Single 5-volt supply, NMOS technology, 40-pin DIP and 44-pin PLCC.

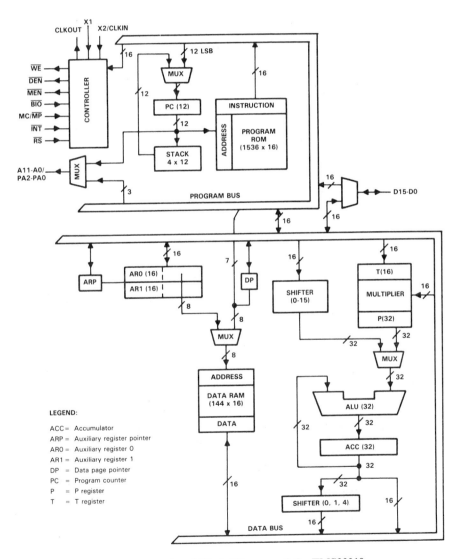

Figure 2. Functional Block Diagram of the TMS32010

TMS320C10 Digital Signal Processor

The TMS320C10[2] is essentially a CMOS replica of the TMS32010. The device is both plug-in and object-code compatible to its NMOS counterpart, the TMS32010. The TMS320C10 also has an instruction cycle time of 200 ns. The device is fabricated in a 2μ CMOS technology with power consumption of 100 mW (typically) at 20-MHz operation. Because of the low-power consumption, the CMOS TMS32010 is very useful for power-sensitive applications, such as digital telephony and portable consumer products.

TMS32011 Digital Signal Processor

The TMS32011[3] is a dedicated microcomputer with 1.5K words of on-chip program ROM (and no external memory expansion) intended for high-volume applications. The TMS32011 is essentially a TMS32010 with the address bus stripped off and two serial ports integrated on-chip. In addition, PCM companding (μ/A-law to and from linear PCM) functions and a timer have been implemented in hardware to reduce the program memory size and increase the CPU utilization for applications using codecs. The device has all the TMS32010 hardware features, such as a 200-ns instruction cycle time, 16 x 16-bit multiplier, and 32-bit ALU/accumulator. The instruction sets are fully compatible, enabling existing TMS32010 development tools to be used in TMS32011 applications.

The following key features distinguish the TMS32011 microcomputer:

- Dual-channel serial port for full-duplex serial communication
- Direct interface to combo-codec and PCM highway systems
- Serial-port timer
- Internal framing-pulse generation
- On-chip companding hardware for μ-law and A-law PCM conversions
- Object-code compatible with the TMS32010 instruction set
- Compatible with TMS32010 development support tools
- Peripheral mode to TMS32010 for application development
- 1.5K words (3K bytes) of program ROM
- 144 words (288 bytes) of data RAM.

TMS32010-25 Digital Signal Processor

The TMS32010-25 is a 160-ns instruction cycle time version of the TMS32010. This microprocessor is intended for higher performance applications using off-chip memory, which require 25 percent greater processor throughput (6.25 million instructions per second) than the TMS32010. Existing TMS32010 designs can simply speed up the system clock to 25 MHz to take advantage of the increased processor throughput without any software redesign. The device is capable of accessing 4K words of external program memory. Other key features of the TMS32010-25 are the same as the ones in the 20-MHz version of the TMS32010.

TMS32020 Digital Signal Processor

The TMS32020[4] digital signal processor is the second-generation member of the TMS320 family of VLSI processors. The TMS32020 architecture is based upon that of the TMS32010, the first member of the TMS320 family. The TMS32020 greatly enhances the memory spaces of the processor, providing 544 words of on-chip data and program memory. Increased throughput is accomplished by means of single-cycle multiply/accumulate instructions with a data move option, five auxiliary registers with a dedicated arithmetic unit, and faster I/O necessary for data-intensive applications. The TMS32020 has a comprehensive instruction set of 109 instructions to increase software throughput and ease of development. The processor contains special repeat instructions for streamlining program space and execution time.

The architectural design of the TMS32020 emphasizes overall system speed, communication, and flexibility in processor configuration. Control signals and software instructions provide block memory transfer, communication to slower off-chip devices, multiprocessing implementations, and floating-point support. Peripheral functions, such as the hardware timer and serial port, have been integrated on-chip to reduce overall system cost. The combination of increased memory (both on-chip and off-chip), expanded instruction set (for example, single-cycle multiply/accumulate), and additional hardware features give the TMS32020 two to three times the performance over the TMS32010 in DSP applications.

Figure 3 shows the functional block diagram of the TMS32020. The TMS32010 source code is upward-compatible with the TMS32020 source code, and can be assembled using the TMS32020 Macro Assembler. The TMS32020 is fabricated in a /4μ NMOS technology and has a chip area of 119K square mil. It is produced in a 68-pin grid array package and has a typical power consumption of 1.2 W. The maximum clock frequency is 20.5 MHz for an instruction rate of five million instructions per second.

Some of the key features of the TMS32020 are:

- 544 words of on-chip data RAM (256 words configurable as either data or program memory)
- 128K words of memory space (64K words program and 64K words data)
- Single-cycle multiply/accumulate instructions
- Repeat instructions
- 200-ns instruction cycle
- Serial port for multiprocessing or codec interface
- Sixteen input and sixteen output channels
- 16-bit parallel interface
- Directly accessible external data memory space
- Global data memory interface
- Block moves for data/program memory
- TMS32010 software-upward compatibility
- Instruction set support for floating-point operations
- On-chip clock
- Single 5-volt supply, NMOS technology, 68-pin grid array package.

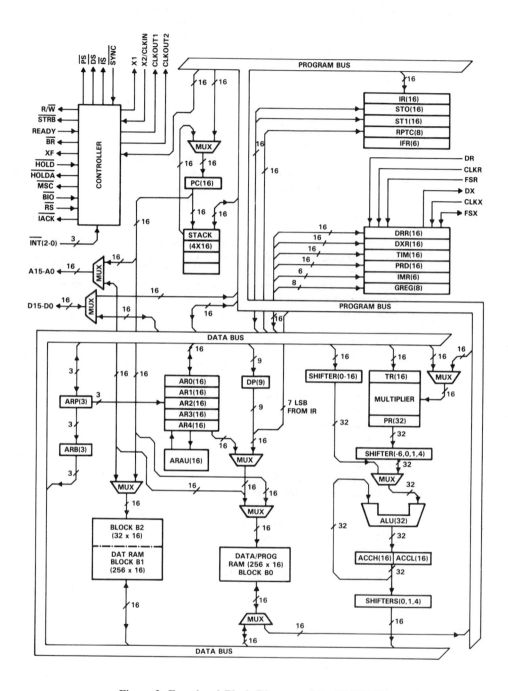

Figure 3. Functional Block Diagram of the TMS32020

TMS320C25 Digital Signal Processor

The TMS320C25 is a pin-compatible CMOS version of the TMS32020 with additional features to further enhance the processor speed, system integration, and ease of application development. The TMS320C25's faster instruction cycle time of 100 ns, additional instructions, and additional on-chip hardware such as an 8-deep hardware stack and 8 auxiliary registers result in two to three times the throughput of its predecessor, the TMS32020. The inclusion of a large on-chip masked ROM (4K words) makes the TMS320C25 ideal for single-chip DSP applications, thus reducing the power, cost, and board space of many applications. All 109 TMS32020 instructions are implemented on the device with object-code compatibility. Two new instructions (MPYA and ZALR) on the device allow an LMS adaptive filter tap and update to be performed in 4 machine cycles. A 256-tap adaptive filter can thus be sampled at 8 kHz to be executed on a single chip (with no external memory) in real time. The device is fabricated in 1.8μ CMOS technology and is packaged in a 68-pin PLCC. Figure 4 shows the functional block diagram of the TMS320C25.

The following outlines the major enhancements of the TMS320C25 over the TMS32020:

- Two versions: 100-ns instruction cycle time
- 125-ns instruction cycle time
- 4K x 16-bit on-chip masked ROM
- Object-code compatible with TMS32020
- MAC/MACD operation with external program memory
- Double-buffered static serial port
- T1/G.711 transmission interface
- Unsigned multiply and carry bit for complex arithmetic
- Bit-reversed addressing for FFTs
- Powerdown mode
- Eight auxiliary registers
- Eight-deep hardware stack
- MPYA and rounding for adaptive filtering
- Concurrent DMA via redefined HOLD mode
- 1.8μ CMOS, 68-pin PLCC.

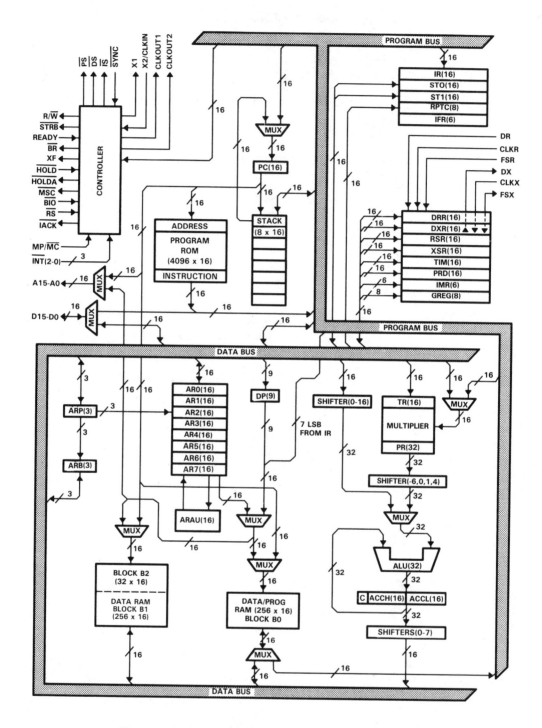

Figure 4. Functional Block Diagram of the TMS320C25

The TMS320 family's unique versatility, computational power, and high I/O throughput give the DSP designer a new powerful solution to a variety of complicated applications. Some of these applications are discussed in Part III of this book.

Development and Support Tools

In developing an application, problems are usually encountered, and questions must be answered before funding to start the project, or to continue it, is granted. Oftentimes the tools and vendor support provided to the designer are the difference between the success and failure of the project. This is especially true when using state-of-the-art electronic devices, such as digital signal processing ICs.

The TMS320 family of digital signal processors has a wide range of development tools available (see Figure 5). These tools range from very inexpensive evaluation modules for application evaluation and benchmarking purposes, assembler/linkers, and software simulators to full-capability hardware emulators. For a complete listing of the support available from TI, DSP text/reference books, design services and support offered by TI Regional Technology Centers and third parties, please refer to the Development Support Reference Guide.[5] A brief summary of them is provided in the succeeding subsections.

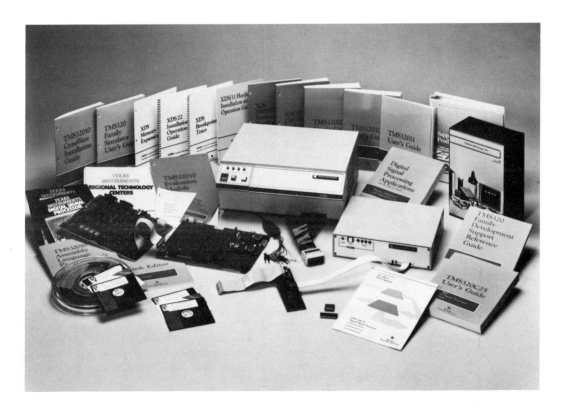

Figure 5. TMS320 Product Development Tools

Software Tools

Assembler/linkers and software simulators are available on PC and VAX for users to develop and debug TMS320 DSP algorithms. Their features are described as follows:

ASSEMBLER/LINKER[4-9]: The TMS320 Macro Assembler translates TMS320 assembly language source code into executable object code. The assembler allows the programmer to work with mnemonics rather than hexadecimal machine instructions and to reference memory locations with symbolic addresses. The macro assembler supports macro calls and definitions along with conditional assembly. The TMS320 Linker permits a program to be designed and implemented in separate modules that will later be linked together to form the complete program. The linker resolves external definitions and references for relocatable code, creating an object file that can be executed by the TMS320 Simulator, TMS320 Emulator, or the TMS320 processor. The TMS320 Macro Assembler/Linker is currently available for the VAX/VMS, TI PC/MS-DOS, and IBM PC/PC-DOS operating systems for both generations of DSP processors.

SIMULATOR[6,9]: The TMS320 Simulator is a software program that simulates operations of the TMS320 digital signal processor to allow program verification. The debug mode enables the user to monitor the state of the simulated TMS320 while the program is executing. The simulator uses the TMS320 object code produced by the TMS320 Macro Assembler/Linker. During program execution, the internal registers and memory of the simulated TMS320 are modified as each instruction is interpreted by the host computer. Once program execution is suspended, the internal registers and both program and data memories can be inspected and/or modified. The TMS320 Simulator is currently available for the VAX/VMS, TI PC/MS-DOS, and IBM PC/PC-DOS operating systems for both generations of DSP devices.

Hardware Tools

Powerful TMS320 evaluation and emulation tools provide in-circuit emulation and hardware program debugging (such as hardware breakpoint/trace) for developing and testing DSP algorithms in a real product environment. The following paragraphs provide a brief description of these tools.

EVALUATION MODULE (EVM)[11]: The EVM allows the designer to determine at minimal cost if the TMS320 meets the timing requirements of an application. It is a standalone single-board module that contains all of the tools necessary to evaluate the device as well as provide in-circuit emulation. The powerful firmware package on the EVM contains a debug monitor, editor, assembler, reverse assembler, EPROM programmer, software communication to two EIA ports, and an audio cassette interface. Communication to a host computer and several peripherals is provided on the TMS320 EVM. Dual EIA ports allow the EVM to be connected to a terminal and either a host computer or a line printer. The EVM accepts either source or object code downloaded from the host computer. The resident assembler converts incoming source text into executable code in just one pass by automatically resolving labels after the first assembly pass is completed. When a session is finished, code is saved via the host computer, audio cassette recorder, or EPROM programmer. The EVM is currently available for the evaluation of the TMS32010.

EMULATOR (XDS)[12]: The TMS320 XDS Emulator is a powerful, sophisticated development tool providing full-speed in-circuit emulation with real-time hardware breakpoint/trace and program execution capability from target memory. The XDS allows integration of the hardware and software modules in the debug mode. By setting breakpoints based on internal conditions or external events, execution of the program can be suspended and control given to the debug mode. In the debug mode, all registers and memory locations can be inspected and modified. Single-step execution is available. Full-trace capabilities at full speed and a reverse assembler that translates machine code back into assembly instructions also increase debugging productivity. The XDS system is designed to interface with either a terminal or a host computer. The object code generated by the assembler/linker can be downloaded to the XDS and then controlled through a terminal. The XDS is available for both the TMS32010 and TMS32020.

ANALOG INTERFACE BOARD (AIB)[13]: The AIB is an analog-to-digital and digital-to-analog conversion board that can be used in conjunction with the EVM or XDS. It can also be used in an educational environment to aid in familiarizing the student with digital signal processing techniques. Analog-to-digital and digital-to-analog converters with 12-bit resolution and anti-aliasing and smoothing filters with cutoff frequency programmable from 4.7 kHz to 20 kHz are included on-board. The AIB is designed to adapt to either the TMS32010 or the TMS32020.

A summary of these tools and their TI part numbers is shown in Tables 2 and 3 for the TMS32010 and TMS32020, respectively.

Table 2. TMS32010 Family Hardware and Software Support Tools

HOST COMPUTER	OPERATING SYSTEM	PART NUMBER
MACRO ASSEMBLERS/LINKERS		
DEC VAX	VMS	TMDS3240210-08
TI/IBM PC	MS/PC-DOS	TMDS3240810-02
SIMULATORS		
DEC VAX	VMS	TMDS3240211-08
TI/IBM PC	MS/PC-DOS	TMDS3240811-02
HARDWARE TOOL		PART NUMBER
Evaluation Module (EVM)		RTC/EVM320A-03
Analog Interface Board (AIB)		RTC/EVM320C-06
Emulator (XDS/22)		TMDS3262210

Table 3. TMS32020 Family Hardware and Software Support Tools

HOST COMPUTER	OPERATING SYSTEM	PART NUMBER
MACRO ASSEMBLERS/LINKERS		
DEC VAX	VMS	TMDS3241210-08
TI/IBM PC	MS/PC-DOS	TMDS3241810-02
SIMULATORS		
DEC VAX	VMS	TMDS3241211-08
TI/IBM PC	MS/PC-DOS	TMDS3241811-02
HARDWARE TOOL		**PART NUMBER**
Analog Interface Board (AIB) Adaptor		RTC/ADP320A-06
Emulator (XDS/11)		TMDS3261120
Emulator (XDS/22)		TMDS3262220

Digital Signal Processing Software Library

A software library is available, which includes the major DSP routines and applications in this book. Table 4 gives the TI part numbers for ordering this library.

Table 4. Digital Signal Processing Software Library

HOST	OPERATING SYSTEM	MEDIUM	PART NUMBER
DEC VAX	VMS	1600 BPI Mag Tape	TMDC3240212-18
TI/IBM PC	MS/PC-DOS	Multiple 5¼″ Floppies	TMDC3240812-12

Digital Filter Design Package

In addition to the above design tools, Texas Instruments has also made available a Digital Filter Design Package (DFDP)[5], developed by Atlanta Signal Processors Inc. (ASPI). The package runs on both TI and IBM PCs. The DFDP allows the user to design a digital filter (lowpass, highpass, bandpass, and bandstop types) using a menu-driven approach. A filter can be designed by simply specifying its characteristics. After the filter design, the DFDP can automatically generate the TMS320 code, which later can be integrated into a designer's application. The DFDP can be ordered through TI or ASPI using TI part number DFDP-IBM001 for the IBM PC version and DFDP-TI001 for the TI PC.

RTC and Third-Party Support

The TI Regional Technology Centers (RTC) are staffed with qualified engineers to provide technical support to the TMS320 customers. TMS320 hands-on workshops are also offered by professional instructors in the RTC to give engineers a quick start designing with the TMS320 digital signal processors. Hands-on exercises are included in the workshop to familiarize students with various hardware and software design tools. Many design examples are given throughout the course, which allows students to learn and practice essential TMS320/DSP design skills. For further information, please contact the nearest RTC through the following phone numbers:

ATLANTA RTC: (404) 662-7945
BOSTON RTC: (617) 890-4271
CHICAGO RTC: (312) 228-6008
DALLAS RTC: (214) 680-5096
NORTHERN CALIFORNIA RTC: (408) 980-0305
SOUTHERN CALIFORNIA RTC: (714) 660-8164

TI also provides a team of third parties[5] with DSP expertise to support customer design. An extensive list of their names, phone numbers, and areas of support is given as follows:

COMPANY	PHONE NUMBER	TMS320 SUPPORT
Alembic Computer System Inc.	(213) 306-2865	IBM PC Speech Development
Allen Ashley	(818) 793-5748	MS-DOS/PC-DOS, CP/M-80 Assemblers
American Automation	(714) 731-1661	EZ-PRO In-Circuit Emulator
Atlanta Signal Processors Inc. (ASPI)	(404) 892-7265	Digital Filter Design Package, Algorithm Development Package
Bedford Research	(617) 275-7246	PDP and VAX Interactive Signal Processing Software
Burr Brown	(602) 746-1111	VMEBUS Board
Computalker	(213) 828-6546	IBM PC, CP/M, Apple II Assemblers, Simulators
Digital Signal Processing Software Inc.	(613) 825-5476	PDP-11, VAX, TI/IBM PC
Digital Sound Corp.	(805) 569-0700	Audio Data Conversion System for Speech Development
DSP Technology	(214) 247-8831	Four-Channel DTMF Detectors
Gas Light Software	(713) 729-1257	PC Digital Filter Software
Hewlett-Packard	Contact the Local HP Sales Office	HP Logic Development System Assembler/Linker
Kontron Electronics	(800) 227-8834	Logic Analyzer, Disassembler
Lawrence Livermore Lab.	Livermore, CA	Signal Processing Software

COMPANY	PHONE NUMBER	TMS320 SUPPORT
Microstuf, Inc.	(404) 952-0267	TI/IBM PC Data Communications Software (CROSSTALK XVI)
Pacific Microcircuits	(604) 536-1886	TMS32010 Peripheral Interface Chip
Pratika SRL	Torino, Italy	Olivetti PC Assembler, Linker Loader
PH Associates	(703) 281-5762	IBM PC, CP/M, PDP-11, UNIX, TEKTRONIX Assemblers
Signal Technology, Inc.	(800) 235-5787	Speech/DSP Software (ILS)
Signix Corporation	(617) 358-5955	IBM PC Digital Filter Design/Evaluation
SKY Computers, Inc.	(617) 454-6200	PDP-11 Q-Bus/IBM PC-Bus Real-Time Processor Boards, Assembler, C Compiler
Televic	Belgium	SPL Compiler for VAX and PDP-11
Texas Instruments Inc.	(512) 250-7474	TI/IBM PC Speech Cards for Speech Vocoding and Recognition
Thorn EMI Elec.	Australia	S/W Support for Eclipse
TIAC Corp.	(604) 461-0120	DSP Hardware for IBM PC
Voice Control Systems Inc.	(214) 248-8244	Speech Recognition
Votan	(415) 490-7600	IBM PC Speech Cards Speech Chip Sets
Whitman Engineering	(305) 628-4516	DSP Development System

DSP Text/Reference Books

A series of DSP books have been or are being written by experts in the field to help both university students and practicing engineers. These books feature DSP theory, algorithms, applications, and TMS320 implementations, and are available from publishers, John Wiley & Sons and Prentice-Hall. The first book in the series, DFT/FFT and Convolution Algorithms, is now available from John Wiley & Sons. Since these books cover both the theoretical and practical aspects of DSP, they can be used by universities as textbooks or DSP engineers as reference guides. The titles, authors, publishers, and availability of these books are provided as follows:

TITLE	AUTHOR	PUBLISHER	AVAILABILITY
DFT/FFT and Convolution Algorithms	C. S. Burrus and T. W. Parks	John Wiley & Sons	now
Digital Filter Design	T. W. Parks and C. S. Burrus	John Wiley & Sons	1986
Adaptive Filter Design	R. Johnson, Jr. et al	John Wiley & Sons	1986
Digital Control	W. Kohn	John Wiley & Sons	1986
Practical Approaches to Speech Coding	P. Papamichalis	Prentice-Hall	1986

For ordering information from publishers, please write or call:

John Wiley & Sons, Inc.
605 Third Avenue
New York, NY 10158
1-800-526-5368

Prentice-Hall,Inc.
Route 9W
Englewood Cliffs, NJ 07632
201-767-9520

TMS320 Documentation Support

Extensive publications are produced by Texas Instruments to support the TMS320 digital signal processing family. A list of these publications is given below.

1. TMS32010 User's Guide (SPRU001B).

2. TMS320C10 Data Sheet (SPRS006).

3. TMS32011 User's Guide (SPRU010).

4. TMS32020 User's Guide (SPRU004A).

5. TMS320C25 User's Guide (SPRU012).

6. TMS320 Development Support Reference Guide (SPRU007A).

7. TMS32010 Assembly Language Programmer's Guide (SPRU002B).

8. Link Editor User's Guide (SPDU037C).

9. TMS32010 Crossware Installation Guide (SPDU049).

10. TMS320 Family Simulator User's Guide (SPRU009).

11. TMS32010 Evaluation Module User's Guide (SPRU005A).

12. XDS/22 TMS32010 Emulator User's Guide (SPDU015).

13. TMS32010 Analog Interface Board User's Guide (SPRU006).

PART II
FUNDAMENTAL DIGITAL SIGNAL PROCESSING OPERATIONS

DIGITAL SIGNAL PROCESSING ROUTINES

DSP INTERFACE TECHNIQUES

Some common digital signal processing routines and interface circuits are frequently used in DSP applications. For example, the same structure of a digital filter used for audio signal processing may also be used for a modem in data communications. A Fast Fourier Transform (FFT) routine can be used for analyzing signals both in instrumentations and in speech coding. Another example is the interface of a digital signal processor to external memory devices in a standalone system or in a multiprocessing environment. A collection of application reports containing these routines and interfaces is included in Part II. Theory, block diagrams, algorithms, and TMS320 implementations are provided in these reports. The DSP Software Library includes code for the major DSP routines and applications.

Part II begins with a series of reports containing DSP routines. The first report discusses the implementation of Finite Impulse Response (FIR)/Infinite Impulse Response (IIR) filters using the TMS32010 and TMS32020. Filters designed with digital processors, such as the TMS320, are superior over their analog counterparts for better specifications, stability, performance, and reproducability. The report describes a variety of methods for implementing FIR/IIR filters using the TMS320. The TMS320 algorithm execution time and data memory requirements are considered. Tradeoffs between several different filter structures are also discussed. This application report complements the Digital Filter Design Package (DFDP) discussed in Section 2.

Fast Fourier Transforms (FFT), containing a class of computationally efficient algorithms implementing the Discrete Fourier Transforms (DFT), are widely used in DSP applications. In the report on FFT, the development of the FFT from the continuous Fourier Transform and DFT is first presented. Issues regarding the implementation of the FFT with the TMS32020 processor are then discussed, such as scaling, special FFT structures, and system memory and I/O considerations. The report also includes the TMS32020 code for 256-point and 1024-point FFT algorithms. The next report discusses companding routines. Companding is required for applications that use codec devices, such as in public and private telephone networks. With the speed and the versatility of the TMS320, companding can be performed in either software or hardware. The report describes both the A-law and μ-law software companding methods. Programs are also provided to show how the software companding can be performed using the computational power of the TMS32010 and the TMS32020. An example of the hardware companding is presented in Section 14 of Part III.

Although The TMS32010 and TMS32020 are fixed-point 16/32-bit digital signal processors, they can also perform floating-point computations at a speed comparable to dedicated floating-point processors. The next two reports present algorithms and code implementing floating-point addition, subtraction, multiplication, and division with the TMS320. The support of floating-point operations by the TI processors has made possible some applications, such as the implementation of the CCITT Adaptive Differential Pulse Code Modulation (ADPCM) algorithm and image/graphics operations.

Sine-wave or waveform generations are used in instrumentations and communications. Both speed and accuracy are major concerns for these applications. The report in Section 8 presents two methods of sine-wave generation. The first method is a fast direct table lookup scheme suitable for applications where speed is critical. The second approach, an enhancement of the first, includes linear interpolation to provide higher accurate waveforms. The last report in the DSP routines portion of Part II is on matrix multiplication with the TMS32010 and TMS32020. Matrix multiplication is useful in applications, such as graphics, numerical analysis, or high-speed control. Because of the high speed of the multiply/accumulate operations and fast data I/O, both processors can multiply in microseconds large matrices with their sizes only limited by the internal data memory. Programs are included in the report to illustrate matrix multiplication on both processors.

The second half of Part II encompasses various hardware interface techniques useful for integrating systems using the TMS320 digital signal processors. The first two reports describe interface circuits for the TMS32010 to asynchronous inputs and to external memory devices, such as external ROM or RAM. A description of a hardware peripheral interface device produced by Pacific Microcircuits is also included, which eases the TMS32010 interface to both external memory and codec devices. The report in Section 12 suggests hardware design techniques for interfacing memory devices and peripherals to the TMS32020. Examples of PROM, EPROM, static RAM, and dynamic RAM circuits built around the TMS32020 are demonstrated, with the timing requirements given for the processor and external devices. Interfaces to a combo-codec and a host computer through UART are also presented. The last report of Part II shows a scheme where the TMS32020 can be used as a numeric coprocessor to a host processor for numeric-intensive applications, such as image/graphics processing. The host processor selected for the example is the MC68000. The interface and communication techniques presented in the report are generic and directly applicable to other host processors.

3. Implementation of FIR/IIR Filters with the TMS32010/TMS32020

Al Lovrich and Ray Simar, Jr.
Digital Signal Processing - Semiconductor Group
Texas Instruments

INTRODUCTION

In many signal processing applications, it is advantageous to use digital filters in place of analog filters. Digital filters can meet tight specifications on magnitude and phase characteristics and eliminate voltage drift, temperature drift, and noise problems associated with analog filter components.

This application report describes a variety of methods for implementing Finite Impulse Response (FIR) and Infinite Impulse Response (IIR) digital filters with the TMS320 family of digital signal processors. Emphasis is on minimizing both the execution time and the number of data memory locations required. Tradeoffs between several different structures of the two classes of digital filters are also discussed.

In this report, TMS320 source code examples are included for the implementation of two FIR filters and three IIR filters based on the techniques presented. Plots of magnitude response, log-magnitude response, unit-sample response, and other pertinent data accompany each of the filter implementations. Important performance considerations in digital filter design are also included. The methods presented for implementing the different types of filters can be readily extended to any desired order of filters.

Readers are assumed to have some familiarity with the basic concepts of digital signal processing theory.[1] The notation used in this report is consistent with that used in reference [1].

FILTERING WITH THE TMS320 FAMILY

Almost every field of science and engineering, such as acoustics, physics, telecommunications, data communications, control systems, and radar, deal with signals. In many applications, it is desirable that the frequency spectrum of a signal be modified, reshaped, or manipulated according to a desired specification. The process may include attenuating a range of frequency components and rejecting or isolating one specific frequency component.

Any system or network that exhibits such frequency-selective characteristics is called a filter. Several types of filters can be identified: lowpass filter (LPF) that passes only "low" frequencies, highpass filter (HPF) that passes "high" frequencies, bandpass filter (BPF) that passes a "band" of frequencies, and band-reject filter that rejects certain frequencies. Filters are used in a variety of applications, such as removing noise from a signal, removing signal distortion due to the transmission channel, separating two or more distinct signals that were mixed in order to maximize communication channel utilization, demodulating signals, and converting discrete-time signals into continuous-time signals.

Advantages of Digital Filtering

The term "digital filter" refers to the computational process or algorithm by which a digital signal or sequence of numbers (acting as input) is transformed into a second sequence of numbers termed the output digital signal. Digital filters involve signals in the digital domain (discrete-time signals), whereas analog filters relate signals in the analog domain (continuous-time signals). Digital filters are used extensively in applications, such as digital image processing, pattern recognition, and spectrum analysis. A band-limited continuous-time signal can be converted to a discrete-time signal by means of sampling. After processing, the discrete-time signal can be converted back to a continuous-time signal. Some of the advantages of using digital filters over their analog counterparts are:

1. High reliability
2. High accuracy
3. No effect of component drift on system performance
4. Component tolerances not critical.

Another important advantage of digital filters when implemented with a programmable processor such as the TMS320 is the ease of changing filter parameters to modify the filter characteristics. This feature allows the design engineer to effectively and easily upgrade or update the characteristics of the designed filter due to changes in the application environment.

Design of Digital Filters

The design of digital filters involves execution of the following steps:

1. Approximation
2. Realization
3. Study of arithmetic errors
4. Implementation.

Approximation is the process of generating a transfer function that satisfies a set of desired specifications, which may involve the time-domain response, frequency-domain response, or some combination of both responses of the filter.

Realization consists of the conversion of the desired transfer function into filter networks. Realization can be accomplished by using several network structures,[2,3] as listed below. Some of these structures are covered in detail in this report.

1. Direct
2. Direct canonic (direct-form II)
3. Cascade
4. Parallel
5. Wave[4]
6. Ladder.

Approximation and realization assume an infinite-precision device for implementation. However, implementation is concerned with the actual hardware circuit or software coding of the filter using a programmable processor. Since practical devices are of finite precision, it is necessary to study the effects of arithmetic errors on the filter response.

TMS320 Digital Signal Processors

Digital Signal Processing (DSP) is concerned with the representation of signals (and the information they contain) by sequences of numbers and with the transformation or processing of such signal representations by numeric-computational procedures. In the past, digital filters were implemented in software using mini- or main-frame computers for non-realtime operation or on specialized dedicated digital hardware for realtime processing of signals.

The recent advances in VLSI technology have resulted in the integration of these digital signal processing systems into small integrated circuits (ICs), such as the TMS320 family of digital signal processors from Texas Instruments. The TMS320 implementation of digital filters allows the filter to operate on realtime signals. This method combines the ease and flexibility of the software implementation of filters with reliable digital hardware. To further ease the design task, it is now possible for engineers to design and test filters using any one of the commercially available filter design packages, some of which create TMS320 code and decrease the design time.

The Texas Instruments TMS320 digital signal processing family contains two generations of digital signal processors. The TMS32010, the first-generation digital signal processor,[5] implements in hardware many functions that other processors typically perform in software. Some of the key features of the TMS32010 are:

- 200-ns instruction cycle
- 1.5K words (3K bytes) program ROM
- 144 words (288 bytes) data RAM
- External memory expansion to 4K words (8K bytes) at full speed
- 16 x 16-bit parallel multiplier
- Interrupt with context save
- Two parallel shifters
- On-chip clock
- Single 5-volt supply, NMOS technology, 40-pin DIP.

The TMS32020 is the second-generation processor[6] in the TMS320 DSP family. To maintain device compatibility, the TMS32020 architecture is based upon that of the TMS32010, the first member of the family, with emphasis on overall speed, communication, and flexibility in processor configuration. Some of the key features of the TMS32020 are:

- 544 words of on-chip data RAM, 256 words of which may be programmed as either data or program memory
- 128K words of data/program space
- Single-cycle multiply/accumulate instructions

- TMS32010 software upward compatibility
- 200-ns instruction cycle
- Sixteen input and sixteen output channels
- 16-bit parallel interface
- Directly accessible external data memory space
- Global data memory interface for multiprocessing
- Instruction set support for floating-point operations
- Block moves for data/program memory
- Serial port for multiprocessing or codec interface
- On-chip clock
- Single 5-volt supply, NMOS technology, 68-pin grid array package.

Because of their computational power, high I/O throughput, and realtime programming, the TMS320 processors have been widely adapted in telecommunication, data communication, and computer applications. In addition to the above features, the TMS320 has efficient DSP-oriented instructions and complete hardware/software development tools, thus making the TMS320 highly suitable for DSP applications.

DIGITAL FILTER IMPLEMENTATION ON THE TMS320

For a large variety of applications, digital filters are usually based on the following relationship between the filter input sequence $x(n)$ and the filter output sequence $y(n)$:

$$y(n) = \sum_{k=0}^{N} a_k \, y(n-k) \; + \; \sum_{k=0}^{M} b_k \, x(n-k) \qquad (1)$$

Equation (1) is referred to as a linear constant-coefficient difference equation. Two classes of filters can be represented by linear constant-coefficient difference equations:

1. Finite Impulse Response (FIR) filters, and
2. Infinite Impulse Response (IIR) filters.

The following sections describe the implementation of these classes of filters on the TMS32010 and TMS32020.

FIR Filters

For FIR filters, all of the a_k in (1) are zero. Therefore, (1) reduces to

$$y(n) = \sum_{k=0}^{M} b_k \, x(n-k) \qquad (2)$$

where $(M + 1)$ is the length of the filter.

As a result, the output of the FIR filter is simply a finite-length weighted sum of the present and previous inputs to the filter. If the unit-sample response of the filter is denoted as h(n), then from (2), it is seen that h(n) = b(n). Therefore, (2) is sometimes written as

$$y(n) = \sum_{k=0}^{M} h(k)x(n-k) \qquad (3)$$

From (3), it can be seen that an FIR filter has, as the name implies, a finite-length response to a unit sample. Denoting the z transforms of x(n), y(n), and h(n) as X(z), Y(z), and H(z), respectively, then

$$H(z) = \frac{Y(z)}{X(z)} = \sum_{k=0}^{M} b_k z^{-1} = \sum_{k=0}^{M} h(k)z^{-k} \qquad (4)$$

Equations (3) and (4) may also be represented by the network structure shown in Figure 1. This structure is referred to as a direct-form realization of an FIR filter, because the filter coefficients can be identified directly from the difference equation (3). The branches labeled with z^{-1} in Figure 1 correspond to the delays in (3) and the multiplications by z^{-1} in (4). Equation (3) may be implemented in a straightforward and efficient manner on a TMS320 processor.

TMS32010 Implementation of FIR Filters

Figure 2 gives an example of a length-5 direct-form FIR filter, and Figure 3 shows a portion of the TMS32010 code for implementing this filter.

The notation developed in this section will be used throughout this application report. XN corresponds to x(n), XNM1 corresponds to x(n−1), etc.

In the above implementation, the following three basic and important concepts for the implementation of FIR filters on the TMS320 should be understood:
1. The relationship between the unit-sample response of an FIR filter and the filter structure,
2. The power of the LTD and MPY instruction pair for this implementation, and
3. The ordering of the input samples in the data memory of the TMS320, which is critical for realtime signal processing.

The input sequence x(n) is stored as shown in Figure 4. In general, each of the multiplies and shifts of x(n) in (3) is implemented with an instruction pair of the form

```
LTD    XNM1
MPY    H1
```

The instruction LTD XNM1 loads the T register with the contents of address XNM1, adds the result of the previous multiply to the accumulator, and shifts the data at address XNM1 to the next higher address in data memory. Using the storage scheme in Figure 4, this corresponds to shifting the data at address XNM1 to address XNM2. The instruction MPY H1 multiplies the contents of the T register with the contents of address H1. The shifting is the reason for the storage scheme used in Figure 4. This scheme, critical for realtime digital signal processing, makes certain that the input sequence x(n) is in the correct location for the next pass through the filter.

By comparing (3) with the code in Figure 3, the reason for the ordering of the data and the importance of the shift implemented by the LTD instruction can be seen. To better

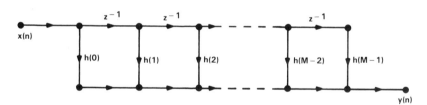

Figure 1. Direct-Form FIR Filter

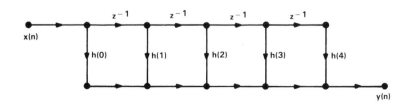

Figure 2. Length-5 Direct-Form FIR Filter

```
* THIS SECTION OF CODE IMPLEMENTS THE FOLLOWING EQUATION:            *
* x(n-4)h(4) + x(n-3)h(3) + x(n-2)h(2) + x(n-1)h(1) + x(n)h(0) = y(n) *
*
NXTPT    IN XN,PA2         * GET THE NEW INPUT VALUE XN FROM PORT PA0 *
*
         ZAC               * ZERO THE ACCUMULATOR *
*
         LT XNM4           * x(n-4)h(4) *
         MPY H4
*
         LTD XNM3          * x(n-4)h(4) + x(n-3)h(3) *
         MPY H3
*
         LTD XNM2          * SIMILAR TO THE PREVIOUS STEPS *
         MPY H2
*
         LTD XNM1
         MPY H1
*
         LTD XN
         MPY H0
*
         APAC              * ADD THE RESULT OF THE LAST MULTIPLY TO *
*                          * THE ACCUMULATOR                        *
*
         SACH YN,1         * STORE THE RESULT IN YN *
*
         OUT YN,PA2        * OUTPUT THE RESPONSE TO PORT PA1 *
*
         B NXTPT           * GO GET THE NEXT POINT *
```

Figure 3. TMS32010 Code for Implementing a Length-5 FIR Filter

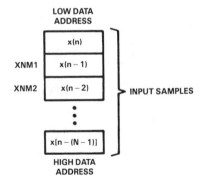

Figure 4. TMS32010 Input Sample Storage for a Length-N FIR Filter

understand the algorithm, the relationship between the input and output of the filter must be considered. Evaluating (3) for a particular value of n, for example, n_0, yields

$$y(n_0) = \sum_{k=0}^{N-1} h(k)\ x(n_0 - k) \qquad (5)$$

If the next sample of the filter response $y(n_0 + 1)$ is needed, it is seen from (3) that

$$y(n_0 + 1) = \sum_{k=0}^{N-1} h(k)\ x(n_0 + 1 - k) \qquad (6)$$

3. Implementation of FIR/IIR Filters with the TMS32010/TMS32020

Equations (5) and (6) show that the samples of $x(n)$ associated with particular values of $h(k)$ in (5) have been shifted to the left (i.e., to a higher data address) by one in (6). This shifting of the input data, illustrated in Figure 5, corresponds to the shifting of the flipped input sequence in relation to the unit-sample response.

Depending on the system constraints, the designer may choose to reduce program memory size by taking advantage of indirect addressing capability provided by the TMS32010. Using either of the auxiliary registers along with the autoincrement or autodecrement feature, the FIR filter program can be rewritten in looped form as shown in Figure 6.

The input sequence $x(n)$ is stored as shown in Figure 4, and the impulse response $h(n)$ is stored as shown in Figure 7. In the looped version, the indirect addressing mode is used with the autodecrement feature and BANZ instruction to control the looping and address generation for data access. While the looped code requires less program memory than the straightline version, the straightline version runs more quickly than the looped code because of the overhead associated with loop control. This design tradeoff should be carefully considered by the design engineer.

It is also possible to use the LTD/MPYK instruction pair to implement each filter tap in straightline code. The MPYK instruction is used to multiply the contents of the T register by a signed 13-bit constant stored in the MPYK instruction word. For many applications, a 13-bit coefficient can adequately implement the filter without significant changes to the filter response. An advantage of using this approach is that the coefficients are stored in program memory and there is no need to transfer them to data memory. This reduces the amount of data memory locations required per filter tap from two to one.

The length-80 FIR filter program in Appendix A implements a linear-phase FIR filter in straightline code. The unit-sample response of the filter is symmetric in order to achieve linear phase. Because of the symmetry, it is necessary to store only 40 (rather than 80) of the samples of the impulse response. This symmetry can often be used to a designer's advantage since it significantly reduces the amount of storage space required to implement the filter.

In summary, by taking advantage of the TMS32010 features, a designer can implement a direct-form FIR filter, optimized for execution time, data memory, or program memory.

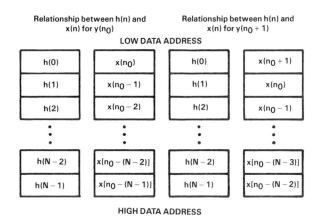

Figure 5. Relationship Between the Contents of Data Registers

```
* THIS SECTION OF CODE IMPLEMENTS THE EQUTION:              *
* x(n-(N-1))h(N-1) + x(n-(N-2))h(N-2) + ... + x(n)h(0) = y(n) *
*
          LARP AR0          * AUXILIARY REGISTER POINTER SET TO AR0 *
*
NXTPT     IN XN,PA2         * PULL IN NEW INPUT FROM PORT PA0 *
*
          LARK AR0,XNMNM1   * AR0 POINTS TO X(n-(N-1)) *
          LARK AR1,HNM1     * AR1 POINTS TO H(N-1) *
*
          ZAC               * ZERO THE ACCUMULATOR *
*
          LT *-,AR1         * x(n-(N-1))h(N-1) *
          MPY *-,AR0
*
LOOP      LTD *,AR1         * x(n-(N-1))h(N-1)+x(n-(N-2))h(N-2)+...+x(n)h(0)=y(n)*
          MPY *-,AR0
*
          BANZ LOOP         * IF AR0 DOES NOT EQUAL ZERO,              *
*                           * THEN DECREMENT AR0 AND BRANCH TO LOOP *
*
          APAC              * ADD THE P REGISTER TO THE ACCUMULATOR *
*
          SACH YN,1         * STORE THE RESULT IN YN *
*
          OUT YN,PA2        * OUTPUT THE RESPONSE TO PORT PA1 *
*
          B NXTPT           * GO GET THE NEXT INPUT POINT *
```

Figure 6. TMS32010 Code for Implementing a Looped FIR Filter

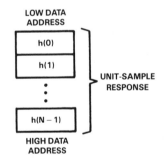

**Figure 7. TMS32010 Unit-Sample Response
Storage for a Looped FIR Filter**

TMS32020 Implementation of FIR Filters

In many DSP applications, realtime processing of signals is very critical. Important choices must be made in selecting a DSP device capable of realtime filtering, For example, in a speech application, a sampling rate of 8 kHz is common, which corresponds to an interval of 125 μs between consecutive samples. This interval is the maximum allowable time for realtime operation, corresponding to 625 cycles on the TMS32010. In order to perform the required signal processing tasks in that interval, it is essential to reduce filter execution time. This can be accomplished by a single-cycle multiply/accumulate instruction. The TMS32020, the second-generation DSP device, is a processor with such a capability. A single-cycle multiply/accumulate with data-move instruction and larger on-chip RAM make it possible to implement each filter tap in approximately 200 ns.

The TMS32020 provides a total of 544 16-bit words of on-chip RAM, divided into three separate blocks of B0, B1, and B2. Of the 544 words, 288 words (blocks B1 and B2) are always data memory, and 256 words (block B0) are programmable as either data or program memory. The CNFD (configure block B0 as data memory) and CNFP (configure block B0 as program memory) instructions allow dynamic configuration of the memory maps through software, as illustrated in Figure 8. After execution of the CNFP instruction, block B0 is mapped into program memory, beginning with address 65280. To take advantage of the MACD (multiply and accumulate with data move) instruction, block B0 must be configured as program memory using the CNFP instruction. MACD only works with on-chip RAM. The use of the MACD instruction helps to speed

3. Implementation of FIR/IIR Filters with the TMS32010/TMS32020

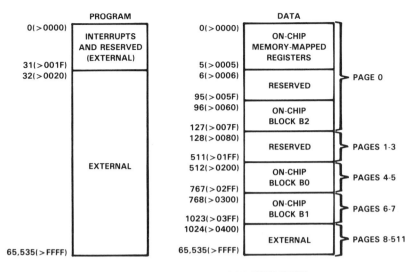

(a) ADDRESS MAPS AFTER A CNFD INSTRUCTION

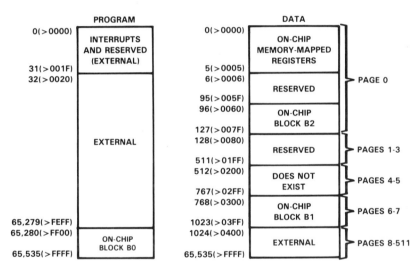

(b) ADDRESS MAPS AFTER A CNFP INSTRUCTION

Figure 8. TMS32020 Memory Maps

the filter execution and allows the size of the FIR filter to expand to 256 taps.[6]

The TMS32020 implementation of (3) is made even more efficient with a repeat instruction, RPTK. It forms a useful instruction pair with MACD, such as

```
RPTK    NM1
MACD    (PMA),(DMA)
```

The RPTK NM1 instruction loads an immediate 8-bit value N-1 into the repeat counter. This causes the next instruction to be executed N times (N = the length of the filter). The instruction MACD (PMA),(DMA) performs the following functions:

1. Loads the program counter with PMA,
2. Multiplies the value in data memory location DMA (on-chip, block B1) by the

value in program memory location PMA (on-chip, block B0),

3. Adds the previous product to the accumulator,
4. Copies the data memory value (block B0) to the next higher on-chip RAM location. The data move is the mechanism by which the z^{-1} delay can be implemented, and
5. Increments the program counter with each multiply/accumulate to point to the next sample of the unit-sample response.

In other words, the MACD instruction combines the LTD/MPY instruction pair into one. With the proper storage of the input samples and the filter unit-sample response, one can take advantage of the power of the MACD instruction. Figure 9 is a data storage scheme that provides the correct sequence of inputs for the next pass through the filter.

In the TMS32020 code example of Figure 10, data memory values are accessed indirectly through auxiliary register 1 (AR1) when the MACD instruction is implemented. For low-order filters (second-order), using the MACD instruction in conjunction with the RPTK instruction is less effective due to the overhead associated with the MACD instruction in setting up the repeat construct. To take advantage of the MACD instruction, the filter order must be greater than three. For lower-order filters, it is recommended to use the LTD/MPY instruction pair in place of RPT/MACD.

Writing looped code for the TMS32020 implementation of an FIR filter gives no further advantage. Since the MACD instruction already uses less program memory, looped code in this case does not reduce program memory size. Implementing FIR filters of length-3 or higher requires the same amount of program memory (excluding coefficient storage). For example, an FIR filter of length-256 takes the same amount of program memory space as a FIR filter of length-4.

Since the TMS32020 instruction set is upward-compatible with the TMS32010 instruction set, it is possible to use the LTD/MPYK instruction pair to implement the filter. With the TMS32020, the designer can use either RPTK/MACD or LTD/MPY(K) where appropriate. Depending on the application and the data memory constraints, the use of the LTD/MPYK instruction pair results in less data memory usage at the cost of increasing the program memory storage.

The FIR filter program of Appendix A is an implementation of the same length-80 FIR filter used in the TMS32010 example. In this implementation, it can be seen that the TMS32020 uses less program memory than the TMS32010 with the tradeoff of using more data memory words. The increase in data memory size is indirectly related to the MACD instruction; i.e., in order to take full advantage of the instruction, it is necessary to keep the multiplier pipeline as busy as possible. Therefore, the filter will execute faster when all 80 coefficients are provided in block B0.

The TMS32020 provides a solution for the faster execution of FIR filters. The combination of the RPTK/MACD instructions provides for a minimum program memory and high-speed execution of an FIR filter. If data memory is a concern, the designer can use the LTD/MPYK instruction pair at the cost of increasing program memory and using 13-bit filter coefficients.

IIR Filters

The concepts introduced for the implementation of FIR filters can be extended to the implementation of IIR filters. However, for an IIR filter, at least one of the a_k in (1) is

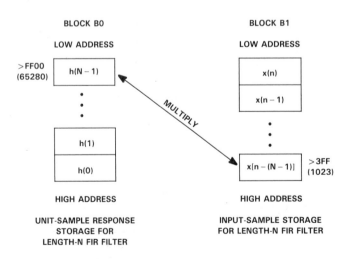

Figure 9. TMS32020 Memory Storage Scheme

3. Implementation of FIR/IIR Filters with the TMS32010/TMS32020

```
* THIS SECTION OF CODE IMPLEMENTS THE EQUATION:
* x(n-(N-1))h(N-1) + x(n-(N-2))h(N-2) + ... + x(n)h(0) = y(n)
*
        CNFP                    * USE BLOCK B0 AS PROGRAM AREA
*
NXTPT   IN      XN,PA0          * BRING IN THE NEW SAMPLE XN
*
        LRLK    AR1,>3FF        * POINT TO THE BOTTOM OF BLOCK B1
        LARP    AR1
*
        MPYK    0               * SET P REGISTER TO ZERO
        ZAC                     * CLEAR THE ACCUMULATOR
*
        RPTK    NM1             * REPEAT N-1 TIMES
        MACD    >FF00,*-        * MULTIPLY/ACCUMULATE
*
        APAC
        SACH    YN,1
*
        OUT     YN,PA1          * OUTPUT THE FILTER RESPONSE y(n)
*
        B       NXTPNT          * GET THE NEXT POINT
```

Figure 10. TMS32020 Code for Implementing a Length-5 FIR Filter

nonzero. It has been shown[1] that the z transform of the unit-sample response of an IIR filter corresponding to (1) is

$$H(z) = \frac{Y(z)}{X(z)} = \frac{\displaystyle\sum_{k=0}^{M} b_k z^{-k}}{1 - \displaystyle\sum_{k=1}^{N} a_k z^{-k}} \qquad (7)$$

where $H(z)$, $Y(z)$, and $X(z)$ are the z transforms of $h(n)$, $y(n)$, and $x(n)$, respectively. Three different network structures often used to implement (7) are the direct form, the cascade form, and the parallel form. Implementation of these structures is discussed in the following sections.

Direct-Form IIR Filter

Equations (1) and (7) may also be represented by the network structure shown in Figure 11. For convenience, it is assumed that $M = N$. This network structure is referred to as the direct-form I realization of an Nth-order difference equation. As was the case for the direct-form FIR filter, the structure in Figure 11 is called direct-form since the coefficients of the network can be obtained directly from the difference equation describing the network. Again, the branches associated with the z^{-1} correspond to the delays in (1) and the multiplications in (7).

The following difference equation:

$$y(n) = \sum_{k=1}^{N} a_k \, y(n-k) + \sum_{k=0}^{M} b_k \, x(n-k) \qquad (8)$$

shows that the output of the filter is a weighted sum of past values of the input to the filter and of the output of the filter. Using techniques similar to those for an FIR filter, this realization can be implemented in a straightforward and efficient way on the TMS32010 and TMS32020.

A network flowgraph equivalent to that in Figure 11 is shown in Figure 12. This system is referred to as the direct-form II structure. Since the direct-form II has the minimum number of delays (branches labeled z^{-1}), it requires the minimum number of storage registers for computation. This structure is advantageous for minimizing the amount of data memory used in the implementation of IIR filters.

In Figures 13 through 17, a second-order direct-form II IIR filter is used as an example for the TMS320 implementation of the IIR filter. The network structure is shown in Figure 13.

The difference equation for this network is

$$d(n) = x(n) + a_1 \, d(n-1) + a_2 \, d(n-2) \qquad (9)$$
$$y(n) = b_0 \, d(n) + b_1 \, d(n-1) + b_2 \, d(n-2)$$

In this case, $d(n)$, shown in (9) and Figure 13, corresponds to the network value at the different delay nodes. The zero-delay register corresponds to $d(n)$; $d(n-1)$ is the register for the delay of one; and $d(n-2)$ is the register for the delay of two. A portion of the TMS32010 code necessary to implement (9) is shown in Figure 14. Initially all $d(n-i)$ for $i=0,1,2$ are set to zero.

The delay-node values of the filter are stored in data memory as shown in Figure 15. At each major step of the algorithm, a multiply is done, and the result from the previous multiply is added to the accumulator. Also, the past delay-node values are shifted to the next higher location in

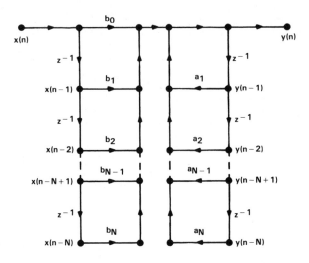

Figure 11. Direct-Form I IIR Filter

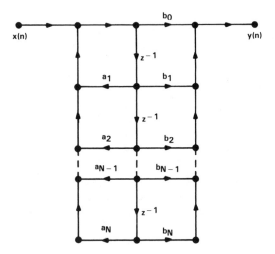

Figure 12. Direct-Form II IIR Filter

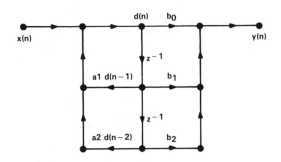

Figure 13. Second-Order Direct-Form II IIR Filter

data memory, thus placing them in the correct position for the next pass through the filter. All of these operations are carried out with instruction pairs, such as

```
LTD     DNM1
MPY     B1
```

where DNM1 corresponds to $d(n-1)$ and B1 corresponds to b_1 as in (9).

When the last multiplication is performed and the result is added to the accumulator, the accumulator contains the result of (9), which is $y(n)$. From (9) and Figure 13, it is evident that the delay-node value $d(n)$ depends on several of the previous delay-node values. This feedback is illustrated by the instruction

```
SACH    DN,1
```

and the use of the statements

```
LTD     DNM1
          .
          .
          .
LTD     DN
```

The ordering of the delay-node values, shown in Figure 15, allows for a simple program structure with minimal computations and minimal data locations. It also accommodates the shifting of the delay-node values in a straightforward way. The feedback of DN makes apparent the underlying structure of the direct-form II filter and (10). This form of the algorithm is flexible and can be extended to higher-order direct-form filters in a straightforward way.

3. Implementation of FIR/IIR Filters with the TMS32010/TMS32020

```
* THIS SECTION OF CODE IMPLEMENTS THE EQUATIONS:   *
* d(n) = x(n) + d(n-1)a  + d(n-2)a                 *
*                       1          2               *
* y(n) = d(n)b  + d(n-1)b  + d(n-2)b               *
*            0          1          2               *
*                                                  *
*
         IN XN,PA0        * NEW INPUT VALUE XN *
*
         LAC XN,15        * LOAD ACCUMULATOR WITH XN *
*
         LT DNM1
         MPY A1
*
         LTA DNM2
         MPY A2
*
         APAC
*
         SACH DN,1        * d(n) = x(n) + d(n-1)a  + d(n-2)a  *
*                         *                      1          2 *
         ZAC
*
         MPY B2
*
         LTD DNM1
         MPY B1
*
         LTD DN
         MPY B0
*
         APAC
*
         SACH YN,1        * y(n) = d(n)b  + d(n-1)b  + d(n-2)b  *
*                         *           0          1          2   *
*
         OUT YN,PA1       * YN IS THE OUTPUT OF THE FILTER *
```

Figure 14. TMS32010 Code for Implementing a Second-Order Direct-Form II IIR Filter

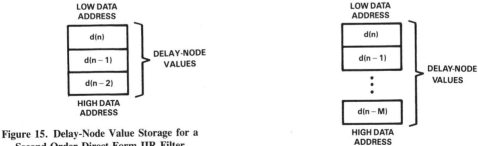

**Figure 15. Delay-Node Value Storage for a
Second-Order Direct-Form IIR Filter**

**Figure 16. Delay-Node Value Storage for a
Direct-Form II IIR Filter**

Figure 16 shows the necessary ordering of the delay-node values for a general direct-form II structure for the case $M \geq N$. Filter order is determined by M or N, whichever is greater.

Figure 17 shows a portion of the TMS32020 code for implementing the same second-order direct-form II IIR filter using the MACD instruction. As discussed in the section on FIR filters, using the RPTK/MACD instruction pair is most effective when the filter order is three or higher. The use of the MACD instruction allows the designer to save one word of program memory over the LTD/MPY implementation. The TMS32020 code in Figure 17 is provided only as an example. For a biquad implementation (second-order direct-form II IIR filter), the TMS32010 code and TMS32020 code for the filter implementation are identical. Note that due to larger on-chip RAM of the TMS32020, higher-order IIR filters or sections of IIR filters can be implemented. For the rest of the IIR filter structures, the same discussion applies to both processors.

An example of a TMS32010/TMS32020 program implementing a fourth-order direct-form II structure can be found in Appendix C.

Cascade-Form IIR Filter

In this section, the realization and implementation of cascade-form IIR filters are discussed. The implementation of a cascade-form IIR filter is an extension of the results of the implementation of the direct-form IIR filter.

The z transform of the unit-sample response of an IIR filter

$$
H(z) = \frac{\displaystyle\sum_{k=0}^{M} b_k z^{-k}}{1 - \displaystyle\sum_{k=1}^{N} a_k z^{-k}} \tag{10}
$$

```
*  THIS SECTION OF CODE IMPLEMENTS A SECOND-ORDER DIRECT-FORM II IIR FILTER
*  d(n) = x(n) + d(n-1)a    + d(n-2)a
*                        1           2
*  y(n) = d(n)b   + d(n-1)b   + d(n-2)b
*             0           1           2
*
NEXT      IN     XN,PA2          * NEW INPUT VALUE XN
*
          LAC    XN
          MPYK   0               * CLEAR P REGISTER
*
          LARP   AR1
          LRLK   AR1,>03FF
          CNFP                   * USE BLOCK B0 AS PROGRAM AREA
*
*  d(n) = x(n) + d(n-1)a    + d(n-2)a
*                        1           2
*
          RPTK   1               * REPEAT 2 TIMES
          MACD   >FF00,*+
*
          APAC
          SACH   DN,1            * d(n)
*
*  y(n) = d(n)b   + d(n-1)b   + d(n-10)b
*             0           1            2
*
          ZAC
          MPYK   0               * CLEAR P REGISTER
*
          MPY    >FF02
*
          RPTK   1
          MACD   >FF03,*-
*
          APAC
          SACH   YN,1            * SAVE FILTERED OUTPUT
*
          OUT    YN,PA2          * YN IS THE OUTPUT OF THE FILTER
          B      NEXT
```

Figure 17. TMS32020 Code for Implementing a Second-Order Direct-Form IIR Filter with MACD

3. Implementation of FIR/IIR Filters with the TMS32010/TMS32020

may also be written in the equivalent form

$$H(z) = \prod_{k=1}^{N/2} \frac{\beta_{0k} + \beta_{1k}z^{-1} + \beta_{2k}z^{-2}}{1 - \alpha_{1k}z^{-1} - \alpha_{2k}z^{-2}} \qquad (11)$$

where the filter is realized as a series of biquads. Therefore, this realization is referred to as the cascade form. Figure 18 shows a fourth-order IIR filter implemented in cascade structure, where the subsections are implemented as direct-form II sections. Each subsection corresponds to one of the terms in the product in (11). Note that any single cascade section is identical to the second-order direct-form II IIR filter described previously.

The difference equation for cascade section i can be written as

$$d_i(n) = y_{i-1}(n) + \alpha_{1i}\, d_i(n-1) + \alpha_{2i}\, d_i(n-2) \qquad (12)$$

$$y_i(n) = \beta_{0i}\, d_i(n) + \beta_{1i}\, d_i(n-1) + \beta_{2i}\, d_i(n-2)$$

where

i	=	$1,2,\ldots,N/2$.
$y_{i-1}(n)$	=	input to section i.
$d_i(n)$	=	value at a particular delay node in section 1.
$y_i(n)$	=	output of section i.
$y_0(n) = x(n)$	=	sample input to the filter.
$y_{N/2} = y(n)$	=	output of the filter.

For the IIR filter consisting of the two cascaded sections shown in Figure 18, there are two sets of equations describing the relationship between the input and output of the filter. The delay-node values for each section are stored as shown in Figure 19. The same indexing scheme used previously

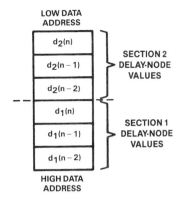

Figure 19. Delay-Node Storage for Cascaded IIR Filter Subsections

is used here (i.e., from the higher address in data memory to the lower address in data memory). In this case, the algorithm can be structured so that the 32-bit accumulator of the TMS320 acts as a storage register and carries the output of one of the second-order subsections to the input of the next second-order subsection. This avoids unnecessary truncation of the intermediate filter values into 16-bit words, and therefore provides better accuracy in the final output.

The implementation of the cascaded fourth-order IIR filter can be summarized as follows:

1. Load the new input value x(n).
2. Operate on the first section as outlined in Figure 12.
3. Leave the output of the first section in the accumulator (i.e., the SACH YN can be omitted for the first-section implementation since the accumulator links the output of one section to the input of the following section).
4. Operate on the second section in the same way as the first section, remembering that

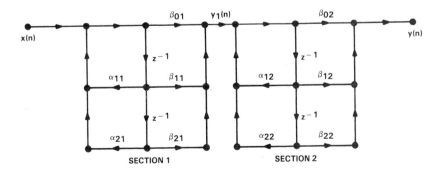

Figure 18. Fourth-Order Cascaded IIR Filter

the accumulator already contains the output of the previous section.

5. The output of the second section is the filter output y(n).

The above procedures can be applied to the IIR filter implementation of higher orders. It can be shown[3] that with proper ordering of the second-order cascades, the resulting filter has better immunity to quantization noise than the direct-form implementation, as will be discussed later.

An example of a TMS32010/TMS32020 program that implements a fourth-order IIR cascaded structure is contained in Appendix C.

Parallel-Form IIR Filter

The third form of an IIR filter is referred to as the parallel form. In this case, H(z) is written as

$$H(z) = \sum_{k=0}^{M-N} C_k z^{-k} + \sum_{k=1}^{N/2} \frac{\gamma_{0k} + \gamma_{1k} z^{-1}}{1 - \alpha_{1k} z^{-1} - \alpha_{2k} z^{-2}} \tag{13}$$

If $M < N$, then the term $(C_k z^{-k}) = 0$. The network form is shown in Figure 20, where it is assumed that $M = N = 4$. The multiplication of the input by C (a constant) is trivial. However, for one of the parallel branches of this structure, the difference equation is

$$d_i(n) = x(n) + \alpha_{1i} d_i(n-1) + \alpha_{2i} d_i(n-2) \tag{14}$$

$$p_i(n) = \gamma_{0i} d_i(n) + \gamma_{1i} d_i(n-1)$$

where $i = 1,2,...,N/2$, and $p_i(n) =$ the present output of a parallel branch.

The similarity to the second-order direct-form II network and the single parallel section is apparent. However, in this case, the outputs of all sections are summed to give the output y(n), i.e.,

$$y(n) = Cx(n) + \sum_{i=1}^{N/2} p_i(n) \tag{15}$$

if $M = N$. For the parallel implementation, the delay-node values are also structured in data memory, as shown in Figure 21, thus allowing for an implementation similar to that used previously. After the output of each section stored in the 32-bit accumulator is determined, these outputs are summed to yield the filter output y(n). An example of a TMS32010/TMS32020 program to implement a parallel structure can be found in Appendix C.

PERFORMANCE CONSIDERATIONS IN DIGITAL FILTER DESIGN

In the previous sections, different realizations of the FIR and IIR digital filters were discussed. This section is mainly concerned with the effects of finite wordlength on filter performance.

Some features of FIR and IIR filters, which distinguish them from each other and need special considerations when they are implemented, include phase characteristics, stability, and coefficient quantization effects.

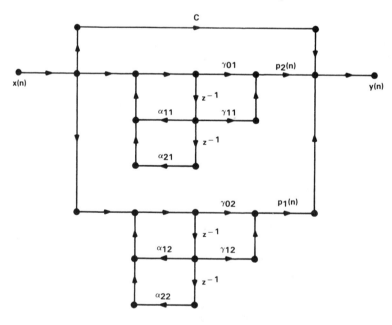

Figure 20. Parallel-Form IIR Filter

3. Implementation of FIR/IIR Filters with the TMS32010/TMS32020

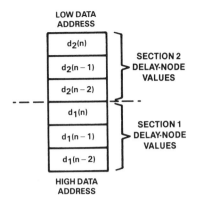

$d_2(n)$

$d_2(n-1)$ — SECTION 2 DELAY-NODE VALUES

$d_2(n-2)$

$d_1(n)$

$d_1(n-1)$ — SECTION 1 DELAY-NODE VALUES

$d_1(n-2)$

HIGH DATA
ADDRESS

**Figure 21. Delay-Node Value Storage
for a Parallel IIR Filter**

Given a set of frequency-response characteristics, typically a higher-order FIR filter is required to match these characteristics to a corresponding IIR filter. However, this does not imply that IIR filters should be used in all cases. In some applications, it is important that the filter have linear phase, and only FIR filters can be designed to have linear phase.

Another important consideration is the stability of the filter. Since the unit-sample response of an FIR filter is of finite length, FIR filters are inherently stable (i.e., a bounded input always produces a bounded output). This can be seen from (5) where the output of an FIR filter is a weighted finite sum of previous inputs. On the other hand, IIR filters may or may not be stable, depending on the locations of the poles of the filter.

Digital filters are designed with the assumption that the filter will be implemented on an infinite precision device. However, since all processors are of finite precision, it is necessary to approximate the "ideal" filter coefficients. This approximation introduces coefficient quantization error. The net result due to imprecise coefficient representations is a deviation of the resultant filter frequency response from the ideal one. For narrowband IIR filters with poles close to the unit circle, longer wordlengths may be required. The worst effect of coefficient quantization is instability resulting from poles being moved outside the unit circle.

The effect of coefficient quantization is highly dependent on the structure of the filter and the wordlength of the implementation hardware. Since the poles and zeroes for a filter implemented with finite wordlength arithmetic are not necessarily the same as the poles and zeroes of a filter implemented on an infinite precision device, the difference may affect the performance of the filter.

In the IIR filter, the cascade and parallel forms implement each pair of complex-conjugate poles separately. As a result, the coefficient quantization effect for each pair of complex-conjugate poles is independent of the other pairs

of complex-conjugate poles. This is generally not true for direct-form filters. Therefore, the cascade and parallel forms of IIR filters are more commonly used than the direct form.

Another problem in implementing a digital filter is the quantization error due to the finite wordlength effect in the hardware. Sources of error arising from the use of finite wordlength include the following:

1. I/O signal quantization
2. Filter coefficient quantization
3. Uncorrelated roundoff (or truncation) noise
4. Correlated roundoff (or truncation) noise
5. Dynamic range constraints.

These problems are addressed in the following paragraphs in more detail.

Representing instantaneous values of a continuous-time signal in digital form introduces errors that are associated with I/O quantization. Input signals are subjected to A/D quantization noise while output signals are subjected to D/A quantization noise. Although output D/A noise is less detrimental, input A/D quantization noise is the more dominant factor in most systems. This is due to the fact that input noise "circulates" within IIR filters and can be "regenerative" while output noise normally just "propagates" off-stage.

The filter coefficients in all of the routines described in this report are initially stored in program memory, and then moved to data memory. These coefficients are represented in Q15 format; i.e., the binary point (represented in two's-complement form) is assumed to follow the most-significant bit. This gives a coefficient range of 0.999969 to -1.0 with increments of 0.000031. The input is also in Q15 format so that when two Q15 numbers are multiplied, the result is a number in Q30 format. When the Q30 number resides in the 32-bit accumulator of the TMS320, the binary point follows the second most-significant bit. Since the output of the filter is assumed to be in Q15 format, the Q30 number must be adjusted by left-shifting by one while maintaining the most-significant 16 bits of the result. This is accomplished with the step SACH YN,1, which shifts the Q30 number to the left by one and stores the upper sixteen bits of the accumulator following the shift. The result YN is in Q15 format. Note that it is important to keep intermediate values in the accumulator as long as possible to maintain the 32-bit accuracy.

Uncorrelated roundoff (or truncation) noise may occur in multiplications. Even though the input to the digital filter is represented with finite wordlength, the result of processing leads to values requiring additional bits for their representation. For example, a b-bit data sample, multiplied by a b-bit coefficient, results in a product that is 2b bits long. In a recursive filter realization, 2b bits are required after the first iteration, 3b bits after the second iteration, and so on. The fact that multiplication results have to be truncated means that every "multiplier" in a digital structure can be regarded as a noise source. The combined effects of various noise sources degrade system performance.

Truncation or rounding off the products formed within the digital filter is referred to as correlated roundoff noise. The result of correlated roundoff (or truncation) noise, including overflow oscillations, is that filters suffer from "limit-cycle effect" (small-amplitude oscillations). For systems with adequate coefficient wordlength and dynamic range, this problem is usually negligible. Overflows are generated by additions resulting in undesirable large-amplitude oscillations. Both limit cycles and overflow oscillations force the digital filter into nonlinear operations. Although limit cycles are difficult to eliminate, saturation arithmetic can be used to reduce overflow oscillations. The overflow mode of operation on the TMS320 family is accomplished with the SOVM (set overflow mode) instruction, which sets the accumulator to the largest representable 32-bit positive ($>$7FFFFFFF hex) or negative ($>$80000000 hex) number according to the direction of overflow.

Dynamic range constraints, such as scaling of parameters, can be used to prevent overflows and underflows of the finite wordlength registers. The dynamic range is the ratio between the largest and smallest signals that can be represented in a filter. For an FIR filter, an overflow of the output results in an error in the output sample. If the input sample has a maximum magnitude of unity, then the worstcase output is

$$y(n) = \sum_{n=0}^{N-1} h(n) = s \qquad (16)$$

To guarantee y(n) to be a fraction, either the filter gain or the input x(n) has to be scaled down by a factor "s". Reducing the filter gain implies scaling down the filter coefficients so that the 16-bit coefficient is no longer used effectively. An implication of this scaling is a degradation of the filter frequency response due to higher quantization errors. As an alternative, the input signal may be scaled, resulting in a reduction in signal-to-noise ratio (SNR). In practice, the second approach is preferred since the scaling factor is normally less than two and does not change the SNR drastically. The required scaling on a TMS32020 is achieved by using the SPM (set P register output shift mode) instruction to invoke a right-shift by six bits to implement up to 128 multiply/accumulates without overflow occurring.

For an IIR filter, an overflow can cause an oscillation with full-scale amplitude, thus rendering the filter useless. In general, if the input signal x(n) is sinusoidal, the reciprocal of the gain "s" of the IIR filter is used to prevent output overflows.

For the TMS320 implementation with its double-precision accumulator and P register, scaling down the input sequence by the scaling factor "s" while maintaining a 16-bit accuracy for the coefficients can accomplish the task. For this reason, use of the MPYK instruction for IIR filter implementation is not recommended. Scaling the input signal by a factor "s" results in a degradation in the overall system SNR. Therefore, for IIR filters, it is important to keep the coefficient quantization errors as small as possible since less accurate coefficients may cause an unstable filter if the poles are moved outside the unit circle. The LAC (load accumulator with shift) instruction on the TMS320 processors easily accomplishes input signal scaling.

In the previous paragraphs, finite wordlength problems associated with digital filter implementation on programmable devices were discussed. The 16-bit coefficients and the 32-bit accumulator of the TMS320 processor help minimize the quantization effects. Special instructions also help overcome problems in the accumulator. These features, in addition to a powerful instruction set, make the TMS32010 and TMS32020 ideal programmable processors for filtering applications.

SOURCE CODE USING THE TMS320

Examples of TMS320 source code for the implementation of two FIR filters and three IIR filters, based on the techniques described in this application report, are contained in the appendixes. Plots of the magnitude response, log-magnitude response, unit-sample response, and other pertinent data precede the filter programs.

Five filter types are presented in the three appendixes as follows:

Appendix A   Length-80 bandpass FIR filter (TMS32010 and TMS32020)

Appendix B   Length-60 FIR differentiator (TMS32010/TMS32020)

Appendix C   Fourth-order lowpass IIR filters: direct-form, cascade, and parallel types (TMS32010/TMS32020)

The purpose of the source code is to further illustrate the use of the TMS320 devices for filtering applications and to allow implemention and analysis of these filters. The code is based on the programming techniques discussed earlier in this report.

TMS32020 source code is listed in the appendix for a length-80 FIR filter. The TMS32020 source code for the rest of the filter programs is identical to the TMS32010 code, as explained earlier. TMS32010 and TMS32020 instructions are compatible only at the mnemonic level. TMS32010 source programs should be reassembled using a TMS32020 assembler before execution. For more detail about code migration, refer to the TMS32020 User's Guide appendix, "TMS32010/TMS32020 System Migration," for detailed information.[6]

These filters were designed using the Digital Filter Design Package (DFDP) developed by Atlanta Signal Processors Incorporated (ASPI).[7] This package runs on either a Texas Instruments Professional Computer or an IBM Personal Computer and can generate TMS320 code for the filter designed. DFDP was used to design the FIR filters with the Remez exchange algorithm developed by Parks and McClellan, and to design the IIR filters by bilinear transformation of an elliptic analog prototype. All plots supplied with the filter programs were produced by DFDP.

Filter design packages, such as DFDP, make the design

and implementation of digital filters straightforward. They allow the DSP engineer to quickly examine a variety of filters and understand the tradeoffs involved in varying the characteristics of the filters. Several digital filter design packages and other useful software support from third parties are described in the TMS32010 Development Support Reference Guide.[8]

All of the TMS320 source code examples have several features in common that depend on the implementation and application. These features include the moving of filter coefficients from storage in program memory to data memory, their representation in Q15 format, and the instructions that control the analog interface used for testing.

The hardware configuration that was used to test these filters included a Texas Instruments analog interface board (AIB) to provide an analog-to-digital and digital-to-analog interface. The sampling rate was 10 kHz in all cases. The filters were driven by a white-noise source, and the frequency response was estimated by a spectrum analyzer. Each filter routine contains several lines of code to initialize the analog interface board. The AIB signals the TMS320 that another input sample is available by pulling the $\overline{\text{BIO}}$ pin low. The TMS320 polls this pin using the BIOZ instruction. The AIB houses a TMS32010 device. In order to use the TMS32020 with the AIB (PN: RTC/EVM320C-06), a specially designed adaptor (PN: RTC/ADP320A-06) must be inserted to convert TMS32020 signals to TMS32010 signals. All of these implementation- and application-dependent sections of code are labeled.

Appendix A provides programs for the implementation of a length-80 linear-phase bandpass FIR filter on the TMS32010 and the TMS32020. The filter has been designed using the Parks-McClellan algorithm. Pertinent data for this filter is as follows:

Passband	1.375 - 3.625 kHz
Stopbands	0.0 - 1.0 kHz
	4.0 - 5.0 kHz
Attenuation in stopbands	-68.4 dB
Transition regions	1.0 - 1.375 kHz
	3.625 - 4.0 kHz

The figures preceding the program show the magnitude response using a linear scale, the log-magnitude response, and the unit-sample response. Both the magnitude response and the log-magnitude response illustrate the equiripple response expected from using the Parks-McClellan algorithm. The unit-sample response possesses the symmetry that is characteristic of linear-phase FIR filters.

A length-60 FIR differentiator, shown in Appendix B, is also designed using the Parks-McClellan algorithm. Characteristics for the FIR differentiator are listed below.

Lower band edge	0.0 kHz
Upper band edge	5.0 kHz

Desired slope	0.4800
Maximum deviation	0.3172 percent

The log-magnitude resonse is illustrated as well as the unit-sample response, which is antisymmetric for an FIR differentiator. Because the code is written in looped form, there is a dramatic reduction in the amount of program space necessary to implement this filter.

The three filters in Appendix C are fourth-order lowpass IIR filters, designed using the bilinear-transform technique. The first filter is based on a direct-form II structure, the second filter is based on a cascade structure with two second-order direct-form II subsections, and the third filter is based on a parallel structure. These three IIR filters are identical in terms of their frequency response and have the following characteristics:

Passband	0.0 - 2.5 kHz
Transition region	2.5 - 2.75 kHz
Stopband	2.75 - 5.0 kHz
Attenuation in stopband	-25.17 dB

The figures that show the magnitude response, log-magnitude response, phase response, group delay, and the unit-sample response for the three IIR filters are treated as a group and precede the three programs for filter implementation.

Table 1 is a summary of information about the five digital filters that are implemented in the appendixes.

An examination of the length-80 FIR filter implementation reveals the advantages of using a TMS32020 over the TMS32010. The program memory size is reduced by a factor of 15 (11 words vs. 163 words) while execution speed is improved by a factor of 1.8. Since the other filter types do not take advantage of the RPTK/MACD instruction pair, the performance results are the same. For example, a fourth-order cascade-form IIR filter executes at 5.4 μs using only 27 program memory words.

When implementing linear-phase FIR filters, the designer must choose the right device for the application. If fast execution time and less program memory are essential, then the TMS32020 is the right choice.

The IIR filters are direct transformations of analog filters, exhibiting the same amplitude and phase characteristics as their analog counterparts. IIR filters tend to be more efficient than FIR filters with respect to transitionband sharpness and filter orders required. Although they require less code for implementation than the FIR filters (TMS32010 straightline code), they show great nonlinearity in phase, which limits their use in some applications.

By far the most commonly used IIR structure is the cascade-form realization. It has been shown that proper ordering of the poles and zeroes results in less sensitivity to quantization noise. The Digital Filter Design Package designs IIR filters in cascade form only.

By using a TMS32020 for both FIR and IIR filter implementations, it is possible to design a higher-order filter

Table 1. Summary Table of Filter Programs

LENGTH-80 LINEAR-PHASE BANDPASS FIR (STRAIGHT-LINE CODE)				
CODE	CYCLES	EXECUTION TIME (MICROSECONDS)	PROGRAM MEMORY (WORDS)	DATA MEMORY (WORDS)
Straight Line:				
TMS32010	163	32.6	163	120
TMS32020	90	18	11	161
(with RPTK)				

LENGTH-60 FIR DIFFERENTIATOR (LOOPED CODE)				
CODE	CYCLES	EXECUTION TIME (MICROSECONDS)	PROGRAM MEMORY (WORDS)	DATA MEMORY (WORDS)
Looped:				
TMS32010/20	243	48.6	11	120

FOURTH-ORDER LOWPASS IIR FILTERS				
STRUCTURE	CYCLES	EXECUTION TIME (MICROSECONDS)	PROGRAM MEMORY (WORDS)	DATA MEMORY (WORDS)
Direct-Form II:				
TMS32010/20	24	4.8	24	16
Cascade:				
TMS32010/20	27	5.4	27	18
Parallel:				
TMS3210/20	28	5.6	28	18

NOTE: The above performance figures are only given as a reference. They should not be taken as benchmarks since programs can always be improved for better speed and memory efficiency.

than with the TMS32010. The TMS32020 is also ideal for higher-order FIR filters that require single-cycle multiply/accumulate operations.

SUMMARY

A brief review of FIR and IIR digital filters has been given to assist in understanding the fundamentals of digital filter structure and their implementations using a digital signal processor. Many design examples have also been included to show the tradeoffs between FIR and IIR structures.

This application report has also described methods for implementing FIR and IIR filters with the TMS32010 and TMS32020. The design engineer can now choose between the two devices, depending on the application.

3. Implementation of FIR/IIR Filters with the TMS32010/TMS32020

REFERENCES

1. A.V. Oppenheim and R.W. Schafer, *Digital Signal Processing*, Prentice-Hall (1975).

2. Andreas Antoniou, *Digital Filters: Analysis and Design*, McGraw-Hill (1979).

3. C.S. Burrus and T.W. Parks, *Digital Filter Design*, John Wiley & Sons (1986).

4. U. Kaiser, "Wave Digital Filters and Their Significance for Customized Digital Signal Processing," *Texas Instruments Engineering Journal*, **Vol 2**, No. 5, 29-44 (September - October 1985).

5. *TMS32010 User's Guide* (SPRU001B), Texas Instruments (1985).

6. *TMS32020 User's Guide* (SPRU004A), Texas Instruments (1985).

7. *Digital Filter Design Package (DFDP)*, Atlanta Signal Processors Inc. (ASPI), 770 Spring St. NW, Suite 208, Atlanta, GA 30308, 404/892-7265 (1984).

8. *TMS32010 Development Support Reference Guide* (SPRU007), Texas Instruments (1984).

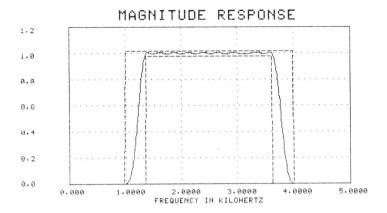

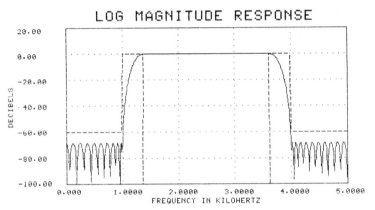

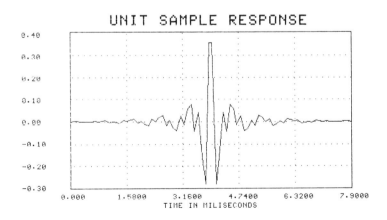

```
0001   ****************************************************
0002   *                                                  *
0003   *            LINEAR-PHASE FIR FILTER               *
0004   *            LENGTH-80 BANDPASS FILTER             *
0005   *                                                  *
0006   *        SAMPLING FREQUENCY = 10 KHZ               *
0007   *                                                  *
0008   *            FILTER CHARACTERISTICS                *
0009   *                                                  *
0010   *               BAND 1     BAND 2     BAND 3       *
0011   *                                                  *
0012   *  LOWER BAND EDGE   0.0000    1.3750     4.0000   *
0013   *  UPPER BAND EDGE   1.0000    3.6250     5.0000   *
0014   *  NOMINAL GAIN      0.0000    1.0000     0.0000   *
0015   *  NOMINAL RIPPLE    0.0010    0.0200     0.0010   *
0016   *  MAXIMUM RIPPLE    0.0004    0.0076     0.0004   *
0017   *  RIPPLE IN DB    -68.3965    0.0657   -68.3997   *
0018   *                                                  *
0019   *               FILTER STRUCTURE                   *
0020   *                                                  *
0021   *              -1              -1                   *
0022   *         z               z                        *
0023   *  o--->---o--->---o--->---o--- ... ---o           *
0024   *  x(n)  h(0)    h(1)    h(2)       h(N-2)  h(N-1)  *
0025   *                                                  *
0026   *  o--->---o--->---o--->---o--- ... ---o---->---o  *
0027   *                                             y(n) *
0028   *                                                  *
0029   ****************************************************
0030   *                                                  *
0031   * CYCLES | EXECUTION TIME | PROGRAM MEMORY | DATA MEMORY *
0032   *        | (MICROSECONDS) |    (WORDS)     |   (WORDS)   *
0033   * -------+----------------+----------------+----------- *
0034   *   90   |      18        |      10        |    161     *
0035   *                                                  *
0036   * (EXCLUDING INITIALIZATION AND I/O)               *
0037   ****************************************************
```

```
0045                    IDT   'FIRBPASS'
0046   002D   YN     EQU   45
0047   002E   MODE   EQU   46
0048   002F   CLOCK  EQU   47
0049   0030   XN     EQU   48
0050          *
0051   0000          AORG  0
0052   0000 FF80     B     START
       0001 0072
0053          *
0054   0020   CTABLE AORG  32
0055   0020 FFDC CH0   DATA  >FFDC  * -0.107251E-02 *
0056   0021 001F CH1   DATA  >001F  *  0.973976E-03 *
```

```
0057   0022 0051 CH2   DATA  >0051  *  0.249065E-02 *
0058   0023 FFE9 CH3   DATA  >FFE9  * -0.675043E-03 *
0059   0024 FFE6 CH4   DATA  >FFE6  * -0.771385E-03 *
0060   0025 FFBA CH5   DATA  >FFBA  * -0.212256E-02 *
0061   0026 FFB4 CH6   DATA  >FFB4  * -0.229530E-02 *
0062   0027 004B CH7   DATA  >004B  *  0.231021E-02 *
0063   0028 FFF9 CH8   DATA  >FFF9  * -0.194902E-03 *
0064   0029 0069 CH9   DATA  >0069  *  0.322896E-02 *
0065   002A 00A2 CH10  DATA  >00A2  *  0.496452E-02 *
0066   002B FF6F CH11  DATA  >FF6F  * -0.440419E-02 *
0067   002C FFFE CH12  DATA  >FFFE  * -0.314831E-04 *
0068   002D FF70 CH13  DATA  >FF70  * -0.438169E-02 *
0069   002E FEF4 CH14  DATA  >FEF4  * -0.815474E-02 *
0070   002F 00CB CH15  DATA  >00CB  *  0.621682E-02 *
0071   0030 000B CH16  DATA  >000B  *  0.342216E-03 *
0072   0031 00E6 CH17  DATA  >00E6  *  0.704627E-02 *
0073   0032 0187 CH18  DATA  >0187  *  0.119391E-01 *
0074   0033 FEE5 CH19  DATA  >FEE5  * -0.860811E-02 *
0075   0034 FFF5 CH20  DATA  >FFF5  * -0.346738E-03 *
0076   0035 FE7F CH21  DATA  >FE7F  * -0.117293E-01 *
0077   0036 FDBF CH22  DATA  >FDBF  * -0.175964E-01 *
0078   0037 0192 CH23  DATA  >0192  *  0.122947E-01 *
0079   0038 026A CH24  DATA  >026A  *  0.188796E-01 *
0080   0039 FC98 CH25  DATA  >FC98  * -0.266148E-01 *
0081   003A FDC2 CH26  DATA  >FDC2  * -0.175126E-01 *
0082   003B FF40 CH27  DATA  >FF40  * -0.586574E-02 *
0083   003C FC0A CH28  DATA  >FC0A  * -0.309240E-01 *
0084   003D FAA3 CH29  DATA  >FAA3  * -0.418954E-01 *
0085   003E 0347 CH30  DATA  >0347  *  0.256315E-01 *
0086   003F FE3D CH31  DATA  >FE3D  * -0.137498E-01 *
0087   0040 0747 CH32  DATA  >0747  *  0.568720E-01 *
0088   0041 09BB CH33  DATA  >09BB  *  0.760286E-01 *
0089   0042 FA3D CH34  DATA  >FA3D  * -0.450011E-01 *
0090   0043 052B CH35  DATA  >052B  *  0.403853E-01 *
0091   0044 EB59 CH36  DATA  >EB59  * -0.161339E+00 *
0092   0045 DC2A CH37  DATA  >DC2A  * -0.279963E+00 *
0093   0046 2D57 CH38  DATA  >2D57  *  0.352454E+00 *
0094   0047 2D57 CH39  DATA  >2D57  *  0.352454E+00 *
0095   0048 DC2A CH40  DATA  >DC2A  * -0.279963E+00 *
0096   0049 EB59 CH41  DATA  >EB59  * -0.161339E+00 *
0097   004A 052B CH42  DATA  >052B  *  0.403853E-01 *
0098   004B FA3D CH43  DATA  >FA3D  * -0.450011E-01 *
0099   004C 09BB CH44  DATA  >09BB  *  0.760286E-01 *
0100   004D 0747 CH45  DATA  >0747  *  0.568720E-01 *
0101   004E FE3D CH46  DATA  >FE3D  * -0.137498E-01 *
0102   004F 0347 CH47  DATA  >0347  *  0.256315E-01 *
0103   0050 FAA3 CH48  DATA  >FAA3  * -0.418954E-01 *
0104   0051 FC0A CH49  DATA  >FC0A  * -0.309240E-01 *
0105   0052 FF40 CH50  DATA  >FF40  * -0.586574E-02 *
0106   0053 FDC2 CH51  DATA  >FDC2  * -0.175126E-01 *
0107   0054 FC98 CH52  DATA  >FC98  * -0.266148E-01 *
0108   0055 026A CH53  DATA  >026A  *  0.188796E-01 *
0109   0056 0192 CH54  DATA  >0192  *  0.122947E-01 *
0110   0057 FDBF CH55  DATA  >FDBF  * -0.175964E-01 *
0111   0058 FE7F CH56  DATA  >FE7F  * -0.117293E-01 *
0112   0059 FFF5 CH57  DATA  >FFF5  * -0.346738E-03 *
0113   005A FEE5 CH58  DATA  >FEE5  * -0.860811E-02 *
```

3. Implementation of FIR/IIR Filters with the TMS32010/TMS32020

```
0114 005B 000B  CH59   DATA  >000B    *  0.347338E-03  *
0115 005C FEE5  CH60   DATA  >FEE5    * -0.860811E-02  *
0116 005D 0187  CH61   DATA  >0187    *  0.119391E-01  *
0117 005E 00E6  CH62   DATA  >00E6    *  0.704627E-02  *
0118 005F 000B  CH63   DATA  >000B    *  0.342216E-03  *
0119 0060 00CB  CH64   DATA  >00CB    *  0.621682E-02  *
0120 0061 FEF4  CH65   DATA  >FEF4    * -0.815474E-02  *
0121 0062 FF70  CH66   DATA  >FF70    * -0.438169E-02  *
0122 0063 FFFE  CH67   DATA  >FFFE    * -0.314831E-04  *
0123 0064 FF6F  CH68   DATA  >FF6F    * -0.440419E-02  *
0124 0065 00A2  CH69   DATA  >00A2    *  0.496452E-02  *
0125 0066 0069  CH70   DATA  >0069    *  0.322896E-02  *
0126 0067 FFF9  CH71   DATA  >FFF9    * -0.194902E-03  *
0127 0068 004B  CH72   DATA  >004B    *  0.231021E-02  *
0128 0069 FFB4  CH73   DATA  >FFB4    * -0.229530E-02  *
0129 006A FFBA  CH74   DATA  >FFBA    * -0.212256E-02  *
0130 006B FFE6  CH75   DATA  >FFE6    * -0.771385E-03  *
0131 006C FFE9  CH76   DATA  >FFE9    * -0.675043E-03  *
0132 006D 0051  CH77   DATA  >0051    *  0.249065E-02  *
0133 006E 001F  CH78   DATA  >001F    *  0.973976E-03  *
0134 006F FFDC  CH79   DATA  >FFDC    * -0.107251E-02  *
0135                *
0136 0070 000A  MD     DATA  >000A
0137 0071 01F3  SMP    DATA  >01F3         ; SAMPLING RATE OF 10 KHZ *
0138                *
0139 0072        START EQU   $
0140                *
0141                * INITIALIZATION OF THE ANALOG INTERFACE BOARD
0142                *
0143 0072 C807        LDPK  7
0144 0073 CA70        LACK  MD
0145 0074 582E        TBLR  MODE
0146 0075 E02E        OUT   MODE,PA0
0147 0076 CA71        LACK  SMP
0148 0077 582F        TBLR  CLOCK
0149 0078 E12F        OUT   CLOCK,PA1
0150                *
0151                ** LOAD FILTER COEFFICIENTS
0152                *
0153 0079 5588        LARP  AR0
0154 007A D000        LRLK  AR0,>200   ; USE AR0 FOR INDIRECT ADDRESSING
     007B 0200                         ; POINT TO BLOCK B0
0155 007C CB4F  RPTK        >4F
0156 007D FCA0  BLKP        CTABLE,*+  ; 80 COEFFICIENTS
     007E 0020
0157                *
0158 007F CE05        CNFP             ; USE BLOCK B0 AS PROGRAM AREA
0159                *
0160 0080 FA80  WAIT  BIOZ  NXTPT      ; BIO PIN GOES LOW WHEN A
     0081 0084
0161 0082 FF80        B     WAIT       ; NEW SAMPLE IS AVAILABLE
     0083 0080
0162                *
0163 0084 8230  NXTPT IN    XN,PA2     ; BRING IN THE NEW SAMPLE XN
0164                *
0165 0085 D100        LRLK  AR1,>3FF   ; POINT TO THE BOTTOM OF BLOCK B1
     0086 03FF
```

```
0166 0087 5589        LARP  AR1
0167                *
0168 0088 A000        MPYK  0
0169 0089 CA00        ZAC
0170                *
0171 008A CB4F        RPTK  >4F
0172 008B 5C90        MACD  >FF00,*-
     008C FF00
0173                *
0174 008D CE15        APAC
0175 008E 692D        SACH  YN,1
0176                *
0177 008F E22D        OUT   YN,PA2     ; OUTPUT THE FILTER RESPONSE y(n)
0178                *
0179 0090 FF80        B     WAIT       ; GO GET THE NEXT POINT
     0091 0080
0180                *
0181                  END

NO ERRORS, NO WARNINGS
```

```
****************************************************
*                                                  *
*            LINEAR-PHASE FIR FILTER               *
*            LENGTH-80 BANDPASS FILTER             *
*                                                  *
*       SAMPLING FREQUENCY = 10 KHZ                *
*                                                  *
*            FILTER CHARACTERISTICS                *
*                                                  *
*                  BAND 1     BAND 2     BAND 3     *
*                                                  *
*  LOWER BAND EDGE  0.0000     1.3750     4.0000    *
*  UPPER BAND EDGE  1.0000     3.6250     5.0000    *
*  NOMINAL GAIN     0.0000     1.0000     0.0000    *
*  NOMINAL RIPPLE   0.0010     0.0200     0.0010    *
*  MAXIMUM RIPPLE   0.0004     0.0076     0.0004    *
*  RIPPLE IN DB   -68.3965     0.0657   -68.3997    *
*                                                  *
*                FILTER STRUCTURE                  *
*                                                  *
*          z^-1        z^-1                  z^-1   *
* o---->---o---->---o---->---o---->---o  -  -o---->---o   *
* x(n)     |        |        |              |        |   *
*          v h(0)   v h(1)   v h(2)   v h(N-2)  v h(N-1) *
*          |        |        |              |        |   *
* o---->---o---->---o---->---o---->---o  -  -o---->---o---->---o *
*                                                       y(n)  *
*                                                  *
****************************************************
* CYCLES | EXECUTION TIME | PROGRAM MEMORY | DATA MEMORY *
*        | (MICROSECONDS) |    (WORDS)     |   (WORDS)   *
* -----------------------------------------------  *
*  163   |      32.6      |      163       |     120     *
*                                                  *
* (EXCLUDING INITIALIZATION AND I/O)               *
****************************************************
```

```
0001                                          IDT 'FIRBPASS'
0002   0000   0000   XN       EQU 0
0003   0001   0001   XNM1     EQU 1
0004   0002   0002   XNM2     EQU 2
0005   0003   0003   XNM3     EQU 3
0006   0004   0004   XNM4     EQU 4
0007   0005   0005   XNM5     EQU 5
0008   0006   0006   XNM6     EQU 6
0009   0007   0007   XNM7     EQU 7
       0008   0008   XNM8     EQU 8
       0009   0009   XNM9     EQU 9
```

```
0058   000A   XNM10    EQU 10
0059   000B   XNM11    EQU 11
0060   000C   XNM12    EQU 12
0061   000D   XNM13    EQU 13
0062   000E   XNM14    EQU 14
0063   000F   XNM15    EQU 15
0064   0010   XNM16    EQU 16
0065   0011   XNM17    EQU 17
0066   0012   XNM18    EQU 18
0067   0013   XNM19    EQU 19
0068   0014   XNM20    EQU 20
0069   0015   XNM21    EQU 21
0070   0016   XNM22    EQU 22
0071   0017   XNM23    EQU 23
0072   0018   XNM24    EQU 24
0073   0019   XNM25    EQU 25
0074   001A   XNM26    EQU 26
0075   001B   XNM27    EQU 27
0076   001C   XNM28    EQU 28
0077   001D   XNM29    EQU 29
0078   001E   XNM30    EQU 30
0079   001F   XNM31    EQU 31
0080   0020   XNM32    EQU 32
0081   0021   XNM33    EQU 33
0082   0022   XNM34    EQU 34
0083   0023   XNM35    EQU 35
0084   0024   XNM36    EQU 36
0085   0025   XNM37    EQU 37
0086   0026   XNM38    EQU 38
0087   0027   XNM39    EQU 39
0088   0028   XNM40    EQU 40
0089   0029   XNM41    EQU 41
0090   002A   XNM42    EQU 42
0091   002B   XNM43    EQU 43
0092   002C   XNM44    EQU 44
0093   002D   XNM45    EQU 45
0094   002E   XNM46    EQU 46
0095   002F   XNM47    EQU 47
0096   0030   XNM48    EQU 48
0097   0031   XNM49    EQU 49
0098   0032   XNM50    EQU 50
0099   0033   XNM51    EQU 51
0100   0034   XNM52    EQU 52
0101   0035   XNM53    EQU 53
0102   0036   XNM54    EQU 54
0103   0037   XNM55    EQU 55
0104   0038   XNM56    EQU 56
0105   0039   XNM57    EQU 57
0106   003A   XNM58    EQU 58
0107   003B   XNM59    EQU 59
0108   003C   XNM60    EQU 60
0109   003D   XNM61    EQU 61
0110   003E   XNM62    EQU 62
0111   003F   XNM63    EQU 63
0112   0040   XNM64    EQU 64
0113   0041   XNM65    EQU 65
0114   0042   XNM66    EQU 66
```

3. Implementation of FIR/IIR Filters with the TMS32010/TMS32020

```
0115  0043        XNM67   EQU  67
0116  0044        XNM68   EQU  68
0117  0045        XNM69   EQU  69
0118  0046        XNM70   EQU  70
0119  0047        XNM71   EQU  71
0120  0048        XNM72   EQU  72
0121  0049        XNM73   EQU  73
0122  004A        XNM74   EQU  74
0123  004B        XNM75   EQU  75
0124  004C        XNM76   EQU  76
0125  004D        XNM77   EQU  77
0126  004E        XNM78   EQU  78
0127  004F        XNM79   EQU  79
0128              *
0129  0050        H0      EQU  80
0130  0051        H1      EQU  81
0131  0052        H2      EQU  82
0132  0053        H3      EQU  83
0133  0054        H4      EQU  84
0134  0055        H5      EQU  85
0135  0056        H6      EQU  86
0136  0057        H7      EQU  87
0137  0058        H8      EQU  88
0138  0059        H9      EQU  89
0139  005A        H10     EQU  90
0140  005B        H11     EQU  91
0141  005C        H12     EQU  92
0142  005D        H13     EQU  93
0143  005E        H14     EQU  94
0144  005F        H15     EQU  95
0145  0060        H16     EQU  96
0146  0061        H17     EQU  97
0147  0062        H18     EQU  98
0148  0063        H19     EQU  99
0149  0064        H20     EQU  100
0150  0065        H21     EQU  101
0151  0066        H22     EQU  102
0152  0067        H23     EQU  103
0153  0068        H24     EQU  104
0154  0069        H25     EQU  105
0155  006A        H26     EQU  106
0156  006B        H27     EQU  107
0157  006C        H28     EQU  108
0158  006D        H29     EQU  109
0159  006E        H30     EQU  110
0160  006F        H31     EQU  111
0161  0070        H32     EQU  112
0162  0071        H33     EQU  113
0163  0072        H34     EQU  114
0164  0073        H35     EQU  115
0165  0074        H36     EQU  116
0166  0075        H37     EQU  117
0167  0076        H38     EQU  118
0168  0077        H39     EQU  119
0169              *
0170  0078        MODE    EQU  120
0171  0079        CLOCK   EQU  121
```

```
0172  007A        YN      EQU  122
0173  007B        ONE     EQU  123
0174              *
0175  0000                AORG 0
0176              *
0177  0000 F900           B    START
      0001 002C
0178              * COEFFICIENTS ARE INITIALLY           *
0179              * STORED IN PROGRAM MEMORY             *
0180              *
0181              * DUE TO THE SYMMETRY OF THE IMPULSE RESPONSE *
0182              * ONLY HALF OF THE SAMPLES OF THE IMPULSE     *
0183              * RESPONSE ARE STORED. THIS MEANS THAT        *
0184              * h(N-1-n) = h(n).                     *
0185              *
0186              *
0187  0002 FFDC   CH0     DATA >FFDC  *  -0.107251E-02 *
0188  0003 001F   CH1     DATA >001F  *   0.973976E-03 *
0189  0004 0051   CH2     DATA >0051  *   0.249065E-02 *
0190  0005 FFE9   CH3     DATA >FFE9  *  -0.675043E-03 *
0191  0006 FFE6   CH4     DATA >FFE6  *  -0.771385E-03 *
0192  0007 FFBA   CH5     DATA >FFBA  *  -0.212256E-02 *
0193  0008 FFB4   CH6     DATA >FFB4  *  -0.229530E-02 *
0194  0009 004B   CH7     DATA >004B  *   0.231021E-02 *
0195  000A FFF9   CH8     DATA >FFF9  *  -0.199002E-03 *
0196  000B 0069   CH9     DATA >0069  *   0.322896E-02 *
0197  000C 00A2   CH10    DATA >00A2  *   0.496452E-02 *
0198  000D FF6F   CH11    DATA >FF6F  *  -0.440419E-02 *
0199  000E FFFE   CH12    DATA >FFFE  *  -0.314831E-04 *
0200  000F FF70   CH13    DATA >FF70  *  -0.438169E-02 *
0201  0010 FEF4   CH14    DATA >FEF4  *  -0.815474E-02 *
0202  0011 00CB   CH15    DATA >00CB  *   0.621682E-02 *
0203  0012 000B   CH16    DATA >000B  *   0.342216E-03 *
0204  0013 00E6   CH17    DATA >00E6  *   0.704627E-02 *
0205  0014 0187   CH18    DATA >0187  *   0.119391E-01 *
0206  0015 FEE5   CH19    DATA >FEE5  *  -0.860811E-02 *
0207  0016 000B   CH20    DATA >000B  *   0.346738E-03 *
0208  0017 FE7F   CH21    DATA >FE7F  *  -0.117293E-01 *
0209  0018 FDBF   CH22    DATA >FDBF  *  -0.175964E-01 *
0210  0019 0192   CH23    DATA >0192  *   0.122947E-01 *
0211  001A FFB5   CH24    DATA >FFB5  *  -0.227426E-02 *
0212  001B 026A   CH25    DATA >026A  *   0.188796E-01 *
0213  001C 0368   CH26    DATA >0368  *   0.266148E-01 *
0214  001D FDC2   CH27    DATA >FDC2  *  -0.175126E-01 *
0215  001E 00C0   CH28    DATA >00C0  *   0.586574E-02 *
0216  001F FC0A   CH29    DATA >FC0A  *  -0.309240E-01 *
0217  0020 FAA3   CH30    DATA >FAA3  *  -0.418954E-01 *
0218  0021 0347   CH31    DATA >0347  *   0.256315E-01 *
0219  0022 FE3D   CH32    DATA >FE3D  *  -0.137498E-01 *
0220  0023 0747   CH33    DATA >0747  *   0.568720E-01 *
0221  0024 09BB   CH34    DATA >09BB  *   0.760286E-01 *
0222  0025 FA3D   CH35    DATA >FA3D  *  -0.450011E-01 *
0223  0026 052B   CH36    DATA >052B  *   0.403853E-01 *
0224  0027 EB59   CH37    DATA >EB59  *  -0.161139E+00 *
0225  0028 DC2A   CH38    DATA >DC2A  *  -0.279963E+00 *
0226  0029 2D57   CH39    DATA >2D57  *   0.352454E+00 *
```

3. Implementation of FIR/IIR Filters with the TMS32010/TMS32020

```
0228  002A 000A   MD      DATA    >000A      * SAMPLING RATE OF 10 KHZ *
0229  002B 01F3   SMP     DATA    >01F3
0230                      *
0231  002C 6E00   START   LDPK    0
0232                      *
0233  002D 7E01           LACK    1
0234  002E 507B           SACL    ONE        * CONTENT OF ONE IS 1 *
0235                      *
0236  002F 7079           LARK    AR0,CLOCK  * THIS SECTION OF CODE LOADS   *
0237  0030 7129           LARK    AR1,>29    * THE FILTER COEFFICIENTS AND  *
0238  0031 7E2B           LACK    SMP        * OTHER VALUES FROM PROGRAM    *
0239  0032 6880           LARP    AR0        * MEMORY TO DATA MEMORY        *
0240  0033 6791   LOAD    TBLR    *-,AR1
0241  0034 107B           SUB     ONE
0242  0035 F400           BANZ    LOAD
      0036 0032
0243                      *
0244  0037 4878           OUT     MODE,PA0   * INITIALIZATION OF ANALOG *
0245  0038 4979           OUT     CLOCK,PA1  * INTERFACE BOARD          *
0246                      *
0247  0039 F600   WAIT    BIOZ    NXTPT      * BIO PIN GOES LOW WHEN A  *
      003A 003D
0248  003B F900           B       WAIT       * NEW SAMPLE IS AVAILABLE  *
      003C 0039
0249                      *
0250  003D 4200   NXTPT   IN      XN,PA2     * BRING IN THE NEW SAMPLE XN *
0251                      *
0252  003E 7F89           ZAC
0253                      *
0254  003F 6A4F           LT      XNM79      * DUE TO SYMMETRY h(0) = h(79) *
0255  0040 6D50           MPY     H0         * x(n-79) * h(79) *
0256                      *
0257  0041 6B4E           LTD     XNM78
0258  0042 6D51           MPY     H1         * h(1) = h(78) *
0259                      *
0260  0043 6B4D           LTD     XNM77
0261  0044 6D52           MPY     H2
0262                      *
0263  0045 6B4C           LTD     XNM76
0264  0046 6D53           MPY     H3
0265                      *
0266  0047 6B4B           LTD     XNM75
0267  0048 6D54           MPY     H4
0268                      *
0269  0049 6B4A           LTD     XNM74
0270  004A 6D55           MPY     H5
0271                      *
0272  004B 6B49           LTD     XNM73
0273  004C 6D56           MPY     H6
0274                      *
0275  004D 6B48           LTD     XNM72
0276  004E 6D57           MPY     H7
0277                      *
0278  004F 6B47           LTD     XNM71
0279  0050 6D58           MPY     H8
0280                      *
0281  0051 6B46           LTD     XNM70
```

```
0282  0052 6D59           MPY     H9
0283                      *
0284  0053 6B45           LTD     XNM69
0285  0054 6D5A           MPY     H10
0286                      *
0287  0055 6B44           LTD     XNM68
0288  0056 6D5B           MPY     H11
0289                      *
0290  0057 6B43           LTD     XNM67
0291  0058 6D5C           MPY     H12
0292                      *
0293  0059 6B42           LTD     XNM66
0294  005A 6D5D           MPY     H13
0295                      *
0296  005B 6B41           LTD     XNM65
0297  005C 6D5E           MPY     H14
0298                      *
0299  005D 6B40           LTD     XNM64
0300  005E 6D5F           MPY     H15
0301                      *
0302  005F 6B3F           LTD     XNM63
0303  0060 6D60           MPY     H16
0304                      *
0305  0061 6B3E           LTD     XNM62
0306  0062 6D61           MPY     H17
0307                      *
0308  0063 6B3D           LTD     XNM61
0309  0064 6D62           MPY     H18
0310                      *
0311  0065 6B3C           LTD     XNM60
0312  0066 6D63           MPY     H19
0313                      *
0314  0067 6B3B           LTD     XNM59
0315  0068 6D64           MPY     H20
0316                      *
0317  0069 6B3A           LTD     XNM58
0318  006A 6D65           MPY     H21
0319                      *
0320  006B 6B39           LTD     XNM57
0321  006C 6D66           MPY     H22
0322                      *
0323  006D 6B38           LTD     XNM56
0324  006E 6D67           MPY     H23
0325                      *
0326  006F 6B37           LTD     XNM55
0327  0070 6D68           MPY     H24
0328                      *
0329  0071 6B36           LTD     XNM54
0330  0072 6D69           MPY     H25
0331                      *
0332  0073 6B35           LTD     XNM53
0333  0074 6D6A           MPY     H26
0334                      *
0335  0075 6B34           LTD     XNM52
0336  0076 6D6B           MPY     H27
0337                      *
0338  0077 6B33           LTD     XNM51
```

3. Implementation of FIR/IIR Filters with the TMS32010/TMS32020

```
0339 0078 6D6C      MPY  H28
0340                *
0341 0079 6B32      LTD  XNM50
0342 007A 6D6D      MPY  H29
0343                *
0344 007B 6B31      LTD  XNM49
0345 007C 6D6E      MPY  H30
0346                *
0347 007D 6B30      LTD  XNM48
0348 007E 6D6F      MPY  H31
0349                *
0350 007F 6B2F      LTD  XNM47
0351 0080 6D70      MPY  H32
0352                *
0353 0081 6B2E      LTD  XNM46
0354 0082 6D71      MPY  H33
0355                *
0356 0083 6B2D      LTD  XNM45
0357 0084 6D72      MPY  H34
0358                *
0359 0085 6B2C      LTD  XNM44
0360 0086 6D73      MPY  H35
0361                *
0362 0087 6B2B      LTD  XNM43
0363 0088 6D74      MPY  H36
0364                *
0365 0089 6B2A      LTD  XNM42
0366 008A 6D75      MPY  H37
0367                *
0368 008B 6B29      LTD  XNM41
0369 008C 6D76      MPY  H38
0370                *
0371 008D 6B28      LTD  XNM40
0372 008E 6D77      MPY  H39
0373                *
0374 008F 6B27      LTD  XNM39
0375 0090 6D77      MPY  H39
0376                *
0377 0091 6B26      LTD  XNM38
0378 0092 6D76      MPY  H38
0379                *
0380 0093 6B25      LTD  XNM37
0381 0094 6D75      MPY  H37
0382                *
0383 0095 6B24      LTD  XNM36
0384 0096 6D74      MPY  H36
0385                *
0386 0097 6B23      LTD  XNM35
0387 0098 6D73      MPY  H35
0388                *
0389 0099 6B22      LTD  XNM34
0390 009A 6D72      MPY  H34
0391                *
0392 009B 6B21      LTD  XNM33
0393 009C 6D71      MPY  H33
0394                *
0395 009D 6B20      LTD  XNM32
```

```
0396 009E 6D70      MPY  H32
0397                *
0398 009F 6B1F      LTD  XNM31
0399 00A0 6D6F      MPY  H31
0400                *
0401 00A1 6B1E      LTD  XNM30
0402 00A2 6D6E      MPY  H30
0403                *
0404 00A3 6B1D      LTD  XNM29
0405 00A4 6D6D      MPY  H29
0406                *
0407 00A5 6B1C      LTD  XNM28
0408 00A6 6D6C      MPY  H28
0409                *
0410 00A7 6B1B      LTD  XNM27
0411 00A8 6D6B      MPY  H27
0412                *
0413 00A9 6B1A      LTD  XNM26
0414 00AA 6D6A      MPY  H26
0415                *
0416 00AB 6B19      LTD  XNM25
0417 00AC 6D69      MPY  H25
0418                *
0419 00AD 6B18      LTD  XNM24
0420 00AE 6D68      MPY  H24
0421                *
0422 00AF 6B17      LTD  XNM23
0423 00B0 6D67      MPY  H23
0424                *
0425 00B1 6B16      LTD  XNM22
0426 00B2 6D66      MPY  H22
0427                *
0428 00B3 6B15      LTD  XNM21
0429 00B4 6D65      MPY  H21
0430                *
0431 00B5 6B14      LTD  XNM20
0432 00B6 6D64      MPY  H20
0433                *
0434 00B7 6B13      LTD  XNM19
0435 00B8 6D63      MPY  H19
0436                *
0437 00B9 6B12      LTD  XNM18
0438 00BA 6D62      MPY  H18
0439                *
0440 00BB 6B11      LTD  XNM17
0441 00BC 6D61      MPY  H17
0442                *
0443 00BD 6B10      LTD  XNM16
0444 00BE 6D60      MPY  H16
0445                *
0446 00BF 6B0F      LTD  XNM15
0447 00C0 6D5F      MPY  H15
0448                *
0449 00C1 6B0E      LTD  XNM14
0450 00C2 6D5E      MPY  H14
0451                *
0452 00C3 6B0D      LTD  XNM13
```

```
0453  00C4 6D5D          MPY H13
0454             *
0455  00C5 6B0C          LTD XNM12
0456  00C6 6D5C          MPY H12
0457             *
0458  00C7 680B          LTD XNM11
0459  00C8 6D5B          MPY H11
0460             *
0461  00C9 6B0A          LTD XNM10
0462  00CA 6D5A          MPY H10
0463             *
0464  00CB 6B09          LTD XNM9
0465  00CC 6D59          MPY H9
0466             *
0467  00CD 6B08          LTD XNM8
0468  00CE 6D58          MPY H8
0469             *
0470  00CF 6B07          LTD XNM7
0471  00D0 6D57          MPY H7
0472             *
0473  00D1 6B06          LTD XNM6
0474  00D2 6D56          MPY H6
0475             *
0476  00D3 6B05          LTD XNM5
0477  00D4 6D55          MPY H5
0478             *
0479  00D5 6B04          LTD XNM4
0480  00D6 6D54          MPY H4
0481             *
0482  00D7 6B03          LTD XNM3
0483  00D8 6D53          MPY H3
0484             *
0485  00D9 6B02          LTD XNM2
0486  00DA 6D52          MPY H2
0487             *
0488  00DB 6B01          LTD XNM1
0489  00DC 6D51          MPY H1
0490             *
0491  00DD 6B00          LTD XN
0492  00DE 6D50          MPY H0
0493             *
0494  00DF 7F8F          APAC
0495             *
0496  00E0 597A          SACH YN,1
0497             *
0498  00E1 4A7A          OUT YN,PA2    * OUTPUT THE FILTER RESPONSE y(n) *
0499             *
0500  00E2 F900          B WAIT        * GO GET THE NEXT POINT *
      00E3 0039
0501
0502
NO ERRORS, NO WARNINGS
```

APPENDIX B
LENGTH-60 FIR DIFFERENTIATOR

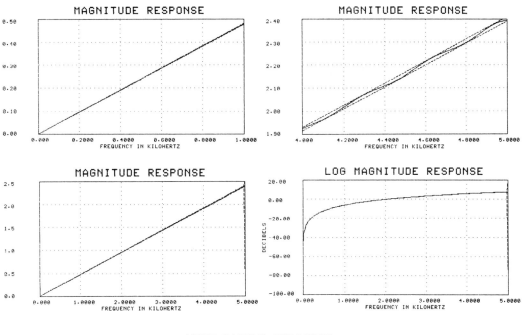

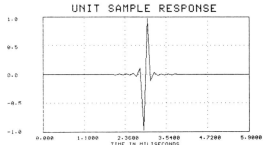

```
0001   *****************************************
0002   *
0003   *                FIR FILTER
0004   *         LENGTH-60 DIFFERENTIATOR
0005   *
0006   *         FILTER CHARACTERISTICS
0007   *
0008   *   SAMPLING FREQUENCY = 10 KHZ
0009   *
0010   *   LOWER BAND EDGE        0.0000
0011   *   UPPER BAND EDGE        5.0000
0012   *   DESIRED SLOPE          0.4800
0013   *   MAX % DEVIATION        0.3171
0014   *
0015   *            FILTER STRUCTURE
0016   *
```

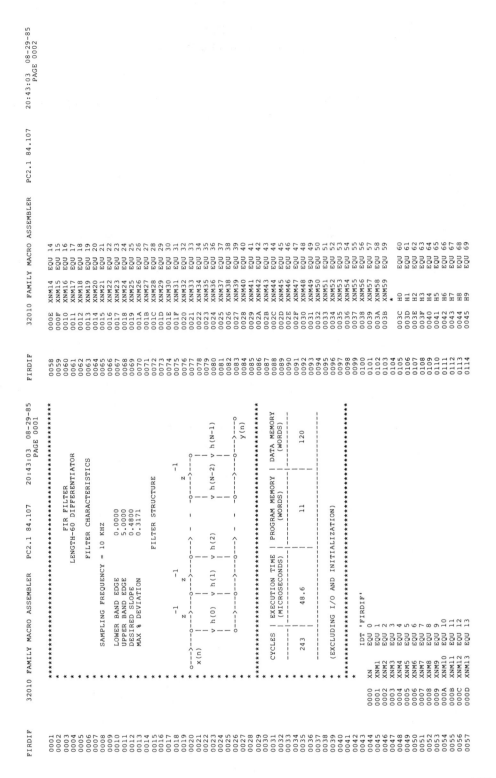

```
0029   *****************************************
0030   *
0031   * CYCLES | EXECUTION TIME | PROGRAM MEMORY | DATA MEMORY
0032   *        | (MICROSECONDS) |    (WORDS)     |   (WORDS)
0033   *
0034   *  243   |     48.6       |      11        |    120
0035   *
0036   *
0037   *
0038   *
0039   *
0040   * (EXCLUDING I/O AND INITIALIZATION)
0041   *
0042   *****************************************
0043   *
0044          IDT 'FIRDIF'
0045   0000   XN      EQU  0
0046   0001   XNM1    EQU  1
0047   0002   XNM2    EQU  2
0048   0003   XNM3    EQU  3
0049   0004   XNM4    EQU  4
0050   0005   XNM5    EQU  5
0051   0006   XNM6    EQU  6
0052   0007   XNM7    EQU  7
0053   0008   XNM8    EQU  8
0054   0009   XNM9    EQU  9
0055   000A   XNM10   EQU  10
0056   000B   XNM11   EQU  11
0057   000C   XNM12   EQU  12
       000D   XNM13   EQU  13
```

```
0058   000E   XNM14   EQU  14
0059   000F   XNM15   EQU  15
0060   0010   XNM16   EQU  16
0061   0011   XNM17   EQU  17
0062   0012   XNM18   EQU  18
0063   0013   XNM19   EQU  19
0064   0014   XNM20   EQU  20
0065   0015   XNM21   EQU  21
0066   0016   XNM22   EQU  22
0067   0017   XNM23   EQU  23
0068   0018   XNM24   EQU  24
0069   0019   XNM25   EQU  25
0070   001A   XNM26   EQU  26
0071   001B   XNM27   EQU  27
0072   001C   XNM28   EQU  28
0073   001D   XNM29   EQU  29
0074   001E   XNM30   EQU  30
0075   001F   XNM31   EQU  31
0076   0020   XNM32   EQU  32
0077   0021   XNM33   EQU  33
0078   0022   XNM34   EQU  34
0079   0023   XNM35   EQU  35
0080   0024   XNM36   EQU  36
0081   0025   XNM37   EQU  37
0082   0026   XNM38   EQU  38
0083   0027   XNM39   EQU  39
0084   0028   XNM40   EQU  40
0085   0029   XNM41   EQU  41
0086   002A   XNM42   EQU  42
0087   002B   XNM43   EQU  43
0088   002C   XNM44   EQU  44
0089   002D   XNM45   EQU  45
0090   002E   XNM46   EQU  46
0091   002F   XNM47   EQU  47
0092   0030   XNM48   EQU  48
0093   0031   XNM49   EQU  49
0094   0032   XNM50   EQU  50
0095   0033   XNM51   EQU  51
0096   0034   XNM52   EQU  52
0097   0035   XNM53   EQU  53
0098   0036   XNM54   EQU  54
0099   0037   XNM55   EQU  55
0100   0038   XNM56   EQU  56
0101   0039   XNM57   EQU  57
0102   003A   XNM58   EQU  58
0103   003B   XNM59   EQU  59
0104          *
0105   003C   H0      EQU  60
0106   003D   H1      EQU  61
0107   003E   H2      EQU  62
0108   003F   H3      EQU  63
0109   0040   H4      EQU  64
0110   0041   H5      EQU  65
0111   0042   H6      EQU  66
0112   0043   H7      EQU  67
0113   0044   H8      EQU  68
0114   0045   H9      EQU  69
```

3. Implementation of FIR/IIR Filters with the TMS32010/TMS32020

```
0115   0046            H10      EQU   70
0116   0047            H11      EQU   71
0117   0048            H12      EQU   72
0118   0049            H13      EQU   73
0119   004A            H14      EQU   74
0120   004B            H15      EQU   75
0121   004C            H16      EQU   76
0122   004D            H17      EQU   77
0123   004E            H18      EQU   78
0124   004F            H19      EQU   79
0125   0050            H20      EQU   80
0126   0051            H21      EQU   81
0127   0052            H22      EQU   82
0128   0053            H23      EQU   83
0129   0054            H24      EQU   84
0130   0055            H25      EQU   85
0131   0056            H26      EQU   86
0132   0057            H27      EQU   87
0133   0058            H28      EQU   88
0134   0059            H29      EQU   89
0135   005A            H30      EQU   90
0136   005B            H31      EQU   91
0137   005C            H32      EQU   92
0138   005D            H33      EQU   93
0139   005E            H34      EQU   94
0140   005F            H35      EQU   95
0141   0060            H36      EQU   96
0142   0061            H37      EQU   97
0143   0062            H38      EQU   98
0144   0063            H39      EQU   99
0145   0064            H40      EQU   100
0146   0065            H41      EQU   101
0147   0066            H42      EQU   102
0148   0067            H43      EQU   103
0149   0068            H44      EQU   104
0150   0069            H45      EQU   105
0151   006A            H46      EQU   106
0152   006B            H47      EQU   107
0153   006C            H48      EQU   108
0154   006D            H49      EQU   109
0155   006E            H50      EQU   110
0156   006F            H51      EQU   111
0157   0070            H52      EQU   112
0158   0071            H53      EQU   113
0159   0072            H54      EQU   114
0160   0073            H55      EQU   115
0161   0074            H56      EQU   116
0162   0075            H57      EQU   117
0163   0076            H58      EQU   118
0164   0077            H59      EQU   119
0165                   *
0166   0078            MODE     EQU   120
0167   0079            CLOCK    EQU   121
0168   007A            YN       EQU   122
0169   007B            ONE      EQU   123
0170                   *
0171   0000            AORG  0
```

```
0172   0000   F900            B     START
0173   0001   0040
0174                   *
0175                   *  COEFFICIENTS ARE INITIALLY *
0176                   *  STORED IN PROGRAM MEMORY  *
0177                   *
0178   0002   0030     CH0      DATA  >0030   *  0.146547E-02 *
0179   0003   FFC2     CH1      DATA  >FFC2   * -0.186717E-02 *
0180   0004   0015     CH2      DATA  >0015   *  0.670857E-03 *
0181   0005   FFEF     CH3      DATA  >FFEF   * -0.507893E-03 *
0182   0006   000F     CH4      DATA  >000F   *  0.479907E-03 *
0183   0007   FFF0     CH5      DATA  >FFF0   * -0.482679E-03 *
0184   0008   0010     CH6      DATA  >0010   *  0.505055E-03 *
0185   0009   FFEE     CH7      DATA  >FFEE   * -0.536998E-03 *
0186   000A   0012     CH8      DATA  >0012   *  0.576256E-03 *
0187   000B   FFEB     CH9      DATA  >FFEB   * -0.624602E-03 *
0188   000C   0016     CH10     DATA  >0016   *  0.681939E-03 *
0189   000D   FFE7     CH11     DATA  >FFE7   * -0.750338E-03 *
0190   000E   001B     CH12     DATA  >001B   *  0.831878E-03 *
0191   000F   FFE1     CH13     DATA  >FFE1   * -0.929373E-03 *
0192   0010   0022     CH14     DATA  >0022   *  0.104702E-02 *
0193   0011   FFD8     CH15     DATA  >FFD8   * -0.119041E-02 *
0194   0012   002C     CH16     DATA  >002C   *  0.136731E-02 *
0195   0013   FFCB     CH17     DATA  >FFCB   * -0.158880E-02 *
0196   0014   003D     CH18     DATA  >003D   *  0.187070E-02 *
0197   0015   FFB6     CH19     DATA  >FFB6   * -0.223732E-02 *
0198   0016   0059     CH20     DATA  >0059   *  0.275799E-02 *
0199   0017   FF90     CH21     DATA  >FF90   * -0.339682E-02 *
0200   0018   008E     CH22     DATA  >008E   *  0.434422E-02 *
0201   0019   FF42     CH23     DATA  >FF42   * -0.578642E-02 *
0202   001A   0108     CH24     DATA  >0108   *  0.806880E-02 *
0203   001B   FE75     CH25     DATA  >FE75   * -0.120382E-01 *
0204   001C   028B     CH26     DATA  >028B   *  0.198777E-01 *
0205   001D   FB04     CH27     DATA  >FB04   * -0.389339E-01 *
0206   001E   0DD6     CH28     DATA  >0DD6   *  0.108105E+00 *
0207   001F   837E     CH29     DATA  >837E   * -0.972714E+00 *
0208   0020   7C81     CH30     DATA  >7C81   * -CH29 *
0209   0021   F229     CH31     DATA  >F229   * -CH28 *
0210   0022   04FB     CH32     DATA  >04FB   * -CH27 *
0211   0023   FD74     CH33     DATA  >FD74   * -CH26 *
0212   0024   018A     CH34     DATA  >018A   * -CH25 *
0213   0025   FEF7     CH35     DATA  >FEF7   * -CH24 *
0214   0026   00BD     CH36     DATA  >00BD   * -CH23 *
0215   0027   FF71     CH37     DATA  >FF71   * -CH22 *
0216   0028   006F     CH38     DATA  >006F   * -CH21 *
0217   0029   FFA6     CH39     DATA  >FFA6   * -CH20 *
0218   002A   0049     CH40     DATA  >0049   * -CH19 *
0219   002B   FFC2     CH41     DATA  >FFC2   * -CH18 *
0220   002C   0034     CH42     DATA  >0034   * -CH17 *
0221   002D   FFD3     CH43     DATA  >FFD3   * -CH16 *
0222   002E   0027     CH44     DATA  >0027   * -CH15 *
0223   002F   FFDD     CH45     DATA  >FFDD   * -CH14 *
0224   0030   001E     CH46     DATA  >001E   * -CH13 *
0225   0031   FFE4     CH47     DATA  >FFE4   * -CH12 *
0226   0032   0018     CH48     DATA  >0018   * -CH11 *
0227   0033   FFE9     CH49     DATA  >FFE9   * -CH10 *
```

3. Implementation of FIR/IIR Filters with the TMS32010/TMS32020

```
0228 0034 0014  CH50   DATA  >0014      * -CH9 *
0229 0035 FFED  CH51   DATA  >FFED      * -CH8 *
0230 0036 0011  CH52   DATA  >0011      * -CH7 *
0231 0037 FFEF  CH53   DATA  >FFEF      * -CH6 *
0232 0038 000F  CH54   DATA  >000F      * -CH5 *
0233 0039 FFF0  CH55   DATA  >FFF0      * -CH4 *
0234 003A 0010  CH56   DATA  >0010      * -CH3 *
0235 003B FFEA  CH57   DATA  >FFEA      * -CH2 *
0236 003C 003D  CH58   DATA  >003D      * -CH1 *
0237 003D FFCF  CH59   DATA  >FFCF      * -CH0 *
0238                   *
0239 003E 000A  MD     DATA  >000A      * SAMPLING RATE OF 10 KHZ *
0240 003F 01F3  SMP    DATA  499
0241                   *
0242 0040 6E00  START  LDPK 0
0243                   *
0244 0041 7E01         LACK 1
0245 0042 507B         SACL ONE         * CONTENT OF ONE IS 1 *
0246                   *
0247 0043 7079         LARK AR0,CLOCK   * THIS SECTION OF CODE LOADS *
0248 0044 713C         LARK AR1,60      * THE FILTER COEFFICIENTS AND *
0249 0045 7E3F         LACK SMP         * OTHER VALUES FROM PROGRAM *
0250 0046 6880         LARP AR0         * MEMORY TO DATA MEMORY *
0251 0047 6791  LOAD   TBLR *-,AR1
0252 0048 107B         SUB ONE
0253 0049 F400         BANZ LOAD
     004A 0046
0254                   *
0255 004B 4878         OUT MODE,PA0     * INITIALIZATION OF ANALOG *
0256 004C 4979         OUT CLOCK,PA1    * INTERFACE BOARD *
0257                   *
0258 004D 6880         LARP AR0         * SET ARP TO AR0 *
0259                   *
0260 004E F600  WAIT   BIOZ NXTPT       * BIO PIN GOES LOW WHEN A *
     004F 0052
0261 0050 F900         B WAIT           * NEW SAMPLE IS AVAILABLE *
     0051 004E
0262                   *
0263 0052 4200  NXTPT  IN XN,PA2        * BRING IN THE NEW SAMPLE XN *
0264                   *
0265 0053 703B         LARK AR0,XNM59   * AR0 POINTS TO THE INPUT SEQUENCE *
0266 0054 7177         LARK AR1,H59     * AR1 POINTS TO THE IMPULSE RESPONSE *
0267                   *
0268 0055 7F89         ZAC
0269                   *
0270 0056 6A91         LT *-,AR1
0271 0057 6D90         MPY *,AR0
0272                   *
0273 0058 6B81  LOOP   LTD *,AR1
0274 0059 6D90         MPY *-,AR0
0275                   *
0276 005A F400         BANZ LOOP
     005B 0058
0277                   *
0278 005C 7F8F         APAC             * ACCUMULATE LAST MULTIPLY *
0279                   *
0280 005D 597A         SACH YN,1
```

```
0281                   *
0282 005E 4A7A         OUT YN,PA2       * OUTPUT THE FILTER RESPONSE y(n) *
0283                   *
0284 005F F900         B WAIT           * GO GET THE NEXT POINT *
     0060 004E
0285                   *
0286                   END
NO ERRORS, NO WARNINGS
```

3. Implementation of FIR/IIR Filters with the TMS32010/TMS32020

APPENDIX C
FOURTH-ORDER LOWPASS IIR FILTERS

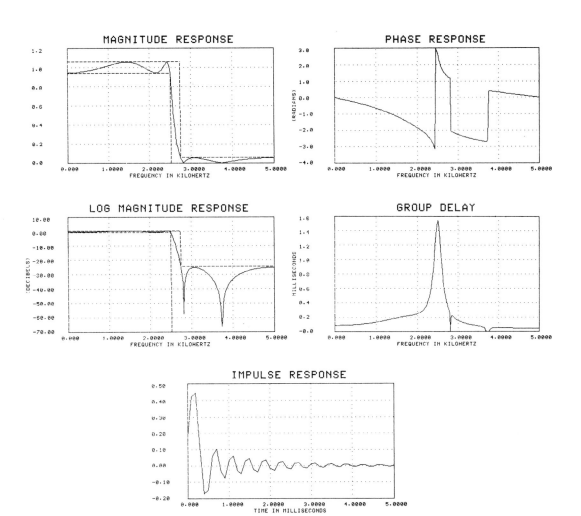

```
*********************************************************************
*                                                                   *
*                     FOURTH-ORDER IIR                              *
*                  ELLIPTIC LOWPASS FILTER                         *
*                                                                   *
*                  DIRECT-FORM II STRUCTURE                         *
*                                                                   *
*                  FILTER CHARACTERISTICS                          *
*                                                                   *
*     SAMPLING FREQUENCY = 10 KHZ                                  *
*                                                                   *
*                         BAND 1        BAND 2                      *
*                                                                   *
*     LOWER BAND EDGE     0.00000       2.75000                    *
*     UPPER BAND EDGE     2.50000       5.00000                    *
*     NOMINAL GAIN        1.00000       0.00000                    *
*     NOMINAL RIPPLE      0.06000       0.06000                    *
*     MAXIMUM RIPPLE      0.05617       0.05514                    *
*     RIPPLE IN DB        0.47469      -25.17089                   *
*                                                                   *
*                  FILTER STRUCTURE                                *
*                                                                   *
*                                  b                                *
*                                   0                               *
*           o---->--o---->--o---->--o---->--o---->--o  y(n)         *
*          x(n)     ( a      v -1   b  )                            *
*                    (  1     z      1 )                            *
*           o--->--o--<---o---->--o---->--o                         *
*                    ( a      v -1   b  )                           *
*                    (  2     z      2 )                            *
*                    o--<---o---->--o---->--o                       *
*                    ( a      v -1   b  )                           *
*                    (  3     z      3 )                            *
*                    o--<---o---->--o---->--o                       *
*                    ( a      v -1   b  )                           *
*                    (  4     z      4 )                            *
*                    o--<---o---->--o---->--o                       *
*                                                                   *
*********************************************************************
```

CYCLES	EXECUTION TIME (MICROSECONDS)	PROGRAM MEMORY (WORDS)	DATA MEMORY (WORDS)
24	4.8	24	16

```
* (EXCLUDING I/O AND INITIALIZATION)
```

```
0058
0059
0060
0061  0000                    *
0062  0000                    *
0063                          *
                              IDT 'IIR4DIR'
0062  0000        DN     EQU 0
0063  0001        DNM1   EQU 1
0064  0002        DNM2   EQU 2
0065  0003        DNM3   EQU 3
0066  0004        DNM4   EQU 4
0067
0068  0005        A1     EQU 5
0069  0006        A2     EQU 6
0070  0007        A3     EQU 7
0071  0008        A4     EQU 8
0072                    *
0073  0009        B0     EQU 9
0074  000A        B1     EQU 10
0075  000B        B2     EQU 11
0076  000C        B3     EQU 12
0077  000D        B4     EQU 13
0078                    *
0079  000E        MODE   EQU 14
0080  000F        CLOCK  EQU 15
0081  0010        YN     EQU 16
0082  0011        XN     EQU 17
0083  0012        ONE    EQU 18
0084                    *
0085  0000                AORG 0
0086
0087  0000 F900            B START
      0001 000D
0088                    *
0089                    *  COEFFICIENTS ARE INITIALLY *
0090                    *  STORED IN PROGRAM MEMORY *
0091                    *
0092  0002 3845   CA1    DATA >3845   * 0.4396070 *
0093                                  * A2 = -1.172416.  THE -1.0 TERM *
0094                                  * IS IMPLEMENTED WITH A SUB AND *
0095  0003 E9EE   CA2    DATA >E9EE   * A2 CONTAINS -0.172416 = >E9EE. *
0096
0097
0098  0004 3167   CA3    DATA >3167   * 0.3859772 *
0099  0005 DDC1   CA4    DATA >DDC1   * -0.2675277 *
0100
0101  0006 17FA   CB0    DATA >17FA   * 0.1873279 *
0102  0007 2B02   CB1    DATA >2B02   * 0.3360168 *
0103  0008 3D1F   CB2    DATA >3D1F   * 0.4775291 *
0104  0009 2AFF   CB3    DATA >2AFF   * 0.3359135 *
0105  000A 17F3   CB4    DATA >17F3   * 0.1871291 *
0106
0107  000B 000A   MD     DATA >000A   * SAMPLING RATE OF 10 KHZ *
0108  000C 01F3   SMP    DATA 499
0109
0110  000D 6E00   START  LDPK 0
0111
0112  000E 7E01          LACK 1
0113  000F 5012          SACL ONE     * CONTENT OF ONE IS 1 *
```

3. Implementation of FIR/IIR Filters with the TMS32010/TMS32020

```
0114                        *
0115  0010  700F    LARK AR0,CLOCK    * THIS SECTION OF CODE LOADS    *
0116  0011  710A    LARK AR1,10       * THE FILTER COEFFICIENTS AND   *
0117  0012  7E0C    LACK SMP          * OTHER VALUES FROM PROGRAM     *
0118  0013  6880 LOAD LARP AR0        * MEMORY TO DATA MEMORY         *
0119  0014  6791    TBLR *-,AR1
0120  0015  1012    SUB ONE
0121  0016  F400    BANZ LOAD
      0017  0013
0122                *
0123  0018  7F89    ZAC               * THIS SECTION SETS THE    *
0124  0019  5000    SACL DN           * INITIAL STATE OF THE     *
0125  001A  5001    SACL DNM1         * FILTER TO ZERO           *
0126  001B  5002    SACL DNM2
0127  001C  5003    SACL DNM3
0128  001D  5004    SACL DNM4
0129                *
0130  001E  480E    OUT MODE,PA0      * INITIALIZATION OF ANALOG *
0131  001F  490F    OUT CLOCK,PA1     * INTERFACE BOARD          *
0132                *
0133  0020  F600 WAIT BIOZ NXTPT      * BIO PIN GOES LOW WHEN A  *
      0021  0024
0134  0022  F900    B WAIT            * NEW SAMPLE IS AVAILABLE   *
      0023  0020
0135                *
0136  0024  4211 NXTPT IN XN,PA2      * BRING IN THE NEW SAMPLE XN *
0137                *
0138                * IMPLEMENTATION OF SYSTEM POLES *
0139  0025  2F11    LAC XN,15
0140                *
0141  0026  6A01    LT DNM1           * d(n-1) * a  *
0142  0027  6D05    MPY A1            *           1
0143                *
0144  0028  6C02    LTA DNM2          * THIS SECTION IS EQUIVALENT TO *
0145  0029  6D06    MPY A2            * -1.172416 * DNM2. THE -1.0    *
0146  002A  1F02    SUB DNM2,15       * TERM IS IMPLEMENTED WITH THE  *
0147                *                 * SUB DNM2,15 AND A2 CONTAINS   *
0148                *                 * -0.172416 = >E9EE.            *
0149                *
0150  002B  6C03    LTA DNM3
0151  002C  6D07    MPY A3
0152                *
0153  002D  6C04    LTA DNM4
0154  002E  6D08    MPY A4
0155                *
0156  002F  7F8F    APAC
0157                *
0158  0030  5900    SACH DN,1
0159                *
0160  0031  7F89    ZAC
0161                *
0162  0032  6D0D    MPY B4            * IMPLEMENTATION OF SYSTEM ZEROES *
0163                *
0164  0033  6803    LTD DNM3          * d(n-3) * b  *
0165  0034  6D0C    MPY B3            *           3
0166                *
0167  0035  6B02    LTD DNM2
```

```
0168  0036  6D0B    MPY B2
0169                *
0170  0037  6801    LTD DNM1
0171  0038  6D0A    MPY B1
0172                *
0173  0039  6B00    LTD DN
0174  003A  6D09    MPY B0
0175                *
0176  003B  7F8F    APAC              * FINISHED FILTER *
0177                *
0178  003C  5910    SACH YN,1
0179                *
0180  003D  4A10    OUT YN,PA2        * OUTPUT THE FILTER RESPONSE y(n) *
0181                *
0182  003E  F900    B WAIT            * GO GET THE NEXT POINT *
      003F  0020
0183                *
0184              END

NO ERRORS, NO WARNINGS
```

```
****************************************
*                                      *
*          FOURTH-ORDER IIR            *
*       ELLIPTIC LOWPASS FILTER        *
*                                      *
*      CASCADE STRUCTURE WITH          *
*  SECOND-ORDER DIRECT-FORM II SUBSECTIONS *
*                                      *
*        FILTER CHARACTERISTICS        *
*                                      *
*  SAMPLING FREQUENCY = 10 KHZ         *
*                                      *
*                        BAND 1      BAND 2
*                                      *
*  LOWER BAND EDGE       0.00000     2.75000
*  UPPER BAND EDGE       2.50000     5.00000
*  NOMINAL GAIN          1.00000     0.00000
*  NOMINAL RIPPLE        0.06000     0.06000
*  MAXIMUM RIPPLE        0.05517     0.05514
*  RIPPLE IN DB          0.47469   -25.17089
*
*            FILTER STRUCTURE
```

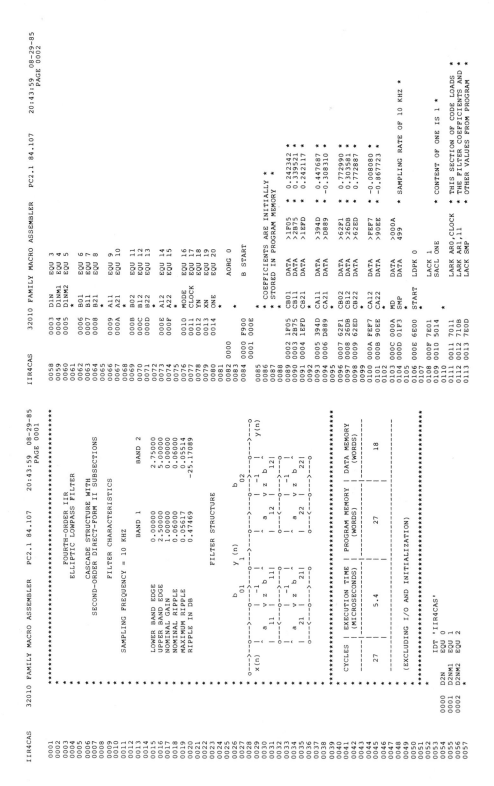

```
*  ---------------------------------------------------------------
*  | CYCLES | EXECUTION TIME | PROGRAM MEMORY | DATA MEMORY |
*  |        | (MICROSECONDS) |    (WORDS)     |   (WORDS)   |
*  ---------------------------------------------------------------
*  |   27   |      5.4       |      27        |     18      |
*  ---------------------------------------------------------------
*
*  (EXCLUDING I/O AND INITIALIZATION)
****************************************

0000             IDT 'IIR4CAS'
0000 0000  D2N    EQU 0
0001 0001  D2NM1  EQU 1
0002 0002  D2NM2  EQU 2
*
```

32010 FAMILY MACRO ASSEMBLER   PC2.1 84.107   20:43:59 08-29-85
                                                        PAGE 0002

```
0058  0003       DIN    EQU 3
0059  0004       DINM1  EQU 4
0060  0005       DINM2  EQU 5
0061             *
0062  0006       B01    EQU 6
0063  0007       B11    EQU 7
0064  0008       B21    EQU 8
0065             *
0066  0009       A11    EQU 9
0067  000A       A21    EQU 10
0068             *
0069  000B       B02    EQU 11
0070  000C       B12    EQU 12
0071  000D       B22    EQU 13
0072             *
0073  000E       A12    EQU 14
0074  000F       A22    EQU 15
0075             *
0076  0010       MODE   EQU 16
0077  0011       CLOCK  EQU 17
0078  0012       YN     EQU 18
0079  0013       XN     EQU 19
0080  0014       ONE    EQU 20
0081             *
0082  0000       AORG 0
0083
0084  0000 F900  START  B START
      0001 000E
0085             *
0086             ** COEFFICIENTS ARE INITIALLY *
0087             ** STORED IN PROGRAM MEMORY *
0088             *
0089  0002 1F05  CB01   DATA >1F05 *  0.242342 *
0090  0003 2B75  CB11   DATA >2B75 *  0.339521 *
0091  0004 1EFD  CB21   DATA >1EFD *  0.242117 *
0092             *
0093  0005 394D  CA11   DATA >394D *  0.447687 *
0094  0006 D889  CA21   DATA >D889 * -0.308310 *
0095             *
0096  0007 62F1  CB02   DATA >62F1 *  0.772990 *
0097  0008 26DB  CB12   DATA >26DB *  0.303581 *
0098  0009 62ED  CB22   DATA >62ED *  0.772887 *
0099             *
0100  000A FEF7  CA12   DATA >FEF7 * -0.008080 *
0101  000B 90EE  CA22   DATA >90EE * -0.867723 *
0102             *
0103  000C 000A  MD     DATA >000A *  SAMPLING RATE OF 10 KHZ *
0104  000D 01F3  SMP    DATA 499
0105             *
0106  000E 6E00  START  LDPK 0
0107             *
0108  000F 7E01         LACK 1
0109  0010 5014         SACL ONE       * CONTENT OF ONE IS 1 *
0110             *
0111  0011 7011         LARK AR0,CLOCK  * THIS SECTION OF CODE LOADS *
0112  0012 710B         LARK AR1,11     * THE FILTER COEFFICIENTS AND *
0113  0013 7E0D         LACK SMP        * OTHER VALUES FROM PROGRAM *
```

```
0114 0014 6880    LOAD    LARP AR0        * MEMORY TO DATA MEMORY         *
0115 0015 6791            TBLR *,AR1
0116 0016 1014            SUB ONE
0117 0017 F400            BANZ LOAD
     0018 0014
0118                *
0119 0019 7F89            ZAC             * THIS SECTION SETS THE         *
0120 001A 5000            SACL D2N        * INITIAL STATE OF THE          *
0121 001B 5001            SACL D2NM1      * FILTER TO ZERO                *
0122 001C 5002            SACL D2NM2
0123 001D 5003            SACL D1N
0124 001E 5004            SACL D1NM1
0125 001F 5005            SACL D1NM2
0126                *
0127 0020 4810            OUT MODE,PA0    * INITIALIZATION OF ANALOG      *
0128 0021 4911            OUT CLOCK,PA1   * INTERFACE BOARD               *
0129                *
0130 0022 F600    WAIT    BIOZ NXTPT      * BIO PIN GOES LOW WHEN A       *
     0023 0026
0131 0024 F900            B WAIT          * NEW SAMPLE IS AVAILABLE        *
     0025 0022
0132                *
0133 0026 4213    NXTPT   IN XN,PA2       * BRING IN THE NEW SAMPLE XN    *
0134                *
0135 0027 2F13            LAC XN,15       * START FIRST CASCADE SECTION   *
0136                *
0137 0028 6A04            LT D1NM1
0138 0029 6D09            MPY A11         * d (n-1) * a                   *
                                          *  1          11
0139                *
0140 002A 6C05            LTA D1NM2
0141 002B 6D0A            MPY A21
0142                *
0143 002C 7F8F            APAC
0144                *
0145 002D 5903            SACH D1N,1
0146                *
0147 002E 7F89            ZAC
0148                *
0149 002F 6D08            MPY B21
0150                *
0151 0030 6B04            LTD D1NM1
0152 0031 6D07            MPY B11
0153                *
0154 0032 6B03            LTD D1N
0155 0033 6D06            MPY B01         * FINISHED FIRST CASCADE SECTION *
0156                *
0157                *                     * START SECOND CASCADE SECTION  *
0158                *
0159 0034 6C01            LTA D2NM1
0160 0035 6D0E            MPY A12         * d (n-1) * a                   *
                                          *  2          12
0161                *
0162 0036 6C02            LTA D2NM2
0163 0037 6D0F            MPY A22
0164                *
0165 0038 7F8F            APAC
0166                *
0167 0039 5900            SACH D2N,1
```

```
0168                *
0169 003A 7F89            ZAC
0170                *
0171 003B 6D0D            MPY B22
0172                *
0173 003C 6B01            LTD D2NM1
0174 003D 6D0C            MPY B12
0175                *
0176 003E 6800            LTD D2N
0177 003F 6D0B            MPY B02
0178                *
0179 0040 7F8F            APAC
0180                *
0181 0041 5912            SACH YN,1       * FINISHED SECOND CASCADE SECTION *
0182                *                     * AND FILTER                      *
0183                *
0184 0042 4A12            OUT YN,PA2      * OUTPUT THE FILTER RESPONSE y(n) *
0185                *
0186 0043 F900            B WAIT          * GO GET THE NEXT POINT           *
     0044 0022
0187                *
0188                     END

NO ERRORS, NO WARNINGS
```

```
0001   ****************************************************
0002   *  *
0003   *  *                  FOURTH-ORDER IIR
0004   *  *              ELLIPTIC LOWPASS FILTER
0005   *  *
0006   *  *                 PARALLEL STRUCTURE
0007   *  *
0008   *  *               FILTER CHARACTERISTICS
0009   *  *
0010   *  *
0011   *  *   SAMPLING FREQUENCY = 10 KHZ
0012   *  *
0013   *  *                      BAND 1        BAND 2
0014   *  *   LOWER BAND EDGE    0.00000      2.75000
0015   *  *   UPPER BAND EDGE    2.50000      5.00000
0016   *  *   NOMINAL GAIN       1.00000      0.00000
0017   *  *   NOMINAL RIPPLE     0.06000      0.06000
0018   *  *   MAXIMUM RIPPLE     0.05617      0.05514
0019   *  *   RIPPLE IN DB       0.47469    -25.17089
0020   *  *
0021   *  *
0022   *  *               FILTER STRUCTURE
0023   *  *
```

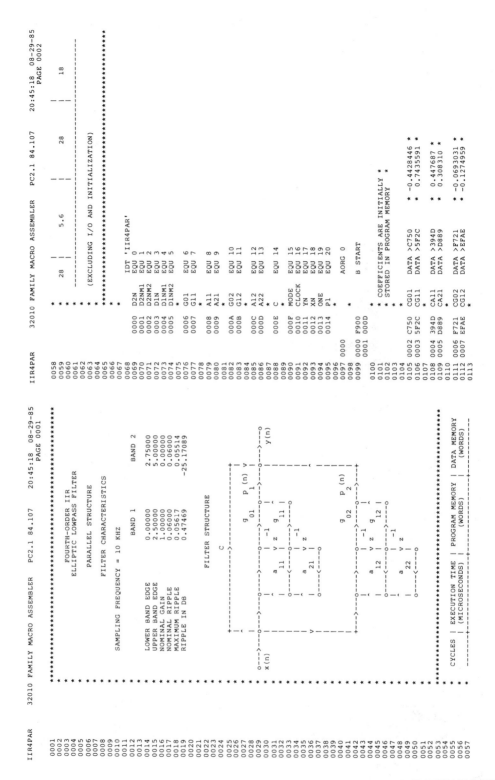

```
0055   *  *   CYCLES | EXECUTION TIME | PROGRAM MEMORY | DATA MEMORY
0056   *  *          | (MICROSECONDS) | (WORDS)        | (WORDS)
0057   *  *
```

```
0058   *  *      28    |      5.6       |      28        |    18
0059   *  *
0060   *  *
0061   *  *            (EXCLUDING I/O AND INITIALIZATION)
0062   *  *
0063   *  ****************************************************
0064
0065
0066
0067   *
0068                  IDT 'IIR4PAR'
0069   *
0070   0000   D2N     EQU   0
0071   0001   D2NM1   EQU   1
0072   0002   D2NM2   EQU   2
0073   0003   D1N     EQU   3
0074   0004   D1NM1   EQU   4
0075   0005   D1NM2   EQU   5
0076   *
0077   0006   G01     EQU   6
0078   0007   G11     EQU   7
0079   *
0080   0008   A11     EQU   8
0081   0009   A21     EQU   9
0082   *
0083   000A   G02     EQU   10
0084   000B   G12     EQU   11
0085   *
0086   000C   A12     EQU   12
0087   000D   A22     EQU   13
0088   *
0089   000E   C       EQU   14
0090   *
0091   000F   MODE    EQU   15
0092   0010   CLOCK   EQU   16
0093   0011   YN      EQU   17
0094   0012   XN      EQU   18
0095   0013   ONE     EQU   19
0096   0014   P1      EQU   20
0097   *
0098          AORG  0
0099   0000   F900            B     START
0100   0001   000D
0101   *
0102   *  COEFFICIENTS ARE INITIALLY *
0103   *  STORED IN PROGRAM MEMORY   *
0104   *
0105   0002   C750   CG01     DATA  >C750    * -0.4428446 *
0106   0003   5F2C   CG11     DATA  >5F2C    *  0.7435591 *
0107
0108   0004   394D   CA11     DATA  >394D    *  0.447687 *
0109   0005   D889   CA21     DATA  >D889    *  0.308310 *
0110
0111   0006   F721   CG02     DATA  >F721    * -0.0693031 *
0112   0007   EFAE   CG12     DATA  >EFAE    * -0.1274959 *
0113
```

3. Implementation of FIR/IIR Filters with the TMS32010/TMS32020

```
IIR4PAR     32010 FAMILY MACRO ASSEMBLER     PC2.1 84.107     20:45:18  08-29-85
                                                                        PAGE 0003

0114 0008 FEF7  CA12    DATA >FEF7      * -0.008080 *
0115 0009 90EE  CA22    DATA >90EE      * -0.867723 *
0116                    *
0117 000A 5988  CC      DATA >5988      * 0.699476 *
0118                    * SAMPLING RATE OF 10 KHZ *
0119 000B 000A  MD      DATA >000A
0120 000C 01F3  SMP     DATA >01F3
0121
0122 000D 6E00  START   LDPK 0
0123                    *
0124 000E 7E01          LACK 1          * CONTENT OF ONE IS 1 *
0125 000F 5013          SACL ONE
0126                    *
0127                    *  THIS SECTION OF CODE LOADS  *
0128 0010 7010          LARK AR0,CLOCK  * THE FILTER COEFFICIENTS AND *
0129 0011 710A          LARK AR1,10     * OTHER VALUES FROM PROGRAM *
0130 0012 7E0C          LACK SMP        * MEMORY TO DATA MEMORY *
0131 0013 6880  LOAD    LARP AR0
0132 0014 6791          TBLR *-,AR1
0133 0015 1013          SUB ONE
0134 0016 F400          BANZ LOAD
     0017 0013
0135                    *
0136 0018 7F89          ZAC             *  THIS SECTION SETS THE *
0137 0019 5000          SACL D2N        *  INITIAL STATE OF THE  *
0138 001A 5001          SACL D2NM1      *  FILTER TO ZERO        *
0139 001B 5002          SACL D2NM2
0140 001C 5003          SACL D1N
0141 001D 5004          SACL D1NM1
0142 001E 5005          SACL D1NM2
0143                    *
0144 001F 480F          OUT MODE,PA0    * INITIALIZATION OF ANALOG *
0145 0020 4910          OUT CLOCK,PA1   * INTERFACE BOARD *
0146                    *
0147 0021 F600  WAIT    BIOZ NXTPT      * BIO PIN GOES LOW WHEN A *
     0022 0025
0148 0023 F900          B WAIT          * NEW SAMPLE IS AVAILABLE *
     0024 0021
0149                    *
0150 0025 4212  NXTPT   IN XN,PA2       * BRING IN THE NEW SAMPLE XN *
0151                    *
0152 0026 2F12          LAC XN,15       * START FIRST PARALLEL SECTION *
0153                    *
0154 0027 6A05          LT D1NM2        * d (n-2) * a  *
0155 0028 6D09          MPY A21         *  1        21 *
0156                    *
0157 0029 6804          LTD D1NM1
0158 002A 6D08          MPY A11
0159                    *
0160 002B 7F8F          APAC
0161                    *
0162 002C 5903          SACH D1N,1
0163                    *
0164 002D 7F89          ZAC
0165                    *
0166 002E 6D07          MPY G11
0167 002F 6803          LTD D1N
```

```
IIR4PAR     32010 FAMILY MACRO ASSEMBLER     PC2.1 84.107     20:45:18  08-29-85
                                                                        PAGE 0004

0168 0030 6D06          MPY G01
0169 0031 7F8F          APAC
0170                    *
0171 0032 5914          SACH P1,1       * FINISHED FIRST PARALLEL SECTION *
0172                    *
0173 0033 2F12          LAC XN,15       * START SECOND PARALLEL SECTION *
0174                    *
0175 0034 6A02          LT D2NM2        * d (n-2) * a  *
0176 0035 6D0D          MPY A22         *  2        22 *
0177                    *
0178 0036 6B01          LTD D2NM1
0179 0037 6D0C          MPY A12
0180                    *
0181 0038 7F8F          APAC
0182                    *
0183 0039 5900          SACH D2N,1
0184                    *
0185 003A 2F14          LAC P1,15
0186                    *
0187 003B 6D0B          MPY G12
0188                    *
0189 003C 6B00          LTD D2N
0190 003D 6D0A          MPY G02
0191                    *
0192 003E 6C0E          LTA C
0193 003F 6D12          MPY XN
0194                    *
0195 0040 7F8F          APAC
0196                    *
0197 0041 5911          SACH YN,1       * FINISHED SECOND PARALLEL SECTION *
0198                    *               * AND FINISHED FILTER *
0199                    *
0200 0042 4A11          OUT YN,PA2      * OUTPUT THE FILTER RESPONSE y(n) *
0201                    *
0202 0043 F900          B WAIT          * GO GET THE NEXT POINT *
     0044 0021
0203                    *
0204 0045              END
0205
NO ERRORS, NO WARNINGS
```

4. Implementation of Fast Fourier Transform Algorithms with the TMS32020

Panos Papamichalis
Digital Signal Processing - Semiconductor Group
Texas Instruments

John So
Atlanta Regional Technology Center
Texas Instruments

INTRODUCTION

The Fourier transform converts information from the time domain into the frequency domain. It is an important analytical tool in such diverse fields as acoustics, optics, seismology, telecommunications, speech, signal processing, and image processing. In discrete-time systems, the Discrete Fourier Transform (DFT) is the counterpart of the continuous-time Fourier transform. Since the DFT is computation-intensive, it had relatively few applications, even with modern computers. The Fast Fourier Transform (FFT) is the generic name for a class of computationally efficient algorithms that implement the DFT and are widely used in the field of Digital Signal Processing (DSP).

Recent advances in VLSI hardware, such as the Texas Instruments TMS320 family of digital signal processors, have further enhanced the popularity of the FFT. This application report describes the implementation of FFT algorithms using the TMS32020 processor, which has features particularly suited to digital signal processing. This report begins with a discussion of the development of the DFT algorithm, leading to the derivation of the FFT algorithm. Special attention is given to various FFT implementation aspects, such as scaling. Although this report refers to radix-2 and radix-4 FFT only, the implementation techniques described are applicable to all FFT algorithms in general.[1-3] Specific examples of FFT implementations on the TMS32010 processor are contained in the book by Burrus and Parks.[4] To expedite TMS32020 FFT code development, two macro libraries are included in the appendices for both the direct and indirect memory addressing modes. TMS32020 source code examples are also given for a 256-point (both radix-2 and radix-4) and a 1024-point complex FFT, along with some system memory considerations for implementing large FFTs. The FFT source code can be found in Appendices A through G.

DEVELOPMENT OF THE DFT ALGORITHM

The Discrete Fourier Transform (DFT) is the discrete-time version of the continuous-time Fourier transform. The continuous-time Fourier transform or frequency spectrum of an analog signal x(t) is

$$X(w) = \int_{-\infty}^{\infty} x(t)e^{-jwt}dt \qquad (1)$$

where, in general, both x(t) and X(w) are complex functions of the continuous-time variable t and the frequency variable w, respectively. The continuous-time signal x(t) is converted to a discrete-time signal x(nT) by sampling it every T seconds. When there is no ambiguity, the sampling period

T notation is dropped and the discrete signal is represented by x(n). The Fourier transform of the discrete signal is given by

$$X(w) = \sum_{n=-\infty}^{\infty} x(n)e^{-jwn} \qquad (2)$$

where w represents normalized frequency and takes on values between 0 and 2π. X(w) is periodic with period 2π and, as a result, it is sufficient to consider its values only between 0 and 2π.[2] The periodicity of X(w) is a direct result of the sampled nature of x(n). In general, sampling in the time domain is associated with periodicity in the frequency domain and, conversely, sampling in the frequency domain is associated with periodicity in the time domain. This property is a basic result in Fourier theory, and forms the foundation of the DFT.

Assume that a signal x(n) consists of N samples. Since no restriction is imposed on what happens outside the interval of N points, it is convenient to assume that the signal is periodically repeated. Under this assumption, and because of the above correspondence of sampling and periodicity, the Fourier transform becomes discrete with the distance between successive samples equal to the fundamental frequency of the signal in the time domain. This distance is $2\pi/N$ in normalized frequency units. The result is the DFT, given by

$$X(k) = \sum_{n=0}^{N-1} x(n) \, W_N^{nk} \qquad k = 0,1,...,N-1 \qquad (3)$$

where $W_N = e^{-j2\pi/N}$, and W_N is known as the phase or twiddle factor. Equation (3) is generally referred to as an N-point DFT. Because the number of complex multiplications and additions required is approximately N^2 for large N, the total number of arithmetic operations required for a given N increases rapidly with the value of N. In fact, the excessively large amount of computations required to compute the DFT directly when N is large has directly prompted alternative methods for computing the DFT efficiently. Most of these methods make use of the inherent symmetry and periodicity of the above twiddle factor, as shown in Figure 1 for the case where N = 8.

Figure 1 shows that the following symmetry and periodicity relationships are true:

Symmetry Property: $\quad W_N^k = -W_N^{k+(N/2)} \qquad (4)$

Periodicity Property: $\quad W_N^k = W_N^{N+k} \qquad (5)$

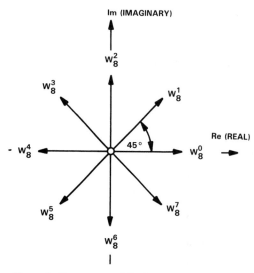

Im (IMAGINARY)

W_8^2

W_8^3 W_8^1

Re (REAL)

$-W_8^4$ 45° W_8^0 →

W_8^5 W_8^7

W_8^6

Figure 1. Symmetry and Periodicity of the Twiddle Factor for $N = 8$

In the next section, these relationships are utilized in the derivation of the radix-2 FFT algorithm.

DERIVATION OF THE FFT ALGORITHM

A more efficient method of computing the DFT that significantly reduces the number of required arithmetic operations is the so-called decimation-in-time (DIT) FFT algorithm.[2] With the FFT, N is a factorable number that allows the overall N-point DFT to be decomposed into successively smaller and smaller transforms. The size of the smallest transform thus derived is known as the radix of the FFT algorithm. Thus, for a radix-2 FFT algorithm, the smallest transform or "butterfly" (basic computational unit) used is the 2-point DFT. Generally, for an N-point FFT,

there are N resultant frequency samples corresponding to N time samples of the input signal x(n). For a radix-2 FFT, N is a power of 2.

The number of arithmetic operations can be reduced initially by decomposing the N-point DFT into two N/2-point DFTs. This means that the input time sequence x(n) is decomposed into two N/2-point subsequences (hence the name, decimation-in-time), which consist of its even-numbered and odd-numbered samples with time indices expressed mathematically as 2n and 2n + 1, respectively. Substituting these time indices into the original DFT equation gives

$$X(k) = \sum_{n=0}^{N/2-1} x(2n) \, W_N^{2nk} + \sum_{n=0}^{N/2-1} x(2n+1) \, W_N^{(2n+1)k} \tag{6}$$

$$= \sum_{n=0}^{N/2-1} x(2n) \, W_N^{2nk} + W_N^k \sum_{n=0}^{N/2-1} x(2n+1) \, W_N^{2nk}$$

Since

$$W_N^2 = [e^{-j(2\pi/N)}]^2 = [e^{-j\pi/(N/2)}]^2 = W_{N/2}$$

equation (6) can be written as

$$X(k) = \sum_{n=0}^{N/2-1} x(2n) \, W_{N/2}^{nk}$$

$$+ W_N^k \sum_{n=0}^{N/2-1} x(2n+1) \, W_{N/2}^{nk} \tag{7}$$

$$= Y(k) + W_N^k \, Z(k) \qquad\qquad k = 0,1,\dots,N-1$$

where $Y(k)$ is the first summation term and $Z(k)$ is the second summation term.

$Y(k)$ and $Z(k)$ are further seen to be the N/2-point DFTs of the even-numbered and odd-numbered time samples, respectively. In this case, the number of complex multiplications and additions is approximately $N + 2(N/2)^2$ because, according to (7), the N-point DFT is split in two N/2-point DFTs, which are then combined by N complex multiplications and additions. Thus, by splitting the original N-point DFT into two N/2-point DFTs, the total number of arithmetic operations has been reduced. This reduction is illustrated in Figure 2.

Implicit in the above derivation is the periodicity of $X(k)$, $Y(k)$, and $Z(k)$. $X(k)$ is periodic in k with a period N, while $Y(k)$ and $Z(k)$ are both periodic in k with a period N/2. Consequently, despite the fact that the index k ranges over N values from 0 to $N-1$ for $X(k)$, both $Y(k)$ and $Z(k)$ must be computed for k between 0 and $(N/2)-1$ only. The periodicity of $Y(k)$ and $Z(k)$ is also assumed in Figure 2.

Although (7) can be used to evaluate $X(k)$ for $0 \leq k \leq N-1$, further reduction in the amount of computation is possible when the symmetry property (4) and periodicity (5) of the twiddle factor are utilized to compute $X(k)$ separately over the following ranges:

1st Half of Frequency Spectrum: $0 \leq k \leq (N/2)-1$

2nd Half of Frequency Spectrum: $(N/2) \leq k \leq (N-1)$

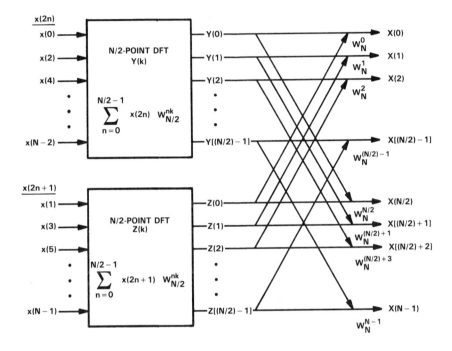

Figure 2. First DIT Decomposition of an N-Point DFT

Equation (7), for $N/2 \leq k \leq N-1$, can be rewritten as

$$X(k+N/2) = \sum_{n=0}^{N/2-1} x(2n)\ W_{N/2}^{n(k+N/2)}$$

$$+ W_N^{k+N/2} \sum_{n=0}^{N/2-1} x(2n+1)\ W_{N/2}^{n(k+N/2)}$$

$$\text{where } 0 \leq k \leq (N/2)-1 \tag{8}$$

Since

$$W_{N/2}^{n(k+N/2)} = W_{N/2}^{n(N/2)}\ W_{N/2}^{nk} = e^{-j2\pi n}\ W_{N/2}^{nk} = W_{N/2}^{nk}$$

and

$$W_N^{k+N/2} = W_N^k\ e^{-j\pi} = -W_N^k$$

equation (8) can be rewritten as

$$X(k+N/2) = \sum_{n=0}^{N/2-1} x(2n)\ W_{N/2}^{nk} \tag{9}$$

$$- W_N^k \sum_{n=0}^{N/2-1} x(2n+1)\ W_{N/2}^{nk}$$

$$= Y(k) - W_N^k\ Z(k) \quad k = 0,1,...,(N/2)-1$$

Therefore, (7) can be used to compute the first half of the frequency spectrum $X(k)$ for the index range $0 \leq k \leq (N/2)-1$, while equation (9) can be used to compute the second half of the frequency spectrum $X(k+N/2)$.

Figure 3 depicts the situation when the symmetry property of the twiddle factor is used to compute $X(k)$. The above decimation process and symmetry exploitation can reduce the DFT computation tremendously. By further decimating the odd-numbered and even-numbered time samples in a similar fashion, four N/4-point DFTs can be obtained, resulting in a further reduction in the DFT computation. Consequently, to arrive at the final radix-2 DIT FFT algorithm, this decimation process is repetitively carried out until eventually the N-point DFT can be evaluated as a collection of 2-point DFTs or butterflies.

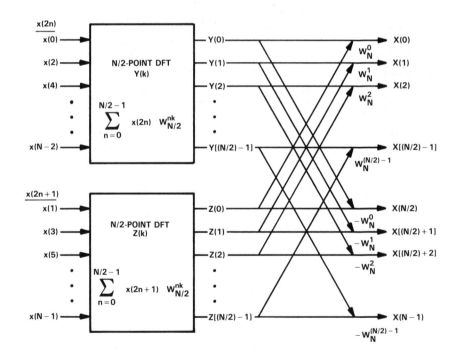

Figure 3. Decomposition of a DFT Using the Symmetry Property

4. Implementation of Fast Fourier Transform Algorithms with the TMS32020

RADIX-2 DECIMATION-IN-TIME (DIT) FFT BUTTERFLY

In the radix-2 DIT FFT algorithm, the time decimation process passes through a total of M stages where $N = 2^M$ with N/2 2-point FFTs or butterflies per stage, giving a total of $(N/2)\log_2 N$ butterflies per N-point FFT.

For the case of an 8-point DFT implemented using the radix-2 DIT FFT algorithm discussed in the previous pages, the input samples are processed through three stages. Four butterflies are required per stage, giving a total of twelve butterflies in the radix-2 implementation. Each butterfly is a 2-point DFT of the form depicted in Figure 4. P and Q are the inputs to the radix-2 DIT FFT butterfly. In general, the inputs to each butterfly are complex as is also the twiddle factor.

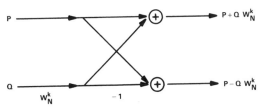

Figure 4. A Radix-2 DIT FFT Butterfly Flowgraph

As shown in Figure 4, the outputs P' and Q' of the radix-2 butterfly are given by

$$P' = P + Q\ W_N^k$$
$$Q' = P - Q\ W_N^k$$

(10)

While Figure 4 actually uses signal flowgraph nomenclature, another commonly used symbol for a radix-2 butterfly is shown in Figure 5.

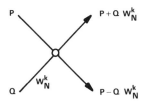

Figure 5. A Simplified Radix-2, DIT FFT Butterfly Symbol

For an explanation of the various notational conventions in use, the reader is referred to reference [3]. Both the flowgraph nomenclature and the butterfly symbol are used interchangeably in this report.

Implementation of the FFT Butterfly with Scaling

In the computation of the FFT, scaling of the intermediate results becomes necessary to prevent overflows. The TMS32020 processor has features optimized for digital signal processing and a number of on-chip shifters for scaling. In particular, the input scaling shifter, the 32-bit double-precision ALU and accumulator, and its output shifters are used extensively for scaling.

To see why scaling is necessary, observe that from the general equation of an N-point DFT (3), application of Parseval's theorem gives

$$\sum_{n=0}^{N-1} x^2(n) = \frac{1}{N} \sum_{k=0}^{N-1} |X(k)|^2$$

(11)

or

$$N\left[\frac{1}{N}\sum_{n=0}^{N-1} x^2(n)\right] = \left[\frac{1}{N}\sum_{k=0}^{N-1} |X(k)|^2\right]$$

(12)

i.e., the mean-squared value of X(k) is N times that of input x(n). Consequently, in computing the DFT of the input sequence x(n), overflows may occur when fixed-point arithmetic is employed without appropriate scaling. To see how overflows can actually occur in FFT computations, consider the general radix-2 butterfly in the mth stage of an N-point DIT FFT as shown in Figure 6.

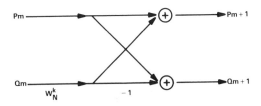

Figure 6. Signal Flowgraph of a Butterfly at the mth Stage

From Figure 6, the final form of the FFT can be written as

$$Pm+1 = Pm + W_N^k\ Qm$$
$$Qm+1 = Pm - W_N^k\ Qm$$

(13)

where Pm and Qm are the inputs, and Pm+1 and Qm+1 are the outputs of the mth stage of the N-point FFT, respectively. In general, Pm, Qm, Pm+1, and Qm+1 are complex as is the twiddle factor. The twiddle factor can be expressed as

$$W_N^k = e^{-j(2\pi/N)k} = \cos(X) - j\sin(X) \qquad (14)$$

where $X = (2\pi/N)k$ and $j = \sqrt{-1}$.

The inputs Pm and Qm can be expressed in terms of their real and imaginary parts by

$$\begin{aligned} Pm &= PR + j\,PI \\ Qm &= QR + j\,QI \end{aligned} \qquad (15)$$

By substituting the values from (14) and (15), equation (13) becomes

$$\begin{aligned} Pm+1 &= PR + j\,PI + (QR\cos(X) + QI\sin(X) \\ &\quad + j\,(QI\cos(X) - QR\sin(X)) \\ &= (PR + QR\cos(X) + QI\sin(X)) \\ &\quad + j\,(PI + QI\cos(X) - QR\sin(X)) \end{aligned}$$

$$\qquad (16)$$

$$\begin{aligned} Qm+1 &= PR + j\,PI - (QR\cos(X) + QI\sin(X)) \\ &\quad - j\,(QI\cos(X) - QR\sin(X)) \\ &= (PR - QR\cos(X) - QI\sin(X)) \\ &\quad + j\,(PI - QI\cos(X) + QR\sin(X)) \end{aligned}$$

Although the inputs of each butterfly stage have real and imaginary parts with magnitudes less than one, the real and imaginary parts of the outputs from (15) can have a maximum magnitude of

$$1 + 1\sin(45) + 1\cos(45) = 2.414213562$$

To avoid the possibility of overflow, each stage of the FFT is scaled down by a factor of 2. In this way, if an FFT consists of M stages, the output is scaled down by $2^M = N$, where N is the length of the FFT. Even with scaling, overflow is possible because of the maximum magnitude value for complex input data. This possibility is avoided by scaling down the input signal by a factor of 1.207106781, and then scaling up the output of the last FFT stage by the same factor. This additional scaling is not implemented in the code of the appendices, because the input

signal is assumed real (i.e., the imaginary part is zero), and the above maximum value cannot be attained. The maximum value for a real input is 2.

Using (15), the TMS32020 butterfly code is given in Figure 7. It is assumed that all input and output data values are in Q15 format; i.e., they are expressed in two's-complement fractional arithmetic with the binary point immediately to the right of the sign bit (15 bits after the binary point). This code incorporates one stage of scaling (i.e., scaling by two) for the implementation of the general radix-2 DIT FFT butterfly with the 16-bit sine and cosine values of the twiddle factor also stored in Q15 format. Note that in performing fractional multiplications, the product of two 16-bit Q15 fractions is a 32-bit double-precision fraction in Q30 format with two sign bits. This result is illustrated in Figure 8, where S stands for sign bit.

The code for a general radix-2 DIT FFT is given in Figure 7. In the comment section, ACC, P-REGISTER, and T-REGISTER represent the on-chip 32-bit accumulator, 32-bit product register, and 16-bit temporary register of the TMS32020 processor, respectively. For more information about the TMS32020 processor and its architecture, see the TMS32020 User's Guide.[5]

The first block in the butterfly code of Figure 7 (starting with the label INIT) is for general system initialization. The second block of code (starting with the label BTRFLY) takes advantage of the double sign bits to provide a "free" divide-by-2 scaling in calculating the term (1/2)(QR COS(X) + QI SIN(X)), which is the scaled real part of the product of the twiddle factor and Qm. In addition, since the current contents of memory location QR are no longer required for subsequent calculations, QR is also used as a temporary storage for this term.

The third block of code calculates the term (1/2) (QI COS(X) − QR SIN(X)), which is the scaled imaginary part of the product of the twiddle factor and Qm. By completing this calculation, QI is also freed as a temporary storage for this term.

The fourth block of code calculates the real parts of Pm + 1 and Qm + 1 and provides the divide-by-2-per-stage scaling function to avoid signal overflows. To perform this function, the input binary scaling shifter of the TMS32020 is used.

```
*****************************************************************
* TMS32020 CODE FOR A GENERAL RADIX-2 DIT FFT BUTTERFLY         *
*****************************************************************
*
* EQUATES FOR THE REAL AND IMAGINARY PARTS OF Xm(P) AND Xm(Q).
* THE LOCATIONS PR, PI, QR, AND QI ARE USED BOTH FOR THE INPUT
* AND THE OUTPUT DATA.
*
PR      EQU     0       Re(Pm) STORED IN LOCATION 0 IN DATA MEMORY
PI      EQU     1       Im(Pm) STORED IN LOCATION 1 IN DATA MEMORY
QR      EQU     2       Re(Qm) STORED IN LOCATION 2 IN DATA MEMORY
QI      EQU     3       Im(Qm) STORED IN LOCATION 3 IN DATA MEMORY
*
* EQUATES FOR THE REAL AND IMAGINARY PARTS OF THE TWIDDLE FACTOR.
*
COSX    EQU     4       COS(X) STORED IN LOCATION 4 IN DATA MEMORY
SINX    EQU     5       SIN(X) STORED IN LOCATION 5 IN DATA MEMORY
*
* INITIALIZE SYSTEM.
*
        AORG    >20
INIT    SPM     0       NO SHIFT AT OUTPUTS OF P-REGISTER
        SSXM            SELECT SIGN-EXTENSION MODE
        ROVM            RESET OVERFLOW MODE
        LDPK    4       CHOOSE DATA PAGE 4
*
* CALCULATE (QR COS(X) + QI SIN(X)); STORE RESULT IN QR.
*
BTRFLY  LT      QR      LOAD T-REGISTER WITH QR
        MPY     COSX    P-REGISTER = (1/2) QR COSX
        LTP     QI      ACC= (1/2) QR COSX ; LOAD T-REGISTER WITH QI
        MPY     SINX    P-REGISTER = (1/2) QI SINX
        APAC            ACC= (1/2)(QR COSX+QI SINX)
        MPY     COSX    P-REGISTER = (1/2) QI COSX
        LT      QR      LOAD T-REGISTER WITH QR
        SACH    QR      QR = (1/2)(QR COSX+QI SINX)
*
* CALCULATE (QI COS(X) - QR SIN(X)); STORE RESULT IN QI.
*
        PAC             ACC= (1/2) QI COSX
        MPY     SINX    P-REGISTER = (1/2) QR SINX
        SPAC            ACC= (1/2)(QI COSX - QR SINX)
        SACH    QI      QI = (1/2)(QI COSX - QR SINX)
*
* CALCULATE Re(Pm+1) AND Re(Qm+1); STORE RESULTS IN PR AND QR.
*
        LAC     PR,14   ACC= (1/4)PR
        ADD     QR,15   ACC= (1/4)(PR + QR COSX + QI SINX)
        SACH    PR,1    PR = (1/2)(PR + QR COSX + QI SINX)
        SUBH    QR      ACC= (1/4)(PR - QR COSX - QI SINX)
        SACH    QR,1    QR = (1/2)(PR - QR COSX - QI SINX)
*
```

Figure 7. TMS32020 code for a General Radix-2 DIT FFT Butterfly

```
* CALCULATE Im[Pm+1] AND Im[Qm+1]; STORE RESULTS IN PI AND QI.
*
      LAC    PI,14     ACC= (1/4)PI
      ADD    QI,15     ACC= (1/4)(PI + QI COSX - QR SINX)
      SACH   PI,1      PI = (1/2)(PI + QI COSX - QR SINX)
      SUBH   QI        ACC= (1/4)(PI - QI COSX + QR SINX)
      SACH   QI,1      QI = (1/2)(PI - QI COSX + QR SINX)
*
```

Figure 7. TMS32020 Code for a General Radix-2 DIT FFT Butterfly (concluded)

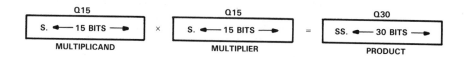

Figure 8. Multiplication of Two Q15 Numbers

Initially, the contents of PR are scaled down by a factor of 4 (equivalent to a 14-bit left-shift). Note that the shift or scaling function is being performed while the contents of PR are being loaded into the 32-bit accumulator. Since the TMS32020 has a 32-bit double-length accumulator, no accuracy is lost in this binary scaling process. To generate the final result of Re(Pm + 1), the contents of QR must be added to the contents of the accumulator with a 1-bit right-shift (equivalent to a 15-bit left-shift). This means adding (1/4)(QR COS(X) + QI SIN(X)) to (1/4)PR, which is the current value held in the accumulator. The upper-half of the accumulator is then stored in PR with a 1-bit left-shift to yield the term (1/2)(PR + QR COS(X) + QI SIN(X)), which is precisely Re(Pm + 1) scaled down by 2. This shift or scaling function is being performed while the contents of the upper half of the accumulator are loaded into PR. At this point, the accumulator still has a value equal to (1/4)(PR + QR COS(X) + QI SIN(X)). Hence, to obtain the final result of Re(Qm + 1), the unscaled contents of QR must be subtracted from the accumulator. The upper-half of the accumulator is again stored in QR with a 1-bit left-shift to yield the term (1/2)(PR + QR COS(X) + QI SIN(X)), which is precisely Re(Qm + 1) scaled down by 2.

In a similar fashion, the fifth block of code calculates the imaginary parts of Pm + 1 and Qm + 1. Note that all the scaling functions performed so far have come "free" with the architecture of the TMS32020.[5]

In summary, the data values are scaled down by right-shifting the 16-bit words as they are loaded into the 32-bit accumulator. In this way, full precision is still maintained in all calculations. The right-shifts are implemented by a corresponding number of left-shifts into the upper half of the accumulator. On the other hand, if the accumulator had

been single precision or 16 bits wide, all scaling operations would have resulted in a loss of accuracy.

In-Place FFT Computations

In the butterfly implementation, the set of input registers in data memory (PR, PI, QR, and QI) for the two complex inputs Pm and Qm are used for holding the two complex outputs Pm + 1 and Qm + 1, respectively. When the same set of input registers is used as output registers for holding the FFT results, the FFT computation is said to be performed in-place. Therefore, FFTs implemented on the TMS32020 using the general butterfly routine are performed in-place.

As a general rule, an in-place FFT computation means that a total of 2N memory locations are required for an N-point FFT since the inputs to the FFT can be complex. On the other hand, a total of up to 4N memory locations is required for not-in-place computations.

Another attractive feature of the butterfly routine is that temporary or scratch-pad registers are not needed for intermediate results or calculations. Where coefficient quantization and other finite wordlength effects are not critical, 13-bit sine and cosine values can be used instead of 16-bit values addressed by the MPY instruction. In this way, the registers COSX and SINX for the twiddle factors can be dispensed with altogether. For this purpose, the MPYK instruction, which has a 13-bit signed constant embedded in its opcode, can be employed instead of the MPY instruction in the butterfly code. In Appendix A, two FFT macros (NORM1 and NORM2) illustrate the use of the MPY and MPYK instructions, respectively. Appendices A and B contain macro libraries that perform the same tasks, but in Appendix A they use direct addressing while in Appendix B they use indirect addressing.

4. Implementation of Fast Fourier Transform Algorithms with the TMS32020

Bit-Reversal/Data Scrambling

As shown in Figure 9, the input time samples x(n) are not in order, i.e., they are scrambled. Such data scrambling or bit reversal is a direct result of the radix-2 FFT derivation. On closer inspection, it is seen that the index of each input sample is actually bit-reversed, as shown in Table 1.

Therefore, the input data sequence must be prescrambled prior to executing the FFT in order to produce in-order outputs. To perform bit reversal on the 8-point FFT, shown in Figure 9, the pairs of input samples, [x(1) and x(4)] and [x(3) and x(6)], must be swapped. On the other hand, Figure 10 has in-order input samples by rearranging the ordering of all the butterflies. However, the outputs are now bit-reversed.

Table 1. Bit-Reversal Algorithm for an 8-Point Radix-2 DIT FFT

INDEX	BIT PATTERN	BIT-REVERSED PATTERN	BIT-REVERSED INDEX
0	000	000	0
1	001	100	4
2	010	010	2
3	011	110	6
4	100	001	1
5	101	101	5
6	110	011	3
7	111	111	7

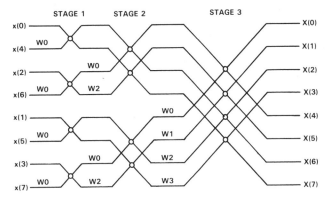

LEGEND FOR TWIDDLE FACTOR: $W0 = W_8^0$ $W1 = W_8^1$ $W2 = W_8^2$ $W3 = W_8^3$

Figure 9. An In-Place DIT FFT with In-Order Outputs and Bit-reversed Inputs

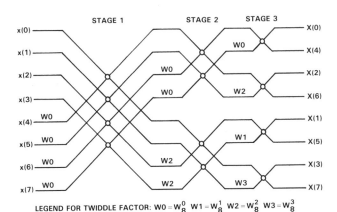

LEGEND FOR TWIDDLE FACTOR: $W0 = W_8^0$ $W1 = W_8^1$ $W2 = W_8^2$ $W3 = W_8^3$

Figure 10. An In-Place DIT FFT with In-Order Inputs but Bit-Reversed Outputs

In general, bit reversal or data scrambling must be performed either at the input stage on the time samples (Figure 9) or at the output stage on the frequency samples (Figure 10). Bit reversal can be performed in-place. Such a process generally requires the use of one temporary data memory location.

Because of its double-precision accumulator and its versatile instruction set, the TMS32020 processor can perform in-place bit reversal or data scrambling without the use of a temporary data memory location. For example, the TMS32020 code for swapping input data locations x(1) and x(4) is given in Figure 11.

Although bit-reversal can be regarded as a separate task performed either at the input or output stage of an FFT implementation, some FFT algorithms exist with bit reversal as an integral part.[5] Such algorithms are said to be in-place and in-order, and they tend to have higher execution speeds than that of the FFT and bit-reversal algorithms executed separately.

A Numerical Example: 8-Point DIT FFT

To illustrate the concept of the FFT, a numerical example of an 8-point, decimation-in-time FFT is presented. The input signal is a square pulse with four samples equal to 0.5 and four samples equal to zero, as shown in Figure 12(a). The broken line in Figure 12 represents the envelope of the plotted signal. Figure 12(b) plots the magnitude of the computed FFT, where the number next to each sample indicates its magnitude. The choice of the amplitude for this example is arbitrary, but it is restricted to be less than 1 since it assumed that the numbers handled by the processor are in Q15 format.

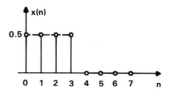

(a) TIME-DOMAIN SIGNAL

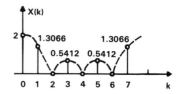

(b) FOURIER TRANSFORM MAGNITUDE

Figure 12. Time-Domain Signal and the Magnitude of Its FFT

The FFT of this time signal is computed by an 8-point DIT FFT as shown in Figure 13. On the left side, the samples x(n) of the time signal are arranged in their normal order. On the right side, the computed samples X(k) of the FFT are in bit-reversed order. Since the computation produces complex numbers, all the numerical values are presented as (R, I), where R is the real part and I is the imaginary part of the complex number. Figure 13 shows also the numerical values computed in the intermediate stages.

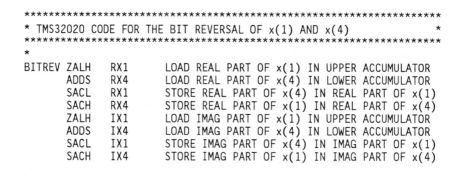

```
***************************************************************
*  TMS32020 CODE FOR THE BIT REVERSAL OF x(1) AND x(4)       *
***************************************************************
*
BITREV ZALH   RX1    LOAD REAL PART OF x(1) IN UPPER ACCUMULATOR
       ADDS   RX4    LOAD REAL PART OF x(4) IN LOWER ACCUMULATOR
       SACL   RX1    STORE REAL PART OF x(4) IN REAL PART OF x(1)
       SACH   RX4    STORE REAL PART OF x(1) IN REAL PART OF x(4)
       ZALH   IX1    LOAD IMAG PART OF x(1) IN UPPER ACCUMULATOR
       ADDS   IX4    LOAD IMAG PART OF x(4) IN LOWER ACCUMULATOR
       SACL   IX1    STORE IMAG PART OF x(4) IN IMAG PART OF x(1)
       SACH   IX4    STORE IMAG PART OF x(1) IN IMAG PART OF x(4)
```

Figure 11. TMS32020 Code for the Bit Reversal of x(1) and x(4)

4. Implementation of Fast Fourier Transform Algorithms with the TMS32020

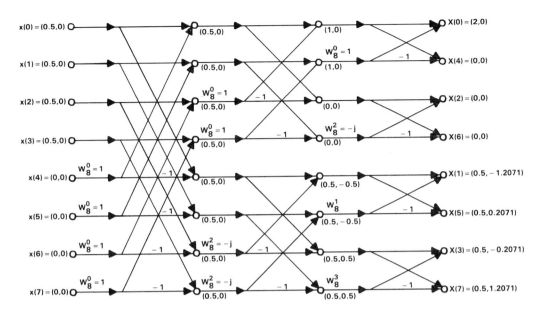

Figure 13. Numerical Example of an 8-Point DIT FFT without Scaling

Table 2 shows the values of the twiddle factor W_8^i for $i = 0,1,...,7$. Of these factors, only the first four are used in the FFT. The other four are related to them through the symmetry property (see equation (4)).

Table 2. Numerical Values of W_8^i, where $i = 0, 1, ..., 7$

TWIDDLE FACTOR	VALUE
W_8^0	1
$W_8^1 = e^{-\pi/4}$	$0.7071 - j\,0.7071$
$W_8^2 = e^{-\pi/2}$	$-j$
$W_8^3 = e^{-3\pi/4}$	$-0.7071 - j\,0.7071$
$W_8^4 = e^{-\pi} = -W_8^0$	-1
$W_8^5 = e^{-5\pi/4} = -W_8^1$	$-0.7071 + j\,0.7071$
$W_8^6 = e^{-3\pi/2} = -W_8^2$	j
$W_8^7 = e^{-7\pi/4} = -W_8^3$	$0.7071 + j\,0.7071$

Figure 13 has demonstrated the need for scaling. Without scaling, the intermediate results can attain values greater than or equal to 1. This would cause overflows in an implementation that uses Q15 numbers. Therefore, scaling is applied as mentioned earlier. Figure 14 shows exactly the same example, but now every stage is scaled by 1/2. No overflows occur with this implementation. The final output is the same as in Figure 13 but scaled by 1/8.

Special Butterflies

Although any N-point FFT (where N is a power of 2) can be directly implemented with the general butterfly only, special butterflies are normally used in order to increase the FFT execution speed.

Special butterflies can be coded by taking advantage of certain sine and cosine values of the twiddle factor. For instance, when the angle X takes on values such as 0, 90, 180, and 270 degrees, butterflies require much less code. Other special butterflies can also be coded for angles such as 45, 135, 225, and 315 degrees. Examples of these special butterflies can be found in nine macros located in Appendix A.

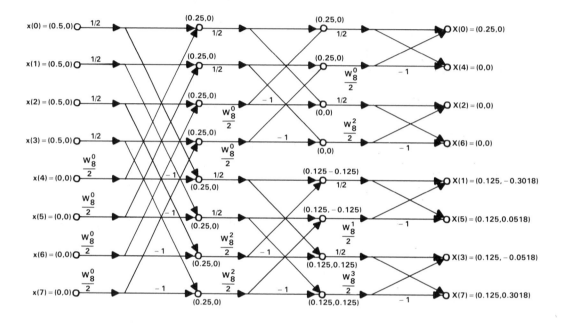

Figure 14. Numerical Example of an 8-Point DIT FFT with Scaling

4. Implementation of Fast Fourier Transform Algorithms with the TMS32020

An interesting point to be noted is that the first two stages of an N-point radix-2 FFT can be performed simultaneously with a special radix-4 butterfly to enhance execution speed. This special radix-4 butterfly is depicted in Figure 15 with the corresponding code (MACRO 9) listed in Appendix A.

The special radix-4 butterfly actually consists of four separate radix-2 butterflies. The radix-4 butterfly is further seen to be a 4-point DFT.

Together with the general butterfly, these special butterflies greatly improve the execution speed of an FFT algorithm. An example of the use of such butterflies for an 8-point DIT FFT is given in Figure 16. Since the FFT implementation is, in general, highly modular, the code in Figure 16 has been structured into a number of macro calls, including a macro for bit reversal.

During assembly time, the TMS32020 Macro Assembler fully expands these macros into in-line code.[5] The first two stages of the 8-point DIT FFT are implemented by the special radix-4 DIT FFT macro COMBO. The last stage consists of the special radix-2 DIT FFT macros ZERO, PIBY4, PIBY2, and PI3BY4. These macros can be found in Appendix A. The difference from the general radix-2 DIT butterfly is that the angle X of the twiddle factor takes on the values 0, 45, 90, and 135, respectively.

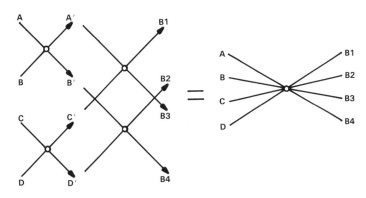

Figure 15. The Equivalence of Four Radix-2 Butterflies to one Radix-4 Butterfly

```
************************************************************
*                 AN 8-POINT DIT FFT                      *
************************************************************
*
X0R     EQU     00
X0I     EQU     01
X1R     EQU     02
X1I     EQU     03
X2R     EQU     04
X2I     EQU     05
X3R     EQU     06
X3I     EQU     07
X4R     EQU     08
X4I     EQU     09
X5R     EQU     10
X5I     EQU     11
X6R     EQU     12
X6I     EQU     13
X7R     EQU     14
X7I     EQU     15
W       EQU     16
        AORG    >20
WTABLE  DATA    >5A82       VALUE FOR SIN(45) OR COS(45)
```

Figure 16. TMS32020 Code for an 8-Point DIT FFT Implementation

```
*
* INITIALIZE SYSTEM
*
INIT    SPM     0       NO SHIFT AT OUTPUTS OF PR
        SSXM            SELECT SIGN-EXTENSION MODE
        ROVM            RESET OVERFLOW MODE
        LDPK    4       CHOOSE DATA PAGE 4
        LALK    WTABLE  GET TWIDDLE FACTOR ADDRESS
        TBLR    W       STORE SIN(45) OR COS(45) IN W
*
* MACRO FOR INPUT BIT REVERSAL
*
BITREV $MACRO PR,PI,QR,QI
        ZALH    :PR:
        ADDS    :QR:
        SACL    :PR:
        SACH    :QR:
        ZALH    :
        ADDS    :QI:
        SACL    :
        SACH    :QI:
        $END
*
* FFT CODE WITH BIT-REVERSED INPUT SAMPLES
*
FFT8PT BITREV 2,3,8,9
        BITREV 6,7,12,13
*
* FIRST & SECOND STAGES COMBINED WITH DIVIDE-BY-4 INTERSTAGE SCALING
*
        COMBO   X0R,X0I,X1R,X1I,X2R,X2I,X3R,X3I
        COMBO   X4R,X4I,X5R,X5I,X6R,X6I,X7R,X7I
*
* THIRD STAGE WITH DIVIDE-BY-2 INTERSTAGE SCALING
*
        ZERO    X0R,X0I,X4R,X4I
        PIBY4   X1R,X1I,X5R,X5I,W
        PIBY2   X2R,X2I,X6R,X6I
        PI3BY4  X3R,X3I,X7R,X7I,W
```

Figure 16. TMS32020 Code for an 8-Point DIT FFT Implementation (concluded)

RADIX-4 DECIMATION-IN-FREQUENCY (DIF) FFT

The implementation described thus far is that of a radix-2 FFT using Decimation In Time (DIT). The decimation-in-time FFT is calculated by breaking the input sequence x(n) into smaller and smaller sequences and computing their FFTs. In an alternate approach, the output sequence X(k), which represents the Fourier transform of x(n), can be broken down into smaller subsequences that are computed from x(n). This method is called Decimation In Frequency (DIF). Computationally, there is no real difference between the two approaches. DIF is introduced here for two reasons: (1) to give the reader a broader understanding of the different methods used for the computation of the FFT, and (2) to allow a comparison of this implementation with the FORTRAN programs provided in the book by Burrus and Parks.[4] The programs from that book were the basis for the development of the radix-4 FFT code on the TMS32020.

In a radix-4 FFT, each butterfly has four inputs and four outputs instead of two as in the case of radix-2 FFT. As shown in the following equations, this is advantageous because the twiddle factor W has special values when the exponent corresponds to multiples of $\pi/2$. The end result is that the computational load of the FFT is reduced, and the radix-4 FFT is computed faster than the radix-2 FFT.

4. Implementation of Fast Fourier Transform Algorithms with the TMS32020

To introduce the radix-4 DIF FFT, equation (3) is broken into four summations. These four summations correspond to the four components in radix-4. The choice of having N/4 consecutive samples of x(n) in each sum is dictated by the choice of Decimation In Frequency (DIF).

$$X(k) = \sum_{n=0}^{N-1} x(n) \, W_N^{nk} = \sum_{n=0}^{(N/4)-1} x(n) \, W_N^{nk}$$

$$+ \sum_{n=N/4}^{(N/2)-1} x(n) \, W_N^{nk} + \sum_{n=N/2}^{(3N/4)-1} x(n) \, W_N^{nk}$$

$$+ \sum_{n=3N/4}^{N-1} x(n) \, W_N^{nk} = \sum_{n=0}^{(N/4)-1} x(n) \, W_N^{nk}$$

$$+ W_N^{Nk/4} \sum_{n=0}^{(N/4)-1} x(n+N/4) \, W_N^{nk}$$

$$\tag{17}$$

$$+ W_N^{Nk/2} \sum_{n=0}^{(N/4)-1} x(n+N/2) \, W_N^{nk}$$

$$+ W_N^{3Nk/4} \sum_{n=0}^{(N/4)-1} x(n+3N/4) \, W_N^{nk}$$

$$k = 0,1,...,N-1$$

From the definition of the twiddle factor, it can be shown that

$$W_N^{Nk/4} = (-j)^k, \quad W_N^{Nk/2} = (-1)^k, \quad \text{and} \quad W_N^{3Nk/4} = (j)^k$$

where j is the square root of -1. With this substitution, (17) can be rewritten as

$$X(k) = \sum_{n=0}^{(N/4)-1} [x(n)+(-j)^k \, x(n+N/4) \tag{18}$$

$$+(-1)^k \, x(n+N/2)+(j)^k \, x(n+3N/4)] \, W_N^{nk}$$

Equation (18) is not yet an FFT of length N/4, because the twiddle factor depends on N and not on N/4. To make it an N/4-point FFT, the sequence X(k) is broken into four sequences (decimation in frequency) for the cases where k = 4r, 4r+1, 4r+2, and 4r+3.

Introducing this segmentation, and remembering that

$$W_N^{4nr} = W_{N/4}^{nr}$$

the following four equations (19) are derived from (18)

$$X(4r) = \sum_{n=0}^{(N/4)-1} [x(n) + x(n+N/4)$$

$$+ x(n+N/2) + x(n+3N/4)] \, W_N^0 \, W_{N/4}^{nr}$$

$$X(4r+1) = \sum_{n=0}^{(N/4)-1} [x(n) - j \, x(n+N/4)$$

$$- x(n+N/2) + j \, x(n+3N/4)] \, W_N^n \, W_{N/4}^{nr}$$

$$X(4r+2) = \sum_{n=0}^{(N/4)-1} [x(n) - x(n+N/4) \tag{19}$$

$$+ x(n+N/2) - x(n+3N/4)] \, W_N^{2n} \, W_{N/4}^{nr}$$

$$X(4r+3) = \sum_{n=0}^{(N/4)-1} [x(n) + j \, x(n+N/4)$$

$$- x(n+N/2) - j \, x(n+3N/4)] \, W_N^{3n} \, W_{N/4}^{nr}$$

Each one of these equations is now an N/4-point FFT that can be computed by repeating the above procedure until $N = 4$. Note that the factors W_N^0, W_N^n, W_N^{2n}, and W_N^{3n} are considered part of the signal. In general, an N-point FFT (where N is a power of 4) can be reduced to the computation of four N/4-point FFTs by transforming the input signal $x(n)$ into an intermediate signal $y(n)$, as suggested by (19). Figure 17 shows the corresponding radix-4 DIF butterfly, which generates one term for each sum in (19).

For simplicity, the notation of Figure 18 is often used instead of that of Figure 17 for the butterfly of radix-4 DIF FFT.

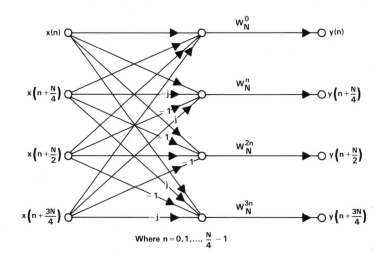

Figure 17. Radix-4 DIF Butterfly

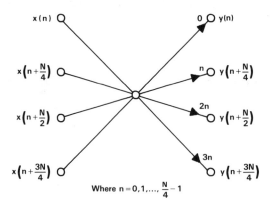

Figure 18. Alternate Form of the Radix-4 DIF Butterfly

Figure 19 shows an example of a 64-point, radix-4 DIF FFT.

Note that the inputs are normally ordered while the outputs are presented in a digit-reversed order. The principle of digit reversal is the same as in radix-2 FFT, but now the digits are 0, 1, 2, and 3 (quaternary system) instead of 0 and 1 (binary system). The code for digit reversal is the same as that shown in Figure 11. For example, the datapoint occupying location 132 (quaternary number corresponding to decimal 30) exchanges positions with the datapoint at location 231 (corresponding to the decimal 45).

Another important point of the radix-4 algorithm regards scaling. Since each stage of the radix-4 algorithm corresponds to two stages of the radix-2 algorithm, equivalent results are obtained by dividing the output of each stage of the radix-4 algorithm by 4.

Appendix E contains the implementation of a 256-point, radix-4 DIF FFT on the TMS32020. This implementation follows the one described in FORTRAN code in the book by Burrus and Parks.[4]

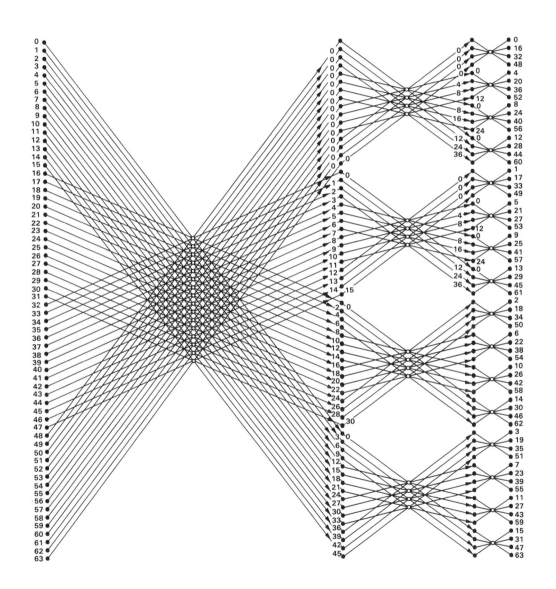

Figure 19. A 64-Point, Radix-4 DIF FFT

4. Implementation of Fast Fourier Transform Algorithms with the TMS32020

SYSTEM MEMORY AND I/O CONSIDERATIONS

Unlike non-realtime FFT applications where data samples to be transformed are assumed to already be in data memory, realtime FFT applications demand careful considerations of data input/output and system memory utilization.

The TMS32020 has 544 words of on-chip data RAM, organized into two 256-word blocks (B0 and B1) and one 32-word block (B2) that can be used as scratch-pad locations.[5] In non-realtime applications, this memory configuration allows a 256-point complex FFT to be easily performed (see Appendix C). However, for realtime FFT applications, input/output data buffering is generally required.

For small transform sizes, up to 128-point complex (or 256-point real) FFTs, the double-buffering technique, shown in Figure 20, can be used for realtime applications without the need of any external data memory. The on-chip RAM blocks B0 and B1 are organized into Buffer A and Buffer B, respectively.

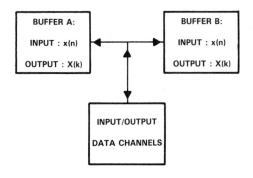

Figure 20. Input/Output Double-Buffering

Consider a 128-point complex FFT. Realtime input data to be transformed can be grouped into "frames" of 256 words (128 complex inputs) read into either Buffer A or Buffer B, depending on which one is not currently being used by the FFT program. The idea is to use the two on-chip RAM blocks B0 and B1 alternatively as I/O and transform buffers.

Assuming that the frame of data in Buffer A, the current transform buffer, is being transformed in-place, a software flag is then set to indicate that Buffer B can now be used as the current I/O buffer. This means that while time-domain data is read into Buffer B, the current I/O buffer, previous transformed data in Buffer B must be transferred out at the same time to make room for the incoming data. This can be acomplished efficiently if the I/O transfers are sequential and organized in a back-to-back manner (i.e., an output operation followed by an input operation).

Resetting the flag indicates that the roles of Buffer A and Buffer B are now reversed. In this case, Buffer B now has a full frame of input data ready to be transformed while Buffer A has a full frame of transformed data (spectral samples) ready to be transferred out to make room for more incoming time-domain data. The setting of the software flag is often implemented as an I/O device service routine (DSR) or as an interrupt handler in the case of interrupt-driven I/O.

Although this double-buffering technique is also applicable to larger transforms with the use of external memory, the actual memory required can be optimized if the transform time for an N-point FFT is shorter than the time to assemble a frame of N complex input data samples. For this purpose, the circular-buffer technique, shown in Figure 21, can be used.

Instead of a double-buffer size of 2N, a circular-buffer size of $N + M < 2N$ can be used where $M < N$ and M depends on the system input data rate in general. For example, M is chosen to be no less than 8T for an 8-kHz input sampling frequency and an N-point complex FFT with a transform time of T ms.

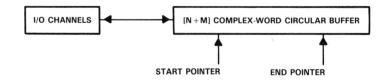

Figure 21. Input/Output Circular-Buffering for Large FFTs

4. Implementation of Fast Fourier Transform Algorithms with the TMS32020

A set of pointers is used to manage the data in the circular buffer. The start pointer is set at the beginning of the current frame, whereas the end pointer always indicates the current input data position in the circular buffer. Both pointers "wrap around" at the end of the circular buffer.

When a complete frame of input data has been collected, the set of pointer values is passed to the FFT program to transform the frame of data. For the next frame of input data, the start pointer points to the location immediately following the last location of the previous frame. As before, the end pointer for the current frame tracks the location of the next input data, and the whole process is repeated.

To decrease execution time, a large N-point FFT can be divided into smaller 256-point complex FFTs and executed 256 complex points at a time utilizing the on-chip RAM, as shown in Figure 22. Note that the system is still collecting incoming time-domain data samples and storing them in the external circular buffer while the FFT program is executing with internal data RAM. When 256 complex points have been processed, the FFT program returns them to the external

buffer while fetching the next set of 256 samples for execution.

This scheme takes advantage of the fact that off-chip data accesses take two cycles each while on-chip data accesses take one cycle each. Certain instructions (e.g., SACL and SACH) even take three cycles to execute when operating on external RAM. To speed execution, off-chip data blocks can be efficiently moved into on-chip data memory via the BLKD (block move from data memory to data memory) instruction, which executes in a single cycle when used in the repeat mode with the repeat counter having a maximum count of 256.

IMPLEMENTING LARGE FFT'S

Figure 23 shows the memory configurations and transfers for a 1024-point complex FFT computed as four 256-point complex FFTs. A kernel 256-point complex FFT can operate on a group of 256 complex points at one time using on-chip RAM. Data transfers between on-chip and off-chip RAM are efficiently performed via the RPTK and BLKD instructions.

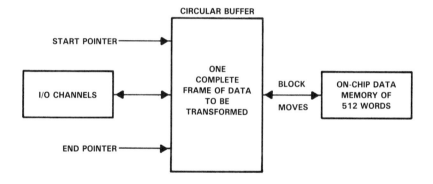

Figure 22. Use of On-Chip Memory to Speed FFT Execution

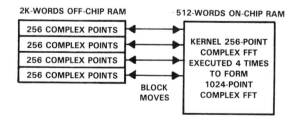

Figure 23. Execution of a 1024-Point Complex FFT with On-Chip RAM

Figure 24 shows a more detailed block diagram of a 1024-point radix-2 complex FFT. It can be seen that 512 butterflies must be performed at each stage. The first eight stages have a total of 4096 butterflies computed by four 256-point FFTs. The 256-point FFT in Appendix C is used as a subroutine for this purpose. Appendix D contains a listing for the 1024-point complex FFT performed with the help of on-chip RAM. However, due to the size of the 1024-point FFT program, the user may find it necessary to subdivide the code into smaller sections prior to assembly.

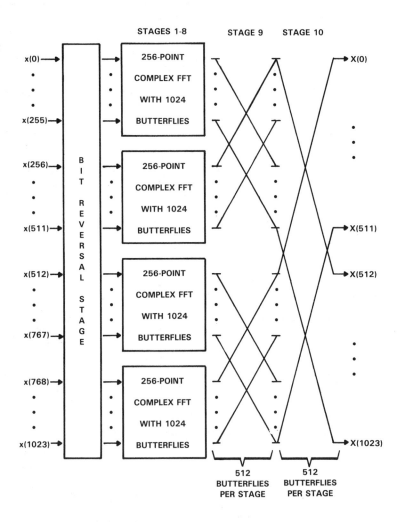

Figure 24. A 1024-Point Complex FFT Using a 256-Point Kernel for Stages 1-8

4. Implementation of Fast Fourier Transform Algorithms with the TMS32020

HIGHER-RADIX FFT'S

The same decomposition principle as for radix-2 FFT algorithms applies to higher radices as well. Table 3 shows the computation requirements[7] for a 4096-point FFT using various radices that are a power of two.

The main benefit of using higher-radix algorithms is the reduced amount of arithmetic operations required for the FFT computation. Beyond the use of radix-8 algorithms, however, the point of diminishing returns rapidly approaches. Memory addressing, data scaling, and program control become more and more complicated. On the other hand, a suitable combination of the radix-2, radix-4, and radix-8 algorithms becomes a flexible and efficient "mixed-radix" algorithm for most FFT applications. Reference [4] contains some useful FORTRAN routines for higher-radix FFT algorithms.

REAL TRANSFORMS VIA COMPLEX FFT'S

In practice, many signals are real functions of time, whereas the FFT algorithm has been derived for complex signals. This means that for real inputs, the imaginary parts of the complex entries are simply set to zero. This results in a certain amount of redundancy. To utilize the bandwidth of the FFT algorithm more effectively, one can use the fact that the frequency spectrum of a real signal is a hermitian function (i.e., the real part is an even function while the imaginary part is an odd function).[11] For example, two N-point real FFTs can be computed simultaneously with a single N-point complex FFT.[8,9,10] On the other hand, a single N-point complex FFT can also be utilized effectively to perform a 2N-point real FFT.[10] Such algorithms substantially reduce the amount of computation required for real FFTs.

Other efficient algorithms exist that further utilize the properties of real signals. In particular, the FFTs of four N-point real symmetric (even) and antisymmetric (odd) sequences can be computed with just one N-point complex FFT.[2] Alternatively, the same N-point complex FFT can be used to compute the FFTs of four N-point real symmetric (even) sequences simultaneously.[2]

INVERSE FFT

The inverse FFT is given by the equation

$$x(n) = (1/N) \sum_{k=0}^{N-1} X(k) \, W_N^{-nk} \quad n = 0,1,\ldots,N-1 \quad (20)$$

where $X(k)$ is the Fourier transform of the time-domain signal $x(n)$. Note that (20) is essentially the same equation as (1), which represents the "forward" FFT, with two important differences: the scaling factor is $(1/N)$, and the exponent of the twiddle factor is the negative of the one in equation (1). Because of the similarity between (20) and (1), the implementation of the inverse FFT is very straightforward. The code can be derived from that given in the appendices by applying the following modifications.

If the forward FFT is implemented with scaling, the resulting values in the frequency domain are $(1/N)X(k)$ and not $X(k)$. Hence, for the inverse FFT, no scaling must be applied in order to get back the original signal. On the other hand, if the forward FFT has not been scaled, the inverse FFT must be scaled. This scaling can be performed all at one point as suggested by (20), or at every stage as described earlier in the forward FFT.

The negative exponent of the twiddle factors implies that the values of $\sin(X)$ will have the opposite sign from that in the forward FFT. Therefore, one way to implement the inverse FFT is to have an additional table with the negatives of $\sin(X)$. Another method is possible if the complex conjugate of (20) is considered.

$$x^*(n) = (1/N) \sum_{k=0}^{N-1} X^*(k) \, W_N^{nk} \quad n = 0,1,\ldots,N-1 \quad (21)$$

In (21), the asterisk indicates complex conjugate. In this form, there is no need to have an additional table for $\sin(X)$. Instead, the inverse FFT is implemented by applying the forward FFT on the complex conjugate of $X(k)$ (with appropriate scaling). The complex conjugate of the resulting time signal is the desired result. Note that if $x(n)$ is real, this last step is not necessary since, in this case, $x^*(n) = x(n)$.

Table 3. Computational Requirements for Higher-Radix FFT Algorithms

ALGORITHM	NUMBER OF REAL MULTIPLICATIONS	NUMBER OF REAL ADDITIONS
RADIX-2 $\left(N = 2^{12}\right)$	81,924	139,266
RADIX-4 $\left(N = 4^6\right)$	57,348	126,978
RADIX-8 $\left(N = 8^4\right)$	49,156	126,978
RADIX-16 $\left(N = 16^3\right)$	48,132	125,442

Table 4. FFT Performance for a TMS32020 Implementation

FFT			EXECUTION		
ALGORITHM	SIZE	TYPE	CYCLES	CLOCK	TIME
RADIX-2	128-Pt	Looped	21,879	5 MHz	4.375 ms
RADIX-2	256-Pt	Looped	42,416	5 MHz	8.483 ms
RADIX-2	256-Pt	Straight-Line	22,595	5 MHz	4.519 ms
RADIX-2	1024-Pt	Straight-Line	159,099	5 MHz	31.8198 ms
RADIX-4	256-Pt	Straight-Line	15,551	5 MHz	3.1102 ms

FFT PERFORMANCE TIMING

Table 4 provides the FFT timing performance for the TMS32020 code in the appendices. The source code examples included in Appendices C through G are not optimized for any specific application since they have been designed to emphasize clarity rather than code optimization. The key feature of these codes is that they do not require any scratch-pad (temporary) memory locations. Consequently, these codes should be useful in memory-critical applications. For time-critical applications, the codes can be optimized for better execution time. Higher execution speed is achieved by using straightline instead of looped code. The tradeoff for this optimization is the larger program memory requirements of the straightline code.

DESCRIPTION OF THE APPENDICES

At the end of this report, there are five appendices with TMS32020 code implementing several FFTs. The contents of the appendices are the following:

Appendix A: N-point, radix-2, DIT FFT (9 macros)
Appendix B: N-point, radix-2, DIT FFT using indirect addressing (7 macros)
Appendix C: 256-point, radix-2 DIT FFT
Appendix D: 1024-point, radix-2 DIT FFT
Appendix E: 256-point, radix-4 DIF FFT
Appendix F: 128-point, radix-2 DIF FFT (looped code)
Appendix G: 256-point, radix-2 DIF FFT (looped code)

SUMMARY

The purpose of this report has been to develop an understanding of the underlying principles in FFT implementations with the TMS32020 processor. The book by Burrus and Parks[4] contains examples of FFT implementations on the TMS32010 processor, the first member of the TMS320 family.

This report has discussed the development of the DFT algorithm, leading to the derivation of the FFT algorithm. The implementation of the radix-2 DIT FFT algorithm was covered in detail, and the radix-4 DIF FFT algorithm was also explained. Special attention was given to various FFT implementation aspects, such as scaling, system memory, and input/output considerations.

The TMS32020 digital signal processor offers many advantages for the implementation of FFT algorithms. Its 200-ns cycle time and special features, such as the single-cycle multiplication, allow high execution speed. The 544 16-bit words of on-chip memory permit the implementation of a 256-point complex FFT without access to external memory, thus further reducing execution time. Furthermore, special instructions, such as RPTK and BLKD, allow the quick transfer of data from external to internal memory, so that portions of large FFTs can be implemented with the on-chip RAM. Due to the flexibility of the TMS32020, the designer can trade-off program memory with execution speed.

REFERENCES

1. B. Gold and C.M. Rader, *Digital Processing of Signals*, McGraw-Hill (1969).
2. A.V. Oppenheim and R.W. Schafer, *Digital Signal Processing*, Prentice-Hall (1975).
3. L.R. Rabiner and B. Gold, *Theory and Applications of Digital Signal Processing*, Prentice-Hall (1975).
4. C.S. Burrus and T.W. Parks, *DFT/FFT and Convolution Algorithms - Theory and Implementation*, John Wiley & Sons (1985).
5. *TMS32020 User's Guide*, Texas Instruments (1985).
6. H.W. Johnson and C.S. Burrus, "An In-order, In-place Radix-2 FFT," *1984 IEEE ICASSP Proceedings*, 28A.2.1-2.4 (March 1984).
7. G.D. Bergland, "A Fast Fourier Transform Algorithm using Base-8 Iterations," *Mathematics of Computation*, **Vol 22**, No. 102, 275-279 (April 1968).
8. J.W. Cooley, P.A.W. Lewis, and P.D. Welch, "The Fast Fourier Transform Algorithm - Programming Considerations in the Calculation of Sine, Cosine, and Laplace Transforms," *Journal of Sound Vibration*, **Vol 12**, 315-337 (July 1970).
9. G.D. Bergland, "A Radix-8 Fast Fourier Transform Subroutine for Real Valued Series," *IEEE Transcriptions of Audio and Electroactivity*, **Vol AU-17**, No. 2 (June 1969).
10. E.O. Brigham, *The Fourier Transform*, Prentice-Hall (1974).
11. R Bracewell, *The Fourier Transform and Its Applications*, McGraw-Hill (1965).

APPENDIX A
FFT MACRO LIBRARY (DIRECT ADDRESSING)

APPENDIX A

```
**********************************************************
* THIS TMS32020 FFT MACRO LIBRARY CONTAINS A TOTAL OF 9   *
* MACROS FOR IMPLEMENTING A GENERAL RADIX-2 DIT N-POINT   *
* FFT. ALL DIT BUTTERFLIES ARE IMPLEMENTED WITH DYNAMIC   *
* SCALING TO AVOID ARITHMETIC OVERFLOWS. TWO'S-COMPLEMENT *
* FIXED-POINT FRACTIONAL ARITHMETIC IS USED THROUGHOUT.   *
* WHILE THESE MACROS ARE NOT NECESSARILY OPTIMIZED FOR    *
* SPEED, THEY ARE SO STRUCTURED THAT NO INTERMEDIATE OR   *
* TEMPORARY REGISTERS ARE REQUIRED FOR THEIR EXECUTION.   *
* CONSEQUENTLY, THESE MACROS ARE PARTICULARLY USEFUL IN   *
* APPLICATIONS WHERE MEMORY SPACE PROVES TO BE CRITICAL.  *
* IN ADDITION, ALL FFT COMPUTATIONS ARE PERFORMED         *
* IN-PLACE AND THE REAL AND IMAGINARY PARTS OF ALL COMPLEX*
* INPUTS ARE ASSUMED TO BE IN CONSECUTIVE DATA MEMORY     *
* LOCATIONS.                                              *
**********************************************************
```

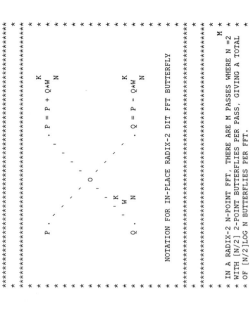

NOTATION FOR IN-PLACE RADIX-2 DIT FFT BUTTERFLY

```
**********************************************************
*                                                      M *
* IN A RADIX-2 N-POINT FFT, THERE ARE M PASSES WHERE N =2 *
* WITH [N/2] 2-POINT BUTTERFLIES PER PASS, GIVING A TOTAL *
* OF [N/2]LOG N BUTTERFLIES PER FFT.                      *
*            2                                           *
**********************************************************
```

```
**********************************************************
* MACRO 1: W =1, k=0.                                     *
*           N                                             *
*                                                        *
*  P=(PR+jPI)       P+Q=(PR+QR)+j(PI+QI)                  *
*             \    /                                      *
*              \  /                                       *
*              /  \                                       *
*             /    \                                      *
*  Q=(QR+jQI)       P-Q=(PR-QR)+j(PI-QI)                  *
*                                                        *
*          -j(2(pi)/N)k                                   *
*  W =e              =COS[(2(pi)/N)k]-jSIN[(2(pi)/N)k]    *
*   N                                                     *
*                   =WR+jWI                               *
*                                                        *
*                   =1                                    *
*                                                        *
* ALL OUTPUT SAMPLES ARE SCALED DOWN BY 2 TO ACCOMMODATE  *
* A 1-BIT OVERFLOW. HOWEVER, NO OVERFLOWS WILL OCCUR      *
* FOR FRACTIONAL INPUTS OF THE FORM X : -1 <= X < 1.  A   *
* TOTAL OF 10 INSTRUCTIONS IS USED.  EXECUTION            *
* TIME IS EQUAL TO 10 MACHINE CYCLES.                     *
**********************************************************
ZERO    $MACRO   PR,PI,QR,QI
*
*       CALCULATE Re[P+Q] AND Re[P-Q]
*
        LAC    :PR:,15    ACC := (1/2)(PR)
        ADD    :QR:,15    ACC := (1/2)(PR+QR)
        SACH   :PR:       PR  := (1/2)(PR+QR)
        SUBH   :QR:       ACC := (1/2)(PR+QR)-(QR)
        SACH   :QR:       QR  := (1/2)(PR-QR)
*
*       CALCULATE Im[P+Q] AND Im[P-Q]
*
        LAC    :PI:,15    ACC := (1/2)(PI)
        ADD    :QI:,15    ACC := (1/2)(PI+QI)
        SACH   :PI:       PI  := (1/2)(PI+QI)
        SUBH   :QI:       ACC := (1/2)(PI+QI)-(QI)
        SACH   :QI:       QI  := (1/2)(PI-QI)
        $END
*
```

```
********************************************
*                                          *
*       MACRO 2: W =-1, k=N/2.             *
*                   N                       *
*                                          *
*    P=(PR+jPI)      P-Q=(PR-QR)+j(PI-QI)   *
*                 \ /                       *
*                  X                        *
*                 / \                       *
*    Q=(QR+jQI)      P+Q=(PR+QR)+j(PI+QI)   *
*                                          *
*          -j(2(pi)/N)k                     *
*    W =e           =COS[(2(pi)/N)k]-jSIN[(2(pi)/N)k] *
*     N                                     *
*                   =WR+jWI                 *
*                                          *
*                   =-1                     *
*                                          *
*    ALL OUTPUT SAMPLES ARE SCALED DOWN BY 2 TO ACCOMMODATE *
*    A 1-BIT OVERFLOW. HOWEVER, NO OVERFLOWS WILL OCCUR *
*    FOR FRACTIONAL INPUTS OF THE FORM X : -1 <= X < 1.  A *
*    TOTAL OF 10 INSTRUCTIONS IS USED. EXECUTION *
*    TIME IS EQUAL TO 10 MACHINE CYCLES.    *
*                                          *
********************************************
PI      SMACRO   PR,PI,QR,QI
*
*       CALCULATE Re[P+Q] AND Re[P-Q]
*
        LAC     :PR.,15  ACC  := (1/2)(PR)
        SUB     :QR.,15  ACC  := (1/2)(PR-QR)
        SACH    :PR:      PR   := (1/2)(PR-QR)
        ADDH    :QR:      ACC  := (1/2)(PR-QR)+(QR)
        SACH    :QR:      QR   := (1/2)(PR+QR)
*
*       CALCULATE Im[P+Q] AND Im[P-Q]
*
        LAC     :PI.,15  ACC  := (1/2)(PI)
        SUB     :QI.,15  ACC  := (1/2)(PI-QI)
        SACH    :PI:      PI   := (1/2)(PI-QI)
        ADDH    :QI:      ACC  := (1/2)(PI-QI)+(QI)
        SACH    :QI:      QI   := (1/2)(PI+QI)
        SEND
*
```

```
********************************************
*                                          *
*       MACRO 3: W =-j, k=N/4.             *
*                   N                       *
*                                          *
*    P=(PR+jPI)      P-jQ=(PR+QI)+j(PI-QR)  *
*                 \ /                       *
*                  X                        *
*                 / \                       *
*    Q=(QR+jQI)      P+jQ=(PR-QI)+j(PI+QR)  *
*                                          *
*          -j(2(pi)/N)k                     *
*    W =e           =COS[(2(pi)/N)k]-jSIN[(2(pi)/N)k] *
*     N                                     *
*                   =WR+jWI                 *
*                                          *
*                   =-j                     *
*                                          *
*    ALL OUTPUT SAMPLES ARE SCALED DOWN BY 2 TO ACCOMMODATE *
*    A 1-BIT OVERFLOW. HOWEVER, NO OVERFLOWS WILL OCCUR *
*    FOR FRACTIONAL INPUTS OF THE FORM X : -1 <= X < 1.  A *
*    TOTAL OF 11 INSTRUCTIONS IS USED. EXECUTION *
*    TIME IS EQUAL TO 11 MACHINE CYCLES.    *
*                                          *
*    WARNING: THIS MACRO REQUIRES THE INPUT SAMPLES QR AND *
*             QI TO BE IN CONSECUTIVE DATA MEMORY LOCATIONS *
*             OF ASCENDING ORDER. THE FOLLOWING STEPS ARE *
*             USED TO IMPLEMENT THIS MACRO:  *
*                                          *
*             (1) [PI-QR] -----> [PI]       *
*             (2) [PI+QR] -----> [QR]       *
*             (3) [PR+QI] -----> [PR]       *
*             (4) [PR-QI] -----> [ACC]      *
*             (5) [QR]    -----> [QI]       *
*             (6) [ACC]   -----> [QR]       *
*                                          *
********************************************
PIBY2   SMACRO   PR,PI,QR,QI
*
*       CALCULATE Re[P+jQ] AND Re[P-jQ]
*
        LAC     :PI.,15   ACC  := (1/2)(PI)
        SUB     :QR.,15   ACC  := (1/2)(PI-QR)
        SACH    :PI:       PI   := (1/2)(PI-QR)
        ADDH    :QR:       ACC  := (1/2)(PI-QR)+(QR)
        SACH    :QR:       QR   := (1/2)(PI+QR)
*
*       CALCULATE Im[P+jQ] AND Im[P-jQ]
*
        LAC     :PR.,15   ACC  := (1/2)(PR)
        ADD     :QI.,15   ACC  := (1/2)(PR+QI)
```

```
*        CALCULATE Im[P+jQ] AND Im[P-jQ]
*
         LAC    :PR:,15   ACC := (1/2)(PR)
         SUB    :QI:,15   ACC := (1/2)(PR-QI)
         SACH   :PR:      PR  := (1/2)(PR-QI)
         ADDH   :QI:      ACC := (1/2)(PR-QI)+(QI)
         DMOV   :QR:      QR  -> QI
         SACH   :QR:      QR  := (1/2)(PR+QI)
         $END
*
```

```
*************************************************************
*                                                           *
*     MACRO 5:   k=N/8.                                      *
*                                                           *
*     P=(PR+jPI)   P+Q*W=(PR+Re[Q*W])+j(PI+Im[Q*W])          *
*                        \  /                                *
*     Q=(QR+jQI)   P-Q*W=(PR-Re[Q*W])+j(PI-Im[Q*W])          *
*                        /  \                                *
*          -j(2(pi)/N)k                                      *
*     W =e            =COS((pi)/4)-jSIN((pi)/4)=WR+jWI        *
*     N                                                      *
*                                                            *
*     LET          W=|COS((pi)/4)|=|SIN((pi)/4)|             *
*                                                            *
*     THEN         [Q*W]=(QR+QI)*W+j(QI-QR)*W                *
*                                                            *
*                  Re[Q*W]=(QI+QR)*W                         *
*                                                            *
*                  Im[Q*W]=(QI-QR)*W                         *
*                                                            *
*     ALL OUTPUT SAMPLES ARE SCALED DOWN BY 2 TO ACCOMMODATE *
*     A 1-BIT OVERFLOW. HOWEVER, NO OVERFLOWS WILL OCCUR     *
*     FOR FRACTIONAL INPUTS OF THE FORM X : -1 <= X < 1. A   *
*     TOTAL OF 20 INSTRUCTIONS ARE USED SUCH THAT EXECUTION  *
*     TIME IS EQUAL TO 20 MACHINE CYCLES.  THIS MACRO        *
*     REQUIRES W TO BE THE ABSOLUTE VALUE (MAGNITUDE) OF     *
*     COS((pi)/4) AND SIN((pi)/4). THE SIGNS OF THESE TRIG   *
*     FUNCTIONS HAVE BEEN TAKEN CARE OF IN THE CODE.         *
*************************************************************
PIBY4   $MACRO    PR,PI,QR,QI,W
*
         LT     :W:       T-REGISTER :=W=COS(PI/4)=SIN(PI/4)
         LAC    :QI:,14   ACC := (1/4)(QI)
         SUB    :QR:,14   ACC := (1/4)(QI-QR)
         SACH   :QI:,15   QI  := (1/2)(QI-QR)
         ADD    :QR:,15   ACC := (1/4)(QI+QR)
         SACH   :QR:,1    QR  := (1/2)(QI+QR)
         LAC    :PR:,14   ACC := (1/4)(PR)
         MPY    :QR:      P-REGISTER := (1/4)(QI+QR)*W
         APAC             ACC := (1/4)[PR+(QI+QR)*W]
         SACH   :PR:,1    PR  := (1/2)[PR+(QI+QR)*W]
```

```
         SACH   :PR:      PR  := (1/2)(PR+QI)
         SUBH   :QI:      ACC := (1/2)(PR+QI)-(QI)
         DMOV   :QR:      QR  -> QI
         SACH   :QR:      QR  := (1/2)(PR-QI)
         $END
*
```

```
*************************************************************
*                                                           *
*     MACRO 4: W =j, k=3N/4.                                 *
*              N                                             *
*                                                            *
*     P=(PR+jPI)   P+jQ=(PR-QI)+j(PI+QR)                     *
*                        \  /                                *
*     Q=(QR+jQI)   P-jQ=(PR+QI)+j(PI-QR)                     *
*                        /  \                                *
*          -j(2(pi)/N)k                                      *
*     W =e            =COS[(2(pi)/N)k]-jSIN[(2(pi)/N)k]       *
*     N                                                      *
*                  =WR+jWI                                   *
*                                                            *
*                  =j                                        *
*                                                            *
*     ALL OUTPUT SAMPLES ARE SCALED DOWN BY 2 TO ACCOMMODATE *
*     A 1-BIT OVERFLOW. HOWEVER, NO OVERFLOWS WILL OCCUR     *
*     FOR FRACTIONAL INPUTS OF THE FORM X : -1 <= X < 1. A   *
*     TOTAL OF 10 INSTRUCTIONS ARE USED SUCH THAT EXECUTION  *
*     TIME IS EQUAL TO 10 MACHINE CYCLES.                    *
*                                                            *
*     WARNING: THIS MACRO REQUIRES THE INPUT SAMPLES QR AND  *
*     QI TO BE IN CONSECUTIVE DATA MEMORY LOCATIONS          *
*     ASCENDING ORDER. THE FOLLOWING STEPS ARE USED          *
*     TO IMPLEMENT THIS MACRO:                               *
*                                                            *
*     (1) [PI-QR]  ----->  [PI]                              *
*     (2) [PI+QR]  ----->  [QR]                              *
*     (3) [PR+QI]  ----->  [PR]                              *
*     (4) [PR-QI]  ----->  [ACC]                             *
*     (5) [QR]     ----->  [QI]                              *
*     (6) [ACC]    ----->  [QR]                              *
*************************************************************
PI3BY2  $MACRO    PR,PI,QR,QI
*
*        CALCULATE Re[P+jQ] AND Re[P-jQ]
*
         LAC    :PI:,15   ACC := (1/2)(PI)
         ADD    :QR:,15   ACC := (1/2)(PI+QR)
         SACH   :PI:      PI  := (1/2)(PI+QR)
         SUBH   :QR:      ACC := (1/2)(PI+QR)-(QR)
         SACH   :QR:      QR  := (1/2)(PI-QR)
*
```

```
MACRO 6 (continued):

        SPAC                ACC := (1/4)(PR)
        SPAC                ACC := (1/4)[PR-(QI+QR)*W]
        SACH  :QR,1         QR := (1/2)[PR-(QI+QR)*W]
        LAC   :PI,14        ACC := (1/4)(PI)
        MPY   :QI:          P-REGISTER := (1/4)(QI+QR)*W
        APAC                ACC := (1/4)[PI+(QI+QR)*W]
        SACH  :PI,1         PI := (1/2)[PI+(QI+QR)*W]
        SPAC                ACC := (1/4)[PI-(QI-QR)*W]
        SACH  :QI,1         QI := (1/2)[PI-(QI-QR)*W]
        $END
*
***********************************************************
*                                                         *
*   MACRO 6:   k=3N/8.                                     *
*                                                         *
*   P=(XR+jXI)   X+Q*W=(XR+Re[Q*W])+j(XI-Im[Q*W])          *
*                       \   /                             *
*   Q=(QR+jQI)   X-Q*W=(XR-Re[Q*W])+j(XI+Im[Q*W])          *
*                                                         *
*        -j(2(pi)/N)k                                      *
*   W =e          =COS(3(pi)/4)-jSIN(3(pi)/4)              *
*   N                                                     *
*                =WR+jWI                                   *
*                                                         *
*   LET    W=|COS(3(pi)/4)|=|SIN(3(pi)/4)|                 *
*                                                         *
*   THEN   [Q*W]=(QI-QR)*W-j(QI+QR)*W                      *
*                                                         *
*          Re[Q*W]=(QI-QR)*W                               *
*                                                         *
*          Im[Q*W]=(QI+QR)*W                               *
*                                                         *
*   ALL OUTPUT SAMPLES ARE SCALED DOWN BY 2 TO ACCOMMODATE *
*   A 1-BIT OVERFLOW. HOWEVER, NO OVERFLOWS WILL OCCUR     *
*   FOR FRACTIONAL INPUTS OF THE FORM X : -1 <= X < 1. A   *
*   TOTAL OF 20 INSTRUCTIONS ARE USED SUCH THAT EXECUTION  *
*   TIME IS EQUAL TO 20 MACHINE CYCLES.  THIS MACRO        *
*   REQUIRES W TO BE THE ABSOLUTE VALUE (MAGNITUDE) OF     *
*   COS(3(pi)/4) AND SIN(3(pi)/4).  THE SIGNS OF THESE TRIG *
*   FUNCTIONS HAVE BEEN TAKEN CARE OF IN THE CODE.         *
***********************************************************
PI3BY4  $MACRO  PR,PI,QR,QI,W
        LT    :W:     T-REGISTER := W=COS(PI/4)=SIN(PI/4)
        LAC   :QI,14         ACC := (1/4)(QI)
        SUB   :QR,14         ACC := (1/4)(QI-QR)
        SACH  :QI,1          QI := (1/2)(QI-QR)
        ADD   :QR,15         ACC := (1/4)(QI+QR)
```

```
MACRO 7 (continued):

        SACH  :QR,1          QR := (1/2)(QI+QR)
        LAC   :PR,14         ACC := (1/4)(PR)
        MPY   :QI:           P-REGISTER := (1/4)(QI-QR)*W
        APAC                 ACC := (1/4)[PR+(QI-QR)*W]
        SACH  :PR,1          PR := (1/2)[PR+(QI-QR)*W]
        SPAC                 ACC := (1/4)(PR)
        SPAC                 ACC := (1/4)[PR-(QI-QR)*W]
        MPY   :QR:           P-REGISTER := (1/4)(QI+QR)*W
        SACH  :QR,1          QR := (1/2)[PR-(QI-QR)*W]
        LAC   :PI,14         ACC := (1/4)(PI)
        SPAC                 ACC := (1/4)[PI-(QI+QR)*W]
        SACH  :PI,1          PI := (1/2)[PI-(QI+QR)*W]
        APAC                 ACC := (1/4)(PI)
        APAC                 ACC := (1/4)[PI+(QI+QR)*W]
        SACH  :QI,1          QI := (1/2)[PI+(QI+QR)*W]
        $END
*
***********************************************************
*                                                         *
*   MACRO 7:  A GENERAL RADIX-2 DIT FFT 'BUTTERFLY'        *
*                                                         *
*   P=(PR+jPI)   P+Q*W=(PR+Re[Q*W])+j(PI-Im[Q*W])          *
*                       \   /                             *
*   Q=(QR+jQI)   P-Q*W=(PR-Re[Q*W])+j(PI+Im[Q*W])          *
*                                                         *
*        -j(2(pi)/N)k                                      *
*   W =e          =COS(O)-jSIN(O)                          *
*   N                                                     *
*                =WR+jWI                                   *
*                                                         *
*   ALL OUTPUT SAMPLES ARE SCALED DOWN BY 2 TO ACCOMMODATE *
*   A 1-BIT OVERFLOW. HOWEVER, NO OVERFLOWS WILL OCCUR     *
*   FOR FRACTIONAL INPUTS OF THE FORM X : -1 <= X < 1. A   *
*   TOTAL OF 23 INSTRUCTIONS ARE USED SUCH THAT EXECUTION  *
*   TIME IS EQUAL TO 23 MACHINE CYCLES.  THIS MACRO        *
*   REQUIRES W TO BE THE ABSOLUTE VALUE (MAGNITUDE) OF     *
*   COS(O) AND SIN(O).  THE SIGNS OF THESE TRIG FUNCTIONS  *
*   HAVE BEEN TAKEN CARE OF IN THE CODE.                   *
***********************************************************
NORM1   $MACRO  PR,PI,QR,QI,WR,WI
*
*   CALCULATE QR*WR + QI*WI AND STORE RESULT IN QI
*
        LT    :QI:    T-REGISTER := QI
        MPYK  :WI:    P-REGISTER := (1/16)(QI*WI)
        LTP   :QR:    ACC := (1/16)(QI*WI); LOAD T WITH QR
        MPYK  :WR:    P-REGISTER := (1/16)(QR*WR)
        APAC          ACC := (1/16)(QR*WR+QI*WI)
        SFR           ACC := (1/32)(QR*WR+QI*WI)
        SACH  :QR,4   QR := (1/2)(QR*WR+QI*WI)
```

```
*
*     CALCULATE QI*WR - QR*WI AND STORE RESULT IN QR
*
      MPYK -:WI:     P-REGISTER := (1/16)(-QR*WI)
      LTP  :QI:      ACC := (1/16)(-QR*WI); T-REGISTER := QI
      MPYK :WR:      P-REGISTER := (1/16)(QI*WR)
      APAC           ACC := (1/16)(QI*WR-QR*WI)
      SFR            ACC := (1/32)(QI*WR-QR*WI)
      SACH :QI:,4    QI := (1/2)(QI*WR-QR*WI)
*
*     CALCULATE Re[P+jQ] & Re[P-jQ] STORE RESULTS in PR & QR
*
      LAC  :PR:,14   ACC := (1/4)PR
      ADD  :QR:,15   ACC := (1/4)[PR+(QR*WR+QI*WI]]
      SACH :PR:,1    PR := (1/2)[PR+(QR*WR+QI*WI]]
      SUBH :QR:      ACC := (1/4)[PR-(QR*WR+QI*WI]]
      SACH :QR:,1    QR := (1/2)[PR-(QR*WR+QI*WI]]
*
*     CALCULATE Im[P+jQ] & Im[P-jQ] STORE RESULTS in PI & QI
*
      LAC  :PI:,14   ACC := (1/4)PI
      ADD  :QI:,15   ACC := (1/4)[PI+(QI*WR-QR*WI]]
      SACH :PI:,1    PI := (1/2)[PI+(QI*WR-QR*WI]]
      SUBH :QI:      ACC := (1/4)[PI-(QI*WR-QR*WI]]
      SACH :QI:,1    QI := (1/2)[PI-(QI*WR-QR*WI]]
      $END

***************************************************
*                                                 *
*    MACRO 8:  A GENERAL RADIX-2 DIT FFT 'BUTTERFLY' *
*                                                 *
*   P=(PR+jPI)  P+Q*W=(PR+Re[Q*W])+j(PI-Im[Q*W])  *
*                      \   /                       *
*   Q=(QR+jQI)         / \                         *
*                 P-Q*W=(PR-Re[Q*W])+j(PI+Im[Q*W]) *
*                                                 *
*        -j(2(pi)/N)k                              *
*   W =e            =COS(O)-jSIN(O)                *
*   N                                             *
*                =WR+jWI                          *
*                                                 *
*   ALL OUTPUT SAMPLES ARE SCALED DOWN BY 2 TO ACCOMMODATE *
*   A 1-BIT OVERFLOW.  HOWEVER, NO OVERFLOWS WILL OCCUR    *
*   FOR FRACTIONAL INPUTS OF THE FORM X : -1 <= X < 1.  A  *
*   TOTAL OF 22 INSTRUCTIONS ARE USED SUCH THAT EXECUTION  *
*   TIME IS EQUAL TO 22 MACHINE CYCLES.  THIS MACRO        *
*   REQUIRES W TO BE THE ABSOLUTE VALUE (MAGNITUDE) OF     *
*   COS(O) AND SIN(O).  THE SIGNS OF THESE TRIG FUNCTIONS  *
*   HAVE BEEN TAKEN CARE OF IN THE CODE.                   *
*                                                 *
***************************************************
NORM2  SMACRO  PR,PI,QR,QI,WR,WI
*
*      CALCULATE QR*WR + QI*WI AND STORE RESULT IN QR
*
       LT   :QR:     T-REGISTER := QR
       MPY  :WR:     P-REGISTER := (1/2)(QR*WR)
       LTP  :QI:     ACC := (1/2)(QR*WR); T-REGISTER := QI
```

```
      MPY  :WI:      P-REGISTER := (1/2)(QI*WI)
      APAC           ACC := (1/2)(QR*WR+QI*WI)
      MPY  :WR:      P-REGISTER := (1/2)(QI*WR)
      LT   :QR:      LOAD T-REGISTER WITH QR
      SACH :QR:      QR := (1/2)(QR*WR+QI*WI)
*
*     CALCULATE QI*WR - QR*WI AND STORE RESULT IN QI
*
      PAC            ACC= (1/2)(QI*WR)
      MPY  :WI:      P-REGISTER := (1/2)(QR*WI)
      SPAC           ACC= (1/2)(QI*WR-QR*WI)
      SACH :QI:      QI = (1/2)(QI*WR-QR*WI)
*
*     CALCULATE Re[Pm+1] & Re[Qm+1] STORE RESULTS in PR & QR
*
      LAC  :PR:,14   ACC= (1/4)PR
      ADD  :QR:,15   ACC= (1/4)[PR+(QR*WR+QI*WI]]
      SACH :PR:,1    PR = (1/2)[PR+(QR*WR+QI*WI]]
      SUBH :QR:      ACC= (1/4)[PR-(QR*WR+QI*WI]]
      SACH :QR:,1    QR = (1/2)[PR-(QR*WR+QI*WI]]
*
*     CALCULATE Im[Pm+1] & Im[Qm+1] STORE RESULTS in PI & QI
*
      LAC  :PI:,14   ACC= (1/4)PI
      ADD  :QI:,15   ACC= (1/4)[PI+(QI*WR-QR*WI]]
      SACH :PI:,1    PI = (1/2)[PI+(QI*WRQR*WI]]
      SUBH :QI:      ACC= (1/4)[PI-(QI*WR-QR*WI]]
      SACH :QI:,1    QI = (1/2)[PI-(QI*WR-QR*WI]]
      $END

*****************************************************
*                                                   *
*   MACRO 9:                                         *
*                                                   *
*                       A ___ A'    B1              *
*                          \ /                      *
*                           X                       *
*                          / \                      *
*                       B  ___ B'                   *
*                       C  ___ C'    B2             *
*                          \ /                      *
*                           X                       *
*                          / \                      *
*                       D  ___ D'    B3             *
*                                    B4             *
*                                                   *
*   A = R1+jI1                                       *
*                                                   *
*   B = R2+jI2                                       *
*                                                   *
*   C = R3+jI3                                       *
*                                                   *
*   D = R4+jI4                                       *
*                                                   *
*   A' = (R1+R2) + j(I1+I2)                          *
*                                                   *
*   B' = (R1-R2) + j(I1-I2)                          *
*                                                   *
*   C' = (R3+R4) + j(R3+R4)                          *
*                                                   *
```

```
*                                                                  *
*    D' = (R3-R4) + j(R3-R4)                                       *
*                                                                  *
*    B1 = A'+C'                                                    *
*                                                                  *
*    B2 = B'-jD'                                                   *
*                                                                  *
*    B3 = A'-C'                                                    *
*                                                                  *
*    B4 = B'+jD'                                                   *
*                                                                  *
*    REAL(B1) = ((R1+R2) + (R3+R4)) /2                             *
*    IMAG(B1) = ((I1+I2) + (I3+I4)) /2                             *
*                                                                  *
*    REAL(B2) = ((R1+R2) + (I3-I4)) /2                             *
*    IMAG(B2) = ((I1-I2) - (R3-R4)) /2                             *
*                                                                  *
*    REAL(B3) = ((R1+R2) - (R3+R4)) /2                             *
*    IMAG(B3) = ((I1+I2) - (I3+I4)) /2                             *
*                                                                  *
*    REAL(B4) = ((R1-R2) - (I3-I4)) /2                             *
*    IMAG(B4) = ((I1-I2) + (R3-R4)) /2                             *
*                                                                  *
* THE FIRST TWO STAGES OF A RADIX-2 N-POINT DIT FFT CAN BE         *
* IMPLEMENTED WITH A SPECIAL RADIX-4 BUTTERFLY WHICH HAS A         *
* UNITY TWIDDLE FACTOR USED TO SPEED UP THE EXECUTION TIME         *
* OF THE FFT WITH THE ABOVE EQUATIONS.  WHEN USING THESE           *
* EQUATIONS, ALL INPUT VALUES MUST BE WITHIN THE RANGE             *
* -1 <= X < 1.0.  TOTAL NUMBER OF INTRUCTIONS IS 37.               *
* EXECUTION TIME IS EQUIVALENT TO 39 MACHINE CYCLES.               *
*                                                                  *
********************************************************************


********************************************************************
*                                                                  *
*   THE FOLLOWING STEPS ARE USED TO IMPLEMENT THE SPECIAL          *
*   RADIX-4 MACRO 'COMBO' FOR THE FIRST TWO STAGES OF AN           *
*   N-POINT RADIX-2 DIT FFT.                                       *
*                                                                  *
*   STEP 1        STEP 2              STEP 3                        *
*   ------        ------              ------                        *
*                                                                  *
*   R1 R1         (R1+R2)/1           [(R1+R2)+(R3+R4)]/2           *
*   I1 I1         (I1+I2)/1           [(I1+I2)+(I3+I4)]/2           *
*                                                                  *
*   R2 R2         [R1-R2)+(I3-I4)]/2  [(R1-R2)+(I3-I4)]/2           *
*   I2 I2         [(I1-I2)-(R3-R4)]/2 [(I1-I2)-(R3-R4)]/2           *
*                                                                  *
*   R3 (R3+R4)/1  (R3+R4)/1           [(R1+R2)-(R3+R4)]/2           *
*   I3 (I3+I4)/1  (I3+I4)/1           [(I1+I2)-(I3+I4)]/2           *
*                                                                  *
*   R4 (R3-R4)/1  [R1-R2)-(I3-I4)]/2  [(R1-R2)-(I3-I4)]/2           *
*   I4 (I3-I4)/1  R4--->I4=(R3-R4)/1                                *
*                 [(I1-I2)+(R3-R4)]/2 [(I1-I2)+(R3-R4)]/2           *
*                                                                  *
********************************************************************
COMBO   $MACRO   R1,I1,R2,I2,R3,I3,R4,I4


*   CALCULATE PARTIAL TERMS FOR R3,R4,I3 AND I4           *
*                                                         *
        LAC     :R3.,14       ACC := (1/4)(R3)
        ADD     :R4.,14       ACC := (1/4)(R3+R4)
        SACH    :R3.,1        R3  := (1/2)(R3+R4)
        SUB     :R4.,15       ACC := (1/4)(R3+R4)-(1/2)(R4)
        SACH    :R4.,1        R4  := (1/2)(R3-R4)
        LAC     :I3.,14       ACC := (1/4)(I3)
        ADD     :I4.,14       ACC := (1/4)(I3+I4)
        SACH    :I3.,1        I3  := (1/2)(I3+I4)
        SUB     :I4.,15       ACC := (1/4)(I3+I4)-(1/2)(I4)
        SACH    :I4.,1        I4  := (1/2)(I3-I4)
*                                                         *
*   CALCULATE PARTIAL TERMS FOR R2,R4,I2 AND I4           *
*                                                         *
        LAC     :R1.,14       ACC := (1/4)(R1)
        ADD     :R2.,14       ACC := (1/4)(R1+R2)
        SACH    :R1.,1        R1  := (1/2)(R1+R2)
        SUB     :R2.,15       ACC := (1/4)(R1+R2)-(1/2)(R2)
        ADD     :I4.,15       ACC := (1/4)[(R1-R2)+(I3-I4)]
        SACH    :R2.,         R2  := (1/4)[(R1-R2)+(I3-I4)]
        SUBH    :I4.          ACC := (1/4)[(R1-R2)-(I3-I4)]
        DMOV    :R4.          R4  = (1/2)(R3-R4)
        SACH    :R4.          R4  := (1/4)[(R1-R2)-(I3-I4)]
        LAC     :I1.,14       ACC := (1/4)(I1)
        ADD     :I2.,14       ACC := (1/4)(I1+I2)
        SACH    :I1.,1        I1  := (1/2)(I1+I2)
        SUB     :I2.,15       ACC := (1/4)(I1+I2)-(1/2)(I2)
        SUB     :I4.,15       ACC := (1/4)[(I1-I2)-(R3-R4)]
        SACH    :I2.          I2  := (1/4)[(I1-I2)-(R3-R4)]
        ADDH    :I4.          ACC := (1/4)[(I1-I2)+(R3-R4)]
        SACH    :I4.          I4  := (1/4)[(I1-I2)+(R3-R4)]
*                                                         *
*   CALCULATE PARTIAL TERMS FOR R1,R3,I1 AND I3           *
*                                                         *
        LAC     :R1.,15       ACC := (1/4)(R1+R2).
        ADD     :R3.,15       ACC := (1/4)[(R1+R2)+(R3+R4)]
        SACH    :R1.          R1  := (1/4)[(R1+R2)+(R3+R4)]
        SUBH    :R3.          ACC := (1/4)[(R1+R2)-(R3+R4)]
        SACH    :R3.          R3  := (1/4)[(R1+R2)-(R3+R4)]
        LAC     :I1.,15       ACC := (1/4)(I1+I2)
        ADD     :I3.,15       ACC := (1/4)[(I1+I2)+(I3+I4)]
        SACH    :I1.          I1  := (1/4)[(I1+I2)+(I3+I4)]
        SUBH    :I3.          ACC := (1/4)[(I1+I2)-(I3+I4)]
        SACH    :I3.          I3  := (1/4)[(I1+I2)-(I3+I4)]
        SEND
```

APPENDIX B
FFT MACRO LIBRARY (INDIRECT ADDRESSING)

```
          A P P E N D I X   B
          -------------------

*********************************************
* THIS TMS32020 FFT MACRO LIBRARY CONTAINS A TOTAL OF 7   *
* MACROS FOR IMPLEMENTING A GENERAL RADIX-2 DIT N-POINT   *
* FFT USING INDIRECT ADDRESS FORMAT OF THE TMS32020. ALL  *
* DIT BUTTERFLIES ARE IMPLEMENTED WITH DYNAMIC SCALING TO *
* AVOID ARITHMETIC OVERFLOWS. 2'S COMPLEMENT FIXED-POINT  *
* FRACTIONAL ARITHMETIC IS USED THROUGHOUT. WHILE THESE   *
* MACROS ARE NOT NECESSARILY OPTIMISED FOR SPEED, THEY    *
* ARE SO STRUCTURED THAT NO INTERMEDIATE OR TEMPORARY     *
* REGISTERS ARE REQUIRED FOR THEIR EXECUTION. HENCE,      *
* THESE MACROS ARE PARTICULARLY USEFUL IN APPLICATIONS    *
* WHERE MEMORY SPACE PROVES TO BE CRITICAL. IN ADDITION,  *
* ALL FFT COMPUTATIONS ARE DONE IN PLACE AND THE REAL AND *
* IMAGINARY PARTS OF ALL COMPLEX INPUTS ARE ASSUMED TO BE *
* IN CONSECUTIVE DATA MEMORY LOCATIONS.                   *
*********************************************

*                                         *
*                                 K       *
*    P  .                   . P = P + Q*W  *
*        ` .             ' '            N  *
*          ` .         ' '                 *
*            ` .     O                     *
*              ' K                         *
*              ' W                         *
*    Q  .      ' N          . Q = P - Q*W  *
*                                      N   *
*                                         *
*        NOTATION FOR RADIX-2 DIT FFT BUTTERFLY  *
*********************************************
* IN A RADIX-2 N-POINT FFT THERE ARE M PASSES WHERE N = 2  *
* WITH [N/2] 2-POINT BUTTERFLIES PER PASS, GIVING A TOTAL  *
* OF [N/2]LOG N BUTTERFLIES PER FFT.                       *
*             2                                            *
*********************************************
```

```
* MACRO 1: W=1, k=0.                                      *
*                                                         *
* P=(PR+jPI)        P+Q=(PR+QR)+j(PI+QI)                  *
*               \ /                                       *
*               / \                                       *
* Q=(QR+jQI)        P-Q=(PR-QR)+j(PI-QI)                  *
*                                                         *
*           -j(2(pi)/N)k                                  *
* W=e             =COS[((2(pi)/N)k]-jSIN[((2(pi)/N)k]     *
*                                                         *
*         =WR+jWI =1                                      *
*                                                         *
* ALL OUTPUT SAMPLES ARE SCALED DOWN BY 2 TO ACCOMMODATE  *
* A 1-BIT OVERFLOW. HOWEVER, NO OVERFLOWS WILL OCCUR      *
* FOR FRACTIONAL INPUTS OF THE FORM X : -1 <= X < 1. A    *
* TOTAL OF 12 INSTRUCTIONS ARE USED SUCH THAT EXECUTION   *
* TIME IS EQUAL TO 14 MACHINE CYCLES.                     *
*                                                         *
* MACRO ENTRY CONDITION: ARP MUST POINT AT AR1            *
* MACRO EXIT CONDITION : ARP MUST POINT AT AR1            *
*                                                         *
* MEMORY ADDRESS BIAS  : BIAS FOR DATA MEMORY PAGE 4      *
*********************************************
ZERO1   $MACRO  PR,QR,BIAS
*
*        INITIALISE AUXILIARY REGISTERS
*
        LRLK   AR1,:PR:+:BIAS:   AR1 POINTS TO PR
        LRLK   AR2,:QR:+:BIAS:   AR2 POINTS TO QR
*
*        CALCULATE Re[P+Q] AND Re[P-Q]
*
        LAC    *,15,AR2    ACC := (1/2)(PR)
        ADD    *,15,AR1    ACC := (1/2)(PR+QR)
        SACH   *+,0,AR2    PR  := (1/2)(PR+QR)
        SUBH   *,0,AR1     ACC := (1/2)(PR+QR)-(QR)
        SACH   *+,0,AR1    QR  := (1/2)(PR-QR)
*
*        CALCULATE Im[P+Q] AND Im[P-Q]
*
        LAC    *,15,AR2    ACC := (1/2)(PI)
        ADD    *,15,AR1    ACC := (1/2)(PI+QI)
        SACH   *,0,AR2     PI  := (1/2)(PI+QI)
        SUBH   *,0,AR1     ACC := (1/2)(PI+QI)-(QI)
        SACH   *-,0,AR1    QI  := (1/2)(PI-QI)
        $END
```

```
        SUBH    *-              ACC := (1/2)(PR+QI)-(QI)
        DMOV    *               QR -> QI
        SACH    *,0,AR1         QR := (1/2)(PR-QI)
        $END
```

```
**********************************************************************
*                                                                    *
*       MACRO 3:    k=N/8.                                            *
*                                                                    *
*   P=(PR+jPI)    P+Q*W=(PR+Re[Q*W])+j(PI+Im[Q*W])                   *
*                         \ /                                         *
*   Q=(QR+jQI)    P-Q*W=(PR-Re[Q*W])+j(PI-Im[Q*W])                   *
*                         / \                                         *
*   W=e       -j(2(pi)/N)k  =COS((pi)/4)-jSIN((pi)/4)=WR+jWI          *
*   N                                                                 *
*                        =WR+jWI                                      *
*                                                                    *
*   LET              W=|COS((pi)/4)|=|SIN((pi)/4)|                    *
*                                                                    *
*   THEN     [Q*W]=(QR+QI)*W+j(QI-QR)*W                               *
*            Re[Q*W]=(QI+QR)*W                                        *
*            Im[Q*W]=(QI-QR)*W                                        *
*                                                                    *
*  ALL OUTPUT SAMPLES ARE SCALED DOWN BY 2 TO ACCOMMODATE            *
*  A 1-BIT OVERFLOW. HOWEVER, NO OVERFLOWS WILL OCCUR                *
*  FOR FRACTIONAL INPUTS OF THE FORM X : -1 <= X < 1. A              *
*  TOTAL OF 22 INSTRUCTIONS ARE USED SUCH THAT EXECUTION             *
*  TIME IS EQUAL TO 24 MACHINE CYCLES. THIS MACRO REQUIRES           *
*  W TO BE THE ABSOLUTE VALUE (MAGNITUDE) OF COS( /4) AND            *
*  SIN( /4). THE SIGNS OF THESE TRIG FUNCTIONS HAVE BEEN             *
*  TAKEN CARE OF IN THE CODE.                                        *
*                                                                    *
*  MACRO ENTRY CONDITION: ARP MUST POINT AT AR1                      *
*  MACRO EXIT CONDITION : ARP MUST POINT AT AR1                      *
*                                                                    *
*  MEMORY ADDRESS BIAS  : BIAS FOR DATA MEMORY PAGE 4                *
**********************************************************************
PBY4I   SMACRO   PR,QR,BIAS
*
        LRLK    AR1,:PR:+:BIAS:+1   ARI POINTS TO QI
        LRLK    AR2,:PR:+:BIAS:     AR2 POINTS TO PR
*
        LAC     *-,14           ACC := (1/4)(QI)
        SUB     *+,14           ACC := (1/4)(QI-QR)
        SACH    *-,1            QI  := (1/2)(QI-QR)
        ADD     *,15            ACC := (1/4)(QI+QR)
        SACH    *,1,AR2         QR  := (1/2)(QI+QR)
        LAC     *,14,AR0        ACC := (1/4)(PR)
        LT      *,AR1           T   :=-W=COS(PI/4)=SIN(PI/4)
        MPY     *,AR2           P   := (1/4)(QI+QR)*W
        APAC                    ACC := (1/4)[PR+(QI+QR)*W]
        SACH    *+,1,AR1        PR  := (1/2)[PR+(QI+QR)*W]
        SPAC                    ACC := (1/4)(PR)
        SPAC                    ACC := (1/4)[PR-(QI+QR)*W]
        SACH    *+,1,AR2        QR  := (1/2)[PR-(QI+QR)*W]
        LAC     *,14,AR1        ACC := (1/4)(PI)
```

```
**********************************************************
*                                                        *
*       MACRO 2: W=-j, k=N/4.                             *
*                                                        *
*   P=(PR+jPI)     P-jQ=(PR+QI)+j(PI-QR)                  *
*                         \ /                             *
*   Q=(QR+jQI)     P+jQ=(PR-QI)+j(PI+QR)                  *
*                         / \                             *
*   W=e       -j(2(pi)/N)k  =COS[(2(pi)/N)k]-jSIN[(2(pi)/N)k] *
*                        =WR+jWI  =-j                     *
*                                                        *
*  ALL OUTPUT SAMPLES ARE SCALED DOWN BY 2 TO ACCOMMODATE *
*  A 1-BIT OVERFLOW. HOWEVER, NO OVERFLOWS WILL OCCUR     *
*  FOR FRACTIONAL INPUTS OF THE FORM X : -1 <= X < 1. A   *
*  TOTAL OF 13 INSTRUCTIONS ARE USED SUCH THAT EXECUTION  *
*  TIME IS EQUAL TO 15 MACHINE CYCLES.                    *
*                                                        *
*  WARNING: THIS MACRO REQUIRES THE INPUT SAMPLES QR AND  *
*           QI TO BE IN CONSECUTIVE DATA MEMORY LOCATIONS *
*           ASCENDING ORDER. THE FOLLOWING STEPS ARE USED *
*           TO IMPLEMENT THIS MACRO:                      *
*                                                        *
*        (1) [PI-QR] -----> [PI]                          *
*        (2) [PI+QR] -----> [QR]                          *
*        (3) [PR+QI] -----> [PR]                          *
*        (4) [PR-QI] -----> [ACC]                         *
*        (5) [QI]    -----> [QR]                          *
*        (6) [ACC]   -----> [QR]                          *
*                                                        *
*  MACRO ENTRY CONDITION: ARP MUST POINT AT AR1           *
*  MACRO EXIT CONDITION : ARP MUST POINT AT AR1           *
*                                                        *
*  MEMORY ADDRESS BIAS  : BIAS FOR DATA MEMORY PAGE 4     *
**********************************************************
PBY2I   SMACRO   PR,QR,BIAS
*
*       INITIALISE AUXILIARY REGISTERS
*
        LRLK    AR1,:PR:+:BIAS:+1   ARI POINTS TO PI
        LRLK    AR2,:QR:+:BIAS:     AR2 POINTS TO QR
*
*       CALCULATE Re[P+jQ] AND Re[P-jQ]
*
        LAC     *,15,AR2        ACC := (1/2)(PI)
        SUB     *,15,AR1        ACC := (1/2)(PI-QR)
        SACH    *,0,AR2         PI  := (1/2)(PI-QR)
        ADDH    *               ACC := (1/2)(PI-QR)+(QR)
        SACH    *+,0,AR1        QR  := (1/2)(PI+QR)
*
*       CALCULATE Im[P+jQ] AND Im[P-jQ]
*
        LAC     *,15,AR2        ACC := (1/2)(PR)
        ADD     *,15,AR1        ACC := (1/2)(PR+QI)
        SACH    *,0,AR2         PR  := (1/2)(PR+QI)
```

```
MPY    *,AR2        P   := (1/4)(QI-QR)*W
APAC               ACC := (1/4)[PI+(QI-QR)*W]
SACH   *-,1,AR1    PI  := (1/2)[PI+(QI-QR)*W]
SPAC               ACC := (1/4)(PI)
SACH   *-,1,AR1    QI  := (1/2)[PI-(QI-QR)*W]
SEND
```

```
*********************************************************
*                                                       *
* MACRO 4:    k=3N/8.                                    *
*                                                        *
* P=(PR+jPI)    P+Q*W=(PR+Re[Q*W])+j(PI+Im[Q*W])         *
*                          \ /                           *
*                          / \                           *
* Q=(QR+jQI)    P-Q*W=(PR-Re[Q*W])+j(PI+Im[Q*W])         *
*                                                        *
*              -j(2(pi)/N)k                              *
* W =e               =COS(3(pi)/4)-jSIN(3(pi)/4)         *
* N                                                      *
*                                                        *
*               =WR+jWI                                  *
*                                                        *
* LET           W=|COS(3(pi)/4)|=|SIN(3(pi)/4)|          *
*                                                        *
* THEN    [Q*W]=(QI-QR)*W-j(QI+QR)*W                     *
*         RE[Q*W]=(QI-QR)*W                              *
*         IM[Q*W]=(QI+QR)*W                              *
*                                                        *
* ALL OUTPUT SAMPLES ARE SCALED DOWN BY 2 TO ACCOMMODATE *
* A 1-BIT OVERFLOW. HOWEVER, NO OVERFLOWS WILL OCCUR     *
* FOR FRACTIONAL INPUTS OF THE FORM X : -1 <= X < 1. A   *
* TOTAL OF 22 INSTRUCTIONS ARE USED SUCH THAT EXECUTION  *
* TIME IS EQUAL TO 24 MACHINE CYCLES. THIS MACRO REQUIRES*
* W TO BE THE ABSOLUTE VALUE (MAGNITUDE) OF COS(3 /4) AND*
* SIN(3 /4).  THE SIGNS OF THESE TRIG FUNCTIONS HAVE BEEN*
* TAKEN CARE OF IN THE CODE.                             *
*                                                        *
* MACRO ENTRY CONDITION: ARP MUST POINT AT AR1           *
* MACRO EXIT CONDITION : ARP MUST POINT AT AR1           *
*                                                        *
* MEMORY ADDRESS BIAS  : BIAS FOR DATA MEMORY PAGE 4     *
*                                                        *
*********************************************************
P3BY4I  $MACRO  PR,QR,BIAS
*
        LRLK    AR1,:QR:+:BIAS:+1  AR1 POINTS TO QI
        LRLK    AR2,:PR:+:BIAS:    AR2 POINTS TO PR
*
        LAC     *-,14     ACC := (1/4)(QI)
        SUB     *+,14     ACC := (1/4)(QI-QR)
        SACH    *-,1      QI  := (1/2)(QI-QR)
        ADD     *,15      ACC := (1/4)(QI+QR)
        SACH    *+,1,AR2  QR  := (1/2)(QI+QR)
        LAC     *,14,AR0  ACC := (1/4)(PR)
        LT      *,AR1     T   := W=COS( /4)=SIN( /4)
        MPY     *-,AR2    P   := (1/4)(QI-QR)*W
        APAC              ACC := (1/4)[PR+(QI-QR)*W]
        SACH    *+,1,AR1  PR  := (1/2)[PR+(QI-QR)*W]
```

```
SPAC              ACC := (1/4)(PR)
SPAC              ACC := (1/4)[PR-(QI-QR)*W]
MPY    *          P   := (1/4)(QI+QR)*W
SACH   *+,1,AR2   QR  := (1/2)[PR-(QI-QR)*W]
LAC    *,14       ACC := (1/4)(PI)
SPAC              ACC := (1/4)[PI-(QI+QR)*W]
SACH   *-,1,AR1   PI  := (1/2)[PI-(QI+QR)*W]
APAC              ACC := (1/4)(PI)
SACH   *-,1,AR1   QI  := (1/2)[PI+(QI+QR)*W]
$END
```

```
*********************************************************
*                                                       *
* MACRO 5:                                               *
*                                                        *
*                                                        *
*              A           A'            B1              *
*                \       /    \        /                 *
*                  \   /        \    /                    *
*              B     X      B'     X      B2             *
*              C     X      C'     X      B3             *
*                  /   \        /    \                    *
*                /       \    /        \                 *
*              D           D'            B4              *
*                                                        *
* A = R1+jI1                                             *
*                                                        *
* B = R2+jI2                                             *
*                                                        *
* C = R3+jI3                                             *
*                                                        *
* D = R4+jI4                                             *
*                                                        *
* A' = (R1+R2) + j(I1+I2)                                *
*                                                        *
* B' = (R1-R2) + j(I1-I2)                                *
*                                                        *
* C' = (R3+R4) + j(R3+R4)                                *
*                                                        *
* D' = (R3-R4) + j(R3-R4)                                *
*                                                        *
* B1 = A'+C'                                             *
*                                                        *
* B2 = B'-jD'                                            *
*                                                        *
* B3 = A'-C'                                             *
*                                                        *
* B4 = B'+jD'                                            *
*                                                        *
* REAL(B1) = ((R1+R2) + (R3+R4)) /2                      *
* IMAG(B1) = ((I1+I2) + (I3+I4)) /2                      *
*                                                        *
* REAL(B2) = ((R1-R2) + (I3-I4)) /2                      *
* IMAG(B2) = ((I1-I2) - (R3-R4)) /2                      *
*                                                        *
* REAL(B3) = ((R1+R2) - (R3+R4)) /2                      *
* IMAG(B3) = ((I1+I2) - (I3+I4)) /2                      *
```

4. Implementation of Fast Fourier Transform Algorithms with the TMS32020

```
*        REAL(B4) = ((R1-R2) - (I3-I4)) /2             *
*        IMAG(B4) = ((I1-I2) + (R3-R4)) /2             *
*                                                       *
*  THE FIRST TWO STAGES OF A RADIX-2 N-POINT DIT FFT CAN BE  *
*  IMPLEMENTED WITH A SPECIAL RADIX-4 BUTTERFLY WHICH HAS A  *
*  UNITY TWIDDLE FACTOR USED TO SPEED UP THE EXECUTION TIME  *
*  OF THE FFT WITH THE ABOVE EQUATIONS. WHEN USING THESE     *
*  EQUATIONS, ALL INPUT VALUES MUST BE WITHIN THE RANGE      *
*  -1 <= X < 1.0.  TOTAL NUMBER OF INSTRUCTIONS IS 39.       *
*  EXECUTION TIME IS EQUIVALENT TO 41 MACHINE CYCLES.        *
*********************************************************


*********************************************************
*  THE FOLLOWING STEPS ARE USED TO IMPLEMENT THE SPECIAL *
*  RADIX-4 MACRO 'COMBO' FOR THE FIRST TWO STAGES OF AN  *
*  N-POINT RADIX-2 DIT FFT.                              *
*                                                       *
*  STEP 1        STEP 2             STEP 3               *
*  ------        ------             ------               *
*                                                       *
*  R1 R1         (R1+R2)/1          [(R1+R2)+(R3+R4)]/2  *
*  I1 I1         (I1+I2)/1          [(I1+I2)+(I3+I4)]/2  *
*                                                       *
*  R2 R2         [(R1-R2)+(I3-I4)]/2   [(R1-R2)-(I3-I4)]/2 *
*  I2 I2         [(I1-I2)-(R3-R4)]/2   [(I1-I2)+(R3-R4)]/2 *
*                                                       *
*  R3 (R3+R4)/1  (R3+R4)/1          [(R1+R2)-(R3+R4)]/2  *
*  I3 (I3+I4)/1  (I3+I4)/1          [(I1+I2)-(I3+I4)]/2  *
*                                                       *
*  R4 (R3-R4)/1  [(R1-R2)-(I3-I4)]/2   [(R1-R2)+(I3-I4)]/2 *
*  I4 (I3-I4)/1  [(I1-I2)+(R3-R4)]/2   [(I1-I2)-(R3-R4)]/2 *
*********************************************************


*********************************************************
*  IN THE FOLLOWING CODE, ALL IMAGINARY TERMS (I'S) ARE  *
*  ASSUMED TO IN LOCATIONS CONSECUTIVE TO THOSE OF THE   *
*  REAL TERMS (R'S).  THEREFORE, THE ADDRESS OF EACH     *
*  IMAGINARY TERM (I) IS REPRESENTED AS (R+1).           *
*********************************************************
COMBOI  $MACRO  R1,R2,R3,R4,BIAS
*
*       CALCULATE PARTIAL TERMS FOR R3,R4,I3 AND I4
*
        LAC     :R3:-:BIAS:,14      ACC := (1/4)(R3)
        ADD     :R4:-:BIAS:,14      ACC := (1/4)(R3+R4)
        SACH    :R3:-:BIAS:,1       R3  := (1/2)(R3+R4)
        SUB     :R4:-:BIAS:,15      ACC := (1/4)(R3+R4)-(1/2)(R4)
        SACH    :R4:-:BIAS:,1       R4  := (1/2)(R3-R4)
        LAC     :R3:+1-:BIAS:,14    ACC := (1/4)(I3)
        ADD     :R4:+1-:BIAS:,14    ACC := (1/4)(I3+I4)
        SACH    :R3:+1-:BIAS:,1     I3  := (1/2)(I3+I4)
        SUB     :R4:+1-:BIAS:,15    ACC := (1/4)(I3+I4)-(1/2)(I4)
        SACH    :R4:+1-:BIAS:,1     I4  := (1/2)(I3-I4)
```

```
*                                                                      *
*       CALCULATE PARTIAL TERMS FOR R2,R4,I2 AND I4
*
        LAC     :R1:-:BIAS:,14      ACC := (1/4)(R1)
        ADD     :R2:-:BIAS:,14      ACC := (1/4)(R1+R2)
        SACH    :R1:-:BIAS:,1       R1  := (1/2)(R1+R2)
        SUB     :R2:-:BIAS:,15      ACC := (1/4)(R1+R2)-(1/2)(R2)
        ADD     :R4:+1-:BIAS:,15    ACC := (1/4)[(R1-R2)+(I3-I4)]
        SUBH    :R2:-:BIAS:         R2  := (1/4)[(R1-R2)+(I3-I4)]
        SUBH    :R4:+1-:BIAS:       ACC := (1/4)[(R1-R2)-(I3-I4)]
        DMOV    :R4:-:BIAS:         I4  := R4 = (1/2)(R3-R4)
        SACH    :R4:-:BIAS:         R4  := (1/4)[(R1-R2)-(I3-I4)]
        LAC     :R1:+1-:BIAS:,14    ACC := (1/4)(I1)
        ADD     :R2:+1-:BIAS:,14    ACC := (1/4)(I1+I2)
        SACH    :R1:+1-:BIAS:,1     I1  := (1/2)(I1+I2)
        SUB     :R2:+1-:BIAS:,15    ACC := (1/4)(I1+I2)-(1/2)(I2)
        SUB     :R4:+1-:BIAS:,15    ACC := (1/4)[(I1-I2)-(R3-R4)]
        SACH    :R2:+1-:BIAS:       I2  := (1/4)[(I1-I2)-(R3-R4)]
        ADDH    :R4:+1-:BIAS:       ACC := (1/4)[(I1-I2)+(R3-R4)]
        SACH    :R4:+1-:BIAS:       I4  := (1/4)[(I1-I2)+(R3-R4)]
```

```
*                                                                      *
*       CALCULATE PARTIAL TERMS FOR R1,R3,I1 AND I3
*
        LAC     :R1:-:BIAS:,15      ACC := (1/4)(R1+R2),
        ADD     :R3:-:BIAS:,15      ACC := (1/4)[(R1+R2)+(R3+R4)]
        SACH    :R1:-:BIAS:         R1  := (1/4)[(R1+R2)+(R3+R4)]
        SUBH    :R3:-:BIAS:         ACC := (1/4)[(R1+R2)-(R3+R4)]
        SACH    :R3:-:BIAS:         R3  := (1/4)[(R1+R2)-(R3+R4)]
        LAC     :R1:+1-:BIAS:,15    ACC := (1/4)(I1+I2)
        ADD     :R3:+1-:BIAS:,15    ACC := (1/4)[(I1+I2)+(I3+I4)]
        SACH    :R1:+1-:BIAS:       I1  := (1/4)[(I1+I2)+(I3+I4)]
        SUBH    :R3:+1-:BIAS:       ACC := (1/4)[(I1+I2)-(I3+I4)]
        SACH    :R3:+1-:BIAS:       I3  := (1/4)[(I1+I2)-(I3+I4)]
        SEND
```

```
*************************************************************
*                                                           *
*  MACRO 6:  A GENERAL RADIX-2 DIT FFT 'BUTTERFLY'          *
*                                                           *
*  P=(PR+jQI)    P+Q*W=(PR+Re[Q*W])+j(PI+Im[Q*W])           *
*                        \   /                              *
*                         \ /                               *
*                          X                                *
*  Q=(QR+jQI)    P-Q*W=(PR-Re[Q*W])+j(PI+Im[Q*W])           *
*                                                           *
*              -j(2(pi)/N)k                                 *
*  W =e                      =cos(o)-jsin(o)                *
*                                                           *
*  N          =WR+jWI                                       *
*                                                           *
*  ALL OUTPUT SAMPLES ARE SCALED DOWN' BY 2 TO ACCOMMODATE  *
*  A 1-BIT OVERFLOW. HOWEVER, NO OVERFLOWS WILL OCCUR       *
*  FOR FRACTIONAL INPUTS OF THE FORM X : -1 <= X < 1. A     *
*  TOTAL OF 25 INSTRUCTIONS ARE USED SUCH THAT EXECUTION    *
*  TIME IS EQUAL TO 27 MACHINE CYCLES.  THIS MACRO          *
*  REQUIRES W TO BE THE ABSOLUTE VALUE (MAGNITUDE) OF       *
*  COS(O) AND SIN(O).  THE SIGNS OF THESE TRIG FUNCTIONS    *
*  HAVE BEEN TAKEN CARE OF IN THE CODE.                     *
*                                                           *
```

```
*    MACRO ENTRY CONDITION: ARP MUST POINT AT AR1          *
*    MACRO EXIT CONDITION: ARP MUST POINT AT AR1           *
*                                                          *
*    MEMORY ADDRESS BIAS    : BIAS FOR DATA MEMORY PAGE 4  *
*                                                          *
***********************************************************
BTRFLI  $MACRO  PR,QR,WR,WI,BIAS
*
*       INITIALISE AUXILIARY REGISTERS
*
        LRLK  AR1,;QR:+:BIAS:+1  AR1 POINTS TO QI
        LRLK  AR2,;PR:+:BIAS:    AR2 POINTS TO PR
*
*       CALCULATE QR*WR + QI*WI AND STORE RESULT IN QI
*
        LT    *-          T-REGISTER := QI
        MPYK  :WI:        P-REGISTER := (1/16)(QI*WI)
        LTP   *           ACC := (1/16)(QI*WI); T-REGISTER=QR
        MPYK  :WR:        P-REGISTER := (1/16)(QR*WR)
        APAC              ACC := (1/16)(QR*WR+QI*WI)
        SFR               ACC := (1/32)(QR*WR+QI*WI)
        SACH  *+,4        QR  := (1/2)(QR*WR+QI*WI)


*
*       CALCULATE QI*WR - QR*WI AND STORE RESULT IN QR
*
        MPYK  -:WI:       P-REGISTER := (1/16)(-QR*WI)
        LTP   *           ACC := (1/16)(-QR*WI); T-REGISTER=QI
        MPYK  :WR:        P-REGISTER := (1/16)(QI*WR)
        APAC              ACC := (1/16)(QI*WR-QR*WI)
        SFR               ACC := (1/2)(QI*WR-QR*WI)
        SACH  *-,4,AR2    QR  := (1/2)(QI*WR-QR*WI)

*
*       CALCULATE Re[P+jQ] & Re[P-jQ] STORE RESULTS in PR & QR
*
        LAC   *,14,AR1    ACC := (1/4)PR
        ADD   *,15,AR2    ACC := (1/4)[PR+(QR*WR+QI*WI)]
        SACH  *+,1,AR1    PR  := (1/2)[PR+(QR*WR+QI*WI)]
        SUBH  *           ACC := (1/4)[PR-(QR*WR+QI*WI)]
        SACH  *+,1,AR2    QR  := (1/2)[PR-(QR*WR+QI*WI)]
*
*       CALCULATE Im[P+jQ] & Im[P-jQ] STORE RESULTS in PI & QI
*
        LAC   *,14,AR1    ACC := (1/4)PI
        ADD   *,15,AR2    ACC := (1/4)[PI+(QI*WR-QR*WI)]
        SACH  *-,1,AR1    PI  := (1/2)[PI+(QI*WR-QR*WI)]
        SUBH  *           ACC := (1/4)[PI-(QI*WR-QR*WI)]
        SACH  *-,1,AR1    QI  := (1/2)[PI-(QI*WR-QR*WI)]
        $END


***********************************************************
*                                                         *
*       MACRO 7: RADIX-2 INPUT BIT REVERSAL               *
*                                                         *
*       ENTRY CONDITION: ARP MUST POINT AT AR1            *
*                                                         *
```

```
*    EXIT CONDITION: ARP MUST POINT AT AR1               *
*                                                        *
*    A TOTAL OF 10 INSTRUCTIONS ARE USED SUCH THAT       *
*    EXECUTION TIME IS EQUAL TO 12 MACHINE CYCLES.       *
*                                                        *
**********************************************************
BITRVI  $MACRO  PR,QR,BIAS
*
*       INITIALISE AUXILIARY REGISTERS
*
        LRLK  AR1,;PR:+:BIAS:    AR1 POINTS TO PR
        LRLK  AR2,;QR:+:BIAS:    AR2 POINTS TO QR
*
        ZALH  *,AR2
        ADDS  *,AR1
        SACL  *+,0,AR2
        SACH  *+,0,AR1
        ZALH  *,AR2
        ADDS  *,AR1
        SACL  *-,0,AR2
        SACH  *-,0,AR1
        $END
```

APPENDIX C
A 256-POINT, RADIX-2 DIT FFT IMPLEMENTATION

```
A P P E N D I X   C
-------------------
        IDT    'FFT256'
*********************************************************
*                                                       *
*    A 256-POINT RADIX-2 DIT COMPLEX FFT FOR THE TMS32020 *
*                                                       *
*    THE FOLLOWING FILE RAD2FFT.MAC CONSISTS OF ALL THE  *
*    MACROS LISTED IN APPENDIX B                         *
*                                                       *
*********************************************************
*
*        COPY    RAD2FFT.MAC
*
*********************************************************
*                                                       *
*    DATA MEMORY MAP FOR PAGES 4, 5, 6 AND 7 (BLOCKS B0,B1) *
*                                                       *
*********************************************************
*
        DORG    0
*
*    DATA MEMORY PAGE 4 (STARTING ADDRESS 512 OR >200)
*
X000    DATA    0,0
X001    DATA    0,0
X002    DATA    0,0
X003    DATA    0,0
X004    DATA    0,0
X005    DATA    0,0
X006    DATA    0,0
X007    DATA    0,0
X008    DATA    0,0
X009    DATA    0,0
X010    DATA    0,0
X011    DATA    0,0
X012    DATA    0,0
X013    DATA    0,0
X014    DATA    0,0
X015    DATA    0,0
X016    DATA    0,0
X017    DATA    0,0
X018    DATA    0,0
X019    DATA    0,0
X020    DATA    0,0
X021    DATA    0,0
X022    DATA    0,0
X023    DATA    0,0
X024    DATA    0,0
X025    DATA    0,0
X026    DATA    0,0
X027    DATA    0,0
X028    DATA    0,0
X029    DATA    0,0
X030    DATA    0,0
X031    DATA    0,0
X032    DATA    0,0
X033    DATA    0,0
X034    DATA    0,0
X035    DATA    0,0
X036    DATA    0,0
X037    DATA    0,0
X038    DATA    0,0
X039    DATA    0,0
X040    DATA    0,0
X041    DATA    0,0
X042    DATA    0,0
X043    DATA    0,0
X044    DATA    0,0
X045    DATA    0,0
X046    DATA    0,0
X047    DATA    0,0
X048    DATA    0,0
X049    DATA    0,0
X050    DATA    0,0
X051    DATA    0,0
X052    DATA    0,0
X053    DATA    0,0
X054    DATA    0,0
X055    DATA    0,0
X056    DATA    0,0
X057    DATA    0,0
X058    DATA    0,0
X059    DATA    0,0
X060    DATA    0,0
X061    DATA    0,0
X062    DATA    0,0
X063    DATA    0,0
*
*    DATA MEMORY PAGE 5 (STARTING ADDRESS 640 OR >280)
*
X064    DATA    0,0
X065    DATA    0,0
X066    DATA    0,0
X067    DATA    0,0
X068    DATA    0,0
X069    DATA    0,0
X070    DATA    0,0
X071    DATA    0,0
X072    DATA    0,0
X073    DATA    0,0
X074    DATA    0,0
X075    DATA    0,0
X076    DATA    0,0
X077    DATA    0,0
X078    DATA    0,0
X079    DATA    0,0
X080    DATA    0,0
X081    DATA    0,0
X082    DATA    0,0
X083    DATA    0,0
X084    DATA    0,0
X085    DATA    0,0
X086    DATA    0,0
X087    DATA    0,0
X088    DATA    0,0
X089    DATA    0,0
X090    DATA    0,0
```

```
X091    DATA    0,0
X092    DATA    0,0
X093    DATA    0,0
X094    DATA    0,0
X095    DATA    0,0
X096    DATA    0,0
X097    DATA    0,0
X098    DATA    0,0
X099    DATA    0,0
X100    DATA    0,0
X101    DATA    0,0
X102    DATA    0,0
X103    DATA    0,0
X104    DATA    0,0
X105    DATA    0,0
X106    DATA    0,0
X107    DATA    0,0
X108    DATA    0,0
X109    DATA    0,0
X110    DATA    0,0
X111    DATA    0,0
X112    DATA    0,0
X113    DATA    0,0
X114    DATA    0,0
X115    DATA    0,0
X116    DATA    0,0
X117    DATA    0,0
X118    DATA    0,0
X119    DATA    0,0
X120    DATA    0,0
X121    DATA    0,0
X122    DATA    0,0
X123    DATA    0,0
X124    DATA    0,0
X125    DATA    0,0
X126    DATA    0,0
X127    DATA    0,0
*
*
*               DATA MEMORY PAGE 6 (STARTING ADDRESS 768 OR >300)
X128    DATA    0,0
X129    DATA    0,0
X130    DATA    0,0
X131    DATA    0,0
X132    DATA    0,0
X133    DATA    0,0
X134    DATA    0,0
X135    DATA    0,0
X136    DATA    0,0
X137    DATA    0,0
X138    DATA    0,0
X139    DATA    0,0
X140    DATA    0,0
X141    DATA    0,0
X142    DATA    0,0
X143    DATA    0,0
X144    DATA    0,0
X145    DATA    0,0
X146    DATA    0,0
X147    DATA    0,0

X148    DATA    0,0
X149    DATA    0,0
X150    DATA    0,0
X151    DATA    0,0
X152    DATA    0,0
X153    DATA    0,0
X154    DATA    0,0
X155    DATA    0,0
X156    DATA    0,0
X157    DATA    0,0
X158    DATA    0,0
X159    DATA    0,0
X160    DATA    0,0
X161    DATA    0,0
X162    DATA    0,0
X163    DATA    0,0
X164    DATA    0,0
X165    DATA    0,0
X166    DATA    0,0
X167    DATA    0,0
X168    DATA    0,0
X169    DATA    0,0
X170    DATA    0,0
X171    DATA    0,0
X172    DATA    0,0
X173    DATA    0,0
X174    DATA    0,0
X175    DATA    0,0
X176    DATA    0,0
X177    DATA    0,0
X178    DATA    0,0
X179    DATA    0,0
X180    DATA    0,0
X181    DATA    0,0
X182    DATA    0,0
X183    DATA    0,0
X184    DATA    0,0
X185    DATA    0,0
X186    DATA    0,0
X187    DATA    0,0
X188    DATA    0,0
X189    DATA    0,0
X190    DATA    0,0
X191    DATA    0,0
*
*               DATA MEMORY PAGE 7 (STARTING ADDRESS 896 OR >380)
*
X192    DATA    0,0
X193    DATA    0,0
X194    DATA    0,0
X195    DATA    0,0
X196    DATA    0,0
X197    DATA    0,0
X198    DATA    0,0
X199    DATA    0,0
X200    DATA    0,0
X201    DATA    0,0
X202    DATA    0,0
X203    DATA    0,0
X204    DATA    0,0
```

```
*************************************************
*                                               *
*   13-BIT TWIDDLE FACTORS FOR 256-POINT COMPLEX FFT *
*                                               *
*************************************************
*
C000    EQU     4095
C001    EQU     4094
C002    EQU     4091
C003    EQU     4085
C004    EQU     4076
C005    EQU     4065
C006    EQU     4052
C007    EQU     4036
C008    EQU     4017
C009    EQU     3996
C010    EQU     3973
C011    EQU     3948
C012    EQU     3920
C013    EQU     3889
C014    EQU     3857
C015    EQU     3822
C016    EQU     3784
C017    EQU     3745
C018    EQU     3703
C019    EQU     3659
C020    EQU     3612
C021    EQU     3564
C022    EQU     3513
C023    EQU     3461
C024    EQU     3406
C025    EQU     3349
C026    EQU     3290
C027    EQU     3229
C028    EQU     3166
C029    EQU     3102
C030    EQU     3035
C031    EQU     2967
C032    EQU     2896
C033    EQU     2824
C034    EQU     2751
C035    EQU     2675
C036    EQU     2598
C037    EQU     2520
C038    EQU     2440
C039    EQU     2359
C040    EQU     2276
C041    EQU     2191
C042    EQU     2106
C043    EQU     2019
C044    EQU     1931
C045    EQU     1842
C046    EQU     1751
C047    EQU     1660
C048    EQU     1567
C049    EQU     1474
C050    EQU     1380
C051    EQU     1285
C052    EQU     1189
```

```
X205    DATA    0,0
X206    DATA    0,0
X207    DATA    0,0
X208    DATA    0,0
X209    DATA    0,0
X210    DATA    0,0
X211    DATA    0,0
X212    DATA    0,0
X213    DATA    0,0
X214    DATA    0,0
X215    DATA    0,0
X216    DATA    0,0
X217    DATA    0,0
X218    DATA    0,0
X219    DATA    0,0
X220    DATA    0,0
X221    DATA    0,0
X222    DATA    0,0
X223    DATA    0,0
X224    DATA    0,0
X225    DATA    0,0
X226    DATA    0,0
X227    DATA    0,0
X228    DATA    0,0
X229    DATA    0,0
X230    DATA    0,0
X231    DATA    0,0
X232    DATA    0,0
X233    DATA    0,0
X234    DATA    0,0
X235    DATA    0,0
X236    DATA    0,0
X237    DATA    0,0
X238    DATA    0,0
X239    DATA    0,0
X240    DATA    0,0
X241    DATA    0,0
X242    DATA    0,0
X243    DATA    0,0
X244    DATA    0,0
X245    DATA    0,0
X246    DATA    0,0
X247    DATA    0,0
X248    DATA    0,0
X249    DATA    0,0
X250    DATA    0,0
X251    DATA    0,0
X252    DATA    0,0
X253    DATA    0,0
X254    DATA    0,0
X255    DATA    0,0
*
***************************************************
*                                                 *
*   DATA LOCATION IN BLOCK B2 FOR W=COS(45) OR SIN(45) *
*                                                 *
***************************************************
*
        DORG    96
W       DATA    0
```

4. Implementation of Fast Fourier Transform Algorithms with the TMS32020

```
C053    EQU     1092
C054    EQU     995
C055    EQU     897
C056    EQU     799
C057    EQU     700
C058    EQU     601
C059    EQU     501
C060    EQU     401
C061    EQU     301
C062    EQU     201
C063    EQU     101
C064    EQU     0
C065    EQU     -101
C066    EQU     -201
C067    EQU     -301
C068    EQU     -401
C069    EQU     -501
C070    EQU     -601
C071    EQU     -700
C072    EQU     -799
C073    EQU     -897
C074    EQU     -995
C075    EQU     -1092
C076    EQU     -1189
C077    EQU     -1285
C078    EQU     -1380
C079    EQU     -1474
C080    EQU     -1567
C081    EQU     -1660
C082    EQU     -1751
C083    EQU     -1842
C084    EQU     -1931
C085    EQU     -2019
C086    EQU     -2106
C087    EQU     -2191
C088    EQU     -2276
C089    EQU     -2359
C090    EQU     -2440
C091    EQU     -2520
C092    EQU     -2598
C093    EQU     -2675
C094    EQU     -2751
C095    EQU     -2824
C096    EQU     -2896
C097    EQU     -2967
C098    EQU     -3035
C099    EQU     -3102
C100    EQU     -3166
C101    EQU     -3229
C102    EQU     -3290
C103    EQU     -3349
C104    EQU     -3406
C105    EQU     -3461
C106    EQU     -3513
C107    EQU     -3564
C108    EQU     -3612
C109    EQU     -3659
C110    EQU     -3703
C111    EQU     -3745
C112    EQU     -3784

C113    EQU     -3822
C114    EQU     -3857
C115    EQU     -3889
C116    EQU     -3920
C117    EQU     -3948
C118    EQU     -3973
C119    EQU     -3996
C120    EQU     -4017
C121    EQU     -4036
C122    EQU     -4052
C123    EQU     -4065
C124    EQU     -4076
C125    EQU     -4085
C126    EQU     -4091
C127    EQU     -4094
*
S000    EQU     0
S001    EQU     101
S002    EQU     201
S003    EQU     301
S004    EQU     401
S005    EQU     501
S006    EQU     601
S007    EQU     700
S008    EQU     799
S009    EQU     897
S010    EQU     995
S011    EQU     1092
S012    EQU     1189
S013    EQU     1285
S014    EQU     1380
S015    EQU     1474
S016    EQU     1567
S017    EQU     1660
S018    EQU     1751
S019    EQU     1842
S020    EQU     1931
S021    EQU     2019
S022    EQU     2106
S023    EQU     2191
S024    EQU     2276
S025    EQU     2359
S026    EQU     2440
S027    EQU     2520
S028    EQU     2598
S029    EQU     2675
S030    EQU     2751
S031    EQU     2824
S032    EQU     2896
S033    EQU     2967
S034    EQU     3035
S035    EQU     3102
S036    EQU     3166
S037    EQU     3229
S038    EQU     3290
S039    EQU     3349
S040    EQU     3406
S041    EQU     3461
S042    EQU     3513
S043    EQU     3564
```

```
S044    EQU     3612
S045    EQU     3659
S046    EQU     3703
S047    EQU     3745
S048    EQU     3784
S049    EQU     3822
S050    EQU     3857
S051    EQU     3889
S052    EQU     3920
S053    EQU     3948
S054    EQU     3973
S055    EQU     3996
S056    EQU     4017
S057    EQU     4036
S058    EQU     4052
S059    EQU     4065
S060    EQU     4076
S061    EQU     4085
S062    EQU     4091
S063    EQU     4094
S064    EQU     4095
S065    EQU     4094
S066    EQU     4091
S067    EQU     4085
S068    EQU     4076
S069    EQU     4065
S070    EQU     4052
S071    EQU     4036
S072    EQU     4017
S073    EQU     3996
S074    EQU     3973
S075    EQU     3948
S076    EQU     3920
S077    EQU     3889
S078    EQU     3857
S079    EQU     3822
S080    EQU     3784
S081    EQU     3745
S082    EQU     3703
S083    EQU     3659
S084    EQU     3612
S085    EQU     3564
S086    EQU     3513
S087    EQU     3461
S088    EQU     3406
S089    EQU     3349
S090    EQU     3290
S091    EQU     3229
S092    EQU     3166
S093    EQU     3102
S094    EQU     3035
S095    EQU     2967
S096    EQU     2896
S097    EQU     2824
S098    EQU     2751
S099    EQU     2675
S100    EQU     2598
S101    EQU     2520
S102    EQU     2440
S103    EQU     2359
S104    EQU     2276
S105    EQU     2191
S106    EQU     2106
S107    EQU     2019
S108    EQU     1931
S109    EQU     1842
S110    EQU     1751
S111    EQU     1660
S112    EQU     1567
S113    EQU     1474
S114    EQU     1380
S115    EQU     1285
S116    EQU     1189
S117    EQU     1092
S118    EQU     995
S119    EQU     897
S120    EQU     799
S121    EQU     700
S122    EQU     601
S123    EQU     501
S124    EQU     401
S125    EQU     301
S126    EQU     201
S127    EQU     101
*
        AORG    0
        B       INIT
*
* SYSTEM INITIALIZTION
*
        AORG    >20
*
* 16-BIT TWIDDLE FACTOR FOR SPECIAL MACROS
*
WVAL    DATA    >5A82
*
INIT    SPM     0
        CNFD
        ROVM
        SSXM
        LARP    ARO
        LRLK    ARO,W
        LALK    WVAL
        TBLR    *,AR1
*
* FFT CODE WITH BIT-REVERSED INPUT SAMPLES
*
FFT256  BITRVI  X001,X128,512
        BITRVI  X002,X064,512
        BITRVI  X003,X192,512
        BITRVI  X004,X032,512
        BITRVI  X005,X160,512
        BITRVI  X006,X096,512
        BITRVI  X007,X224,512
        BITRVI  X008,X016,512
        BITRVI  X009,X144,512
        BITRVI  X010,X080,512
        BITRVI  X011,X208,512
        BITRVI  X012,X048,512
        BITRVI  X013,X176,512
```

4. Implementation of Fast Fourier Transform Algorithms with the TMS32020

```
BITRVI  X014,X112,512
BITRVI  X015,X240,512
BITRVI  X017,X136,512
BITRVI  X018,X072,512
BITRVI  X019,X200,512
BITRVI  X020,X040,512
BITRVI  X021,X168,512
BITRVI  X022,X104,512
BITRVI  X023,X232,512
BITRVI  X025,X152,512
BITRVI  X026,X088,512
BITRVI  X027,X216,512
BITRVI  X028,X056,512
BITRVI  X029,X184,512
BITRVI  X030,X120,512
BITRVI  X031,X248,512
BITRVI  X033,X132,512
BITRVI  X034,X068,512
BITRVI  X035,X196,512
BITRVI  X037,X164,512
BITRVI  X038,X100,512
BITRVI  X039,X228,512
BITRVI  X041,X148,512
BITRVI  X042,X084,512
BITRVI  X043,X212,512
BITRVI  X044,X052,512
BITRVI  X045,X180,512
BITRVI  X046,X116,512
BITRVI  X047,X244,512
BITRVI  X049,X140,512
BITRVI  X050,X076,512
BITRVI  X051,X204,512
BITRVI  X053,X172,512
BITRVI  X054,X108,512
BITRVI  X055,X236,512
BITRVI  X057,X156,512
BITRVI  X058,X092,512
BITRVI  X059,X220,512
BITRVI  X061,X188,512
BITRVI  X062,X124,512
BITRVI  X063,X252,512
BITRVI  X065,X130,512
BITRVI  X067,X194,512
BITRVI  X069,X162,512
BITRVI  X070,X098,512
BITRVI  X071,X226,512
BITRVI  X073,X146,512
BITRVI  X074,X082,512
BITRVI  X075,X210,512
BITRVI  X077,X178,512
BITRVI  X078,X114,512
BITRVI  X079,X242,512
BITRVI  X081,X138,512
BITRVI  X083,X202,512
BITRVI  X085,X170,512
BITRVI  X086,X106,512
BITRVI  X087,X234,512
BITRVI  X089,X154,512
BITRVI  X091,X218,512
BITRVI  X093,X186,512
BITRVI  X094,X122,512
BITRVI  X095,X250,512
BITRVI  X097,X134,512
BITRVI  X099,X198,512
BITRVI  X101,X166,512
BITRVI  X103,X230,512
BITRVI  X105,X150,512
BITRVI  X107,X214,512
BITRVI  X109,X182,512
BITRVI  X110,X118,512
BITRVI  X111,X246,512
BITRVI  X113,X142,512
BITRVI  X115,X206,512
BITRVI  X117,X174,512
BITRVI  X119,X238,512
BITRVI  X121,X158,512
BITRVI  X123,X222,512
BITRVI  X125,X190,512
BITRVI  X127,X254,512
BITRVI  X131,X193,512
BITRVI  X133,X161,512
BITRVI  X135,X225,512
BITRVI  X137,X145,512
BITRVI  X139,X209,512
BITRVI  X141,X177,512
BITRVI  X143,X241,512
BITRVI  X147,X201,512
BITRVI  X149,X169,512
BITRVI  X151,X233,512
BITRVI  X155,X217,512
BITRVI  X157,X185,512
BITRVI  X159,X249,512
BITRVI  X163,X197,512
BITRVI  X167,X229,512
BITRVI  X171,X213,512
BITRVI  X173,X181,512
BITRVI  X175,X245,512
BITRVI  X179,X205,512
BITRVI  X183,X237,512
BITRVI  X187,X221,512
BITRVI  X191,X253,512
BITRVI  X199,X227,512
BITRVI  X203,X211,512
BITRVI  X207,X243,512
BITRVI  X215,X235,512
BITRVI  X223,X251,512
BITRVI  X239,X247,512

FFT CODE FOR STAGES 1 AND 2

LDPK    4
COMBOI  X000,X001,X002,X003,0
COMBOI  X004,X005,X006,X007,0
COMBOI  X008,X009,X010,X011,0
COMBOI  X012,X013,X014,X015,0
COMBOI  X016,X017,X018,X019,0
COMBOI  X020,X021,X022,X023,0
COMBOI  X024,X025,X026,X027,0
COMBOI  X028,X029,X030,X031,0
COMBOI  X032,X033,X034,X035,0
```

* * *

```
ZEROI   X000,X004,512
PBY4I   X001,X005,512
PBY2I   X002,X006,512
P3BY4I  X003,X007,512
ZEROI   X008,X012,512
PBY4I   X009,X013,512
PBY2I   X010,X014,512
P3BY4I  X011,X015,512
ZEROI   X016,X020,512
PBY4I   X017,X021,512
PBY2I   X018,X022,512
P3BY4I  X019,X023,512
ZEROI   X024,X028,512
PBY4I   X025,X029,512
PBY2I   X026,X030,512
P3BY4I  X027,X031,512
ZEROI   X032,X036,512
PBY4I   X033,X037,512
PBY2I   X034,X038,512
P3BY4I  X035,X039,512
ZEROI   X040,X044,512
PBY4I   X041,X045,512
PBY2I   X042,X046,512
P3BY4I  X043,X047,512
ZEROI   X048,X052,512
PBY4I   X049,X053,512
PBY2I   X050,X054,512
P3BY4I  X051,X055,512
ZEROI   X056,X060,512
PBY4I   X057,X061,512
PBY2I   X058,X062,512
P3BY4I  X059,X063,512
ZEROI   X064,X068,512
PBY4I   X065,X069,512
PBY2I   X066,X070,512
P3BY4I  X067,X071,512
ZEROI   X072,X076,512
PBY4I   X073,X077,512
PBY2I   X074,X078,512
P3BY4I  X075,X079,512
ZEROI   X080,X084,512
PBY4I   X081,X085,512
PBY2I   X082,X086,512
P3BY4I  X083,X087,512
ZEROI   X088,X092,512
PBY4I   X089,X093,512
PBY2I   X090,X094,512
P3BY4I  X091,X095,512
ZEROI   X096,X100,512
PBY4I   X097,X101,512
PBY2I   X098,X102,512
P3BY4I  X099,X103,512
ZEROI   X104,X108,512
PBY4I   X105,X109,512
PBY2I   X106,X110,512
P3BY4I  X107,X111,512
ZEROI   X112,X116,512
PBY4I   X113,X117,512
PBY2I   X114,X118,512
```

```
COMBOI  X036,X037,X038,X039,0
COMBOI  X040,X041,X042,X043,0
COMBOI  X044,X045,X046,X047,0
COMBOI  X048,X049,X050,X051,0
COMBOI  X052,X053,X054,X055,0
COMBOI  X056,X057,X058,X059,0
COMBOI  X060,X061,X062,X063,0
LDPK    5
COMBOI  X064,X065,X066,X067,128
COMBOI  X068,X069,X070,X071,128
COMBOI  X072,X073,X074,X075,128
COMBOI  X076,X077,X078,X079,128
COMBOI  X080,X081,X082,X083,128
COMBOI  X084,X085,X086,X087,128
COMBOI  X088,X089,X090,X091,128
COMBOI  X092,X093,X094,X095,128
COMBOI  X096,X097,X098,X099,128
COMBOI  X100,X101,X102,X103,128
COMBOI  X104,X105,X106,X107,128
COMBOI  X108,X109,X110,X111,128
COMBOI  X112,X113,X114,X115,128
COMBOI  X116,X117,X118,X119,128
COMBOI  X120,X121,X122,X123,128
COMBOI  X124,X125,X126,X127,128
LDPK    6
COMBOI  X128,X129,X130,X131,256
COMBOI  X132,X133,X134,X135,256
COMBOI  X136,X137,X138,X139,256
COMBOI  X140,X141,X142,X143,256
COMBOI  X144,X145,X146,X147,256
COMBOI  X148,X149,X150,X151,256
COMBOI  X152,X153,X154,X155,256
COMBOI  X156,X157,X158,X159,256
COMBOI  X160,X161,X162,X163,256
COMBOI  X164,X165,X166,X167,256
COMBOI  X168,X169,X170,X171,256
COMBOI  X172,X173,X174,X175,256
COMBOI  X176,X177,X178,X179,256
COMBOI  X180,X181,X182,X183,256
COMBOI  X184,X185,X186,X187,256
COMBOI  X188,X189,X190,X191,256
LDPK    7
COMBOI  X192,X193,X194,X195,384
COMBOI  X196,X197,X198,X199,384
COMBOI  X200,X201,X202,X203,384
COMBOI  X204,X205,X206,X207,384
COMBOI  X208,X209,X210,X211,384
COMBOI  X212,X213,X214,X215,384
COMBOI  X216,X217,X218,X219,384
COMBOI  X220,X221,X222,X223,384
COMBOI  X224,X225,X226,X227,384
COMBOI  X228,X229,X230,X231,384
COMBOI  X232,X233,X234,X235,384
COMBOI  X236,X237,X238,X239,384
COMBOI  X240,X241,X242,X243,384
COMBOI  X244,X245,X246,X247,384
COMBOI  X248,X249,X250,X251,384
COMBOI  X252,X253,X254,X255,384
```

FFT CODE FOR STAGE 3

* *

4. Implementation of Fast Fourier Transform Algorithms with the TMS32020

```
P3BY4I    X115,X119,512
ZEROI     X120,X124,512
PBY4I     X121,X125,512
PBY2I     X122,X126,512
P3BY4I    X123,X127,512
ZEROI     X128,X132,512
PBY4I     X129,X133,512
PBY2I     X130,X134,512
P3BY4I    X131,X135,512
ZEROI     X136,X140,512
PBY4I     X137,X141,512
PBY2I     X138,X142,512
P3BY4I    X139,X143,512
ZEROI     X144,X148,512
PBY4I     X145,X149,512
PBY2I     X146,X150,512
P3BY4I    X147,X151,512
ZEROI     X152,X156,512
PBY4I     X153,X157,512
PBY2I     X154,X158,512
P3BY4I    X155,X159,512
ZEROI     X160,X164,512
PBY4I     X161,X165,512
PBY2I     X162,X166,512
P3BY4I    X163,X167,512
ZEROI     X168,X172,512
PBY4I     X169,X173,512
PBY2I     X170,X174,512
P3BY4I    X171,X175,512
ZEROI     X176,X180,512
PBY4I     X177,X181,512
PBY2I     X178,X182,512
P3BY4I    X179,X183,512
ZEROI     X184,X188,512
PBY4I     X185,X189,512
PBY2I     X186,X190,512
P3BY4I    X187,X191,512
ZEROI     X192,X196,512
PBY4I     X193,X197,512
PBY2I     X194,X198,512
P3BY4I    X195,X199,512
ZEROI     X200,X204,512
PBY4I     X201,X205,512
PBY2I     X202,X206,512
P3BY4I    X203,X207,512
ZEROI     X208,X212,512
PBY4I     X209,X213,512
PBY2I     X210,X214,512
P3BY4I    X211,X215,512
ZEROI     X216,X220,512
PBY4I     X217,X221,512
PBY2I     X218,X222,512
P3BY4I    X219,X223,512
ZEROI     X224,X228,512
PBY4I     X225,X229,512
PBY2I     X226,X230,512
P3BY4I    X227,X231,512
ZEROI     X232,X236,512
PBY4I     X233,X237,512
PBY2I     X234,X238,512
P3BY4I    X235,X239,512
ZEROI     X240,X244,512
PBY4I     X241,X245,512
PBY2I     X242,X246,512
P3BY4I    X243,X247,512
ZEROI     X248,X252,512
PBY4I     X249,X253,512
PBY2I     X250,X254,512
P3BY4I    X251,X255,512
```

* * *

```
FFT CODE FOR STAGE 4

ZEROI     X000,X008,512
BTRFLI    X001,X009,C016,S016,512
PBY4I     X002,X010,512
BTRFLI    X003,X011,C048,S048,512
PBY2I     X004,X012,512
BTRFLI    X005,X013,C080,S080,512
P3BY4I    X006,X014,512
BTRFLI    X007,X015,C112,S112,512
ZEROI     X016,X024,512
BTRFLI    X017,X025,C016,S016,512
PBY4I     X018,X026,512
BTRFLI    X019,X027,C048,S048,512
PBY2I     X020,X028,512
BTRFLI    X021,X029,C080,S080,512
P3BY4I    X022,X030,512
BTRFLI    X023,X031,C112,S112,512
ZEROI     X032,X040,512
BTRFLI    X033,X041,C016,S016,512
PBY4I     X034,X042,512
BTRFLI    X035,X043,C048,S048,512
PBY2I     X036,X044,512
BTRFLI    X037,X045,C080,S080,512
P3BY4I    X038,X046,512
BTRFLI    X039,X047,C112,S112,512
ZEROI     X048,X056,512
BTRFLI    X049,X057,C016,S016,512
PBY4I     X050,X058,512
BTRFLI    X051,X059,C048,S048,512
PBY2I     X052,X060,512
BTRFLI    X053,X061,C080,S080,512
P3BY4I    X054,X062,512
BTRFLI    X055,X063,C112,S112,512
ZEROI     X064,X072,512
BTRFLI    X065,X073,C016,S016,512
PBY4I     X066,X074,512
BTRFLI    X067,X075,C048,S048,512
PBY2I     X068,X076,512
BTRFLI    X069,X077,C080,S080,512
P3BY4I    X070,X078,512
BTRFLI    X071,X079,C112,S112,512
ZEROI     X080,X088,512
BTRFLI    X081,X089,C016,S016,512
PBY4I     X082,X090,512
BTRFLI    X083,X091,C048,S048,512
PBY2I     X084,X092,512
BTRFLI    X085,X093,C080,S080,512
P3BY4I    X086,X094,512
BTRFLI    X087,X095,C112,S112,512
```

```
ZEROI    X096,X104,512
BTRFLI   X097,X105,C016,S016,512
PBY4I    X098,X106,512
BTRFLI   X099,X107,C048,S048,512
PBY2I    X100,X108,512
BTRFLI   X101,X109,C080,S080,512
P3BY4I   X102,X110,512
BTRFLI   X103,X111,C112,S112,512
ZEROI    X112,X120,512
BTRFLI   X113,X121,C016,S016,512
PBY4I    X114,X122,512
BTRFLI   X115,X123,C048,S048,512
PBY2I    X116,X124,512
BTRFLI   X117,X125,C080,S080,512
P3BY4I   X118,X126,512
BTRFLI   X119,X127,C112,S112,512
ZEROI    X128,X136,512
BTRFLI   X129,X137,C016,S016,512
PBY4I    X130,X138,512
BTRFLI   X131,X139,C048,S048,512
PBY2I    X132,X140,512
P3BY4I   X133,X141,C080,S080,512
BTRFLI   X134,X142,512
BTRFLI   X135,X143,C112,S112,512
ZEROI    X144,X152,512
BTRFLI   X145,X153,C016,S016,512
PBY4I    X146,X154,512
BTRFLI   X147,X155,C048,S048,512
PBY2I    X148,X156,512
BTRFLI   X149,X157,C080,S080,512
P3BY4I   X150,X158,512
BTRFLI   X151,X159,C112,S112,512
ZEROI    X160,X168,512
BTRFLI   X161,X169,C016,S016,512
PBY4I    X162,X170,512
BTRFLI   X163,X171,C048,S048,512
PBY2I    X164,X172,512
BTRFLI   X165,X173,C080,S080,512
P3BY4I   X166,X174,512
BTRFLI   X167,X175,C112,S112,512
ZEROI    X176,X184,512
BTRFLI   X177,X185,C016,S016,512
PBY4I    X178,X186,512
BTRFLI   X179,X187,C048,S048,512
PBY2I    X180,X188,512
BTRFLI   X181,X189,C080,S080,512
P3BY4I   X182,X190,512
BTRFLI   X183,X191,C112,S112,512
ZEROI    X192,X200,512
BTRFLI   X193,X201,C016,S016,512
PBY4I    X194,X202,512
BTRFLI   X195,X203,C048,S048,512
PBY2I    X196,X204,512
BTRFLI   X197,X205,C080,S080,512
P3BY4I   X198,X206,512
BTRFLI   X199,X207,C112,S112,512
ZEROI    X208,X216,512
BTRFLI   X209,X217,C016,S016,512
PBY4I    X210,X218,512
BTRFLI   X211,X219,C048,S048,512
PBY2I    X212,X220,512
BTRFLI   X213,X221,C080,S080,512
P3BY4I   X214,X222,512
BTRFLI   X215,X223,C112,S112,512
ZEROI    X224,X232,512
BTRFLI   X225,X233,C016,S016,512
PBY4I    X226,X234,512
BTRFLI   X227,X235,C048,S048,512
PBY2I    X228,X236,512
BTRFLI   X229,X237,C080,S080,512
P3BY4I   X230,X238,512
BTRFLI   X231,X239,C112,S112,512
ZEROI    X240,X248,512
BTRFLI   X241,X249,C016,S016,512
PBY4I    X242,X250,512
BTRFLI   X243,X251,C048,S048,512
PBY2I    X244,X252,512
BTRFLI   X245,X253,C080,S080,512
P3BY4I   X246,X254,512
BTRFLI   X247,X255,C112,S112,512

FFT CODE FOR STAGE 5

ZEROI    X000,X016,512
BTRFLI   X001,X017,C008,S008,512
BTRFLI   X002,X018,C016,S016,512
BTRFLI   X003,X019,C024,S024,512
PBY4I    X004,X020,512
BTRFLI   X005,X021,C040,S040,512
BTRFLI   X006,X022,C048,S048,512
BTRFLI   X007,X023,C056,S056,512
PBY2I    X008,X024,512
BTRFLI   X009,X025,C072,S072,512
BTRFLI   X010,X026,C080,S080,512
BTRFLI   X011,X027,C088,S088,512
P3BY4I   X012,X028,512
BTRFLI   X013,X029,C104,S104,512
BTRFLI   X014,X030,C112,S112,512
BTRFLI   X015,X031,C120,S120,512
ZEROI    X032,X048,512
BTRFLI   X033,X049,C008,S008,512
BTRFLI   X034,X050,C016,S016,512
BTRFLI   X035,X051,C024,S024,512
PBY4I    X036,X052,512
BTRFLI   X037,X053,C040,S040,512
BTRFLI   X038,X054,C048,S048,512
BTRFLI   X039,X055,C056,S056,512
PBY2I    X040,X056,512
BTRFLI   X041,X057,C072,S072,512
BTRFLI   X042,X058,C080,S080,512
BTRFLI   X043,X059,C088,S088,512
P3BY4I   X044,X060,512
BTRFLI   X045,X061,C104,S104,512
BTRFLI   X046,X062,C112,S112,512
BTRFLI   X047,X063,C120,S120,512
ZEROI    X064,X080,512
BTRFLI   X065,X081,C008,S008,512
BTRFLI   X066,X082,C016,S016,512
BTRFLI   X067,X083,C024,S024,512
PBY4I    X068,X084,512
```

* * *

4. Implementation of Fast Fourier Transform Algorithms with the TMS32020

```
BTRFLI  X193,X209,C008,S008,512
BTRFLI  X194,X210,C016,S016,512
BTRFLI  X195,X211,C024,S024,512
PBY4I   X196,X212,512
BTRFLI  X197,X213,C040,S040,512
BTRFLI  X198,X214,C048,S048,512
BTRFLI  X199,X215,C056,S056,512
PBY2I   X200,X216,512
BTRFLI  X201,X217,C072,S072,512
BTRFLI  X202,X218,C080,S080,512
BTRFLI  X203,X219,C088,S088,512
P3BY4I  X204,X220,512
BTRFLI  X205,X221,C104,S104,512
BTRFLI  X206,X222,C112,S112,512
BTRFLI  X207,X223,C120,S120,512
ZEROI   X224,X240,512
BTRFLI  X225,X241,C008,S008,512
BTRFLI  X226,X242,C016,S016,512
BTRFLI  X227,X243,C024,S024,512
PBY4I   X228,X244,512
BTRFLI  X229,X245,C040,S040,512
BTRFLI  X230,X246,C048,S048,512
BTRFLI  X231,X247,C056,S056,512
PBY2I   X232,X248,512
BTRFLI  X233,X249,C072,S072,512
BTRFLI  X234,X250,C080,S080,512
BTRFLI  X235,X251,C088,S088,512
P3BY4I  X236,X252,512
BTRFLI  X237,X253,C104,S104,512
BTRFLI  X238,X254,C112,S112,512
BTRFLI  X239,X255,C120,S120,512


FFT CODE FOR STAGE 6

ZEROI   X000,X032,512
BTRFLI  X001,X033,C004,S004,512
BTRFLI  X002,X034,C008,S008,512
BTRFLI  X003,X035,C012,S012,512
BTRFLI  X004,X036,C016,S016,512
BTRFLI  X005,X037,C020,S020,512
BTRFLI  X006,X038,C024,S024,512
BTRFLI  X007,X039,C028,S028,512
PBY4I   X008,X040,512
BTRFLI  X009,X041,C036,S036,512
BTRFLI  X010,X042,C040,S040,512
BTRFLI  X011,X043,C044,S044,512
BTRFLI  X012,X044,C048,S048,512
BTRFLI  X013,X045,C052,S052,512
BTRFLI  X014,X046,C056,S056,512
BTRFLI  X015,X047,C060,S060,512
PBY2I   X016,X048,512
BTRFLI  X017,X049,C068,S068,512
BTRFLI  X018,X050,C072,S072,512
BTRFLI  X019,X051,C076,S076,512
BTRFLI  X020,X052,C080,S080,512
BTRFLI  X021,X053,C084,S084,512
BTRFLI  X022,X054,C088,S088,512
BTRFLI  X023,X055,C092,S092,512
P3BY4I  X024,X056,512
BTRFLI  X025,X057,C100,S100,512
```

* * *

```
BTRFLI  X069,X085,C040,S040,512
BTRFLI  X070,X086,C048,S048,512
BTRFLI  X071,X087,C056,S056,512
PBY2I   X072,X088,512
BTRFLI  X073,X089,C072,S072,512
BTRFLI  X074,X090,C080,S080,512
BTRFLI  X075,X091,C088,S088,512
P3BY4I  X076,X092,512
BTRFLI  X077,X093,C104,S104,512
BTRFLI  X078,X094,C112,S112,512
BTRFLI  X079,X095,C120,S120,512
ZEROI   X096,X112,512
BTRFLI  X097,X113,C008,S008,512
BTRFLI  X098,X114,C016,S016,512
BTRFLI  X099,X115,C024,S024,512
PBY4I   X100,X116,512
BTRFLI  X101,X117,C040,S040,512
BTRFLI  X102,X118,C048,S048,512
BTRFLI  X103,X119,C056,S056,512
PBY2I   X104,X120,512
BTRFLI  X105,X121,C072,S072,512
BTRFLI  X106,X122,C080,S080,512
BTRFLI  X107,X123,C088,S088,512
P3BY4I  X108,X124,512
BTRFLI  X109,X125,C104,S104,512
BTRFLI  X110,X126,C112,S112,512
BTRFLI  X111,X127,C120,S120,512
ZEROI   X128,X144,512
BTRFLI  X129,X145,C008,S008,512
BTRFLI  X130,X146,C016,S016,512
BTRFLI  X131,X147,C024,S024,512
PBY4I   X132,X148,512
BTRFLI  X133,X149,C040,S040,512
BTRFLI  X134,X150,C048,S048,512
BTRFLI  X135,X151,C056,S056,512
PBY2I   X136,X152,512
BTRFLI  X137,X153,C072,S072,512
BTRFLI  X138,X154,C080,S080,512
BTRFLI  X139,X155,C088,S088,512
P3BY4I  X140,X156,512
BTRFLI  X141,X157,C104,S104,512
BTRFLI  X142,X158,C112,S112,512
BTRFLI  X143,X159,C120,S120,512
ZEROI   X160,X176,512
BTRFLI  X161,X177,C008,S008,512
BTRFLI  X162,X178,C016,S016,512
BTRFLI  X163,X179,C024,S024,512
PBY4I   X164,X180,512
BTRFLI  X165,X181,C040,S040,512
BTRFLI  X166,X182,C048,S048,512
BTRFLI  X167,X183,C056,S056,512
PBY2I   X168,X184,512
BTRFLI  X169,X185,C072,S072,512
BTRFLI  X170,X186,C080,S080,512
BTRFLI  X171,X187,C088,S088,512
P3BY4I  X172,X188,512
BTRFLI  X173,X189,C104,S104,512
BTRFLI  X174,X190,C112,S112,512
BTRFLI  X175,X191,C120,S120,512
ZEROI   X192,X208,512
```

```
BTRFLI  X026,X058,C104,S104,512
BTRFLI  X027,X059,C108,S108,512
BTRFLI  X028,X060,C112,S112,512
BTRFLI  X029,X061,C116,S116,512
BTRFLI  X030,X062,C120,S120,512
BTRFLI  X031,X063,C124,S124,512
ZEROI   X064,X096,512
BTRFLI  X065,X097,C004,S004,512
BTRFLI  X066,X098,C008,S008,512
BTRFLI  X067,X099,C012,S012,512
BTRFLI  X068,X100,C016,S016,512
BTRFLI  X069,X101,C020,S020,512
BTRFLI  X070,X102,C024,S024,512
BTRFLI  X071,X103,C028,S028,512
PBY4I   X072,X104,512
BTRFLI  X073,X105,C036,S036,512
BTRFLI  X074,X106,C040,S040,512
BTRFLI  X075,X107,C044,S044,512
BTRFLI  X076,X108,C048,S048,512
BTRFLI  X077,X109,C052,S052,512
BTRFLI  X078,X110,C056,S056,512
BTRFLI  X079,X111,C060,S060,512
PBY2I   X080,X112,512
BTRFLI  X081,X113,C068,S068,512
BTRFLI  X082,X114,C072,S072,512
BTRFLI  X083,X115,C076,S076,512
BTRFLI  X084,X116,C080,S080,512
BTRFLI  X085,X117,C084,S084,512
BTRFLI  X086,X118,C088,S088,512
BTRFLI  X087,X119,C092,S092,512
P3BY4I  X088,X120,512
BTRFLI  X089,X121,C100,S100,512
BTRFLI  X090,X122,C104,S104,512
BTRFLI  X091,X123,C108,S108,512
BTRFLI  X092,X124,C112,S112,512
BTRFLI  X093,X125,C116,S116,512
BTRFLI  X094,X126,C120,S120,512
BTRFLI  X095,X127,C124,S124,512
ZEROI   X128,X160,512
BTRFLI  X129,X161,C004,S004,512
BTRFLI  X130,X162,C008,S008,512
BTRFLI  X131,X163,C012,S012,512
BTRFLI  X132,X164,C016,S016,512
BTRFLI  X133,X165,C020,S020,512
BTRFLI  X134,X166,C024,S024,512
BTRFLI  X135,X167,C028,S028,512
PBY4I   X136,X168,512
BTRFLI  X137,X169,C036,S036,512
BTRFLI  X138,X170,C040,S040,512
BTRFLI  X139,X171,C044,S044,512
BTRFLI  X140,X172,C048,S048,512
BTRFLI  X141,X173,C052,S052,512
BTRFLI  X142,X174,C056,S056,512
BTRFLI  X143,X175,C060,S060,512
PBY2I   X144,X176,512
BTRFLI  X145,X177,C068,S068,512
BTRFLI  X146,X178,C072,S072,512
BTRFLI  X147,X179,C076,S076,512
BTRFLI  X148,X180,C080,S080,512
BTRFLI  X149,X181,C084,S084,512

BTRFLI  X150,X182,C088,S088,512
BTRFLI  X151,X183,C092,S092,512
P3BY4I  X152,X184,512
BTRFLI  X153,X185,C100,S100,512
BTRFLI  X154,X186,C104,S104,512
BTRFLI  X155,X187,C108,S108,512
BTRFLI  X156,X188,C112,S112,512
BTRFLI  X157,X189,C116,S116,512
BTRFLI  X158,X190,C120,S120,512
BTRFLI  X159,X191,C124,S124,512
ZEROI   X192,X224,512
BTRFLI  X193,X225,C004,S004,512
BTRFLI  X194,X226,C008,S008,512
BTRFLI  X195,X227,C012,S012,512
BTRFLI  X196,X228,C016,S016,512
BTRFLI  X197,X229,C020,S020,512
BTRFLI  X198,X230,C024,S024,512
BTRFLI  X199,X231,C028,S028,512
PBY4I   X200,X232,512
BTRFLI  X201,X233,C036,S036,512
BTRFLI  X202,X234,C040,S040,512
BTRFLI  X203,X235,C044,S044,512
BTRFLI  X204,X236,C048,S048,512
BTRFLI  X205,X237,C052,S052,512
BTRFLI  X206,X238,C056,S056,512
BTRFLI  X207,X239,C060,S060,512
PBY2I   X208,X240,512
BTRFLI  X209,X241,C068,S068,512
BTRFLI  X210,X242,C072,S072,512
BTRFLI  X211,X243,C076,S076,512
BTRFLI  X212,X244,C080,S080,512
BTRFLI  X213,X245,C084,S084,512
BTRFLI  X214,X246,C088,S088,512
BTRFLI  X215,X247,C092,S092,512
P3BY4I  X216,X248,512
BTRFLI  X217,X249,C100,S100,512
BTRFLI  X218,X250,C104,S104,512
BTRFLI  X219,X251,C108,S108,512
BTRFLI  X220,X252,C112,S112,512
BTRFLI  X221,X253,C116,S116,512
BTRFLI  X222,X254,C120,S120,512
BTRFLI  X223,X255,C124,S124,512

FFT CODE FOR STAGE 7

ZEROI   X000,X064,512
BTRFLI  X001,X065,C002,S002,512
BTRFLI  X002,X066,C004,S004,512
BTRFLI  X003,X067,C006,S006,512
BTRFLI  X004,X068,C008,S008,512
BTRFLI  X005,X069,C010,S010,512
BTRFLI  X006,X070,C012,S012,512
BTRFLI  X007,X071,C014,S014,512
BTRFLI  X008,X072,C016,S016,512
BTRFLI  X009,X073,C018,S018,512
BTRFLI  X010,X074,C020,S020,512
BTRFLI  X011,X075,C022,S022,512
BTRFLI  X012,X076,C024,S024,512
BTRFLI  X013,X077,C026,S026,512
BTRFLI  X014,X078,C028,S028,512
```

* * *

4. Implementation of Fast Fourier Transform Algorithms with the TMS32020

```
BTRFLI    X015,X079,C030,S030,512
PBY4I     X016,X080,512
BTRFLI    X017,X081,C034,S034,512
BTRFLI    X018,X082,C036,S036,512
BTRFLI    X019,X083,C038,S038,512
BTRFLI    X020,X084,C040,S040,512
BTRFLI    X021,X085,C042,S042,512
BTRFLI    X022,X086,C044,S044,512
BTRFLI    X023,X087,C046,S046,512
BTRFLI    X024,X088,C048,S048,512
BTRFLI    X025,X089,C050,S050,512
BTRFLI    X026,X090,C052,S052,512
BTRFLI    X027,X091,C054,S054,512
BTRFLI    X028,X092,C056,S056,512
BTRFLI    X029,X093,C058,S058,512
BTRFLI    X030,X094,C060,S060,512
BTRFLI    X031,X095,C062,S062,512
PBY2I     X032,X096,512
BTRFLI    X033,X097,C066,S066,512
BTRFLI    X034,X098,C068,S068,512
BTRFLI    X035,X099,C070,S070,512
BTRFLI    X036,X100,C072,S072,512
BTRFLI    X037,X101,C074,S074,512
BTRFLI    X038,X102,C076,S076,512
BTRFLI    X039,X103,C078,S078,512
BTRFLI    X040,X104,C080,S080,512
BTRFLI    X041,X105,C082,S082,512
BTRFLI    X042,X106,C084,S084,512
BTRFLI    X043,X107,C086,S086,512
BTRFLI    X044,X108,C088,S088,512
BTRFLI    X045,X109,C090,S090,512
BTRFLI    X046,X110,C092,S092,512
BTRFLI    X047,X111,C094,S094,512
P3BY4I    X048,X112,512
BTRFLI    X049,X113,C098,S098,512
BTRFLI    X050,X114,C100,S100,512
BTRFLI    X051,X115,C102,S102,512
BTRFLI    X052,X116,C104,S104,512
BTRFLI    X053,X117,C106,S106,512
BTRFLI    X054,X118,C108,S108,512
BTRFLI    X055,X119,C110,S110,512
BTRFLI    X056,X120,C112,S112,512
BTRFLI    X057,X121,C114,S114,512
BTRFLI    X058,X122,C116,S116,512
BTRFLI    X059,X123,C118,S118,512
BTRFLI    X060,X124,C120,S120,512
BTRFLI    X061,X125,C122,S122,512
BTRFLI    X062,X126,C124,S124,512
BTRFLI    X063,X127,C126,S126,512
ZEROI     X128,X192,512
BTRFLI    X129,X193,C002,S002,512
BTRFLI    X130,X194,C004,S004,512
BTRFLI    X131,X195,C006,S006,512
BTRFLI    X132,X196,C008,S008,512
BTRFLI    X133,X197,C010,S010,512
BTRFLI    X134,X198,C012,S012,512
BTRFLI    X135,X199,C014,S014,512
BTRFLI    X136,X200,C016,S016,512
BTRFLI    X137,X201,C018,S018,512
BTRFLI    X138,X202,C020,S020,512

BTRFLI    X139,X203,C022,S022,512
BTRFLI    X140,X204,C024,S024,512
BTRFLI    X141,X205,C026,S026,512
BTRFLI    X142,X206,C028,S028,512
BTRFLI    X143,X207,C030,S030,512
PBY4I     X144,X208,512
BTRFLI    X145,X209,C034,S034,512
BTRFLI    X146,X210,C036,S036,512
BTRFLI    X147,X211,C038,S038,512
BTRFLI    X148,X212,C040,S040,512
BTRFLI    X149,X213,C042,S042,512
BTRFLI    X150,X214,C044,S044,512
BTRFLI    X151,X215,C046,S046,512
BTRFLI    X152,X216,C048,S048,512
BTRFLI    X153,X217,C050,S050,512
BTRFLI    X154,X218,C052,S052,512
BTRFLI    X155,X219,C054,S054,512
BTRFLI    X156,X220,C056,S056,512
BTRFLI    X157,X221,C058,S058,512
BTRFLI    X158,X222,C060,S060,512
BTRFLI    X159,X223,C062,S062,512
PBY2I     X160,X224,512
BTRFLI    X161,X225,C066,S066,512
BTRFLI    X162,X226,C068,S068,512
BTRFLI    X163,X227,C070,S070,512
BTRFLI    X164,X228,C072,S072,512
BTRFLI    X165,X229,C074,S074,512
BTRFLI    X166,X230,C076,S076,512
BTRFLI    X167,X231,C078,S078,512
BTRFLI    X168,X232,C080,S080,512
BTRFLI    X169,X233,C082,S082,512
BTRFLI    X170,X234,C084,S084,512
BTRFLI    X171,X235,C086,S086,512
BTRFLI    X172,X236,C088,S088,512
BTRFLI    X173,X237,C090,S090,512
BTRFLI    X174,X238,C092,S092,512
BTRFLI    X175,X239,C094,S094,512
P3BY4I    X176,X240,512
BTRFLI    X177,X241,C098,S098,512
BTRFLI    X178,X242,C100,S100,512
BTRFLI    X179,X243,C102,S102,512
BTRFLI    X180,X244,C104,S104,512
BTRFLI    X181,X245,C106,S106,512
BTRFLI    X182,X246,C108,S108,512
BTRFLI    X183,X247,C110,S110,512
BTRFLI    X184,X248,C112,S112,512
BTRFLI    X185,X249,C114,S114,512
BTRFLI    X186,X250,C116,S116,512
BTRFLI    X187,X251,C118,S118,512
BTRFLI    X188,X252,C120,S120,512
BTRFLI    X189,X253,C122,S122,512
BTRFLI    X190,X254,C124,S124,512
BTRFLI    X191,X255,C126,S126,512

FFT CODE FOR STAGE 8

ZEROI     X000,X128,512
BTRFLI    X001,X129,C001,S001,512
BTRFLI    X002,X130,C002,S002,512
BTRFLI    X003,X131,C003,S003,512
```

* * *

```
BTRFLI  X004,X132,C004,S004,512
BTRFLI  X005,X133,C005,S005,512
BTRFLI  X006,X134,C006,S006,512
BTRFLI  X007,X135,C007,S007,512
BTRFLI  X008,X136,C008,S008,512
BTRFLI  X009,X137,C009,S009,512
BTRFLI  X010,X138,C010,S010,512
BTRFLI  X011,X139,C011,S011,512
BTRFLI  X012,X140,C012,S012,512
BTRFLI  X013,X141,C013,S013,512
BTRFLI  X014,X142,C014,S014,512
BTRFLI  X015,X143,C015,S015,512
BTRFLI  X016,X144,C016,S016,512
BTRFLI  X017,X145,C017,S017,512
BTRFLI  X018,X146,C018,S018,512
BTRFLI  X019,X147,C019,S019,512
BTRFLI  X020,X148,C020,S020,512
BTRFLI  X021,X149,C021,S021,512
BTRFLI  X022,X150,C022,S022,512
BTRFLI  X023,X151,C023,S023,512
BTRFLI  X024,X152,C024,S024,512
BTRFLI  X025,X153,C025,S025,512
BTRFLI  X026,X154,C026,S026,512
BTRFLI  X027,X155,C027,S027,512
BTRFLI  X028,X156,C028,S028,512
BTRFLI  X029,X157,C029,S029,512
BTRFLI  X030,X158,C030,S030,512
BTRFLI  X031,X159,C031,S031,512
PBY4I   X032,X160,512
BTRFLI  X033,X161,C033,S033,512
BTRFLI  X034,X162,C034,S034,512
BTRFLI  X035,X163,C035,S035,512
BTRFLI  X036,X164,C036,S036,512
BTRFLI  X037,X165,C037,S037,512
BTRFLI  X038,X166,C038,S038,512
BTRFLI  X039,X167,C039,S039,512
BTRFLI  X040,X168,C040,S040,512
BTRFLI  X041,X169,C041,S041,512
BTRFLI  X042,X170,C042,S042,512
BTRFLI  X043,X171,C043,S043,512
BTRFLI  X044,X172,C044,S044,512
BTRFLI  X045,X173,C045,S045,512
BTRFLI  X046,X174,C046,S046,512
BTRFLI  X047,X175,C047,S047,512
BTRFLI  X048,X176,C048,S048,512
BTRFLI  X049,X177,C049,S049,512
BTRFLI  X050,X178,C050,S050,512
BTRFLI  X051,X179,C051,S051,512
BTRFLI  X052,X180,C052,S052,512
BTRFLI  X053,X181,C053,S053,512
BTRFLI  X054,X182,C054,S054,512
BTRFLI  X055,X183,C055,S055,512
BTRFLI  X056,X184,C056,S056,512
BTRFLI  X057,X185,C057,S057,512
BTRFLI  X058,X186,C058,S058,512
BTRFLI  X059,X187,C059,S059,512
BTRFLI  X060,X188,C060,S060,512
BTRFLI  X061,X189,C061,S061,512
BTRFLI  X062,X190,C062,S062,512
BTRFLI  X063,X191,C063,S063,512

PBY2I   X064,X192,512
BTRFLI  X065,X193,C065,S065,512
BTRFLI  X066,X194,C066,S066,512
BTRFLI  X067,X195,C067,S067,512
BTRFLI  X068,X196,C068,S068,512
BTRFLI  X069,X197,C069,S069,512
BTRFLI  X070,X198,C070,S070,512
BTRFLI  X071,X199,C071,S071,512
BTRFLI  X072,X200,C072,S072,512
BTRFLI  X073,X201,C073,S073,512
BTRFLI  X074,X202,C074,S074,512
BTRFLI  X075,X203,C075,S075,512
BTRFLI  X076,X204,C076,S076,512
BTRFLI  X077,X205,C077,S077,512
BTRFLI  X078,X206,C078,S078,512
BTRFLI  X079,X207,C079,S079,512
BTRFLI  X080,X208,C080,S080,512
BTRFLI  X081,X209,C081,S081,512
BTRFLI  X082,X210,C082,S082,512
BTRFLI  X083,X211,C083,S083,512
BTRFLI  X084,X212,C084,S084,512
BTRFLI  X085,X213,C085,S085,512
BTRFLI  X086,X214,C086,S086,512
BTRFLI  X087,X215,C087,S087,512
BTRFLI  X088,X216,C088,S088,512
BTRFLI  X089,X217,C089,S089,512
BTRFLI  X090,X218,C090,S090,512
BTRFLI  X091,X219,C091,S091,512
BTRFLI  X092,X220,C092,S092,512
BTRFLI  X093,X221,C093,S093,512
BTRFLI  X094,X222,C094,S094,512
BTRFLI  X095,X223,C095,S095,512
P3BY4I  X096,X224,512
BTRFLI  X097,X225,C097,S097,512
BTRFLI  X098,X226,C098,S098,512
BTRFLI  X099,X227,C099,S099,512
BTRFLI  X100,X228,C100,S100,512
BTRFLI  X101,X229,C101,S101,512
BTRFLI  X102,X230,C102,S102,512
BTRFLI  X103,X231,C103,S103,512
BTRFLI  X104,X232,C104,S104,512
BTRFLI  X105,X233,C105,S105,512
BTRFLI  X106,X234,C106,S106,512
BTRFLI  X107,X235,C107,S107,512
BTRFLI  X108,X236,C108,S108,512
BTRFLI  X109,X237,C109,S109,512
BTRFLI  X110,X238,C110,S110,512
BTRFLI  X111,X239,C111,S111,512
BTRFLI  X112,X240,C112,S112,512
BTRFLI  X113,X241,C113,S113,512
BTRFLI  X114,X242,C114,S114,512
BTRFLI  X115,X243,C115,S115,512
BTRFLI  X116,X244,C116,S116,512
BTRFLI  X117,X245,C117,S117,512
BTRFLI  X118,X246,C118,S118,512
BTRFLI  X119,X247,C119,S119,512
BTRFLI  X120,X248,C120,S120,512
BTRFLI  X121,X249,C121,S121,512
BTRFLI  X122,X250,C122,S122,512
BTRFLI  X123,X251,C123,S123,512

BTRFLI  X124,X252,C124,S124,512
BTRFLI  X125,X253,C125,S125,512
BTRFLI  X126,X254,C126,S126,512
BTRFLI  X127,X255,C127,S127,512
END
```

4. Implementation of Fast Fourier Transform Algorithms with the TMS32020

APPENDIX D
A 1024-POINT, RADIX-2 DIT FFT IMPLEMENTATION

```
        A P P E N D I X   D
        ---------------------

        IDT       'FT1024'
*****************************************************************
*                                                              *
*       A 1024-POINT RADIX-2 DIT COMPLEX FFT FOR THE TMS32020  *
*       ------------------------------------------------       *
*                                                              *
*       THE FOLLOWING FILE RAD2FFT.MAC CONSISTS OF ALL THE     *
*       MACROS LISTED IN APPENDIX B                            *
*                                                              *
*****************************************************************
*
*       COPY      RAD2FFT.MAC
*
*****************************************************************
*                                                              *
*       DATA MEMORY MAP FOR PAGES 4, 5, 6 AND 7 (BLOCKS B0,B1) *
*                                                              *
*****************************************************************
*
*       DORG      0
*
*       DATA MEMORY PAGE 4 (STARTING ADDRESS 512 OR >200)
*
X000    DATA      0,0
X001    DATA      0,0
X002    DATA      0,0
X003    DATA      0,0
X004    DATA      0,0
X005    DATA      0,0
X006    DATA      0,0
X007    DATA      0,0
X008    DATA      0,0
X009    DATA      0,0
X010    DATA      0,0
X011    DATA      0,0
X012    DATA      0,0
X013    DATA      0,0
X014    DATA      0,0
X015    DATA      0,0
X016    DATA      0,0
X017    DATA      0,0
X018    DATA      0,0
X019    DATA      0,0
X020    DATA      0,0
X021    DATA      0,0
X022    DATA      0,0
X023    DATA      0,0
X024    DATA      0,0
X025    DATA      0,0
X026    DATA      0,0
X027    DATA      0,0
X028    DATA      0,0
X029    DATA      0,0
X030    DATA      0,0
X031    DATA      0,0
X032    DATA      0,0
X033    DATA      0,0
```

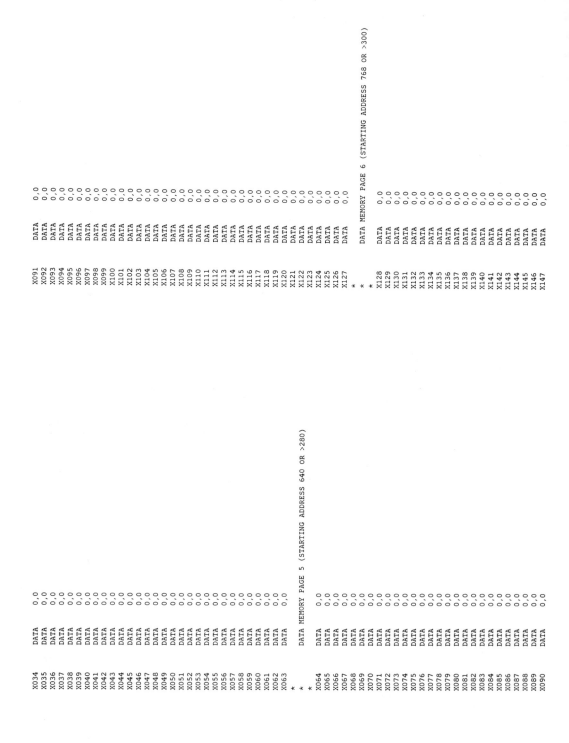

```
X034      DATA    0,0
X035      DATA    0,0
X036      DATA    0,0
X037      DATA    0,0
X038      DATA    0,0
X039      DATA    0,0
X040      DATA    0,0
X041      DATA    0,0
X042      DATA    0,0
X043      DATA    0,0
X044      DATA    0,0
X045      DATA    0,0
X046      DATA    0,0
X047      DATA    0,0
X048      DATA    0,0
X049      DATA    0,0
X050      DATA    0,0
X051      DATA    0,0
X052      DATA    0,0
X053      DATA    0,0
X054      DATA    0,0
X055      DATA    0,0
X056      DATA    0,0
X057      DATA    0,0
X058      DATA    0,0
X059      DATA    0,0
X060      DATA    0,0
X061      DATA    0,0
X062      DATA    0,0
X063      DATA    0,0
*
* DATA MEMORY PAGE 5 (STARTING ADDRESS 640 OR >280)
X064      DATA    0,0
X065      DATA    0,0
X066      DATA    0,0
X067      DATA    0,0
X068      DATA    0,0
X069      DATA    0,0
X070      DATA    0,0
X071      DATA    0,0
X072      DATA    0,0
X073      DATA    0,0
X074      DATA    0,0
X075      DATA    0,0
X076      DATA    0,0
X077      DATA    0,0
X078      DATA    0,0
X079      DATA    0,0
X080      DATA    0,0
X081      DATA    0,0
X082      DATA    0,0
X083      DATA    0,0
X084      DATA    0,0
X085      DATA    0,0
X086      DATA    0,0
X087      DATA    0,0
X088      DATA    0,0
X089      DATA    0,0
X090      DATA    0,0

X091      DATA    0,0
X092      DATA    0,0
X093      DATA    0,0
X094      DATA    0,0
X095      DATA    0,0
X096      DATA    0,0
X097      DATA    0,0
X098      DATA    0,0
X099      DATA    0,0
X100      DATA    0,0
X101      DATA    0,0
X102      DATA    0,0
X103      DATA    0,0
X104      DATA    0,0
X105      DATA    0,0
X106      DATA    0,0
X107      DATA    0,0
X108      DATA    0,0
X109      DATA    0,0
X110      DATA    0,0
X111      DATA    0,0
X112      DATA    0,0
X113      DATA    0,0
X114      DATA    0,0
X115      DATA    0,0
X116      DATA    0,0
X117      DATA    0,0
X118      DATA    0,0
X119      DATA    0,0
X120      DATA    0,0
X121      DATA    0,0
X122      DATA    0,0
X123      DATA    0,0
X124      DATA    0,0
X125      DATA    0,0
X126      DATA    0,0
X127      DATA    0,0
*
* DATA MEMORY PAGE 6 (STARTING ADDRESS 768 OR >300)
X128      DATA    0,0
X129      DATA    0,0
X130      DATA    0,0
X131      DATA    0,0
X132      DATA    0,0
X133      DATA    0,0
X134      DATA    0,0
X135      DATA    0,0
X136      DATA    0,0
X137      DATA    0,0
X138      DATA    0,0
X139      DATA    0,0
X140      DATA    0,0
X141      DATA    0,0
X142      DATA    0,0
X143      DATA    0,0
X144      DATA    0,0
X145      DATA    0,0
X146      DATA    0,0
X147      DATA    0,0
```

```
X148    DATA    0,0
X149    DATA    0,0
X150    DATA    0,0
X151    DATA    0,0
X152    DATA    0,0
X153    DATA    0,0
X154    DATA    0,0
X155    DATA    0,0
X156    DATA    0,0
X157    DATA    0,0
X158    DATA    0,0
X159    DATA    0,0
X160    DATA    0,0
X161    DATA    0,0
X162    DATA    0,0
X163    DATA    0,0
X164    DATA    0,0
X165    DATA    0,0
X166    DATA    0,0
X167    DATA    0,0
X168    DATA    0,0
X169    DATA    0,0
X170    DATA    0,0
X171    DATA    0,0
X172    DATA    0,0
X173    DATA    0,0
X174    DATA    0,0
X175    DATA    0,0
X176    DATA    0,0
X177    DATA    0,0
X178    DATA    0,0
X179    DATA    0,0
X180    DATA    0,0
X181    DATA    0,0
X182    DATA    0,0
X183    DATA    0,0
X184    DATA    0,0
X185    DATA    0,0
X186    DATA    0,0
X187    DATA    0,0
X188    DATA    0,0
X189    DATA    0,0
X190    DATA    0,0
X191    DATA    0,0
*
*       DATA MEMORY PAGE 7 (STARTING ADDRESS 896 OR >380)
*
X192    DATA    0,0
X193    DATA    0,0
X194    DATA    0,0
X195    DATA    0,0
X196    DATA    0,0
X197    DATA    0,0
X198    DATA    0,0
X199    DATA    0,0
X200    DATA    0,0
X201    DATA    0,0
X202    DATA    0,0
X203    DATA    0,0
X204    DATA    0,0

X205    DATA    0,0
X206    DATA    0,0
X207    DATA    0,0
X208    DATA    0,0
X209    DATA    0,0
X210    DATA    0,0
X211    DATA    0,0
X212    DATA    0,0
X213    DATA    0,0
X214    DATA    0,0
X215    DATA    0,0
X216    DATA    0,0
X217    DATA    0,0
X218    DATA    0,0
X219    DATA    0,0
X220    DATA    0,0
X221    DATA    0,0
X222    DATA    0,0
X223    DATA    0,0
X224    DATA    0,0
X225    DATA    0,0
X226    DATA    0,0
X227    DATA    0,0
X228    DATA    0,0
X229    DATA    0,0
X230    DATA    0,0
X231    DATA    0,0
X232    DATA    0,0
X233    DATA    0,0
X234    DATA    0,0
X235    DATA    0,0
X236    DATA    0,0
X237    DATA    0,0
X238    DATA    0,0
X239    DATA    0,0
X240    DATA    0,0
X241    DATA    0,0
X242    DATA    0,0
X243    DATA    0,0
X244    DATA    0,0
X245    DATA    0,0
X246    DATA    0,0
X247    DATA    0,0
X248    DATA    0,0
X249    DATA    0,0
X250    DATA    0,0
X251    DATA    0,0
X252    DATA    0,0
X253    DATA    0,0
X254    DATA    0,0
X255    DATA    0,0
*
*****************************************************
*                                                   *
*       DATA LOCATION IN BLOCK B2 FOR W=COS(45) OR SIN(45)  *
*                                                   *
*****************************************************
*
W       DORG    96
        DATA    0
```

```
*
***********************************************
*                                             *
*   13-BIT TWIDDLE FACTORS FOR 256-POINT COMPLEX FFT   *
*                                             *
***********************************************
*
C000    EQU    4095          C053    EQU    1092
C001    EQU    4094          C054    EQU    995
C002    EQU    4091          C055    EQU    897
C003    EQU    4085          C056    EQU    799
C004    EQU    4076          C057    EQU    700
C005    EQU    4065          C058    EQU    601
C006    EQU    4052          C059    EQU    501
C007    EQU    4036          C060    EQU    401
C008    EQU    4017          C061    EQU    301
C009    EQU    3996          C062    EQU    201
C010    EQU    3973          C063    EQU    101
C011    EQU    3948          C064    EQU    0
C012    EQU    3920          C065    EQU    -101
C013    EQU    3889          C066    EQU    -201
C014    EQU    3857          C067    EQU    -301
C015    EQU    3822          C068    EQU    -401
C016    EQU    3784          C069    EQU    -501
C017    EQU    3745          C070    EQU    -601
C018    EQU    3703          C071    EQU    -700
C019    EQU    3659          C072    EQU    -799
C020    EQU    3612          C073    EQU    -897
C021    EQU    3564          C074    EQU    -995
C022    EQU    3513          C075    EQU    -1092
C023    EQU    3461          C076    EQU    -1189
C024    EQU    3406          C077    EQU    -1285
C025    EQU    3349          C078    EQU    -1380
C026    EQU    3290          C079    EQU    -1474
C027    EQU    3229          C080    EQU    -1567
C028    EQU    3166          C081    EQU    -1660
C029    EQU    3102          C082    EQU    -1751
C030    EQU    3035          C083    EQU    -1842
C031    EQU    2967          C084    EQU    -1931
C032    EQU    2896          C085    EQU    -2019
C033    EQU    2824          C086    EQU    -2106
C034    EQU    2751          C087    EQU    -2191
C035    EQU    2675          C088    EQU    -2276
C036    EQU    2598          C089    EQU    -2359
C037    EQU    2520          C090    EQU    -2440
C038    EQU    2440          C091    EQU    -2520
C039    EQU    2359          C092    EQU    -2598
C040    EQU    2276          C093    EQU    -2675
C041    EQU    2191          C094    EQU    -2751
C042    EQU    2106          C095    EQU    -2824
C043    EQU    2019          C096    EQU    -2896
C044    EQU    1931          C097    EQU    -2967
C045    EQU    1842          C098    EQU    -3035
C046    EQU    1751          C099    EQU    -3102
C047    EQU    1660          C100    EQU    -3166
C048    EQU    1567          C101    EQU    -3229
C049    EQU    1474          C102    EQU    -3290
C050    EQU    1380          C103    EQU    -3349
C051    EQU    1285          C104    EQU    -3406
C052    EQU    1189          C105    EQU    -3461
                            C106    EQU    -3513
                            C107    EQU    -3564
                            C108    EQU    -3612
                            C109    EQU    -3659
                            C110    EQU    -3703
                            C111    EQU    -3745
                            C112    EQU    -3784
```

```
C113  EQU  -3822
C114  EQU  -3857
C115  EQU  -3889
C116  EQU  -3920
C117  EQU  -3948
C118  EQU  -3973
C119  EQU  -3996
C120  EQU  -4017
C121  EQU  -4036
C122  EQU  -4052
C123  EQU  -4065
C124  EQU  -4076
C125  EQU  -4085
C126  EQU  -4091
C127  EQU  -4094
*
S000  EQU  0
S001  EQU  101
S002  EQU  201
S003  EQU  301
S004  EQU  401
S005  EQU  501
S006  EQU  601
S007  EQU  700
S008  EQU  799
S009  EQU  897
S010  EQU  995
S011  EQU  1092
S012  EQU  1189
S013  EQU  1285
S014  EQU  1380
S015  EQU  1474
S016  EQU  1567
S017  EQU  1660
S018  EQU  1751
S019  EQU  1842
S020  EQU  1931
S021  EQU  2019
S022  EQU  2106
S023  EQU  2191
S024  EQU  2276
S025  EQU  2359
S026  EQU  2440
S027  EQU  2520
S028  EQU  2598
S029  EQU  2675
S030  EQU  2751
S031  EQU  2824
S032  EQU  2896
S033  EQU  2967
S034  EQU  3035
S035  EQU  3102
S036  EQU  3166
S037  EQU  3229
S038  EQU  3290
S039  EQU  3349
S040  EQU  3406
S041  EQU  3461
S042  EQU  3513
S043  EQU  3564

S044  EQU  3612
S045  EQU  3659
S046  EQU  3703
S047  EQU  3745
S048  EQU  3784
S049  EQU  3822
S050  EQU  3857
S051  EQU  3889
S052  EQU  3920
S053  EQU  3948
S054  EQU  3973
S055  EQU  3996
S056  EQU  4017
S057  EQU  4036
S058  EQU  4052
S059  EQU  4065
S060  EQU  4076
S061  EQU  4085
S062  EQU  4091
S063  EQU  4094
S064  EQU  4095
S065  EQU  4094
S066  EQU  4091
S067  EQU  4085
S068  EQU  4076
S069  EQU  4065
S070  EQU  4052
S071  EQU  4036
S072  EQU  4017
S073  EQU  3996
S074  EQU  3973
S075  EQU  3948
S076  EQU  3920
S077  EQU  3889
S078  EQU  3857
S079  EQU  3822
S080  EQU  3784
S081  EQU  3745
S082  EQU  3703
S083  EQU  3659
S084  EQU  3612
S085  EQU  3564
S086  EQU  3513
S087  EQU  3461
S088  EQU  3406
S089  EQU  3349
S090  EQU  3290
S091  EQU  3229
S092  EQU  3166
S093  EQU  3102
S094  EQU  3035
S095  EQU  2967
S096  EQU  2896
S097  EQU  2824
S098  EQU  2751
S099  EQU  2675
S100  EQU  2598
S101  EQU  2520
S102  EQU  2440
S103  EQU  2359
```

```
S104    EQU     2276
S105    EQU     2191
S106    EQU     2106
S107    EQU     2019
S108    EQU     1931
S109    EQU     1842
S110    EQU     1751
S111    EQU     1660
S112    EQU     1567
S113    EQU     1474
S114    EQU     1380
S115    EQU     1285
S116    EQU     1189
S117    EQU     1092
S118    EQU     995
S119    EQU     897
S120    EQU     799
S121    EQU     700
S122    EQU     601
S123    EQU     501
S124    EQU     401
S125    EQU     301
S126    EQU     201
S127    EQU     101
*
        LIST
        AORG    0
        B       INIT
*
        AORG    >20
*
WVAL    DATA    >5A82
*
*       SYSTEM INITIALIZATION
*
INIT    SPM     0
        CNFD
        ROVM
        SSXM
        LARP    AR0
        LRLK    AR0,W
        LALK    WVAL
        TBLR    *,AR1
        B       FT1024
*
****************************************************************
*                                                              *
*       256-POINT FFT KERNEL - STAGES 1 AND 2                  *
*                                                              *
****************************************************************
*
KNL256  LDPK    4
        COMBOI  X000,X001,X002,X003,0
        COMBOI  X004,X005,X006,X007,0
        COMBOI  X008,X009,X010,X011,0
        COMBOI  X012,X013,X014,X015,0
        COMBOI  X016,X017,X018,X019,0
        COMBOI  X020,X021,X022,X023,0
        COMBOI  X024,X025,X026,X027,0
        COMBOI  X028,X029,X030,X031,0
        COMBOI  X032,X033,X034,X035,0
        COMBOI  X036,X037,X038,X039,0
        COMBOI  X040,X041,X042,X043,0
        COMBOI  X044,X045,X046,X047,0
        COMBOI  X048,X049,X050,X051,0
        COMBOI  X052,X053,X054,X055,0
        COMBOI  X056,X057,X058,X059,0
        COMBOI  X060,X061,X062,X063,0
        LDPK    5
        COMBOI  X064,X065,X066,X067,128
        COMBOI  X068,X069,X070,X071,128
        COMBOI  X072,X073,X074,X075,128
        COMBOI  X076,X077,X078,X079,128
        COMBOI  X080,X081,X082,X083,128
        COMBOI  X084,X085,X086,X087,128
        COMBOI  X088,X089,X090,X091,128
        COMBOI  X092,X093,X094,X095,128
        COMBOI  X096,X097,X098,X099,128
        COMBOI  X100,X101,X102,X103,128
        COMBOI  X104,X105,X106,X107,128
        COMBOI  X108,X109,X110,X111,128
        COMBOI  X112,X113,X114,X115,128
        COMBOI  X116,X117,X118,X119,128
        COMBOI  X120,X121,X122,X123,128
        COMBOI  X124,X125,X126,X127,128
        LDPK    6
        COMBOI  X128,X129,X130,X131,256
        COMBOI  X132,X133,X134,X135,256
        COMBOI  X136,X137,X138,X139,256
        COMBOI  X140,X141,X142,X143,256
        COMBOI  X144,X145,X146,X147,256
        COMBOI  X148,X149,X150,X151,256
        COMBOI  X152,X153,X154,X155,256
        COMBOI  X156,X157,X158,X159,256
        COMBOI  X160,X161,X162,X163,256
        COMBOI  X164,X165,X166,X167,256
        COMBOI  X168,X169,X170,X171,256
        COMBOI  X172,X173,X174,X175,256
        COMBOI  X176,X177,X178,X179,256
        COMBOI  X180,X181,X182,X183,256
        COMBOI  X184,X185,X186,X187,256
        COMBOI  X188,X189,X190,X191,256
        LDPK    7
        COMBOI  X192,X193,X194,X195,384
        COMBOI  X196,X197,X198,X199,384
        COMBOI  X200,X201,X202,X203,384
        COMBOI  X204,X205,X206,X207,384
        COMBOI  X208,X209,X210,X211,384
        COMBOI  X212,X213,X214,X215,384
        COMBOI  X216,X217,X218,X219,384
        COMBOI  X220,X221,X222,X223,384
        COMBOI  X224,X225,X226,X227,384
        COMBOI  X228,X229,X230,X231,384
        COMBOI  X232,X233,X234,X235,384
        COMBOI  X236,X237,X238,X239,384
        COMBOI  X240,X241,X242,X243,384
        COMBOI  X244,X245,X246,X247,384
        COMBOI  X248,X249,X250,X251,384
        COMBOI  X252,X253,X254,X255,384
```

```
***************************************************
*                                                 *
*     256-POINT FFT KERNEL  -  STAGE 3            *
*                                                 *
***************************************************
*

        ZEROI   X000,X004,512
        PBY4I   X001,X005,512
        PBY2I   X002,X006,512
        P3BY4I  X003,X007,512
        ZEROI   X008,X012,512
        PBY4I   X009,X013,512
        PBY2I   X010,X014,512
        P3BY4I  X011,X015,512
        ZEROI   X016,X020,512
        PBY4I   X017,X021,512
        PBY2I   X018,X022,512
        P3BY4I  X019,X023,512
        ZEROI   X024,X028,512
        PBY4I   X025,X029,512
        PBY2I   X026,X030,512
        P3BY4I  X027,X031,512
        ZEROI   X032,X036,512
        PBY4I   X033,X037,512
        PBY2I   X034,X038,512
        P3BY4I  X035,X039,512
        ZEROI   X040,X044,512
        PBY4I   X041,X045,512
        PBY2I   X042,X046,512
        P3BY4I  X043,X047,512
        ZEROI   X048,X052,512
        PBY4I   X049,X053,512
        PBY2I   X050,X054,512
        P3BY4I  X051,X055,512
        ZEROI   X056,X060,512
        PBY4I   X057,X061,512
        PBY2I   X058,X062,512
        P3BY4I  X059,X063,512
        ZEROI   X064,X068,512
        PBY4I   X065,X069,512
        PBY2I   X066,X070,512
        P3BY4I  X067,X071,512
        ZEROI   X072,X076,512
        PBY4I   X073,X077,512
        PBY2I   X074,X078,512
        P3BY4I  X075,X079,512
        ZEROI   X080,X084,512
        PBY4I   X081,X085,512
        PBY2I   X082,X086,512
        P3BY4I  X083,X087,512
        ZEROI   X088,X092,512
        PBY4I   X089,X093,512
        PBY2I   X090,X094,512
        P3BY4I  X091,X095,512
        ZEROI   X096,X100,512
        PBY4I   X097,X101,512
        PBY2I   X098,X102,512
        P3BY4I  X099,X103,512
        ZEROI   X104,X108,512
        PBY4I   X105,X109,512
        PBY2I   X106,X110,512
        P3BY4I  X107,X111,512
        ZEROI   X112,X116,512
        PBY4I   X113,X117,512
        PBY2I   X114,X118,512
        P3BY4I  X115,X119,512
        ZEROI   X120,X124,512
        PBY4I   X121,X125,512
        PBY2I   X122,X126,512
        P3BY4I  X123,X127,512
        ZEROI   X128,X132,512
        PBY4I   X129,X133,512
        PBY2I   X130,X134,512
        P3BY4I  X131,X135,512
        ZEROI   X136,X140,512
        PBY4I   X137,X141,512
        PBY2I   X138,X142,512
        P3BY4I  X139,X143,512
        ZEROI   X144,X148,512
        PBY4I   X145,X149,512
        PBY2I   X146,X150,512
        P3BY4I  X147,X151,512
        ZEROI   X152,X156,512
        PBY4I   X153,X157,512
        PBY2I   X154,X158,512
        P3BY4I  X155,X159,512
        ZEROI   X160,X164,512
        PBY4I   X161,X165,512
        PBY2I   X162,X166,512
        P3BY4I  X163,X167,512
        ZEROI   X168,X172,512
        PBY4I   X169,X173,512
        PBY2I   X170,X174,512
        P3BY4I  X171,X175,512
        ZEROI   X176,X180,512
        PBY4I   X177,X181,512
        PBY2I   X178,X182,512
        P3BY4I  X179,X183,512
        ZEROI   X184,X188,512
        PBY4I   X185,X189,512
        PBY2I   X186,X190,512
        P3BY4I  X187,X191,512
        ZEROI   X192,X196,512
        PBY4I   X193,X197,512
        PBY2I   X194,X198,512
        P3BY4I  X195,X199,512
        ZEROI   X200,X204,512
        PBY4I   X201,X205,512
        PBY2I   X202,X206,512
        P3BY4I  X203,X207,512
        ZEROI   X208,X212,512
        PBY4I   X209,X213,512
        PBY2I   X210,X214,512
        P3BY4I  X211,X215,512
        ZEROI   X216,X220,512
        PBY4I   X217,X221,512
        PBY2I   X218,X222,512
        P3BY4I  X219,X223,512
        ZEROI   X224,X228,512
        PBY4I   X225,X229,512
```

```
*
******************************************
*                                        *
*   256-POINT FFT KERNEL - STAGE 4        *
*                                        *
******************************************
*
        ZEROI   X000,X008,512
        BTRFLI  X001,X009,C016,S016,512
        PBY4I   X002,X010,512
        BTRFLI  X003,X011,C048,S048,512
        PBY2I   X004,X012,512
        BTRFLI  X005,X013,C080,S080,512
        P3BY4I  X006,X014,512
        BTRFLI  X007,X015,C112,S112,512
        ZEROI   X016,X024,512
        BTRFLI  X017,X025,C016,S016,512
        PBY4I   X018,X026,512
        BTRFLI  X019,X027,C048,S048,512
        PBY2I   X020,X028,512
        BTRFLI  X021,X029,C080,S080,512
        P3BY4I  X022,X030,512
        BTRFLI  X023,X031,C112,S112,512
        ZEROI   X032,X040,512
        BTRFLI  X033,X041,C016,S016,512
        PBY4I   X034,X042,512
        BTRFLI  X035,X043,C048,S048,512
        PBY2I   X036,X044,512
        BTRFLI  X037,X045,C080,S080,512
        P3BY4I  X038,X046,512
        BTRFLI  X039,X047,C112,S112,512
        ZEROI   X048,X056,512
        BTRFLI  X049,X057,C016,S016,512
        PBY4I   X050,X058,512
        BTRFLI  X051,X059,C048,S048,512
        PBY2I   X052,X060,512
        BTRFLI  X053,X061,C080,S080,512
        P3BY4I  X054,X062,512
        BTRFLI  X055,X063,C112,S112,512
        ZEROI   X064,X072,512
        BTRFLI  X065,X073,C016,S016,512
        PBY4I   X066,X074,512
        BTRFLI  X067,X075,C048,S048,512
        PBY2I   X068,X076,512
        BTRFLI  X069,X077,C080,S080,512
        P3BY4I  X070,X078,512
        BTRFLI  X071,X079,C112,S112,512
        ZEROI   X080,X088,512
        BTRFLI  X081,X089,C016,S016,512
        PBY4I   X082,X090,512
        BTRFLI  X083,X091,C048,S048,512
        PBY2I   X084,X092,512
        BTRFLI  X085,X093,C080,S080,512
        P3BY4I  X086,X094,512
        BTRFLI  X087,X095,C112,S112,512
        ZEROI   X096,X104,512
        BTRFLI  X097,X105,C016,S016,512
        PBY4I   X098,X106,512
        BTRFLI  X099,X107,C048,S048,512
        PBY2I   X100,X108,512
        BTRFLI  X101,X109,C080,S080,512
        P3BY4I  X102,X110,512
        BTRFLI  X103,X111,C112,S112,512
        ZEROI   X112,X120,512
        BTRFLI  X113,X121,C016,S016,512
        PBY4I   X114,X122,512
        BTRFLI  X115,X123,C048,S048,512
        PBY2I   X116,X124,512
        BTRFLI  X117,X125,C080,S080,512
        P3BY4I  X118,X126,512
        BTRFLI  X119,X127,C112,S112,512
        ZEROI   X128,X136,512
        BTRFLI  X129,X137,C016,S016,512
        PBY4I   X130,X138,512
        BTRFLI  X131,X139,C048,S048,512
        PBY2I   X132,X140,512
        BTRFLI  X133,X141,C080,S080,512
        P3BY4I  X134,X142,512
        BTRFLI  X135,X143,C112,S112,512
        ZEROI   X144,X152,512
        BTRFLI  X145,X153,C016,S016,512
        PBY4I   X146,X154,512
        BTRFLI  X147,X155,C048,S048,512
        PBY2I   X148,X156,512
        BTRFLI  X149,X157,C080,S080,512
        P3BY4I  X150,X158,512
        BTRFLI  X151,X159,C112,S112,512
        ZEROI   X160,X168,512
        BTRFLI  X161,X169,C016,S016,512
        PBY4I   X162,X170,512
        BTRFLI  X163,X171,C048,S048,512
        PBY2I   X164,X172,512
        BTRFLI  X165,X173,C080,S080,512
        P3BY4I  X166,X174,512
        BTRFLI  X167,X175,C112,S112,512
        ZEROI   X176,X184,512
        BTRFLI  X177,X185,C016,S016,512
        PBY4I   X178,X186,512
        BTRFLI  X179,X187,C048,S048,512
        PBY2I   X180,X188,512
        BTRFLI  X181,X189,C080,S080,512
        P3BY4I  X182,X190,512
        BTRFLI  X183,X191,C112,S112,512
        ZEROI   X192,X200,512
        BTRFLI  X193,X201,C016,S016,512
        PBY4I   X194,X202,512
        PBY2I   X226,X230,512
        P3BY4I  X227,X231,512
        ZEROI   X232,X236,512
        PBY4I   X233,X237,512
        PBY2I   X234,X238,512
        P3BY4I  X235,X239,512
        ZEROI   X240,X244,512
        PBY4I   X241,X245,512
        PBY2I   X242,X246,512
        P3BY4I  X243,X247,512
        ZEROI   X248,X252,512
        PBY4I   X249,X253,512
        PBY2I   X250,X254,512
        P3BY4I  X251,X255,512
```

```
BTRFLI   X195,X203,C048,S048,512
PBY2I    X196,X204,512
BTRFLI   X197,X205,C080,S080,512
P3BY4I   X198,X206,512
BTRFLI   X199,X207,C112,S112,512
ZEROI    X208,X216,512
BTRFLI   X209,X217,C016,S016,512
PBY4I    X210,X218,512
BTRFLI   X211,X219,C048,S048,512
PBY2I    X212,X220,512
BTRFLI   X213,X221,C080,S080,512
P3BY4I   X214,X222,512
BTRFLI   X215,X223,C112,S112,512
ZEROI    X224,X232,512
BTRFLI   X225,X233,C016,S016,512
PBY4I    X226,X234,512
BTRFLI   X227,X235,C048,S048,512
PBY2I    X228,X236,512
BTRFLI   X229,X237,C080,S080,512
P3BY4I   X230,X238,512
BTRFLI   X231,X239,C112,S112,512
ZEROI    X240,X248,512
BTRFLI   X241,X249,C016,S016,512
PBY4I    X242,X250,512
BTRFLI   X243,X251,C048,S048,512
PBY2I    X244,X252,512
BTRFLI   X245,X253,C080,S080,512
P3BY4I   X246,X254,512
BTRFLI   X247,X255,C112,S112,512
```

```
PBY2I    X040,X056,512
BTRFLI   X041,X057,C072,S072,512
BTRFLI   X042,X058,C080,S080,512
BTRFLI   X043,X059,C088,S088,512
P3BY4I   X044,X060,512
BTRFLI   X045,X061,C104,S104,512
BTRFLI   X046,X062,C112,S112,512
BTRFLI   X047,X063,C120,S120,512
ZEROI    X064,X080,512
BTRFLI   X065,X081,C008,S008,512
BTRFLI   X066,X082,C016,S016,512
BTRFLI   X067,X083,C024,S024,512
PBY4I    X068,X084,512
BTRFLI   X069,X085,C040,S040,512
BTRFLI   X070,X086,C048,S048,512
BTRFLI   X071,X087,C056,S056,512
PBY2I    X072,X088,512
BTRFLI   X073,X089,C072,S072,512
BTRFLI   X074,X090,C080,S080,512
BTRFLI   X075,X091,C088,S088,512
P3BY4I   X076,X092,512
BTRFLI   X077,X093,C104,S104,512
BTRFLI   X078,X094,C112,S112,512
BTRFLI   X079,X095,C120,S120,512
ZEROI    X096,X112,512
BTRFLI   X097,X113,C008,S008,512
BTRFLI   X098,X114,C016,S016,512
BTRFLI   X099,X115,C024,S024,512
PBY4I    X100,X116,512
BTRFLI   X101,X117,C040,S040,512
BTRFLI   X102,X118,C048,S048,512
BTRFLI   X103,X119,C056,S056,512
PBY2I    X104,X120,512
BTRFLI   X105,X121,C072,S072,512
BTRFLI   X106,X122,C080,S080,512
BTRFLI   X107,X123,C088,S088,512
P3BY4I   X108,X124,512
BTRFLI   X109,X125,C104,S104,512
BTRFLI   X110,X126,C112,S112,512
BTRFLI   X111,X127,C120,S120,512
ZEROI    X128,X144,512
BTRFLI   X129,X145,C008,S008,512
BTRFLI   X130,X146,C016,S016,512
BTRFLI   X131,X147,C024,S024,512
PBY4I    X132,X148,512
BTRFLI   X133,X149,C040,S040,512
BTRFLI   X134,X150,C048,S048,512
BTRFLI   X135,X151,C056,S056,512
PBY2I    X136,X152,512
BTRFLI   X137,X153,C072,S072,512
BTRFLI   X138,X154,C080,S080,512
BTRFLI   X139,X155,C088,S088,512
P3BY4I   X140,X156,512
BTRFLI   X141,X157,C104,S104,512
BTRFLI   X142,X158,C112,S112,512
BTRFLI   X143,X159,C120,S120,512
ZEROI    X160,X176,512
BTRFLI   X161,X177,C008,S008,512
BTRFLI   X162,X178,C016,S016,512
BTRFLI   X163,X179,C024,S024,512
```

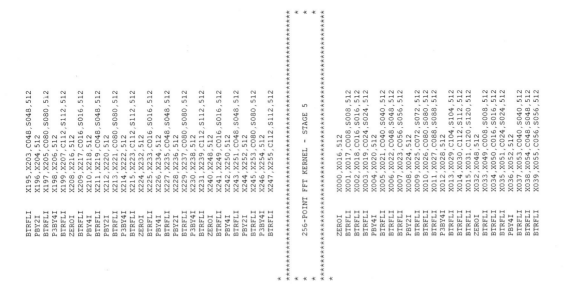

```
*
****************************************************
*           256-POINT FFT KERNEL - STAGE 5         *
*                                                  *
****************************************************
*
ZEROI    X000,X016,512
BTRFLI   X001,X017,C008,S008,512
BTRFLI   X002,X018,C016,S016,512
BTRFLI   X003,X019,C024,S024,512
PBY4I    X004,X020,512
BTRFLI   X005,X021,C040,S040,512
BTRFLI   X006,X022,C048,S048,512
BTRFLI   X007,X023,C056,S056,512
PBY2I    X008,X024,512
BTRFLI   X009,X025,C072,S072,512
BTRFLI   X010,X026,C080,S080,512
BTRFLI   X011,X027,C088,S088,512
P3BY4I   X012,X028,512
BTRFLI   X013,X029,C104,S104,512
BTRFLI   X014,X030,C112,S112,512
BTRFLI   X015,X031,C120,S120,512
ZEROI    X032,X048,512
BTRFLI   X033,X049,C008,S008,512
BTRFLI   X034,X050,C016,S016,512
BTRFLI   X035,X051,C024,S024,512
PBY4I    X036,X052,512
BTRFLI   X037,X053,C040,S040,512
BTRFLI   X038,X054,C048,S048,512
BTRFLI   X039,X055,C056,S056,512
```

```
BTRFLI  X009,X041,C036,S036,512
BTRFLI  X010,X042,C040,S040,512
BTRFLI  X011,X043,C044,S044,512
BTRFLI  X012,X044,C048,S048,512
BTRFLI  X013,X045,C052,S052,512
BTRFLI  X014,X046,C056,S056,512
BTRFLI  X015,X047,C060,S060,512
PBY2I   X016,X048,512
BTRFLI  X017,X049,C068,S068,512
BTRFLI  X018,X050,C072,S072,512
BTRFLI  X019,X051,C076,S076,512
BTRFLI  X020,X052,C080,S080,512
BTRFLI  X021,X053,C084,S084,512
BTRFLI  X022,X054,C088,S088,512
BTRFLI  X023,X055,C092,S092,512
P3BY4I  X024,X056,512
BTRFLI  X025,X057,C100,S100,512
BTRFLI  X026,X058,C104,S104,512
BTRFLI  X027,X059,C108,S108,512
BTRFLI  X028,X060,C112,S112,512
BTRFLI  X029,X061,C116,S116,512
BTRFLI  X030,X062,C120,S120,512
BTRFLI  X031,X063,C124,S124,512
ZEROI   X064,X096,512
BTRFLI  X065,X097,C004,S004,512
BTRFLI  X066,X098,C008,S008,512
BTRFLI  X067,X099,C012,S012,512
BTRFLI  X068,X100,C016,S016,512
BTRFLI  X069,X101,C020,S020,512
BTRFLI  X070,X102,C024,S024,512
BTRFLI  X071,X103,C028,S028,512
PBY4I   X072,X104,512
BTRFLI  X073,X105,C036,S036,512
BTRFLI  X074,X106,C040,S040,512
BTRFLI  X075,X107,C044,S044,512
BTRFLI  X076,X108,C048,S048,512
BTRFLI  X077,X109,C052,S052,512
BTRFLI  X078,X110,C056,S056,512
BTRFLI  X079,X111,C060,S060,512
PBY2I   X080,X112,512
BTRFLI  X081,X113,C068,S068,512
BTRFLI  X082,X114,C072,S072,512
BTRFLI  X083,X115,C076,S076,512
BTRFLI  X084,X116,C080,S080,512
BTRFLI  X085,X117,C084,S084,512
BTRFLI  X086,X118,C088,S088,512
BTRFLI  X087,X119,C092,S092,512
P3BY4I  X088,X120,512
BTRFLI  X089,X121,C100,S100,512
BTRFLI  X090,X122,C104,S104,512
BTRFLI  X091,X123,C108,S108,512
BTRFLI  X092,X124,C112,S112,512
BTRFLI  X093,X125,C116,S116,512
BTRFLI  X094,X126,C120,S120,512
BTRFLI  X095,X127,C124,S124,512
ZEROI   X128,X160,512
BTRFLI  X129,X161,C004,S004,512
BTRFLI  X130,X162,C008,S008,512
BTRFLI  X131,X163,C012,S012,512
BTRFLI  X132,X164,C016,S016,512

PBY4I   X164,X180,512
BTRFLI  X165,X181,C040,S040,512
BTRFLI  X166,X182,C048,S048,512
BTRFLI  X167,X183,C056,S056,512
PBY2I   X168,X184,512
BTRFLI  X169,X185,C072,S072,512
BTRFLI  X170,X186,C080,S080,512
BTRFLI  X171,X187,C088,S088,512
P3BY4I  X172,X188,512
BTRFLI  X173,X189,C104,S104,512
BTRFLI  X174,X190,C112,S112,512
BTRFLI  X175,X191,C120,S120,512
ZEROI   X192,X208,512
BTRFLI  X193,X209,C008,S008,512
BTRFLI  X194,X210,C016,S016,512
BTRFLI  X195,X211,C024,S024,512
PBY4I   X196,X212,512
BTRFLI  X197,X213,C040,S040,512
BTRFLI  X198,X214,C048,S048,512
BTRFLI  X199,X215,C056,S056,512
PBY2I   X200,X216,512
BTRFLI  X201,X217,C072,S072,512
BTRFLI  X202,X218,C080,S080,512
BTRFLI  X203,X219,C088,S088,512
P3BY4I  X204,X220,512
BTRFLI  X205,X221,C104,S104,512
BTRFLI  X206,X222,C112,S112,512
BTRFLI  X207,X223,C120,S120,512
ZEROI   X224,X240,512
BTRFLI  X225,X241,C008,S008,512
BTRFLI  X226,X242,C016,S016,512
BTRFLI  X227,X243,C024,S024,512
PBY4I   X228,X244,512
BTRFLI  X229,X245,C040,S040,512
BTRFLI  X230,X246,C048,S048,512
BTRFLI  X231,X247,C056,S056,512
PBY2I   X232,X248,512
BTRFLI  X233,X249,C072,S072,512
BTRFLI  X234,X250,C080,S080,512
BTRFLI  X235,X251,C088,S088,512
P3BY4I  X236,X252,512
BTRFLI  X237,X253,C104,S104,512
BTRFLI  X238,X254,C112,S112,512
BTRFLI  X239,X255,C120,S120,512

****************************************************
*                                                  *
*         256-POINT FFT KERNEL - STAGE 6           *
*                                                  *
****************************************************
*
ZEROI   X000,X032,512
BTRFLI  X001,X033,C004,S004,512
BTRFLI  X002,X034,C008,S008,512
BTRFLI  X003,X035,C012,S012,512
BTRFLI  X004,X036,C016,S016,512
BTRFLI  X005,X037,C020,S020,512
BTRFLI  X006,X038,C024,S024,512
BTRFLI  X007,X039,C028,S028,512
PBY4I   X008,X040,512
```

4. Implementation of Fast Fourier Transform Algorithms with the TMS32020

```
******************************************************************
*                                                                *
*         256-POINT FFT KERNEL - STAGE 7                         *
*                                                                *
******************************************************************
*
        ZEROI   X000,X064,512
        BTRFLI  X001,X065,C002,S002,512
        BTRFLI  X002,X066,C004,S004,512
        BTRFLI  X003,X067,C006,S006,512
        BTRFLI  X004,X068,C008,S008,512
        BTRFLI  X005,X069,C010,S010,512
        BTRFLI  X006,X070,C012,S012,512
        BTRFLI  X007,X071,C014,S014,512
        BTRFLI  X008,X072,C016,S016,512
        BTRFLI  X009,X073,C018,S018,512
        BTRFLI  X010,X074,C020,S020,512
        BTRFLI  X011,X075,C022,S022,512
        BTRFLI  X012,X076,C024,S024,512
        BTRFLI  X013,X077,C026,S026,512
        BTRFLI  X014,X078,C028,S028,512
        BTRFLI  X015,X079,C030,S030,512
        PBY4I   X016,X080,512
        BTRFLI  X017,X081,C034,S034,512
        BTRFLI  X018,X082,C036,S036,512
        BTRFLI  X019,X083,C038,S038,512
        BTRFLI  X020,X084,C040,S040,512
        BTRFLI  X021,X085,C042,S042,512
        BTRFLI  X022,X086,C044,S044,512
        BTRFLI  X023,X087,C046,S046,512
        BTRFLI  X024,X088,C048,S048,512
        BTRFLI  X025,X089,C050,S050,512
        BTRFLI  X026,X090,C052,S052,512
        BTRFLI  X027,X091,C054,S054,512
        BTRFLI  X028,X092,C056,S056,512
        BTRFLI  X029,X093,C058,S058,512
        BTRFLI  X030,X094,C060,S060,512
        BTRFLI  X031,X095,C062,S062,512
        PBY2I   X032,X096,512
        BTRFLI  X033,X097,C066,S066,512
        BTRFLI  X034,X098,C068,S068,512
        BTRFLI  X035,X099,C070,S070,512
        BTRFLI  X036,X100,C072,S072,512
        BTRFLI  X037,X101,C074,S074,512
        BTRFLI  X038,X102,C076,S076,512
        BTRFLI  X039,X103,C078,S078,512
        BTRFLI  X040,X104,C080,S080,512
        BTRFLI  X041,X105,C082,S082,512
        BTRFLI  X042,X106,C084,S084,512
        BTRFLI  X043,X107,C086,S086,512
        BTRFLI  X044,X108,C088,S088,512
        BTRFLI  X045,X109,C090,S090,512
        BTRFLI  X046,X110,C092,S092,512
        BTRFLI  X047,X111,C094,S094,512
        P3BY4I  X048,X112,512
        BTRFLI  X049,X113,C098,S098,512
        BTRFLI  X050,X114,C100,S100,512
        BTRFLI  X051,X115,C102,S102,512
        BTRFLI  X052,X116,C104,S104,512
        BTRFLI  X053,X117,C106,S106,512

        BTRFLI  X133,X165,C020,S020,512
        BTRFLI  X134,X166,C024,S024,512
        BTRFLI  X135,X167,C028,S028,512
        PBY4I   X136,X168,512
        BTRFLI  X137,X169,C036,S036,512
        BTRFLI  X138,X170,C040,S040,512
        BTRFLI  X139,X171,C044,S044,512
        BTRFLI  X140,X172,C048,S048,512
        BTRFLI  X141,X173,C052,S052,512
        BTRFLI  X142,X174,C056,S056,512
        BTRFLI  X143,X175,C060,S060,512
        PBY2I   X144,X176,512
        BTRFLI  X145,X177,C068,S068,512
        BTRFLI  X146,X178,C072,S072,512
        BTRFLI  X147,X179,C076,S076,512
        BTRFLI  X148,X180,C080,S080,512
        BTRFLI  X149,X181,C084,S084,512
        BTRFLI  X150,X182,C088,S088,512
        BTRFLI  X151,X183,C092,S092,512
        P3BY4I  X152,X184,512
        BTRFLI  X153,X185,C100,S100,512
        BTRFLI  X154,X186,C104,S104,512
        BTRFLI  X155,X187,C108,S108,512
        BTRFLI  X156,X188,C112,S112,512
        BTRFLI  X157,X189,C116,S116,512
        BTRFLI  X158,X190,C120,S120,512
        BTRFLI  X159,X191,C124,S124,512
        ZEROI   X192,X224,512
        BTRFLI  X193,X225,C004,S004,512
        BTRFLI  X194,X226,C008,S008,512
        BTRFLI  X195,X227,C012,S012,512
        BTRFLI  X196,X228,C016,S016,512
        BTRFLI  X197,X229,C020,S020,512
        BTRFLI  X198,X230,C024,S024,512
        BTRFLI  X199,X231,C028,S028,512
        PBY4I   X200,X232,512
        BTRFLI  X201,X233,C036,S036,512
        BTRFLI  X202,X234,C040,S040,512
        BTRFLI  X203,X235,C044,S044,512
        BTRFLI  X204,X236,C048,S048,512
        BTRFLI  X205,X237,C052,S052,512
        BTRFLI  X206,X238,C056,S056,512
        BTRFLI  X207,X239,C060,S060,512
        PBY2I   X208,X240,512
        BTRFLI  X209,X241,C068,S068,512
        BTRFLI  X210,X242,C072,S072,512
        BTRFLI  X211,X243,C076,S076,512
        BTRFLI  X212,X244,C080,S080,512
        BTRFLI  X213,X245,C084,S084,512
        BTRFLI  X214,X246,C088,S088,512
        BTRFLI  X215,X247,C092,S092,512
        P3BY4I  X216,X248,512
        BTRFLI  X217,X249,C100,S100,512
        BTRFLI  X218,X250,C104,S104,512
        BTRFLI  X219,X251,C108,S108,512
        BTRFLI  X220,X252,C112,S112,512
        BTRFLI  X221,X253,C116,S116,512
        BTRFLI  X222,X254,C120,S120,512
        BTRFLI  X223,X255,C124,S124,512
*
```

```
BTRFLI   X178,X242,C100,S100,512
BTRFLI   X179,X243,C102,S102,512
BTRFLI   X180,X244,C104,S104,512
BTRFLI   X181,X245,C106,S106,512
BTRFLI   X182,X246,C108,S108,512
BTRFLI   X183,X247,C110,S110,512
BTRFLI   X184,X248,C112,S112,512
BTRFLI   X185,X249,C114,S114,512
BTRFLI   X186,X250,C116,S116,512
BTRFLI   X187,X251,C118,S118,512
BTRFLI   X188,X252,C120,S120,512
BTRFLI   X189,X253,C122,S122,512
BTRFLI   X190,X254,C124,S124,512
BTRFLI   X191,X255,C126,S126,512

*************************************************
*        *        *
*        *        *     256-POINT FFT KERNEL - STAGE 8
*        *        *
*************************************************
*

ZEROI    X000,X128,512
BTRFLI   X001,X129,C001,S001,512
BTRFLI   X002,X130,C002,S002,512
BTRFLI   X003,X131,C003,S003,512
BTRFLI   X004,X132,C004,S004,512
BTRFLI   X005,X133,C005,S005,512
BTRFLI   X006,X134,C006,S006,512
BTRFLI   X007,X135,C007,S007,512
BTRFLI   X008,X136,C008,S008,512
BTRFLI   X009,X137,C009,S009,512
BTRFLI   X010,X138,C010,S010,512
BTRFLI   X011,X139,C011,S011,512
BTRFLI   X012,X140,C012,S012,512
BTRFLI   X013,X141,C013,S013,512
BTRFLI   X014,X142,C014,S014,512
BTRFLI   X015,X143,C015,S015,512
BTRFLI   X016,X144,C016,S016,512
BTRFLI   X017,X145,C017,S017,512
BTRFLI   X018,X146,C018,S018,512
BTRFLI   X019,X147,C019,S019,512
BTRFLI   X020,X148,C020,S020,512
BTRFLI   X021,X149,C021,S021,512
BTRFLI   X022,X150,C022,S022,512
BTRFLI   X023,X151,C023,S023,512
BTRFLI   X024,X152,C024,S024,512
BTRFLI   X025,X153,C025,S025,512
BTRFLI   X026,X154,C026,S026,512
BTRFLI   X027,X155,C027,S027,512
BTRFLI   X028,X156,C028,S028,512
BTRFLI   X029,X157,C029,S029,512
BTRFLI   X030,X158,C030,S030,512
BTRFLI   X031,X159,C031,S031,512
PBY4I    X032,X160,512
BTRFLI   X033,X161,C033,S033,512
BTRFLI   X034,X162,C034,S034,512
BTRFLI   X035,X163,C035,S035,512
BTRFLI   X036,X164,C036,S036,512
BTRFLI   X037,X165,C037,S037,512
BTRFLI   X038,X166,C038,S038,512
```

```
BTRFLI   X054,X118,C108,S108,512
BTRFLI   X055,X119,C110,S110,512
BTRFLI   X056,X120,C112,S112,512
BTRFLI   X057,X121,C114,S114,512
BTRFLI   X058,X122,C116,S116,512
BTRFLI   X059,X123,C118,S118,512
BTRFLI   X060,X124,C120,S120,512
BTRFLI   X061,X125,C122,S122,512
BTRFLI   X062,X126,C124,S124,512
BTRFLI   X063,X127,C126,S126,512
ZEROI    X128,X192,512
BTRFLI   X129,X193,C002,S002,512
BTRFLI   X130,X194,C004,S004,512
BTRFLI   X131,X195,C006,S006,512
BTRFLI   X132,X196,C008,S008,512
BTRFLI   X133,X197,C010,S010,512
BTRFLI   X134,X198,C012,S012,512
BTRFLI   X135,X199,C014,S014,512
BTRFLI   X136,X200,C016,S016,512
BTRFLI   X137,X201,C018,S018,512
BTRFLI   X138,X202,C020,S020,512
BTRFLI   X139,X203,C022,S022,512
BTRFLI   X140,X204,C024,S024,512
BTRFLI   X141,X205,C026,S026,512
BTRFLI   X142,X206,C028,S028,512
BTRFLI   X143,X207,C030,S030,512
PBY4I    X144,X208,512
BTRFLI   X145,X209,C034,S034,512
BTRFLI   X146,X210,C036,S036,512
BTRFLI   X147,X211,C038,S038,512
BTRFLI   X148,X212,C040,S040,512
BTRFLI   X149,X213,C042,S042,512
BTRFLI   X150,X214,C044,S044,512
BTRFLI   X151,X215,C046,S046,512
BTRFLI   X152,X216,C048,S048,512
BTRFLI   X153,X217,C050,S050,512
BTRFLI   X154,X218,C052,S052,512
BTRFLI   X155,X219,C054,S054,512
BTRFLI   X156,X220,C056,S056,512
BTRFLI   X157,X221,C058,S058,512
BTRFLI   X158,X222,C060,S060,512
BTRFLI   X159,X223,C062,S062,512
PBY2I    X160,X224,512
BTRFLI   X161,X225,C066,S066,512
BTRFLI   X162,X226,C068,S068,512
BTRFLI   X163,X227,C070,S070,512
BTRFLI   X164,X228,C072,S072,512
BTRFLI   X165,X229,C074,S074,512
BTRFLI   X166,X230,C076,S076,512
BTRFLI   X167,X231,C078,S078,512
BTRFLI   X168,X232,C080,S080,512
BTRFLI   X169,X233,C082,S082,512
BTRFLI   X170,X234,C084,S084,512
BTRFLI   X171,X235,C086,S086,512
BTRFLI   X172,X236,C088,S088,512
BTRFLI   X173,X237,C090,S090,512
BTRFLI   X174,X238,C092,S092,512
BTRFLI   X175,X239,C094,S094,512
P3BY4I   X176,X240,512
BTRFLI   X177,X241,C098,S098,512
```

4. Implementation of Fast Fourier Transform Algorithms with the TMS32020

```
BTRFLI   X039,X167,C039,S039,512
BTRFLI   X040,X168,C040,S040,512
BTRFLI   X041,X169,C041,S041,512
BTRFLI   X042,X170,C042,S042,512
BTRFLI   X043,X171,C043,S043,512
BTRFLI   X044,X172,C044,S044,512
BTRFLI   X045,X173,C045,S045,512
BTRFLI   X046,X174,C046,S046,512
BTRFLI   X047,X175,C047,S047,512
BTRFLI   X048,X176,C048,S048,512
BTRFLI   X049,X177,C049,S049,512
BTRFLI   X050,X178,C050,S050,512
BTRFLI   X051,X179,C051,S051,512
BTRFLI   X052,X180,C052,S052,512
BTRFLI   X053,X181,C053,S053,512
BTRFLI   X054,X182,C054,S054,512
BTRFLI   X055,X183,C055,S055,512
BTRFLI   X056,X184,C056,S056,512
BTRFLI   X057,X185,C057,S057,512
BTRFLI   X058,X186,C058,S058,512
BTRFLI   X059,X187,C059,S059,512
BTRFLI   X060,X188,C060,S060,512
BTRFLI   X061,X189,C061,S061,512
BTRFLI   X062,X190,C062,S062,512
BTRFLI   X063,X191,C063,S063,512
PBY2I    X064,X192,512
BTRFLI   X065,X193,C065,S065,512
BTRFLI   X066,X194,C066,S066,512
BTRFLI   X067,X195,C067,S067,512
BTRFLI   X068,X196,C068,S068,512
BTRFLI   X069,X197,C069,S069,512
BTRFLI   X070,X198,C070,S070,512
BTRFLI   X071,X199,C071,S071,512
BTRFLI   X072,X200,C072,S072,512
BTRFLI   X073,X201,C073,S073,512
BTRFLI   X074,X202,C074,S074,512
BTRFLI   X075,X203,C075,S075,512
BTRFLI   X076,X204,C076,S076,512
BTRFLI   X077,X205,C077,S077,512
BTRFLI   X078,X206,C078,S078,512
BTRFLI   X079,X207,C079,S079,512
BTRFLI   X080,X208,C080,S080,512
BTRFLI   X081,X209,C081,S081,512
BTRFLI   X082,X210,C082,S082,512
BTRFLI   X083,X211,C083,S083,512
BTRFLI   X084,X212,C084,S084,512
BTRFLI   X085,X213,C085,S085,512
BTRFLI   X086,X214,C086,S086,512
BTRFLI   X087,X215,C087,S087,512
BTRFLI   X088,X216,C088,S088,512
BTRFLI   X089,X217,C089,S089,512
BTRFLI   X090,X218,C090,S090,512
BTRFLI   X091,X219,C091,S091,512
BTRFLI   X092,X220,C092,S092,512
BTRFLI   X093,X221,C093,S093,512
BTRFLI   X094,X222,C094,S094,512
BTRFLI   X095,X223,C095,S095,512
P3BY4I   X096,X224,512
BTRFLI   X097,X225,C097,S097,512
BTRFLI   X098,X226,C098,S098,512

BTRFLI   X099,X227,C099,S099,512
BTRFLI   X100,X228,C100,S100,512
BTRFLI   X101,X229,C101,S101,512
BTRFLI   X102,X230,C102,S102,512
BTRFLI   X103,X231,C103,S103,512
BTRFLI   X104,X232,C104,S104,512
BTRFLI   X105,X233,C105,S105,512
BTRFLI   X106,X234,C106,S106,512
BTRFLI   X107,X235,C107,S107,512
BTRFLI   X108,X236,C108,S108,512
BTRFLI   X109,X237,C109,S109,512
BTRFLI   X110,X238,C110,S110,512
BTRFLI   X111,X239,C111,S111,512
BTRFLI   X112,X240,C112,S112,512
BTRFLI   X113,X241,C113,S113,512
BTRFLI   X114,X242,C114,S114,512
BTRFLI   X115,X243,C115,S115,512
BTRFLI   X116,X244,C116,S116,512
BTRFLI   X117,X245,C117,S117,512
BTRFLI   X118,X246,C118,S118,512
BTRFLI   X119,X247,C119,S119,512
BTRFLI   X120,X248,C120,S120,512
BTRFLI   X121,X249,C121,S121,512
BTRFLI   X122,X250,C122,S122,512
BTRFLI   X123,X251,C123,S123,512
BTRFLI   X124,X252,C124,S124,512
BTRFLI   X125,X253,C125,S125,512
BTRFLI   X126,X254,C126,S126,512
BTRFLI   X127,X255,C127,S127,512
RET

*****************************************************************
*                                                               *
*   1024-POINT FFT CODE WITH BIT-REVERSED INPUT SAMPLES         *
*                                                               *
*   ALL INPUT REAL AND IMAGINARY DATA POINTS ARE ASSUMED        *
*   TO BE IN CONSECUTIVE LOCATIONS (A TOTAL OF 2048 ) IN        *
*   EXTERNAL DATA MEMORY STARTING FROM LOCATION 1024 IN         *
*   PAGE 8.  OUT OF THE 1024 COMPLEX POINTS, THERE ARE          *
*   ALTOGETHER 496 PAIRS OF INPUT DATA WHICH NEED TO BE         *
*   SCRAMBLED AS SHOWN BELOW.                                   *
*                                                               *
*****************************************************************
*
FFT1024  BITRVI   2,1024,1024
         BITRVI   4,512,1024
         BITRVI   6,1536,1024
         BITRVI   8,256,1024
         BITRVI   10,1280,1024
         BITRVI   12,768,1024
         BITRVI   14,1792,1024
         BITRVI   16,128,1024
         BITRVI   18,1152,1024
         BITRVI   20,640,1024
         BITRVI   22,1664,1024
         BITRVI   24,384,1024
         BITRVI   26,1408,1024
         BITRVI   28,896,1024
         BITRVI   30,1920,1024
         BITRVI   32,64,1024
```

```
        BITRVI  34,1088,1024
        BITRVI  36,576,1024
        BITRVI  38,1600,1024
        BITRVI  40,320,1024
        BITRVI  42,1344,1024
        BITRVI  44,832,1024
        BITRVI  46,1856,1024
        BITRVI  48,192,1024
        BITRVI  50,1216,1024
        BITRVI  52,704,1024
        BITRVI  54,1728,1024
        BITRVI  56,448,1024
        BITRVI  58,1472,1024
        BITRVI  60,960,1024
        BITRVI  62,1984,1024
        BITRVI  66,1056,1024
        BITRVI  68,544,1024
        BITRVI  70,1568,1024
        BITRVI  72,288,1024
        BITRVI  74,1312,1024
        BITRVI  76,800,1024
        BITRVI  78,1824,1024
        BITRVI  80,160,1024
        BITRVI  82,1184,1024
        BITRVI  84,672,1024
        BITRVI  86,1696,1024
        BITRVI  88,416,1024
        BITRVI  90,1440,1024
        BITRVI  92,928,1024
        BITRVI  94,1952,1024
        BITRVI  98,1120,1024
        BITRVI  100,608,1024
        BITRVI  102,1632,1024
        BITRVI  104,352,1024
        BITRVI  106,1376,1024
        BITRVI  108,864,1024
        BITRVI  110,1888,1024
        BITRVI  112,224,1024
        BITRVI  114,1248,1024
        BITRVI  116,736,1024
        BITRVI  118,1760,1024
        BITRVI  120,480,1024
        BITRVI  122,1504,1024
        BITRVI  124,992,1024
        BITRVI  126,2016,1024
        BITRVI  130,1040,1024
        BITRVI  132,528,1024
        BITRVI  134,1552,1024
        BITRVI  136,272,1024
        BITRVI  138,1296,1024
        BITRVI  140,784,1024
        BITRVI  142,1808,1024
        BITRVI  146,1168,1024
        BITRVI  148,656,1024
        BITRVI  150,1680,1024
        BITRVI  152,400,1024
        BITRVI  154,1424,1024
        BITRVI  156,912,1024
        BITRVI  158,1936,1024
        BITRVI  162,1104,1024
        BITRVI  164,592,1024
        BITRVI  166,1616,1024
        BITRVI  168,336,1024
        BITRVI  170,1360,1024
        BITRVI  172,848,1024
        BITRVI  174,1872,1024
        BITRVI  176,208,1024
        BITRVI  178,1232,1024
        BITRVI  180,720,1024
        BITRVI  182,1744,1024
        BITRVI  184,464,1024
        BITRVI  186,1488,1024
        BITRVI  188,976,1024
        BITRVI  190,2000,1024
        BITRVI  194,1072,1024
        BITRVI  196,560,1024
        BITRVI  198,1584,1024
        BITRVI  200,304,1024
        BITRVI  202,1328,1024
        BITRVI  204,816,1024
        BITRVI  206,1840,1024
        BITRVI  210,1200,1024
        BITRVI  212,688,1024
        BITRVI  214,1712,1024
        BITRVI  216,432,1024
        BITRVI  218,1456,1024
        BITRVI  220,944,1024
        BITRVI  222,1968,1024
        BITRVI  226,1136,1024
        BITRVI  228,624,1024
        BITRVI  230,1648,1024
        BITRVI  232,368,1024
        BITRVI  234,1392,1024
        BITRVI  236,880,1024
        BITRVI  238,1904,1024
        BITRVI  242,1264,1024
        BITRVI  244,752,1024
        BITRVI  246,1776,1024
        BITRVI  248,496,1024
        BITRVI  250,1520,1024
        BITRVI  252,1008,1024
        BITRVI  254,2032,1024
        BITRVI  258,1032,1024
        BITRVI  260,520,1024
        BITRVI  262,1544,1024
        BITRVI  266,1288,1024
        BITRVI  268,776,1024
        BITRVI  270,1800,1024
        BITRVI  274,1160,1024
        BITRVI  276,648,1024
        BITRVI  278,1672,1024
        BITRVI  280,392,1024
        BITRVI  282,1416,1024
        BITRVI  284,904,1024
        BITRVI  286,1928,1024
        BITRVI  290,1096,1024
        BITRVI  292,584,1024
        BITRVI  294,1608,1024
        BITRVI  296,328,1024
        BITRVI  298,1352,1024
```

4. Implementation of Fast Fourier Transform Algorithms with the TMS32020

```
BITRVI 300,840,1024
BITRVI 302,1864,1024
BITRVI 306,1224,1024
BITRVI 308,712,1024
BITRVI 310,1736,1024
BITRVI 312,456,1024
BITRVI 314,1480,1024
BITRVI 316,968,1024
BITRVI 318,1992,1024
BITRVI 322,1064,1024
BITRVI 324,552,1024
BITRVI 326,1576,1024
BITRVI 330,1320,1024
BITRVI 332,808,1024
BITRVI 334,1832,1024
BITRVI 338,1192,1024
BITRVI 340,680,1024
BITRVI 342,1704,1024
BITRVI 344,424,1024
BITRVI 346,1448,1024
BITRVI 348,936,1024
BITRVI 350,1960,1024
BITRVI 354,1128,1024
BITRVI 356,616,1024
BITRVI 358,1640,1024
BITRVI 362,1384,1024
BITRVI 364,872,1024
BITRVI 366,1896,1024
BITRVI 370,1256,1024
BITRVI 372,744,1024
BITRVI 374,1768,1024
BITRVI 376,488,1024
BITRVI 378,1512,1024
BITRVI 380,1000,1024
BITRVI 382,2024,1024
BITRVI 386,1048,1024
BITRVI 388,536,1024
BITRVI 390,1560,1024
BITRVI 394,1304,1024
BITRVI 396,792,1024
BITRVI 398,1816,1024
BITRVI 402,1176,1024
BITRVI 404,664,1024
BITRVI 406,1688,1024
BITRVI 410,1432,1024
BITRVI 412,920,1024
BITRVI 414,1944,1024
BITRVI 418,1112,1024
BITRVI 420,600,1024
BITRVI 422,1624,1024
BITRVI 426,1368,1024
BITRVI 428,856,1024
BITRVI 430,1880,1024
BITRVI 434,1240,1024
BITRVI 436,728,1024
BITRVI 438,1752,1024
BITRVI 440,472,1024
BITRVI 442,1496,1024
BITRVI 444,984,1024
BITRVI 446,2008,1024

BITRVI 450,1080,1024
BITRVI 452,568,1024
BITRVI 454,1592,1024
BITRVI 458,1336,1024
BITRVI 460,824,1024
BITRVI 462,1848,1024
BITRVI 466,1208,1024
BITRVI 468,696,1024
BITRVI 470,1720,1024
BITRVI 474,1464,1024
BITRVI 476,952,1024
BITRVI 478,1976,1024
BITRVI 482,1144,1024
BITRVI 484,632,1024
BITRVI 486,1656,1024
BITRVI 490,1400,1024
BITRVI 492,888,1024
BITRVI 494,1912,1024
BITRVI 498,1272,1024
BITRVI 500,760,1024
BITRVI 502,1784,1024
BITRVI 506,1528,1024
BITRVI 508,1016,1024
BITRVI 510,2040,1024
BITRVI 514,1028,1024
BITRVI 518,1540,1024
BITRVI 522,1284,1024
BITRVI 524,772,1024
BITRVI 526,1796,1024
BITRVI 530,1156,1024
BITRVI 532,644,1024
BITRVI 534,1668,1024
BITRVI 538,1412,1024
BITRVI 540,900,1024
BITRVI 542,1924,1024
BITRVI 546,1092,1024
BITRVI 548,580,1024
BITRVI 550,1604,1024
BITRVI 554,1348,1024
BITRVI 556,836,1024
BITRVI 558,1860,1024
BITRVI 562,1220,1024
BITRVI 564,708,1024
BITRVI 566,1732,1024
BITRVI 570,1476,1024
BITRVI 572,964,1024
BITRVI 574,1988,1024
BITRVI 578,1060,1024
BITRVI 582,1572,1024
BITRVI 586,1316,1024
BITRVI 588,804,1024
BITRVI 590,1828,1024
BITRVI 594,1188,1024
BITRVI 596,676,1024
BITRVI 598,1700,1024
BITRVI 602,1444,1024
BITRVI 604,932,1024
BITRVI 606,1956,1024
BITRVI 610,1124,1024
BITRVI 614,1636,1024
```

```
BITRVI  618,1380,1024
BITRVI  620,868,1024
BITRVI  622,1892,1024
BITRVI  626,1252,1024
BITRVI  628,740,1024
BITRVI  630,1764,1024
BITRVI  634,1508,1024
BITRVI  636,996,1024
BITRVI  638,2020,1024
BITRVI  642,1044,1024
BITRVI  646,1556,1024
BITRVI  650,1300,1024
BITRVI  652,788,1024
BITRVI  654,1812,1024
BITRVI  658,1172,1024
BITRVI  662,1684,1024
BITRVI  666,1428,1024
BITRVI  668,916,1024
BITRVI  670,1940,1024
BITRVI  674,1108,1024
BITRVI  678,1620,1024
BITRVI  682,1364,1024
BITRVI  684,852,1024
BITRVI  686,1876,1024
BITRVI  690,1236,1024
BITRVI  692,724,1024
BITRVI  694,1748,1024
BITRVI  698,1492,1024
BITRVI  700,980,1024
BITRVI  702,2004,1024
BITRVI  706,1076,1024
BITRVI  710,1588,1024
BITRVI  714,1332,1024
BITRVI  716,820,1024
BITRVI  718,1844,1024
BITRVI  722,1204,1024
BITRVI  726,1716,1024
BITRVI  730,1460,1024
BITRVI  732,948,1024
BITRVI  734,1972,1024
BITRVI  738,1140,1024
BITRVI  742,1652,1024
BITRVI  746,1396,1024
BITRVI  748,884,1024
BITRVI  750,1908,1024
BITRVI  754,1268,1024
BITRVI  758,1780,1024
BITRVI  762,1524,1024
BITRVI  764,1012,1024
BITRVI  766,2036,1024
BITRVI  770,1036,1024
BITRVI  774,1548,1024
BITRVI  778,1292,1024
BITRVI  782,1804,1024
BITRVI  786,1164,1024
BITRVI  790,1676,1024
BITRVI  794,1420,1024
BITRVI  796,908,1024
BITRVI  798,1932,1024
BITRVI  802,1100,1024

BITRVI  806,1612,1024
BITRVI  810,1356,1024
BITRVI  812,844,1024
BITRVI  814,1868,1024
BITRVI  818,1228,1024
BITRVI  822,1740,1024
BITRVI  826,1484,1024
BITRVI  828,972,1024
BITRVI  830,1996,1024
BITRVI  834,1068,1024
BITRVI  838,1580,1024
BITRVI  842,1324,1024
BITRVI  846,1836,1024
BITRVI  850,1196,1024
BITRVI  854,1708,1024
BITRVI  858,1452,1024
BITRVI  860,940,1024
BITRVI  862,1964,1024
BITRVI  866,1132,1024
BITRVI  870,1644,1024
BITRVI  874,1388,1024
BITRVI  878,1900,1024
BITRVI  882,1260,1024
BITRVI  886,1772,1024
BITRVI  890,1516,1024
BITRVI  892,1004,1024
BITRVI  894,2028,1024
BITRVI  898,1052,1024
BITRVI  902,1564,1024
BITRVI  906,1308,1024
BITRVI  910,1820,1024
BITRVI  914,1180,1024
BITRVI  918,1692,1024
BITRVI  922,1436,1024
BITRVI  926,1948,1024
BITRVI  930,1116,1024
BITRVI  934,1628,1024
BITRVI  938,1372,1024
BITRVI  942,1884,1024
BITRVI  946,1244,1024
BITRVI  950,1756,1024
BITRVI  954,1500,1024
BITRVI  956,988,1024
BITRVI  958,2012,1024
BITRVI  962,1084,1024
BITRVI  966,1596,1024
BITRVI  970,1340,1024
BITRVI  974,1852,1024
BITRVI  978,1212,1024
BITRVI  982,1724,1024
BITRVI  986,1468,1024
BITRVI  990,1980,1024
BITRVI  994,1148,1024
BITRVI  998,1660,1024
BITRVI  1002,1404,1024
BITRVI  1006,1916,1024
BITRVI  1010,1276,1024
BITRVI  1014,1788,1024
BITRVI  1018,1532,1024
BITRVI  1022,2044,1024
```

4. Implementation of Fast Fourier Transform Algorithms with the TMS32020

```
BITRVI  1030,1538,1024
BITRVI  1034,1282,1024
BITRVI  1038,1794,1024
BITRVI  1042,1154,1024
BITRVI  1046,1666,1024
BITRVI  1050,1410,1024
BITRVI  1054,1922,1024
BITRVI  1058,1090,1024
BITRVI  1062,1602,1024
BITRVI  1066,1346,1024
BITRVI  1070,1858,1024
BITRVI  1074,1218,1024
BITRVI  1078,1730,1024
BITRVI  1082,1474,1024
BITRVI  1086,1986,1024
BITRVI  1094,1570,1024
BITRVI  1098,1314,1024
BITRVI  1102,1826,1024
BITRVI  1106,1186,1024
BITRVI  1110,1698,1024
BITRVI  1114,1442,1024
BITRVI  1118,1954,1024
BITRVI  1126,1634,1024
BITRVI  1130,1378,1024
BITRVI  1134,1890,1024
BITRVI  1138,1250,1024
BITRVI  1142,1762,1024
BITRVI  1146,1506,1024
BITRVI  1150,2018,1024
BITRVI  1158,1554,1024
BITRVI  1162,1298,1024
BITRVI  1166,1810,1024
BITRVI  1174,1682,1024
BITRVI  1178,1426,1024
BITRVI  1182,1938,1024
BITRVI  1190,1618,1024
BITRVI  1194,1362,1024
BITRVI  1198,1874,1024
BITRVI  1202,1234,1024
BITRVI  1206,1746,1024
BITRVI  1210,1490,1024
BITRVI  1214,2002,1024
BITRVI  1222,1586,1024
BITRVI  1226,1330,1024
BITRVI  1230,1842,1024
BITRVI  1238,1714,1024
BITRVI  1242,1458,1024
BITRVI  1246,1970,1024
BITRVI  1254,1650,1024
BITRVI  1258,1394,1024
BITRVI  1262,1906,1024
BITRVI  1270,1778,1024
BITRVI  1274,1522,1024
BITRVI  1278,2034,1024
BITRVI  1286,1546,1024
BITRVI  1294,1802,1024
BITRVI  1302,1674,1024
BITRVI  1306,1418,1024
BITRVI  1310,1930,1024
BITRVI  1318,1610,1024
BITRVI  1322,1354,1024
BITRVI  1326,1866,1024
BITRVI  1334,1738,1024
BITRVI  1338,1482,1024
BITRVI  1342,1994,1024
BITRVI  1350,1578,1024
BITRVI  1358,1834,1024
BITRVI  1366,1706,1024
BITRVI  1370,1450,1024
BITRVI  1374,1962,1024
BITRVI  1382,1642,1024
BITRVI  1390,1898,1024
BITRVI  1398,1770,1024
BITRVI  1402,1514,1024
BITRVI  1406,2026,1024
BITRVI  1414,1562,1024
BITRVI  1422,1818,1024
BITRVI  1430,1690,1024
BITRVI  1438,1946,1024
BITRVI  1446,1626,1024
BITRVI  1454,1882,1024
BITRVI  1462,1754,1024
BITRVI  1466,1498,1024
BITRVI  1470,2010,1024
BITRVI  1478,1594,1024
BITRVI  1486,1850,1024
BITRVI  1494,1722,1024
BITRVI  1502,1978,1024
BITRVI  1510,1658,1024
BITRVI  1518,1914,1024
BITRVI  1526,1786,1024
BITRVI  1534,2042,1024
BITRVI  1550,1798,1024
BITRVI  1558,1670,1024
BITRVI  1566,1926,1024
BITRVI  1574,1606,1024
BITRVI  1582,1862,1024
BITRVI  1590,1734,1024
BITRVI  1598,1990,1024
BITRVI  1614,1830,1024
BITRVI  1622,1702,1024
BITRVI  1630,1958,1024
BITRVI  1646,1894,1024
BITRVI  1654,1766,1024
BITRVI  1662,2022,1024
BITRVI  1678,1814,1024
BITRVI  1694,1942,1024
BITRVI  1710,1878,1024
BITRVI  1718,1750,1024
BITRVI  1726,2006,1024
BITRVI  1742,1846,1024
BITRVI  1758,1974,1024
BITRVI  1774,1910,1024
BITRVI  1790,2038,1024
BITRVI  1822,1934,1024
BITRVI  1838,1870,1024
BITRVI  1854,1998,1024
BITRVI  1886,1966,1024
BITRVI  1918,2030,1024
BITRVI  1982,2014,1024
```

```
*****************************************************
*                                                   *
*   PERFORM STAGE 9 OF THE 1024-POINT FFT -- TWIDDLE *
*   FACTOR VALUES ARE TABLE-READ FROM EXTERNAL PROGRAM *
*   MEMORY TO ON-CHIP RAM AND THE GENERAL 'BUTTERFLY' *
*   SUBROUTINE IS EXECUTED 512 TIMES WITH ALL DATA IN *
*   EXTERNAL DATA MEMORY.  A SPEED IMPROVEMENT CAN BE *
*   ACHIEVED BY MOVING FFT DATA ON-CHIP INSTEAD OF TWIDDLE *
*   FACTOR VALUES FOR EXECUTION.  HOWEVER, THE TWIDDLE *
*   FACTOR VALUES WILL HAVE TO BE STORED IN EXTERNAL DATA *
*   RAM.                                             *
*****************************************************

STAGE9  LRLK  AR0,512        INITIALISE TWIDDLE FACTORS
        LRLK  AR1,255        SET UP TWIDDLE TABLE SIZE
        LARP  AR0            USE AR0 TO POINT AT TABLE
        LALK  W000           SET UP TWIDDLE TABLE ADDRESS
        TBLR  *+             AND STORE IN INTERNAL RAM
        ADLK  1
LOOP    TBLR  *+,AR1         AND STORE IN INTERNAL RAM
        ADLK  3
        BANZ  LOOP,*-,AR0
*
        LRLK  AR0,512        INITIALISE STEP SIZE
        LRLK  AR1,512        AR1 POINTS AT TWIDDLE FACTORS
        LRLK  AR2,1024+512   AR2 POINTS AT REAL DATA
        LRLK  AR3,255        INITIALISE LOOP COUNT
        LARP  AR2
LOOP1   CALL  BTRFLY         PERFORM LOOPED FFT FOR STAGE 9
        LARP  AR3
        BANZ  LOOP1,*-,AR2
*
        LRLK  AR1,512        AR1 POINTS AT TWIDDLE FACTORS
        LRLK  AR2,2048+512   AR2 POINTS AT REAL DATA
        LRLK  AR3,255        INITIALISE LOOP COUNT
        LARP  AR2
LOOP2   CALL  BTRFLY         PERFORM LOOPED FFT FOR STAGE 9
        LARP  AR3
        BANZ  LOOP2,*-,AR2
*
*****************************************************
*                                                   *
*   PERFORM STAGE 10 OF THE 1024-POINT FFT -- TWIDDLE *
*   FACTOR VALUES ARE TABLE-READ FROM EXTERNAL PROGRAM *
*   MEMORY TO ON-CHIP RAM AND THE GENERAL 'BUTTERFLY' *
*   SUBROUTINE IS EXECUTED 512 TIMES WITH ALL DATA IN *
*   EXTERNAL DATA MEMORY.  A SPEED IMPROVEMENT CAN BE *
*   ACHIEVED BY MOVING FFT DATA ON-CHIP INSTEAD OF TWIDDLE *
*   FACTOR VALUES FOR EXECUTION.  HOWEVER, THE TWIDDLE *
*   FACTOR VALUES WILL HAVE TO BE STORED IN EXTERNAL DATA *
*   RAM.                                             *
*****************************************************

STGE10  LARP  AR0            INITIALISE TWIDDLE FACTORS
        LRLK  AR0,512        AND STORE IN INTERNAL RAM.
        RPTK  255
        BLKP  W000,*+
```

```
*****************************************************
*                                                   *
*   THE FIRST 8 STAGES OF THE 10-STAGE 1024-POINT COMPLEX *
*   FFT WILL BE PERFORMED AS 4 SEPARATE 256-POINT COMPLEX *
*   FFT'S USING THE ON-CHIP DATA BLOCKS B0 AND B1.  THE *
*   KERNEL CODE BELOW WILL THEREFORE BE CALLED 4 TIMES TO *
*   ACCOMPLISH THIS WHILE EXTERNAL DATA MEMORY WILL HAVE *
*   MOVE ON-CHIP USING THE BLKD INSTRUCTION.         *
*                                                   *
*****************************************************

        LARP  AR1            BLOCK MOVE FIRST GROUP OF 256
        LRLK  AR1,512        COMPLEX POINTS (1024-1535) FROM
        RPTK  255            EXTERNAL RAM TO ON-CHIP RAM
        BLKD  1024,*+
        RPTK  255
        BLKD  1280,*+
        CALL  KNL256         EXECUTE 256-POINT KERNEL FFT
        LRLK  AR1,1024       AND RETURN RESULTS TO
        RPTK  255            EXTERNAL RAM (1024-1535)
        BLKD  512,*+
        RPTK  255
        BLKD  768,*+
*
        LRLK  AR1,512        BLOCK MOVE SECOND GROUP OF 256
        RPTK  255            COMPLEX POINTS (1536-2047) FROM
        BLKD  1536,*+        EXTERNAL RAM TO ON-CHIP RAM
        RPTK  255
        BLKD  1792,*+
        CALL  KNL256         EXECUTE 256-POINT KERNEL FFT
        LRLK  AR1,1536       AND RETURN RESULTS TO
        RPTK  255            EXTERNAL RAM (1536-2047)
        BLKD  512,*+
        RPTK  255
        BLKD  768,*+
*
        LRLK  AR1,512        BLOCK MOVE THIRD GROUP OF 256
        RPTK  255            COMPLEX POINTS (2048-2559) FROM
        BLKD  2048,*+        EXTERNAL RAM TO ON-CHIP RAM
        RPTK  255
        BLKD  2304,*+
        CALL  KNL256         EXECUTE 256-POINT KERNEL FFT
        LRLK  AR1,2048       AND RETURN RESULTS TO
        RPTK  255            EXTERNAL RAM (2048-2559)
        BLKD  512,*+
        RPTK  255
        BLKD  768,*+
*
        LRLK  AR1,512        BLOCK MOVE FOURTH GROUP OF 256
        RPTK  255            COMPLEX POINTS (2560-3071) FROM
        BLKD  2560,*+        EXTERNAL RAM TO ON-CHIP RAM
        RPTK  255
        BLKD  2816,*+
        CALL  KNL256         EXECUTE 256-POINT KERNEL FFT
        LRLK  AR1,2560       AND RETURN RESULTS TO
        RPTK  255            EXTERNAL RAM (2560-3071)
        BLKD  512,*+
        RPTK  255
        BLKD  768,*+
```

```
        RPTK   255
        BLKP   W128,*+
*
        LARP   AR0
        LRLK   AR0,1024             INITIALISE STEP SIZE
        LRLK   AR1,512              AR1 POINTS AT TWIDDLE FACTORS
        LRLK   AR2,2048             AR2 POINTS AT REAL DATA
        LRLK   AR3,255              INITIALISE LOOP COUNT
        LARP   AR2
LOOP3   CALL   BTRFLY               PERFORM LOOPED FFT FOR STAGE 10
        LARP   AR3
        BANZ   LOOP3,*-,AR2
*
        LARP   AR0
        LRLK   AR0,512              INITIALISE TWIDDLE FACTORS
        RPTK   255                  AND STORE IN INTERNAL RAM.
        BLKP   W256,*+
        RPTK   255
        BLKP   W384,*+
*
        LARP   AR0
        LRLK   AR0,1024             INITIALISE STEP SIZE
        LRLK   AR1,512              AR1 POINTS AT TWIDDLE FACTORS
        LRLK   AR2,2048+512         AR2 POINTS AT REAL DATA
        LRLK   AR3,255              INITIALISE LOOP COUNT
        LARP   AR2
LOOP4   CALL   BTRFLY               PERFORM LOOPED FFT FOR STAGE 10
        LARP   AR3
        BANZ   LOOP4,*-,AR2
        B      DONE
*
***********************************************************
*                                                         *
*       RADIX-2 GENERAL BUTTERFLY SUBROUTINE              *
*                                                         *
*                                                         *
*       P=(PR+jQI)    P+Q*W=(PR+Re[Q*W])+j(PI+Im[Q*W])    *
*                          \   /                          *
*                           \ /                           *
*                           / \                           *
*       Q=(QR+jQI)    P-Q*W=(PR-Re[Q*W])+j(PI+Im[Q*W])    *
*                                                         *
*            -j(2(pi)/N)k                                 *
*       W =e              =COS(X)-jSIN(X)                 *
*                                                         *
*       N                          =WR+jWI               *
*                                                         *
***********************************************************
*       CALCULATE QR*COS(X) + QI*SIN(X) AND STORE RESULT IN QR    *
***********************************************************
BTRFLY  LT   *+,AR1        LOAD T-REGISTER WITH QR
        MPY  *+,AR2        P-REGISTER = (1/2)(QR*COSX)
        LTP  *-,AR1        ACC= (1/2)(QR*COSX); LOAD TR WITH QI
        MPY  *-            P-REGISTER = (1/2)(QI*SINX)
        APAC               ACC= (1/2)(QI*COSX+QI*SINX)
        MPY  *+,AR2        P-REGISTER = (1/2)(QI*COSX)
        LT   *             LOAD T-REGISTER WITH QR
        SACH *+,0,AR1      QR = (1/2)(QR*COSX+QI*SINX)
*
*       CALCULATE QI*COS(X) - QR*SIN(X) AND STORE RESULT IN QI
*
        PAC                          ACC= (1/2)(QI*COSX)
        MPY  *+,AR2                  P-REGISTER = (1/2)(QR*SINX)
        SPAC                         ACC= (1/2)(QI*COSX-QR*SINX)
        SACH *0-                     QI = (1/2)(QI*COSX-QR*SINX)
*
*       CALCULATE Re[Pm+1] & Re[Qm+1] STORE RESULTS in PR & QR
*
        MAR  *-
        LAC  *0+,14                  ACC= (1/4)PR
        ADD  *0-,15                  ACC= (1/4)[PR+(QR*COSX+QI*SINX)]
        SACH *0+,1                   PR = (1/2)[PR+(QR*COSX+QI*SINX)]
        SUBH *                       ACC= (1/4)[PR-(QR*COSX+QI*SINX)]
        SACH *0-,1                   QR = (1/2)[PR-(QR*COSX+QI*SINX)]
*
*       CALCULATE Im[Pm+1] & Im[Qm+1] STORE RESULTS in PI & QI
*
        MAR  *+
        LAC  *0+,14                  ACC= (1/4)PI
        ADD  *0-,15                  ACC= (1/4)[PI+(QI*COSX-QR*SINX)]
        SACH *0+,1                   PI = (1/2)[PI+(QI*COSX-QR*SINX)]
        SUBH *                       ACC= (1/4)[PI-(QI*COSX-QR*SINX)]
        SACH *+,1                    QI = (1/2)[PI-(QI*COSX-QR*SINX)]
        RET
*
***********************************************************
*                                                         *
*       16-BIT TWIDDLE FACTORS FOR 1024-POINT COMPLEX FFT *
*                                                         *
***********************************************************
*
W000    DATA   32767,0
W001    DATA   32766,201
W002    DATA   32765,402
W003    DATA   32762,603
W004    DATA   32758,804
W005    DATA   32752,1005
W006    DATA   32746,1206
W007    DATA   32738,1407
W008    DATA   32728,1608
W009    DATA   32718,1809
W010    DATA   32706,2009
W011    DATA   32693,2210
W012    DATA   32679,2410
W013    DATA   32664,2611
W014    DATA   32647,2811
W015    DATA   32629,3012
W016    DATA   32610,3212
W017    DATA   32590,3412
W018    DATA   32568,3612
W019    DATA   32545,3811
W020    DATA   32521,4011
W021    DATA   32496,4210
W022    DATA   32470,4410
W023    DATA   32442,4609
W024    DATA   32413,4808
W025    DATA   32383,5007
W026    DATA   32352,5205
W027    DATA   32319,5404
W028    DATA   32285,5602
```

W029	DATA	32250,5800
W030	DATA	32214,5998
W031	DATA	32177,6195
W032	DATA	32138,6393
W033	DATA	32098,6590
W034	DATA	32057,6786
W035	DATA	32015,6983
W036	DATA	31972,7179
W037	DATA	31927,7375
W038	DATA	31881,7571
W039	DATA	31834,7767
W040	DATA	31786,7962
W041	DATA	31736,8157
W042	DATA	31686,8351
W043	DATA	31634,8546
W044	DATA	31581,8739
W045	DATA	31527,8933
W046	DATA	31471,9126
W047	DATA	31415,9319
W048	DATA	31357,9512
W049	DATA	31298,9704
W050	DATA	31238,9896
W051	DATA	31176,10088
W052	DATA	31114,10279
W053	DATA	31050,10469
W054	DATA	30985,10660
W055	DATA	30919,10850
W056	DATA	30852,11039
W057	DATA	30784,11228
W058	DATA	30715,11417
W059	DATA	30644,11605
W060	DATA	30572,11793
W061	DATA	30499,11980
W062	DATA	30425,12167
W063	DATA	30350,12354
W064	DATA	30274,12540
W065	DATA	30196,12725
W066	DATA	30117,12910
W067	DATA	30038,13095
W068	DATA	29957,13279
W069	DATA	29875,13462
W070	DATA	29791,13645
W071	DATA	29707,13828
W072	DATA	29622,14010
W073	DATA	29535,14191
W074	DATA	29448,14372
W075	DATA	29359,14553
W076	DATA	29269,14733
W077	DATA	29178,14912
W078	DATA	29086,15091
W079	DATA	28993,15269
W080	DATA	28899,15447
W081	DATA	28803,15624
W082	DATA	28707,15800
W083	DATA	28609,15976
W084	DATA	28511,16151
W085	DATA	28411,16326
W086	DATA	28310,16500
W087	DATA	28209,16673
W088	DATA	28106,16846
W089	DATA	28002,17018
W090	DATA	27897,17190
W091	DATA	27791,17360
W092	DATA	27684,17531
W093	DATA	27576,17700
W094	DATA	27467,17869
W095	DATA	27357,18037
W096	DATA	27245,18205
W097	DATA	27133,18372
W098	DATA	27020,18538
W099	DATA	26906,18703
W100	DATA	26790,18868
W101	DATA	26674,19032
W102	DATA	26557,19195
W103	DATA	26439,19358
W104	DATA	26319,19520
W105	DATA	26199,19681
W106	DATA	26078,19841
W107	DATA	25956,20001
W108	DATA	25832,20160
W109	DATA	25708,20318
W110	DATA	25583,20475
W111	DATA	25457,20632
W112	DATA	25330,20788
W113	DATA	25202,20943
W114	DATA	25073,21097
W115	DATA	24943,21250
W116	DATA	24812,21403
W117	DATA	24680,21555
W118	DATA	24548,21706
W119	DATA	24414,21856
W120	DATA	24279,22005
W121	DATA	24144,22154
W122	DATA	24007,22302
W123	DATA	23870,22449
W124	DATA	23732,22595
W125	DATA	23593,22740
W126	DATA	23453,22884
W127	DATA	23312,23028
W128	DATA	23170,23170
W129	DATA	23028,23312
W130	DATA	22884,23453
W131	DATA	22740,23593
W132	DATA	22595,23732
W133	DATA	22449,23870
W134	DATA	22302,24007
W135	DATA	22154,24144
W136	DATA	22005,24279
W137	DATA	21856,24414
W138	DATA	21706,24548
W139	DATA	21555,24680
W140	DATA	21403,24812
W141	DATA	21250,24943
W142	DATA	21097,25073
W143	DATA	20943,25202
W144	DATA	20788,25330
W145	DATA	20632,25457
W146	DATA	20475,25583
W147	DATA	20318,25708
W148	DATA	20160,25832

4. Implementation of Fast Fourier Transform Algorithms with the TMS32020

W149	DATA	20001,25956	W209	DATA	9319,31415
W150	DATA	19841,26078	W210	DATA	9126,31471
W151	DATA	19681,26199	W211	DATA	8933,31527
W152	DATA	19520,26319	W212	DATA	8739,31581
W153	DATA	19358,26439	W213	DATA	8546,31634
W154	DATA	19195,26557	W214	DATA	8351,31686
W155	DATA	19032,26674	W215	DATA	8157,31736
W156	DATA	18868,26790	W216	DATA	7962,31786
W157	DATA	18703,26906	W217	DATA	7767,31834
W158	DATA	18538,27020	W218	DATA	7571,31881
W159	DATA	18372,27133	W219	DATA	7375,31927
W160	DATA	18205,27245	W220	DATA	7179,31972
W161	DATA	18037,27357	W221	DATA	6983,32015
W162	DATA	17869,27467	W222	DATA	6786,32057
W163	DATA	17700,27576	W223	DATA	6590,32098
W164	DATA	17531,27684	W224	DATA	6393,32138
W165	DATA	17360,27791	W225	DATA	6195,32177
W166	DATA	17190,27897	W226	DATA	5998,32214
W167	DATA	17018,28002	W227	DATA	5800,32250
W168	DATA	16846,28106	W228	DATA	5602,32285
W169	DATA	16673,28209	W229	DATA	5404,32319
W170	DATA	16500,28311	W230	DATA	5205,32352
W171	DATA	16326,28411	W231	DATA	5007,32383
W172	DATA	16151,28511	W232	DATA	4808,32413
W173	DATA	15976,28609	W233	DATA	4609,32442
W174	DATA	15800,28707	W234	DATA	4410,32470
W175	DATA	15624,28803	W235	DATA	4211,32496
W176	DATA	15447,28899	W236	DATA	4011,32521
W177	DATA	15269,28993	W237	DATA	3811,32545
W178	DATA	15091,29086	W238	DATA	3612,32568
W179	DATA	14912,29178	W239	DATA	3412,32590
W180	DATA	14733,29269	W240	DATA	3212,32610
W181	DATA	14553,29359	W241	DATA	3012,32629
W182	DATA	14372,29448	W242	DATA	2811,32647
W183	DATA	14191,29535	W243	DATA	2611,32664
W184	DATA	14010,29622	W244	DATA	2410,32679
W185	DATA	13828,29707	W245	DATA	2210,32693
W186	DATA	13645,29791	W246	DATA	2009,32706
W187	DATA	13462,29875	W247	DATA	1809,32718
W188	DATA	13279,29957	W248	DATA	1608,32728
W189	DATA	13095,30038	W249	DATA	1407,32738
W190	DATA	12910,30117	W250	DATA	1206,32746
W191	DATA	12725,30196	W251	DATA	1005,32752
W192	DATA	12540,30274	W252	DATA	804,32758
W193	DATA	12354,30350	W253	DATA	603,32762
W194	DATA	12167,30425	W254	DATA	402,32765
W195	DATA	11980,30499	W255	DATA	201,32766
W196	DATA	11793,30572	W256	DATA	0,32767
W197	DATA	11605,30644	W257	DATA	-201,32766
W198	DATA	11417,30715	W258	DATA	-402,32765
W199	DATA	11228,30784	W259	DATA	-603,32762
W200	DATA	11039,30852	W260	DATA	-804,32758
W201	DATA	10850,30920	W261	DATA	-1005,32752
W202	DATA	10660,30986	W262	DATA	-1206,32746
W203	DATA	10469,31050	W263	DATA	-1407,32738
W204	DATA	10279,31114	W264	DATA	-1608,32728
W205	DATA	10087,31176	W265	DATA	-1809,32718
W206	DATA	9896,31238	W266	DATA	-2009,32706
W207	DATA	9704,31298	W267	DATA	-2210,32693
W208	DATA	9512,31357	W268	DATA	-2410,32679

```
W269    DATA    -2611,32663        W329    DATA    -14191,29535
W270    DATA    -2811,32647        W330    DATA    -14372,29448
W271    DATA    -3012,32629        W331    DATA    -14553,29359
W272    DATA    -3212,32610        W332    DATA    -14733,29269
W273    DATA    -3412,32590        W333    DATA    -14912,29178
W274    DATA    -3612,32568        W334    DATA    -15091,29086
W275    DATA    -3811,32545        W335    DATA    -15269,28993
W276    DATA    -4011,32521        W336    DATA    -15447,28899
W277    DATA    -4210,32496        W337    DATA    -15624,28803
W278    DATA    -4410,32470        W338    DATA    -15800,28707
W279    DATA    -4609,32442        W339    DATA    -15976,28609
W280    DATA    -4808,32413        W340    DATA    -16151,28511
W281    DATA    -5007,32383        W341    DATA    -16326,28411
W282    DATA    -5205,32352        W342    DATA    -16500,28310
W283    DATA    -5404,32319        W343    DATA    -16673,28209
W284    DATA    -5602,32285        W344    DATA    -16846,28106
W285    DATA    -5800,32250        W345    DATA    -17018,28002
W286    DATA    -5998,32214        W346    DATA    -17190,27897
W287    DATA    -6195,32177        W347    DATA    -17360,27791
W288    DATA    -6393,32138        W348    DATA    -17531,27684
W289    DATA    -6590,32098        W349    DATA    -17700,27576
W290    DATA    -6786,32057        W350    DATA    -17869,27467
W291    DATA    -6983,32015        W351    DATA    -18037,27357
W292    DATA    -7179,31972        W352    DATA    -18205,27245
W293    DATA    -7375,31927        W353    DATA    -18372,27133
W294    DATA    -7571,31881        W354    DATA    -18538,27020
W295    DATA    -7767,31834        W355    DATA    -18703,26906
W296    DATA    -7962,31786        W356    DATA    -18868,26790
W297    DATA    -8157,31736        W357    DATA    -19032,26674
W298    DATA    -8351,31686        W358    DATA    -19195,26557
W299    DATA    -8546,31634        W359    DATA    -19358,26439
W300    DATA    -8739,31581        W360    DATA    -19520,26319
W301    DATA    -8933,31527        W361    DATA    -19681,26199
W302    DATA    -9126,31471        W362    DATA    -19841,26078
W303    DATA    -9319,31415        W363    DATA    -20001,25956
W304    DATA    -9512,31357        W364    DATA    -20160,25832
W305    DATA    -9704,31298        W365    DATA    -20318,25708
W306    DATA    -9896,31238        W366    DATA    -20475,25583
W307    DATA    -10087,31176       W367    DATA    -20632,25457
W308    DATA    -10279,31114       W368    DATA    -20788,25330
W309    DATA    -10469,31050       W369    DATA    -20943,25202
W310    DATA    -10660,30986       W370    DATA    -21097,25073
W311    DATA    -10850,30920       W371    DATA    -21250,24943
W312    DATA    -11039,30852       W372    DATA    -21403,24812
W313    DATA    -11228,30784       W373    DATA    -21555,24680
W314    DATA    -11417,30715       W374    DATA    -21706,24548
W315    DATA    -11605,30644       W375    DATA    -21856,24414
W316    DATA    -11793,30572       W376    DATA    -22005,24279
W317    DATA    -11980,30499       W377    DATA    -22154,24144
W318    DATA    -12167,30425       W378    DATA    -22302,24007
W319    DATA    -12354,30350       W379    DATA    -22449,23870
W320    DATA    -12540,30274       W380    DATA    -22595,23732
W321    DATA    -12725,30196       W381    DATA    -22740,23593
W322    DATA    -12910,30117       W382    DATA    -22884,23453
W323    DATA    -13095,30037       W383    DATA    -23028,23312
W324    DATA    -13279,29957       W384    DATA    -23170,23170
W325    DATA    -13462,29875       W385    DATA    -23312,23028
W326    DATA    -13645,29791       W386    DATA    -23453,22884
W327    DATA    -13828,29707       W387    DATA    -23593,22740
W328    DATA    -14010,29622       W388    DATA    -23732,22595
```

4. Implementation of Fast Fourier Transform Algorithms with the TMS32020

```
W389  DATA  -23870,22449
W390  DATA  -24007,22302
W391  DATA  -24144,22154
W392  DATA  -24279,22005
W393  DATA  -24414,21856
W394  DATA  -24548,21706
W395  DATA  -24680,21555
W396  DATA  -24812,21403
W397  DATA  -24943,21250
W398  DATA  -25073,21097
W399  DATA  -25202,20943
W400  DATA  -25330,20788
W401  DATA  -25457,20632
W402  DATA  -25583,20475
W403  DATA  -25708,20318
W404  DATA  -25832,20160
W405  DATA  -25956,20001
W406  DATA  -26078,19841
W407  DATA  -26199,19681
W408  DATA  -26319,19520
W409  DATA  -26439,19358
W410  DATA  -26557,19195
W411  DATA  -26674,19032
W412  DATA  -26790,18868
W413  DATA  -26906,18703
W414  DATA  -27020,18538
W415  DATA  -27133,18372
W416  DATA  -27245,18205
W417  DATA  -27357,18037
W418  DATA  -27467,17869
W419  DATA  -27576,17700
W420  DATA  -27684,17530
W421  DATA  -27791,17360
W422  DATA  -27897,17190
W423  DATA  -28002,17018
W424  DATA  -28106,16846
W425  DATA  -28209,16673
W426  DATA  -28311,16500
W427  DATA  -28411,16326
W428  DATA  -28511,16151
W429  DATA  -28609,15976
W430  DATA  -28707,15800
W431  DATA  -28803,15624
W432  DATA  -28899,15447
W433  DATA  -28993,15269
W434  DATA  -29086,15091
W435  DATA  -29178,14912
W436  DATA  -29269,14733
W437  DATA  -29359,14553
W438  DATA  -29448,14372
W439  DATA  -29535,14191
W440  DATA  -29622,14010
W441  DATA  -29707,13828
W442  DATA  -29791,13645
W443  DATA  -29875,13462
W444  DATA  -29957,13279
W445  DATA  -30038,13095
W446  DATA  -30117,12910
W447  DATA  -30196,12725
W448  DATA  -30274,12540

W449  DATA  -30350,12354
W450  DATA  -30425,12167
W451  DATA  -30499,11980
W452  DATA  -30572,11793
W453  DATA  -30644,11605
W454  DATA  -30715,11417
W455  DATA  -30784,11228
W456  DATA  -30852,11039
W457  DATA  -30920,10850
W458  DATA  -30985,10660
W459  DATA  -31050,10469
W460  DATA  -31114,10279
W461  DATA  -31176,10087
W462  DATA  -31238,9896
W463  DATA  -31298,9704
W464  DATA  -31357,9512
W465  DATA  -31415,9319
W466  DATA  -31471,9126
W467  DATA  -31527,8933
W468  DATA  -31581,8739
W469  DATA  -31634,8546
W470  DATA  -31686,8351
W471  DATA  -31736,8157
W472  DATA  -31786,7962
W473  DATA  -31834,7767
W474  DATA  -31881,7571
W475  DATA  -31927,7375
W476  DATA  -31972,7179
W477  DATA  -32015,6983
W478  DATA  -32057,6786
W479  DATA  -32098,6589
W480  DATA  -32138,6393
W481  DATA  -32177,6195
W482  DATA  -32214,5997
W483  DATA  -32250,5800
W484  DATA  -32285,5602
W485  DATA  -32319,5404
W486  DATA  -32352,5205
W487  DATA  -32383,5007
W488  DATA  -32413,4808
W489  DATA  -32442,4609
W490  DATA  -32470,4410
W491  DATA  -32496,4210
W492  DATA  -32521,4011
W493  DATA  -32545,3811
W494  DATA  -32568,3612
W495  DATA  -32590,3412
W496  DATA  -32610,3212
W497  DATA  -32629,3012
W498  DATA  -32647,2811
W499  DATA  -32664,2611
W500  DATA  -32679,2410
W501  DATA  -32693,2210
W502  DATA  -32706,2009
W503  DATA  -32718,1809
W504  DATA  -32728,1608
W505  DATA  -32738,1407
W506  DATA  -32746,1206
W507  DATA  -32752,1005
W508  DATA  -32758,804

W509  DATA  -32762,603
W510  DATA  -32765,402
W511  DATA  -32766,201
DONE  NOP
      END
```

APPENDIX E
A 256-POINT, RADIX-4 DIF FFT IMPLEMENTATION

```
        A P P E N D I X   E
        -------------------

        IDT 'S4FFT256'

*  COOLEY-TUKEY RADIX-4, DIF FFT PROGRAM - 256-POINT STRAIGHT-LINE.
*
*       FOUR STAGES OF THREE TYPES RADIX-4 BUTTERFLIES -
*       ZERO, SPECIAL, AND NORMAL - IMPLEMENTED WITH MACROS.
*       COMPLEX INPUT DATA LOCATED ON PAGES 4-7 OF DATA MEMORY.
*       RESULTS ARE LEFT IN DATA RAM.
*       USES 13-BIT COEFFICIENTS FROM MPYK INSTRUCTIONS.
*       INTERMEDIATE VALUES ARE SCALED BY 1/4 AT EACH STAGE TO PREVENT
*       OVERFLOW.
*
*****************************************************************************
*
*  DATA MEMORY ALLOCATION.
*
*****************************************************************************
N       EQU  256    * FFT LENGTH
*
T1      EQU  96     * TEMPORARY LOCATIONS ADDRESSED
T2      EQU  97     * BY AUXILIARY REGISTERS.
*
X0      EQU  512
X1      EQU  514
X2      EQU  516
X3      EQU  518
X4      EQU  520
X5      EQU  522
X6      EQU  524
X7      EQU  526
X8      EQU  528
X9      EQU  530
X10     EQU  532
X11     EQU  534
X12     EQU  536
X13     EQU  538
X14     EQU  540
X15     EQU  542
X16     EQU  544
X17     EQU  546
X18     EQU  548
X19     EQU  550
X20     EQU  552
X21     EQU  554
X22     EQU  556
X23     EQU  558
X24     EQU  560
X25     EQU  562
X26     EQU  564
X27     EQU  566
X28     EQU  568
X29     EQU  570
```

```
X30  EQU  572
X31  EQU  574
X32  EQU  576
X33  EQU  578
X34  EQU  580
X35  EQU  582
X36  EQU  584
X37  EQU  586
X38  EQU  588
X39  EQU  590
X40  EQU  592
X41  EQU  594
X42  EQU  596
X43  EQU  598
X44  EQU  600
X45  EQU  602
X46  EQU  604
X47  EQU  606
X48  EQU  608
X49  EQU  610
X50  EQU  612
X51  EQU  614
X52  EQU  616
X53  EQU  618
X54  EQU  620
X55  EQU  622
X56  EQU  624
X57  EQU  626
X58  EQU  628
X59  EQU  630
X60  EQU  632
X61  EQU  634
X62  EQU  636
X63  EQU  638
X64  EQU  640
X65  EQU  642
X66  EQU  644
X67  EQU  646
X68  EQU  648
X69  EQU  650
X70  EQU  652
X71  EQU  654
X72  EQU  656
X73  EQU  658
X74  EQU  660
X75  EQU  662
X76  EQU  664
X77  EQU  666
X78  EQU  668
X79  EQU  670
X80  EQU  672
X81  EQU  674
X82  EQU  676
X83  EQU  678
X84  EQU  680
X85  EQU  682
X86  EQU  684
X87  EQU  686
X88  EQU  688
X89  EQU  690
X90  EQU  692
X91  EQU  694
X92  EQU  696
X93  EQU  698
X94  EQU  700
X95  EQU  702
X96  EQU  704
X97  EQU  706
X98  EQU  708
X99  EQU  710
X100 EQU  712
X101 EQU  714
X102 EQU  716
X103 EQU  718
X104 EQU  720
X105 EQU  722
X106 EQU  724
X107 EQU  726
X108 EQU  728
X109 EQU  730
X110 EQU  732
X111 EQU  734
X112 EQU  736
X113 EQU  738
X114 EQU  740
X115 EQU  742
X116 EQU  744
X117 EQU  746
X118 EQU  748
X119 EQU  750
X120 EQU  752
X121 EQU  754
X122 EQU  756
X123 EQU  758
X124 EQU  760
X125 EQU  762
X126 EQU  764
X127 EQU  766
X128 EQU  768
X129 EQU  770
X130 EQU  772
X131 EQU  774
X132 EQU  776
X133 EQU  778
X134 EQU  780
X135 EQU  782
X136 EQU  784
X137 EQU  786
X138 EQU  788
X139 EQU  790
X140 EQU  792
X141 EQU  794
X142 EQU  796
X143 EQU  798
X144 EQU  800
X145 EQU  802
X146 EQU  804
X147 EQU  806
X148 EQU  808
X149 EQU  810
```

4. Implementation of Fast Fourier Transform Algorithms with the TMS32020

```
X150  EQU  812
X151  EQU  814
X152  EQU  816
X153  EQU  818
X154  EQU  820
X155  EQU  822
X156  EQU  824
X157  EQU  826
X158  EQU  828
X159  EQU  830
X160  EQU  832
X161  EQU  834
X162  EQU  836
X163  EQU  838
X164  EQU  840
X165  EQU  842
X166  EQU  844
X167  EQU  846
X168  EQU  848
X169  EQU  850
X170  EQU  852
X171  EQU  854
X172  EQU  856
X173  EQU  858
X174  EQU  860
X175  EQU  862
X176  EQU  864
X177  EQU  866
X178  EQU  868
X179  EQU  870
X180  EQU  872
X181  EQU  874
X182  EQU  876
X183  EQU  878
X184  EQU  880
X185  EQU  882
X186  EQU  884
X187  EQU  886
X188  EQU  888
X189  EQU  890
X190  EQU  892
X191  EQU  894
X192  EQU  896
X193  EQU  898
X194  EQU  900
X195  EQU  902
X196  EQU  904
X197  EQU  906
X198  EQU  908
X199  EQU  910
X200  EQU  912
X201  EQU  914
X202  EQU  916
X203  EQU  918
X204  EQU  920
X205  EQU  922
X206  EQU  924
X207  EQU  926
X208  EQU  928
X209  EQU  930

X210  EQU  932
X211  EQU  934
X212  EQU  936
X213  EQU  938
X214  EQU  940
X215  EQU  942
X216  EQU  944
X217  EQU  946
X218  EQU  948
X219  EQU  950
X220  EQU  952
X221  EQU  954
X222  EQU  956
X223  EQU  958
X224  EQU  960
X225  EQU  962
X226  EQU  964
X227  EQU  966
X228  EQU  968
X229  EQU  970
X230  EQU  972
X231  EQU  974
X232  EQU  976
X233  EQU  978
X234  EQU  980
X235  EQU  982
X236  EQU  984
X237  EQU  986
X238  EQU  988
X239  EQU  990
X240  EQU  992
X241  EQU  994
X242  EQU  996
X243  EQU  998
X244  EQU  1000
X245  EQU  1002
X246  EQU  1004
X247  EQU  1006
X248  EQU  1008
X249  EQU  1010
X250  EQU  1012
X251  EQU  1014
X252  EQU  1016
X253  EQU  1018
X254  EQU  1020
X255  EQU  1022
*
* TABLE WITH COSINES
*
C0    EQU  4095
C1    EQU  4094
C2    EQU  4090
C3    EQU  4084
C4    EQU  4075
C5    EQU  4064
C6    EQU  4051
C7    EQU  4035
C8    EQU  4016
C9    EQU  3996
C10   EQU  3972
```

```
C11   EQU   3947
C12   EQU   3919
C13   EQU   3888
C14   EQU   3856
C15   EQU   3821
C16   EQU   3783
C17   EQU   3744
C18   EQU   3702
C19   EQU   3658
C20   EQU   3611
C21   EQU   3563
C22   EQU   3512
C23   EQU   3460
C24   EQU   3405
C25   EQU   3348
C26   EQU   3289
C27   EQU   3228
C28   EQU   3165
C29   EQU   3101
C30   EQU   3034
C31   EQU   2966
C32   EQU   2896
C33   EQU   2824
C34   EQU   2750
C35   EQU   2675
C36   EQU   2598
C37   EQU   2519
C38   EQU   2439
C39   EQU   2358
C40   EQU   2275
C41   EQU   2191
C42   EQU   2105
C43   EQU   2018
C44   EQU   1930
C45   EQU   1841
C46   EQU   1751
C47   EQU   1659
C48   EQU   1567
C49   EQU   1474
C50   EQU   1380
C51   EQU   1285
C52   EQU   1189
C53   EQU   1092
C54   EQU   995
C55   EQU   897
C56   EQU   799
C57   EQU   700
C58   EQU   601
C59   EQU   501
C60   EQU   401
C61   EQU   301
C62   EQU   201
C63   EQU   100
C64   EQU   0
C65   EQU   -99
C66   EQU   -200
C67   EQU   -300
C68   EQU   -400
C69   EQU   -500
C70   EQU   -600

C71   EQU   -699
C72   EQU   -798
C73   EQU   -896
C74   EQU   -994
C75   EQU   -1091
C76   EQU   -1188
C77   EQU   -1284
C78   EQU   -1379
C79   EQU   -1473
C80   EQU   -1566
C81   EQU   -1658
C82   EQU   -1750
C83   EQU   -1840
C84   EQU   -1929
C85   EQU   -2017
C86   EQU   -2104
C87   EQU   -2190
C88   EQU   -2274
C89   EQU   -2357
C90   EQU   -2438
C91   EQU   -2518
C92   EQU   -2597
C93   EQU   -2674
C94   EQU   -2749
C95   EQU   -2823
C96   EQU   -2895
C97   EQU   -2965
C98   EQU   -3033
C99   EQU   -3100
C100  EQU   -3164
C101  EQU   -3227
C102  EQU   -3288
C103  EQU   -3347
C104  EQU   -3404
C105  EQU   -3459
C106  EQU   -3511
C107  EQU   -3562
C108  EQU   -3610
C109  EQU   -3657
C110  EQU   -3701
C111  EQU   -3743
C112  EQU   -3782
C113  EQU   -3820
C114  EQU   -3855
C115  EQU   -3887
C116  EQU   -3918
C117  EQU   -3946
C118  EQU   -3971
C119  EQU   -3995
C120  EQU   -4015
C121  EQU   -4034
C122  EQU   -4050
C123  EQU   -4063
C124  EQU   -4074
C125  EQU   -4083
C126  EQU   -4089
C127  EQU   -4093
C128  EQU   -4094
C129  EQU   -4093
C130  EQU   -4089
```

```
C131    EQU     -4083          C191    EQU     -99
C132    EQU     -4074          C192    EQU     0
C133    EQU     -4063          C193    EQU     100
C134    EQU     -4050          C194    EQU     201
C135    EQU     -4034          C195    EQU     301
C136    EQU     -4015          C196    EQU     401
C137    EQU     -3995          C197    EQU     501
C138    EQU     -3971          C198    EQU     601
C139    EQU     -3946          C199    EQU     700
C140    EQU     -3918          C200    EQU     799
C141    EQU     -3887          C201    EQU     897
C142    EQU     -3855          C202    EQU     995
C143    EQU     -3820          C203    EQU     1092
C144    EQU     -3782          C204    EQU     1189
C145    EQU     -3743          C205    EQU     1285
C146    EQU     -3701          C206    EQU     1380
C147    EQU     -3657          C207    EQU     1474
C148    EQU     -3610          C208    EQU     1567
C149    EQU     -3562          C209    EQU     1659
C150    EQU     -3511          C210    EQU     1751
C151    EQU     -3459          C211    EQU     1841
C152    EQU     -3404          C212    EQU     1930
C153    EQU     -3347          C213    EQU     2018
C154    EQU     -3288          C214    EQU     2105
C155    EQU     -3227          C215    EQU     2191
C156    EQU     -3164          C216    EQU     2275
C157    EQU     -3100          C217    EQU     2358
C158    EQU     -3033          C218    EQU     2439
C159    EQU     -2965          C219    EQU     2519
C160    EQU     -2895          C220    EQU     2598
C161    EQU     -2823          C221    EQU     2675
C162    EQU     -2749          C222    EQU     2750
C163    EQU     -2674          C223    EQU     2824
C164    EQU     -2597          C224    EQU     2896
C165    EQU     -2518          C225    EQU     2966
C166    EQU     -2438          C226    EQU     3034
C167    EQU     -2357          C227    EQU     3101
C168    EQU     -2274          C228    EQU     3165
C169    EQU     -2190          C229    EQU     3228
C170    EQU     -2104          C230    EQU     3289
C171    EQU     -2017          C231    EQU     3348
C172    EQU     -1929          C232    EQU     3405
C173    EQU     -1840          C233    EQU     3460
C174    EQU     -1750          C234    EQU     3512
C175    EQU     -1658          C235    EQU     3563
C176    EQU     -1566          C236    EQU     3611
C177    EQU     -1473          C237    EQU     3658
C178    EQU     -1379          C238    EQU     3702
C179    EQU     -1284          C239    EQU     3744
C180    EQU     -1188          C240    EQU     3783
C181    EQU     -1091          C241    EQU     3821
C182    EQU     -994           C242    EQU     3856
C183    EQU     -896           C243    EQU     3888
C184    EQU     -798           C244    EQU     3919
C185    EQU     -699           C245    EQU     3947
C186    EQU     -600           C246    EQU     3972
C187    EQU     -500           C247    EQU     3996
C188    EQU     -400           C248    EQU     4016
C189    EQU     -300           C249    EQU     4035
C190    EQU     -200           C250    EQU     4051
```

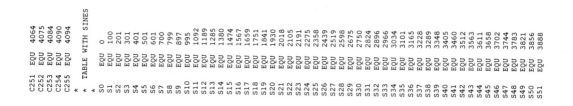

S52 EQU 3919
S53 EQU 3947
S54 EQU 3972
S55 EQU 3996
S56 EQU 4016
S57 EQU 4035
S58 EQU 4051
S59 EQU 4064
S60 EQU 4075
S61 EQU 4084
S62 EQU 4090
S63 EQU 4094
S64 EQU 4095
S65 EQU 4094
S66 EQU 4090
S67 EQU 4084
S68 EQU 4075
S69 EQU 4064
S70 EQU 4051
S71 EQU 4035
S72 EQU 4016
S73 EQU 3996
S74 EQU 3972
S75 EQU 3947
S76 EQU 3919
S77 EQU 3888
S78 EQU 3856
S79 EQU 3821
S80 EQU 3783
S81 EQU 3744
S82 EQU 3702
S83 EQU 3658
S84 EQU 3611
S85 EQU 3563
S86 EQU 3512
S87 EQU 3460
S88 EQU 3405
S89 EQU 3348
S90 EQU 3289
S91 EQU 3228
S92 EQU 3165
S93 EQU 3101
S94 EQU 3034
S95 EQU 2966
S96 EQU 2896
S97 EQU 2824
S98 EQU 2750
S99 EQU 2675
S100 EQU 2598
S101 EQU 2519
S102 EQU 2439
S103 EQU 2358
S104 EQU 2275
S105 EQU 2191
S106 EQU 2105
S107 EQU 2018
S108 EQU 1930
S109 EQU 1841
S110 EQU 1751
S111 EQU 1659

C251 EQU 4064
C252 EQU 4075
C253 EQU 4084
C254 EQU 4090
C255 EQU 4094
*
*   TABLE WITH SINES
*
S0   EQU 0
S1   EQU 100
S2   EQU 201
S3   EQU 301
S4   EQU 401
S5   EQU 501
S6   EQU 601
S7   EQU 700
S8   EQU 799
S9   EQU 897
S10 EQU 995
S11 EQU 1092
S12 EQU 1189
S13 EQU 1285
S14 EQU 1380
S15 EQU 1474
S16 EQU 1567
S17 EQU 1659
S18 EQU 1751
S19 EQU 1841
S20 EQU 1930
S21 EQU 2018
S22 EQU 2105
S23 EQU 2191
S24 EQU 2275
S25 EQU 2358
S26 EQU 2439
S27 EQU 2519
S28 EQU 2598
S29 EQU 2675
S30 EQU 2750
S31 EQU 2824
S32 EQU 2896
S33 EQU 2966
S34 EQU 3034
S35 EQU 3101
S36 EQU 3165
S37 EQU 3228
S38 EQU 3289
S39 EQU 3348
S40 EQU 3405
S41 EQU 3460
S42 EQU 3512
S43 EQU 3563
S44 EQU 3611
S45 EQU 3658
S46 EQU 3702
S47 EQU 3744
S48 EQU 3783
S49 EQU 3821
S50 EQU 3856
S51 EQU 3888

4. Implementation of Fast Fourier Transform Algorithms with the TMS32020

```
S112    EQU     1567
S113    EQU     1474
S114    EQU     1380
S115    EQU     1285
S116    EQU     1189
S117    EQU     1092
S118    EQU     995
S119    EQU     897
S120    EQU     799
S121    EQU     700
S122    EQU     601
S123    EQU     501
S124    EQU     401
S125    EQU     301
S126    EQU     201
S127    EQU     100
S128    EQU     0
S129    EQU     -99
S130    EQU     -200
S131    EQU     -300
S132    EQU     -400
S133    EQU     -500
S134    EQU     -600
S135    EQU     -699
S136    EQU     -798
S137    EQU     -896
S138    EQU     -994
S139    EQU     -1091
S140    EQU     -1188
S141    EQU     -1284
S142    EQU     -1379
S143    EQU     -1473
S144    EQU     -1566
S145    EQU     -1658
S146    EQU     -1750
S147    EQU     -1840
S148    EQU     -1929
S149    EQU     -2017
S150    EQU     -2104
S151    EQU     -2190
S152    EQU     -2274
S153    EQU     -2357
S154    EQU     -2438
S155    EQU     -2518
S156    EQU     -2597
S157    EQU     -2674
S158    EQU     -2749
S159    EQU     -2823
S160    EQU     -2895
S161    EQU     -2965
S162    EQU     -3033
S163    EQU     -3100
S164    EQU     -3164
S165    EQU     -3227
S166    EQU     -3288
S167    EQU     -3347
S168    EQU     -3404
S169    EQU     -3459
S170    EQU     -3511
S171    EQU     -3562

S172    EQU     -3610
S173    EQU     -3657
S174    EQU     -3701
S175    EQU     -3743
S176    EQU     -3782
S177    EQU     -3820
S178    EQU     -3855
S179    EQU     -3887
S180    EQU     -3918
S181    EQU     -3946
S182    EQU     -3971
S183    EQU     -3995
S184    EQU     -4015
S185    EQU     -4034
S186    EQU     -4050
S187    EQU     -4063
S188    EQU     -4074
S189    EQU     -4083
S190    EQU     -4089
S191    EQU     -4093
S192    EQU     -4094
S193    EQU     -4093
S194    EQU     -4089
S195    EQU     -4083
S196    EQU     -4074
S197    EQU     -4063
S198    EQU     -4050
S199    EQU     -4034
S200    EQU     -4015
S201    EQU     -3995
S202    EQU     -3971
S203    EQU     -3946
S204    EQU     -3918
S205    EQU     -3887
S206    EQU     -3855
S207    EQU     -3820
S208    EQU     -3782
S209    EQU     -3743
S210    EQU     -3701
S211    EQU     -3657
S212    EQU     -3610
S213    EQU     -3562
S214    EQU     -3511
S215    EQU     -3459
S216    EQU     -3404
S217    EQU     -3347
S218    EQU     -3288
S219    EQU     -3227
S220    EQU     -3164
S221    EQU     -3100
S222    EQU     -3033
S223    EQU     -2965
S224    EQU     -2895
S225    EQU     -2823
S226    EQU     -2749
S227    EQU     -2674
S228    EQU     -2597
S229    EQU     -2518
S230    EQU     -2438
S231    EQU     -2357
```

```
S232    EQU     -2274
S233    EQU     -2190
S234    EQU     -2104
S235    EQU     -2017
S236    EQU     -1929
S237    EQU     -1840
S238    EQU     -1750
S239    EQU     -1658
S240    EQU     -1566
S241    EQU     -1473
S242    EQU     -1379
S243    EQU     -1284
S244    EQU     -1188
S245    EQU     -1091
S246    EQU     -994
S247    EQU     -896
S248    EQU     -798
S249    EQU     -699
S250    EQU     -600
S251    EQU     -500
S252    EQU     -400
S253    EQU     -300
S254    EQU     -200
S255    EQU     -99

        PAGE

*
* COOLEY-TUKEY RADIX-4, DIF FFT PROGRAM - 256-POINT STRAIGHT-LINE.
*
* FOUR STAGES OF THREE TYPES RADIX-4 BUTTERFLIES -
* ZERO, SPCIAL, AND NORMAL - IMPLEMENTED WITH MACROS.
* COMPLEX INPUT DATA LOCATED ON PAGES 4-7 OF DATA MEMORY.
* RESULTS ARE LEFT IN DATA RAM.
* USES 13-BIT COEFFICIENTS FROM MPYK INSTRUCTIONS.
* INTERMEDIATE VALUES ARE SCALED BY 1/4 AT EACH STAGE TO PREVENT
* OVERFLOW.
*
*****************************************************************
*
* MACROS TO PRODUCE STRAIGHT-LINE 256-POINT COMPLEX FFT.
*
*****************************************************************
*
* ZERO FOR CASE W = 1 (THETA = 0).
*
* X'S AND Y'S ARE INPUT AND OUTPUT LOCATIONS FOR BUTTERFLY.
* ENTER WITH ARP = 1, AR1 --> XI1, AR2 --> XI, AR0 = [XI3]-[XI1]
* EXIT WITH ARP = 1, AR1 --> NEXT XI1, AR2 --> NEXT XI
*
ZERO    $MACRO
*
        LAC     *0+,15          * (1/2)XI1
        ADD     *0-,15          * (1/2)XI3
        SACH    *0+             * XI1 = (1/2)(XI1 + XI3)
        SUBH    *               * XI3
        SACH    *0-,0,AR2       * XI3 = (1/2)(XI1 - XI3)
        LAC     *0+,14          * (1/4)XI
        SUB     *,14            * (1/4)XI2
        SACH    T1              * T1 = (1/4)(XI - XI2)
        ADD     *,15,AR1        * R1 (ACC) = (1/4)(XI + XI2)
        SUB     *,15,AR2        * (1/2)XI1
        SACH    *0-,0,AR1       * XI2 = R1 - (1/2)XI1
        ADDH    *+,AR2          * XI1
        SACH    *+,0,AR1        * XI = R1 + (1/2)XI1

        LAC     *0+,15          * (1/2)YI1
        ADD     *0-,15          * (1/2)YI3
        SUBH    *0+             * YI1 = (1/2)(YI1 + YI3)
        SACH    *               * YI3
        SACH    *0-,0,AR2       * YI3 = (1/2)(YI1 - YI3)
        LAC     *0+,14          * (1/4)YI
        SUB     *,14            * (1/4)YI2
        SACH    T2              * T2 = (1/4)(YI - YI2)
        ADD     *,15,AR1        * S1 (ACC) = (1/4)(YI + YI2)
        SUB     *,15,AR2        * (1/2)YI1
        SACH    *0-,0,AR1       * YI2 = S1 - (1/2)YI1
        ADDH    *0+,AR2         * YI1
        SACH    *+,0,AR1        * YI = S1 + (1/2)YI1, POINT TO NEXT XI

        ZALH    T1              * (1/2)YI3
        ADD     *0-,15          * POINT TO XII
        MAR     *-              * XII = T1 + (1/2)YI3
        MAR     *0+             * POINT TO YI3
        SUBH    *-              * YI3
        SACH    T1              * T1 = T1 - (1/2)YI3
        ZALH    T2              * (1/2)XI3
        ADD     *+,15           * YI3 = T2 + (1/2)XI3
        SUBH    *-              * XI3
        LAR     AR3,T1          * AR3 = T1
        SAR     AR3,*+          * XI3 = T1
        MAR     *0-             * POINT TO YI1
        SACH    *+              * YII = T2 - (1/2)XI3 , POINT TO NEXT XI1

        $END

*****************************************************************
*
* NORMAL - STANDARD RADIX-4 BUTTERFLY.
*
* X'S AND Y'S ARE INPUT AND OUTPUT LOCATIONS FOR BUTTERFLY.
* IA'S SPECIFY TWIDDLE FACTOR LOCATIONS.
* ENTER WITH ARP = 1, AR1 --> XI1, AR2 --> XI, AR0 = [XI3] - [XI1]
* EXIT WITH ARP = 1, AR1 --> NEXT XI1, AR2 --> NEXT XI
*
NORMAL  $MACRO  IA1,IA2,IA3
*
        LAC     *0+,15          * (1/2)XI1
        ADD     *0-,15          * (1/2)XI3
        SACH    *0+             * XI1 = (1/2)(XI1 + XI3)
        SUBH    *               * XI3
        SACH    *0-,0,AR2       * XI3 = (1/2)(XI1 - XI3)
        LAC     *0+,14          * (1/4)XI
        SUB     *,14            * (1/4)XI2
        SACH    T1              * T1 = (1/4)(XI - XI2)
        ADD     *,15,AR1        * R1 (ACC) = (1/4)(XI + XI2)
        SUB     *,15,AR2        * (1/2)XI1
        SACH    *0-,0,AR1       * XI2 = R1 - (1/2)XI1
```

4. Implementation of Fast Fourier Transform Algorithms with the TMS32020

```
        ADDH  *+,AR2        * XII
        SACH  *+,0,AR1      * XI  = RI + (1/2)XII
*
        LAC   *0+,15        * (1/2)YII
        ADD   *0-,15        * (1/2)YI3
        SACH  *0+           * YI1 = (1/2)(YI1 + YI3)
        SUBH  *             * YI3
        SACH  *-,0,AR2      * YI3 = (1/2)(YI1 - YI3)
        LAC   *0+,14        * (1/4)YI
        SUB   *,14          * (1/4)YI2
        SACH  T2            * T2 = (1/4)(YI - YI2)
        ADD   *,15,AR1      * S1 (ACC) = (1/4)(YI + YI2)
        SUB   *,15,AR2      * (1/2)YII
        ADDH  *0+,AR2       * YI2 = S1 - (1/2)YII
        SACH  *0+,0,AR1     * YI = S1 + (1/2)YII, POINT TO YI2
*
        ZALH  T1            * (1/2)YI3
        ADD   *0-,15        * POINT TO XI1
        MAR   *-            * XI1 = T1 + (1/2)YI3
        SACH  *+            * POINT TO YI3
        MAR   *0+           * YI3
        SUBH  *-            * T1 = T1 - (1/2)YI3
        SACH  T1
        ZALH  T2            * (1/2)XI3
        ADD   *,15          * T2 = T2 + (1/2)XI3
        SACH  T2            * XI3, POINT TO YI1
        SUBH  *+            * POINT TO YI1
        MAR   *0-           * YI1 = T2 - (1/2)XI3
        SACH  *
*
        LT    *-            * YI1
        MPYK  C:IA1:        * CO1 * YI1
        LTP   *+            * XI1
        MPYK  S:IA1:        * SI1 * XI1
        LTS   *             * YI1
        SACH  *-,4          * YI1 = YI1*CO1 - XI1*SI1
        MPYK  S:IA1:        * SI1 * YI1
        LTP   *,AR2         * XI1
        MPYK  C:IA1:        * CO1 * XI1
        LTA   *-,AR1        * XI1 = YI1*SI1 + XI1*CO1, POINT TO XI3
        SACH  *0+,4,AR2     *
*
        MPYK  C:IA2:        * CO2 * YI2
        LTP   *+            * XI2
        MPYK  S:IA2:        * SI2 * XI2
        LTS   *             * YI2
        SACH  *-,4          * YI2 = YI2*CO2 - XI2*SI2
        MPYK  S:IA2:        * SI2 * YI2
        LTP   *             * XI2
        MPYK  C:IA2:        * CO2 * XI2
        LTA   T1            * T1
        SACH  *0-,4         * XI2 = YI2*SI2 + XI2*CO2
        MAR   *+,AR1
*
        MPYK  C:IA3:        * CO3 * T1
        LTP   T2            * T2
        MPYK  S:IA3:        * SI3 * T2
        APAC
        SACH  *+,4,AR2      * XI3 = T1*CO3 + T2*SI3
```

```
        MPYK  C:IA3:        * CO3 * T2
        LTP   T1            * T1
        MPYK  S:IA3:        * SI3 * T1
        LTS   *+,AR1        * SPAC, POINT TO NEXT XI, (LT = DUMMY OP)
        SACH  *+,4          * YI3 = T2*CO3 - T1*SI3
        MAR   *0-           * POINT TO NEXT XI1

              $END

*********************************************************************
*
* SPCIAL FOR CASE THETA = pi/4
*
* X'S AND Y'S ARE INPUT AND OUTPUT LOCATIONS FOR BUTTERFLY.
* ENTER WITH ARP = 1, AR1 --> XI1, AR2 --> XI, AR0 = [XI3] - [XI1]
* EXIT WITH ARP = 1, AR1 --> NEXT XI1, AR2 --> NEXT XI
*
SPCIAL  $MACRO
*
        LAC   *0+,14        * (1/4)XII
        ADD   *0-,14        * (1/4)XI3
        SACH  *0+           * XI1 = (1/4)(XI1 + XI3)
        SUB   *,15          * (1/2)XI3
        SACH  *+            * XI3 = (1/4)(XI1 - XI3)
*
        LAC   *0-,14        * (1/4)YI3
        ADD   *0+,14        * (1/4)YI1
        SACH  *0+           * YI1 = (1/4)(YI1 + YI3)
        SUB   *,15          * (1/2)YI3
        SACH  *0-,0,AR2     * YI3 = (1/4)(YI1 - YI3), POINT TO YI1
*
        LAC   *0+,14        * (1/4)XI
        ADD   *0-,14        * (1/4)XI2
        SACH  *0+           * XI = (1/4)(XI + XI2)
        SUB   *,15          * (1/2)XI2
        SACH  *+            * XI2 = (1/4)(XI - XI2)
*
        LAC   *0-,14        * (1/4)YI2
        ADD   *0+,14        * (1/4)YI
        SACH  *0+           * YI = (1/4)(YI + YI2)
        SUB   *,15          * (1/2)YI2
        SACH  *0-           * YI2 = (1/4)(YI - YI2), POINT TO YI
*
        LAC   *,0,AR1       * YI
        ADD   *,0,AR2       * YI1
        SACL  *-,0,AR1      * YI = YI + YI1
        SUB   *,-,1         * 2 YI1
        SACL  T2            * T2 = YI - YI1
        LAC   *0+,0,AR2     * XI1, POINT TO XI3
        SUB   *             * XI
        SACL  T1            * T1 = XI1 - XI
        ADD   *,1           * 2 XI
        SACL  *0+           * XI = XI1 + XI, POINT TO XI2
*
        LT    *+,AR1        * XI2
        MPYK  C32
        MAR   *+            *
        LTP   *             * YI3
        MPYK  C32
        SPAC
        SACH  *-,4          * YI3 = (XI2-YI3)*C32
```

```
                AORG 0
        B    32
*
                AORG 32
FF256   LDPK 0              * STORE ALL TEMPORARY VARIABLES IN B2.
        SOVM
        SSXM
        SPM  0
        LARP 1
*
*                INPUT DATA (256 REAL AND 256 IMAGINARY VALUES)
*
        LRLK AR1,X0
        RPTK 255
        IN   *+,PA0
        RPTK 255
        IN   *+,PA0
*
* PASS 1
*
        LRLK AR0,256       * STEP VALUE BETWEEN BUTTERFLY "LEGS"
        LRLK AR1,X64
        LRLK AR2,X0
        ZERO               * X0,X64,X128,X192,Y0,Y64,Y128,Y192
*
        NORMAL 1,2,3       * X1,X65,X129,X193,Y1,Y65,Y129,Y193
        NORMAL 2,4,6       * X2,X66,X130,X194,...
        NORMAL 3,6,9
        NORMAL 4,8,12
        NORMAL 5,10,15
        NORMAL 6,12,18
        NORMAL 7,14,21
        NORMAL 8,16,24
        NORMAL 9,18,27
        NORMAL 10,20,30
        NORMAL 11,22,33
        NORMAL 12,24,36
        NORMAL 13,26,39
        NORMAL 14,28,42
        NORMAL 15,30,45
        NORMAL 16,32,48
        NORMAL 17,34,51
        NORMAL 18,36,54
        NORMAL 19,38,57
        NORMAL 20,40,60
        NORMAL 21,42,63
        NORMAL 22,44,66
        NORMAL 23,46,69
        NORMAL 24,48,72
        NORMAL 25,50,75
        NORMAL 26,52,78
```

```
        APAC
        LTA  *0-           * XI3
        SACH *0+,4,AR2     * XI1 = (XI2+YI3)*C32
*
        MPYK C32
        LTP  *,AR1         * YI2
        MPYK C32
        APAC
        SACH *0-,4         * XI3 = (YI2+XI3)*C32
        SPAC
        SPAC
        NEG
        MAR  *+
        SACH *,4,AR2       * YI1 = (YI2-XI3)*C32
*
        LAC  T1
        SACL *-            * YI2 = T1
        LAC  T2
        SACL *0-           * XI2 = T2
        MAR  *+
        MAR  *+,AR1        * POINT TO NEXT XI
*
        LAC  *-            * YI1
        SUB  *+            * XI1
        SACL *-            * YI1 = YI1 - XI1
        ADD  *,1           * 2 XI1
        SACL *0+           * XI1 = YI1 + XI1
        LAC  *+            * XI3
        SUB  *-            * YI3
        SACL *+            * XI3 = XI3 - YI3
        ADD  *,1           * 2 YI3
        NEG
        SACL *0-           * YI3 = -(YI3 + XI3)
        MAR  *+            * POINT TO NEXT XI1
*
        $END
*****************************************************
* DIGREV MACRO TO DO EXCHANGE OF LOCATIONS FOR DIGIT REVERSAL.
*
DIGREV $MACRO I,J
*
        LRLK AR1,X:I:      * AR1 POINTS TO XI
        LRLK AR2,X:J:      * AR1 POINTS TO XJ
*
        ZALH *,AR2
        ADDS *,AR1
        SACL *+,0,AR2
        SACH *+,0,AR1
        ZALH *,AR2
        ADDS *,AR1
        SACL *,0,AR2
        SACH *,0,AR1
*
        $END
*****************************************************
* * * *   MAIN ROUTINE TO CALL ABOVE MACROS WITH APPROPRIATE PARAMETERS.
```

4. Implementation of Fast Fourier Transform Algorithms with the TMS32020

```
NORMAL  27,54,81
NORMAL  28,56,84
NORMAL  29,58,87
NORMAL  30,60,90
NORMAL  31,62,93
*
SPCIAL                  * X31,X95,X159,X223,Y31,Y95,Y159,Y223
*
NORMAL  33,66,99        * X32,X96,X160,X224,Y32,Y96,Y160,Y224
NORMAL  34,68,102       * X33,X97,X161,X225,Y33,Y97,Y161,Y225
NORMAL  35,70,105       * X34,X98,X162,X226,...
NORMAL  36,72,108
NORMAL  37,74,111
NORMAL  38,76,114
NORMAL  39,78,117
NORMAL  40,80,120
NORMAL  41,82,123
NORMAL  42,84,126
NORMAL  43,86,129
NORMAL  44,88,132
NORMAL  45,90,135
NORMAL  46,92,138
NORMAL  47,94,141
NORMAL  48,96,144
NORMAL  49,98,147
NORMAL  50,100,150
NORMAL  51,102,153
NORMAL  52,104,156
NORMAL  53,106,159
NORMAL  54,108,162
NORMAL  55,110,165
NORMAL  56,112,168
NORMAL  57,114,171
NORMAL  58,116,174
NORMAL  59,118,177
NORMAL  60,120,180
NORMAL  61,122,183
NORMAL  62,124,186
NORMAL  63,126,189      * X63,X127,X191,X255,Y63,Y127,Y191,Y255
*
*************************************************************
* PASS 2
*
*************************************************************
*
FIRST SET OF BUTTERFLIES
*
LARK    AR0,64          * STEP VALUE BETWEEN BUTTERFLY "LEGS"
LRLK    AR1,X16
LRLK    AR2,X0
ZERO                    * X0,X16,X32,X48,Y0,Y16,Y32,Y48
*
NORMAL  4,8,12          * X1,X17,X33,X49,Y1,Y17,Y33,Y49
NORMAL  8,16,24         * X2,X18,X34,X50,...
NORMAL  12,24,36
NORMAL  16,32,48
NORMAL  20,40,60
NORMAL  24,48,72
NORMAL  28,56,84        * X7,X23,X39,X55,Y7,Y23,Y39,Y55
```

```
*                       * X8,X24,X40,X56,Y8,Y24,Y40,Y56
*
SPCIAL                  * X9,X25,X41,X57,Y9,Y25,Y41,Y57
                        * X10,X26,X42,X58,...
NORMAL  36,72,108
NORMAL  40,80,120
NORMAL  44,88,132
NORMAL  48,96,144
NORMAL  52,104,156
NORMAL  56,112,168
NORMAL  60,120,180      * X15,X31,X47,X63,X15,X31,X47,Y63
*
*
SECOND SET OF BUTTERFLIES
*
*
LRLK    AR1,X80
LRLK    AR2,X64
ZERO                    * X64,X80,X96,X112,Y64,Y80,Y96,Y112
*
NORMAL  4,8,12          * X65,X81,X97,X113,Y65,Y81,Y97,Y113
NORMAL  8,16,24         * X66,X82,X98,X114,...
NORMAL  12,24,36
NORMAL  16,32,48
NORMAL  20,40,60
NORMAL  24,48,72
NORMAL  28,56,84        * X71,X87,X103,X119,Y71,Y87,Y103,Y119
*
SPCIAL                  * X72,X88,X104,X120,Y72,Y88,Y104,Y120
*
NORMAL  36,72,108       * X73,X89,X105,X121,Y73,Y89,Y105,Y121
NORMAL  40,80,120       * X74,X90,X106,X122,...
NORMAL  44,88,132
NORMAL  48,96,144
NORMAL  52,104,156
NORMAL  56,112,168
NORMAL  60,120,180      * X79,X95,X111,X127,Y79,Y95,Y111,Y127
*
*
THIRD SET OF BUTTERFLIES
*
*
LRLK    AR1,X144
LRLK    AR2,X128
ZERO                    * X128,X144,X160,X176,X128,Y144,Y160,Y176
*
NORMAL  4,8,12          * X129,X145,X161,X177,Y129,Y145,Y161,Y177
NORMAL  8,16,24         * X130,X146,X162,X178,....
NORMAL  12,24,36
NORMAL  16,32,48
NORMAL  20,40,60
NORMAL  24,48,72
NORMAL  28,56,84        * X135,X151,X167,X183,Y135,Y151,Y167,Y183
*
SPCIAL                  * X136,X152,X168,X184,Y136,Y152,Y168,Y184
*
NORMAL  36,72,108       * X137,X153,X169,X185,Y137,Y153,Y169,Y185
NORMAL  40,80,120       * X138,X154,X170,X186,...
NORMAL  44,88,132
NORMAL  48,96,144
NORMAL  52,104,156
NORMAL  56,112,168
NORMAL  60,120,180      * X143,X159,X175,X191,Y143,Y159,Y175,Y191
*
FOURTH SET OF BUTTERFLIES
```

```
*
        LRLK    AR1,X208
        LRLK    AR2,X192
        ZERO
*                       * X192,X208,X224,X240,Y192,Y208,Y224,Y240
        NORMAL  4,8,12   * X193,X209,X225,X241,Y193,Y209,Y225,Y241
        NORMAL  8,16,24   * X194,X210,X226,X242,...
        NORMAL  12,24,36
        NORMAL  16,32,48
        NORMAL  20,40,60
        NORMAL  24,48,72
        NORMAL  28,56,84   * X199,X215,X231,X247,Y199,Y215,Y231,Y247
*
        SPCIAL             * X200,X216,X232,X248,Y200,Y216,Y232,Y248
*                          * X201,X217,X233,X249,Y201,Y217,Y233,Y249
        NORMAL  36,72,108   * X202,X218,X234,X250,....
        NORMAL  40,80,120
        NORMAL  44,88,132
        NORMAL  48,96,144
        NORMAL  52,104,156
        NORMAL  56,112,168
        NORMAL  60,120,180  * X207,X223,X239,X255,Y207,Y223,Y239,Y255
*
* ***********************************************************
* PASS 3
* ***********************************************************
*
        LARK    AR0,16             * STEP VALUE BETWEEN BUTTERFLY "LEGS"
*
        LRLK    AR1,X4
        LRLK    AR2,X0
*                                  * X0,X4,X8,X12,Y0,Y4,Y8,Y12
        ZERO                       * X1,X5,X9,X13,Y1,Y5,Y9,Y13
        NORMAL  16,32,48           * X2,X6,X10,X14,Y2,Y6,Y10,Y14
        SPCIAL
        NORMAL  48,96,144          * X3,X7,X11,X15,Y3,Y7,Y11,Y15
*
        LRLK    AR1,X20
        LRLK    AR2,X16
*                                  * X16,X20,X24,X28,Y16,Y20,Y24,Y28
        ZERO                       * X17,X21,X25,X29,Y17,Y21,Y25,Y29
        NORMAL  16,32,48           * X18,X22,X26,X30,Y18,Y22,Y26,Y30
        SPCIAL
        NORMAL  48,96,144          * X19,X23,X27,X31,Y19,Y23,Y27,Y31
*
        LRLK    AR1,X36
        LRLK    AR2,X32
*                                  * X32,X36,X40,X44,Y32,Y36,Y40,Y44
        ZERO                       * X33,X37,X41,X45,Y33,Y37,Y41,Y45
        NORMAL  16,32,48           * X34,X38,X42,X46,Y34,Y38,Y42,Y46
        SPCIAL
        NORMAL  48,96,144          * X35,X39,X43,X47,Y35,Y39,Y43,Y47
*
        LRLK    AR1,X52
        LRLK    AR2,X48
*                                  * X48,X52,X56,X60,Y48,Y52,Y56,Y60
        ZERO
        NORMAL  16,32,48           * X49,X53,X57,X61,Y49,Y53,Y57,Y61
        SPCIAL                     * X50,X54,X58,X62,Y50,Y54,Y58,Y62
        NORMAL  48,96,144          * X51,X55,X59,X63,Y51,Y55,Y59,Y63
*
        LRLK    AR1,X68
        LRLK    AR2,X64
*                                  * X64,X68,X72,X76,Y64,Y68,Y72,Y76
        ZERO                       * X65,X69,X73,X77,Y65,Y69,Y73,Y77
        NORMAL  16,32,48           * X66,X70,X74,X78,Y66,Y70,Y74,Y78
        SPCIAL                     * X67,X71,X75,X79,Y67,Y71,Y75,Y79
        NORMAL  48,96,144
*
        LRLK    AR1,X84
        LRLK    AR2,X80
*                                  * X80,X84,X88,X92,Y80,Y84,Y88,Y92
        ZERO                       * X81,X85,X89,X93,Y81,Y85,Y89,Y93
        NORMAL  16,32,48           * X82,X86,X90,X94,Y82,Y86,Y90,Y94
        SPCIAL                     * X83,X87,X91,X95,Y83,Y87,Y91,Y95
        NORMAL  48,96,144
*
        LRLK    AR1,X100
        LRLK    AR2,X96
*                                  * X96,X100,X104,X108,Y96,Y100,Y104,Y108
        ZERO                       * X97,X101,X105,X109,Y97,Y101,Y105,Y109
        NORMAL  16,32,48           * X98,X102,X106,X110,Y98,Y102,Y106,Y110
        SPCIAL                     * X99,X103,X107,X111,Y99,Y103,Y107,Y111
        NORMAL  48,96,144
*
        LRLK    AR1,X116
        LRLK    AR2,X112
*                                  * X112,X116,X120,X124,Y112,Y116,Y120,Y124
        ZERO                       * X113,X117,X121,X125,Y113,Y117,Y121,Y125
        NORMAL  16,32,48           * X114,X118,X122,X126,Y114,Y118,Y122,Y126
        SPCIAL                     * X115,X119,X123,X127,Y115,Y119,Y123,Y127
        NORMAL  48,96,144
*
        LRLK    AR1,X132
        LRLK    AR2,X128
*                                  * X128,X132,X136,X140,Y128,Y132,Y136,Y140
        ZERO                       * X129,X133,X137,X141,Y129,Y133,Y137,Y141
        NORMAL  16,32,48           * X130,X134,X138,X142,Y130,Y134,Y138,Y142
        SPCIAL                     * X131,X135,X139,X143,Y131,Y135,Y139,Y143
        NORMAL  48,96,144
*
        LRLK    AR1,X148
        LRLK    AR2,X144
*                                  * X144,X148,X152,X156,Y144,Y148,Y152,Y156
        ZERO                       * X145,X149,X153,X157,Y145,Y149,Y153,Y157
        NORMAL  16,32,48           * X146,X150,X154,X158,Y146,Y150,Y154,Y158
        SPCIAL                     * X147,X151,X155,X159,Y147,Y151,Y155,Y159
        NORMAL  48,96,144
*
        LRLK    AR1,X164
        LRLK    AR2,X160
*                                  * X160,X164,X168,X172,Y160,Y164,Y168,Y172
        ZERO                       * X161,X165,X169,X173,Y161,Y165,Y169,Y173
        NORMAL  16,32,48           * X162,X166,X170,X174,Y162,Y166,Y170,Y174
        SPCIAL                     * X163,X167,X171,X175,Y163,Y167,Y171,Y175
        NORMAL  48,96,144
```

4. Implementation of Fast Fourier Transform Algorithms with the TMS32020

```
        ZERO
        LRLK    ARI,X17
        LRLK    AR2,X16         * X12,X13,X14,X15,Y12,Y13,Y14,Y15
        ZERO
        LRLK    ARI,X21
        LRLK    AR2,X20         * X16,X17,X18,X19,Y16,Y17,Y18,Y19
        ZERO
        LRLK    ARI,X25
        LRLK    AR2,X24         * X20,X21,X22,X23,Y20,Y21,Y22,Y23
        ZERO
        LRLK    ARI,X29
        LRLK    AR2,X28         * X24,X25,X26,X27,Y24,Y25,Y26,Y27
        ZERO
        LRLK    ARI,X33
        LRLK    AR2,X32         * X28,X29,X30,X31,Y28,Y29,Y30,Y31
        ZERO
        LRLK    ARI,X37
        LRLK    AR2,X36         * X32,X33,X34,X35,Y32,Y33,Y34,Y35
        ZERO
        LRLK    ARI,X41
        LRLK    AR2,X40         * X36,X37,X38,X39,Y36,Y37,Y38,Y39
        ZERO
        LRLK    ARI,X45
        LRLK    AR2,X44         * X40,X41,X42,X43,Y40,Y41,Y42,Y43
        ZERO
        LRLK    ARI,X49
        LRLK    AR2,X48         * X44,X45,X46,X47,Y44,Y45,Y46,Y47
        ZERO
        LRLK    ARI,X53
        LRLK    AR2,X52         * X48,X49,X50,X51,Y48,Y49,Y50,Y51
        ZERO
        LRLK    ARI,X57
        LRLK    AR2,X56         * X52,X53,X54,X55,Y52,Y53,Y54,Y55
        ZERO
        LRLK    ARI,X61
        LRLK    AR2,X60         * X56,X57,X58,X59,Y56,Y57,Y58,Y59
        ZERO
        LRLK    ARI,X65
        LRLK    AR2,X64         * X60,X61,X62,X63,Y60,Y61,Y62,Y63
        ZERO
        LRLK    ARI,X69
        LRLK    AR2,X68         * X64,X65,X66,X67,Y64,Y65,Y66,Y67
        ZERO
        LRLK    ARI,X73
        LRLK    AR2,X72         * X68,X69,X70,X71,Y68,Y69,Y70,Y71
        ZERO
        LRLK    ARI,X77
        LRLK    AR2,X76         * X72,X73,X74,X75,Y72,Y73,Y74,Y75
        ZERO
        LRLK    ARI,X81
        LRLK    AR2,X80         * X76,X77,X78,X79,Y76,Y77,Y78,Y79
        ZERO
        LRLK    ARI,X85
        LRLK    AR2,X84         * X80,X81,X82,X83,Y80,Y81,Y82,Y83
        ZERO
        LRLK    ARI,X89
        LRLK    AR2,X88         * X84,X85,X86,X87,Y84,Y85,Y86,Y87
        ZERO
        LRLK    ARI,X93         * X88,X89,X90,X91,Y88,Y89,Y90,Y91


        LRLK    ARI,X180
        LRLK    AR2,X176
        ZERO                    * X176,X180,X184,X188,Y176,Y180,Y184,Y188
        NORMAL  16,32,48        * X177,X181,X185,X189,Y177,Y181,Y185,Y189
        SPCIAL                  * X178,X182,X186,X190,Y178,Y182,Y186,Y190
        NORMAL  48,96,144       * X179,X183,X187,X191,Y179,Y183,Y187,Y191
        LRLK    ARI,X196
        LRLK    AR2,X192
        ZERO                    * X192,X196,X200,X204,Y192,Y196,Y200,Y204
        NORMAL  16,32,48        * X193,X197,X201,X205,Y193,Y197,Y201,Y205
        SPCIAL                  * X194,X198,X202,X206,Y194,Y198,Y202,Y206
        NORMAL  48,96,144       * X195,X199,X203,X207,Y195,Y199,Y203,Y207
        LRLK    ARI,X212
        LRLK    AR2,X208
        ZERO                    * X208,X212,X216,X220,Y208,Y212,Y216,Y220
        NORMAL  16,32,48        * X209,X213,X217,X221,Y209,Y213,Y217,Y221
        SPCIAL                  * X210,X214,X218,X222,Y210,Y214,Y218,Y222
        NORMAL  48,96,144       * X211,X215,X219,X223,Y211,Y215,Y219,Y223
        LRLK    ARI,X228
        LRLK    AR2,X224
        ZERO                    * X224,X228,X232,X236,Y224,Y228,Y232,Y236
        NORMAL  16,32,48        * X225,X229,X233,X237,Y225,Y229,Y233,Y237
        SPCIAL                  * X226,X230,X234,X238,Y226,Y230,Y234,Y238
        NORMAL  48,96,144       * X227,X231,X235,X239,Y227,Y231,Y235,Y239
        LRLK    ARI,X244
        LRLK    AR2,X240
        ZERO                    * X240,X244,X248,X252,Y240,Y244,Y248,Y252
        NORMAL  16,32,48        * X241,X245,X249,X253,Y241,Y245,Y249,Y253
        SPCIAL                  * X242,X246,X250,X254,Y242,Y246,Y250,Y254
        NORMAL  48,96,144       * X243,X247,X251,X255,Y243,Y247,Y251,Y255
*
**************************************************************
*
* PASS 4
*
**************************************************************
*                               * STEP VALUE BETWEEN BUTTERFLY "LEGS"
*
        LARK    ARO,4
        LRLK    ARI,X1
        LRLK    AR2,X0          * X0,X1,X2,X3,Y0,Y1,Y2,Y3
        ZERO
        LRLK    ARI,X5
        LRLK    AR2,X4          * X4,X5,X6,X7,Y4,Y5,Y6,Y7
        ZERO
        LRLK    ARI,X9
        LRLK    AR2,X8          * X8,X9,X10,X11,Y8,Y9,Y10,Y11
        ZERO
        LRLK    ARI,X13
        LRLK    AR2,X12
```

```
        LRLK  AR2,X92
        ZERO
        LRLK  AR1,X97     *  X92,X93,X94,X95,Y92,Y93,Y94,Y95
        LRLK  AR2,X96
        ZERO
        LRLK  AR1,X101    *  X96,X97,X98,X99,Y96,Y97,Y98,Y99
        LRLK  AR2,X100
        ZERO
        LRLK  AR1,X105    *  X100,X101,X102,X103,Y100,Y101,Y102,Y103
        LRLK  AR2,X104
        ZERO
        LRLK  AR1,X109    *  X104,X105,X106,X107,Y104,Y105,Y106,Y107
        LRLK  AR2,X108
        ZERO
        LRLK  AR1,X113    *  X108,X109,X110,X111,Y108,Y109,Y110,Y111
        LRLK  AR2,X112
        ZERO
        LRLK  AR1,X117    *  X112,X113,X114,X115,Y112,Y113,Y114,Y115
        LRLK  AR2,X116
        ZERO
        LRLK  AR1,X121    *  X116,X117,X118,X119,Y116,Y117,Y118,Y119
        LRLK  AR2,X120
        ZERO
        LRLK  AR1,X125    *  X120,X121,X122,X123,Y120,Y121,Y122,Y123
        LRLK  AR2,X124
        ZERO
        LRLK  AR1,X129    *  X124,X125,X126,X127,Y124,Y125,Y126,Y127
        LRLK  AR2,X128
        ZERO
        LRLK  AR1,X133    *  X128,X129,X130,X131,Y128,Y129,Y130,Y131
        LRLK  AR2,X132
        ZERO
        LRLK  AR1,X137    *  X132,X133,X134,X135,Y132,Y133,Y134,Y135
        LRLK  AR2,X136
        ZERO
        LRLK  AR1,X141    *  X136,X137,X138,X139,Y136,Y137,Y138,Y139
        LRLK  AR2,X140
        ZERO
        LRLK  AR1,X145    *  X140,X141,X142,X143,Y140,Y141,Y142,Y143
        LRLK  AR2,X144
        ZERO
        LRLK  AR1,X149    *  X144,X145,X146,X147,Y144,Y145,Y146,Y147
        LRLK  AR2,X148
        ZERO
        LRLK  AR1,X153    *  X148,X149,X150,X151,Y148,Y149,Y150,Y151
        LRLK  AR2,X152
        ZERO
        LRLK  AR1,X157    *  X152,X153,X154,X155,Y152,Y153,Y154,Y155
        LRLK  AR2,X156
        ZERO
        LRLK  AR1,X161    *  X156,X157,X158,X159,Y156,Y157,Y158,Y159
        LRLK  AR2,X160
        ZERO
        LRLK  AR1,X165    *  X160,X161,X162,X163,Y160,Y161,Y162,Y163
        LRLK  AR2,X164
        ZERO
        LRLK  AR1,X169    *  X164,X165,X166,X167,Y164,Y165,Y166,Y167
        LRLK  AR2,X168
        ZERO
                          *  X168,X169,X170,X171,Y168,Y169,Y170,Y171

        LRLK  AR1,X173    *  X172,X173,X174,X175,Y172,Y173,Y174,Y175
        LRLK  AR2,X172
        ZERO
        LRLK  AR1,X177    *  X176,X177,X178,X179,Y176,Y177,Y178,Y179
        LRLK  AR2,X176
        ZERO
        LRLK  AR1,X181    *  X180,X181,X182,X183,Y180,Y181,Y182,Y183
        LRLK  AR2,X180
        ZERO
        LRLK  AR1,X185    *  X184,X185,X186,X187,Y184,Y185,Y186,Y187
        LRLK  AR2,X184
        ZERO
        LRLK  AR1,X189    *  X188,X189,X190,X191,Y188,Y189,Y190,Y191
        LRLK  AR2,X188
        ZERO
        LRLK  AR1,X193    *  X192,X193,X194,X195,Y192,Y193,Y194,Y195
        LRLK  AR2,X192
        ZERO
        LRLK  AR1,X197    *  X196,X197,X198,X199,Y196,Y197,Y198,Y199
        LRLK  AR2,X196
        ZERO
        LRLK  AR1,X201    *  X200,X201,X202,X203,Y200,Y201,Y202,Y203
        LRLK  AR2,X200
        ZERO
        LRLK  AR1,X205    *  X204,X205,X206,X207,Y204,Y205,Y206,Y207
        LRLK  AR2,X204
        ZERO
        LRLK  AR1,X209    *  X208,X209,X210,X211,Y208,Y209,Y210,Y211
        LRLK  AR2,X208
        ZERO
        LRLK  AR1,X213    *  X212,X213,X214,X215,Y212,Y213,Y214,Y215
        LRLK  AR2,X212
        ZERO
        LRLK  AR1,X217    *  X216,X217,X218,X219,Y216,Y217,Y218,Y219
        LRLK  AR2,X216
        ZERO
        LRLK  AR1,X221    *  X220,X221,X222,X223,Y220,Y221,Y222,Y223
        LRLK  AR2,X220
        ZERO
        LRLK  AR1,X225    *  X224,X225,X226,X227,Y224,Y225,Y226,Y227
        LRLK  AR2,X224
        ZERO
        LRLK  AR1,X229    *  X228,X229,X230,X231,Y228,Y229,Y230,Y231
        LRLK  AR2,X228
        ZERO
        LRLK  AR1,X233    *  X232,X233,X234,X235,Y232,Y233,Y234,Y235
        LRLK  AR2,X232
        ZERO
        LRLK  AR1,X237    *  X236,X237,X238,X239,Y236,Y237,Y238,Y239
        LRLK  AR2,X236
        ZERO
        LRLK  AR1,X241    *  X240,X241,X242,X243,Y240,Y241,Y242,Y243
        LRLK  AR2,X240
        ZERO
        LRLK  AR1,X245    *  X244,X245,X246,X247,Y244,Y245,Y246,Y247
        LRLK  AR2,X244
        ZERO
        LRLK  AR1,X249
        LRLK  AR2,X248
        LRLK
```

4. Implementation of Fast Fourier Transform Algorithms with the TMS32020

```
                        * X248,X249,X250,X251,Y248,Y249,Y250,Y251

                        * X252,X253,X254,X255,Y252,Y253,Y254,Y255

        ZERO
        LRLK    AR1,X253
        LRLK    AR2,X252
        ZERO
*
*
* DIGIT REVERSE COUNTER FOR RADIX-4 FFT COMPUTATION.
*
        LARP    1
        DIGREV  1,64          DIGREV  61,124        DIGREV  183,222
        DIGREV  2,128         DIGREV  62,188        DIGREV  187,238
        DIGREV  3,192         DIGREV  63,252        DIGREV  191,254
        DIGREV  4,16          DIGREV  66,129        DIGREV  199,211
        DIGREV  5,80          DIGREV  67,193        DIGREV  203,227
        DIGREV  6,144         DIGREV  69,81         DIGREV  207,243
        DIGREV  7,208         DIGREV  70,145        DIGREV  219,231
        DIGREV  8,32          DIGREV  71,209        DIGREV  223,247
        DIGREV  9,96          DIGREV  73,97         DIGREV  239,251
        DIGREV  10,160        DIGREV  74,161      *
        DIGREV  11,224        DIGREV  75,225      *     OUTPUT FFT DATA
        DIGREV  12,48         DIGREV  77,113      *
        DIGREV  13,112        DIGREV  78,177        LARP    1
        DIGREV  14,176        DIGREV  79,241        LRLK    AR1,X0
        DIGREV  15,240        DIGREV  82,133        RPTK    255
        DIGREV  17,68         DIGREV  83,197        OUT     *+,PA1
        DIGREV  18,132        DIGREV  86,149        RPTK    255
        DIGREV  19,196        DIGREV  87,213        OUT     *+,PA1
        DIGREV  21,84         DIGREV  89,101      *
        DIGREV  22,148        DIGREV  90,165      * FFT COMPLETE.
        DIGREV  23,212        DIGREV  91,229      *
        DIGREV  24,36         DIGREV  93,117      WHOA    B       WHOA
        DIGREV  25,100        DIGREV  94,181              END
        DIGREV  26,164        DIGREV  95,245
        DIGREV  27,228        DIGREV  98,137
        DIGREV  28,52         DIGREV  99,201
        DIGREV  29,116        DIGREV  102,153
        DIGREV  30,180        DIGREV  103,217
        DIGREV  31,244        DIGREV  106,169
        DIGREV  33,72         DIGREV  107,233
        DIGREV  34,136        DIGREV  109,121
        DIGREV  35,200        DIGREV  110,185
        DIGREV  37,88         DIGREV  111,249
        DIGREV  38,152        DIGREV  114,141
        DIGREV  39,216        DIGREV  115,205
        DIGREV  41,104        DIGREV  118,157
        DIGREV  42,168        DIGREV  119,221
        DIGREV  43,232        DIGREV  122,173
        DIGREV  44,56         DIGREV  123,237
        DIGREV  45,120        DIGREV  126,189
        DIGREV  46,184        DIGREV  127,253
        DIGREV  47,248        DIGREV  131,194
        DIGREV  49,76         DIGREV  134,146
        DIGREV  50,140        DIGREV  135,210
        DIGREV  51,204        DIGREV  138,162
        DIGREV  53,92         DIGREV  139,226
        DIGREV  54,156        DIGREV  142,178
        DIGREV  55,220        DIGREV  143,242
        DIGREV  57,108        DIGREV  147,198
        DIGREV  58,172        DIGREV  151,214
        DIGREV  59,236        DIGREV  154,166
                              DIGREV  155,230
                              DIGREV  158,182
                              DIGREV  159,246
                              DIGREV  163,202
                              DIGREV  167,218
                              DIGREV  171,234
                              DIGREV  174,186
                              DIGREV  175,250
                              DIGREV  179,206
```

APPENDIX F
A 128-POINT, RADIX-2 DIF FFT IMPLEMENTATION (LOOPED CODE)

```
*
*
        IDT 'FFT12'
*
*       COOLEY-TUKEY 128-POINT, RADIX-2, DIF FFT PROGRAM FOR THE TMS32020.
*
*       SINGLE FFT BUTTERFLY.
*       REAL INPUT DATA, STORED IN PAGE 4 OF BLOCK B0.
*       FFT COMPUTATION IS DONE IN PAGES 6, 7 OF BLOCK B1 (COMPLEX NUMBERS).
*       USES TABLE LOOKUP (FROM PROGRAM MEMORY) OF THE TWIDDLE FACTORS.
*       INTERMEDIATE VALUES ARE SCALED BY .5 AT EACH STAGE SO AS TO PREVENT
*       THE POSSIBILITY OF OVERFLOW.
*       NO EXTERNAL RAM IS USED.
*       THE MAGNITUDE OF THE FFT IS COMPUTED AND NORMALIZED SO THAT ITS MAXIMUM
*       VALUE HAS A NONZERO BIT AFTER THE BINARY POINT.
*
**********************************************************************
*
*       N IS THE SIZE OF THE TRANSFORM.  N = 2*M.
*
N       EQU 128
M       EQU 7
*
XIN     EQU 512          * LOCATION OF REAL INPUT DATA
XOUT    EQU 640          * LOCATION OF REAL OUTPUT DATA
XFFT    EQU 768          * LOCATION OF COMPLEX DATA FOR THE FFT
*
*       BLOCK B2 DATA MEMORY ALLOCATION (DP = 0 WILL ALWAYS POINT TO B2).
*
XT      EQU 96           * TEMPORARY - REAL PART
YT      EQU 97           * TEMPORARY - IMAGINARY PART
I       EQU 98           * 1ST INDEX
IE      EQU 99           * INDEX TO TWIDDLE FACTORS
HOLDN   EQU 100          * INCREMENT TO IA
QUARTN  EQU 101          * CONTAINS VALUE N
N1      EQU 102          * CONTAINS VALUE N/4
N2      EQU 103          * INCREMENT TO I.
J       EQU 104          * SEPARATION OF I AND L
K       EQU 105          * LOOP COUNTER
ONE     EQU 106          * BIT REVERSAL INDEX COUNTER
IADDR   EQU 107          * CONTAINS VALUE 1
SINTBL  EQU 108          * CONTAINS INPUT
SIN     EQU 109          * SINE TABLE POINTER
COS     EQU 110          * SINE LOCATION
MAX     EQU 111          * COSINE LOCATION
        EQU 112          * MAX VALUE OF THE SQR MAGNITUDE OF FFT
*
*       BEGIN PROGRAM MEMORY SECTION.
**********************************************************************
RSVECT  AORG 0
B       32
        AORG 32
INIT    LDPK 0
        CNFD
        SOVM             * ALWAYS POINT TO B2 FOR TEMP STORAGE
        SSXM
        SPM 1            * 32010 ARITHMETIC
        LACK 1           * SHIFT PRODUCT LEFT BY 1
        SACL ONE
        SACL IE          * INITIALIZE IE = 1
        LALK N
        SACL HOLDN       * HOLDN = N

        SACL N2          * INITIALIZE N2 = N
        LAC HOLDN,14
        SACH QUARTN      * QUARTN = N/4
        LRLK AR3,XFFT    * ADDRESS OF COMPLEX INPUT DATA
        SAR AR3,IADDR    * STORE ON PAGE 0
        LARP 2
        LARK AR1,M-1     * AR1 CONTAINS K COUNTER
*
*       READ IN 128 REAL POINTS
*
        LRLK AR2,XIN
        RPTK 127
        IN *+,PA0
*
*       INITIALIZE FFT RAM
*
        LRLK AR2,XFFT
        ZAC
        SACL MAX
        RPTK 255
        SACL *+
*
*       MOVE REAL DATA FROM INPUT LOCATIONS TO COMPLEX FFT LOCATIONS
*
        LRLK AR2,XFFT
        LARK AR0,2
        RPTK 127
        BLKD XIN,*0+
*
*       MOVE SQUARED MAGNITUDE OF PREVIOUS COMPUTATION TO OUTPUT LOCATIONS
*
        LRLK AR2,XOUT
        RPTK 127
        BLKD XFFT,*+
*
*       FFT COMPUTATION
*
KLOOP   LAC N2,15
        SACH N1,1        * N1 = N2
        SACH N2          * N2 = N2/2
        ZAC
        SACL IA          * IA = 0
        SACL J           * J = 0
        LAR AR2,N2       * AR2 CONTAINS J VALUE
        MAR *-,AR3       * START AT N2-1
*
        LALK SINE
        SACL SINTBL      * SINE TABLE BASE ADDRESS
*
JLOOP   LAC J,1          * I = J (DATA ORGANIZED AS REAL VALUE FOLLOWED
        SACL I           *     BY IMAGINARY SO THAT ADDRESS I IS 2 TIMES J).
*
ILOOP   LAR AR0,I
        LAR AR3,IADDR    * LOAD INPUT BASE ADDRESS
        MAR *0+          * AR3 = I + IADDR
        LAR AR0,N1       * ADD N2*2 (N1 = N2*2)
        LAC *0+,15       * LOAD (1/2)(XI), POINT TO XL
                         *     (L = I + N2)
        SUB *,15         * XT = (1/2)(XI - XL)
        SACH XT          * STORE XT ON PAGE 0
        ADDH *0-         * XI = (1/2)(XI + XL), POINT TO XI
        SACH *+          * STORE XI, POINT TO YI
```

4. Implementation of Fast Fourier Transform Algorithms with the TMS32020

```
        LAC    *0+,15        * LOAD (1/2)YI, POINT TO YL
        SUB    *,15
        SACH   YT            * YT = (1/2)(YI - YL)
        ADDH   *0-
        SACH   *0+           * YI = (1/2)(YI + YL), POINT TO YI
                             * STORE YI, POINT TO YL
*
        LAC    SINTBL
        TBLR   SIN           * READ IN SINE
        ADD    QUARTN
        TBLR   COS           * READ IN COSINE
*
        LT     COS
        MPY    YT            * YT*COS-->P
        LTP    SIN
        MPY    XT            * XT*SIN-->P
        SPAC
        SACH   *-            * YL = XT*COS - XT*SIN
        MPY    YT            * YT*SIN-->P
        LTP    COS
        MPY    XT            * XT*COS-->P
        APAC
        SACH   *             * XL = XT*COS + YT*SIN
*
* ADD INCREMENT FOR NEXT LOOP.
*
        LAC    I
        ADD    N1,1          * I = I + N1
        SACL   I
        SUB    HOLDN,1
        BLZ    ILOOP         * WHILE I < N
*
        LAC    SINTBL        * INDEX SINTBL POINTER BY IE
        ADD    IE
        SACL   SINTBL
*
        LAC    J
        ADD    ONE           * J = J + 1
        SACL   J
        LARP   2
        BANZ   JLOOP,*-,AR3
*
        LAC    IE,1
        SACL   IE            * IE = 2 * IE
        LARP   1
        BANZ   KLOOP,*-,AR2
*
* DIGIT REVERSE COUNTER FOR RADIX-2 FFT COMPUTATION.
*
DRC2    ZAC
        SACL   I
        SACL   J             * FOR I = 0 TO N-2
        LAR    AR0,IADDR
        LAR    AR4,HOLDN
        LARP   4
        MAR    *-
        MAR    *-,AR2
DRLOOP  SUB    J
        BGEZ   NOSWAP        * IF I < J, THEN SWAP
*
        LAR    AR2,J
        MAR    *0+,AR3       * J = J + IADDR
        LAR    AR3,I
```

```
        MAR    *0+,AR2       * I = I + IADDR
* SWAP I AND L VALUES.
        ZALH   *,AR3
        ADDS   *
        SACH   *+,0,AR2      * X(I) = X(J)
        SACL   *+            * X(J) = X(I)
*
        ZALH   *,AR3
        ADDS   *
        SACH   *,0,AR2       * Y(I) = Y(J)
        SACL   *,0,AR3       * Y(J) = Y(I)
NOSWAP  LAC    HOLDN
        SACL   K             * K = N
INLOOP  LAC    J
        SUB    K
        BLZ    OUTLOP,*,AR4  * IF J >= K THEN
        SACL   J             * J = J - K
        LAC    K,15
        SACH   K             * K = K/2.
        B      INLOOP
OUTLOP  ADD    K,1
        SACL   J             * L = L + J
        LAC    I
        ADD    ONE,1
        SACL   I             * INCREMENT I
        BANZ   DRLOOP,*-,AR2
*
*
* COMPUTE THE MAXIMUM VALUE OF THE SQUARED MAGNITUDE OF FFT
*
        LRLK   AR2,XFFT
        LARK   AR3,127
*
LOOP1   ZAC
        MPYK   0
        SQRA   *+            * X(I)**2
        SQRA   *+,AR3        * Y(I)**2
        APAC
        SACH   IADDR         * IADDR = X(I)**2 + Y(I)**2
        SUBH   MAX
        BLEZ   CONT
        LAC    IADDR
        SACL   MAX
CONT    BANZ   LOOP1,*-,AR2
*
* NORMALIZE THE MAX VALUE TO FIND EXPONENT
*
        LARK   AR2,0
        ZALH   MAX
        RPTK   14
        NORM                 * AR2 CONTAINS EXPONENT
        SAR    AR2,I         * STORE EXPONENT IN I
*
* COMPUTE SQUARED MAGNITUDE OF FFT
*
LOOP2   ZAC
        MPYK   0
        SQRA   *+            * X(I)**2
        SQRA   *+,AR1        * Y(I)**2
```

```
        APAC
        SFR         I
        RPT
        SFL         *+,0,AR3
        SACH        LOOP2,*-,AR2
        BANZ
*
* OUTPUT SQUARED MAGNITUDE
*
        LRLK        AR2,XFFT
        RPTK        127
        OUT         *+,PA2
*
* FFT COMPLETE.
*
WHOA    B   WHOA
*
* COEFFICIENT TABLE (SIZE OF TABLE IS 3N/4).
*
SINE EQU $
        DATA    >0
        DATA    >648
        DATA    >C8C
        DATA    >12C8
        DATA    >18F9
        DATA    >1F1A
        DATA    >2528
        DATA    >2B1F
        DATA    >30FC
        DATA    >36BA
        DATA    >3C57
        DATA    >41CE
        DATA    >471D
        DATA    >4C40
        DATA    >5134
        DATA    >55F6
        DATA    >5A82
        DATA    >5ED7
        DATA    >62F2
        DATA    >66D0
        DATA    >6A6E
        DATA    >6DCA
        DATA    >70E3
        DATA    >73B6
        DATA    >7642
        DATA    >7885
        DATA    >7A7D
        DATA    >7C2A
        DATA    >7D8A
        DATA    >7E9D
        DATA    >7F62
        DATA    >7FD9
COSINE EQU $
        DATA    >7FFF
        DATA    >7FD9
        DATA    >7F62
        DATA    >7E9D
        DATA    >7D8A
        DATA    >7C2A
        DATA    >7A7D
        DATA    >7885
        DATA    >7642
        DATA    >73B6
```

```
* NORMALIZE RESULT
* XOUT(I) = X(I)**2 + Y(I)**2
```

```
        DATA    >70E3
        DATA    >6DCA
        DATA    >6A6E
        DATA    >66D0
        DATA    >62F2
        DATA    >5ED7
        DATA    >5A82
        DATA    >5134
        DATA    >4C40
        DATA    >471D
        DATA    >41CE
        DATA    >3C57
        DATA    >36BA
        DATA    >30FC
        DATA    >2B1F
        DATA    >2528
        DATA    >1F1A
        DATA    >18F9
        DATA    >12C8
        DATA    >C8C
        DATA    >648
        DATA    >0
        DATA    >F9B8
        DATA    >F374
        DATA    >ED38
        DATA    >E707
        DATA    >E0E6
        DATA    >DAD8
        DATA    >D4E1
        DATA    >CF04
        DATA    >C946
        DATA    >C3A9
        DATA    >BE32
        DATA    >B8E3
        DATA    >B3C0
        DATA    >AECC
        DATA    >AA0A
        DATA    >A57E
        DATA    >A129
        DATA    >9D0E
        DATA    >9930
        DATA    >9592
        DATA    >9236
        DATA    >8F1D
        DATA    >8C4A
        DATA    >89BE
        DATA    >877B
        DATA    >8583
        DATA    >83D6
        DATA    >8276
        DATA    >8163
        DATA    >809E
        DATA    >8027
        END
```

APPENDIX G
A 256-POINT, RADIX-2 DIF FFT IMPLEMENTATION (LOOPED CODED)

```
        IDT   'FFT2'
*
*   COOLEY-TUKEY 256-POINT, RADIX-2, DIF FFT PROGRAM FOR THE TMS32020.
*
*   SINGLE FFT BUTTERFLY.
*   COMPLEX INPUT DATA, STORED AS X(I), Y(I), X(I+1), Y(I+1), ...
*   USES TABLE LOOKUP (FROM EXTERNAL DATA MEMORY) OF THE TWIDDLE FACTORS.
*   INTERMEDIATE VALUES ARE SCALED BY .5 AT EACH STAGE SO AS TO PREVENT
*   THE POSSIBILITY OF OVERFLOW.
*
************************************************************************
*
* N IS THE SIZE OF THE TRANSFORM.  N = 2**M.
*
N       EQU 256
M       EQU 8
*
INPUT   EQU 512    * LOCATION OF COMPLEX INPUT DATA IN INTERNAL DATA MEM
TABLE   EQU 1024   * LOCATION OF COEFFICIENT TABLE IN EXTERNAL DATA MEM
*
* BLOCK B2 DATA MEMORY ALLOCATION (DP = 0 WILL ALWAYS POINT TO B2).
*
XT      EQU 96     * TEMPORARY - REAL PART
YT      EQU 97     * TEMPORARY - IMAGINARY PART
I       EQU 98     * 1ST INDEX
IA      EQU 99     * INDEX TO TWIDDLE FACTORS
IE      EQU 100    * INCREMENT TO IA
HOLDN   EQU 101    * CONTAINS VALUE N
QUARTN  EQU 102    * CONTAINS VALUE N/4
N1      EQU 103    * INCREMENT TO I.
N2      EQU 104    * SEPARATION OF I AND L
J       EQU 105    * LOOP COUNTER
K       EQU 106    * BIT REVERSAL INDEX COUNTER
ONE     EQU 107    * CONTAINS VALUE 1
IADDR   EQU 108    * CONTAINS INPUT
COSTBL  EQU 109    * COSTBL = TABLE + N/4
*
* BEGIN PROGRAM MEMORY SECTION.
*
************************************************************************
*
RSVECT  AORG 0
        B   32
INIT    AORG 32
        SOVM
        SSXM
        SPM   1         * 32010 ARITHMETIC
        LACK  1         * SHIFT PRODUCT LEFT BY 1
        SACL  ONE
        LACK  IE
        SACL  IE        * INITIALIZE IE = 1
        LALK  N.
        SACL  HOLDN     * HOLDN = N
        SACL  N2        * INITIALIZE N2 = N
        LAC   HOLDN,14
        SACH  QUARTN    * QUARTN = N/4
        LRLK  AR3,INPUT * ADDRESS OF COMPLEX INPUT DATA
        SAR   AR3,IADDR * STORE ON PAGE 0
        LRLK  AR4,TABLE * ADDRESS OF SINE TABLE
        LARP  4
        RPTK  191       * MOVE 192 COEFFICIENTS

        BLKP  SINE,*+
        LRLK  AR4,TABLE    * ADDRESS OF SINE TABLE
        LAR   AR0,QUARTN   * COSTBL = TABLE + N/4, point to J counter
        MAR   *0+,AR2
        SAR   AR4,COSTBL
        LARK  AR1,M-1      * AR1 CONTAINS K COUNTER
*
*   READ IN 256 COMPLEX POINTS
*
        LRLK  AR2,INPUT
        RPTK  255
        IN    *+,PA0
        RPTK  255
        IN    *+,PA0
*
* FFT COMPUTATION
*
KLOOP   LAC   N2,15
        SACH  N1,1         * N1 = N2
        SACH  N2           * N2 = N2/2
        ZAC
        SACL  IA           * IA = 0
        SACL  J            * J = 0
        LAR   AR2,N2       * AR2 CONTAINS J VALUE
        MAR   *-,AR3       * START AT N2-1
*
        LAR   AR4,COSTBL   * COSINE TABLE BASE ADDRESS
*
JLOOP   LAC   J,1          * I = J (DATA ORGANIZED AS REAL VALUE FOLLOWED
        SACL  I            BY IMAGINARY SO THAT ADDRESS I IS 2 TIMES J).
*
* ILOOP
*
ILOOP   LAR   AR0,I        * LOAD INPUT BASE ADDRESS
        LAR   AR3,IADDR    * AR3 = I + IADDR
        MAR   *0+          * ADD N2*2 (N1 = N2*2)
        LAR   AR0,N1       * LOAD (1/2)XI, POINT TO XL
        LAC   *0+,15          (L = I + N2)
        SUB   *,15         * XT = (1/2)(XI - XL)
        SACH  XT           * STORE XT ON PAGE 0
        ADDH  *0-          * XI = (1/2)(XI + XL), POINT TO XI
        SACH  *+           * STORE XI, POINT TO YI
        LAC   *0+,15       * LOAD (1/2)YI, POINT TO YL
        SUB   *,15         * YT = (1/2)(YI - YL)
        ADDH  *0-          * YI = (1/2)(YI + YL), POINT TO YI
        SACH  *0+,0,AR4    * STORE YI, POINT TO YL
*                            AR4 POINTS TO COSTBL
        LAR   AR0,QUARTN
        LT    *0-          * LOAD T WITH COS, POINT TO SIN
        MPY   YT
        LTP   *0+,AR3      * ACC <-- C*YT, POINT TO YL
        MPY   XT
        SPAC
        SACH  *-,0,AR4     * YL = C*YT, POINT TO YL
        MPY   YT           * STORE YL
        LTP   *,AR3
        MPY   XT           * ACC <-- S*YT, POINT TO XL
        APAC
        SACH  *            * XL = C*XT + S*YT, POINT TO XL
*                            STORE XL
* ADD INCREMENT FOR NEXT LOOP.
```

```
        LAC   I
        ADD   N1,1          * I = I + N1
        SACL  I
        SUB   HOLDN,1       * WHILE I < N
        BLZ   ILOOP
*
*                           * INDEX COSTBL POINTER BY IE
        LAR   AR0,IE
        LARP  4
        MAR   *0+,AR2
        LAC   J
        ADD   ONE           * J = J + 1
        SACL  J
        BANZ  JLOOP,*-,AR3
*
        LAC   IE,1
        SACL  IE            * IE = 2 * IE
        LARP  1
        BANZ  KLOOP,*-,AR2
*
* DIGIT REVERSE COUNTER FOR RADIX-2 FFT COMPUTATION.
*
DRC2    ZAC
        SACL  I
        SACL  J
        LAR   AR0,IADDR
        LARP  4
        LAR   AR4,HOLDN
        MAR   *-
        MAR   *-,AR2        * FOR I = 0 TO N-2
DRLOOP  SUB   J
        BGEZ  NOSWAP        * IF I < J, THEN SWAP
*
        LAR   AR2,J
        MAR   *0+,AR3       * J = J + IADDR
        LAR   AR3,I
        MAR   *0+,AR2       * I = I + IADDR
* SWAP I AND L VALUES.
        ZALH  *,AR3
        ADDS  *
        SACH  *++,0,AR2     * X(I) = X(J)
        SACL  *++           * X(J) = X(I)
*
        ZALH  *,AR3
        ADDS  *
        SACH  *,0,AR2       * Y(I) = Y(J)
        SACL  *,0,AR3       * Y(J) = Y(I)
*
NOSWAP  LAC   HOLDN
        SACL  K             * K = N
INLOOP  LAC   J
        SUB   K             * IF J >= K THEN
        BLZ   OUTLOP,*,AR4
        SACL  J             * J = J - K
        LAC   K,15
        SACH  K             * K = K/2.
        B     INLOOP
*
OUTLOP  ADD   K,1
        SACL  J             * L = L + J
        LAC   I
        ADD   ONE,1         * INCREMENT I
        SACL  I
        BANZ  DRLOOP,*-,AR2
*
```

```
* OUTPUT FFT VALUES
*
        LRLK  AR2,INPUT
        RPTK  255
        OUT   *+,PA1
        RPTK  255
        OUT   *+,PA1
*
* FFT COMPLETE.
*
WHOA    B     WHOA
*
* COEFFICIENT TABLE (SIZE OF TABLE IS 3N/4).
*
SINE    EQU   $
        DATA  >0
        DATA  >324
        DATA  >648
        DATA  >96B
        DATA  >C8C
        DATA  >FAB
        DATA  >12C8
        DATA  >15E2
        DATA  >18F9
        DATA  >1C0C
        DATA  >1F1A
        DATA  >2224
        DATA  >2528
        DATA  >2827
        DATA  >2B1F
        DATA  >2E11
        DATA  >30FC
        DATA  >33DF
        DATA  >36BA
        DATA  >398D
        DATA  >3C57
        DATA  >3F17
        DATA  >41CE
        DATA  >447B
        DATA  >471D
        DATA  >49B4
        DATA  >4C40
        DATA  >4EC0
        DATA  >5134
        DATA  >539B
        DATA  >55F6
        DATA  >5843
        DATA  >5A82
        DATA  >5CB4
        DATA  >5ED7
        DATA  >60EC
        DATA  >62F2
        DATA  >64E9
        DATA  >66D0
        DATA  >68A7
        DATA  >6A6E
        DATA  >6C24
        DATA  >6DCA
        DATA  >6F5F
        DATA  >70E3
        DATA  >7255
        DATA  >73B6
        DATA  >7505
```

```
       DATA >7642
       DATA >776C
       DATA >7885
       DATA >798A
       DATA >7A7D
       DATA >7B5D
       DATA >7C2A
       DATA >7CE4
       DATA >7D8A
       DATA >7E1E
       DATA >7E9D
       DATA >7F0A
       DATA >7F62
       DATA >7FA7
       DATA >7FD9
       DATA >7FF6
COSINE EQU  $
       DATA >7FFF
       DATA >7FF6
       DATA >7FD9
       DATA >7FA7
       DATA >7F62
       DATA >7F0A
       DATA >7E9D
       DATA >7E1E
       DATA >7D8A
       DATA >7CE4
       DATA >7C2A
       DATA >7B5D
       DATA >798A
       DATA >7885
       DATA >776C
       DATA >7642
       DATA >7505
       DATA >73B6
       DATA >7255
       DATA >70E3
       DATA >6F5F
       DATA >6DCA
       DATA >6C24
       DATA >6A6E
       DATA >68A7
       DATA >66D0
       DATA >64E9
       DATA >62F2
       DATA >60EC
       DATA >5ED7
       DATA >5CB4
       DATA >5A82
       DATA >5843
       DATA >55F6
       DATA >539B
       DATA >5134
       DATA >4EC0
       DATA >4C40
       DATA >49B4
       DATA >471D
       DATA >447B
       DATA >41CE
       DATA >3F17
       DATA >3C57
       DATA >398D

       DATA >36BA
       DATA >33DF
       DATA >30FC
       DATA >2E11
       DATA >2B1F
       DATA >2827
       DATA >2528
       DATA >2224
       DATA >1F1A
       DATA >1C0C
       DATA >18F9
       DATA >15E2
       DATA >12C8
       DATA >FAB
       DATA >C8C
       DATA >96B
       DATA >648
       DATA >324
       DATA >0
       DATA >FCDC
       DATA >F9B8
       DATA >F695
       DATA >F374
       DATA >F055
       DATA >ED38
       DATA >EA1E
       DATA >E707
       DATA >E3F4
       DATA >E0E6
       DATA >DDDC
       DATA >DAD8
       DATA >D7D9
       DATA >D4E1
       DATA >D1EF
       DATA >CF04
       DATA >CC21
       DATA >C946
       DATA >C673
       DATA >C3A9
       DATA >C0E9
       DATA >BE32
       DATA >BB85
       DATA >B8E3
       DATA >B64C
       DATA >B3C0
       DATA >B140
       DATA >AECC
       DATA >AC65
       DATA >AA0A
       DATA >A7BD
       DATA >A57E
       DATA >A34C
       DATA >A129
       DATA >9F14
       DATA >9D0E
       DATA >9B17
       DATA >9930
       DATA >9759
       DATA >9592
       DATA >93DC
       DATA >9236
       DATA >90A1
       DATA >8F1D

       DATA >8DAB
       DATA >8C4A
       DATA >8AFB
       DATA >89BE
       DATA >8894
       DATA >877B
       DATA >8676
       DATA >8583
       DATA >84A3
       DATA >83D6
       DATA >831C
       DATA >8276
       DATA >81E2
       DATA >8163
       DATA >80F6
       DATA >809E
       DATA >8059
       DATA >8027
       DATA >800A
       END
```

4. Implementation of Fast Fourier Transform Algorithms with the TMS32020

5. Companding Routines for the TMS32010/TMS32020

Lou Pagnucco and Cole Erskine
Digital Signal Processing - Semiconductor Group
Texas Instruments

INTRODUCTION

In Pulse Code Modulation (PCM) systems, which are commonly used in public and private (PBX) telephone networks, samples of an analog speech waveform are encoded as binary words and transmitted serially usually at a rate of 8000 samples per second. This digitized data is communicated most efficiently if the amplitude of the waveform is compressed to logarithmic scale before transmission (reducing the number of bits required for their representation), and then expanded at the receiver.

The conversion to logarithmic scale insures that low-amplitude signals are digitized with a minimal loss of fidelity. This procedure of first compressing and then expanding the signal is known as "companding" (COMpressing and exPANDING). Figures 1(a) and 1(b) show the procedures involved in companding, typically accomplished by a hardware device called a codec or combo-codec (the combined PCM codec and filter).[1] Since codecs are inexpensive, they have been widely used as input/output (I/O) devices for analog signals in many digital signal processing applications, such as digital telephony.

In a digital signal processing system that incorporates codecs, a reversed companding process is required as shown in Figure 1(c). The compressed PCM data is first converted to linear PCM to be processed by the digital signal processor. After the digital signal processing, the processed linear PCM is then compressed before sending it to the codec to produce an analog output signal.

The TMS320 family of digital signal processors is designed for numeric-intensive applications. Because of the processor's high speed, the companding (actually an expand-and-compress procedure) described in Figure 1(c) can be performed with minimum execution time and program requirements. This allows the processor to dedicate most of its resources for real-time digital signal processing applications, such as filtering, tone generation/detection, transcoding, vocoding, and echo cancellation. Companding can be performed in software in two ways: (1) by calculating the companding algorithm in real-time, or (2) by looking up a table pregenerated using the algorithm. The lookup-table approach naturally requires more storage, but it provides faster execution than the algorithmic approach. The tradeoff must be made by the digital signal processing designer between memory and speed requirements. In addition, companding can be accomplished externally in hardware (see the application report, "Telecommunications Interfacing to the TMS32010").[3]

The main portion of this report presents four TMS32010 programs that implement the standard companding algorithms. The TMS32010 is the first generation of the TMS320 family of digital signal processors. Three programs for companding, which use the TMS32020, the second-generation digital signal processor, are included in the appendix. One of these programs uses the lookup-table approach. Note that no special effort is taken to further optimize these programs for any particular application. The purpose of this report is to show how companding can be

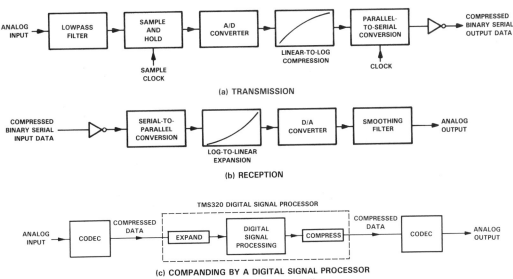

(a) TRANSMISSION

(b) RECEPTION

(c) COMPANDING BY A DIGITAL SIGNAL PROCESSOR

Figure 1. PCM Companding

performed by both generations of digital signal processors. Other application reports are available, which show how the companding routines can be optimized for special applications, such as Adaptive Differential Pulse Code Modulation (ADPCM)[2] and telecommunication interfaces[3] using the TMS32010, and echo cancellation[4] using the TMS32020.

COMPANDING

In any sampled data system, the analog-to-digital (A/D) conversion process introduces quantization noise. For the usual linear A/D encoding scheme, the digitized code word is a truncated binary representation of the analog sample. The effect of this truncation is most pronounced for small signals. For voice transmission, this is undesirable since most information in speech signals resides in the lower amplitudes even though speech signals typically require a wide dynamic range. This can be remedied by adjusting the size of the quantization interval so that it is proportional to the input signal level. In this case, the quantization interval is small for small amplitude signals and larger for larger signals. Consequently, lower amplitudes are represented with more quantization levels and, therefore, with greater resolution.

The resulting encoding scheme is logarithmic in nature and has the property of yielding the greatest dynamic range for a given signal-to-noise ratio and word length. Companding is defined by two international standards based on this relation — both compress the equivalent of 13 bits of dynamic range into 7. The standard employed in the United States and Japan is known as the μ-255 law companding characteristic and is given by the equation

$$F(x) = sgn(x) \ \frac{\ln(1 + \mu|x|)}{\ln(1 + \mu)}$$

where:

F(x) is the compressed output value
x is the normalized input signal (between -1 and 1)
μ is the compression parameter (= 255 in North America)
sgn(x) is the sign ($\pm$) of x

The European standard is referred to as A-law companding and is defined by the equation

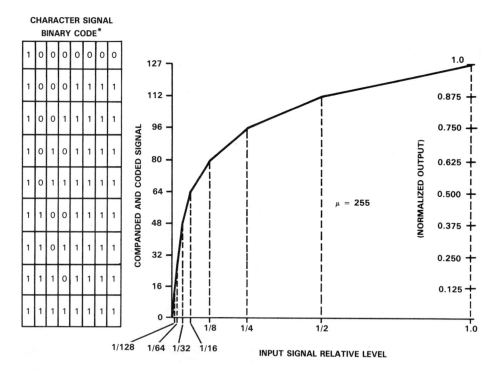

* This is the bit pattern transmitted for positive input values. The left-most bit is a 0 for negative input values.

Figure 2. Companding Curve of the μ-Law Compander (from Reference 5)

5. Companding Routines for the TMS32010/TMS32020

$$F(x) = \begin{cases} \text{sgn}(x) \; \dfrac{A|x|}{1 + \ln(A)} & \text{for } 0 \le |x| < \dfrac{1}{A} \\[3mm] \text{sgn}(x) \; \dfrac{(1 + \ln A|x|)}{1 + \ln(A)} & \text{for } \dfrac{1}{A} \le |x| \le 1 \end{cases}$$

where:

F(x) is the compressed output value
x is the normalized input signal (between -1 and 1)
A is the compression parameter ($= 87.6$ in Europe)
sgn(x) is the sign ($\pm$) of x

In practice, the code word is actually inverted before transmission. Low amplitude signals tend to be more numerous than large amplitude samples. Consequently, inverting the bits increases the density of positive pulses on the transmission line which improves the performance of timing and clock recovery circuits.

μ-255 COMPANDING

Eight-bit sign-magnitude words can represent 255 different code words. This made 255 the most convenient choice for the μ-law companding parameter. This companding characteristic exhibits the valuable property of being closely approximated by a set of eight straight-line segments, as shown in Figure 2.5 This figure illustrates how the input sample values of successively larger intervals are compressed into intervals of uniform size. The slope of each segment is exactly one-half that of the preceding one. The step size between adjacent code words is doubled in each succeeding segment. This property allows the conversion to and from a linear format to be done very efficiently.

TMS32010 Algorithm

Uniformly quantized 14-bit sign-magnitude numbers are compressed into 8-bit signed μ-255 code words by the program 'MULAWCMP' and expanded to their original amplitude by the program 'MULAWEXP', both listed and flowcharted in the Program Listings section. The code word Y, formed by MULAWCMP, has the format Y = PSSSQQQQ composed of:

Polarity bit: P
3-bit segment number: SSS
4-bit quantization bin number: QQQQ

The encoding algorithm is best understood by examining the segment endpoints of Table 1, which begin with the values 31, 95, 223,..., 4063.

Note that

$$\begin{aligned} 31 &= 2^6 - 33 \\ 95 &= 2^7 - 33 \\ 223 &= 2^8 - 33 \end{aligned}$$

$$\cdot$$
$$\cdot$$
$$\cdot$$

$$4063 = 2^{12} - 33$$

so that if 33 is added to each value in the table, the end points become powers of two.

This means that the segment number corresponding to a number N (which is to be encoded) can be determined by finding the most significant '1' bit in the binary representation of N + 33. Furthermore, as Table 2 indicates, the following four bits make up the quantization bin number. The remaining bits are discarded.

On expansion, these lost bits are assumed to have been the median of the possible numbers which these lost bits could have represented — a one followed by zeroes (see Table 3). This rounding limits the loss in accuracy.

Table 1. Encoding/Decoding Table for μ-255 PCM*
(Courtesy of John Wiley & Sons, see Reference 6)

Input Amplitude Range	Step Size	Segment Code S	Quantization Code Q	Code Value	Decoder Amplitude
0-1	1		0000	0	0
1-3			0001	1	2
3-5	2	000	0010	2	4
.			.	.	.
.			.	.	.
.			.	.	.
29-31			1111	15	30
31-35			0000	16	33
.			.	.	.
.	4	001	.	.	.
.			.	.	.
91-95			1111	31	93
95-103			0000	32	99
.			.	.	.
.	8	010	.	.	.
.			.	.	.
215-223			1111	47	219
223-239			0000	48	231
.			.	.	.
.	16	011	.	.	.
.			.	.	.
463-479			1111	63	471
479-511			0000	64	495
.			.	.	.
.	32	100	.	.	.
.			.	.	.
959-991			1111	79	975
991-1055			0000	80	1023
.			.	.	.
.	64	101	.	.	.
.			.	.	.
1951-2015			1111	95	1983
2015-2143			0000	96	2079
.			.	.	.
.	128	110	.	.	.
.			.	.	.
3935-4063			1111	111	3999
4063-4319			0000	112	4191
.			.	.	.
.	256	111	.	.	.
.			.	.	.
7903-8159			1111	127	8031

*This table displays magnitude encoding only. Polarity bits are assigned as "0" for positive and "1" for negative. In transmission, all bits are inverted.

5. Companding Routines for the TMS32010/TMS32020

Table 2. μ-255 Binary Encoding Table[*]

Biased Input Values													Compressed Code Word						
Bit: 12	11	10	9	8	7	6	5	4	3	2	1	0	Bit: 6	5	4	3	2	1	0
0	0	0	0	0	0	0	1	Q_3	Q_2	Q_1	Q_0	x	0	0	0	Q_3	Q_2	Q_1	Q_0
0	0	0	0	0	0	1	Q_3	Q_2	Q_1	Q_0	x	x	0	0	1	Q_3	Q_2	Q_1	Q_0
0	0	0	0	0	1	Q_3	Q_2	Q_1	Q_0	x	x	x	0	1	0	Q_3	Q_2	Q_1	Q_0
0	0	0	0	1	Q_3	Q_2	Q_1	Q_0	x	x	x	x	0	1	1	Q_3	Q_2	Q_1	Q_0
0	0	0	1	Q_3	Q_2	Q_1	Q_0	x	x	x	x	x	1	0	0	Q_3	Q_2	Q_1	Q_0
0	0	1	Q_3	Q_2	Q_1	Q_0	x	x	x	x	x	x	1	0	1	Q_3	Q_2	Q_1	Q_0
0	1	Q_3	Q_2	Q_1	Q_0	x	x	x	x	x	x	x	1	1	0	Q_3	Q_2	Q_1	Q_0
1	Q_3	Q_2	Q_1	Q_0	x	x	x	x	x	x	x	x	1	1	1	Q_3	Q_2	Q_1	Q_0

[*] The polarity is not shown in this table.

NOTE: The leading bit is the sign bit.

EXAMPLES:

(1) $+865_{10} \xrightarrow{\text{BIAS}} +865_{10}+33_{10} = +898_{10} = +382_{16} = (0)0\ 0011\ 1000\ 0010_2 \rightarrow (0)100\ 1100_2$

(2) $-2513_{10} \xrightarrow{\text{BIAS}} -2513_{10}-33_{10} = -2546_{10} = -9F2_{16} = (1)0\ 1001\ 1111\ 0010_2 \rightarrow (1)110\ 0011_2$

Table 3. μ-255 Binary Decoding Table[*]

Compressed Code Word							Biased Output Values												
Bit: 6	5	4	3	2	1	0	Bit: 12	11	10	9	8	7	6	5	4	3	2	1	0
0	0	0	Q_3	Q_2	Q_1	Q_0	0	0	0	0	0	0	0	1	Q_3	Q_2	Q_1	Q_0	1
0	0	1	Q_3	Q_2	Q_1	Q_0	0	0	0	0	0	0	1	Q_3	Q_2	Q_1	Q_0	1	0
0	1	0	Q_3	Q_2	Q_1	Q_0	0	0	0	0	0	1	Q_3	Q_2	Q_1	Q_0	1	0	0
0	1	1	Q_3	Q_2	Q_1	Q_0	0	0	0	0	1	Q_3	Q_2	Q_1	Q_0	1	0	0	0
1	0	0	Q_3	Q_2	Q_1	Q_0	0	0	0	1	Q_3	Q_2	Q_1	Q_0	1	0	0	0	0
1	0	1	Q_3	Q_2	Q_1	Q_0	0	0	1	Q_3	Q_2	Q_1	Q_0	1	0	0	0	0	0
1	1	0	Q_3	Q_2	Q_1	Q_0	0	1	Q_3	Q_2	Q_1	Q_0	1	0	0	0	0	0	0
1	1	1	Q_3	Q_2	Q_1	Q_0	1	Q_3	Q_2	Q_1	Q_0	1	0	0	0	0	0	0	0

[*] The polarity is not shown in this table.

NOTE: The leading bit is the sign bit.

EXAMPLES:

(1) $3C_{16} = (0)011\ 1100_2 \rightarrow (0)0\ 0001\ 1100\ 1000_2 = +01C8_{16} = +456_{10} \xrightarrow{\text{REMOVE BIAS}} 456_{10}-33_{10} = +423_{10}$

(2) $E3_{16} = (1)110\ 0011_2 \rightarrow (1)0\ 1001\ 1100\ 0000_2 = -09C0_{16} = -2496_{10} \xrightarrow{\text{REMOVE BIAS}} -2496_{10}+33_{10} = -2463_{10}$

Performance

Analysis of the PCM μ-255 companding system of Figure 1 shows that the approximated digital values approach the original inputs closely, with a signal-to-quantization noise ratio of 39.3 dB for a full-range sinusoid.[6] In general, voice signals have smaller quantization errors but lower signal power, so μ-255 performance in voice transmission is approximately the same. The signal-to-quantization noise ratio for μ-255 law encoding is given in Figure 3 for sinusoid inputs. The algorithm space and time requirements are given in Table 4.

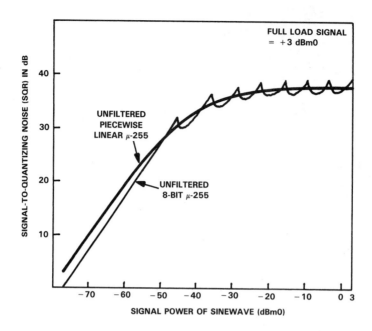

Figure 3. Signal-to-Quantizing Noise of μ-Law Coding with Sinewave Inputs
(Courtesy of John Wiley & Sons, see Reference 6)

Table 4. Summary of μ-Law Program Space and Time Requirements

Function	Words of Memory		Program Cycles		Time Required[†]
	Program	Data	Initialization	Loop[‡]	μsec
Compress	105	13	17	40	8.0
Expand	46	8	6	23	4.6

[†] Assuming initialization
[‡] Worst case

5. Companding Routines for the TMS32010/TMS32020

A-LAW COMPANDING

The companding characteristic recommended by CCITT and adopted in Europe is the 'A-law' standard. Not only can this characteristic be approximated with linear segments, as with the μ-law approximation, but a portion of the rule is linear by definition.

The A-law programs 'ALAWCOMP' and 'ALAWEXP' are listed and flowcharted in the Program Listings section. They differ from the μ-law routines in the handling of the first segment. This segment is defined to be exactly linear for the A-law. Also, biasing is not required before conversion. The inputs should be scaled to a maximum value of 4096 for representation, as opposed to 8158 for the μ-law case. Since this allows for a minimum step less refined than the μ-law, the A-law characteristic provides less fidelity for small signals but is superior in terms of dynamic range.

Figures 4 and 5 and Tables 5, 6, and 7 presented below are analogous to those given above for μ-law.

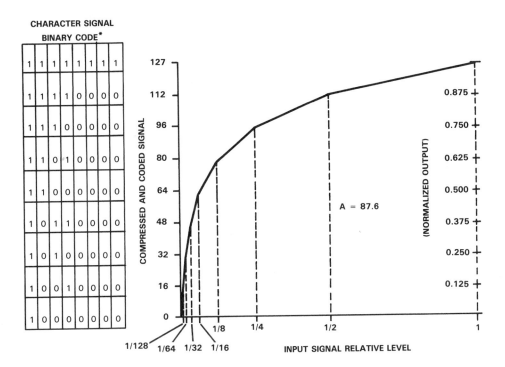

*For positive input values. The left-most bit is a 0 for negative input values. Even bits (beginning with 1 at the left) are inverted before transmission.

Figure 4. Companding Curve of the A-Law Compander (from Reference 5)

Table 5. Segmented A-Law Encoding/Decoding Table
(Courtesy of John Wiley & Sons, see Reference 6)

Input Amplitude Range	Step Size	Segment Code S	Quantization Code Q	Code Value	Decoder Amplitude
0-2			0000	0	1
2-4		000	0001	1	3
.			.	.	.
.			.	.	.
.			.	.	.
30-32	2		1111	15	31
32-34			0000	16	33
.			.	.	.
.		001	.	.	.
.			.	.	.
62-64			1111	31	63
64-68			0000	32	66
.			.	.	.
.	4	010	.	.	.
.			.	.	.
124-128			1111	47	126
128-136			0000	48	132
.			.	.	.
.	8	011	.	.	.
.			.	.	.
248-256			1111	63	252
256-272			0000	64	264
.			.	.	.
.	16	100	.	.	.
.			.	.	.
496-512			1111	79	504
512-544			0000	80	528
.			.	.	.
.	32	101	.	.	.
.			.	.	.
992-1024			1111	95	1008
1024-1088			0000	96	1056
.			.	.	.
.	64	110	.	.	.
.			.	.	.
1984-2048			1111	111	2016
2048-2176			0000	112	2112
.			.	.	.
.	128	111	.	.	.
.			.	.	.
3968-4096			1111	127	4032

Table 6. A-Law Binary Encoding Table[*]

Input Values												Compressed Code Word						
Bit: 11	10	9	8	7	6	5	4	3	2	1	0	Bit: 6	5	4	3	2	1	0
0	0	0	0	0	0	0	Q_3	Q_2	Q_1	Q_0	x	0	0	0	Q_3	Q_2	Q_1	Q_0
0	0	0	0	0	0	1	Q_3	Q_2	Q_1	Q_0	x	0	0	1	Q_3	Q_2	Q_1	Q_0
0	0	0	0	0	1	Q_3	Q_2	Q_1	Q_0	x	x	0	1	0	Q_3	Q_2	Q_1	Q_0
0	0	0	0	1	Q_3	Q_2	Q_1	Q_0	x	x	x	0	1	1	Q_3	Q_2	Q_1	Q_0
0	0	0	1	Q_3	Q_2	Q_1	Q_0	x	x	x	x	1	0	0	Q_3	Q_2	Q_1	Q_0
0	0	1	Q_3	Q_2	Q_1	Q_0	x	x	x	x	x	1	0	1	Q_3	Q_2	Q_1	Q_0
0	1	Q_3	Q_2	Q_1	Q_0	x	x	x	x	x	x	1	1	0	Q_3	Q_2	Q_1	Q_0
1	Q_3	Q_2	Q_1	Q_0	x	x	x	x	x	x	x	1	1	1	Q_3	Q_2	Q_1	Q_0

[*] The polarity is not shown in this table.
NOTE: The leading bit is the sign bit.
EXAMPLES:
(1) $+3221_{10} = +C95_{16} = $ (0) 1100 1001 0101$_2 \rightarrow$ (0) 111 1001$_2$
(2) $-199_{10} = -C7_{16} = $ (1) 0000 1100 0111$_2 \rightarrow$ (1) 011 1000$_2$

Table 7. A-Law Binary Decoding Table[*]

Compressed Code Word							Output Values											
Bit: 6	5	4	3	2	1	0	Bit: 11	10	9	8	7	6	5	4	3	2	1	0
0	0	0	Q_3	Q_2	Q_1	Q_0	0	0	0	0	0	0	0	Q_3	Q_2	Q_1	Q_0	1
0	0	1	Q_3	Q_2	Q_1	Q_0	0	0	0	0	0	0	1	Q_3	Q_2	Q_1	Q_0	1
0	1	0	Q_3	Q_2	Q_1	Q_0	0	0	0	0	0	1	Q_3	Q_2	Q_1	Q_0	1	0
0	1	1	Q_3	Q_2	Q_1	Q_0	0	0	0	0	1	Q_3	Q_2	Q_1	Q_0	1	0	0
1	0	0	Q_3	Q_2	Q_1	Q_0	0	0	0	1	Q_3	Q_2	Q_1	Q_0	1	0	0	0
1	0	1	Q_3	Q_2	Q_1	Q_0	0	0	1	Q_3	Q_2	Q_1	Q_0	1	0	0	0	0
1	1	0	Q_3	Q_2	Q_1	Q_0	0	1	Q_3	Q_2	Q_1	Q_0	1	0	0	0	0	0
1	1	1	Q_3	Q_2	Q_1	Q_0	1	Q_3	Q_2	Q_1	Q_0	1	0	0	0	0	0	0

[*] The polarity is not shown in this table.
NOTE: The leading bit is the sign bit.
EXAMPLES:
(1) (0) 001 1101$_2 \rightarrow$ (0) 0000 0011 1011$_2 = +3B_{16} = +59_{10}$
(2) (1) 110 0100$_2 \rightarrow$ (1) 0101 0010 0000$_2 \rightarrow -520_{16} = -1312_{10}$

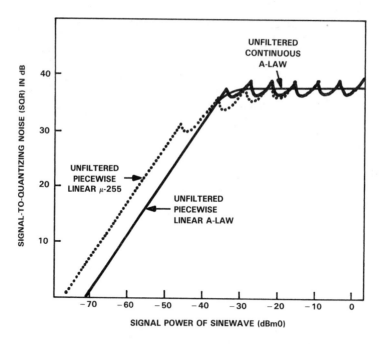

Figure 5. Signal-to-Quantizing Noise of A-Law PCM Coding with Sinewave Inputs (Courtesy of John Wiley & Sons, see Reference 6)

Table 8. Summary of A-Law Program Space and Time Requirements

Function	Words of Memory		Program Cycles		Time Required[†]
	Program	Data	Initialization	Loop[‡]	μsec
Compress	97	11	14	36	7.2
Expand	48	7	4	25	5.0

[†] Assuming initialization
[‡] Worst case

SUMMARY

The programs, listed in the next section, have been designed to reduce both memory space used and "loop time," i.e., that time required to complete the calculations after initializations have been made. Of course, other space/time tradeoffs are possible. Dedicated to companding, the TMS32010 can compress 125,000 or expand 200,000 words in a second using these routines. This speed and the versatility of the TMS32010 allow one device to compand a PCM data stream while simultaneously performing related functions such as filtering, vocoding, and tone generation/recognition.

PROGRAM LISTINGS

The following TMS32010 program flowcharts and assembly language routines are listed below:

'MULAWCMP': μ-LAW COMPRESSION
'MULAWEXP': μ-LAW EXPANSION
'ALAWCOMP': A-LAW COMPRESSION
'ALAWEXP':   A-LAW EXPANSION

FLOWCHART: MULAWCMP

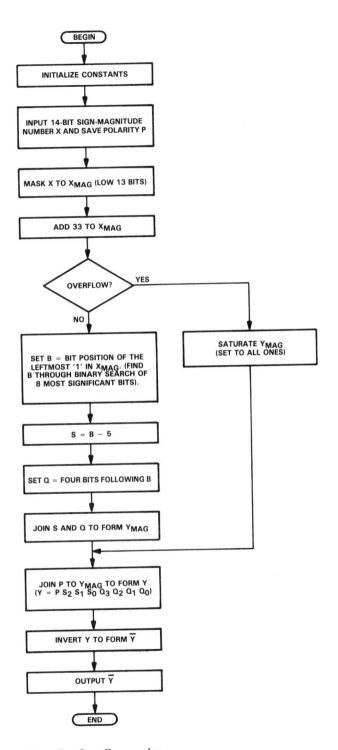

Figure 6. μ-Law Compression

5. Companding Routines for the TMS32010/TMS32020

```
0001                         IDT     'MULAWCMP'
0002                 ***
0003                 *       'MULAWCMP' PERFORMS A MU-255 COMPRESSION. THE
0004                 *       14-BIT SIGN-MAGNITUDE INPUT X,
0005                 *
0006                 *       X = P X12 X11 ... X2 X1 X0
0007                 *
0008                 *       IS ENCODED AS AN 8-BIT SIGN-MAGNITUDE NUMBER Y,
0009                 *
0010                 *       Y = P S2 S1 S0 Q3 Q2 Q1 Q0 consisting of
0011                 *
0012                 *       POLARITY BIT: P,
0013                 *       3-BIT SEGMENT NUMBER: S = S2 S1 S0
0014                 *       4-BIT QUANTIZATION BIN NUMBER: Q = Q3 Q2 Q1 Q0
0015                 *
0016                 *       Y IS INVERTED BEFORE TRANSMISSION.
0017                 *       PORT 0 IS USED FOR I/O.
0018                 *
0019                 *       WORST-CASE TIMING IN CYCLES: 17 INIT / 40 LOOP
0020                 *       SPACE REQUIREMENTS IN WORDS: 13 DATA / 105 PROGRAM
0021                 *
0022                 *       CONSTANTS:
0023                 *
0024                 *
0025      0001  ONE    EQU    1      =1
0026      0002  BIT4   EQU    2      =>0010  (ONE IN BIT 4)
0027      0003  BIT13  EQU    3      =>2000  (ONE IN BIT 13)
0028      0004  MASK13 EQU    4      =>1FFF  (13 ONES)
0029      0005  MASK8  EQU    5      =>00FF  (8 ONES)
0030      0006  MASK4  EQU    6      =>000F  (4 ONES)
0031      0007  MASK2  EQU    7      =>0003  (2 ONES)
0032      0008  BIAS   EQU    8      =33
0033            *
0034            *
0035            *       VARIABLES:
0036            *
0037      0009  X      EQU    9      DATA INPUT (14 BITS)
0038      000A  Y      EQU    10     ENCODED DATA OUTPUT (8 BITS)
0039      000B  P      EQU    11     POLARITY OF DATA (0 FOR POS)
0040      000C  S      EQU    12     3-BIT SEGMENT NUMBER
0041      000D  Q      EQU    13     4-BIT QUANTIZATION BIN NUMBER
0042            *
0043 0000              AORG   0
0044            *
0045 0000 7E01  INIT   LACK   1
0046 0001 5001         SACL   ONE
0047 0002 2401         LAC    ONE,4
0048 0003 5002         SACL   BIT4
0049 0004 2D01         LAC    ONE,13
0050 0005 5003         SACL   BIT13
0051 0006 1001         SUB    ONE
0052 0007 5004         SACL   MASK13
0053 0008 7EFF         LACK   >00FF
0054 0009 5005         SACL   MASK8
```

```
0055 000A 7E0F        LACK     >000F
0056 000B 5006        SACL     MASK4
0057 000C 7E03        LACK     >0003
0058 000D 5007        SACL     MASK2
0059 000E 7E21        LACK     33
0060 000F 5008        SACL     BIAS
0061              *
0062              *    GET INPUT, SAVE POLARITY, AND MASK TO MAGITUDE
0063              *
0064 0010 4009  START IN       X,0       INPUT DATA
0065 0011 2003        LAC      BIT13     POLARITY BIT MASK
0066 0012 7909        AND      X
0067 0013 5C0B        SACH     P,4       0 FOR POS; 2 FOR NEG.
0068 0014 2009        LAC      X
0069 0015 7904        AND MASK13
0070              *
0071              *    BIAS MAGNITUDE AND SATURATE IF OVERFLOW OCCURS
0072              *
0073 0016 0008        ADD      BIAS      BIAS INPUT X BY 33
0074 0017 5009        SACL     X
0075 0018 7903        AND      BIT13     CHECK FOR OVERFLOW INTO P
0076 0019 FF00        BZ       LMOST     NO OVERFLOW IF ZERO
     001A 001E
0077              *
0078              *    SATURATION: ENCODE LARGEST CODE WORD
0079              *
0080 001B 7E7F        LACK     >7F       7 ONES
0081 001C F900        B        SIGN
     001D 0063
0082              *
0083              *    COMPUTE THE THREE BITS S(= S2 S1 S0) OF THE
0084              *    COMPRESSED WORD
0085              *
0086              *         S = BP - 5
0087              *
0088              *    WHERE BP IS THE BIT POSITION OF THE LEFTMOST '1'
0089              *    IN X.  THE FOUR FOLLOWING BITS ARE IN THE HIGH
0090              *    HALF OF THE ACCUMULATOR. THESE FOUR BITS ARE
0091              *    Q (=Q3 Q2 Q1 Q0).
0092              *
0093              *    SEARCH BITS X12 THRU X5. (SEE TABLE 1 OF TEXT.)
0094              *
0095              *
0096 001E 2906  LMOST LAC      MASK4,9       1111 0000
0097 001F 7909        AND      X
0098 0020 FF00        BZ       EEE
     0021 0042
0099 0022 2B07        LAC      MASK2,11      1100 0000
0100 0023 7909        AND      X
0101 0024 FF00        BZ       CC
     0025 0034
0102 0026 2C01        LAC      ONE,12        1000 0000
0103 0027 7909        AND      X
```

```
0104 0028 FF00        BZ      B
     0029 002F
0105 002A 7E07        LACK    7                    1... ....
0106 002B 500C        SACL    S
0107 002C 2809        LAC     X,8
0108 002D F900        B       XDONE
     002E 005F
0109 002F 7E06   B    LACK    6                    01.. ....
0110 0030 500C        SACL    S
0111 0031 2909        LAC     X,9
0112 0032 F900        B       XDONE
     0033 005F
0113 0034 2A01   CC   LAC     ONE,10          0100 0000
0114 0035 7909        AND     X
0115 0036 FF00        BZ      D
     0037 003D
0116 0038 7E05        LACK    5                    001. ....
0117 0039 500C        SACL    S
0118 003A 2A09        LAC     X,10
0119 003B F900        B       XDONE
     003C 005F
0120 003D 7E04   D    LACK    4               0001 ....
0121 003E 500C        SACL    S
0122 003F 2B09        LAC     X,11
0123 0040 F900        B       XDONE
     0041 005F
0124 0042 2707   EEE  LAC     MASK2,7         0000 1100
0125 0043 7909        AND     X
0126 0044 FF00        BZ      GG
     0045 0054
0127 0046 2801        LAC     ONE,8           0000 1000
0128 0047 7909        AND     X
0129 0048 FF00        BZ      F
     0049 004F
0130                *
0131 004A 7E03        LACK    3               0000 1...
0132 004B 500C        SACL    S
0133 004C 2C09        LAC     X,12
0134 004D F900        B       XDONE
     004E 005F
0135                *
0136 004F 7E02   F    LACK    2               0000 01..
0137 0050 500C        SACL    S
0138 0051 2D09        LAC     X,13
0139 0052 F900        B       XDONE
     0053 005F
0140                *
0141 0054 2601   GG   LAC     ONE,6      0000 0010
0142 0055 7909        AND     X
0143 0056 FF00        BZ      H
     0057 005D
0144                *
0145 0058 7E01        LACK    1          0000 001.
0146 0059 500C        SACL    S
```

5. Companding Routines for the TMS32010/TMS32020

```
0147 005A 2E09           LAC     X,14
0148 005B F900           B       XDONE
     005C 005F
0149              *
0150 005D 500C  H         SACL    S        0000 0001 (ACC = 0)
0151 005E 2F09           LAC     X,15
0152              *
0153              * REMOVE LEFTMOST '1' AND STORE Q
0154              *
0155 005F 6202  XDONE     SUBH    BIT4
0156 0060 580D           SACH    Q
0157              *
0158              * FORM 8-BIT COMPRESSED WORD FROM Q, S, AND P.
0159              *
0160 0061 200D           LAC     Q        Q: BITS 0-3   ____QQQQ
0161 0062 040C           ADD     S,4      S: BITS 4-6   _SSSQQQQ
0162 0063 060B  SIGN      ADD     P,6      P: BIT 7      PSSSQQQQ
0163              *
0164              * COMPLEMENT FOR TRANSMISSION AND OUTPUT
0165              *
0166 0064 7805           XOR     MASK8
0167 0065 500A           SACL    Y
0168 0066 480A           OUT     Y,0      PORT 0
0169 0067 F900  FIN       B       FIN
     0068 0067
0170              *
0171                     END
NO ERRORS, NO WARNINGS
```

FLOWCHART: MULAWEXP

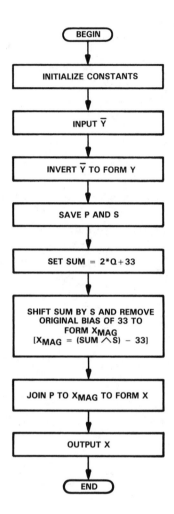

Figure 7. μ-Law Expansion

```
0001                            IDT     'MULAWEXP'
0002                  ***
0003                  *     'MULAWEXP' PERFORM A MU-LAW EXPANSION. THE
0004                  *     8-BIT DATA INPUT IS
0005                  *
0006                  *     Y = P S2 S1 S0 Q3 Q2 Q1 Q0  WHICH CONSISTS OF
0007                  *
0008                  *       POLARITY BIT: P
0009                  *       3-BIT SEGMENT NUMBER: S = S2 S1 S0
0010                  *       4-BIT QUANTIZATION NUMBER: Q = Q3 Q2 Q1 Q0
0011                  *
0012                  *     THE INPUT Y IS EXPANDED INTO A 14-BIT OUTPUT
0013                  *
0014                  *     X = P X12 X11 X10 ... X2 X1 X0 CONSISTING OF
0015                  *
0016                  *       POLARITY BIT:P
0017                  *       AND A 13-BIT MAGNITUDE
0018                  *                                      S
0019                  *       (X12...X0) = (33 + 2Q) X 2  - 33
0020                  *
0021                  *     PORT 0 IS USED FOR I/O.
0022                  *     WORST-CASE TIMING IN CYCLES: 6 INIT / 23 LOOP
0023                  *     SPACE REQUIREMENTS IN WORDS: 8 DATA / 46 PROGRAM
0024                  *
0025                  * CONSTANTS:
0026                  *
0027       0001  ONE     EQU     1        = 1
0028       0002  BIT7    EQU     2        = >0080
0029       0003  BIAS    EQU     3        = 33
0030                  *
0031                  * VARIABLES:
0032                  *
0033       0004  Y       EQU     4        MU-LAW COMPRESSED 8-BIT DATA INPUT
0034       0005  X       EQU     5        DECODED (EXPANDED) 14-BIT OUTPUT
0035       0006  P       EQU     6        POLARITY OF DATA (0 FOR POS)
0036       0007  S       EQU     7        3-BIT SEGMENT NUMBER
0037       0008  SUM     EQU     8        VALUE TO BE SHIFTED
0038                  *
0039 0000                    AORG    0
0040                  *
0041 0000 7E01  INIT    LACK    1
0042 0001 5001          SACL    ONE
0043 0002 2701          LAC     ONE,7
0044 0003 5002          SACL    BIT7
0045 0004 7E21          LACK    33
0046 0005 5003          SACL    BIAS
0047                  *
0048                  * INVERT INPUT
0049                  *
0050 0006 4004  START   IN      Y,0
0051 0007 7EFF          LACK    >00FF
0052 0008 7804          XOR     Y
0053 0009 5004          SACL    Y
0054                  *
```

```
0055                   * SAVE POLARITY AND STRIP TO LOW 7 BITS
0056 000A 7902                 AND      BIT7
0057 000B 5006                 SACL     P         0000 FOR POS; 0080 FOR NEG
0058 000C 7E7F                 LACK     >007F
0059 000D 7904                 AND      Y
0060 000E 5004                 SACL     Y
0061                   *
0062                   * MAGNITUDE IS CORRECT. STRIP Y OF S AND Q.
0063 000F 2C04                 LAC      Y,12      SHIFT S INTO HIGH HALF OF ACC
0064 0010 5807                 SACH     S
0065 0011 2104                 LAC      Y,1       DOUBLE
0066 0012 1507                 SUB      S,5       REMOVE S BITS (STRIP TO 2Q)
0067 0013 0003                 ADD      BIAS
0068 0014 5008                 SACL     SUM       SUM = 2Q + BIAS
0069                   *
0070                   * SHIFT SUM BY S AND REMOVE BIAS
0071 0015 7E1E                 LACK     SBASE     OFFSET FOR SHIFT ROUTINE
0072 0016 0107                 ADD      S,1       DOUBLE S (2 WORDS/SHIFT SEGMENT)
0073 0017 7F8C                 CALA               SHIFT SUM BY S
0074 0018 1003                 SUB      BIAS
0075                   *
0076                   * ACC = MAGNITUDE, ADD POLARITY TO BIT 13
0077 0019 0606                 ADD      P,6       SHIFT P TO BIT 13
0078 001A 5005                 SACL     X
0079 001B 4805                 OUT      X,0       OUTPUT RESULT TO PORT 0
0080 001C F900   FIN   B        FIN
     001D 001C
0081                   *
0082                   *
0083                   * LOAD SUM SHIFTED 0:7
0084 001E 2008   SBASE LAC      SUM,0
0085 001F 7F8D                 RET
0086 0020 2108                 LAC      SUM,1
0087 0021 7F8D                 RET
0088 0022 2208                 LAC      SUM,2
0089 0023 7F8D                 RET
0090 0024 2308                 LAC      SUM,3
0091 0025 7F8D                 RET
0092 0026 2408                 LAC      SUM,4
0093 0027 7F8D                 RET
0094 0028 2508                 LAC      SUM,5
0095 0029 7F8D                 RET
0096 002A 2608                 LAC      SUM,6
0097 002B 7F8D                 RET
0098 002C 2708                 LAC      SUM,7
0099 002D 7F8D                 RET
0100                   *
0101                          END
NO ERRORS, NO WARNINGS
```

5. Companding Routines for the TMS32010/TMS32020

FLOWCHART: ALAWCOMP

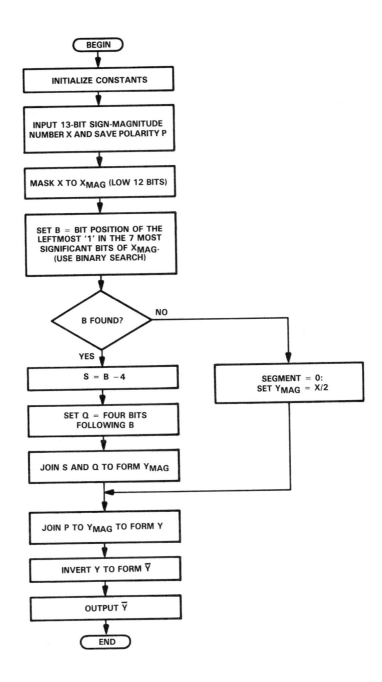

Figure 8. A-Law Compression

```
0001                         IDT      'ALAWCOMP'
0002                ***
0003                *        'ALAWCOMP' PERFORMS AN A-LAW COMPRESSION.
0004                *         THE 13-BIT SIGN-MAGNITUDE INPUT X,
0005                *
0006                *        X = P X11 X10 ... X2 X1 X0
0007                *
0008                *        IS ENCODED AS AN 8-BIT SIGN-MAGNITUDE NUMBER Y,
0009                *
0010                *        Y = P S2 S1 S0 Q3 Q2 Q1 Q0 consisting of
0011                *
0012                *        POLARITY BIT: P,
0013                *        3-BIT SEGMENT NUMBER: S = S2 S1 S0
0014                *        4-BIT QUANTIZATION BIN NUMBER: Q = Q3 Q2 Q1 Q0
0015                *
0016                *        Y IS INVERTED BEFORE TRANSMISSION.
0017                *        PORT 0 IS USED FOR I/O.
0018                *
0019                *        WORST-CASE TIMING IN CYCLES: 14 INIT / 36 LOOP
0020                *        SPACE REQUIREMENTS IN WORDS: 11 DATA / 97 PROG
0021                *
0022                *   .    CONSTANTS:
0023                *   :
0024                *   .
0025      0001  ONE    EQU     1        =1
0026      0002  BIT4   EQU     2        = >0010 (ONE IN BIT 4)
0027      0003  MASK12 EQU     3        = >0FFF (12 ONES)
0028      0004  MASK8  EQU     4        = >00FF ( 8 ONES)
0029      0005  MASK4  EQU     5        = >000F ( 4 ONES)
0030      0006  MASK2  EQU     6        = >0003 ( 2 ONES)
0031                *
0032                *VARIABLES:
0033      0007  X      EQU     7        DATA INPUT (13 BITS)
0034      0008  Y      EQU     8        ENCODED DATA OUTPUT (8 BITS)
0035      0009  P      EQU     9        POLARITY OF DATA (0 FOR POS)
0036      000A  S      EQU     10       3-BIT SEGMENT NUMBER
0037      000B  Q      EQU     11       4-BIT QUANTIZATION BIN NUMBER
0038 0000
0039                *
0040                *
0041 0000                    AORG     0
0042                *
0043 0000 7E01  INIT  LACK     1
0044 0001 5001        SACL     ONE
0045 0002 2401        LAC      ONE,4
0046 0003 5002        SACL     BIT4
0047 0004 2C01        LAC      ONE,12
0048 0005 1001        SUB      ONE
0049 0006 5003        SACL     MASK12
0050 0007 7EFF        LACK     >00FF
0051 0008 5004        SACL     MASK8
0052 0009 7E0F        LACK     >000F
0053 000A 5005        SACL     MASK4
0054 000B 7E03        LACK     >0003
```

5. Companding Routines for the TMS32010/TMS32020

```
0055 000C 5006              SACL    MASK2
0056                   *
0057                   * GET INPUT AND SAVE POLARITY
0058 000D 4007   START  IN      X,0      INPUT DATA THRU PORT 0
0059 000E 2C01          LAC     ONE,12   POLARITY BIT MASK
0060 000F 7907          AND     X
0061 0010 5C09          SACH    P,4      0 FOR POS; 1 FOR NEG.
0062                   *
0063                   * STRIP TO LOW 12 BITS
0064 0011 2007          LAC     X
0065 0012 7903          AND     MASK12
0066 0013 5007          SACL    X
0067                   *
0068                   * S =BP -4 WHERE BP = BIT POSITION OF THE LEFTMOST '1'
0069                   * IN X. TIND THE '1' THROUGH A BINARY SEARCH OF 8 MSB'S
0070                   * OF X. STORE S AND LOAD X SHIFTED LEFT BY 16-S SO
0071                   * THAT THE '1' AND FOUR FOLLOWING BITS ARE IN THE HIGH
0072                   * HALF OF THE ACCUMULATOR. SEARCH BITS 4 THRU 11.
0073                   *
0074 0014 2805   LMOST  LAC     MASK4,8
0075 0015 7907          AND     X
0076 0016 FF00          BZ      EEE
     0017 0038
0077                   *
0078 0018 2A06          LAC     MASK2,10          1100 0000
0079 0019 7907          AND     X
0080 001A FF00          BZ      CC
     001B 002A
0081                   *
0082 001C 2B01          LAC     ONE,11            1000 0000
0083 001D 7907          AND     X
0084 001E FF00          BZ      B
     001F 0025
0085                   *
0086 0020 7E07          LACK    7                 1... ....
0087 0021 500A          SACL    S
0088 0022 2907          LAC     X,9
0089 0023 F900          B       XDONE
     0024 0058
0090                   *
0091 0025 7E06   B      LACK    6                 01.. ....
0092 0026 500A          SACL    S
0093 0027 2A07          LAC     X,10
0094 0028 F900          B       XDONE
     0029 0058
0095                   *
0096 002A 2901   CC     LAC     ONE,9             0010 0000
0097 002B 7907          AND     X
0098 002C FF00          BZ      D
     002D 0033
0099                   *
0100 002E 7E05          LACK    5                 001. ....
0101 002F 500A          SACL    S
0102 0030 2B07          LAC     X,11
```

```
0103 0031 F900          B       XDONE
     0032 0058
0104                *
0105 0033 7E04    D      LACK    4                   0001 ....
0106 0034 500A           SACL    S
0107 0035 2C07           LAC     X,12
0108 0036 F900           B       XDONE
     0037 0058
0109                *
0110 0038 2606    EEE    LAC     MASK2,6             0000 1100
0111 0039 7907           AND     X
0112 003A FF00           BZ      GG
     003B 004A
0113                *
0114 003C 2701           LAC     ONE,7               0000 100
0115 003D 7907           AND     X
0116 003E FF00           BZ      F
     003F 0045
0117                *
0118 0040 7E03           LACK    3                   0000 1...
0119 0041 500A           SACL    S
0120 0042 2D07.          LAC     X,13
0121 0043 F900           B       XDONE
     0044 0058
0122                *
0123 0045 7E02    F      LACK    2                   0000 01..
0124 0046 500A           SACL    S
0125 0047 2E07           LAC     X,14
0126 0048 F900           B       XDONE
     0049 0058
0127                *
0128 004A 2501    GG     LAC     ONE,5               0000 0010
0129 004B 7907           AND     X
0130 004C FF00           BZ      SEGZ
     004D 0053
0131                *
0132 004E 7E01           LACK    1                   0000 001.
0133 004F 500A           SACL    S
0134 0050 2F07           LAC     X,15
0135 0051 F900           B       XDONE
     0052 0058
0136                * SEGMENT 0: _SSSQQQQ =X/2
0137 0053 2F07    SEGZ   LAC     X,15
0138 0054 5807           SACH    X
0139 0055 2007           LAC     X
0140 0056 F900           B       SIGN
     0057 005C
0141                *
0142                * REMOVE LEFTMOST '1' AND STORE Q
0143 0058 6202    XDONE  SUBH    BIT4
0144 0059 580B           SACH    Q
0145                *
0146                * FORM 8-BIT COMPRESSED WORS FIR Q, S, AND P.
0147 005A 200B           LAC     Q        Q:BITS 0-3 ____QQQQ
```

```
0148 005B 040A           ADD     S,4       S:BITS 4-6  _SSSQQQQ
0149 005C 0709  SIGN     ADD     P,7       PSSSQQQQ
0150                *
0151                * COMPLEMENT FOR TRANSMISSION AND OUTPUT
0152 005D 7804           XOR     MASK8
0153 005E 5008           SACL    Y
0154 005F 4808           OUT     Y,0       PORT 0
0155 0060 F900  FIN      B       FIN
     0061 0060
0156                *
0157                     END
NO ERRORS, NO WARNINGS
```

FLOW CHART: ALAWEXP

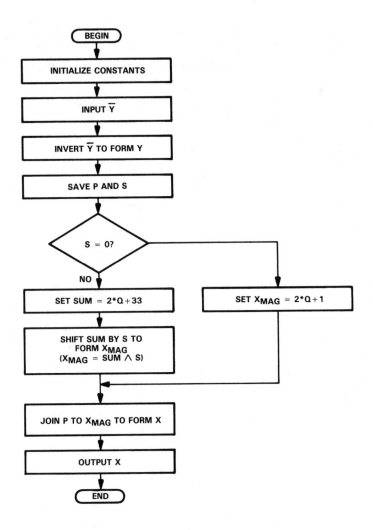

Figure 9. A-Law Expansion

5. Companding Routines for the TMS32010/TMS32020

```
0001                        IDT     'ALAWEXP'
0002                ***
0003                *       'ALAWEXP' PERFORM AN A-LAW EXPANSION. THE 8-BIT
0004                *        DATA INPUT IS
0005                *
0006                *        Y = P S2 S1 S0 Q3 Q2 Q1 Q0   WHICH CONSISTS OF
0007                *
0008                *          POLARITY BIT: P
0009                *          3-BIT SEGMENT NUMBER: S = S2 S1 S0
0010                *          4-BIT QUANTIZATION NUMBER: Q = Q3 Q2 Q1 Q0
0011                *
0012                *        THE INPUT Y IS EXPANDED INTO A 13-BIT OUTPUT
0013                *
0014                *          X = P X11 X10 X9 ... X2 X1 X0 CONSISTING OF
0015                *
0016                *          POLARITY BIT:P
0017                *          AND A 13-BIT MAGNITUDE (X12...X0)
0018                *
0019                *        PORT 0 IS USED FOR I/O.
0020                *        WORST-CASE TIMING IN CYCLES: 4 INIT / 25 LOOP
0021                *        SPACE REQUIREMENTS IN WORDS: 7 DATA / 48 LOOP
0022                *
0023                * CONSTANTS:
0024                *
0025      0001  ONE   EQU     1       = 1
0026      0002  BIT7  EQU     2       = >0080 (ONE IN BIT 7)
0027                *
0028                * VARIABLES:
0029                *
0030      0003  Y     EQU     3       A-LAW COMPRESSED 8-BIT DATA INPUT
0031      0004  X     EQU     4       DECODED (EXPANDED) 13-BIT OUTPUT
0032      0005  P     EQU     5       POLARITY OF DATA (0 FOR POS)
0033      0006  S     EQU     6       3-BIT SEGMENT NUMBER
0034      0007  SUM   EQU     7       VALUE TO BE SHIFTED
0035                *
0036 0000               AORG    0
0037                *
0038 0000 7E01  INIT    LACK    1
0039 0001 5001          SACL    ONE
0040 0002 2701          LAC     ONE,7
0041 0003 5002          SACL    BIT7
0042                *
0043                * INVERT INPUT
0044                *
0045 0004 4003  START   IN      Y,0
0046 0005 7EFF          LACK    >00FF
0047 0006 7803          XOR     Y
0048 0007 5003          SACL    Y
0049                *
0050                * SAVE POLARITY AND STRIP TO LOW 7 BITS
0051 0008 7902          AND     BIT7
0052 0009 5005          SACL    P       0000 FOR POS; 0080 FOR NEG
0053 000A 7E7F          LACK    >007F
0054 000B 7903          AND     Y
```

```
0055 000C 5003            SACL    Y
0056               *
0057               * MAGNITUDE IS CORRECT. STRIP Y OF S AND Q.
0058 000D 2C03            LAC     Y,12    SHIFT S INTO HIGH HALF OF ACC
0059 000E 5806            SACH    S
0060 000F 2006            LAC     S       CHECK FOR SEGMENT 0
0061 0010 FE00            BNZ     SEGNZ
     0011 0016
0062               * SEGMENT 0: EXPAND X TO 2*Q + 1
0063 0012 2103            LAC     Y,1
0064 0013 0001            ADD     ONE
0065 0014 F900            B       SIGN
     0015 001D
0066               * NONZERO SEGMENT: SUM = 2*Q + 33
0067 0016 7E21     SEGNZ  LACK    33
0068 0017 0103            ADD     Y,1
0069 0018 1506            SUB     S,5     REMOVE S BITS
0070 0019 5007            SACL    SUM
0071               *
0072               * SHIFT SUM BY S USING VARIABLE SHIFT ROUTINE AT SBASE
0073 001A 7E20            LACK    SBASE-2 OFFSET (MINUS 0 CASE)
0074 001B 0106            ADD     S,1     DOUBLE S (2 WDS/SHIFT SEGMENT)
0075 001C 7F8C            CALA            SHIFT SUM BY S
0076               *
0077               * ACC = MAGNITUDE. ADD POLARITY TO BIT 12.
0078 001D 0505     SIGN   ADD     P,5     SHIFT P TO BIY 12
0079 001E 5004            SACL    X
0080 001F 4804            OUT     X,0     OUTPUT RESULT TO PORT 0
0081 0020 F900     FIN    B       FIN
     0021 0020
0082               *
0083               * LOAD SUM SHIFTED 0:6
0084 0022 2007     SBASE  LAC     SUM,0
0085 0023 7F8D            RET
0086 0024 2107            LAC     SUM,1
0087 0025 7F8D            RET
0088 0026 2207            LAC     SUM,2
0089 0027 7F8D            RET
0090 0028 2307            LAC     SUM,3
0091 0029 7F8D            RET
0092 002A 2407            LAC     SUM,4
0093 002B 7F8D            RET
0094 002C 2507            LAC     SUM,5
0095 002D 7F8D            RET
0096 002E 2607            LAC     SUM,6
0097 002F 7F8D            RET
0098               *
0099                      END
NO ERRORS, NO WARNINGS
```

5. Companding Routines for the TMS32010/TMS32020

Tables 9 and 10 are included to aid in verifying particular implementations of the algorithms that have been presented.

Table 9. Segmented μ-255 Companding[*]
(Courtesy of John Wiley & Sons, see Reference 6)

| | Segment S | | | | | | | | Quantization | |
	000	001	010	011	100	101	110	111	BIN	Q
Quantization Endpoints	0	31	95	223	479	991	2015	4063	0000	0
	1	35	103	239	511	1055	2143	4319	0001	1
	3	39	111	255	543	1119	2271	4575	0010	2
	5	43	119	271	575	1183	2399	4831	0011	3
	7	47	127	287	607	1247	2527	5087	0100	4
	9	51	135	303	639	1311	2655	5343	0101	5
	11	55	143	319	671	1375	2783	5599	0110	6
	13	59	151	335	703	1439	2911	5855	0111	7
	15	63	159	351	735	1503	3039	6111	1000	8
	17	67	167	367	767	1567	3167	6367	1001	9
	19	71	175	383	799	1631	3295	6623	1010	10
	21	75	183	399	831	1695	3423	6879	1011	11
	23	79	191	415	863	1759	3551	7135	1100	12
	25	83	199	431	895	1823	3679	7391	1101	13
	27	87	207	447	927	1887	3807	7647	1110	14
	29	91	215	463	959	1951	3935	7903	1111	15
	31	95	223	479	991	2015	4063	8159		

[*] (1) Sample values are referenced to a full-scale value of 8159. (2) Negative samples are encoded in sign-magnitude format with a polarity bit of 1. (3) In actual transmission the codes are inverted to increase the density of 1's when low signal amplitudes are encoded. (4) Analog output samples are decoded as the center of the encoded quantization interval. (5) Quantization error is the difference between the reconstructed output value and the original input sample value.

Table 10. Segmented A-Law Companding
(Courtesy of John Wiley & Sons, see Reference 6)

| | Segment S | | | | | | | | Quantization | |
	000	001	010	011	100	101	110	111	BIN	Q
Quantization Endpoints	0	32	64	128	256	512	1024	2048	0000	0
	2	34	68	136	272	544	1088	2176	0001	1
	4	36	72	144	288	576	1152	2304	0010	2
	6	38	76	152	304	608	1216	2432	0011	3
	8	40	80	160	320	640	1280	2560	0100	4
	10	42	84	168	336	672	1344	2688	0101	5
	12	44	88	176	352	704	1408	2816	0110	6
	14	46	92	184	368	736	1472	2944	0111	7
	16	48	96	192	384	768	1536	3072	1000	8
	18	50	100	200	400	800	1600	3200	1001	9
	20	52	104	208	416	832	1664	3328	1010	10
	22	54	108	216	432	864	1728	3456	1011	11
	24	56	112	224	448	896	1792	3584	1100	12
	26	58	116	232	464	928	1856	3712	1101	13
	28	60	120	240	480	960	1920	3840	1110	14
	30	62	124	248	496	992	1984	3968	1111	15
	32	64	128	256	512	1024	2048	4096		

REFERENCES

1. *TCM2913, TCM2914, TCM2916, TCM2917 Combined Single-Chip PCM Codec and Filter* (Data Sheet SCTS012), Texas Instruments (1983).

2. J.B. Reimer, M.L. McMahan, and M. Arjmand, *32-kbit/s ADPCM with the TMS32010* (Application Report), Texas Instruments (1985).

3. J. Robillard, *Telcommunications Interfacing to the TMS32010* (Application Report), Texas Instruments (1985).

4. D.G. Messerschmitt, D.J. Hedberg, C.R. Cole, A. Haoui, and P. Winship, *Digital Voice Echo Canceller with a TMS32020* (Application Report), Texas Instruments (1985).

5. J.L. Fike and G.E. Friend, *Understanding Telephone Electronics,* Texas Instruments (1983).

6. J.C. Bellamy, *Digital Telephony*, John Wiley & Sons (1982). (Figures and tables reprinted by permission of John Wiley & Sons.)

APPENDIX

COMPANDING ROUTINES FOR THE TMS32020

This appendix provides companding programs for the TMS32020. The basic theory and operation are similar to what is described for the TMS32010 in the major portion of this report.

Programs included in the appendix for μ-law and A-law expansion and compression utilize the serial port of the TMS32020 for 8-bit serial I/O. The routines are interrupt-driven to allow direct interfacing to a codec, such as the Texas Instruments TCM2913 (see the data sheet for further information). The following paragraphs briefly describe each of the programs included in the appendix.

The first program reads a μ-law value from the Data Receive Register (DRR), expands it to a 14-bit linear value, and writes it to location X. This value is then compressed back to an 8-bit μ-law value and written to the Data Transmit Register (DXR). Since μ-law codecs invert all bits of the 8-bit value for transmission, XORK (exclusive-OR immediate with accumulator with shift) instructions are used to perform inversion in the TMS32020 before expansion and after compression. Note also that the LACT (load accumulator with shift specified by T register) instruction is useful for performing the conditional shift implemented in the TMS32020 program by a computed subroutine call (CALA). The compression routine of the program assumes a left-justified 14-bit value within the accumulator. Therefore, the value stored in X is left-shifted twice before the compression routine is entered. The NORM (normalize contents of the accumulator) instruction is then used to find the MSB of the accumulator by performing an in-place accumulator left-shift if the two MSBs are the same. At the same time, the count of the left-shifts is maintained in auxiliary register 0 (AR0) and used to compute the segment number S.

The second program performs A-law expansion and compression, and is similar to the μ-law program. For A-law transmission, however, only the even-order bits of the 8-bit value are inverted. Note that the LACT and NORM instructions are still used to compute the expanded and compressed values.

The third program is an example of a simplified solution to μ-law expansion. The 8-bit μ-law value is used as an index into a 256-word table of 14-bit linear values accessed by the TBLR (table write) instruction. This lookup-table approach may also be utilized for A-law expansion. While this method is obviously the fastest method of performing expansion, it is also the most inefficient in terms of program memory requirements.

```
0001                   ****************************************************************
0002              *                          PROGRAM 1                           *
0003              *                                                               *
0004              *    U-LAW EXPAND/COMPRESS CODEC LOOPBACK PROGRAM. UPON         *
0005              *    RECEIVING AN RINT INSTRUCTION, DRR IS READ, EXPANDED,      *
0006              *    COMPRESSED BACK, AND WRITTEN TO DXR, AT WHICH TIME THE      *
0007              *    PROCESSOR IDLES UNTIL ANOTHER RINT INSTRUCTION IS          *
0008              *    GENERATED. THE FIRST WORD TRANSMITTED IS A ZERO.           *
0009              ****************************************************************
0010              *
0011      0000    DRR      EQU    0
0012      0001    DXR      EQU    1
0013      0004    IMR      EQU    4
0014      0060    S        EQU    96        * U-LAW SEGMENT NUMBER
0015      0061    BIAS     EQU    97        * = 33
0016      0062    X        EQU    98
0017      0063    SUM      EQU    99
0018      0064    SIGN     EQU    100
0019      0065    NEG7     EQU    101       * = >FFF9
0020      0066    Q        EQU    102       * U-LAW QUANTIZATION BIN
0021      0067    BIAS2    EQU    103
0022 0000         AORG   0
0023 0000 FF80    RSVECT   B      INIT
     0001 0400
0024 001A         AORG   26
0025 001A 9800    EXPAND   BIT    DRR,8     * TEST DRR FOR SIGN
0026 001B 2C00             LAC    DRR,12
0027 001C D404             ANDK   >7F00,4   * ZERO SIGN AND OTHER MSBS
     001D 7F00
0028 001E D406             XORK   >7F00,4   * INVERT ALL BITS
     001F 7F00
0029 0020 6860             SACH   S
0030 0021 4460             SUBH   S         * ZERO ACCH
0031 0022 0B61             ADD    BIAS,11
0032 0023 CE18             SFL
0033 0024 6C63             SACH   SUM,4
0034 0025 3C60             LT     S
0035 0026 4263             LACT   SUM
0036 0027 1061             SUB    BIAS
0037 0028 F980             BBNZ   POSVAL    * POSITIVE IF TC = 1
     0029 002B
0038 002A CE23             NEG
0039 002B 6062    POSVAL   SACL   X
0040              *
0041 002C 2262    JSTIFY   LAC    X,2       * LEFT-JUSTIFY 14-BIT NUMBER
0042 002D 6062             SACL   X
0043              *
0044 002E 4062    COMPRS   ZALH   X         * U-LAW COMPRESS # IN ACCH
0045 002F F380             BLZ    NEGCMP
     0030 0040
0046 0031 4867             ADDH   BIAS2     * (# ALREADY LEFT-JUSTIFIED)
0047 0032 3065             LAR    AR0,NEG7
0048 0033 CB06             RPTK   6         * FIND MSB
0049 0034 CEA2             NORM
0050 0035 DE04             ANDK   >F000,14  * ZERO 2 MSBS & ALL LSBS
     0036 F000
0051 0037 6866             SACH   Q
```

```
0052 0038 7060            SAR    AR0,S
0053 0039 4060            ZALH   S
0054 003A CE1B            ABS
0055 003B 0266            ADD    Q,2
0056 003C D406            XORK   >FF00,4        * INVERT ALL BITS
     003D FF00
0057 003E FF80            B      SATCH
     003F 004E
0058           *
0059 0040 CE1B   NEGCMP   ABS
0060 0041 4867            ADDH   BIAS2          * (# ALREADY LEFT-JUSTIFIED)
0061 0042 3065            LAR    AR0,NEG7
0062 0043 CB06            RPTK   6              * FIND MSB
0063 0044 CEA2            NORM
0064 0045 DE04            ANDK   >F000,14       * ZERO 2 MSBS & ALL LSBS
     0046 F000
0065 0047 6866            SACH   Q
0066 0048 7060            SAR    AR0,S
0067 0049 4060            ZALH   S
0068 004A CE1B            ABS
0069 004B 0266            ADD    Q,2
0070 004C D406            XORK   >7F00,4        * INVERT ALL BITS IN Q
     004D 7F00
0071           *                                (P=0 FOR NEGATIVE VALUES)
0072 004E 6C01   SATCH    SACH   DXR,4
0073           *
0074 004F CE00   WAIT     EINT
0075 0050 CE1F            IDLE                  * WAIT FOR RINT
0076           *
0077           *
0078           *          INITIALIZATION ROUTINE
0079           *
0080           *
0081 0400                 AORG   1024
0082 0400 C800   INIT     LDPK   0              * POINT DP TO B2 AND MMRS
0083 0401 CA10            LACK   >10
0084 0402 6004            SACL   IMR            * ENABLE RINT BUT DISABLE
0085           *                                  ALL OTHERS
0086 0403 CE0F            FORT   1              * CONFIGURE SERIAL PORT TO
0087           *                                  BYTE MODE
0088 0404 CE03            SOVM
0089 0405 CE07            SSXM
0090 0406 CE08            SPM    0
0091 0407 CA21            LACK   33
0092 0408 6061            SACL   BIAS
0093 0409 2261            LAC    BIAS,2
0094 040A 6067            SACL   BIAS2
0095 040B D001            LALK   -7
     040C FFF9
0096 040D 6065            SACL   NEG7
0097 040E CA00            LACK   0
0098 040F 6000            SACL   DRR            * ZERO MSBS OF DRR
0099 0410 6001            SACL   DXR            * ZERO DXR FOR FIRST
0100           *                                  TRANSMIT OPERATION
0101 0411 5588            LARP   0              * ZERO ARP
0102 0412 5588            LARP   0              *   AND ARB
0103 0413 CE00            EINT
```

```
0104 0414 CE1F          IDLE              * WAIT FOR RINT
0105                    END
NO ERRORS, NO WARNINGS
```

5. Companding Routines for the TMS32010/TMS32020

```
0001          *********************************************************************
0002          *                                                                   *
0003          *                        PROGRAM 2                                   *
0004          *    A-LAW EXPAND/COMPRESS CODEC LOOPBACK PROGRAM. UPON              *
0005          *    RECEIVING AN RINT INSTRUCTION, DRR IS READ, EXPANDED,           *
0006          *    COMPRESSED BACK, AND WRITTEN TO DXR, AT WHICH TIME THE          *
0007          *    PROCESSOR IDLES UNTIL ANOTHER RINT INSTRUCTION IS               *
0008          *    GENERATED. THE FIRST WORD TRANSMITTED IS A ZERO.                *
0009          *********************************************************************
0010          *
0011    0000  DRR       EQU    0
0012    0001  DXR       EQU    1
0013    0004  IMR       EQU    4
0014    0060  S         EQU    96          * A-LAW SEGMENT NUMBER
0015    0061  BIAS      EQU    97          * = 33
0016    0062  X         EQU    98
0017    0063  SUM       EQU    99
0018    0064  SIGN      EQU    100
0019    0065  NEG7      EQU    101         * = >FFF9
0020    0066  Q         EQU    102         * A-LAW QUANTIZATION BIN
0021    0067  ONE       EQU    103
0022 0000               AORG   0
0023 0000 FF80 RSVECT   B      INIT
     0001 0400
0024 001A               AORG   26
0025 001A 9800 EXPAND   BIT    DRR,8       * TEST DRR FOR SIGN
0026 001B 2C00          LAC    DRR,12
0027 001C D404          ANDK   >7F00,4     * ZERO SIGN AND OTHER MSBS
     001D 7F00
0028 001E D406          XORK   >5500,4     * INVERT EVEN-ORDER BITS
     001F 5500
0029 0020 6860          SACH   S
0030 0021 4460          SUBH   S           * ZERO ACCH
0031 0022 3060          LAR    AR0,S
0032 0023 FB90          BANZ   SEGNZ,*-    * TEST FOR SEGMENT NUMBER 0
     0024 002D
0033 0025 0B67          ADD    ONE,11
0034 0026 CE18          SFL
0035 0027 F880          BBZ    PSVAL1
     0028 002A
0036 0029 CE23          NEG
0037 002A 6C62 PSVAL1   SACH   X,4
0038 002B FF80          B      JSTIFY
     002C 0037
0039          *
0040 002D 7060 SEGNZ    SAR    AR0,S       * STORE DECREMENTED S
0041 002E 0B61          ADD    BIAS,11
0042 002F CE18          SFL
0043 0030 6C63          SACH   SUM,4
0044 0031 3C60          LT     S
0045 0032 4263          LACT   SUM
0046 0033 F880          BBZ    POSVAL
     0034 0036
0047 0035 CE23          NEG
0048 0036 6062 POSVAL   SACL   X
0049          *
0050 0037 2362 JSTIFY   LAC    X,3
```

```
0051 0038 6062              SACL   X              * LEFT-JUSTIFY 13-BIT NUMBER
0052                *
0053 0039 4062   COMPRS     ZALH   X              * A-LAW COMPRESS # IN ACCH
0054 003A F380              BLZ    NEGCMP
     003B 0052
0055 003C 3065              LAR    AR0,NEG7
0056 003D CB06              RPTK   6              * FIND MSB
0057 003E CEA2              NORM
0058 003F FB80              BANZ   SEGNZ1,*
     0040 0047
0059 0041 6866              SACH   Q              * SEGMENT NUMBER = 0
0060 0042 2166              LAC    Q,1
0061 0043 D406              XORK   >5500,4        * INVERT EVEN-ORDER BITS
     0044 5500
0062 0045 FF80              B      SATCH
     0046 0067
0063 0047 DE04   SEGNZ1     ANDK   >F000,14       * ZERO 2 MSBS & ALL LSBS
     0048 F000
0064 0049 6866              SACH   Q
0065 004A 7060              SAR    AR0,S
0066 004B 4060              ZALH   S
0067 004C CE1B              ABS
0068 004D 0266              ADD    Q,2
0069 004E D406              XORK   >5500,4        * INVERT EVEN-ORDER BITS
     004F 5500
0070 0050 FF80              B      SATCH
     0051 0067
0071                *
0072 0052 CE1B   NEGCMP     ABS
0073 0053 3065              LAR    AR0,NEG7
0074 0054 CB06              RPTK   6              * FIND MSB
0075 0055 CEA2              NORM
0076 0056 FB80              BANZ   SEGNZ2,*
     0057 005E
0077 0058 6866              SACH   Q              * SEGMENT NUMBER = 0
0078 0059 2166              LAC    Q,1
0079 005A D406              XORK   >D500,4        * INVERT EVEN-ORDER BITS
     005B D500
0080                *                               AND SET SIGN BIT TO 1
0081 005C FF80              B      SATCH
     005D 0067
0082 005E DE04   SEGNZ2     ANDK   >F000,14       * ZERO 2 MSBS & ALL LSBS
     005F F000
0083 0060 6866              SACH   Q
0084 0061 7060              SAR    AR0,S
0085 0062 4060              ZALH   S
0086 0063 CE1B              ABS
0087 0064 0266              ADD    Q,2
0088 0065 D406              XORK   >D500,4        * INVERT EVEN-ORDER BITS
     0066 D500
0089                *                               AND SET SIGN BIT TO 1
0090 0067 6C01   SATCH      SACH   DXR,4
0091                *
0092 0068 CE00   WAIT       EINT
0093 0069 CE1F              IDLE                  * WAIT FOR RINT
0094                *
0095                *
```

5. Companding Routines for the TMS32010/TMS32020

```
0096                  *          INITIALIZATION ROUTINE
0097                  *
0098                  *
0099 0400                        AORG  1024
0100 0400 C800  INIT             LDPK  0              * POINT DP TO B2 AND MMRS
0101 0401 CA10                   LACK  >10
0102 0402 6004                   SACL  IMR            * ENABLE RINT BUT DISABLE
0103                  *                                 ALL OTHERS
0104 0403 CE0F                   FORT  1              * CONFIGURE SERIAL PORT TO
0105                  *                                 BYTE MODE
0106 0404 CE03                   SOVM
0107 0405 CE07                   SSXM
0108 0406 CE08                   SPM   0
0109 0407 CA01                   LACK  1
0110 0408 6067                   SACL  ONE
0111 0409 CA21                   LACK  33
0112 040A 6061                   SACL  BIAS
0113 040B D001                   LALK  -7
     040C FFF9
0114 040D 6065                   SACL  NEG7
0115 040E CA00                   LACK  0
0116 040F 6000                   SACL  DRR            * ZERO MSBS OF DRR
0117 0410 6001                   SACL  DXR            * ZERO DXR FOR FIRST
0118                  *                                 TRANSMIT OPERATION
0119 0411 5588                   LARP  0              * ZERO ARP
0120 0412 5588                   LARP  0              *   AND ARB
0121 0413 CE00                   EINT
0122 0414 CE1F                   IDLE                 * WAIT FOR RINT
0123                              END
NO ERRORS, NO WARNINGS
```

```
0001             ***********************************************************
0002             *                      PROGRAM 3                          *
0003             *                                                         *
0004             *   U-LAW TABLE LOOKUP EXPANSION PROGRAM. UPON RECEIVING   *
0005             *   AN RINT INSTRUCTION, DRR IS READ, EXPANDED, AND        *
0006             *   WRITTEN TO THE STORAGE LOCATION X, AT WHICH TIME THE    *
0007             *   PROCESSOR IDLES UNTIL ANOTHER RINT INSTRUCTION IS       *
0008             *   GENERATED. DXR IS SET TO ZERO SO THAT ALL WORDS         *
0009             *   TRANSMITTED ARE ZEROES.                                *
0010             ***********************************************************
0011             *
0012     0000  DRR     EQU   0
0013     0001  DXR     EQU   1
0014     0004  IMR     EQU   4
0015     0060  X       EQU   96
0016     0061  BADDR   EQU   97       * CONTAINS BASE ADDR FOR
0017           *                        TABLE
0018 0000            AORG  0
0019 0000 FF80  RSVECT  B     INIT
     0001 001F
0020 001A            AORG  26
0021 001A 2000  EXPAND  LAC   DRR
0022 001B 0061          ADD   BADDR    * ADD TABLE BASE ADDR FOR
0023           *                         U-LAW EXPANSION TABLE
0024           *                         LOOKUP
0025 001C 5860          TBLR  X        * READ INTO LOCATION X
0026           *
0027 001D CE00          EINT
0028 001E CE1F          IDLE           * WAIT FOR RINT
0029           *
0030           *
0031           *        INITIALIZATION ROUTINE
0032           *
0033           *
0034 001F C800  INIT    LDPK  0        * POINT DP TO B2 AND MMRS
0035 0020 CA10          LACK  >10
0036 0021 6004          SACL  IMR      * ENABLE RINT BUT DISABLE
0037           *                         ALL OTHERS
0038 0022 CE0F          FORT  1        * CONFIGURE SERIAL PORT TO
0039           *                         BYTE MODE
0040 0023 CE03          SOVM
0041 0024 CE07          SSXM
0042 0025 CE08          SPM   0
0043 0026 D001          LALK  XTBL
     0027 0300
0044 0028 6061          SACL  BADDR    * BASE ADDR FOR TBL LOOKUP
0045 0029 CA00          LACK  0
0046 002A 6000          SACL  DRR      * ZERO MSBS OF DRR
0047 002B 6001          SACL  DXR      * ZERO DXR FOR FIRST
0048           *                         TRANSMIT OPERATION
0049 002C CE00          EINT
0050 002D CE1F          IDLE           * WAIT FOR RINT
0051           *
0052           *
0053           *        TABLE FOR U-LAW EXPANSION TABLE LOOKUP
0054           *
0055           *
```

```
0056 0300                        AORG   768
0057       0300  XTBL            EQU    $
0058                   *                                * NEGATIVE VALUES FIRST
0059 0300 E0A1                   DATA   >E0A1           * (FF, FE, ETC.)
0060 0301 E1A1                   DATA   >E1A1
0061 0302 E2A1                   DATA   >E2A1
0062 0303 E3A1                   DATA   >E3A1
0063 0304 E4A1                   DATA   >E4A1
0064 0305 E5A1                   DATA   >E5A1
0065 0306 E6A1                   DATA   >E6A1
0066 0307 E7A1                   DATA   >E7A1
0067 0308 E8A1                   DATA   >E8A1
0068 0309 E9A1                   DATA   >E9A1
0069 030A EAA1                   DATA   >EAA1
0070 030B EBA1                   DATA   >EBA1
0071 030C ECA1                   DATA   >ECA1
0072 030D EDA1                   DATA   >EDA1
0073 030E EEA1                   DATA   >EEA1
0074 030F EFA1                   DATA   >EFA1
0075 0310 F061                   DATA   >F061
0076 0311 F0E1                   DATA   >F0E1
0077 0312 F161                   DATA   >F161
0078 0313 F1E1                   DATA   >F1E1
0079 0314 F261                   DATA   >F261
0080 0315 F2E1                   DATA   >F2E1
0081 0316 F361                   DATA   >F361
0082 0317 F3E1                   DATA   >F3E1
0083 0318 F461                   DATA   >F461
0084 0319 F4E1                   DATA   >F4E1
0085 031A F561                   DATA   >F561
0086 031B F5E1                   DATA   >F5E1
0087 031C F661                   DATA   >F661
0088 031D F6E1                   DATA   >F6E1
0089 031E F761                   DATA   >F761
0090 031F F7E1                   DATA   >F7E1
0091 0320 F841                   DATA   >F841
0092 0321 F881                   DATA   >F881
0093 0322 F8C1                   DATA   >F8C1
0094 0323 F901                   DATA   >F901
0095 0324 F941                   DATA   >F941
0096 0325 F981                   DATA   >F981
0097 0326 F9C1                   DATA   >F9C1
0098 0327 FA01                   DATA   >FA01
0099 0328 FA41                   DATA   >FA41
0100 0329 FA81                   DATA   >FA81
0101 032A FAC1                   DATA   >FAC1
0102 032B FB01                   DATA   >FB01
0103 032C FB41                   DATA   >FB41
0104 032D FB81                   DATA   >FB81
0105 032E FBC1                   DATA   >FBC1
0106 032F FC01                   DATA   >FC01
0107 0330 FC31                   DATA   >FC31
0108 0331 FC51                   DATA   >FC51
0109 0332 FC71                   DATA   >FC71
0110 0333 FC91                   DATA   >FC91
0111 0334 FCB1                   DATA   >FCB1
0112 0335 FCD1                   DATA   >FCD1
```

5. Companding Routines for the TMS32010/TMS32020   207

```
0113 0336 FCF1          DATA   >FCF1
0114 0337 FD11          DATA   >FD11
0115 0338 FD31          DATA   >FD31
0116 0339 FD51          DATA   >FD51
0117 033A FD71          DATA   >FD71
0118 033B FD91          DATA   >FD91
0119 033C FDB1          DATA   >FDB1
0120 033D FDD1          DATA   >FDD1
0121 033E FDF1          DATA   >FDF1
0122 033F FE11          DATA   >FE11
0123 0340 FE29          DATA   >FE29
0124 0341 FE39          DATA   >FE39
0125 0342 FE49          DATA   >FE49
0126 0343 FE59          DATA   >FE59
0127 0344 FE69          DATA   >FE69
0128 0345 FE79          DATA   >FE79
0129 0346 FE89          DATA   >FE89
0130 0347 FE99          DATA   >FE99
0131 0348 FEA9          DATA   >FEA9
0132 0349 FEB9          DATA   >FEB9
0133 034A FEC9          DATA   >FEC9
0134 034B FED9          DATA   >FED9
0135 034C FEE9          DATA   >FEE9
0136 034D FEF9          DATA   >FEF9
0137 034E FF09          DATA   >FF09
0138 034F FF19          DATA   >FF19
0139 0350 FF25          DATA   >FF25
0140 0351 FF2D          DATA   >FF2D
0141 0352 FF35          DATA   >FF35
0142 0353 FF3D          DATA   >FF3D
0143 0354 FF45          DATA   >FF45
0144 0355 FF4D          DATA   >FF4D
0145 0356 FF55          DATA   >FF55
0146 0357 FF5D          DATA   >FF5D
0147 0358 FF65          DATA   >FF65
0148 0359 FF6D          DATA   >FF6D
0149 035A FF75          DATA   >FF75
0150 035B FF7D          DATA   >FF7D
0151 035C FF85          DATA   >FF85
0152 035D FF8D          DATA   >FF8D
0153 035E FF95          DATA   >FF95
0154 035F FF9D          DATA   >FF9D
0155 0360 FFA3          DATA   >FFA3
0156 0361 FFA7          DATA   >FFA7
0157 0362 FFAB          DATA   >FFAB
0158 0363 FFAF          DATA   >FFAF
0159 0364 FFB3          DATA   >FFB3
0160 0365 FFB7          DATA   >FFB7
0161 0366 FFBB          DATA   >FFBB
0162 0367 FFBF          DATA   >FFBF
0163 0368 FFC3          DATA   >FFC3
0164 0369 FFC7          DATA   >FFC7
0165 036A FFCB          DATA   >FFCB
0166 036B FFCF          DATA   >FFCF
0167 036C FFD3          DATA   >FFD3
0168 036D FFD7          DATA   >FFD7
0169 036E FFDB          DATA   >FFDB
```

5. Companding Routines for the TMS32010/TMS32020

```
0170  036F  FFDF          DATA    >FFDF
0171  0370  FFE2          DATA    >FFE2
0172  0371  FFE4          DATA    >FFE4
0173  0372  FFE6          DATA    >FFE6
0174  0373  FFE8          DATA    >FFE8
0175  0374  FFEA          DATA    >FFEA
0176  0375  FFEC          DATA    >FFEC
0177  0376  FFEE          DATA    >FFEE
0178  0377  FFF0          DATA    >FFF0
0179  0378  FFF2          DATA    >FFF2
0180  0379  FFF4          DATA    >FFF4
0181  037A  FFF6          DATA    >FFF6
0182  037B  FFF8          DATA    >FFF8
0183  037C  FFFA          DATA    >FFFA
0184  037D  FFFC          DATA    >FFFC
0185  037E  FFFE          DATA    >FFFE
0186  037F  0000          DATA    >0
0187              *
0188  0380  1F5F          DATA    >1F5F        * POSITIVE VALUES NEXT
0189  0381  1E5F          DATA    >1E5F        *   (POLARITY BIT = 1)
0190  0382  1D5F          DATA    >1D5F
0191  0383  1C5F          DATA    >1C5F
0192  0384  1B5F          DATA    >1B5F
0193  0385  1A5F          DATA    >1A5F
0194  0386  195F          DATA    >195F
0195  0387  185F          DATA    >185F
0196  0388  175F          DATA    >175F
0197  0389  165F          DATA    >165F
0198  038A  155F          DATA    >155F
0199  038B  145F          DATA    >145F
0200  038C  135F          DATA    >135F
0201  038D  125F          DATA    >125F
0202  038E  115F          DATA    >115F
0203  038F  105F          DATA    >105F
0204  0390  0F9F          DATA    >F9F
0205  0391  0F1F          DATA    >F1F
0206  0392  0E9F          DATA    >E9F
0207  0393  0E1F          DATA    >E1F
0208  0394  0D9F          DATA    >D9F
0209  0395  0D1F          DATA    >D1F
0210  0396  0C9F          DATA    >C9F
0211  0397  0C1F          DATA    >C1F
0212  0398  0B9F          DATA    >B9F
0213  0399  0B1F          DATA    >B1F
0214  039A  0A9F          DATA    >A9F
0215  039B  0A1F          DATA    >A1F
0216  039C  099F          DATA    >99F
0217  039D  091F          DATA    >91F
0218  039E  089F          DATA    >89F
0219  039F  081F          DATA    >81F
0220  03A0  07BF          DATA    >7BF
0221  03A1  077F          DATA    >77F
0222  03A2  073F          DATA    >73F
0223  03A3  06FF          DATA    >6FF
0224  03A4  06BF          DATA    >6BF
0225  03A5  067F          DATA    >67F
0226  03A6  063F          DATA    >63F
```

5. Companding Routines for the TMS32010/TMS32020

```
0227 03A7 05FF          DATA    >5FF
0228 03A8 05BF          DATA    >5BF
0229 03A9 057F          DATA    >57F
0230 03AA 053F          DATA    >53F
0231 03AB 04FF          DATA    >4FF
0232 03AC 04BF          DATA    >4BF
0233 03AD 047F          DATA    >47F
0234 03AE 043F          DATA    >43F
0235 03AF 03FF          DATA    >3FF
0236 03B0 03CF          DATA    >3CF
0237 03B1 03AF          DATA    >3AF
0238 03B2 038F          DATA    >38F
0239 03B3 036F          DATA    >36F
0240 03B4 034F          DATA    >34F
0241 03B5 032F          DATA    >32F
0242 03B6 030F          DATA    >30F
0243 03B7 02EF          DATA    >2EF
0244 03B8 02CF          DATA    >2CF
0245 03B9 02AF          DATA    >2AF
0246 03BA 028F          DATA    >28F
0247 03BB 026F          DATA    >26F
0248 03BC 024F          DATA    >24F
0249 03BD 022F          DATA    >22F
0250 03BE 020F          DATA    >20F
0251 03BF 01EF          DATA    >1EF
0252 03C0 01D7          DATA    >1D7
0253 03C1 01C7          DATA    >1C7
0254 03C2 01B7          DATA    >1B7
0255 03C3 01A7          DATA    >1A7
0256 03C4 0197          DATA    >197
0257 03C5 0187          DATA    >187
0258 03C6 0177          DATA    >177
0259 03C7 0167          DATA    >167
0260 03C8 0157          DATA    >157
0261 03C9 0147          DATA    >147
0262 03CA 0137          DATA    >137
0263 03CB 0127          DATA    >127
0264 03CC 0117          DATA    >117
0265 03CD 0107          DATA    >107
0266 03CE 00F7          DATA    >F7
0267 03CF 00E7          DATA    >E7
0268 03D0 00DB          DATA    >DB
0269 03D1 00D3          DATA    >D3
0270 03D2 00CB          DATA    >CB
0271 03D3 00C3          DATA    >C3
0272 03D4 00BB          DATA    >BB
0273 03D5 00B3          DATA    >B3
0274 03D6 00AB          DATA    >AB
0275 03D7 00A3          DATA    >A3
0276 03D8 009B          DATA    >9B
0277 03D9 0093          DATA    >93
0278 03DA 008B          DATA    >8B
0279 03DB 0083          DATA    >83
0280 03DC 007B          DATA    >7B
0281 03DD 0073          DATA    >73
0282 03DE 006B          DATA    >6B
0283 03DF 0063          DATA    >63
```

```
0284 03E0 005D         DATA   >5D
0285 03E1 0059         DATA   >59
0286 03E2 0055         DATA   >55
0287 03E3 0051         DATA   >51
0288 03E4 004D         DATA   >4D
0289 03E5 0049         DATA   >49
0290 03E6 0045         DATA   >45
0291 03E7 0041         DATA   >41
0292 03E8 003D         DATA   >3D
0293 03E9 0039         DATA   >39
0294 03EA 0035         DATA   >35
0295 03EB 0031         DATA   >31
0296 03EC 002D         DATA   >2D
0297 03ED 0029         DATA   >29
0298 03EE 0025         DATA   >25
0299 03EF 0021         DATA   >21
0300 03F0 001E         DATA   >1E
0301 03F1 001C         DATA   >1C
0302 03F2 001A         DATA   >1A
0303 03F3 0018         DATA   >18
0304 03F4 0016         DATA   >16
0305 03F5 0014         DATA   >14
0306 03F6 0012         DATA   >12
0307 03F7 0010         DATA   >10
0308 03F8 000E         DATA   >E
0309 03F9 000C         DATA   >C
0310 03FA 000A         DATA   >A
0311 03FB 0008         DATA   >8
0312 03FC 0006         DATA   >6
0313 03FD 0004         DATA   >4
0314 03FE 0002         DATA   >2
0315 03FF 0000         DATA   >0
0316            *
0317                   END
NO ERRORS, NO WARNINGS
```

6. Floating-Point Arithmetic with the TMS32010

Ray Simar, Jr.
Digital Signal Processing - Semiconductor Group
Texas Instruments

INTRODUCTION

The TMS32010 Digital Signal Processor is a fixed-point 16/32-bit microprocessor. However, it can also perform floating-point computations at a speed comparable to dedicated floating-point processors.

The purpose of this application report is to analyze an implementation of floating-point addition and multiplication on the TMS32010. The floating-point single-precision standard proposed by the IEEE will be examined. Using this standard, the TMS32010 performs a floating-point multiplication in 8.4 microseconds and a floating-point addition in 17.2 microseconds.

To illustrate floating-point formats and the tradeoffs involved in making a choice between different floating-point formats, a review of floating-point arithmetic notation and of addition and multiplication algorithms is first presented.

FLOATING-POINT NOTATION

The floating-point number f may be written in floating-point format as

$$f = m \times b^e$$

where

m = mantissa
b = base
e = exponent

For example, 6,789,320 may be written as

$$0.6789320 \times 10^7$$

In this case,

m = 0.6789320
b = 10
e = 7

The two floating-point numbers f_1 and f_2 may be written as

$$f_1 = m_1 \times b^{e1}$$
$$f_2 = m_2 \times b^{e2}$$

Floating-point addition/subtraction, multiplication, and division for f_1 and f_2 are defined as follows:

$$f_1 \pm f_2 = (m_1 \pm m_2 \times b^{-(e1-e2)}) \times b^{e1} \quad \text{if } e_1 \geq e_2 \quad (1)$$

or

$$= (m_1 \times b^{-(e2-e1)}) \pm m_2) \times b^{e2} \quad \text{if } e_1 < e_2$$

$$f_1 \times f_2 = m_1 \times m_2 \times b^{(e1+e2)} \quad (2)$$

$$f_1/f_2 = (m_1/m_2) \times b^{(e1-e2)} \quad (3)$$

A cursory examination of these expressions reveals some of the factors involved in the implementation of floating-point arithmetic. For addition, it is necessary to shift the mantissa of the floating-point number which has the smaller exponent to the right by the difference in the magnitude of the two exponents. This is shown in the multiplication by the terms

$$b^{-(e1-e2)} \quad \text{and} \quad b^{-(e2-e1)}$$

This right shift can result in mantissa underflow. There are also possibilities for mantissa overflow. Addition and subtraction of exponents can lead to exponent underflow and overflow. To alleviate underflow and overflow, it is necessary to decide on some scheme for roundoff. For a detailed description and analysis of underflow and overflow conditions and rounding schemes, see reference 1.

It is desirable to have all numbers normalized, i.e., the mantissas of f_1 and f_2 have the most significant digit in the leftmost position. This provides the representation with the greatest accuracy possible for a fixed mantissa length. The result of any floating-point operation must also be normalized. The factors associated with normalization, overflow, and other characteristics of floating-point implementations are best illustrated with a few examples.

Consider the addition of two binary floating-point numbers f_1 and f_2 where

$$f_1 = 0.10100 \times 2^{011}$$
$$f_2 = 0.11100 \times 2^{001}$$

Both of these numbers are normalized, i.e., the first bit after the binary point is a 1. Addition requires equal exponents, so the fractions are aligned by shifting right the one with the smaller exponent and adjusting the smaller exponent. This yields

$$f_2 = 0.00111 \times 2^{011}$$

Then,

$$f_1 + f_2 = 0.10100 \times 2^{011} + 0.00111 \times 2^{011}$$
$$= 0.11011 \times 2^{011} = f_3$$

The sum may overflow the left end by one digit, thus requiring a postaddition adjustment or renormalization step. Since it is assumed that the register is only of a finite length, this renormalization will result in the loss of the lowest order bit.

Another example illustrates the overflow past the most significant bit. With an assumed register length of five, let

$$f_1 = 0.11100 \times 2^{011}$$
$$f_2 = 0.10101 \times 2^{001}$$

Then,

$$
\begin{array}{r}
0.11100 \times 2^{011} = f_1 \\
+ \quad 0.0010101 \times 2^{011} = f_2 \\
\hline
\underline{1}.000010\underline{1} \times 2^{011} = f_3
\end{array}
$$

The significance of the two digits underlined in the right part of the mantissa is suspect, since it is assumed that the corresponding bits of f_1 are zero. The left underlined digit is the overflow past the most significant bit. To finish the addition, f_3 is shifted to the right and the exponent adjusted accordingly. Thus,

$$1.0000101 \times 2^{011} = f_3$$

The shift of the fraction and the adjustment of the exponent yield

$$0.10000101 \times 2^{100} = f_3$$

The result may be rounded, giving

$$0.10001 \times 2^{100} = f_3$$

or truncated, giving

$$0.10000 \times 2^{100} = f_3$$

FLOATING-POINT ALGORITHMS

Multiplication Algorithm
The algorithm for normalized floating-point multiplication is illustrated in Figure 1. This algorithm is an implementation of Equation 2 in the section on floating-point notation.

The floating-point numbers being multiplied are A and B written as

$$A = m_A \times b^{e_A} \quad \text{and} \quad B = m_B \times b^{e_B}$$

The result is

$$C = m_C \times b^{e_C}$$

For the resulting m_C, there are three special cases. The m_C may be zero, in which case there is a branch to Step 10 to set $C = 0$. If $m_C \neq 0$, then the most significant bit will be in either the first or second leftmost bit. If the most significant bit is in the second leftmost bit, then a left shift of m_C is necessary (see Step 5). Otherwise, C is already in normalized form, and there is a branch to Step 6.

Step 6 implements the desired rounding scheme. After this rounding, it is possible that m_C will overflow (see Step 7). In this case, it is necessary to right-shift m_C one bit (see Step 8). Step 9 checks for special cases of e_C. If there is an overflow or underflow of e_C, it is handled in Step 10. Otherwise, the result is in range, and the calculation is complete.

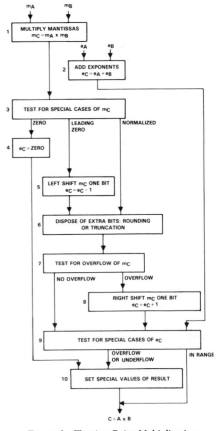

Figure 1. Floating-Point Multiplication

Addition Algorithm

The implementation of normalized floating-point addition is more involved than for multiplication. This addition algorithm, outlined in Figure 2, is an implementation of Equation 1 in the section on floating-point notation.

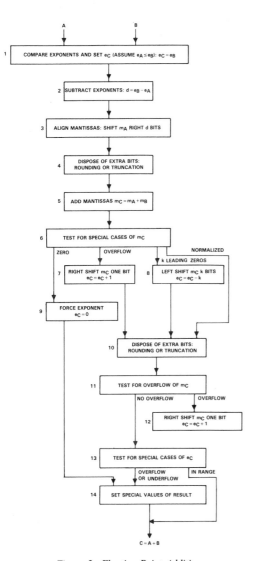

Figure 2. *Floating-Point Addition*

Step 1 compares e_A and e_B for determining e_C. For this illustration of the algorithm, it is assumed that $e_A \le e_B$. Step 2 determines the right shift (d) that is required to align m_A. Step 3 implements this right shift of m_A. Step 4 disposes of the extra bits of m_A by using the desired rounding technique. The mantissas of A and B are then added in Step 5.

Now, things become somewhat more involved. The m_C may be zero, in which case there is a branch to Step 9 which sets $e_C = 0$; a branch to Step 14 sets the special value of the result. The m_C may overflow, in which case a right shift of one is necessary (see Step 7). The m_C may have k leading zeroes, in which case a left shift of k is required. This normalization step is generally the most involved and time-consuming step to perform. Steps 10, 11, and 12 round m_C, test for a possible overflow due to the rounding, and adjust e_C accordingly. Step 13 involves the determination of the special case of e_C. Finally, after Step 14, the sum C = A + B is formed.

IEEE FLOATING-POINT SINGLE-PRECISION FORMAT

Of interest is a set of formats known as the IEEE standard. This IEEE recommended format consists of a variety of precision formats (single, double, single-extended, and double-extended). The IEEE has also proposed several techniques for handling special cases such as overflow, underflow, $\pm \infty$, and rounding. For complete details, the reader is referred to the proposed IEEE standard.[2]

The single-precision format is a 32-bit format consisting of a 1-bit sign field s, an 8-bit biased exponent e, and a 23-bit fraction f (see Figure 3). The value of a binary floating-point number X is determined as follows:

$$X = (-1)^s \times 2^{(e-127)} \times 1.f$$

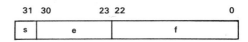

Figure 3. *IEEE Floating-Point Single-Precision Format*

The advantage of this format is that it is structured in such a way as to provide easy storage and straightforward input/output operations on 8-, 16- and 32-bit processors. The disadvantage with this format is that the large mantissa will generally span several words of memory.

FLOATING-POINT IMPLEMENTATION

IEEE Implementation

The IEEE single-precision format is described here as it applies to the addition and multiplication algorithms. In these floating-point routines written for the TMS32010, all results are truncated to 31 bits. This was done so that the user has more flexibility to develop a rounding scheme suitable for his application. The representations of $\pm \infty$ are ignored so that the user can decide how to handle these exceptions in a manner that is appropriate for his particular application.

I/O Considerations

The first consideration is the internal representation of the binary floating-point number. If the number is read into the TMS32010 as two 16-bit words, some processing is then necessary to put the floating-point number into a representation which is easier to process. The representation used in the TMS32010 programs in the Appendix is shown in Figure 4. This internal representation may be arrived at by a simple manipulation of the IEEE bit fields. For this particular algorithm, it is assumed that the floating-point number is input to the TMS32010 as the four 16-bit fields shown in Figure 4. However, the user can easily supply his own routine to arrive at this format from two 16-bit inputs to the TMS32010 where the inputs contain the IEEE single-precision format.

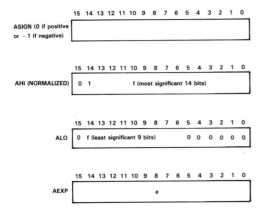

Figure 4. Floating-Point Representation

The format in Figure 4 was chosen to minimize the execution time of the floating-point addition and multiplication routines. The format of the result is shown in Figure 5. Notice that it is identical to the format in Figure 4 except for CLO. CLO has its 16 most significant bits valid for both the multiplication and addition routines.

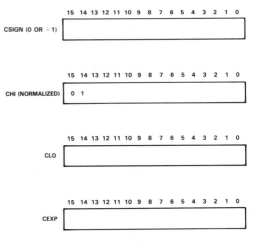

Figure 5. Result Representation

Normalization

Since the floating-point addition involves a normalization, a technique similar to a binary search algorithm is used in the addition routine in the Appendix. To begin the normalization routine, note that with the format used for the result (see Figure 5), all mantissas can be considered to be positive. The binary search for the most significant bit (the leftmost 1 since the mantissa is positive) is illustrated in Figure 6.

The first move is to split C into CHI and CLO. If $CHI \neq 0$, then the most significant bit (MSB) is in CHI; otherwise, it is in CLO. For this example, it is in CHI. The next step is to split CHI into C11 and C12. If $C11 \neq 0$, then the MSB is in the eight bits of C11; otherwise, it is in C12. For this example, the MSB is in C12. Next, split C12 into C23 and C24. Again, if $C24 \neq 0$, then the MSB is in C24; otherwise, it is in C23. Since $C24 \neq 0$, split C24 into C37 and C38. Since $C37 \neq 0$, the MSB is in C37. Finally, splitting C37, a simple bit test shows that the MSB is in position 19. Using this technique, it is possible to find the MSB in a 32-bit field with only five compares.

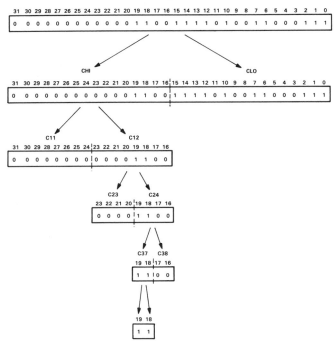

Figure 6. Binary Search

Added Precision

As illustrated in Figure 5, the 16 most significant bits of CLO are valid, i.e., C is valid for 31 places beyond the binary point. Oftentimes the user is not as concerned with the IEEE standard as in being certain that he has enough accuracy for his particular application. Since the TMS32010 uses 16-bit words, the routines in the Appendix implicitly maintain a 30-bit mantissa. They also implicitly use a 16-bit exponent. If the user desires this added accuracy and dynamic range, then it is readily implementable with no additional cost in execution time. The normalization for the addition, as mentioned previously, operates over the entire 32-bit accumulator. For the strict IEEE format, the user will only want to normalize over the 25 most significant bits of the accumulator. The structure of the normalization routine makes this modification simple.

The routines in the Appendix make no provision for the representations of $\pm \infty$ and exponent underflow and overflow. The user of the routines should consider the degree of significance of these results and the way they should be handled for his particular application. Since these routines are written to operate at maximum speed, truncation of results is used. If the user desires to implement a rounding scheme, then he will also need to check for the possibility of overflow due to the rounding scheme. This step is shown in the multiplication and addition flowcharts (see Figures 1 and 2).

SUMMARY

The TMS32010 may be used to perform floating-point operations with great accuracy, wide dynamic range, and high-speed execution. The design engineer has the responsibility of deciding what type of floating-point format is best for his application. To aid in understanding floating-point operations, several examples have been given illustrating the manipulations necessary to implement floating-point addition and multiplication algorithms. Flowcharts for these algorithms are also included. The Appendix contains the TMS32010 code for the IEEE floating-point single-precision format used in multiplication and addition. These same routines may also be used without modification to implement a format with up to a 30-bit mantissa and a 16-bit exponent without any increase in execution time.

REFERENCES

1. Kuck, D.J., THE STRUCTURE OF COMPUTERS AND COMPUTATIONS, Volume 1. New York: John Wiley & Sons, 1978.

2. Coonen, J. et al, "A Proposed Standard for Binary Floating-Point Arithmetic," ACM SIGNUM NEWS-LETTER, October, 1979, 4-12.

```
IEEEMULT      320 FAMILY MACRO ASSEMBLER  2.1 83.076      16:33:40   1/18/84
0001               *****************************************************************
0002               *
0003               *     THIS IS A FLOATING-POINT MULTIPLICATION ROUTINE WHICH
0004               *     IMPLEMENTS THE IEEE PROPOSED FLOATING-POINT FORMAT ON
0005               *     THE TMS32010.
0006               *
0007               *****************************************************************
0008               *
0009               *     INITIAL FORMAT (ALL 16 BIT WORDS)
0010               *     ------------------
0011               *     |   ALL 0 OR 1   |      ASIGN (0 OR -1)
0012               *     ------------------
0013               *
0014               *     ------------------
0015               *     |0|. 15 BITS     |      AHI (NORMALIZED)
0016               *     ------------------
0017               *
0018               *     ------------------
0019               *     |0| 9 BITS |--0--|      ALO
0020               *     ------------------
0021               *
0022               *     ------------------
0023               *     |                |      AEXP (-127 TO 128)
0024               *     ------------------
0025               *
0026               *     TO HAVE THIS CORRESPOND TO IEEE FORMAT, INPUT 0.1F *
0027               *     2 ** (E + 1) INSTEAD OF 1.F * 2 ** E AND SUBTRACT
0028               *     127 FROM E.
0029               *
0030               *     THE FINAL FORMAT IS THE SAME AS THE INITIAL FORMAT
0031               *     EXCEPT THAT FOR CLO WE HAVE:
0032               *
0033               *     ------------------
0034               *     |    16 BITS     |      CLO
0035               *     ------------------
0036               *
0037               *     THE 16 BITS OF CLO ARE VALID.  ANYTHING PAST THESE
0038               *     HAS BEEN TRUNCATED.
0039               *
0040               *****************************************************************
0041               *
0042               *     WORST CASE (EXCLUDING INITIALIZATION AND I/O):
0043               *     8.4 MICROSECONDS
0044               *     WORDS OF PROGRAM MEMORY: 72
0045               *
0046               *****************************************************************
0047               *
0048                     IDT 'IEEEMULT'
0049 0000              DSEG
0050               *
0051 0000       AEXP BSS 1
0052 0001       AHI BSS 1
0053 0002       ALO BSS 1
0054 0003       ASIGN BSS 1
0055 0004       BEXP BSS 1
```

```
0056 0005          BHI BSS 1
0057 0006          BLO BSS 1
0058 0007          BSIGN BSS 1
0059 0008          CEXP BSS 1
0060 0009          CHI BSS 1
0061 000A          CLO BSS 1
0062 000B          CSIGN BSS 1
0063 000C          TLO BSS 1
0064 000D          THI BSS 1
0065 000E          CLOHI BSS 1
0066 000F          M0003 BSS 1
0067 0010          ONE BSS 1
0068 0011          NEGONE BSS 1
0069 0012          TWO BSS 1
0070          *
0071 0013          DEND
0072          *
0073 0000          PSEG      * BEGIN THE PROGRAM SEGMENT *
0074          *
0075 0000 4003          IN ASIGN,PA0      * INPUT *
0076 0001 4000          IN AEXP,PA0
0077 0002 4001          IN AHI,PA0
0078 0003 4002          IN ALO,PA0
0079 0004 4007          IN BSIGN,PA0
0080 0005 4004          IN BEXP,PA0
0081 0006 4005          IN BHI,PA0
0082 0007 4006          IN BLO,PA0
0083          *          * FINISHED THE INPUT ROUTINE *
0084          *
0085          *                    * A LITTLE INITIALIZATION *
0086 0008 6E00 START    LDPK 0
0087 0009 7E01          LACK 1
0088 000A 5010          SACL ONE
0089 000B 7E03          LACK 3
0090 000C 500F          SACL M0003
0091 000D 7F89          ZAC
0092 000E 1010          SUB ONE
0093 000F 5011          SACL NEGONE
0094 0010 7E02          LACK 2
0095 0011 5012          SACL TWO
0096          *          * DONE WITH THE INITIALIZATION *
0097          *          * ADD EXPONENTS *
0098 0012 2000          LAC AEXP
0099 0013 0004          ADD BEXP
0100 0014 5008          SACL CEXP      * CEXP = AEXP + BEXP *
0101          *          * FINISHED ADDING EXPONENTS *
0102          *          * MULTIPLY MANTISSAS *
0103 0015 6A02          LT ALO      * FIRST PRODUCT, (ALO * BHI) *
0104 0016 6D05          MPY BHI
0105 0017 7F8E          PAC
0106 0018 580D          SACH THI
0107 0019 500C          SACL TLO
0108          *
0109 001A 6A01          LT AHI      * SECOND PRODUCT, (AHI * BLO) *
0110 001B 6D06          MPY BLO
0111          *
0112 001C 7F8F          APAC      * (ALO * BHI + AHI * BLO) *
```

```
0113                    *
0114 001D 7F8F          APAC        * HAS THE EFFECT OF
0115                    *             (AHI * BLO + ALO * BHI)
0116                    *             * 2 ** -15 *
0117 001E 600D          ADDH THI
0118 001F 610C          ADDS TLO
0119 0020 580D          SACH THI
0120                    *
0121 0021 6D05          MPY BHI     * (AHI * BHI) *
0122 0022 7F8E          PAC
0123 0023 610D          ADDS THI
0124                    *
0125 0024 5909          SACH CHI,1     * GET RID OF EXTRA SIGN BIT *
0126 0025 500A          SACL CLO
0127                    *           * THE (ALO * BLO * 2 ** -30) IS LOST DUE
0128                    *             TO THE IEEE FORMAT *
0129                    *
0130                    *           * FINISHED MULTIPLYING THE MANTISSAS *
0131                    *
0132                    *           * CHECK SPECIAL CASES AND WRAP THINGS UP
0133 0026 FE00          BNZ OK      * CHI AND CLO ARE STILL IN THE ACC *
     0027 002C'
0134                    *
0135 0028 7F89          ZAC         * IF C IS ZERO LOAD CEXP WITH ZERO *
0136 0029 5008          SACL CEXP
0137 002A F900          B SETSIN    * BRANCH TO SET THE SIGN *
     002B 003A'
0138                    *
0139 002C 210A  OK      LAC CLO,1      * TAKING CARE OF EXTRA SIGN BIT
0140                    *                AS ABOVE *
0141 002D 500A          SACL CLO
0142 002E 2E10          LAC ONE,14     * MASK OFF POSSIBLE MSB *
0143 002F 7909          AND CHI
0144 0030 FE00          BNZ SETSIN     * BRANCH IF NORMALIZATION NOT
     0031 003A'
0145                    *                NECESSARY *
0146                    *
0147 0032 2008  SHIFT1  LAC CEXP       * HERE A LEFT SHIFT OF ONE IS
0148                    *                NECESSARY *
0149 0033 1010          SUB ONE
0150 0034 5008          SACL CEXP
0151                    *
0152 0035 6509          ZALH CHI
0153 0036 610A          ADDS CLO
0154 0037 5909          SACH CHI,1
0155 0038 210A          LAC CLO,1
0156 0039 500A          SACL CLO
0157                    *
0158 003A 6603  SETSIN  ZALS ASIGN
0159 003B 7807          XOR BSIGN
0160 003C FE00          BNZ NEG        * IF ASIGN XOR BSIGN != 0
     003D 0042'
0161                    *                THE PRODUCT IS NEGATIVE *
0162                    *
0163 003E 7F89          ZAC
0164 003F 500B          SACL CSIGN
0165 0040 F900          B OUTPUT
```

```
       0041 0044'
0166                    *
0167 0042 2011   NEG        LAC NEGONE     * NEGONE = -1 *
0168                    *
0169 0043 500B             SACL CSIGN
0170                    *
0171 0044 490B   OUTPUT     OUT CSIGN,PA1
0172 0045 4908             OUT CEXP,PA1
0173 0046 4909             OUT CHI,PA1
0174 0047 490A             OUT CLO,PA1
0175                    *
0176 0048 F900   SELF       B SELF
       0049 0048'
0177                    *
0178                    *            * END THE PROGRAM SEGMENT *
0179                                 END
NO ERRORS, NO WARNINGS
```

```
0001              ********************************************************
0002         *
0003         *    THIS IS A FLOATING POINT ADDITION ROUTINE WHICH
0004         *    IMPLEMENTS THE IEEE PROPOSED FLOATING POINT FORMAT
0005         *    ON THE TMS32010.
0006         *
0007              ********************************************************
0008         *
0009         *    INITIAL FORMAT (ALL 16 BIT WORDS)
0010         *    ------------------
0011         *    | ALL 0 OR ALL 1 |     ASIGN (0 OR -1)
0012         *    ------------------
0013         *
0014         *    ------------------
0015         *    |0|.  15 BITS   |     AHI (NORMALIZED)
0016         *    ------------------
0017         *
0018         *    ------------------
0019         *    |0|  9 BITS |--0-|     ALO
0020         *    ------------------
0021         *
0022         *    ------------------
0023         *    |                |     AEXP (-127 TO 128)
0024         *    ------------------
0025         *
0026         *    TO HAVE THIS CORRESPOND TO IEEE FORMAT, INPUT
0027         *    0.1F * 2 ** (E + 1)
0028         *    INSTEAD OF 1.F * 2 ** E AND SUBTRACT 127 FROM E.
0029         *
0030         *    THE FINAL FORMAT IS THE SAME AS THE INITIAL FORMAT
0031         *    EXCEPT THAT FOR CLO WE HAVE:
0032         *
0033         *    ------------------
0034         *    |  16 BITS  |     CLO
0035         *    ------------------
0036         *
0037         *    ALL 16 BITS OF CLO ARE VALID.  ANYTHING PAST THESE
0038         *    HAS BEEN TRUNCATED.
0039         *
0040              ********************************************************
0041         *
0042         *    WORST CASE (EXCLUDING INITIALIZATION AND I/O):
0043         *    17.2 MICROSECONDS.
0044         *    THIS FIGURE INCLUDES THE NORMALIZATION.
0045         *    WORDS OF PROGRAM MEMORY: 768
0046         *
0047              ********************************************************
0048         *
0049              IDT 'IEEEADD'
0050 0000         DORG 0
0051         *
0052 0000    AEXP BSS 1
0053 0001    AHI BSS 1
0054 0002    ALO BSS 1
```

6. Floating-Point Arithmetic with the TMS32010

```
0055 0003        ASIGN BSS 1
0056 0004        BEXP BSS 1
0057 0005        BHI BSS 1
0058 0006        BLO BSS 1
0059 0007        BSIGN BSS 1
0060 0008        CEXP BSS 1
0061 0009        CHI BSS 1
0062 000A        CLO BSS 1
0063 000B        CSIGN BSS 1
0064 000C        C11 BSS 1
0065 000D        C12 BSS 1
0066 000E        C21 BSS 1
0067 000F        C22 BSS 1
0068 0010        C31 BSS 1
0069 0011        C32 BSS 1
0070 0012        C33 BSS 1
0071 0013        C34 BSS 1
0072 0014        C23 BSS 1
0073 0015        C24 BSS 1
0074 0016        C35 BSS 1
0075 0017        C36 BSS 1
0076 0018        C37 BSS 1
0077 0019        C38 BSS 1
0078 001A        CTEMP BSS 1
0079 001B        D BSS 1
0080 001C        SHIFT1 BSS 1
0081 001D        SHIFT2 BSS 1
0082 001E        AHITL BSS 1
0083 001F        BHITL BSS 1
0084 0020        N1 BSS 1
0085 0021        N2 BSS 1
0086 0022        N3 BSS 1
0087 0023        N4 BSS 1
0088 0024        N5 BSS 1
0089 0025        N6 BSS 1
0090 0026        N7 BSS 1
0091 0027        N8 BSS 1
0092 0028        N9 BSS 1
0093 0029        N10 BSS 1
0094 002A        N11 BSS 1
0095 002B        N12 BSS 1
0096 002C        N13 BSS 1
0097 002D        N14 BSS 1
0098 002E        N15 BSS 1
0099 002F        N16 BSS 1
0100 0030        N17 BSS 1
0101 0031        N18 BSS 1
0102 0032        N19 BSS 1
0103 0033        N20 BSS 1
0104 0034        N21 BSS 1
0105 0035        N22 BSS 1
0106 0036        N23 BSS 1
0107 0037        N24 BSS 1
0108 0038        N25 BSS 1
0109 0039        N26 BSS 1
0110 003A        N27 BSS 1
```

6. Floating-Point Arithmetic with the TMS32010

```
0111 003B        N28 BSS 1
0112 003C        N29 BSS 1
0113 003D        N30 BSS 1
0114 003E        ONE BSS 1
0115 003F        TWO BSS 1
0116 0040        FIFTEN BSS 1
0117 0041        SIXTEN BSS 1
0118 0042        M0002 BSS 1
0119 0043        M0003 BSS 1
0120 0044        M000F BSS 1
0121 0045        M001F BSS 1
0122 0046        M007F BSS 1
0123 0047        M00FF BSS 1
0124 0048        M01FF BSS 1
0125 0049        M03FF BSS 1
0126 004A        M07FF BSS 1
0127 004B        M0FFF BSS 1
0128 004C        M1FFF BSS 1
0129 004D        M3FFF BSS 1
0130 004E        M7FFF BSS 1
0131 004F        M8000 BSS 1
0132 0050        OS BSS 1
0133             *
0134             *
0135             *      BEGIN THE PROGRAM SEGMENT *
0136             *
0137 0000              AORG >0
0138             *
0139 0000 F900         B START
     0001 0011
0140             *
0141 0002 0002  SHIFTS  DATA 2
0142 0003 0004          DATA 4
0143 0004 0008          DATA 8
0144 0005 0010          DATA 16
0145 0006 0020          DATA 32
0146 0007 0040          DATA 64
0147 0008 0080          DATA 128
0148 0009 0100          DATA 256
0149 000A 0200          DATA 512
0150 000B 0400          DATA 1024
0151 000C 0800          DATA 2048
0152 000D 1000          DATA 4096
0153 000E 2000          DATA 8192
0154 000F 4000          DATA 16384
0155 0010 8000          DATA 32768
0156             *
0157             *              * A LITTLE INITIALIZATION *
0158 0011 6E00  START   LDPK 0
0159 0012 7E01          LACK 1
0160 0013 503E          SACL ONE
0161 0014 7020          LARK AR0,N1
0162 0015 711D          LARK AR1,29
0163 0016 7F89          ZAC
0164 0017 103E  LOOP    SUB ONE
0165 0018 6880          LARP AR0
```

6. Floating-Point Arithmetic with the TMS32010

```
0166 0019 50A1          SACL *+,0,AR1
0167 001A F400          BANZ LOOP
     001B 0017
0168 001C 7E02          LACK 2
0169 001D 5042          SACL M0002
0170 001E 503F          SACL TWO
0171 001F 7E03          LACK 3
0172 0020 5043          SACL M0003
0173 0021 7E0F          LACK 15
0174 0022 5044          SACL M000F
0175 0023 5040          SACL FIFTEN
0176 0024 7E10          LACK 16
0177 0025 5041          SACL SIXTEN
0178 0026 7E1F          LACK 31
0179 0027 5045          SACL M001F
0180 0028 7E7F          LACK 127
0181 0029 5046          SACL M007F
0182 002A 2444          LAC M000F,4
0183 002B 0044          ADD M000F
0184 002C 5047          SACL M00FF
0185 002D 083E          ADD ONE,8
0186 002E 5048          SACL M01FF
0187 002F 093E          ADD ONE,9
0188 0030 5049          SACL M03FF
0189 0031 0A3E          ADD ONE,10
0190 0032 504A          SACL M07FF
0191 0033 0B3E          ADD ONE,11
0192 0034 504B          SACL M0FFF
0193 0035 0C3E          ADD ONE,12
0194 0036 504C          SACL M1FFF
0195 0037 0D3E          ADD ONE,13
0196 0038 504D          SACL M3FFF
0197 0039 0E3E          ADD ONE,14
0198 003A 504E          SACL M7FFF
0199 003B 2F3E          LAC ONE,15
0200 003C 504F          SACL M8000
0201 003D 6A3E          LT ONE
0202 003E 8002          MPYK SHIFTS
0203 003F 7F8E          PAC
0204 0040 5050          SACL OS
0205           *
0206           *                        * FINISHED INITIALIZATION *
0207           *
0208 0041 4003          IN ASIGN,PA0
0209 0042 4000          IN AEXP,PA0
0210 0043 4001          IN AHI,PA0
0211 0044 4002          IN ALO,PA0
0212 0045 4007          IN BSIGN,PA0
0213 0046 4004          IN BEXP,PA0
0214 0047 4005          IN BHI,PA0
0215 0048 4006          IN BLO,PA0
0216           *
0217 0049 2000          LAC AEXP
0218 004A 1004          SUB BEXP
0219 004B FA00          BLZ ALTB       * BRANCH IF AEXP < BEXP *
     004C 0091
```

```
0220                *
0221 004D FF00          BZ AEQB        * BRANCH IF AEXP = BEXP *
     004E 00E8
0222                *
0223 004F 501B          SACL D         * D IS THE RIGHT SHIFT NEEDED FOR B *
0224                *
0225 0050 2000          LAC AEXP
0226 0051 5008          SACL CEXP      * THE EXP FOR THE RESULT IS AEXP *
0227                *
0228 0052 2040   AGTB   LAC FIFTEN     * A > B SO SHIFT B TO THE RIGHT D *
0229 0053 101B          SUB D
0230 0054 FB00          BLEZ BHIZER    * IF D >= 15 BHI IS ZERO *
     0055 0069
0231                *
0232 0056 0050          ADD OS         ------------------
0233 0057 671C          TBLR SHIFT1    |0|   15 BITS   |   BHI
0234                *                  ------------------
0235                *                  ------------------
0236                *                  |0| 8 BITS |   |   BLO
0237                *                  ------------------
0238 0058 6A05          LT BHI                    || IS
0239 0059 6D1C          MPY SHIFT1                || CHANGED
0240 005A 7F8E          PAC                       VV TO
0241 005B 7F88          ABS                       VV
0242 005C 5805          SACH BHI                  VV
0243 005D 501F          SACL BHITL     ------------------
0244                *                  | -0- | 15-D  |   BHI
0245 005E 6A06          LT BLO         ------------------
0246 005F 6D1C          MPY SHIFT1     ------------------
0247 0060 7F8E          PAC            | D |   16-D   |   BLO
0248 0061 7F8F          APAC           ------------------
0249 0062 7F88          ABS
0250 0063 601F          ADDH BHITL
0251 0064 5806          SACH BLO
0252 0065 2102          LAC ALO,1
0253 0066 5002          SACL ALO
0254                *
0255 0067 F900          B DUN          * FINISHED SHIFT OF B BY D TO THE RIGHT
     0068 00EE
0256                *
0257 9069 003E   BHIZER ADD ONE        * IF D >= 16 BLO LOSES BITS *
0258 006A FB00          BLEZ BLOLUZ
     006B 0077
0259                *                  * WE ONLY GET HERE IF D = 15 *
0260                *                  ------------------
0261                *                  |0|   15 BITS   |   BHI
0262 006C 2206          LAC BLO,2      ------------------
0263 006D 5806          SACH BLO       ------------------
0264 006E 6606          ZALS BLO       |0| 8 BITS |   |   BLO
0265 006F 0105          ADD BHI,1      ------------------
0266 0070 5006          SACL BLO                 ||
0267 0071 7F89          ZAC                       VV
0268 0072 5005          SACL BHI       ------------------
0269                *                  |        -0-     |   BHI
0270                *                  ------------------
```

6. Floating-Point Arithmetic with the TMS32010

```
0271                *                      -------------------
0272                *                      ¦   15 BITS   ¦?¦  BLO
0273                *                      -------------------
0274 0073 2102               LAC ALO,1
0275 0074 5002               SACL ALO
0276                *                      * FINISHED SHIFT OF B
0277 0075 F900               B DUN           BY D TO THE RIGHT *
     0076 00EE
0278                *
0279 0077 FF00      BLOLUZ   BZ SHB16
     0078 0089
0280 0079 0040               ADD FIFTEN
0281 007A FB00               BLEZ BZERO     IF D >= 31, THEN B IS ZERO.
     007B 00DE
0282                *                      -------------------
0283 007C 0050               ADD OS         ¦0¦  15 BITS    ¦  BHI
0284 007D 671C               TBLR SHIFT1    -------------------
0285                *                      -------------------
0286 007E 6A05               LT BHI         ¦0¦ 8 BITS ¦    ¦  BLO
0287 007F 6D1C               MPY SHIFT1     -------------------
0288 0080 7F8E               PAC                    ¦¦
0289 0081 7F88               ABS                    VV
0290 0082 5806               SACH BLO               VV
0291 0083 7F89               ZAC            -------------------
0292 0084 5005               SACL BHI       ¦0¦  ---0---     ¦  BHI
0293                *                      -------------------
0294                *                      -------------------
0295                *                      ¦ --0-- ¦ 31 - D¦  BLO
0296                *                      -------------------
0297 0085 2102               LAC ALO,1
0298 0086 5002               SACL ALO
0299                *
0300 0087 F900               B DUN
     0088 00EE
0301                *
0302 0089 2005      SHB16    LAC BHI
0303 008A 5006               SACL BLO
0304 008B 7F89               ZAC
0305 008C 5005               SACL BHI
0306                *
0307 008D 2102               LAC ALO,1
0308 008E 5002               SACL ALO
0309                *
0310 008F F900               B DUN
     0090 00EE
0311                *
0312 0091 7F88      ALTB     ABS            * TO SEE WHAT IS GOING ON LOOK AT
0313                *                         THE PREVIOUS CASE FOR AGTB *
0314 0092 501B               SACL D
0315                *
0316 0093 2004               LAC BEXP
0317 0094 5008               SACL CEXP
0318                *
0319 0095 2040               LAC FIFTEN
0320 0096 101B               SUB D
0321 0097 FB00               BLEZ AHIZER
```

6. Floating-Point Arithmetic with the TMS32010

```
          0098 00AC
0322                        *
0323 0099 0050              ADD OS
0324 009A 671C              TBLR SHIFT1
0325                        *
0326 009B 6A01              LT AHI
0327 009C 6D1C              MPY SHIFT1
0328 009D 7F8E              PAC
0329 009E 7F88              ABS
0330 009F 5801              SACH AHI
0331 00A0 501E              SACL AHITL
0332                        *
0333 00A1 6A02              LT ALO
0334 00A2 6D1C              MPY SHIFT1
0335 00A3 7F8E              PAC
0336 00A4 7F8F              APAC
0337 00A5 7F88              ABS
0338 00A6 601E              ADDH AHITL
0339 00A7 5802              SACH ALO
0340 00A8 2106              LAC BLO,1
0341 00A9 5006              SACL BLO
0342                        *
0343 00AA F900              B DUN
     00AB 00EE
0344                        *
0345 00AC 003E    AHIZER    ADD ONE
0346 00AD FB00              BLEZ ALOLUZ
     00AE 00BA
0347                        *
0348 00AF 2202              LAC ALO,2
0349 00B0 5802              SACH ALO
0350 00B1 2002              LAC ALO
0351 00B2 0101              ADD AHI,1
0352 00B3 5002              SACL ALO
0353 00B4 7F89              ZAC
0354 00B5 5001              SACL AHI
0355 00B6 2106              LAC BLO,1
0356 00B7 5006              SACL BLO
0357                        *
0358 00B8 F900              B DUN
     00B9 00EE
0359                        *
0360 00BA FF00    ALOLUZ    BZ SHA16
     00BB 00CC
0361 00BC 0040              ADD FIFTEN
0362 00BD FB00              BLEZ AZERO
     00BE 00D4
0363                        *
0364 00BF 0050              ADD OS
0365 00C0 671C              TBLR SHIFT1
0366                        *
0367 00C1 6A01              LT AHI
0368 00C2 6D1C              MPY SHIFT1
0369 00C3 7F8E              PAC
0370 00C4 7F88              ABS
0371 00C5 5802              SACH ALO
```

6. Floating-Point Arithmetic with the TMS32010

```
0372 00C6 7F89          ZAC
0373 00C7 5001          SACL AHI
0374 00C8 2106          LAC BLO,1
0375 00C9 5006          SACL BLO
0376            *
0377 00CA F900          B DUN
     00CB 00EE
0378            *
0379 00CC 2001  SHA16   LAC AHI
0380 00CD 5002          SACL ALO
0381 00CE 7F89          ZAC
0382 00CF 5001          SACL AHI
0383            *
0384 00D0 2106          LAC BLO,1
0385 00D1 5006          SACL BLO
0386            *
0387 00D2 F900          B DUN
     00D3 00EE
0388            *
0389 00D4 2005  AZERO   LAC BHI
0390 00D5 5009          SACL CHI
0391 00D6 2004          LAC BEXP
0392 00D7 5008          SACL CEXP
0393 00D8 2106          LAC BLO,1
0394 00D9 500A          SACL CLO
0395 00DA 2007          LAC BSIGN
0396 00DB 500B          SACL CSIGN
0397 00DC F900          B OUTPUT
     00DD 02FB
0398            *
0399 00DE 2001  BZERO   LAC AHI
0400 00DF 5009          SACL CHI
0401 00E0 2000          LAC AEXP
0402 00E1 5008          SACL CEXP
0403 00E2 2102          LAC ALO,1
0404 00E3 500A          SACL CLO
0405 00E4 2003          LAC ASIGN
0406 00E5 500B          SACL CSIGN
0407 00E6 F900          B OUTPUT
     00E7 02FB
0408            *
0409 00E8 2102  AEQB    LAC ALO,1
0410 00E9 5002          SACL ALO
0411 00EA 2106          LAC BLO,1
0412 00EB 5006          SACL BLO
0413 00EC 2000          LAC AEXP
0414 00ED 5008          SACL CEXP
0415            *
0416            *                    * GO TO DUN *
0417            *
0418 00EE 2003  DUN     LAC ASIGN
0419 00EF 1007          SUB BSIGN
0420 00F0 FE00          BNZ DIFSIN    * BRANCH IF THERE IS A
     00F1 0113                          SIGN DIFFERENCE *
0421            *
0422 00F2 6501          ZALH AHI
```

6. Floating-Point Arithmetic with the TMS32010

```
0423 00F3 6102          ADDS ALO
0424 00F4 6106          ADDS BLO
0425 00F5 6005          ADDH BHI        ------------------
0426            *                       |0|  15 BITS  |   CHI
0427 00F6 5809          SACH CHI        ------------------
0428 00F7 500A          SACL CLO        ------------------
0429 00F8 FF00          BZ CZERO        |     16 BITS  |  CLO
     00F9 0131                          ------------------
0431            *
0432 00FA FD00          BGEZ NOOV       * CHECKING FOR OVERFLOW DUE TO ADD *
     00FB 010F
0433            *
0434 00FC 2F09          LAC CHI,15      * SHIFT TO RIGHT ONE TO CANCEL OVERFLOW
0435 00FD 5809          SACH CHI
0436 00FE 501A          SACL CTEMP
0437            *
0438 00FF 2009          LAC CHI
0439 0100 794E          AND M7FFF       * CANCEL SIGN EXTENSION *
0440 0101 5009          SACL CHI
0441            *
0442 0102 2F0A          LAC CLO,15
0443 0103 580A          SACH CLO
0444 0104 200A          LAC CLO
0445 0105 794E          AND M7FFF       * CANCEL SIGN EXTENSION *
0446 0106 611A          ADDS CTEMP
0447 0107 500A          SACL CLO
0448            *
0449 0108 2008          LAC CEXP
0450 0109 003E          ADD ONE         * DUE TO RIGHT SHIFT *
0451 010A 5008          SACL CEXP
0452            *
0453 010B 2003          LAC ASIGN
0454 010C 500B          SACL CSIGN
0455            *
0456 010D F900          B OUTPUT        * FINISHED RIGHT SHIFT *
     010E 02FB
0457            *
0458 010F 2003  NOOV    LAC ASIGN
0459 0110 500B          SACL CSIGN
0460            *
0461 0111 F900          B NORM
     0112 013D
0462            *
0463 0113 FA00  DIFSIN  BLZ CHAS        * IF < 0 DO (B - A) *
     0114 0123
0464            *
0465 0115 6501  CHBS    ZALH AHI        * DO (|A| - |B|) SINCE B < 0 AND A > 0
0466 0116 6102          ADDS ALO
0467 0117 6306          SUBS BLO
0468 0118 6205          SUBH BHI
0469            *
0470 0119 FF00          BZ CZERO
     011A 0131
0471 011B FA00          BLZ CNEG
     011C 0138
0472            *
```

6. Floating-Point Arithmetic with the TMS32010

```
0473 011D 5809          SACH CHI
0474 011E 500A          SACL CLO
0475           *
0476 011F 7F89          ZAC
0477 0120 500B          SACL CSIGN
0478           *
0479 0121 F900          B NORM
     0122 013D
0480           *
0481 0123 6505  CHAS    ZALH BHI        * DO (|B| - |A|) SINCE A < 0 AND B > 0
0482 0124 6106          ADDS BLO
0483 0125 6302          SUBS ALO
0484 0126 6201          SUBH AHI
0485           *
0486 0127 FF00          BZ CZERO
     0128 0131
0487 0129 FA00          BLZ CNEG
     012A 0138
0488           *
0489 012B 5809          SACH CHI
0490 012C 500A          SACL CLO
0491           *
0492 012D 7F89          ZAC
0493 012E 500B          SACL CSIGN
0494           *
0495 012F F900          B NORM
     0130 013D
0496           *
0497 0131 7F89  CZERO   ZAC
0498 0132 5008          SACL CEXP
0499 0133 5009          SACL CHI
0500 0134 500A          SACL CLO
0501 0135 500B          SACL CSIGN
0502 0136 F900          B OUTPUT
     0137 02FB
0503           *
0504 0138 7F88  CNEG    ABS
0505 0139 5809          SACH CHI
0506 013A 500A          SACL CLO
0507 013B 2020          LAC N1
0508 013C 500B          SACL CSIGN
0509           *
0510           *                    * GO TO NORM *
0511           *
0512           *
0513           *************************************************************
0514           *
0515           *    NORM DOES THE NORMALIZATION.  THAT IS TO SAY THAT IT
0516           *    FINDS THE MSB OF CHI AND CLO.  IN THIS CASE THE MSB WILL
0517           *    BE THE FIRST ONE (1) FOUND.  THE SEARCH FOR THIS SPECIAL
0518           *    ONE IS DONE WITH A BINARY SEARCH. THE NOTATION USED
0519           *    IS SUMMARIZED HERE:
0520           *
0521           *
0522           *    |------------------- 16 BITS -------------------|
0523           *
```

6. Floating-Point Arithmetic with the TMS32010

233

```
0524       *     ------------------------------------------------
0525       *     |                 CHI OR CLO                |   CHI OR
0526       *     ------------------------------------------------ CLO
0527       *     ------------------------------------------------ 8
0528       *     |         C11           I           C12     |   BITS
0529       *     ------------------------------------------------ EACH
0530       *     ------------------------------------------------ 4
0531       *     |   C21   I   C22   |   C23   I   C24   |   BITS
0532       *     ------------------------------------------------ EACH
0533       *     ------------------------------------------------ 2
0534       *     | C31 I C32 | C33 I C34 | C35 I C36 | C37 I C38 |  BITS
0535       *     ------------------------------------------------ EACH
0536       *     ------------------------------------------------ 1
0537       *     | | | | | | | | | | | | | | | | |   BIT
0538       *     ------------------------------------------------ EACH
0539       *
0540             THE I'S REPRESENT THE BOUNDARY BETWEEN THE ACC HIGH BIT
0541       *     AND THE ACC LOW BITS WHEN THE HIGHER LEVEL IS LOADED INTO
0542       *     THE ACC WITH THE NECESSARY SHIFT TO SPLIT BY A FACTOR OF
0543       *     TWO.
0544       *     WORST CASE: 6.4 MICROSECONDS.
0545       *
0546       ***********************************************************
0547 013D 2809  NORM    LAC CHI,8
0548 013E FF00          BZ CLOB        * BRANCH IF MSB IS IN CLO *
     013F 0228
0549       *
0550 0140 580C          SACH C11       * SPLIT THE 16 BIT WORD INTO TWO
0551 0141 500D          SACL C12         8 BIT WORDS *
0552       *
0553 0142 2C0C          LAC C11,12
0554 0143 FF00          BZ C12B        * IF THERE IS A ONE IN C11 IT WILL
     0144 01A9
0555       *                           BE NONZERO *
0556 0145 580E          SACH C21
0557 0146 500F          SACL C22
0558       *
0559 0147 2E0E          LAC C21,14
0560 0148 FF00          BZ C22B        * IF THERE IS A ONE ETC. *
     0149 016C
0561       *
0562 014A 5810          SACH C31
0563 014B 5011          SACL C32
0564       *
0565 014C 2010          LAC C31
0566 014D FF00          BZ C32B        * IF THERE IS A ONE ETC. *
     014E 0151
0567       *
0568 014F F900          B OUTPUT       * THE MSB CANNOT BE IN BIT ONE BECAUSE
     0150 02FB
0569       *                           THIS WAS HANDLED EARLIER *
0570 0151 2011  C32B    LAC C32
0571 0152 794F          AND M8000
0572 0153 FF00          BZ MSB4        * MSB# MEANS THE MSB IS IN BIT # FROM
     0154 0160
0573       *                           LEFT TO RIGHT *
```

6. Floating-Point Arithmetic with the TMS32010

```
0574              *
0575 0155 2020            LAC N1          * LEFT SHIFT OF 1 *
0576 0156 0008            ADD CEXP
0577 0157 5008            SACL CEXP
0578              *
0579 0158 6509            ZALH CHI
0580 0159 610A            ADDS CLO
0581 015A 6009            ADDH CHI
0582 015B 610A            ADDS CLO
0583 015C 5809            SACH CHI
0584 015D 500A            SACL CLO
0585              *
0586 015E F900            B OUTPUT
     015F 02FB
0587              *
0588 0160 2021   MSB4     LAC N2          * LEFT SHIFT OF 2 *
0589 0161 0008            ADD CEXP
0590 0162 5008            SACL CEXP
0591              *
0592 0163 220A            LAC CLO,2
0593 0164 500A            SACL CLO
0594 0165 581A            SACH CTEMP
0595 0166 201A            LAC CTEMP
0596 0167 7943            AND M0003
0597 0168 0209            ADD CHI,2
0598 0169 5009            SACL CHI
0599              *
0600 016A F900            B OUTPUT
     016B 02FB
0601              *
0602 016C 220F   C22B     LAC C22,2
0603 016D 5812            SACH C33
0604 016E 5013            SACL C34
0605              *
0606 016F 2012            LAC C33
0607 0170 FF00            BZ C34B         * BRANCH IF NO ONE *
     0171 018D
0608              *
0609 0172 7942            AND M0002
0610 0173 FF00            BZ MSB6
     0174 0181
0611              *
0612 0175 2022            LAC N3          * LEFT SHIFT OF THREE *
0613 0176 0008            ADD CEXP
0614 0177 5008            SACL CEXP
0615 0178 230A            LAC CLO,3
0616 0179 500A            SACL CLO
0617 017A 581A            SACH CTEMP
0618 017B 201A            LAC CTEMP
0619 017C 7944            AND M000F
0620 017D 0309            ADD CHI,3
0621 017E 5009            SACL CHI
0622              *
0623 017F F900            B OUTPUT
     0180 02FB
0624              *
```

6. Floating-Point Arithmetic with the TMS32010

```
0625 0181 2023  MSB6    LAC N4          * LEFT SHIFT OF 4 *
0626 0182 0008          ADD CEXP
0627 0183 5008          SACL CEXP
0628              *
0629 0184 240A          LAC CLO,4
0630 0185 500A          SACL CLO
0631 0186 581A          SACH CTEMP
0632 0187 201A          LAC CTEMP
0633 0188 7944          AND M000F
0634 0189 0409          ADD CHI,4
0635 018A 5009          SACL CHI
0636              *
0637 018B F900          B OUTPUT
     018C 02FB
0638              *
0639 018D 2013  C34B    LAC C34
0640              *
0641 018E 794F          AND M8000
0642 018F FF00          BZ MSB8
     0190 019D
0643              *
0644 0191 2024          LAC N5          * LEFT SHIFT OF 5 *
0645 0192 0008          ADD CEXP
0646 0193 5008          SACL CEXP
0647              *
0648 0194 250A          LAC CLO,5
0649 0195 500A          SACL CLO
0650 0196 581A          SACH CTEMP
0651 0197 201A          LAC CTEMP
0652 0198 7945          AND M001F
0653 0199 0509          ADD CHI,5
0654 019A 5009          SACL CHI
0655              *
0656 019B F900          B OUTPUT
     019C 02FB
0657              *
0658 019D 2025  MSB8    LAC N6          * LEFT SHIFT OF 6 *
0659 019E 0008          ADD CEXP
0660 019F 5008          SACL CEXP
0661              *
0662 01A0 260A          LAC CLO,6
0663 01A1 500A          SACL CLO
0664 01A2 581A          SACH CTEMP
0665 01A3 201A          LAC CTEMP
0666 01A4 7945          AND M001F
0667 01A5 0609          ADD CHI,6
0668 01A6 5009          SACL CHI
0669              *
0670 01A7 F900          B OUTPUT        * ^^ COMPLETES TOP 8 BITS ^^ *
     01A8 02FB
0671              *
0672 01A9 240D  C12B    LAC C12,4
0673 01AA 5814          SACH C23
0674 01AB 5015          SACL C24
0675              *
0676 01AC 2E14          LAC C23,14
```

6. Floating-Point Arithmetic with the TMS32010

```
0677 01AD FF00          BZ C24B
     01AE 01EB
0678                *
0679 01AF 5816          SACH C35
0680 01B0 5017          SACL C36
0681                *
0682 01B1 2016          LAC C35
0683 01B2 FF00          BZ C36B
     01B3 01CF
0684                *
0685 01B4 7942          AND M0002
0686 01B5 FF00          BZ MSB10
     01B6 01C3
0687                *
0688 01B7 2026          LAC N7        * LEFT SHIFT OF 7 *
0689 01B8 0008          ADD CEXP
0690 01B9 5008          SACL CEXP
0691                *
0692 01BA 270A          LAC CLO,7
0693 01BB 500A          SACL CLO
0694 01BC 581A          SACH CTEMP
0695 01BD 201A          LAC CTEMP
0696 01BE 7946          AND M007F
0697 01BF 0709          ADD CHI,7
0698 01C0 5009          SACL CHI
0699                *
0700 01C1 F900          B OUTPUT
     01C2 02FB
0701                *
0702 01C3 2027  MSB10   LAC N8        * LEFT SHIFT OF 8 *
0703 01C4 0008          ADD CEXP
0704 01C5 5008          SACL CEXP
0705                *
0706 01C6 280A          LAC CLO,8
0707 01C7 500A          SACL CLO
0708 01C8 581A          SACH CTEMP
0709 01C9 201A          LAC CTEMP
0710 01CA 7947          AND M00FF
0711 01CB 0809          ADD CHI,8
0712 01CC 5009          SACL CHI
0713 01CD F900          B OUTPUT
     01CE 02FB
0714                *
0715 01CF 2017  C36B    LAC C36
0716 01D0 794F          AND M8000
0717 01D1 FF00          BZ MSB12
     01D2 01DF
0718 01D3
0719 01D3 2028          LAC N9        * LEFT SHIFT OF 9 *
0720 01D4 0008          ADD CEXP
0721 01D5 5008          SACL CEXP
0722 01D6 290A          LAC CLO,9
0723 01D7 500A          SACL CLO
0724 01D8 581A          SACH CTEMP
0725 01D9 201A          LAC CTEMP
0726 01DA 7948          AND M01FF
```

```
0727 01DB 0909          ADD CHI,9
0728 01DC 5009          SACL CHI
0729           *
0730 01DD F900          B OUTPUT
     01DE 02FB
0731           *
0732 01DF 2029  MSB12   LAC N10      * LEFT SHIFT OF 10 *
0733 01E0 0008          ADD CEXP
0734 01E1 5008          SACL CEXP
0735           *
0736 01E2 2A0A          LAC CLO,10
0737 01E3 500A          SACL CLO
0738 01E4 581A          SACH CTEMP
0739 01E5 201A          LAC CTEMP
0740 01E6 7949          AND M03FF
0741 01E7 0A09          ADD CHI,10
0742 01E8 5009          SACL CHI
0743           *
0744 01E9 F900          B OUTPUT
     01EA 02FB
0745           *
0746 01EB 2215  C24B    LAC C24,2
0747 01EC 5818          SACH C37
0748 01ED 5019          SACL C38
0749           *
0750 01EE 2018          LAC C37
0751 01EF FF00          BZ C38B
     01F0 020C
0752           *
0753 01F1 7942          AND M0002
0754 01F2 FF00          BZ MSB14
     01F3 0200
0755           *
0756 01F4 202A          LAC N11      * LEFT SHIFT OF 11 *
0757 01F5 0008          ADD CEXP
0758 01F6 5008          SACL CEXP
0759           *
0760 01F7 2B0A          LAC CLO,11
0761 01F8 500A          SACL CLO
0762 01F9 581A          SACH CTEMP
0763 01FA 201A          LAC CTEMP
0764 01FB 794A          AND M07FF
0765 01FC 0B09          ADD CHI,11
0766 01FD 5009          SACL CHI
0767           *
0768 01FE F900          B OUTPUT
     01FF 02FB
0769           *
0770 0200 202B  MSB14   LAC N12      * LEFT SHIFT OF 12 *
0771 0201 0008          ADD CEXP
0772 0202 5008          SACL CEXP
0773           *
0774 0203 2C0A          LAC CLO,12
0775 0204 500A          SACL CLO
0776 0205 581A          SACH CTEMP
```

6. Floating-Point Arithmetic with the TMS32010

```
0777 0206 201A          LAC CTEMP
0778 0207 794B          AND MOFFF
0779 0208 0C09          ADD CHI,12
0780 0209 5009          SACL CHI
0781              *
0782 020A F900          B OUTPUT
     020B 02FB
0783              *
0784 020C 2019   C38B   LAC C38
0785 020D 794F          AND M8000
0786 020E FF00          BZ MSB16
     020F 021C
0787              *
0788 0210 202C          LAC N13          * LEFT SHIFT OF 13 *
0789 0211 0008          ADD CEXP
0790 0212 5008          SACL CEXP
0791              *
0792 0213 2D0A          LAC CLO,13
0793 0214 500A          SACL CLO
0794 0215 581A          SACH CTEMP
0795 0216 201A          LAC CTEMP
0796 0217 794C          AND M1FFF
0797 0218 0D09          ADD CHI,13
0798 0219 5009          SACL CHI
0799              *
0800 021A F900          B OUTPUT
     021B 02FB
0801              *
0802 021C 202D   MSB16  LAC N14          * LEFT SHIFT OF 14 *
0803 021D 0008          ADD CEXP
0804 021E 5008          SACL CEXP
0805              *
0806 021F 2E0A          LAC CLO,14
0807 0220 500A          SACL CLO
0808 0221 581A          SACH CTEMP
0809 0222 201A          LAC CTEMP
0810 0223 794D          AND M3FFF
0811 0224 0E09          ADD CHI,14
0812 0225 5009          SACL CHI
0813              *
0814 0226 F900          B OUTPUT
     0227 02FB
0815              *
0816 0228 280A   CLOB   LAC CLO,8        * CHI IS ZERO *
0817              *
0818 0229 580C          SACH C11         * SPLIT THE 16 BIT WORD INTO TWO
0819 022A 500D          SACL C12           8 BIT PIECES *
0820              *
0821 022B 2C0C          LAC C11,12
0822 022C FF00          BZ C12BP         * IF THERE IS A ONE ETC. *
     022D 0296
0823              *
0824 022E 580E          SACH C21
0825 022F 500F          SACL C22
0826              *
```

```
0827 0230 2E0E              LAC  C21,14
0828 0231 FF00              BZ   C22BP      * IF THERE IS A ONE ETC. *
     0232 0265
0829                *
0830 0233 5810              SACH C31
0831 0234 5011              SACL C32
0832                *
0833 0235 2010              LAC  C31
0834 0236 FF00              BZ   C32BP      * IF THERE IS A ONE ETC. *
     0237 024F
0835                *
0836 0238 7942              AND  M0002
0837 0239 FF00              BZ   MSB18
     023A 0246
0838                *
0839 023B 202E              LAC  N15        * LEFT SHIFT OF 15 *
0840 023C 0008              ADD  CEXP
0841 023D 5008              SACL CEXP
0842                *
0843 023E 2F0A              LAC  CLO,15
0844 023F 5809              SACH CHI
0845 0240 500A              SACL CLO
0846 0241 2009              LAC  CHI
0847 0242 794E              AND  M7FFF
0848 0243 5009              SACL CHI
0849                *
0850 0244 F900              B    OUTPUT
     0245 02FB
0851                *
0852 0246 202F    MSB18     LAC  N16        * LEFT SHIFT OF 16 *
0853 0247 0008              ADD  CEXP
0854 0248 5008              SACL CEXP
0855                *
0856 0249 200A              LAC  CLO
0857 024A 5009              SACL CHI
0858 024B 7F89              ZAC
0859 024C 500A              SACL CLO
0860                *
0861 024D F900              B    OUTPUT
     024E 02FB
0862                *
0863 024F 2011    C32BP     LAC  C32
0864 0250 794F              AND  M8000
0865 0251 FF00              BZ   MSB20
     0252 025C
0866                *
0867 0253 2030              LAC  N17        * LEFT SHIFT OF 17 *
0868 0254 0008              ADD  CEXP
0869 0255 5008              SACL CEXP
0870                *
0871 0256 210A              LAC  CLO,1
0872 0257 5009              SACL CHI
0873 0258 7F89              ZAC
0874 0259 500A              SACL CLO
0875                *
0876 025A F900              B    OUTPUT
```

6. Floating-Point Arithmetic with the TMS32010

```
          025B 02FB
0877                      *
0878 025C 2031  MSB20     LAC N18         * LEFT SHIFT OF 18 *
0879 025D 0008            ADD CEXP
0880 025E 5008            SACL CEXP
0881                      *
0882 025F 220A            LAC CLO,2
0883 0260 5009            SACL CHI
0884 0261 7F89            ZAC
0885 0262 500A            SACL CLO
0886                      *
0887 0263 F900            B OUTPUT
          0264 02FB
0888                      *
0889 0265 220F  C22BP     LAC C22,2
0890 0266 5812            SACH C33
0891 0267 5013            SACL C34
0892                      *
0893 0268 2012            LAC C33
0894 0269 FF00            BZ C34BP        * BRANCH IF NO ONE *
          026A 0280
0895                      *
0896 026B 7942            AND M0002
0897 026C FF00            BZ MSB22
          026D 0277
0898                      *
0899 026E 2032            LAC N19         * LEFT SHIFT OF 19 *
0900 026F 0008            ADD CEXP
0901 0270 5008            SACL CEXP
0902                      *
0903 0271 230A            LAC CLO,3
0904 0272 5009            SACL CHI
0905 0273 7F89            ZAC
0906 0274 500A            SACL CLO
0907                      *
0908 0275 F900            B OUTPUT
          0276 02FB
0909                      *
0910 0277 2033  MSB22     LAC N20         * LEFT SHIFT OF 20 *
0911 0278 0008            ADD CEXP
0912 0279 5008            SACL CEXP
0913                      *
0914 027A 240A            LAC CLO,4
0915 027B 5009            SACL CHI
0916 027C 7F89            ZAC
0917 027D 500A            SACL CLO
0918                      *
0919 027E F900            B OUTPUT
          027F 02FB
0920                      *
0921 0280 2013  C34BP     LAC C34
0922 0281 794F            AND M8000
0923 0282 FF00            BZ MSB24
          0283 028D
0924                      *
0925 0284 2034            LAC N21         * LEFT SHIFT OF 21 *
```

6. Floating-Point Arithmetic with the TMS32010

```
0926 0285 0008          ADD CEXP
0927 0286 5008          SACL CEXP
0928              *
0929 0287 250A          LAC CLO,5
0930 0288 5009          SACL CHI
0931 0289 7F89          ZAC
0932 028A 500A          SACL CLO
0933              *
0934 028B F900          B OUTPUT
     028C 02FB
0935              *
0936 028D 2035  MSB24   LAC N22         * LEFT SHIFT OF 22 *
0937 028E 0008          ADD CEXP
0938 028F 5008          SACL CEXP
0939              *
0940 0290 260A          LAC CLO,6
0941 0291 5009          SACL CHI
0942 0292 7F89          ZAC
0943 0293 500A          SACL CLO
0944              *
0945 0294 F900          B OUTPUT        * ^^ COMPLETES TOP 8 ^^ *
     0295 02FB
0946              *
0947 0296 240D  C12BP   LAC C12,4
0948 0297 5814          SACH C23
0949 0298 5015          SACL C24
0950              *
0951 0299 2E14          LAC C23,14
0952 029A FF00          BZ C24BP
     029B 02CC
0953 029C 5816          SACH C35
0954 029D 5017          SACL C36
0955              *
0956 029E 2016          LAC C35
0957 029F FF00          BZ C36BP
     02A0 02B6
0958              *
0959 02A1 7942          AND M0002
0960 02A2 FF00          BZ MSB26
     02A3 02AD
0961              *
0962 02A4 2036          LAC N23         * LEFT SHIFT OF 23 *
0963 02A5 0008          ADD CEXP
0964 02A6 5008          SACL CEXP
0965              *
0966 02A7 270A          LAC CLO,7
0967 02A8 5009          SACL CHI
0968 02A9 7F89          ZAC
0969 02AA 500A          SACL CLO
0970              *
0971 02AB F900          B OUTPUT
     02AC 02FB
0972              *
0973 02AD 2037  MSB26   LAC N24         * LEFT SHIFT OF 24 *
0974 02AE 0008          ADD CEXP
0975 02AF 5008          SACL CEXP
```

6. Floating-Point Arithmetic with the TMS32010

```
0976                    *
0977 02B0 280A              LAC CLO,8
0978 02B1 5009             SACL CHI
0979 02B2 7F89             ZAC
0980 02B3 500A             SACL CLO
0981                    *
0982 02B4 F900             B OUTPUT
     02B5 02FB
0983                    *
0984 02B6 2017  C36BP      LAC C36
0985 02B7 794F             AND M8000
0986 02B8 FF00             BZ MSB28
     02B9 02C3
0987                    *
0988 02BA 2038             LAC N25        * LEFT SHIFT OF 25 *
0989 02BB 0008             ADD CEXP
0990 02BC 5008             SACL CEXP
0991 02BD 290A             LAC CLO,9
0992 02BE 5009             SACL CHI
0993 02BF 7F89             ZAC
0994 02C0 500A             SACL CLO
0995                    *
0996 02C1 F900             B OUTPUT
     02C2 02FB
0997                    *
0998 02C3 2039  MSB28      LAC N26        * LEFT SHIFT OF 26 *
0999 02C4 0008             ADD CEXP
1000 02C5 5008             SACL CEXP
1001                    *
1002 02C6 2A0A             LAC CLO,10
1003 02C7 5009             SACL CHI
1004 02C8 7F89             ZAC
1005 02C9 500A             SACL CLO
1006                    *
1007 02CA F900             B OUTPUT
     02CB 02FB
1008                    *
1009 02CC 2215  C24BP      LAC C24,2
1010 02CD 5818             SACH C37
1011 02CE 5019             SACL C38
1012                    *
1013 02CF 2018             LAC C37
1014 02D0 FF00             BZ C38BP
     02D1 02E7
1015                    *
1016 02D2 7942             AND M0002
1017 02D3 FF00             BZ MSB30
     02D4 02DE
1018                    *
1019 02D5 203A             LAC N27        * LEFT SHIFT OF 27 *
1020 02D6 0008             ADD CEXP
1021 02D7 5008             SACL CEXP
1022                    *
1023 02D8 2B0A             LAC CLO,11
1024 02D9 5009             SACL CHI
1025 02DA 7F89             ZAC
```

6. Floating-Point Arithmetic with the TMS32010

```
1026 02DB 500A            SACL CLO
1027              *
1028 02DC F900            B OUTPUT
     02DD 02FB
1029              *
1030 02DE 203B   MSB30    LAC N28       * LEFT SHIFT OF 28 *
1031 02DF 0008            ADD CEXP
1032 02E0 5008            SACL CEXP
1033              *
1034 02E1 2C0A            LAC CLO,12
1035 02E2 5009            SACL CHI
1036 02E3 7F89            ZAC
1037 02E4 500A            SACL CLO
1038              *
1039 02E5 F900            B OUTPUT
     02E6 02FB
1040              *
1041 02E7 2019   C38BP    LAC C38
1042 02E8 794F            AND M8000
1043 02E9 FF00            BZ MSB32
     02EA 02F4
1044              *
1045 02EB 203C            LAC N29       * LEFT SHIFT OF 29 *
1046 02EC 0008            ADD CEXP
1047 02ED 5008            SACL CEXP
1048              *
1049 02EE 2D0A            LAC CLO,13
1050 02EF 5009            SACL CHI
1051 02F0 7F89            ZAC
1052 02F1 500A            SACL CLO
1053              *
1054 02F2 F900            B OUTPUT
     02F3 02FB
1055              *
1056 02F4 203D   MSB32    LAC N30       * LEFT SHIFT OF 30 *
1057 02F5 0008            ADD CEXP      * SORRY, BUT THATS ALL THE ACC CAN HOLD
1058 02F6 5008            SACL CEXP
1059              *
1060 02F7 2E0A            LAC CLO,14
1061 02F8 5009            SACL CHI
1062 02F9 7F89            ZAC
1063 02FA 500A            SACL CLO
1064              *
1065              *
1066              *
1067 02FB 490B   OUTPUT   OUT CSIGN,PA1
1068 02FC 4908            OUT CEXP,PA1
1069 02FD 4909            OUT CHI,PA1
1070 02FE 490A            OUT CLO,PA1
1071              *
1072 02FF F900   SELF     B SELF
     0300 02FF
1073              *
1074                      END
NO ERRORS, NO WARNINGS
```

6. Floating-Point Arithmetic with the TMS32010

7. Floating-Point Arithmetic with the TMS32020

Charles Crowell

Digital Signal Processing - Semiconductor Group

Texas Instruments

INTRODUCTION

The TMS32020 Digital Signal Processor is a fixed-point 16/32-bit microprocessor. However, it can also perform floating-point computations at a speed comparable to some dedicated floating-point processors.

The purpose of this application report is to analyze an implementation of floating-point addition, multiplication, and division on the TMS32020. The floating-point single-precision standard proposed by the IEEE will be examined. Using this standard, the TMS32020 performs a floating-point multiplication in 7.8 microseconds, a floating-point addition in 15.4 microseconds, and a floating-point division in 22.8 microseconds.

To illustrate floating-point formats and the tradeoffs involved in making a choice between different floating-point formats, a review of floating-point arithmetic notation and of addition, multiplication, and division algorithms is first presented.

FLOATING-POINT NOTATION

The floating-point number f may be written in floating-point format as

$$f = m \times b^e$$

where

$$m = \text{mantissa}$$
$$b = \text{base}$$
$$e = \text{exponent}$$

For example, 6,789,320 may be written as

$$0.6789320 \times 10^7$$

In this case,

$$m = 0.6789320$$
$$b = 10$$
$$e = 7$$

The two floating-point numbers f_1 and f_2 may be written as

$$f_1 = m_1 \times b^{e_1}$$
$$f_2 = m_2 \times b^{e_2}$$

Floating-point addition/subtraction, multiplication, and division for f_1 and f_2 are defined as follows:

$$f_1 \pm f_2 = (m_1 \pm m_2 \times b^{-(e_1-e_2)}) \times b^{e_1} \text{ if } e_1 \geq e_2 \quad (1)$$

or

$$= (m_1 \times b^{-(e_2-e_1)}) \pm m_2) \times b^{e_2} \text{ if } e_1 < e_2$$

$$f_1 \times f_2 = m_1 \times m_2 \times b^{(e_1+e_2)} \quad (2)$$

$$f_1/f_2 = (m_1/m_2) \times b^{(e_1-e_2)} \quad (3)$$

A cursory examination of these expressions reveals some of the factors involved in the implementation of floating-point arithmetic. For addition, it is necessary to shift the mantissa of the floating-point number which has the smaller exponent to the right by the difference in the magnitude of the two exponents. This is shown in the multiplication by the terms

$$b^{-(e_1-e_2)} \text{ and } b^{-(e_2-e_1)}$$

This right shift can result in mantissa underflow. There are also possibilities for mantissa overflow. Addition and subtraction of exponents can lead to exponent underflow and overflow. To alleviate underflow and overflow, it is necessary to decide on some scheme for roundoff. For a detailed description and analysis of underflow and overflow conditions and rounding schemes, see reference 1.

It is desirable to have all numbers normalized, i.e., the mantissas of f_1 and f_2 have the most significant digit in the leftmost position. This provides the representation with the greatest accuracy possible for a fixed mantissa length. The result of any floating-point operation must also be normalized. The factors associated with normalization, overflow, and other characteristics of floating-point implementations are best illustrated with a few examples.

Consider the addition of two binary floating-point numbers f_1 and f_2 where

$$f_1 = 0.10100 \times 2^{011}$$
$$f_2 = 0.11100 \times 2^{001}$$

Both of these numbers are normalized, i.e., the first bit after the binary point is a 1. Addition requires equal exponents, so the fractions are aligned by shifting right the one with the smaller exponent and adjusting the smaller exponent. This yields

$$f_2 = 0.00111 \times 2^{011}$$

Then,

$$f_1 + f_2 = 0.10100 \times 2^{011} + 0.00111 \times 2^{011}$$
$$= 0.11011 \times 2^{011} = f_3$$

The sum may overflow the left end by one digit, thus requiring a postaddition adjustment or renormalization step. Since it is assumed that the register is only of a finite length, this renormalization will result in the loss of the lowest order bit.

Another example illustrates the overflow past the most significant bit. With an assumed register length of five, let

$$f_1 = 0.11100 \times 2^{011}$$
$$f_2 = 0.10101 \times 2^{001}$$

Then,

$$
\begin{array}{r}
0.11100 \times 2^{011} = f_1 \\
+ 0.0010101 \times 2^{011} = f_2 \\
\hline
1.0000101 \times 2^{011} = f_3
\end{array}
$$

The significance of the two digits underlined in the right part of the mantissa is suspect, since it is assumed that the corresponding bits of f_1 are zero. The left underlined digit is the overflow past the most significant bit. To finish the addition, f_3 is shifted to the right and the exponent adjusted accordingly. Thus,

$$1.0000101 \times 2^{011} = f_3$$

The shift of the fraction and the adjustment of the exponent yield

$$0.10000101 \times 2^{100} = f_3$$

The result may be rounded, giving

$$0.10001 \times 2^{100} = f_3$$

or truncated, giving

$$0.10000 \times 2^{100} = f_3$$

FLOATING-POINT ALGORITHMS

Multiplication Algorithm

The algorithm for normalized floating-point multiplication is illustrated in Figure 1. This algorithm is an implementation of Equation 2 in the section on floating-point notation. The floating-point numbers being multiplied are A and B written as

$$A = m_A \times b^{e_A} \text{ and } B = m_B \times b^{e_B}$$

The result is

$$C = m_C \times b^{e_C}$$

For the resulting m_C, there are three special cases. The m_C may be zero, in which case there is a branch to Step 10 to set $C = 0$. If $m_C \neq 0$, then the most significant bit will

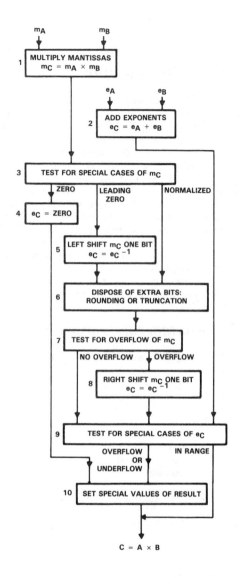

Figure 1. Floating-Point Multiplication

be in either the first or second leftmost bit. If the most significant bit is in the second leftmost bit, then a left shift of m_C is necessary (see Step 5). Otherwise, C is already in normalized form, and there is a branch to Step 6.

In Step 6, the desired rounding scheme is implemented. After this rounding, it is possible that m_C will overflow (see Step 7). In this case, it is necessary to right-shift m_C one bit (see Step 8). Special cases of e_C, are tested for in Step 9.

If there is an overflow or underflow of e_C, it is corrected in Step 10. Otherwise, the result is in range, and the calculation is complete.

Addition Algorithm

The implementation of normalized floating-point addition is more involved than for multiplication. This addition algorithm, outlined in Figure 2, is an implementation of Equation 1 in the section on floating-point notation.

In Step 1, e_A and e_B are compared to determine e_C. For this illustration of the algorithm, it is assumed that $e_A \leq e_B$. The right shift (d) required to align m_A is determined in Step 2. The procedure in Step 3 implements the right shift of m_A. In Step 4, the extra bits of m_A are discarded by using the desired rounding technique. The mantissas of A and B are then added in Step 5.

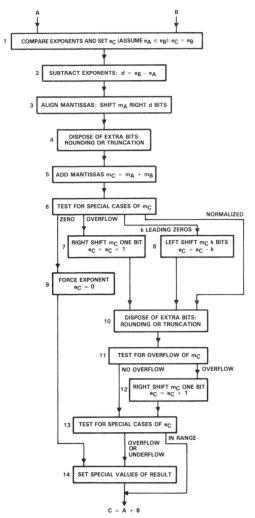

Figure 2. Floating-Point Addition

Now, the procedure becomes somewhat more involved. The m_C may be zero, in which case there is a branch to Step 9 which sets $e_C = 0$; a branch to Step 14 sets the special value of the result. The m_C may overflow, making a right shift of one necessary (see Step 7). The m_C may have k leading zeroes; therefore, a left shift of k is required. This normalization step is generally the most involved and time-consuming step to perform. The procedures in Steps 10, 11, and 12 round the m_C, test for a possible overflow due to the rounding, and adjust e_C accordingly. The special case of e_C is determined in Step 13. Finally, after Step 14, the sum C = A + B is formed.

Division Agorithm

Floating-point division is more sophisticated than multiplication and addition since fixed-point processors such as the TMS32020 are not inherently capable of performing division. For example, 1/3 = 0.3333...; only an approximation can be calculated since 1/3 must be represented in a finite number of terms. Several algorithms can be implemented to find good approximations of such numbers. The algorithm implemented in this report is shown in Figure 3.

Step 1 shows the equivalent of A/B. In Step 2, the latter term is expanded using a power series of 1/(1 + X), where ϵ (BLO/BHI) is X (ϵ simply denotes that the term is right-shifted 16 bits forming the least significant bits of a 32-bit number). The third term in the power series only affects the LSB of a 32-bit result; therefore, this term and all the following terms can be dropped, as shown in Step 3.

The equation in Step 3 can be implemented on the TMS32020 in two steps. Assuming that the result is a 32-bit number Q and that it is composed of a 16-bit QHI and a 16-bit QLO, think of the equation in Step 3 in the following manner: A/B = Q − ϵX. The first term is a fair approximation of the result Q, and the second term is a correction term to obtain a better approximation. With this in mind, it can be shown that (AHI + ϵALO)/BHI will give a 16-bit quotient and a 16-bit remainder. Due to the architecture of the TMS32020, the 16-bit quotient will be in the low word of the accumulator and the remainder will be in the high word of the accumulator after the division. Since it is desirable

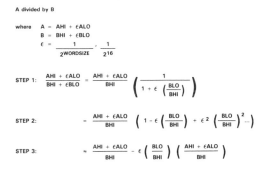

Figure 3. Division Equation

to have a floating-point result, the remainder must be divided by BHI to obtain the low word of the quotient. Now QHI and QLO have been calculated. When placing Q into the correction term (equation in Step 3), note that Q is equal to QHI + QLO. It can be shown that QLO will have no effect on the result since the correction term is multiplied by ϵ. Therefore, to calculate A divided by B, simply implement the following equation:

$$\frac{A}{B} = \frac{A}{BHI} - \epsilon\left(\frac{BLO}{BHI} \times QHI\right)$$

where the division is fixed binary (left-shifts and subtracts).

Figure 4 shows the implementation of the division algorithm that was outlined in Figure 3.

In Step 1, the dividend is right-shifted four times to prevent an overflow. Note that the result is not shifted left to compensate for this shift, because the normalization routine automatically does this. The shift causes the dividend to be limited to 27 significant bits instead of 31. In Step 2, a binary divide (left-shifts and subtracts) is implemented on the dividend by the high 16 bits of the divisor. The 32-bit result contains a quotient in the low 16 bits of the accumulator, and a remainder (R1) in the high 16 bits of the accumulator. R1 is left-shifted fifteen places in Step 3. The new R1 is divided by BHI in Step 4 to calculate the lower 16 bits of the quotient.

The quotient has now been approximated. The 32-bit result is composed of QHI and QLO, as shown in Figure 3. To obtain a better approximation, one term in the power series expansion must be added to the quotient. Therefore, the procedure in Step 5 calculates a 16-bit correction term, which is then added (or subtracted since it is the term following the "1" in the power series) to the 32-bit quotient.

Testing for an overflow of the resulting mantissa is necessary. Since the dividend was left-shifted four places, the resulting quotient will not be negative if an overflow occurred. To detect an overflow, bit 28 in the quotient must be tested. If this bit is a 1, an overflow occurred; if it is a 0, no overflow occurred. If an overflow has occurred, the exponent must be incremented. Finally, it is necessary to normalize the quotient and output the results.

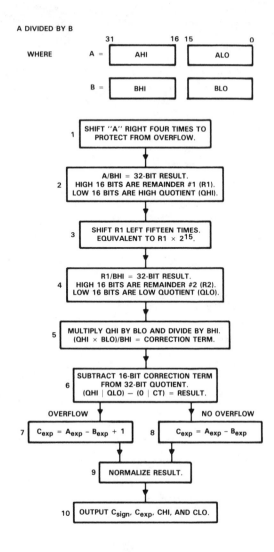

Figure 4. Floating-Point Division

IEEE FLOATING-POINT SINGLE-PRECISION FORMAT

Of interest is a set of formats known as the IEEE standard. This IEEE recommended format consists of a variety of precision formats (single, double, single-extended, and double-extended). The IEEE has also proposed several techniques for handling special cases such as overflow, underflow, ± ∞, and rounding. For complete details, the reader is referred to the proposed IEEE standard.[2]

The single-precision format is a 32-bit format consisting of a 1-bit sign field s, an 8-bit biased exponent e, and a 23-bit fraction f (see Figure 5). The value of a binary floating-point number X is determined as follows:

$$X = (-1)^s \times 2^{(e-127)} \times 1.f$$

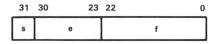

Figure 5. IEEE Floating-Point Single-Precision Format

The advantage of this format is that it is structured in such a way as to provide easy storage and straightforward input/output operations on 8-, 16- and 32-bit processors. The disadvantage with this format is that the large mantissa will generally span several words of memory.

FLOATING-POINT IMPLEMENTATION

IEEE Implementation

The IEEE single-precision format is described here as it applies to the addition, multiplication, and division algorithms. In these floating-point routines written for the TMS32020, all results are truncated to 31 bits to provide more flexibility in the user's development of a rounding scheme suitable for his application. The representations of ± ∞ are ignored so that the user can decide how to handle these exceptions in a manner that is appropriate for his particular application.

I/O Considerations

The first consideration is the internal representation of the binary floating-point number. If the number is read into the TMS32020 as two 16-bit words, some processing is then necessary to put the floating-point number into a representation which is easier to process. The representation used in the TMS32020 programs in the appendices is shown in Figure 6. This internal representation may be arrived at by a simple manipulation of the IEEE bit fields. For this particular algorithm, it is assumed that the floating-point number is input to the TMS32020 as the four 16-bit fields shown in Figure 6. However, the user can easily supply his own routine to arrive at this format from two 16-bit inputs to the TMS32020 where the inputs contain the IEEE single-precision format.

The format in Figure 6 was chosen to minimize the execution time of the floating-point addition, multiplication, and division routines. The format of the result is shown in Figure 7. Notice that it is identical to the format in Figure 5 except for CLO. CLO has its 16 most significant bits valid for both the addition, and multiplication, and division routines.

Normalization

Since the floating-point routines require normalization, a partial binary search algorithm is implemented in the addition and division routines in the appendices. To begin the normalization routine, note that all mantissas can be considered to be positive with the format used for the result shown in Figure 7. The binary search for the most significant bit (the leftmost 1 since the mantissa is positive) is illustrated in Figure 8.

The first move is to split the result into CHI and CLO. If CHI ≠ 0, the most significant bit (MSB) is the CHI; otherwise, it is the CLO. For this example, it is in CLO.

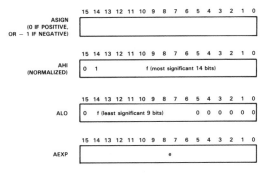

Figure 6. Floating-Point Representation

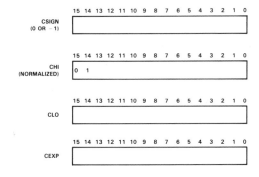

Figure 7. Result Representation

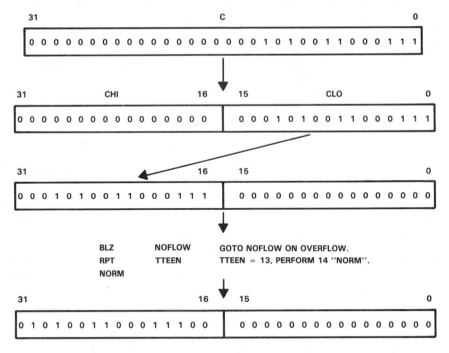

Figure 8. Partial Binary Search

The next step is to form a 32-bit result with CLO in the most significant word position. It is now possible for the MSB to be in the highest bit location since CLO has been left-shifted 16 times. If this is the case, an overflow has occurred, and the result must be right-shifted once. The normalization routine tests this by branching to NOFLOW if the result is negative. If the number is not negative, the normalization can continue.

The NORM instruction is used in the repeat mode to complete the normalization. Note that this whole normalization routine can be replaced by the following two instructions: RPTK 29 and NORM. The RPTK instruction causes the NORM instruction to be repeated 30 times, thus normalizing a 32-bit number. This method is not implemented here due to the timing. These two instructions always take 31 cycles to normalize a 32-bit number. The normalization routine here takes only 22 cycles (worst case) for normalizing a 32-bit number. Therefore, if program space is more important than timing efficiency, it is best to replace the normalization routine with these two instructions.

Added Precision

As illustrated in Figure 7, the 16 most significant bits of CLO are valid, i.e., C is valid for 31 places beyond the

binary point. Oftentimes the user is not as concerned with the IEEE standard as in being certain that he has enough accuracy for his particular application. Since the TMS32020 uses 16-bit words, the routines in the appendices implicitly maintain a 30-bit mantissa. They also implicitly use a 16-bit exponent. If the user desires this added accuracy and dynamic range, then it is readily implementable with no additional cost in execution time. The normalization for the addition, as mentioned previously, operates over the entire 32-bit accumulator. For the strict IEEE format, the user will only want to normalize over the 25 most significant bits of the accumulator. The structure of the normalization routine makes this modification simple.

The routines in the appendices make no provision for the representation of $\pm \infty$ and exponent underflow and overflow. The user of the routines should consider the degree of significance of these results and the way they should be handled for his particular application. Since these routines are written to operate at maximum speed, truncation of results is used. If the user desires to implement a rounding scheme, then he will also need to check for the possibility of overflow due to the rounding scheme. This step is shown in the multiplication, addition, and division flowcharts (see Figures 1, 2, and 3).

SUMMARY

The TMS32020 may be used to perform floating-point operations with great accuracy, wide dynamic range, and high-speed execution. The design engineer has the responsibility of deciding what type of floating-point format is best for his application. To aid in understanding floating-point operations, several examples have been given that illustrate the manipulations necessary to implement floating-point addition, multiplication, and division algorithms. Flowcharts for these algorithms are also included. The appendices contain the TMS32020 code for the IEEE floating-point single-precision format used in addition, multiplication, and division. The addition and multiplication routines may also be used without modification to implement a format with up to a 30-bit mantissa and a 16-bit exponent without any increase in execution time.

ACKNOWLEDGEMENTS

Major portions of this application report were taken from ''Floating-Point Arithmetic with the TMS32010,'' an application report written by Ray Simar, Jr. The author would also like to thank Gwyn Guidy for her assistance with the floating-point division algorithm.

REFERENCES

1. D.J. Kuck, *The Structure of Computers and Computations*, **Vol 1**, John Wiley & Sons (1978).
2. J. Coonen et al, ''A Proposed Standard for Binary Floating-Point Arithmetic,'' *ACM Signum Newsletter*, 4-12 (October 1979).
3. Donald E. Knuth, *Seminumerical Algorithms*, **Vol 2**, 2nd Edition, Addison-Wesley (1981).

APPENDIX A

```
0001              **********************************************************
0002              *                                                        *
0003              *     THIS IS A FLOATING-POINT ADDITION ROUTINE WHICH     *
0004              *     IMPLEMENTS THE IEEE PROPOSED FLOATING-POINT         *
0005              *     FORMAT ON THE TMS32020.                            *
0006              *                                                        *
0007              **********************************************************
0008              *
0009              *     INITIAL FORMAT (ALL 16 BIT WORDS)
0010              *     ------------------
0011              *     |   ALL 0 OR 1  |      ASIGN (0 OR -1)
0012              *     ------------------
0013              *
0014              *     ------------------
0015              *     |0|.  15 BITS   |      AHI (NORMALIZED)
0016              *     ------------------
0017              *
0018              *     ------------------
0019              *     |0|  9 BITS |--0-|      ALO
0020              *     ------------------
0021              *
0022              *     ------------------
0023              *     |                |      AEXP (-127 TO 128)
0024              *     ------------------
0025              *
0026              *     TO CORRESPOND WITH IEEE FORMAT,
0027              *     INPUT 0.1F * 2 ** (E + 1)
0028              *     INSTEAD OF 1.F * 2 **E, AND SUBTRACT 127 FROM E.
0029              *
0030              *     THE FINAL FORMAT IS THE SAME AS THE INITIAL FORMAT
0031              *     EXCEPT THAT FOR CLO WE HAVE:
0032              *
0033              *     ------------------
0034              *     |   16 BITS     |      CLO
0035              *     ------------------
0036              *
0037              *     ALL 16 BITS OF CLO ARE VALID. ANYTHING PAST THESE HAS
0038              *     BEEN TRUNCATED.
0039              *
0040              **********************************************************
0041              *                                                        *
0042              *     WORST CASE (EXCLUDING INITIALIZATION AND I/O):      *
0043              *     15.4 MICROSECONDS.                                 *
0044              *     THIS TIMING INCLUDES THE NORMALIZATION.            *
0045              *     WORDS OF PROGRAM MEMORY: 217                        *
0046              *                                                        *
0047              **********************************************************
0048              *
0049                       IDT      ' FLTADD'
0050 0000                  AORG
0051      0000 ASIGN EQU   0
0052      0001 AEXP  EQU   1
0053      0002 AHI   EQU   2
0054      0003 ALO   EQU   3
0055      0004 BSIGN EQU   4
0056      0005 BEXP  EQU   5
0057      0006 BHI   EQU   6
```

```
0058      0007  BLO     EQU     7
0059      0008  CSIGN   EQU     8
0060      0009  CEXP    EQU     9
0061      000A  CHI     EQU     10
0062      000B  CLO     EQU     11
0063      000C  D       EQU     12
0064      000D  ONE     EQU     13
0065      000E  TEMP    EQU     14
0066      000F  THREE   EQU     15
0067      0010  SIXT    EQU     16
0068      0011  RESID   EQU     17
0069      0012  TTEEN   EQU     18
0070            *
0071            *       INITIALIZATION
0072            *
0073 0000 C804          LDPK    4           BEGIN ON PAGE 4.
0074 0001 CE07          SSXM                SET SIGN EXTENSION.
0075 0002 5589          LARP    1
0076 0003 D100          LRLK    AR1,>200
     0004 0200
0077 0005 CB07          RPTK    7
0078 0006 80A0          IN      *+,PA0
0079 0007 5588          LARP    0
0080 0008 C000          LARK    AR0,0       CLEAR EXPONENT REGISTER.
0081 0009 CA01          LACK    1
0082 000A 600D          SACL    ONE         ONE = 1
0083 000B CA10          LACK    16
0084 000C 6010          SACL    SIXT
0085 000D CA03          LACK    3
0086 000E 600F          SACL    THREE
0087 000F CA0D          LACK    13
0088 0010 6012          SACL    TTEEN
0089            *
0090            *       BEGIN FLOATING POINT ADD
0091            *
0092 0011 2001  UP      LAC     AEXP        FIND LARGEST NUMBER.
0093 0012 1005          SUB     BEXP
0094 0013 F680          BZ      AEQB        IF EXP ARE THE SAME, JUMP TO AEQB.
     0014 0043
0095 0015 F380          BLZ     ALTB        IF A IS LESS THAN B, JUMP TO ALTB.
     0016 004D
0096            *
0097 0017 CE23  AGTB    NEG
0098 0018 0010          ADD     SIXT        D = (16-D)
0099 0019 F380          BLZ     A1          JUNP IF EXP DIFFERENCE IS > 16
     001A 0028
0100            *
0101            *   EXPONENT DIFFERENCE < 16
0102            *
0103 001B 600C          SACL    D
0104 001C 3C0C          LT      D
0105 001D 4206          LACT    BHI         BHI IS SHIFTED RIGHT "D" TIMES.
0106 001E 6806          SACH    BHI
0107 001F 6011          SACL    RESID       RESIDUAL BITS MUST BE MAINTAINED.
0108 0020 4207          LACT    BLO         BLO IS SHIFTED RIGHT "D" TIMES.
0109 0021 CE18          SFL                 MSB (THE 0) IS SHIFTED AWAY.
0110 0022 6807          SACH    BLO
```

7. Floating-Point Arithmetic with the TMS32020

```
0111 0023 2007          LAC     BLO
0112 0024 4D11          OR      RESID       GET BITS THAT WERE SHIFTED FROM BHI.
0113 0025 6007          SACL    BLO
0114 0026 FF80          B       A2
     0027 0031
0115            *
0116            *       EXPONENT DIFFERENCE >16
0117            *
0118 0028 0010  A1      ADD     SIXT
0119 0029 F380          BLZ     A3          JUMP IF EXPONENT DIFF > 32
     002A 0039
0120 002B 600C          SACL    D
0121 002C 3C0C          LT      D
0122 002D 4206          LACT    BHI
0123 002E 6807          SACH    BLO
0124 002F CA00          ZAC
0125 0030 6006          SACL    BHI
0126 0031 2000  A2      LAC     ASIGN       A IS LARGER THAN B.
0127 0032 6008          SACL    CSIGN       THEREFORE, CSIGN = ASIGN.
0128 0033 2001          LAC     AEXP        ALIGN THE B MANTISSA.
0129 0034 6009          SACL    CEXP
0130 0035 2103          LAC     ALO,1       GET RID OF EXTRA BIT.
0131 0036 6003          SACL    ALO
0132 0037 FF80          B       CHKSGN      DO BOTH NUMBERS HAVE THE SAME SIGN?
     0038 0078
0133            *
0134            *       A >> B ,  RESULT = A
0135            *
0136 0039 2002  A3      LAC     AHI
0137 003A 600A          SACL    CHI
0138 003B 2103          LAC     ALO,1
0139 003C 600B          SACL    CLO         A IS LARGER THAN B
0140 003D 2000          LAC     ASIGN       THEREFORE CSIGN = ASIGN
0141 003E 6008          SACL    CSIGN
0142 003F 2001          LAC     AEXP
0143 0040 6009          SACL    CEXP
0144 0041 FF80          B       AROUND
     0042 00D6
0145            *
0146 0043 2000  AEQB    LAC     ASIGN       IF SIGNS ARE THE SAME, CSIGN = ASIGN
0147 0044 6008          SACL    CSIGN
0148 0045 2103          LAC     ALO,1       ALIGN MANTISSAS.
0149 0046 6003          SACL    ALO
0150 0047 2107          LAC     BLO,1
0151 0048 6007          SACL    BLO
0152 0049 2001          LAC     AEXP        SET C EXPONENT = A EXPONENT.
0153 004A 6009          SACL    CEXP
0154 004B FF80          B       CHKSGN      DO BOTH NUMBERS HAVE THE SAME SIGN?
     004C 0078
0155            *
0156 004D 0010  ALTB    ADD     SIXT        D = (16-D)
0157 004E F380          BLZ     B1          JUMP IF EXP DIFF > 16
     004F 005D
0158 0050 600C          SACL    D
0159 0051 3C0C          LT      D
0160 0052 4202          LACT    AHI         AHI GETS SHIFTED "D" TIMES.
0161 0053 6802          SACH    AHI
```

```
0162 0054 6011          SACL    RESID       MAINTAIN EXTRA BITS.
0163 0055 4203          LACT    ALO         ALO GETS SHIFTED "D" TIMES.
0164 0056 CE18          SFL                 MSB (THE 0) IS SHIFTED AWAY.
0165 0057 6803          SACH    ALO
0166 0058 2003          LAC     ALO
0167 0059 4D11          OR      RESID       GET RESIDUAL BITS.
0168 005A 6003          SACL    ALO
0169 005B FF80          B       B2
     005C 0066
0170             *
0171             *    EXPONENENT DIFFERENCE > 16
0172             *
0173 005D 0010  B1      ADD     SIXT
0174 005E F380          BLZ     B3          JUMP IF EXP DIFF > 32
     005F 006E
0175 0060 600C          SACL    D
0176 0061 3C0C          LT      D
0177 0062 4202          LACT    AHI
0178 0063 6803          SACH    ALO
0179 0064 CA00          ZAC
0180 0065 6002          SACL    AHI
0181 0066 2004  B2      LAC     BSIGN       B IS THE BIGGEST NUMBER.
0182 0067 6008          SACL    CSIGN       THEREFORE, LET THE SIGN OF C=BSIGN.
0183 0068 2005          LAC     BEXP        SET C EXPONENT = B EXPONENT.
0184 0069 6009          SACL    CEXP
0185 006A 2107          LAC     BLO,1       GET RID OF EXTRA BIT.
0186 006B 6007          SACL    BLO
0187 006C FF80          B       CHKSGN      DO BOTH NUMBERS HAVE THE SAME SIGN?
     006D 0078
0188             *
0189             *    B >> A ,     RESULT = B
0190             *
0191 006E 2006  B3      LAC     BHI
0192 006F 600A          SACL    CHI
0193 0070 2107          LAC     BLO,1
0194 0071 600B          SACL    CLO         B IS THE BIGGEST NUMBER
0195 0072 2004          LAC     BSIGN       THEREFORE, LET THE SIGN OF C=BSIGN
0196 0073 6008          SACL    CSIGN
0197 0074 2005          LAC     BEXP        SET C EXPONENT = B EXPONENT
0198 0075 6009          SACL    CEXP
0199 0076 FF80          B       AROUND
     0077 00D6
0200             *
0201 0078 2000  CHKSGN  LAC     ASIGN       CHECK THE SIGNS.
0202 0079 1004          SUB     BSIGN
0203 007A F680          BZ      ADNOW       IF THEY ARE THE SAME, JUST ADD.
     007B 00A9
0204 007C F380          BLZ     AISNEG
     007D 008C
0205 007E 4002  BISNEG  ZALH    AHI         DO (|A| - |B|),
0206 007F 4903          ADDS    ALO         SINCE B < 0 AND A > 0.
0207 0080 4507          SUBS    BLO
0208 0081 4406          SUBH    BHI
0209 0082 F680          BZ      CZERO
     0083 009A
0210 0084 F380          BLZ     CNEG
     0085 00A1
```

7. Floating-Point Arithmetic with the TMS32020

```
0211 0086 680A          SACH    CHI
0212 0087 600B          SACL    CLO
0213 0088 CA00          ZAC
0214 0089 6008          SACL    CSIGN
0215 008A FF80          B       NORMAL          GO AND NORMALIZE RESULT.
     008B 00B3
0216 008C 4006  AISNEG  ZALH    BHI             DO (¦B¦ - ¦A¦),
0217 008D 4907          ADDS    BLO             SINCE A < 0  AND  B > 0.
0218 008E 4503          SUBS    ALO
0219 008F 4402          SUBH    AHI
0220 0090 F680          BZ      CZERO
     0091 009A
0221 0092 F380          BLZ     CNEG
     0093 00A1
0222 0094 680A          SACH    CHI
0223 0095 600B          SACL    CLO
0224 0096 CA00          ZAC
0225 0097 6008          SACL    CSIGN
0226 0098 FF80          B       NORMAL          GO AND NORMALIZE RESULTS.
     0099 00B3
0227            *
0228 009A CA00  CZERO   ZAC                     HERE, ONLY IF RESULT = 0.
0229 009B 6009          SACL    CEXP
0230 009C 6008          SACL    CSIGN
0231 009D 600A          SACL    CHI
0232 009E 600B          SACL    CLO
0233 009F FF80          B       AROUND          OUTPUT A ZERO.
     00A0 00D6
0234            *
0235 00A1 CE1B  CNEG    ABS                     HERE, IF RESULT IS NEGATIVE.
0236 00A2 680A          SACH    CHI
0237 00A3 600B          SACL    CLO
0238 00A4 D001          LALK    >FFFF
     00A5 FFFF
0239 00A6 6008          SACL    CSIGN
0240 00A7 FF80          B       NORMAL          GO NORMALIZE RESULT.
     00A8 00B3
0241            *
0242 00A9 4002  ADNOW   ZALH    AHI             IF SIGNS ARE THE SAME, JUST ADD.
0243 00AA 4903          ADDS    ALO
0244 00AB 4907          ADDS    BLO
0245 00AC 4806          ADDH    BHI
0246 00AD 680A          SACH    CHI
0247 00AE 600B          SACL    CLO
0248 00AF F080          BV      OVFLOW          DID AN OVERFLOW OCCUR?
     00B0 00C4
0249 00B1 F680          BZ      CZERO           IS RESULT = 0 ?
     00B2 009A
0250            *
0251            *       NORMALIZE
0252            *
0253 00B3 200A  NORMAL  LAC     CHI             DOES CHI HAVE THE MSB?
0254 00B4 F680          BZ      LO1
     00B5 00BC
0255 00B6 400A          ZALH    CHI             IF YES, NORMALIZE RESULT.
0256 00B7 490B          ADDS    CLO
0257 00B8 4B12          RPT     TTEEN           WILL PERFORM 14 "NORMS"
```

```
0258 00B9 CEA2          NORM
0259 00BA FF80          B       OUTPUT      GO OUTPUT RESULTS.
     00BB 00D0
0260 00BC 400B  LO1     ZALH    CLO         HERE IF CLO HAS MSB.
0261 00BD C010          LARK    AR0,16      OFFSET EXPONENT BY 16.
0262 00BE F380          BLZ     NOFLOW      DID BIT SEARCH CAUSE OVERFLOW?
     00BF 00CD
0263 00C0 4B12          RPT     TTEEN       IF NOT, NORMALIZE RESULT.
0264 00C1 CEA2          NORM
0265 00C2 FF80          B       OUTPUT      GO OUTPUT RESULT.
     00C3 00D0
0266           *
0267           *
0268           *        FINISHED WITH NORMALIZATION
0269           *
0270           *        HERE ONLY IF OVERFLOW OCCURRED DURING ADDITION
0271           *
0272           *
0273 00C4 CE06  OVFLOW  RSXM                RESET SIGN EXTENSION TO SHIFT RIGHT
0274 00C5 CE19          SFR                 SHIFT RIGHT.
0275 00C6 680A          SACH    CHI         STORE NORMALIZED MANTISSA.
0276 00C7 600B          SACL    CLO
0277 00C8 2009          LAC     CEXP        DECREMENT EXPONENT.
0278 00C9 000D          ADD     ONE
0279 00CA 6009          SACL    CEXP
0280 00CB FF80          B       AROUND      GO OUTPUT RESULTS.
     00CC 00D6
0281           *
0282           *        OVERLOW OCCURRED DURING BIT SEARCH
0283           *
0284 00CD 5590  NOFLOW  MAR     *-          DECREMENT EXPONENT.
0285 00CE CE06          RSXM                RSXM FOR LOGICAL RIGHT SHIFT.
0286 00CF CE19          SFR                 PERFORM RIGHT SHIFT.
0287           *
0288           *
0289           *        TAKE CARE OF EXPONENT & NORMALIZED MANTISSA,
0290           *        THEN OUTPUT RESULTS.
0291           *
0292           *
0293 00D0 700E  OUTPUT  SAR     AR0,TEMP    HERE AFTER NORMALIZATION.
0294 00D1 680A          SACH    CHI         SAVE NORMALIZED MANTISSA.
0295 00D2 600B          SACL    CLO
0296 00D3 2009          LAC     CEXP        ADJUST EXPONENT.
0297 00D4 100E          SUB     TEMP
0298 00D5 6009          SACL    CEXP
0299           *
0300 00D6 5589  AROUND  LARP    1           RESET POINTER.
0301 00D7 4B0F          RPT     THREE
0302 00D8 E0A0          OUT     *+,PA0
0303 00D9 CE1F          IDLE                WAIT FOR INTERRUPT.
NO ERRORS, NO WARNINGS
```

7. Floating-Point Arithmetic with the TMS32020

APPENDIX B

```
0001              ************************************************************
0002              *                                                          *
0003              *     THIS IS A FLOATING-POINT MULTIPLICATION ROUTINE WHICH *
0004              *     IMPLEMENTS THE IEEE PROPOSED FLOATING-POINT FORMAT     *
0005              *     ON THE TMS32020.                                       *
0006              *                                                          *
0007              ************************************************************
0008              *
0009              *     INITIAL FORMAT (ALL 16-BIT WORDS)
0010              *     ------------------
0011              *     !   ALL 0 OR 1   !     ASIGN (0 OR -1)
0012              *     ------------------
0013              *
0014              *     ------------------
0015              *     !0!.   15 BITS   !     AHI (NORMALIZED)
0016              *     ------------------
0017              *
0018              *     ------------------
0019              *     !0!  9 BITS !--0-!     ALO
0020              *     ------------------
0021              *
0022              *     ------------------
0023              *     !              !     AEXP (-127 TO 128)
0024              *     ------------------
0025              *
0026              *     TO CORRESPOND WITH IEEE FORMAT,
0027              *     INPUT 0.1F * 2 ** (E + 1)
0028              *     INSTEAD OF 1.F * 2 **E, AND SUBTRACT 127 FROM E.
0029              *
0030              *     THE FINAL FORMAT IS THE SAME AS THE INITIAL FORMAT
0031              *     EXCEPT THAT FOR CLO WE HAVE:
0032              *
0033              *     ------------------
0034              *     !   16 BITS      !     CLO
0035              *     ------------------
0036              *
0037              *     ALL 16 BITS OF CLO ARE VALID. ANYTHING PAST THESE HAS
0038              *     BEEN TRUNCATED.
0039              *
0040              ************************************************************
0041              *                                                          *
0042              *     WORST CASE (EXCLUDING INITIALIZATION AND I/O):         *
0043              *     7.8 MICROSECONDS.                                      *
0044              *     THIS TIMING INCLUDES THE NORMALIZATION.                *
0045              *     WORDS OF PROGRAM MEMORY: 60                            *
0046              *                                                          *
0047              ************************************************************
0048              *
0049 0000              AORG
0050      0000  ASIGN  EQU    0
0051      0001  AEXP   EQU    1
0052      0002  AHI    EQU    2
0053      0003  ALO    EQU    3
0054      0004  BSIGN  EQU    4
0055      0005  BEXP   EQU    5
0056      0006  BHI    EQU    6
```

```
0057        0007  BLO    EQU     7
0058        0008  CSIGN  EQU     8
0059        0009  CEXP   EQU     9
0060        000A  CHI    EQU     10
0061        000B  CLO    EQU     11
0062        000C  THI    EQU     12
0063        000D  NEGONE EQU     13
0064        000E  TLO    EQU     14
0065        000F  TEMP   EQU     15
0066              *
0067              *
0068              *      INITIALIZATION
0069              *
0070              *
0071              *
0072 0000 C804          LDPK    4         BEGIN ON PAGE 4.
0073 0001 CE07          SSXM              SET SIGN EXTENSION.
0074 0002 5589          LARP    1
0075 0003 D100          LRLK    AR1,>200
     0004 0200
0076 0005 CB07          RPTK    7         READ NUMBERS INTO BLOCK B0.
0077 0006 80A0          IN      *+,PA0
0078 0007 C000          LARK    AR0,0     CLEAR EXPONENT REGISTER.
0079 0008 5588          LARP    0
0080 0009 D001          LALK    >FFFF
     000A FFFF
0081 000B 600D          SACL    NEGONE    NEGONE = -1
0082            *
0083            *
0084            *      BEGIN FLOATING-POINT MULTIPLICATION.
0085            *
0086            *
0087 000C 2001  UP      LAC     AEXP      ADD EXPONENTS.
0088 000D 0005          ADD     BEXP
0089 000E 6009          SACL    CEXP
0090            *
0091 000F 3C03          LT      ALO       FIRST PRODUCT (ALO * BHI)
0092 0010 3806          MPY     BHI
0093 0011 CE14          PAC
0094 0012 680C          SACH    THI
0095 0013 600E          SACL    TLO
0096            *
0097 0014 3C02          LT      AHI       SECOND PRODUCT (AHI * BLO)
0098 0015 3807          MPY     BLO
0099            *
0100 0016 CE15          APAC              HAS EFFECT OF (AHI * BLO + ALO * BHI) * 2 ** -15.
0101 0017 CE15          APAC
0102            *
0103 0018 480C          ADDH    .THI
0104 0019 490E          ADDS    TLO
0105 001A 680C          SACH    THI
0106            *
0107 001B 3806          MPY     BHI       (AHI * BHI)
0108 001C CE14          PAC
0109 001D 490C          ADDS    THI
0110            *
```

7. Floating-Point Arithmetic with the TMS32020

```
0111 001E 690A        SACH   CHI,1    GET RID OF EXTRA SIGN BITS.
0112 001F 610B        SACL   CLO,1
0113              *
0114 0020 F580        BNZ    OK        IS RESULT ZERO?
     0021 0026
0115 0022 CA00        ZAC
0116 0023 6009        SACL   CEXP
0117 0024 FF80        B      SETSIN
     0025 002F
0118              *
0119 0026 400A  OK    ZALH   CHI       NORMALIZE AND WRAP UP.
0120 0027 490B        ADDS   CLO
0121 0028 CEA2        NORM
0122 0029 680A        SACH   CHI
0123 002A 600B        SACL   CLO
0124 002B 700F        SAR    AR0,TEMP
0125 002C 2009        LAC    CEXP
0126 002D 100F        SUB    TEMP
0127 002E 6009        SACL   CEXP
0128              *
0129 002F 4100  SETSIN ZALS  ASIGN     WHAT IS SIGN OF RESULT?
0130 0030 4C04        XOR    BSIGN
0131 0031 F580        BNZ    NEG
     0032 0037
0132 0033 CA00        ZAC
0133 0034 6008        SACL   CSIGN
0134 0035 FF80        B      OUTPUT
     0036 0039
0135 0037 200D  NEG   LAC    NEGONE
0136 0038 6008        SACL   CSIGN
0137 0039 5589  OUTPUT LARP  1         OUTPUT RESULTS.
0138 003A CB03        RPTK   3
0139 003B E0A0        OUT    *+,PA0
0140 003C CE1F        IDLE
NO ERRORS, NO WARNINGS
```

APPENDIX C

```
0001          ************************************************************
0002          *                                                         *
0003          *    THIS IS A FLOATING-POINT DIVISION ROUTINE WHICH       *
0004          *    IMPLEMENTS THE IEEE PROPOSED FLOATING-POINT FORMAT    *
0005          *    ON THE TMS32020.                                      *
0006          *                                                         *
0007          ************************************************************
0008          *
0009          *    INITIAL FORMAT (ALL 16-BIT WORDS)
0010          *    -------------------
0011          *    |   ALL 0 OR 1   |     ASIGN (0 OR -1)
0012          *    -------------------
0013          *
0014          *    -------------------
0015          *    |0|.   15 BITS   |     AHI (NORMALIZED)
0016          *    -------------------
0017          *
0018          *    -------------------
0019          *    |0|  9 BITS |--0-|     ALO
0020          *    -------------------
0021          *
0022          *    -------------------
0023          *    |               |     AEXP (-127 TO 128)
0024          *    -------------------
0025          *
0026          *    TO CORRESPOND WITH IEEE FORMAT,
0027          *    INPUT 0.1F * 2 ** (E + 1)
0028          *    INSTEAD OF 1.F * 2 **E, AND SUBTRACT 127 FROM E.
0029          *
0030          *    THE FINAL FORMAT IS THE SAME AS THE INITIAL FORMAT
0031          *    EXCEPT THAT FOR CLO WE HAVE:
0032          *
0033          *    -------------------
0034          *    |   16 BITS      |     CLO
0035          *    -------------------
0036          *
0037          *    ALL 16 BITS OF CLO ARE VALID. ANYTHING PAST THESE HAS
0038          *    BEEN TRUNCATED.
0039          *
0040          ************************************************************
0041          *                                                         *
0042          *    WORST CASE (EXCLUDING INITIALIZATION AND I/O):        *
0043          *    22.8 MICROSECONDS.                                    *
0044          *    THIS TIMING INCLUDES THE NORMALIZATION.               *
0045          *    WORDS OF PROGRAM MEMORY: 92                           *
0046          *                                                         *
0047          ************************************************************
0048          *
0049 0000            AORG    0
0050      0000 ASIGN  EQU     0
0051      0001 AEXP   EQU     1
0052      0002 AHI    EQU     2
0053      0003 ALO    EQU     3
0054      0004 BSIGN  EQU     4
0055      0005 BEXP   EQU     5
0056      0006 BHI    EQU     6
```

```
0057      0007  BLO    EQU    7
0058      0008  CSIGN  EQU    8
0059      0009  CEXP   EQU    9
0060      000A  CHI    EQU    10
0061      000B  CLO    EQU    11
0062      000C  NEGONE EQU    12
0063      000D  TEMP   EQU    13
0064      000E  FOUR   EQU    14
0065      000F  QM     EQU    15
0066      0010  QL     EQU    16
0067      0011  R1     EQU    17
0068      0012  R2     EQU    18
0069      0013  CL     EQU    19
0070      0014  M1000  EQU    20
0071      0015  ONE    EQU    21
0072      0016  THREE  EQU    22
0073      0017  FITEEN EQU    23
0074      0018  THIRTY EQU    24
0075      0019  TTEEN  EQU    25
0076            *
0077            *
0078            *      INITIALIZATION
0079            *
0080            *
0081            *
0082 0000 C804        LDPK   4          BEGIN ON PAGE 4.
0083 0001 CE07        SSXM              SET SIGN EXTENSION.
0084 0002 5589        LARP   1
0085 0003 D100        LRLK   AR1,>200
     0004 0200
0086 0005 CB07        RPTK   7          READ NUMBERS INTO BLOCK B0.
0087 0006 80A0        IN     *+,PA0
0088 0007 5588        LARP   0
0089 0008 C000        LARK   AR0,0      CLEAR EXPONENT REGISTER.
0090 0009 D001        LALK   >FFFF
     000A FFFF
0091 000B 600C        SACL   NEGONE     NEGONE = -1
0092 000C D001        LALK   >1000
     000D 1000
0093 000E 6014        SACL   M1000      M1000 = >1000
0094 000F CA04        LACK   4
0095 0010 600E        SACL   FOUR       FOUR = 4
0096 0011 CA01        LACK   1
0097 0012 6015        SACL   ONE        ONE = 1
0098 0013 CA03        LACK   3
0099 0014 6016        SACL   THREE      THREE = 3
0100 0015 CA0F        LACK   15
0101 0016 6017        SACL   FITEEN     FITEEN = 15
0102 0017 CA1E        LACK   30
0103 0018 6018        SACL   THIRTY     THIRTY = 30
0104 0019 CA0D        LACK   13
0105 001A 6019        SACL   TTEEN      TTEEN = 13
0106 001B CA00        ZAC
0107 001C 6009        SACL   CEXP       CLEAR CEXP
0108           *
0109           *
```

```
0110              *     FINISHED WITH INITIALIZATION
0111              *
0112              *
0113 001D 2000          LAC     ASIGN     CSIGN = ASIGN, IF ASIGN = BSIGN.
0114 001E 6008          SACL    CSIGN
0115 001F 1004          SUB     BSIGN
0116 0020 F680          BZ      OK        ELSE, CSIGN = -1.
     0021 0023
0117 0022 200C          LAC     NEGONE
0118              *     SACL    CSIGN
0119 0023 4002   OK     ZALH    AHI       SHIFT DIVIDEND TO PROTECT FROM OVERFLOW.
0120 0024 4903          ADDS    ALO
0121 0025 4B16          RPT     THREE
0122 0026 CE19          SFR
0123              *
0124 0027 4B17          RPT     FITEEN    QM = AHI!ALO / BHI, R1 = REMAINDER.
0125 0028 4706          SUBC    BHI
0126 0029 6811          SACH    R1        HIGH ACCUMULATOR RETAINS REMAINDER.
0127 002A 600F          SACL    QM
0128 002B 2F11          LAC     R1,15     (R1 * 2**15) / BHI   GIVES QL, AND R2.
0129 002C 4B17          RPT     FITEEN    COMPUTES (R1 * 2**15) / BHI.
0130 002D 4706          SUBC    BHI
0131 002E 6812          SACH    R2        HIGH ACCUMULATOR RETAINS REMAINDER.
0132 002F 6010          SACL    QL
0133              *
0134 0030 3C0F          LT      QM        CORRECTION TERM = (QM * BLO) / BHI.
0135 0031 3807          MPY     BLO       COMPUTES (QM * BLO).
0136 0032 CE14          PAC
0137 0033 4B17          RPT     FITEEN    COMPUTES (QM * BLO) / BHI.
0138 0034 4706          SUBC    BHI
0139 0035 6013          SACL    CL
0140 0036 400F          ZALH    QM        QM!QL - 0!CL = CHI!CLO
0141 0037 4910          ADDS    QL
0142 0038 1013          SUB     CL
0143 0039 600B          SACL    CLO
0144 003A 680A          SACH    CHI
0145 003B 200A          LAC     CHI       DID AN OVERFLOW OCCUR?
0146 003C 4E14          AND     M1000
0147 003D F680          BZ      NOOVF     IF NOT, GOTO NOOVF.
     003E 0041
0148 003F 2015          LAC     ONE       ELSE, INCREMENT CEXP.
0149 0040 6009          SACL    CEXP
0150 0041 2001   NOOVF  LAC     AEXP      COMPUTE RESULTING EXPONENT.
0151 0042 1005          SUB     BEXP
0152 0043 0009          ADD     CEXP
0153 0044 6009          SACL    CEXP
0154              *
0155              *
0156              *     NORMALIZE
0157              *
0158 0045 200A   NORMAL LAC     CHI       DOES CHI HAVE THE MSB?
0159 0046 F680          BZ      LO1
     0047 004E
0160 0048 400A          ZALH    CHI       IF YES, NORMALIZE RESULT.
0161 0049 490B          ADDS    CLO
0162 004A 4B19          RPT     TTEEN     WILL PERFORM 14 "NORMS".
```

```
0163 004B CEA2          NORM
0164 004C FF80          B       OUTPUT      GO OUTPUT RESULTS.
     004D 0057
0165 004E 400B  LO1     ZALH    CLO         HERE, IF CLO HAS MSB.
0166 004F F380          BLZ     NOFLOW      DID BIT SEARCH CAUSE OVERFLOW?
     0050 0055
0167 0051 4B19          RPT     TTEEN       IF NOT, NORMALIZE RESULT.
0168 0052 CEA2          NORM
0169 0053 FF80          B       OUTPUT      GO OUTPUT RESULT.
     0054 0057
0170            *
0171            *
0172            *       FINISHED WITH NORMALIZATION
0173            *
0174            *       OVERFLOW OCCURRED DURING BIT SEARCH
0175            *
0176 0055 CE06  NOFLOW  RSXM                RSXM FOR LOGICAL RIGHT SHIFT.
0177 0056 CE19          SFR                 PERFORM RIGHT SHIFT.
0178            *
0179            *
0180            *       TAKE CARE OF EXPONENT & NORMALIZED MANTISSA,
0181            *       THEN OUTPUT RESULTS.
0182            *
0183            *
0184 0057 680A  OUTPUT  SACH    CHI         SAVE NORMALIZED MANTISSA.
0185 0058 600B          SACL    CLO
0186 0059 5589          LARP    1           RESET POINTER.
0187 005A 4B16          RPT     THREE       OUTPUT RESULTS, CSIGN, CEXP, CHI, AND CLO.
0188 005B E0A0          OUT     *+,PA0
0189 005C CE1F          IDLE                WAIT FOR INTERRUPT.
NO ERRORS, NO WARNINGS
```

8. Precision Digital Sine-Wave Generation with the TMS32010

Domingo Garcia
Digital Signal Processing - Semiconductor Group
Texas Instruments

INTRODUCTION

Sine-wave generators are fundamental building blocks of signal processing systems which are used in diverse applications, such as communication, instrumentation, and control. In the past, engineers usually designed these oscillators with analog circuitry. Now, however, new high-speed digital signal processors like the TMS32010 present designers with an alternative that in many cases is superior. The TMS32010 provides the speed and accuracy to produce stable, low-distortion sine waves over a wide range of frequencies.

This application report describes two different methods for implementing a digital sine wave generator using the TMS32010. The first method is a fast direct table lookup scheme suitable for applications not requiring extreme accuracy. The second approach, an enhancement of the first, includes linear interpolation to provide sine waveforms with a minimum of harmonic distortion.

DIRECT TABLE LOOKUP METHOD

The first algorithm is a simple, fast table lookup scheme. The sine values for N angles which are uniformly spaced around the unit circle are stored in a table which has the following format:

INDEX	ANGLE	SINE TABLE
0	0 X 360°/N	S[0] = sin(0°/N)
1	1 X 360°/N	S[1] = sin(360°/N)
2	2 X 360°/N	S[2] = sin(720°/N)
.	.	.
.		
.	.	.
N-2	(N-2) X 360°/N	S[N-2] = sin((N-2) X 360°/N)
N-1	(N-1) X 360°/N	S[N-1] = sin((N-1) X 360°/N)

A sine wave is generated by stepping through the table at a constant rate (in effect, moving counterclockwise around the unit circle), wrapping around at the end of the table whenever 360° is exceeded. Using the table index as the angle parameter and DELTA as the step size, this lookup method generates the sequence:

S[mod(k X DELTA,N)]   for k = 1, 2, 3, 4, ...

where mod(a,b) =   remainder of the division a/b when this quotient is computed as an integer [e.g., mod(22.34,5) = 2.34)]

The 'mod' operator provides the wraparound at the end of the table. Figure 1 illustrates this algorithm.

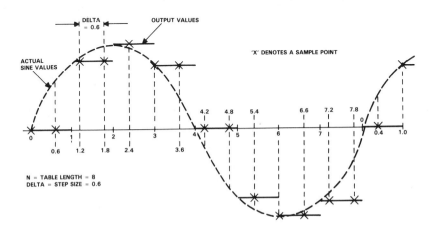

Figure 1. Direct Table Lookup

The sampled waveform generated is only an approximation to a sampled sinusoid. In general, the longer the table is the more resolution it provides, and consequently, the closer the approximation will be.

The frequency, f, of the sine wave depends on two factors:

(1) The time interval between successive samples, i.e., the sampling interval, t

(2) The step size, DELTA

f is given by the equation:

$$f = \frac{DELTA}{t \times N} \text{ [Hz]} \quad \text{where t is expressed in seconds}$$

Note that to satisfy the Nyquist criterion there must be at least two samples generated each sinusoid period. This requires that DELTA ≤ N/2.

In Figure 1, N = 8 and DELTA = 0.6. If, for instance, eight samples are generated each millisecond, then t = 0.000125 seconds and

$$f = \frac{0.6}{8 \times 0.000125} \text{ Hz} = 600 \text{ Hz}$$

TMS32010 Implementation

This section describes the concise TMS32010 subroutine, given in Appendix B, which implements the table lookup scheme based on a sine table with 128 entries. Each time this subroutine is called, the next sample point is calculated. This subroutine uses:

(1) 138 (=128 + 10) words of program memory space (128 words for sine table storage and 10 words for program memory)

(2) 6 words in data memory as working registers

If this program is used as a subroutine, each sample can be computed in 3.0 microseconds. However, if the code is inserted directly in line with the code of a master program, avoiding the overhead of a subroutine, a sample can be computed in 2.2 microseconds.

The values in the sine table are all scaled. The decimal values, +1.0 and −1.0, are represented by the two's complement hexadecimal values 4000 and C000, respectively. All other values are scaled and rounded to the closest hexadecimal number. Rounding is used, rather than truncation, to avoid adding unnecessary distortion.

The 16-bit data memory location 'ALPHA' serves as a modulo 128 counter which cycles through the sine table to select the sample points. ALPHA is regarded as having an integer and fractional part with the format:

Q Q Q Q Q Q Q Q . Q Q Q Q Q Q Q Q
15 14 13 12 11 10 9 8   7 6 5 4 3 2 1 0

The 16-bit data memory locaion 'DELTA' contains the step size. DELTA has the same (integer.fraction) format as ALPHA. Every time the sine wave subroutine is called, the contents of ALPHA are incremented by the contents of DELTA. The integer portion of ALPHA (i.e., the eight MSBs) is the pointer to the sine table. However, because the table starts at address location SINE, this pointer is offset by the value for that address before the table is accessed. The eight most significant bits of ALPHA are masked when ALPHA is updated to insure that they never exceed 127. The routine returns the sine value in the data memory location 'SINA'.

For any given sampling interval, t, the frequencies which can be generated must be of the form

$$f = \frac{DELTA}{t \times 128} \text{ [Hz]} \quad \text{where t is expressed in seconds}$$

Since DELTA has a precision of eight bits to the right of the decimal place, any desired frequency (≤ 1/2t [Hz]) can be approximated with an error of no more than

$$\frac{1/256}{t \times 128} \text{ [Hz]} \quad = \quad \frac{1}{32768 \times t} \text{ [Hz]}$$

For example, if the sampling frequency is 8 kHz, then the frequency resolution is

$$\frac{8000}{32768} \text{ Hz} = 0.25 \text{ Hz}$$

Harmonic Distortion

Due to approximations made in calculating the samples of a sine wave of frequency f, a certain amount of the "energy" of the samples' waveform will fall into other frequencies as well. These frequencies are either:

(1) Harmonic frequencies, nf, where n = 2, 3, 4, ..., or

(2) Subharmonic frequencies, nf/m, where n and m are integers.

This spurious energy results in noise which is referred to as "harmonic distortion." It is usually measured in terms of Total Harmonic Distortion (THD) which is defined as the ratio

$$THD = \frac{\text{spurious harmonic energy}}{\text{total energy of the waveform}}$$

There are two sources of error in the table lookup algorithm which cause harmonic distortion:

(1) Quantization error is introduced by representing the sine table values by 16-bit numbers.
(2) Larger errors are introduced when points between table entries are sampled. This occurs when DELTA is not an integer.

The longer the sine table is, the less significant the second error source will be. Consequently, harmonic distortion decreases with increasing table length. Furthermore, when DELTA is an integer, quantization is the only error source, and THD is extremely small regardless of table size. THD is given for several table lengths and values of DELTA in Figure 2. Note that the figures in this table only represent the THD in the digitized sine wave. If the sine wave is reconstructed using a digital-to-analog converter and analog filters, these analog devices will contribute additional distortion. (The procedure for computing THD is described in Appendix A.)

LINEAR INTERPOLATION METHOD

To decrease the harmonic distortion for a given table size, an interpolation scheme can be used to compute the sine values between table entries more accurately. Linear interpolation is the simplest method to implement. This method uses the values of two consecutive table entries as the end points of a line segment. Sample points for parameter values falling between table entries assume values on the line segment between the points. This algorithm is illustrated in Figure 3.

TABLE LENGTH: 32

DELTA	THD
2.0	0.00000024
2.25	0.00300893
2.50	0.00240751
2.75	0.00300917
3.0	0.00000024
8.25	0.00300924
11.625	0.00315807

TABLE LENGTH: 64

DELTA	THD
2.00	0.00000048
2.25	0.00075269
2.50	0.00060219
2.75	0.00075239
3.00	0.00000018
8.25	0.00075204
11.625	0.00079078

TABLE LENGTH: 128

DELTA	THD
2.00	0.00000054
2.25	0.00018859
2.50	0.00015080
2.75	0.00018835
3.00	0.00000012
8.25	0.00018889
11.625	0.00020128

Figure 2. Total Harmonic Distortion Using Direct Table Lookup

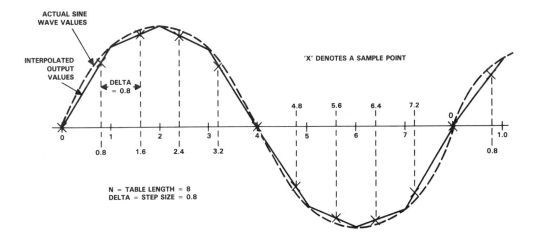

Figure 3. Linear Interpolation

This algorithm is based on the linear approximation

$$\sin(360°(I+D)/N) \cong \sin(360°I/N) \\ + D \times \{\sin(360°(I+1)/N) \\ - \sin(360°I/N)\}$$

$$= S[I] + D \times \{ S[I+1] - S[I] \}$$

where N is the sine table length,

I is an integer such that $0 \leq I \leq N-1$, and

D is a decimal number such that $0 \leq D < 1.0$

The value, $S[I+1] - S[I]$, is the slope of the line segment between the two sample points which bracket the value $I+D$ (i.e., $I \leq I+D < I+1$).

All the values required for this interpolation scheme are stored in the following two tables:

INDEX	ANGLE	SINE TABLE	SLOPE TABLE
0	0 X 360°/N	S[0] = sin(0°/N)	S[1] − S[0]
1	1 X 360°/N	S[1] = sin(360°/N)	S[2] − S[1]
2	2 X 360°/N	S[2] = sin(720°/N)	S[3] − S[2]
.	.	.	.
.	.	.	.
.	.	.	.
N − 2	(N − 2) X 360°/N	S[N − 2] = sin(360°(N − 2)/N)	S[N − 1] − S[N − 2]
N − 1	(N − 1) X 360°/N	S[N − 1] = sin(360°(N − 1)/N)	S[0] − S[N − 1]

TMS32010 Implementation

The sample TMS32010 implementation of this linear interpolation scheme, given in Appendix C, is an enhancement of the table lookup method. This subroutine is based on 128-entry sine and slope tables. Each time this subroutine is called, the next sample point is calculated. This subroutine uses:

(1) 276 (= 128 + 128 + 20) words of program memory space

(128 words for sine table storage,

128 words for slope table storage, and

20 words for program memory)

(2) 9 words in data memory as working registers

If this program is used as a subroutine, each sample can be computed in 5.4 microseconds. However, if the code is inserted directly in line with the code of a master program, avoiding the overhead of a subroutine, a sample can be computed in only 4.6 microseconds.

Just as in the table lookup algorithm, a sine wave is generated by stepping through the sine table at a constant rate, wrapping around at the end of the table whenever 360° is exceeded. The table index is used as the angle parameter, denoted by ALPHA.

DELTA denotes the step size for this routine also. In this case, however, sample points falling between the samples in the sine table are evaluated using the linear approximation formula given above.

The values in both the sine and slope tables are calculated in the same way as they were for the table lookup program. The decimal values, +1.0 and −1.0, are

represented by the two's complement hexadecimal values 4000 and C000, respectively. All hexadecimal values are rounded rather than truncated to the closest 16-bit representations to reduce quantization noise.

Because the method to compute the step size is the same as that used in the table lookup scheme, the frequency resolution will also be the same. However, because of the linear interpolation between table entries, sine values are no longer limited to the values stored in the table. This allows the error between the computed value and the actual value to be less.

Harmonic Distortion

Figure 4 lists the distortion of several sine waves generated using the TMS32010 linear interpolation routine for various table lengths and step sizes. These results clearly show that the distortion for a particular fractional step size decreases if the size of the table is increased just as in the direct table lookup case. However, for the same non-integer step size and the same table length, the distortion for the linear distortion method is much lower than that of direct table lookup.

These values were experimentally determined and the method used to compute them is given in Appendix A.

TABLE LENTH: 32

DELTA	THD
2.0	0.00000024
2.25	0.00169343
2.50	0.00135476
2.75	0.00169379
3.0	0.00000024
8.25	0.00169361
11.625	0.00177808

TABLE LENGTH: 64

DELTA	THD
2.00	0.00000048
2.25	0.00018884
2.50	0.00015055
2.75	0.00018771
3.00	0.00000018
8.25	0.00018806
11.625	0.00019815

TABLE LENGTH: 128

DELTA	THD
2.00	0.00000054
2.25	0.00000054
2.50	0.00000012
2.75	0.00000101
3.00	0.00000012
8.25	0.00000006
11.625	0.00000155

Figure 4. Harmonic Distortion Using Linear Interpolation

8. Precision Digital Sine-Wave Generation with the TMS32010

IMPLEMENTATION TRADE-OFFS

There are three trade-offs that must be considered when implementing the algorithms described above. They are speed, accuracy, and the size of the table ROM.

The direct table lookup method is the fastest implementation. Using a table that ranges from $0°$ to $360°$, the routine needs only to address the table and compute the next angle. However, the table occupies more program memory space than is absolutely required.

To minimize the amount of program memory required for the sine table, one can take advantage of the symmetry of the sine function. By keeping track of the quadrant as ALPHA is increased, a table that ranges from $0°$ to $90°$ will be sufficient. This decreases the size of the table by three-fourths. However, the extra code necessary to keep track of the quadrant will increase the execution time of the routine.

If harmonic distortion is important, then some form of interpolation is needed. One can use the linear interpolation method of the second example or other approximations such as a Taylor Series or Maclaurin Series expansions carried out to the second- or third-order term, or beyond. These schemes will, however, also increase the amount of code as well as the execution time.

APPENDIX A: COMPUTATION OF TOTAL HARMONIC DISTORTION

To determine the Total Harmonic Distortion (THD) of a sampled data sine wave, the amount of energy due to frequency components other than the fundamental is divided by the total energy of the wave. This is computed from the formula:

$$THD = [E(total) - E(fundamental)] / E(total)$$

For the most accurate results, these energy terms should be calculated over a full cycle of the signal. In the case of a sine wave generated by either of the two methods, a full cycle may actually consist of several sinusoid periods. For instance, if N = table length = 128 and if DELTA = step size = 1.5, a cycle will only be completed for the smallest n for which n x 1.5 is evenly divisible by 128. This occurs for n = 256 which marks the end of the second sinusoid period.

In general, if DELTA = A/B where A and B are relatively prime integers, and N = table length, then the sequence x(n), n = 1,2,3, ... of sine-wave samples will cycle after no more than B x N points.

The amount of total "energy" in a cycle of this length is

$$E(total) = \sum_{n=0}^{BN-1} x^2(n)$$

The amount of "energy" in the fundamental frequency over this period is

$$E(fundamental) = 1/BN \ (|X(A)|^2 + |X(BN-A)|^2)$$
$$= 2/BN \ |X(A)|^2 \quad \text{for a real sequence}$$

where the X(k) terms are terms of the Discrete Fourier Transform defined by the equation

$$X(k) = \sum_{n=0}^{BN-1} x(n)exp(-j(2\pi/N)nk)$$

The values given in Figures 2 and 4 are based on actual values computed by the TMS32010 for the two sample sine-wave generator programs. The computation of THD was carried out on a VAX 11/780 using the above formulas with double-precision floating-point arithmetic.

```
GENER1        320 FAMILY MACRO ASSEMBLER  2.1 83.076     17:14:48   1/18/84
                                                                    PAGE 0001

0001                              IDT     'GENER1'
0002                  ********************************************************
0003           *              SINE WAVE GENERATOR                          *
0004           *           DIRECT TABLE LOOKUP METHOD                      *
0005           * THIS PROGRAM USES A LOOKUP TABLE OF SINE VALUES TO        *
0006           * COMPUTE THE SAMPLES OF THE WAVE.  THE FREQUENCY IS        *
0007           * DETERMINED BY THE SIZE BY WHICH ONE STEPS THROUGH THE     *
0008           * TABLE. THE TABLE CONSISTS OF 128 ENTRIES THAT CORRESPOND  *
0009           * TO EQUALLY SPACED ANGLES BETWEEN 0 AND 360 DEGREES.       *
0010                  ********************************************************
0011           *           NOTE:   Q NOTATION                              *
0012           * THE TMS32010 USES FIXED-POINT TWO'S COMPLEMENT NUMBERS.   *
0013           * EACH 16-BIT NUMBER HAS A SIGN BIT, i INTEGER BITS, AND    *
0014           * (15-i) FRACTIONAL BITS.  THE VALUE AFTER THE LETTER Q     *
0015           * REFERS TO THE NUMBER OF FRACTIONAL BITS THAT ARE          *
0016           * REPRESENTED BY THAT NUMBER, i.e., A Q14 NUMBER IS         *
0017           * CONSIDERED TO HAVE 14 FRACTIONAL BITS.                    *
0018                  ********************************************************
0019 0000 F900        B       START
     0001 0083'
0020 0002 0000  SINE  DATA    >0              *THE SINE TABLE
0021 0003 0324        DATA    >324            *VALUES ARE REPRESENTED IN
0022 0004 0646        DATA    >646            *Q14 FORMAT, i.e., THERE
0023 0005 0964        DATA    >964            *ARE 14 BITS AFTER THE
0024 0006 0C7C        DATA    >C7C            *BINARY POINT.
0025 0007 0F8D        DATA    >F8D
0026 0008 1294        DATA    >1294
0027 0009 1590        DATA    >1590
0028 000A 187E        DATA    >187E
0029 000B 1B5D        DATA    >1B5D
0030 000C 1E2B        DATA    >1E2B
0031 000D 20E7        DATA    >20E7
0032 000E 238E        DATA    >238E
0033 000F 2620        DATA    >2620
0034 0010 289A        DATA    >289A
0035 0011 2AFB        DATA    >2AFB
0036 0012 2D41        DATA    >2D41
0037 0013 2F6C        DATA    >2F6C
0038 0014 3179        DATA    >3179
0039 0015 3368        DATA    >3368
0040 0016 3537        DATA    >3537
0041 0017 36E5        DATA    >36E5
0042 0018 3871        DATA    >3871
0043 0019 39DB        DATA    >39DB
0044 001A 3B21        DATA    >3B21
0045 001B 3C42        DATA    >3C42
0046 001C 3D3F        DATA    >3D3F
0047 001D 3E15        DATA    >3E15
0048 001E 3EC5        DATA    >3EC5
0049 001F 3F4F        DATA    >3F4F
0050 0020 3FB1        DATA    >3FB1
0051 0021 3FEC        DATA    >3FEC
0052 0022 4000        DATA    >4000
0053 0023 3FEC        DATA    >3FEC
```

```
0054 0024 3FB1            DATA    >3FB1
0055 0025 3F4F            DATA    >3F4F
0056 0026 3EC5            DATA    >3EC5
0057 0027 3E15            DATA    >3E15
0058 0028 3D3F            DATA    >3D3F
0059 0029 3C42            DATA    >3C42
0060 002A 3B21            DATA    >3B21
0061 002B 39DB            DATA    >39DB
0062 002C 3871            DATA    >3871
0063 002D 36E5            DATA    >36E5
0064 002E 3537            DATA    >3537
0065 002F 3368            DATA    >3368
0066 0030 3179            DATA    >3179
0067 0031 2F6C            DATA    >2F6C
0068 0032 2D41            DATA    >2D41
0069 0033 2AFB            DATA    >2AFB
0070 0034 289A            DATA    >289A
0071 0035 2620            DATA    >2620
0072 0036 238E            DATA    >238E
0073 0037 20E7            DATA    >20E7
0074 0038 1E2B            DATA    >1E2B
0075 0039 1B5D            DATA    >1B5D
0076 003A 187E            DATA    >187E
0077 003B 1590            DATA    >1590
0078 003C 1294            DATA    >1294
0079 003D 0F8D            DATA    >F8D
0080 003E 0C7C            DATA    >C7C
0081 003F 0964            DATA    >964
0082 0040 0646            DATA    >646
0083 0041 0324            DATA    >324
0084 0042 0000            DATA    >0
0085 0043 FCDC            DATA    >FCDC
0086 0044 F9BA            DATA    >F9BA
0087 0045 F69C            DATA    >F69C
0088 0046 F384            DATA    >F384
0089 0047 F073            DATA    >F073
0090 0048 ED6C            DATA    >ED6C
0091 0049 EA70            DATA    >EA70
0092 004A E782            DATA    >E782
0093 004B E4A3            DATA    >E4A3
0094 004C E1D5            DATA    >E1D5
0095 004D DF19            DATA    >DF19
0096 004E DC72            DATA    >DC72
0097 004F D9E0            DATA    >D9E0
0098 0050 D766            DATA    >D766
0099 0051 D505            DATA    >D505
0100 0052 D2BF            DATA    >D2BF
0101 0053 D094            DATA    >D094
0102 0054 CE87            DATA    >CE87
0103 0055 CC98            DATA    >CC98
0104 0056 CAC9            DATA    >CAC9
0105 0057 C91B            DATA    >C91B
0106 0058 C78F            DATA    >C78F
0107 0059 C625            DATA    >C625
```

8. Precision Digital Sine-Wave Generation with the TMS32010

```
0108 005A C4DF          DATA    >C4DF
0109 005B C3BE          DATA    >C3BE
0110 005C C2C1          DATA    >C2C1
0111 005D C1EB          DATA    >C1EB
0112 005E C13B          DATA    >C13B
0113 005F C0B1          DATA    >C0B1
0114 0060 C04F          DATA    >C04F
0115 0061 C014          DATA    >C014
0116 0062 C000          DATA    >C000
0117 0063 C014          DATA    >C014
0118 0064 C04F          DATA    >C04F
0119 0065 C0B1          DATA    >C0B1
0120 0066 C13B          DATA    >C13B
0121 0067 C1EB          DATA    >C1EB
0122 0068 C2C1          DATA    >C2C1
0123 0069 C3BE          DATA    >C3BE
0124 006A C4DF          DATA    >C4DF
0125 006B C625          DATA    >C625
0126 006C C78F          DATA    >C78F
0127 006D C91B          DATA    >C91B
0128 006E CAC9          DATA    >CAC9
0129 006F CC98          DATA    >CC98
0130 0070 CE87          DATA    >CE87
0131 0071 D094          DATA    >D094
0132 0072 D2BF          DATA    >D2BF
0133 0073 D505          DATA    >D505
0134 0074 D766          DATA    >D766
0135 0075 D9E0          DATA    >D9E0
0136 0076 DC72          DATA    >DC72
0137 0077 DF19          DATA    >DF19
0138 0078 E1D5          DATA    >E1D5
0139 0079 E4A3          DATA    >E4A3
0140 007A E782          DATA    >E782
0141 007B EA70          DATA    >EA70
0142 007C ED6C          DATA    >ED6C
0143 007D F073          DATA    >F073
0144 007E F384          DATA    >F384
0145 007F F69C          DATA    >F69C
0146 0080 F9BA          DATA    >F9BA
0147 0081 FCDC          DATA    >FCDC
0148 0082
0149 0082 7FFF  M1      DATA    >7FFF
0150 0083
0151                ***************************
0152                *DATA MEMORY LOCATIONS USED*
0153                ***************************
0154      0000  DELTA   EQU     0
0155      0001  ALPHA   EQU     1
0156      0002  SINA    EQU     2
0157      0003  TEMP    EQU     3          *WORKSPACE REGISTER
0158      0004  MASK    EQU     4
0159      0005  OFSET   EQU     5
0160 0083
```

```
0161              *************************************************************
0162              * NECESSARY INITIALIZATIONS:                                *
0163              * MASK    INITIALIZED TO >7FFF FOR 128 POINT TABLE          *
0164              * OFSET   INITIALIZED TO THE ADDRESS AT THE BEGINNING       *
0165              *           OF TABLE.                                       *
0166              * ALPHA   INITIALLY CLEARED                                 *
0167              * DELTA   INITIALIZED TO INCREMENT VALUE USING Q8 FORMAT    *
0168              *************************************************************
0169 0083 6F00    START  LDP     0               * SET DATA PAGE POINTER
0170 0084 7E82           LACK    M1
0171 0085 6704           TBLR    MASK
0172 0086 7E02           LACK    SINE
0173 0087 5005           SACL    OFSET
0174 0088 7F89           ZAC
0175 0089 5001           SACL    ALPHA
0176 008A 4100           IN      DELTA,PA1              * IN THIS EXAMPLE,
0177 008B F800    L1     CALL    SWAVE1                * DELTA IS INPUT
     008C 008F'
0178              ********************************
0179              *       REST OF PROGRAM        *
0180              ********************************
0181 008D F900           B       L1
     008E 008B'
0182 008F
0183              *************************************************************
0184              * SINE WAVE SUBROUTINE:                                     *
0185              * THIS ROUTINE EXTRACTS THE SINE OF AN ANGLE FROM THE       *
0186              * TABLE AND RETURNS THE VALUE IN THE DATA LOCATION          *
0187              * 'SINA'.  IT USES A FRACTIONAL STEP SIZE TO COMPUTE        *
0188              * THE NEXT POINT OF THE WAVE.  IT TAKES 2.6 microseconds    *
0189              * TO EXECUTE.                                               *
0190              *************************************************************
0191 008F 2801    SWAVE1 LAC     ALPHA,8
0192 0090 5803           SACH    TEMP            *ISOLATE INTEGER PORTION
0193 0091 2003           LAC     TEMP
0194 0092 0005           ADD     OFSET
0195 0093 6702           TBLR    SINA            *SINE VALUE FROM TABLE (Q14)
0196 0094 2001           LAC     ALPHA
0197 0095 0000           ADD     DELTA           *COMPUTE NEXT ADDRRESS
0198 0096 7904           AND     MASK            *MODULO 128 MASK = >7FFF
0199 0097 5001           SACL    ALPHA           *SAVE NEXT ADDRESS
0200 0098 7F8D           RET                     *RETURN TO MAIN PROGRAM
0201                     END
NO ERRORS, NO WARNINGS
```

8. Precision Digital Sine-Wave Generation with the TMS32010

APPENDIX C: TMS32010 LINEAR INTERPOLATION ROUTINE

GENER2   320 FAMILY MACRO ASSEMBLER 2.1 83.076   10:54:48   1/19/84

PAGE 0001

```
0001                         IDT      'GENER2'
0002                ***********************************************
0003                *           SINE WAVE GENERATOR              *
0004                *          LINEAR INTERPOLATION METHOD       *
0005                * THIS PROGRAM USES A LOOKUP TABLE OF SINE VALUES TO *
0006                * COMPUTE THE SAMPLES OF THE WAVE.  THE FREQUENCY IS *
0007                * DETERMINED BY THE SIZE BY WHICH ONE STEPS THROUGH  *
0008                * THE TABLE.  THE TABLE CONSISTS OF 128 ENTRIES THAT *
0009                * CORRESPOND TO EQUALLY SPACED ANGLES BETWEEN 0 AND  *
0010                * 360 DEGREES.  POINTS BETWEEN THE TABLE ENTRIES ARE *
0011                * APPROXIMATED USING A LINEAR APPROXIMATION,         *
0012                * sin(A) ~= sin(INT[A])                             *
0013                *          + {sin(INT[A]+1)-sin(INT[A])}xFRACT[A]   *
0014                * ALL THE POSSIBLE SLOPES BETWEEN ANY TWO CONSECUTIVE*
0015                * SINE TABLE ENTRIES ARE STORED IN A SEPARATE TABLE. *
0016                * THE SLOPE TABLE IS ALSO 128 ENTRIES LONG.          *
0017                ***********************************************
0018                *         NOTE:  Q NOTATION                  *
0019                * THE TMS32010 USES FIXED-POINT TWO'S COMPLEMENT     *
0020                * NUMBERS.  EACH 16-BIT NUMBER HAS A SIGN BIT, i     *
0021                * INTEGER BITS, AND (15-i) FRACTIONAL BITS.  THE     *
0022                * VALUE AFTER THE LETTER Q REFERS TO THE NUMBER OF   *
0023                * FRACTIONAL BITS THAT ARE REPRESENTED BY THAT       *
0024                * NUMBER, i.e., A Q14 NUMBER IS CONSIDERED TO HAVE   *
0025                * 14 FRACTIONAL BITS.                                *
0026                ***********************************************
0027 0000
0028 0000 F900          B        START
     0001 0104'
0029          *
0030 0002 7FFF   M1     DATA     >7FFF              *MASK VALUES
0031 0003 0FFF   M2     DATA     >0FFF
0032          *
0033 0004 0000   SINE   DATA     >0                 *THE SINE TABLE
0034 0005 0324          DATA     >324               *VALUES ARE REPRESENTED
0035 0006 0646          DATA     >646               *IN Q14 FORMAT, i.e.,
0036 0007 0964          DATA     >964               *THERE ARE 14 BITS AFTER
0037 0008 0C7C          DATA     >C7C               *THE BINARY POINT.
0038 0009 0F8D          DATA     >F8D
0039 000A 1294          DATA     >1294
0040 000B 1590          DATA     >1590
0041 000C 187E          DATA     >187E
0042 000D 1B5D          DATA     >1B5D
0043 000E 1E2B          DATA     >1E2B
0044 000F 20E7          DATA     >20E7
0045 0010 238E          DATA     >238E
0046 0011 2620          DATA     >2620
0047 0012 289A          DATA     >289A
0048 0013 2AFB          DATA     >2AFB
0049 0014 2D41          DATA     >2D41
0050 0015 2F6C          DATA     >2F6C
0051 0016 3179          DATA     >3179
0052 0017 3368          DATA     >3368
0053 0018 3537          DATA     >3537
```

```
0054 0019 36E5          DATA    >36E5
0055 001A 3871          DATA    >3871
0056 001B 39DB          DATA    >39DB
0057 001C 3B21          DATA    >3B21
0058 001D 3C42          DATA    >3C42
0059 001E 3D3F          DATA    >3D3F
0060 001F 3E15          DATA    >3E15
0061 0020 3EC5          DATA    >3EC5
0062 0021 3F4F          DATA    >3F4F
0063 0022 3FB1          DATA    >3FB1
0064 0023 3FEC          DATA    >3FEC
0065 0024 4000          DATA    >4000
0066 0025 3FEC          DATA    >3FEC
0067 0026 3FB1          DATA    >3FB1
0068 0027 3F4F          DATA    >3F4F
0069 0028 3EC5          DATA    >3EC5
0070 0029 3E15          DATA    >3E15
0071 002A 3D3F          DATA    >3D3F
0072 002B 3C42          DATA    >3C42
0073 002C 3B21          DATA    >3B21
0074 002D 39DB          DATA    >39DB
0075 002E 3871          DATA    >3871
0076 002F 36E5          DATA    >36E5
0077 0030 3537          DATA    >3537
0078 0031 3368          DATA    >3368
0079 0032 3179          DATA    >3179
0080 0033 2F6C          DATA    >2F6C
0081 0034 2D41          DATA    >2D41
0082 0035 2AFB          DATA    >2AFB
0083 0036 289A          DATA    >289A
0084 0037 2620          DATA    >2620
0085 0038 238E          DATA    >238E
0086 0039 20E7          DATA    >20E7
0087 003A 1E2B          DATA    >1E2B
0088 003B 1B5D          DATA    >1B5D
0089 003C 187E          DATA    >187E
0090 003D 1590          DATA    >1590
0091 003E 1294          DATA    >1294
0092 003F 0F8D          DATA    >F8D
0093 0040 0C7C          DATA    >C7C
0094 0041 0964          DATA    >964
0095 0042 0646          DATA    >646
0096 0043 0324          DATA    >324
0097 0044 0000          DATA    >0
0098 0045 FCDC          DATA    >FCDC
0099 0046 F9BA          DATA    >F9BA
0100 0047 F69C          DATA    >F69C
0101 0048 F384          DATA    >F384
0102 0049 F073          DATA    >F073
0103 004A ED6C          DATA    >ED6C
0104 004B EA70          DATA    >EA70
0105 004C E782          DATA    >E782
0106 004D E4A3          DATA    >E4A3
0107 004E E1D5          DATA    >E1D5
```

8. Precision Digital Sine-Wave Generation with the TMS32010

```
0108 004F DF19          DATA     >DF19
0109 0050 DC72          DATA     >DC72
0110 0051 D9E0          DATA     >D9E0
0111 0052 D766          DATA     >D766
0112 0053 D505          DATA     >D505
0113 0054 D2BF          DATA     >D2BF
0114 0055 D094          DATA     >D094
0115 0056 CE87          DATA     >CE87
0116 0057 CC98          DATA     >CC98
0117 0058 CAC9          DATA     >CAC9
0118 0059 C91B          DATA     >C91B
0119 005A C78F          DATA     >C78F
0120 005B C625          DATA     >C625
0121 005C C4DF          DATA     >C4DF
0122 005D C3BE          DATA     >C3BE
0123 005E C2C1          DATA     >C2C1
0124 005F C1EB          DATA     >C1EB
0125 0060 C13B          DATA     >C13B
0126 0061 C0B1          DATA     >C0B1
0127 0062 C04F          DATA     >C04F
0128 0063 C014          DATA     >C014
0129 0064 C000          DATA     >C000
0130 0065 C014          DATA     >C014
0131 0066 C04F          DATA     >C04F
0132 0067 C0B1          DATA     >C0B1
0133 0068 C13B          DATA     >C13B
0134 0069 C1EB          DATA     >C1EB
0135 006A C2C1          DATA     >C2C1
0136 006B C3BE          DATA     >C3BE
0137 006C C4DF          DATA     >C4DF
0138 006D C625          DATA     >C625
0139 006E C78F          DATA     >C78F
0140 006F C91B          DATA     >C91B
0141 0070 CAC9          DATA     >CAC9
0142 0071 CC98          DATA     >CC98
0143 0072 CE87          DATA     >CE87
0144 0073 D094          DATA     >D094
0145 0074 D2BF          DATA     >D2BF
0146 0075 D505          DATA     >D505
0147 0076 D766          DATA     >D766
0148 0077 D9E0          DATA     >D9E0
0149 0078 DC72          DATA     >DC72
0150 0079 DF19          DATA     >DF19
0151 007A E1D5          DATA     >E1D5
0152 007B E4A3          DATA     >E4A3
0153 007C E782          DATA     >E782
0154 007D EA70          DATA     >EA70
0155 007E ED6C          DATA     >ED6C
0156 007F F073          DATA     >F073
0157 0080 F384          DATA     >F384
0158 0081 F69C          DATA     >F69C
0159 0082 F9BA          DATA     >F9BA
0160 0083 FCDC          DATA     >FCDC
0161 0084 0324 TSLOPE DATA      >324              *SLOPE BETWEEN TWO
```

```
0162 0085 0322          DATA    >322            *SINE ENTRIES (Q14)
0163 0086 031E          DATA    >31E
0164 0087 0318          DATA    >318
0165 0088 0311          DATA    >311
0166 0089 0307          DATA    >307
0167 008A 02FC          DATA    >2FC
0168 008B 02EE          DATA    >2EE
0169 008C 02DF          DATA    >2DF
0170 008D 02CE          DATA    >2CE
0171 008E 02BC          DATA    >2BC
0172 008F 02A7          DATA    >2A7
0173 0090 0291          DATA    >291
0174 0091 027A          DATA    >27A
0175 0092 0261          DATA    >261
0176 0093 0246          DATA    >246
0177 0094 022B          DATA    >22B
0178 0095 020D          DATA    >20D
0179 0096 01EF          DATA    >1EF
0180 0097 01CF          DATA    >1CF
0181 0098 01AE          DATA    >1AE
0182 0099 018C          DATA    >18C
0183 009A 016A          DATA    >16A
0184 009B 0146          DATA    >146
0185 009C 0121          DATA    >121
0186 009D 00FC          DATA    >FC
0187 009E 00D6          DATA    >D6
0188 009F 00B0          DATA    >B0
0189 00A0 0089          DATA    >89
0190 00A1 0062          DATA    >62
0191 00A2 003B          DATA    >3B
0192 00A3 0014          DATA    >14
0193 00A4 FFEC          DATA    >FFEC
0194 00A5 FFC5          DATA    >FFC5
0195 00A6 FF9E          DATA    >FF9E
0196 00A7 FF77          DATA    >FF77
0197 00A8 FF50          DATA    >FF50
0198 00A9 FF2A          DATA    >FF2A
0199 00AA FF04          DATA    >FF04
0200 00AB FEDF          DATA    >FEDF
0201 00AC FEBA          DATA    >FEBA
0202 00AD FE96          DATA    >FE96
0203 00AE FE74          DATA    >FE74
0204 00AF FE52          DATA    >FE52
0205 00B0 FE31          DATA    >FE31
0206 00B1 FE11          DATA    >FE11
0207 00B2 FDF3          DATA    >FDF3
0208 00B3 FDD5          DATA    >FDD5
0209 00B4 FDBA          DATA    >FDBA
0210 00B5 FD9F          DATA    >FD9F
0211 00B6 FD86          DATA    >FD86
0212 00B7 FD6F          DATA    >FD6F
0213 00B8 FD59          DATA    >FD59
0214 00B9 FD44          DATA    >FD44
0215 00BA FD32          DATA    >FD32
```

```
0216 00BB FD21        DATA    >FD21
0217 00BC FD12        DATA    >FD12
0218 00BD FD04        DATA    >FD04
0219 00BE FCF9        DATA    >FCF9
0220 00BF FCEF        DATA    >FCEF
0221 00C0 FCE8        DATA    >FCE8
0222 00C1 FCE2        DATA    >FCE2
0223 00C2 FCDE        DATA    >FCDE
0224 00C3 FCDC        DATA    >FCDC
0225 00C4 FCDC        DATA    >FCDC
0226 00C5 FCDE        DATA    >FCDE
0227 00C6 FCE2        DATA    >FCE2
0228 00C7 FCE8        DATA    >FCE8
0229 00C8 FCEF        DATA    >FCEF
0230 00C9 FCF9        DATA    >FCF9
0231 00CA FD04        DATA    >FD04
0232 00CB FD12        DATA    >FD12
0233 00CC FD21        DATA    >FD21
0234 00CD FD32        DATA    >FD32
0235 00CE FD44        DATA    >FD44
0236 00CF FD59        DATA    >FD59
0237 00D0 FD6F        DATA    >FD6F
0238 00D1 FD86        DATA    >FD86
0239 00D2 FD9F        DATA    >FD9F
0240 00D3 FDBA        DATA    >FDBA
0241 00D4 FDD5        DATA    >FDD5
0242 00D5 FDF3        DATA    >FDF3
0243 00D6 FE11        DATA    >FE11
0244 00D7 FE31        DATA    >FE31
0245 00D8 FE52        DATA    >FE52
0246 00D9 FE74        DATA    >FE74
0247 00DA FE96        DATA    >FE96
0248 00DB FEBA        DATA    >FEBA
0249 00DC FEDF        DATA    >FEDF
0250 00DD FF04        DATA    >FF04
0251 00DE FF2A        DATA    >FF2A
0252 00DF FF50        DATA    >FF50
0253 00E0 FF77        DATA    >FF77
0254 00E1 FF9E        DATA    >FF9E
0255 00E2 FFC5        DATA    >FFC5
0256 00E3 FFEC        DATA    >FFEC
0257 00E4 0014        DATA    >14
0258 00E5 003B        DATA    >3B
0259 00E6 0062        DATA    >62
0260 00E7 0089        DATA    >89
0261 00E8 00B0        DATA    >B0
0262 00E9 00D6        DATA    >D6
0263 00EA 00FC        DATA    >FC
0264 00EB 0121        DATA    >121
0265 00EC 0146        DATA    >146
0266 00ED 016A        DATA    >16A
0267 00EE 018C        DATA    >18C
0268 00EF 01AE        DATA    >1AE
0269 00F0 01CF        DATA    >1CF
```

```
0270 00F1 01EF          DATA     >1EF
0271 00F2 020D          DATA     >20D
0272 00F3 022B          DATA     >22B
0273 00F4 0246          DATA     >246
0274 00F5 0261          DATA     >261
0275 00F6 027A          DATA     >27A
0276 00F7 0291          DATA     >291
0277 00F8 02A7          DATA     >2A7
0278 00F9 02BC          DATA     >2BC
0279 00FA 02CE          DATA     >2CE
0280 00FB 02DF          DATA     >2DF
0281 00FC 02EE          DATA     >2EE
0282 00FD 02FC          DATA     >2FC
0283 00FE 0307          DATA     >307
0284 00FF 0311          DATA     >311
0285 0100 0318          DATA     >318
0286 0101 031E          DATA     >31E
0287 0102 0322          DATA     >322
0288 0103 0324          DATA     >324
0289 0104
0290                    ****************************
0291                    *DATA MEMORY LOCATIONS USED *
0292                    ****************************
0293      0000 DELTA  EQU    0
0294      0001 ALPHA  EQU    1
0295      0002 SLOPE  EQU    2
0296      0003 SINA   EQU    3
0297      0004 TEMP   EQU    4             *SCRATCH PAD LOCATION
0298      0005 MASK1  EQU    5             *MASK FOR MODULO 128 POINTER
0299      0006 MASK2  EQU    6             *ISOLATE FRACTIONAL PART
0300      0007 OFSET1 EQU    7             *ADDRESS OF SINE TABLE
0301      0008 OFSET2 EQU    8             *ADDRESS OF SLOPE TABLE
0302                    ***********************************************
0303                    * NECESSARY INITIALIZATIONS:                  *
0304                    * MASK1   INITIALIZED TO >7FFF FOR 128 POINT TABLE *
0305                    * MASK2   INITIALIZED TO >0FFF                 *
0306                    * OFSET1  SET TO THE ADDRESS AT THE BEGINNING OF *
0307                    *         SINE TABLE                           *
0308                    * OFSET2  SET TO THE ADDRESS AT THE BEGINNING OF *
0309                    *         SLOPE TABLE WITH RESPECT TO SINE TABLE *
0310                    * ALPHA   INITIALLY CLEARED                    *
0311                    * DELTA   INITIALIZED TO INCREMENT VALUE USING *
0312                    *         Q8 FORMAT                            *
0313                    ***********************************************
0314 0104
0315 0104 6F00 START  LDP    0             *SET DATA PAGE POINTER.
0316 0105 7E04        LACK   SINE          *SINE TABLE ADDRESS.
0317 0106 5007        SACL   OFSET1
0318 0107 7E84        LACK   TSLOPE        *SLOPE TABLE ADDRESS.
0319 0108 1007        SUB    OFSET1
0320 0109 5008        SACL   OFSET2
0321 010A 7E02        LACK   M1            *RETRIEVE MASK1.
0322 010B 6705        TBLR   MASK1
0323 010C 7E03        LACK   M2            *RETRIEVE MASK2.
```

8. Precision Digital Sine-Wave Generation with the TMS32010

```
0324 010D 6706          TBLR    MASK2
0325 010E 7F89          ZAC
0326 010F 5001          SACL    ALPHA           *START ANGLE AT ZERO.
0327 0110 4100          IN      DELTA,PA1       *FOR THIS EXAMPLE,
0328 0111 F800   L1     CALL    SWAVE2          *DELTA IS INPUT.
     0112 0115'
0329                    ****************************
0330                    *    REST OF PROGRAM       *
0331                    ****************************
0332 0113 F900          B       L1
     0114 0111'
0333 0115
0334 0115
0335                    **********************************************
0336                    * SINE WAVE SUBROUTINE:                      *
0337                    * THIS ROUTINE COMPUTES THE SINE OF AN ANGLE AND     *
0338                    * RETURNS THE VALUE IN DATA LOCATION 'SINA'.  IT USES *
0339                    * A FRACTIONAL STEP SIZE TO AUTOMATICALLY COMPUTE THE *
0340                    * ADDRESS OF THE NEXT POINT OF THE SINE WAVE. IT     *
0341                    * TAKES 5.0 microseconds TO EXECUTE.          *
0342                    **********************************************
0343 0115 2801   SWAVE2 LAC     ALPHA,8
0344 0116 5804          SACH    TEMP            *ISOLATE INTEGER PORTION
0345 0117 2004          LAC     TEMP
0346 0118 0007          ADD     OFSET1
0347 0119 6703          TBLR    SINA            *GET CLOSEST LOWER SINE-
0348 011A 0008          ADD     OFSET2          *TABLE ENTRY (b).
0349 011B 6702          TBLR    SLOPE           *SLOPE FOR CORRESPONDING INTERVA
0350 011C 2401          LAC     ALPHA,4         *ISOLATE FRACTIONAL PORTION
0351 011D 7906          AND     MASK2           *AND MAKE Q8 NUMBER INTO Q12.
0352 011E 5004          SACL    TEMP
0353 011F 6A04          LT      TEMP
0354 0120 6D02          MPY     SLOPE           * mx
0355 0121 7F8E          PAC
0356 0122 0C03          ADD     SINA,12
0357 0123 5C03          SACH    SINA,4          * y = mx + b.
0358 0124 2001          LAC     ALPHA           *COMPUTE ADDRESS OF NEXT POINT.
0359 0125 0000          ADD     DELTA           *OF WAVE.
0360 0126 7905          AND     MASK1           *CLEAR MOST SIGNIFICANT BIT.
0361 0127 5001          SACL    ALPHA           *SAVE NEXT ADDRESS.
0362 0128 7F8D          RET                     *RETURN TO MAIN PROGRAM.
0363                    END
NO ERRORS, NO WARNINGS
```

8. Precision Digital Sine-Wave Generation with the TMS32010   289

9. Matrix Multiplication with the TMS32010 and TMS32020

Charles Crowell
Digital Signal Processing - Semiconductor Group
Texas Instruments

INTRODUCTION

Matrix multiplication is useful in applications such as graphics, numerical analysis, or high-speed control. The purpose of this application report is to illustrate matrix multiplication on two digital signal processors, the TMS32010 and TMS32020.

Both the TMS32010 and TMS32020 can multiply any two matrices of size $M \times N$ and $N \times P$. The programs for the TMS32010 and TMS32020, included in the appendices, can multiply large matrices and are only limited by the amount of internal data RAM available. Assuming a 200-ns cycle time, the TMS32010 and TMS32020 can calculate $[1 \times 3] \times [3 \times 3]$ in 5.4 microseconds.

Before discussing the two versions of implementing a matrix multiplication algorithm, a brief review of matrix multiplication is presented along with three examples of graphics applications.

MATRIX MULTIPLICATION

The size of a matrix is defined by the number of rows and columns it contains. For example, the following is a 5×3 matrix since it contains five rows and three columns.

$$
A = \begin{bmatrix}
a_{11} & a_{12} & a_{13} \\
a_{21} & a_{22} & a_{23} \\
a_{31} & a_{32} & a_{33} \\
a_{41} & a_{42} & a_{43} \\
a_{51} & a_{52} & a_{53}
\end{bmatrix}
$$

Any two matrices can be multiplied together as long as the second matrix has the same number of rows as the first has of columns. This condition is called conformability. For example, if a matrix A is an $M \times N$ matix and a matrix B is an $N \times P$ matrix, then the two can be multiplied together with the resulting matrix being of size $M \times P$.

$$
A = \begin{bmatrix} 3 & 4 \\ 2 & 7 \end{bmatrix} \qquad B = \begin{bmatrix} 4 \\ 6 \end{bmatrix} \qquad AB = \begin{bmatrix} 36 \\ 50 \end{bmatrix}
$$

$$M \times N = 2 \times 2 \qquad N \times P = 2 \times 1 \qquad M \times P = 2 \times 1$$

Example: $(3)(4) + (4)(6) = 36$

Given the two conformable matrices A and B, the elements of $C = A \times B$ are given by:

$$
C_{ij} = \sum_{k=1}^{N} a_{ik} \times b_{kj}
$$

for $i = 1,\ldots,M$ and $j = 1,\ldots,P$

Q12 FORMAT

Applications often require multiplication of mixed numbers. Since the TMS32010 and TMS32020 implement fixed-point arithmetic, the programs in the appendices assume a Q12 format, i.e., 12 bits follow an assumed binary point. The bits to the right of the assumed binary point represent the fractional part of the number and the four bits to the left represent the integer part of the number. An example of Q12 format is as follows:

$$0\,0\,0\,1.1\,1\,0\,1\,1\,1\,1\,0\,0\,0\,0\,0 = 1.866$$

$\uparrow$

ASSUMED BINARY POINT

$$
\begin{array}{rl}
& 0000.110111100000 = 0.866 \text{ in Q12} \\
\times & 0000.100000000000 = 0.5 \text{ in Q12} \\
\hline
00000000.&011011110000000000000000 = 0.433 \text{ in Q24}
\end{array}
$$

The result of a Q12 by Q12 multiplication is a number in a Q24 format that can easily be converted to Q12 by a logical left-shift of four. The first four bits will be lost as well as the last twelve, but these bits are insignificant for Q12. Note that the programs in the appendices provide no protection against overflow; therefore, the design engineer should implement a format that best fits the application.

GRAPHICS APPLICATIONS

Operations in graphics applications, such as translation, scaling, or rotation, require matrix manipulations to be performed in a limited amount of time. Therefore, the TMS32010 and TMS32020 processors are ideal for these applications. Graphics applications, such as scaling and rotation of points in a coordinate system, require multiplication of matrices. Translation is typically implemented by addition of two matrices. However, when points are represented in a homogeneous coordinate system, translation can be implemented by multiplication. In a homogeneous coordinate system, a point $P(x,y)$ is represented as $P(X,Y,1)$. This type of coordinate system is desirable since it relates translation with scaling and rotation.

Translation can be defined as the moving of a point or points in a coordinate system from one location to another without rotating. This is accomplished by adding a displacement value D_x to the X coordinate of a point and adding a displacement value D_y to the Y coordinate, thus moving the point from one location to another. Figure 1 shows both addition and multiplication methods of translation and an example of each.

Similar to translation, scaling can be implemented by matrix multiplication. Points can be scaled by multiplying

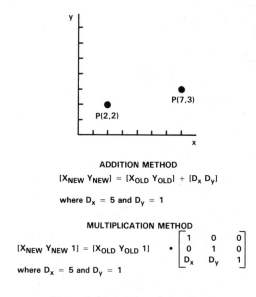

ADDITION METHOD

$$[X_{NEW}\ Y_{NEW}] = [X_{OLD}\ Y_{OLD}] + [D_x\ D_y]$$

where $D_x = 5$ and $D_y = 1$

MULTIPLICATION METHOD

$$[X_{NEW}\ Y_{NEW}\ 1] = [X_{OLD}\ Y_{OLD}\ 1] \bullet \begin{bmatrix} 1 & 0 & 0 \\ 0 & 1 & 0 \\ D_x & D_y & 1 \end{bmatrix}$$

where $D_x = 5$ and $D_y = 1$

Figure 1. Translation of Coordinates

each coordinate of a point (or points) by a scaling value S_x and S_y. Scaling an object is similar to stretching or shrinking an object. The coordinates of each point that makes up the object are multiplied by a scaling value which scales the object to a larger or smaller scale. Figure 2 shows the scaling of an object from one size to another.

Rotation of the coordinates of a point (or points) about an angle theta can also be accomplished by a matrix multiplication. The following set of equations results with the matrix multiplication required to rotate an object about any angle.

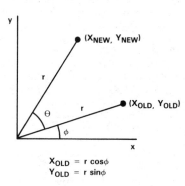

$$X_{OLD} = r\cos\phi$$
$$Y_{OLD} = r\sin\phi$$

$$X_{NEW} = r\cos(\theta+\phi) = r\cos\phi\cos\theta - r\sin\phi\sin\theta$$
$$Y_{NEW} = r\sin(\theta+\phi) = r\cos\phi\sin\theta + r\sin\phi\cos\theta$$

$$X_{NEW} = X_{OLD}\cos\theta - Y_{OLD}\sin\theta$$
$$Y_{NEW} = X_{OLD}\sin\theta + Y_{OLD}\cos\theta$$

OR

$$[X_{NEW}\ Y_{NEW}\ 1] = [X_{OLD}\ Y_{OLD}\ 1] \bullet \begin{bmatrix} \cos\theta & \sin\theta & 0 \\ -\sin\theta & \cos\theta & 0 \\ 0 & 0 & 1 \end{bmatrix}$$

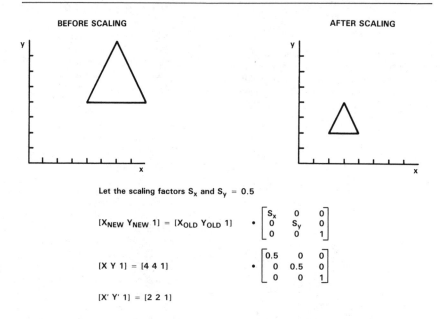

Let the scaling factors S_x and $S_y = 0.5$

$$[X_{NEW}\ Y_{NEW}\ 1] = [X_{OLD}\ Y_{OLD}\ 1] \bullet \begin{bmatrix} S_x & 0 & 0 \\ 0 & S_y & 0 \\ 0 & 0 & 1 \end{bmatrix}$$

$$[X\ Y\ 1] = [4\ 4\ 1] \bullet \begin{bmatrix} 0.5 & 0 & 0 \\ 0 & 0.5 & 0 \\ 0 & 0 & 1 \end{bmatrix}$$

$$[X'\ Y'\ 1] = [2\ 2\ 1]$$

Figure 2. Scaling From One Size To Another

9. Matrix Multiplication with the TMS32010 and TMS32020

Figure 3 shows an implementation of these equations to rotate an object 30 degrees about the origin.

Figures 4 and 5 show a segment of straight-line TMS32010 and TMS32020 code, respectively. These programs calculate the coordinate rotation example using a Q12 format. Note that once the matrices are loaded into memory, the procssors can calculate the results in 5.4 microseconds. The segment of TMS32020 code in Figure 5 implements the MAC instruction. For small matrices, the MAC instruction in conjunction with the RPT instruction gains little due to the overhead timing of the MAC instruction. However, for larger matrices, this method is most efficient since the MAC instruction becomes single-cycle in the repeat mode. For applications that only require translation, scaling, or rotation of coordinates, straight-line code as in Figures 4 and 5 is more efficient than the larger programs in the appendices.

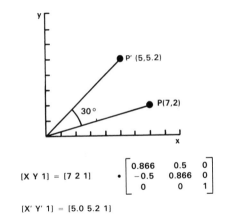

$$[X\ Y\ 1] = [7\ 2\ 1] \cdot \begin{bmatrix} 0.866 & 0.5 & 0 \\ -0.5 & 0.866 & 0 \\ 0 & 0 & 1 \end{bmatrix}$$

$$[X'\ Y'\ 1] = [5.0\ 5.2\ 1]$$

Figure 3. Implementation of Rotation Matrix

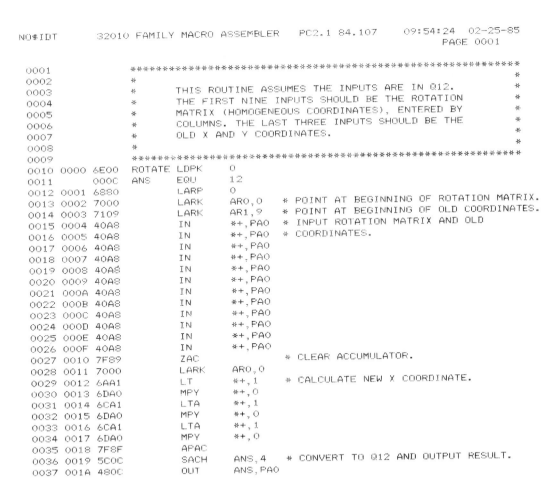

```
NO$IDT        32010 FAMILY MACRO ASSEMBLER    PC2.1 84.107    09:54:24  02-25-85
                                                                        PAGE 0001


0001          ***********************************************************
0002          *                                                        *
0003          *       THIS ROUTINE ASSUMES THE INPUTS ARE IN Q12.      *
0004          *       THE FIRST NINE INPUTS SHOULD BE THE ROTATION     *
0005          *       MATRIX (HOMOGENEOUS COORDINATES), ENTERED BY     *
0006          *       COLUMNS. THE LAST THREE INPUTS SHOULD BE THE     *
0007          *       OLD X AND Y COORDINATES.                         *
0008          *                                                        *
0009          ***********************************************************
0010 0000 6E00  ROTATE LDPK   0
0011      000C  ANS    EQU    12
0012 0001 6880         LARP   0
0013 0002 7000         LARK   AR0,0     * POINT AT BEGINNING OF ROTATION MATRIX.
0014 0003 7109         LARK   AR1,9     * POINT AT BEGINNING OF OLD COORDINATES.
0015 0004 40A8         IN     *+,PA0    * INPUT ROTATION MATRIX AND OLD
0016 0005 40A8         IN     *+,PA0    * COORDINATES.
0017 0006 40A8         IN     *+,PA0
0018 0007 40A8         IN     *+,PA0
0019 0008 40A8         IN     *+,PA0
0020 0009 40A8         IN     *+,PA0
0021 000A 40A8         IN     *+,PA0
0022 000B 40A8         IN     *+,PA0
0023 000C 40A8         IN     *+,PA0
0024 000D 40A8         IN     *+,PA0
0025 000E 40A8         IN     *+,PA0
0026 000F 40A8         IN     *+,PA0
0027 0010 7F89         ZAC              * CLEAR ACCUMULATOR.
0028 0011 7000         LARK   AR0,0
0029 0012 6AA1         LT     *+,1      * CALCULATE NEW X COORDINATE.
0030 0013 6DA0         MPY    *+,0
0031 0014 6CA1         LTA    *+,1
0032 0015 6DA0         MPY    *+,0
0033 0016 6CA1         LTA    *+,1
0034 0017 6DA0         MPY    *+,0
0035 0018 7F8F         APAC
0036 0019 5C0C         SACH   ANS,4     * CONVERT TO Q12 AND OUTPUT RESULT.
0037 001A 480C         OUT    ANS,PA0
```

Figure 4. TMS32010 Code for Rotation

9. Matrix Multiplication with the TMS32010 and TMS32020

```
0038 001B 7F89        ZAC
0039 001C 7109        LARK      AR1,9     * CALCULATE NEW Y COORDINATES.
0040 001D 6AA1        LT        *+,1
0041 001E 6DA0        MPY       *+,0
0042 001F 6CA1        LTA       *+,1
0043 0020 6DA0        MPY       *+,0
0044 0021 6CA1        LTA       *+,1
0045 0022 6DA0        MPY       *+,0
0046 0023 7F8F        APAC
0047 0024 5C0C        SACH      ANS,4     * CONVERT TO Q12 AND OUTPUT RESULT.
0048 0025 480C        OUT       ANS,PAO
0049 0026 7F89        ZAC
0050 0027 7109        LARK      AR1,9     * FINISH HOMOGENEOUS MATRIX.
0051 0028 6AA1        LT        *+,1
0052 0029 6DA0        MPY       *+,0
0053 002A 6CA1        LTA       *+,1
0054 002B 6DA0        MPY       *+,0
0055 002C 6CA1        LTA       *+,1
0056 002D 6DA0        MPY       *+,0
0057 002E 7F8F        APAC
0058 002F 5C0C        SACH      ANS,4
0059 0030 480C        OUT       ANS,PAO
0060 0031 7F8D        RET
NO ERRORS, NO WARNINGS
```

Figure 4. TMS32010 Code for Rotation (Concluded)

```
0001             *****************************************************************
0002             *                                                               *
0003             *         THIS ROUTINE ASSUMES THE INPUTS ARE IN Q12.            *
0004             *         THE FIRST NINE INPUTS SHOULD BE THE ROTATION           *
0005             *         MATRIX (HOMOGENEOUS COORDINATES), ENTERED BY           *
0006             *         COLUMNS. THE LAST THREE INPUTS SHOULD BE THE           *
0007             *         OLD X AND Y COORDINATES.                               *
0008             *                                                               *
0009             *****************************************************************
0010 0000 5589   ROTATE LARP   1            * USE AUXILIARY REGISTER 1.
0011      000C   ANS    EQU    12
0012 0001 CA00          ZAC                 * INITIALIZE ACCUMULATOR.
0013 0002 C806          LDPK   6
0014 0003 D100          LRLK   AR1,>300     * LOAD ROTATION MATRIX INTO B1.
     0004 0300
0015 0005 CB08          RPTK   8
0016 0006 80A0          IN     *+,PA0
0017 0007 D10'          LRLK   AR1,>200     * LOAD COORDINATES INTO BLOCK B0.
     0008 0200
0018 0009 CB02          RPTK   2
0019 000A 80A0          IN     *+,PA0
0020 000B CE05          CNFP                * CONFIGURE B0 AS PROGRAM MEMORY.
0021 000C A000          MPYK   >0           * CLEAR P REGISTER.
0022 000D D100          LRLK   AR1,>300
     000E 0300
0023 000F CB02          RPTK   2
0024 0010 5DA0          MAC    >FF00,*+     * CALCULATE THE NEW X COORDINATE.
     0011 FF00
0025 0012 CE15          APAC
0026 0013 6C0C          SACH   ANS,4
0027 0014 E00C          OUT    ANS,PA0      * OUTPUT NEW X COORDINATE.
0028 0015 A000          MPYK   >0           * CLEAR P REGISTER.
0029 0016 CA00          ZAC
0030 0017 CB02          RPTK   2
0031 0018 5DA0          MAC    >FF00,*+     * CALCULATE NEW Y COORDINATE.
     0019 FF00
0032 001A CE15          APAC
0033 001B 6C0C          SACH   ANS,4
0034 001C E00C          OUT    ANS,PA0      * OUTPUT NEW Y COORDINATE.
0035 001D A000          MPYK   >0           * CLEAR P REGISTER.
0036 001E CA00          ZAC
0037 001F CB02          RPTK   2
0038 0020 5DA0          MAC    >FF00,*+     * FINISH HOMOGENEOUS MATRIX.
     0021 FF00
0039 0022 CE15          APAC
0040 0023 6C0C          SACH   ANS,4
0041 0024 E00C          OUT    ANS,PA0
0042 0025 CE26          RET
NO ERRORS, NO WARNINGS
```

Figure 5. TMS32020 Code for Rotation

To combine translation, scaling, and rotation, a more general matrix can be implemented.

GENERAL MATRIX FOR TWO-DIMENSIONAL SYSTEMS

$$
\begin{bmatrix}
r_{11} & r_{12} & 0 \\
r_{21} & r_{22} & 0 \\
t_x & t_y & 1
\end{bmatrix}
$$

The upper 2×2 matrix is a combination rotation matrix and scaling matrix. The t_x and t_y values are the translation values. A three-dimensional general matrix can be developed similar to the two-dimensional translation, scaling, and rotation matrix.

GENERAL MATRIX FOR THREE-DIMENSIONAL SYSTEMS

$$
\begin{bmatrix}
r_{11} & r_{12} & r_{13} & 0 \\
r_{21} & r_{22} & r_{23} & 0 \\
r_{31} & r_{32} & r_{33} & 0 \\
t_x & t_y & t_z & 1
\end{bmatrix}
$$

IMPLEMENTATION OF THE MATRIX MULTIPLICATION ALGORITHM FOR THE TMS32010

The implementation of the algorithm for the TMS32010 shown in Figure 6 assumes that the two matrices to be multiplied together are of size $M \times N$ and $N \times P$. Three major

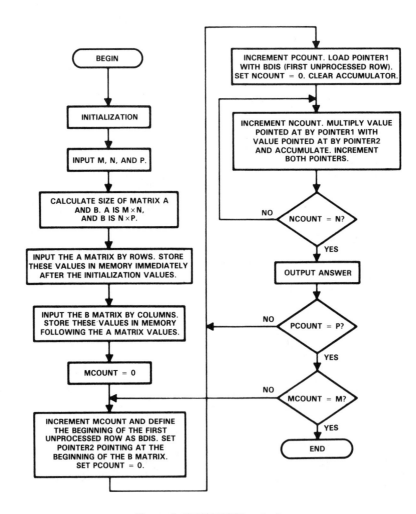

Figure 6. TMS32010 Flowchart

9. Matrix Multiplication with the TMS32010 and TMS32020

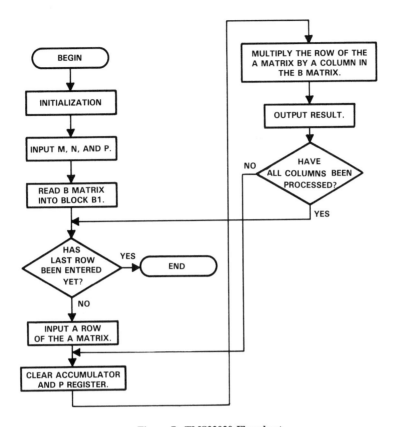

Figure 7. TMS32020 Flowchart

loops are included to multiply the two matrices. The outside loop control is labeled MCOUNT since it controls which row in the A matrix is being referenced during the multiplication. The secondary loop control is labeled PCOUNT because it counts how many columns in the B matrix have been processed. The inside loop control is labeled NCOUNT since it controls the multiplication of the values in the A matrix with the values in the B matrix.

IMPLEMENTATION OF THE MATRIX MULTIPLICATION ALGORITHM FOR THE TMS32020

The implementation of the algorithm for the TMS32020 is somewhat different since its advanced instruction set allows for a more efficient method of computing matrix multiplication. The TMS32020 version in Figure 7 also assumes that the two matrices to be multiplied are of size $M \times N$ and $N \times P$. This program takes a row of the A matrix,

loads it into block B0 of data memory, and then multiplies this row by all columns in the B matrix. The TMS32020 continues this process until all the rows in the A matrix have been multiplied by all the columns in the B matrix. The TMS32020 version is similar to the TMS32010 in that the A matrix must be entered by rows and the B matrix by columns. This allows for a faster execution time. Figure 7 shows the basic implementation of the matrix multiplication algorithm that the TMS32020 uses to multiply two matrices.

Since the programs in the appendices treat the matrices differently, a memory map is included to help in understanding the two versions. Figure 8 shows how the matrices should look in memory after they have been entered. Note that for the TMS32020 version, the A matrix values reside in program memory since the CNFP (configure as program memory) instruction was implemented. Note also that only one row of the A matrix is in this block since the program enters one row at a time.

For the following matrices,

$$A = \begin{bmatrix} a_{11} & a_{12} \\ a_{21} & a_{22} \end{bmatrix} \quad B = \begin{bmatrix} b_{11} & b_{12} & b_{13} \\ b_{21} & b_{22} & b_{23} \end{bmatrix}$$

the memory would be configured in this manner for the TMS32010 and TMS32020.

TMS32010 DATA MEMORY		TMS32020 DATA MEMORY		PROGRAM MEMORY	
LOCATION (IN HEX)	VALUE	LOCATION (IN HEX)	VALUE	LOCATION (IN HEX)	VALUE
>00F	a_{11}	>308	b_{11}	>FF00	a_{i1}
>010	a_{12}	>309	b_{21}	>FF01	a_{i2}
>011	a_{21}	>30A	b_{12}		
>012	a_{22}	>30B	b_{22}		
>013	b_{11}	>30C	b_{13}		
>014	b_{21}	>30D	b_{23}		
>015	b_{12}				
>016	b_{22}				
>017	b_{13}				
>018	b_{23}				

Figure 8. Memory Maps

SUMMARY

The TMS32010 and TMS32020 processors can be used to multiply large matrices efficiently. A brief review of matrix multiplication has been given to assist in the understanding of fundamental matrix multiplication. Three examples of graphics applications have been presented since these applications often require multiplication of matrices.

The TMS320 family has the power and flexibility to cost-effectively implement a wide range of high-speed graphics, numerical analysis, digital signal processing, and control applications. Since the TMS32010 and TMS32020 combine the flexibility of a high-speed controller with the numerical capability of an array processor, a new approach to applications such as graphics can now be considered.

REFERENCES

1. J.D. Foley and A. Van Dam, *Fundamentals of Interactive Commputer Graphics*, Addison-Wesley Publishing Company, Inc. (1982).
2. S.D. Conte and Carl de Boor, *Elementary Numerical Analysis*, McGraw-Hill, Inc. (1980).

Appendix A

```
0001              ********************************************************
0002              *    ALL INPUTS AND OUTPUTS FOR THIS PROGRAM SHOULD  *
0003              *    BE OR ARE IN Q12 FORMAT EXCEPT FOR THE M, N,     *
0004            . *    AND P INPUTS, WHICH SHOULD BE Q0.                *
0005              ********************************************************
0006 0000                AORG      0
0007      0000    M      EQU       >0
0008      0001    N      EQU       >1
0009      0002    P      EQU       >2
0010      0003    C1     EQU       >3
0011      0004    C2     EQU       >4
0012      0005    C3     EQU       >5
0013      0006    ANS    EQU       >6
0014      0007    ADIS   EQU       >7
0015      0008    BDIS   EQU       >8
0016      0009    CDIS   EQU       >9
0017      000A    TEMP   EQU       >A
0018      000B    COI    EQU       >B
0019      000C    COS    EQU       >C
0020      000D    T      EQU       >D
0021      000E    ONE    EQU       >E
0022             *
0023             * INITIALIZATION
0024             *
0025 0000 6E00         LDPK      0
0026 0001 6880         LARP      0
0027 0002 7E0F         LACK      15
0028 0003 500C         SACL      COS
0029 0004 500D         SACL      T
0030 0005 7E01         LACK      1
0031 0006 500E         SACL      ONE
0032             *
0033             * MATRIX A IS M x N AND MATRIX B IS N x P.
0034             * THESE STATEMENTS READ IN THE SIZES OF
0035             * THE TWO MATRICES.
0036             *
0037 0007 4000         IN        M,PAO
0038 0008 4001         IN        N,PAO
0039 0009 4002         IN        P,PAO
0040             *
0041             * CALCULATE THE LENGTH OF THE A MATRIX AND
0042             * STORE THIS VALUE IN ADIS.
0043             *
0044 000A 6A00         LT        M
0045 000B 6D01         MPY       N
0046 000C 7F8E         PAC
0047 000D 5007         SACL      ADIS
0048             *
0049             * CALCULATE THE LENGTH OF THE B MATRIX AND
0050             * STORE THIS VALUE IN BDIS.
0051             *
0052 000E 6A01         LT        N
0053 000F 6D02         MPY       P
0054 0010 7F8E         PAC
0055 0011 5008         SACL      BDIS
0056             *
0057             * POINT AT THE END OF THE INITIAL DATA.
0058             *
0059 0012 380C         LAR ARO,COS
```

9. Matrix Multiplication with the TMS32010 and TMS32020

```
0060              *
0061              * READ THE A MATRIX VALUES INTO DATA RAM.
0062              * THIS MATRIX MUST BE ENTERED BY ROWS.
0063              * THE MATRIX VALUES WILL BE LOCATED IN
0064              * DATA RAM FOLLOWING THE INITIALIZATION
0065              * VALUES.
0066              *
0067 0013 200B  FST   LAC    COI
0068 0014 000E        ADD    ONE
0069 0015 500B        SACL   COI
0070 0016 4088        IN     *,PA0
0071 0017 68A8        MAR    *+
0072 0018 2007        LAC    ADIS
0073 0019 100B        SUB    COI
0074 001A FE00        BNZ    FST
     001B 0013
0075              *
0076              * RESET COUNTER TO READ IN THE B MATRIX VALUES.
0077              *
0078 001C 7F89        ZAC
0079 001D 500B        SACL   COI
0080              *
0081              * READ THE B MATRIX VALUES INTO DATA RAM.
0082              * UNLIKE THE A MATRIX, THESE VALUES MUST BE
0083              * ENTERED BY COLUMNS.  THESE VALUES WILL BE
0084              * LOCATED IN DATA RAM FOLLOWING THE A MATRIX VALUES.
0085              *
0086              *
0087 001E 200B  SND   LAC    COI
0088 001F 000E        ADD    ONE
0089 0020 500B        SACL   COI
0090 0021 4088        IN     *,PA0
0091 0022 68A8        MAR    *+
0092 0023 2008        LAC    BDIS
0093 0024 100B        SUB    COI
0094 0025 FE00        BNZ    SND
     0026 001E
0095              *
0096              * MORE INITIALIZATION
0097              *
0098 0027 200D        LAC    T
0099 0028 1001        SUB    N
0100 0029 5003        SACL   C1
0101 002A 200D        LAC    T
0102 002B 0007        ADD    ADIS
0103 002C 500D        SACL   T
0104 002D 1001        SUB    N
0105 002E 5007        SACL   ADIS
0106              *
0107              * CALCULATE A × B
0108              *
0109              *
0110              *
0111              *
0112              *                      N
0113              *                    _____
0114              *                    \
0115              * OUTPUT(i,) =        \  A(ik) × B(kj)
0116              *                     /
0117              *                    /
0118              *                    -----
0119              *                    k = 1
0120              *
0121 002F 2003  FS    LAC    C1
0122 0030 0001        ADD    N
```

302 9. Matrix Multiplication with the TMS32010 and TMS32020

```
0123 0031 5003          SACL    C1
0124 0032 6881          LARP    1
0125 0033 390D          LAR     AR1,T
0126 0034 6880          LARP    0
0127 0035 7F89          ZAC
0128 0036 5004          SACL    C2
0129 0037 2004    SN    LAC     C2
0130 0038 000E          ADD     ONE
0131 0039 5004          SACL    C2
0132 003A 3803          LAR     AR0,C1
0133 003B 7F89          ZAC
0134 003C 5006          SACL    ANS
0135 003D 5005          SACL    C3
0136 003E 2005    TH    LAC     C3
0137 003F 000E          ADD     ONE
0138 0040 5005          SACL    C3
0139 0041 6506          ZALH    ANS
0140 0042 6AA1          LT      *+,AR1
0141 0043 6DA0          MPY     *+,AR0
0142 0044 7F8F          APAC
0143 0045 5806          SACH    ANS
0144 0046 2005          LAC     C3
0145 0047 1001          SUB     N
0146 0048 FE00          BNZ     TH
     0049 003E
0147               *
0148               * LOAD ACCUMULATOR WITH HIGH WORD OF Q24 RESULT.
0149               * LEFT-SHIFT FOUR TO CONVERT TO Q12.
0150               * NOTE THAT ONLY THE 12 MSB'S ARE SIGNIFICANT.
0151               *
0152 004A 2406          LAC     ANS,4
0153 004B 5006          SACL    ANS
0154 004C 4806          OUT     ANS,PA0
0155 004D 2004          LAC     C2
0156 004E 1002          SUB     P
0157 004F FE00          BNZ     SN
     0050 0037
0158 0051 2003          LAC     C1
0159 0052 1007          SUB     ADIS
0160 0053 FE00          BNZ     FS
     0054 002F
0161 0055 F900    QUIT  B       QUIT
     0056 0055
NO ERRORS, NO WARNINGS
```

Appendix B

```
0001                  ***************************************
0002                  *    ALL INPUTS AND OUTPUTS FOR THIS PROGRAM    *
0003                  *    SHOULD BE OR ARE IN Q12 FORMAT EXCEPT      *
0004                  *    FOR THE M, N, AND P, WHICH SHOULD BE Q0.   *
0005                  ***************************************
0006 0020                     AORG    32
0007       0000   M    EQU     >0
0008       0001   N    EQU     >1
0009       0002   P    EQU     >2
0010       0003   ANS  EQU     >3
0011       0004   BDM1 EQU     >4
0012       0005   ONE  EQU     >5
0013       0006   NM1  EQU     >6
0014       0007   PM1  EQU     >7 .
0015                  *
0016                  * INITIALIZATION
0017                  *
0018 0020 C806                LDPK    6
0019 0021 D100                LRLK    AR1,>300
     0022 0300
0020 0023 5589                LARP    1
0021 0024 CA01                LACK    >1
0022 0025 6005                SACL    ONE
0023                  *
0024                  * READ SIZES OF MATRICES.
0025                  *
0026 0026 CB02                RPTK    2
0027 0027 80A0                IN      *+,PA0
0028                  *
0029                  * MORE INITIALIZATION
0030                  *
0031 0028 2001                LAC     M
0032 0029 0005                ADD     ONE
0033 002A 6001                SACL    M
0034 002B 2000                LAC     N
0035 002C 1005                SUB     ONE
0036 002D 6006                SACL    NM1
0037 002E 3C00                LT      N
0038 002F 3802                MPY     P
0039 0030 CE14                PAC
0040 0031 1005                SUB     ONE
0041 0032 6004                SACL    BDM1
0042 0033 2002                LAC     P
0043 0034 1005                SUB     ONE
0044 0035 6007                SACL    PM1
0045                  *
0046                  * READ IN THE B MATRIX.
0047                  *
0048 0036 D100                LRLK    AR1,>308
     0037 0308
0049 0038 4B04                RPT     BDM1
0050 0039 80A0                IN      *+,PA0
0051 003A 2001   CALLER LAC    M
0052 003B 1005                SUB     ONE
0053 003C 6001                SACL    M
0054 003D F680                BZ      QT
     003E 0052
0055                  *
0056                  * CALL ROUTINE TO READ IN A ROW
```

```
0057                   * OF THE A MATRIX.
0058                   *
0059 003F FE80             CALL    IO
     0040 0053
0060 0041 D100             LRLK    AR1,>308
     0042 0308
0061 0043 5589             LARP    1
0062 0044 3007             LAR     ARO,PM1
0063                   *
0064                   * CLEAR ACCUMULATOR AND P REGISTER.
0065                   *
0066 0045 A000   MUL       MPYK    0
0067 0046 CA00             ZAC
0068                   *
0069                   * MULTIPLY A ROW BY A COLUMN.
0070                   *
0071 0047 4B06             RPT     NM1
0072 0048 5DA0             MAC     >FF00,*+
     0049 FF00
0073 004A CE15             APAC
0074                   *
0075                   * OUTPUT RESULT.
0076                   *
0077 004B 6C03             SACH    ANS,4
0078 004C E003             OUT     ANS,PAO
0079 004D 5588             LARP    0
0080                   *
0081                   * CHECK TO SEE IF ALL COLUMNS HAVE BEEN PROCESSED.
0082                   *
0083 004E FB99             BANZ    MUL,*-,1
     004F 0045
0084                   *
0085                   *     GO GET NEXT ROW.
0086                   *
0087 0050 FF80             B       CALLER
     0051 003A
0088 0052 CE1F   QT        IDLE
0089 0053 CE04   IO        CNFD
0090 0054 5589             LARP    1
0091 0055 D100             LRLK    AR1,>200
     0056 0200
0092 0057 4B06             RPT     NM1
0093 0058 80A0             IN      *+,PAO
0094 0059 CE05             CNFP
0095 005A CE26             RET
NO ERRORS, NO WARNINGS
```

9. Matrix Multiplication with the TMS32010 and TMS32020

10. Interfacing to Asynchronous Inputs with the TMS32010

Jon Bradley
Digital Signal Processing - Semiconductor Group
Texas Instruments

INTRODUCTION

Interrupt ($\overline{\text{INT}}$), reset ($\overline{\text{RS}}$), and branch on I/O ($\overline{\text{BIO}}$) are inputs to the TMS32010 microprocessor that are typically provided from asynchronous sources within a system. These three inputs are subject to certain considerations in addition to those relevant to signals that are synchronized to TMS32010 operation. Observing these considerations is important to insure reliable operation of these three input functions.

INTERRUPT AND $\overline{\text{BIO}}$ INPUTS

Interrupt ($\overline{\text{INT}}$) and $\overline{\text{BIO}}$ are the two inputs on which synchronization is most important.

The $\overline{\text{BIO}}$ input provides a convenient approach to implementing polled I/O on the TMS32010. $\overline{\text{BIO}}$, unlike $\overline{\text{INT}}$, is not latched internally, so the state of the $\overline{\text{BIO}}$ input must be maintained while the BIOZ instruction is executing. The previous state of $\overline{\text{BIO}}$ is irrelevant. When properly synchronized to CLKOUT, $\overline{\text{BIO}}$ is recognized by the TMS32010 on the next CLKOUT cycle. It should also be noted that the cycle during which $\overline{\text{BIO}}$ is sampled by the BIOZ instruction is the first of the two cycles that it takes BIOZ to execute.

Several aspects of interrupt operation are important for proper implementation of the interrupt function within a system. These fall into three basic categories:

1. Interrupt and input synchronization
2. Internal interrupt control logic
3. Interrupt programming.

Interrupt and $\overline{\text{BIO}}$ Input Synchronization

Synchronization of the interrupt and $\overline{\text{BIO}}$ inputs is absolutely critical in maintaining proper operation of these functions. This must be accomplished externally since the TMS32010 does not contain the necessary synchronizing logic. This requirement results from the fact that these inputs are sampled within the TMS32010 on the falling edge of CLKOUT. If the input level is changing at this point, the internal circuitry may receive an invalid logic level, causing unpredictable operation. Specifically, the input must be at a stable, valid logic level at least 50 ns before the falling edge of the CLKOUT signal, and must remain stable for at least one full CLKOUT cycle (200 ns for a 20-MHz TMS32010). These timing relationships are shown graphically in Figure 1.

The issue of synchronization has been addressed at length in numerous publications,[1] so details of the subject are not presented here. It can be shown that adequate synchronization for the TMS32010 may be accomplished by simply passing the interrupt input through a sufficiently fast

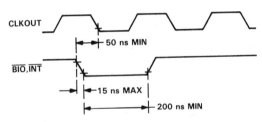

NOTE: Timing measurements are referenced to and from a low voltage of 0.8 V and a high voltage of 2.0 V.

Figure 1. Input Timing Relationships

positive edge-triggered flip-flop clocked by CLKOUT. The flip-flop used must have a clock-rising to output-valid delay of no more than 15 ns. Examples of circuits that accomplish input synchronization using a flip-flop are shown in Figure 2.

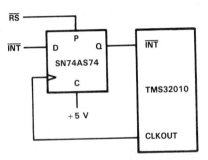

a. $\overline{\text{INT}}$ Synchronization Circuit

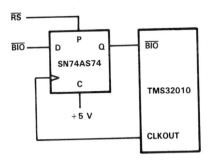

b. $\overline{\text{BIO}}$ Synchronization Circuit

Figure 2. Sample Input Synchronization Circuits

All further references to the interrupt or $\overline{\text{BIO}}$ functions in this application report assume that this synchronization has been provided. These functions should never be implemented without synchronization; behavior of unsynchronized inputs is unpredictable and not guaranteed. For example, the interrupt input signal travels through two different paths with unequal propagation delays. If asynchronous interrupts are allowed, interrupts may become disabled before an interrupt can be accepted. Since the pending interrupt is not accepted, the interrupt service routine is not executed. Therefore, probably no EINT instruction will be executed, since this instruction is unlikely to be included in normal program flow and interrupts will remain disabled unless the system is reset.

Internal Interrupt Control Logic

The interrupt input on the TMS32010 is both edge and level triggered, i.e., a low-going pulse on the interrupt input can generate an interrupt (provided the pulse is at least as long as one CLKOUT cycle). However, if the interrupt input goes low and remains low, the interrupt will remain pending until the input goes high and the interrupt is cleared

(see Figure 3). This results in several distinct situations which can occur with respect to how interrupts are serviced. Three specific cases are presented for illustration.

In the first case, shown in Figure 4, a low pulse occurs on the interrupt input. The falling edge of this pulse causes IFLAG to be set to one, resulting in IREQ and IACT going high (since $\overline{\text{INTM}}$ is high). Then, when the $\overline{\text{INT}}$ input returns to a one, IFLAG (and therefore, IREQ and IACT) remain high until the interrupt is accepted (according to the rules described in the next section). When the interrupt is accepted, the interrupt service routine is entered, and the IACK signal is pulsed, which disables the interrupt by setting the INTM bit. At the end of the interrupt service routine, interrupts are enabled using the EINT instruction, which clears the INTM bit. The important point in this case is that, once the interrupt input goes high and the IFLAG flip-flop is cleared, control returns directly to the mainline program following the interrupt service routine.

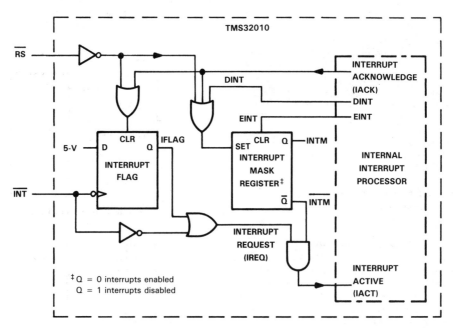

Figure 3. Internal Interrupt Control Logic

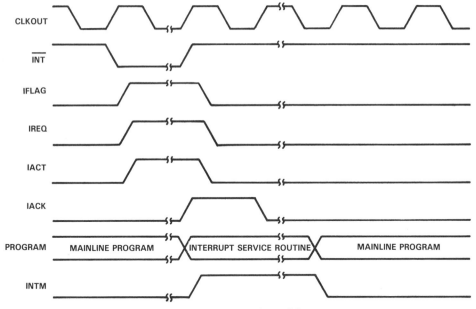

Figure 4. Pulse-Triggered Interrupt

In the second case, $\overline{\text{INT}}$ goes low and stays low for an indefinite period (see Figure 5). Here IFLAG is set to one, and IREQ and IACT go high as they did in the case of a pulsed interrupt. However, when the interrupt service routine is entered and the IACK pulse is generated, IFLAG is cleared, but IREQ remains high since $\overline{\text{INT}}$ is still low.

The result of this is that, when interrupts are reenabled at the end of the interrupt service routine, the interrupt will be accepted again, and the interrupt routine will be reentered without returning to the mainline program. This will continue until the interrupt is removed.

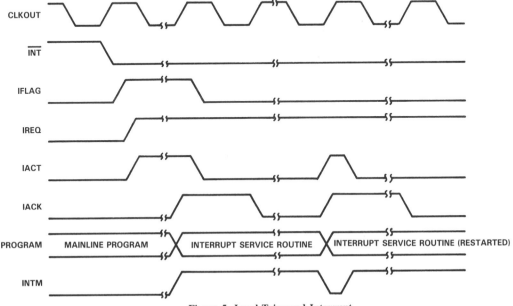

Figure 5. Level-Triggered Interrupt

The third case is a situation where, after a pulse has occurred on $\overline{\text{INT}}$ and the normal interrupt processing has begun, another interrupt occurs before servicing of the first interrupt is complete (see Figure 6). It is evident from the timing diagram that, once the first interrupt is cleared by the IACK pulse, another $\overline{\text{INT}}$ pulse may cause a second interrupt to become pending. This interrupt will be serviced immediately following the return from the first interrupt service routine (the mainline program will not be reentered until the second interrupt is serviced).

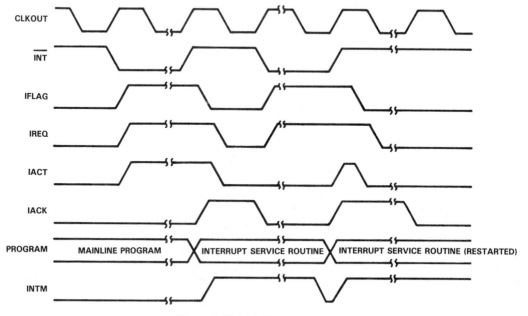

Figure 6. Multiple-Pulse Interrupts

10. Interfacing to Asynchronous Inputs with the TMS32010

The following general rule describes all of these situations: An interrupt may be generated either by the falling edge of a low-going pulse or by a low level on the $\overline{\text{INT}}$ input. It should be noted that this is distinctly different from many other types of systems in which an actual edge is required to produce an interrupt, and a level alone has no effect. Level triggering is provided on the TMS32010 in order to allow multiple external devices to interrupt the processor and be handled in a polled fashion.

Interrupt Programming

From a software standpoint, operation of interrupts on the TMS32010 is quite straightforward and very similar to interrupts on most other microprocessors, with only two exceptions.

Interrupt operation in general can be characterized by the following rules:

1. If an interrupt occurs while interrupts are enabled, it is accepted following completion of the instruction currently executing, even if the current instruction is a multicycle instruction.
2. If an interrupt occurs while interrupts are disabled, the interrupt remains pending and will be accepted as soon as interrupts are enabled, unless cleared in the meantime.
3. Pending interrupts are cleared by $\overline{\text{RS}}$ or by entry into the interrupt service routine unless $\overline{\text{INT}}$ has remained low.

Table I shows a summary of interrupt operation depending upon the circumstances under which the interrupt occurred. It should be emphasized that, although it may not be immediately obvious, most of these cases are a direct consequence of standard interrupt processing. The two exceptions to the general rules are those marked with an asterisk.

As an example of standard interrupt processing, consider the situation in which interrupts are enabled, and an interrupt occurs during the execution of a DINT (disable interrupts) instruction. Even though interrupts are enabled, the general rules state that an interrupt will not be accepted until the execution of the current instruction is completed. In this case, since the completion of this instruction results in interrupts being disabled, the interrupt will not be accepted.

Also, it should be noted that, even though some instructions take more than one cycle to execute, interrupt acceptance will always be delayed until the instruction has fully completed execution.

The first of the two exceptions to the general rules is the situation in which interrupts are enabled and an interrupt occurs during an MPY (or an MPYK) instruction. In this case, acceptance of the interrupt is delayed one more instruction cycle, until after the instruction following the MPY(K) has been executed. The reason for this is that the contents of the P Register (the results of the multiply) are very difficult to recreate. Delaying acceptance of interrupts an additional cycle guarantees that the program may safely store the results of the operation before trapping into the interrupt service routine, which might alter the contents of the P Register. (Thus, the user would normally code an instruction which stores the P Register into the accumulator immediately following a multiply instruction if interrupts are present in the system). However, if the interrupt service routine does not require the use of the P Register, it is not necessary that its contents be stored.

The second exception to the general rules occurs in the situation where interrupts are disabled (such as in an interrupt service routine), and an interrupt occurs during the execution of an EINT (enable interrupts) instruction. Here also, acceptance of the interrupt is delayed one additional instruction cycle, until after the instruction following the EINT has been executed. The reason for this is that, if the RET

Table 1. Interrupt Operation in Software

If Interrupt Occurs:	And Interrupts Are:	
	Enabled	Disabled
During DINT	The interrupt is ignored (but remains pending until cleared or an EINT is executed)	The interrupt is ignored (but remains pending until cleared or an EINT is executed)
During EINT	The interrupt is accepted after the EINT is executed	*The instruction following the EINT is completed and then the interrupt is accepted (unless the instruction is a DINT)
During MPY or MPYK	*The instruction following the MPY(K) is completed and then the interrupt is accepted (unless the instruction is a DINT)	The interrupt is ignored (but remains pending until cleared or an EINT is executed)
During any other instruction	The interrupt is accepted after the instruction executes (even if the instruction is multicycle)	The interrupt is ignored (but remains pending until cleared or an EINT is executed)

*Exception to standard interrupt servicing

instruction normally following an EINT in an interrupt service routine is not allowed to execute before another interrupt is accepted, a stack overflow can eventually result, destroying return linkage to the mainline program.

In these two exceptions to the general rules, note that if the instruction following an EINT (or MPY or MPYK) instruction is a DINT instruction, the interrupt will not be accepted.

The following sequence of instructions illustrates a combination of several of these concepts:

```
DINT
EINT
MPY     VAL1
MPY     VAL2
MPYK    CONST
DINT
```

An interrupt occurring at any point during this sequence of instructions would never be accepted (at least during this sequence). It would, however, remain pending until cleared or accepted in a later section of code following an EINT. The reasons for this are as follows: if the interrupt occurred during the DINT, it would wait until the instruction had executed before being considered for acceptance. Since the completion of this instruction results in interrupts being disabled, it would not be accepted at this point. If the interrupt occurred during the EINT or was still pending at this time, interrupts would still be disabled, and the interrupt would be ignored. After the EINT had executed (because of the two exceptions to the general rules), all of the multiplies would be executed. Then finally, after the MPYK, the DINT would also be executed (since it follows a multiply instruction) and would disable interrupts again.

$\overline{\text{RS}}$ INPUT

$\overline{\text{RS}}$ serves as the master reset for the TMS32010. Asynchronous inputs may be used on $\overline{\text{RS}}$ since this input is provided with an internal synchronizer circuit. However, if it is necessary to maintain synchronization of program execution to the reset input, the 50-ns setup time between $\overline{\text{RS}}$ and CLKOUT must be observed. This may be required, for example, when multiple TMS32010s are used concurrently in a system. Figure 7 shows the timing of the $\overline{\text{RS}}$ input.

It should be noted that reset should only be used to perform system initialization-type functions (such as would be used after powerup). Reset cannot be used to perform hold or interrupt functions, since it does not preserve any linkage to the previous program environment. Also, the state of the internal logic (including data RAM, status and general-purpose registers, accumulator contents, etc.) is not predictable after reset; once $\overline{\text{RS}}$ is driven low and then high, the results are simply that interrupts are disabled, the interrupt flag is cleared, and program execution is forced to location zero.

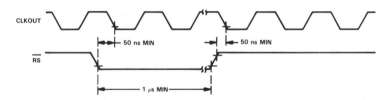

NOTE: Timing measurements are referenced to and from a low voltage of 0.8 V and a high voltage of 2.0 V.

Figure 7. $\overline{\text{RS}}$ Input Timing

SUMMARY

Adherence to the basic considerations described in this application report regarding these three inputs provides efficient and reliable system operation. Briefly restated, these considerations are as follows:

1. The interrupt and $\overline{\text{BIO}}$ inputs must be synchronized.
2. Interrupts are either level or edge triggered, and function according to standard interrupt processing, except for the special cases of the EINT, MPY, and MPYK instructions.
3. On the $\overline{\text{RS}}$ input, no synchronization is required, except in cases where precise timing immediately following reset is necessary.

REFERENCE

1. Stoll, Peter A., "How To Avoid Synchronization Problems," VLSI DESIGN, November/December 1982, pp. 56-59.

10. Interfacing to Asynchronous Inputs with the TMS32010

11. Interfacing External Memory to the TMS32010

Jon Bradley
Digital Signal Processing - Semiconductor Group
Texas Instruments

INTRODUCTION

External ROM or RAM can be interfaced to the memory bus of the TMS32010 microprocessor in applications that require additional RAM or program memory.

The purpose of this application report is to describe two basic low-cost methods for expanding the TMS32010's memory configuration:

1. Direct expansion, utilizing a standard memory cycle for memory access
2. Extended memory interface, utilizing an address (latched using a standard memory cycle) that automatically increments or decrements after each access.

Of the two methods, the first method is very efficient for program and small data memory expansion, whereas the second method is more useful for large data memory expansion.

The design techniques presented here can easily be extended to encompass interface of other devices to the TMS32010. Each of the circuits discussed in this application report has been built and tested to verify its operation.

RAM/ROM PROGRAM MEMORY EXPANSION

For systems requiring additional program memory or small amounts of external data memory, the direct expansion circuit described in this section provides a straightforward approach.

The TMS32010 program memory can be expanded beyond the 1.5K-word internal capacity in two ways:

1. Implementing an additional 2.5K words externally (MC/$\overline{\text{MP}}$ = 1), or
2. Implementing the full 4K-word program memory space externally (MC/$\overline{\text{MP}}$ = 0).

In either case, the memory is accessed in a single memory cycle and appears no different to the TMS32010 than internal program memory. The circuit described in this section uses the full 4K-word program memory space implemented externally using MC/$\overline{\text{MP}}$ = 0. This configuration is useful in applications where, perhaps for cost reasons, using a masked ROM version of the TMS32010 is impractical.

Design Considerations

An important consideration in the design of the direct expansion circuit is the desire to minimize chip count in order to reduce cost. This is an important factor in digital signal processing (DSP) systems since their cost must compete with that of analog approaches. In this circuit, as little additional logic as possible is used without sacrificing performance.

The memories used in the circuit are chosen to provide minimum chip count using currently available devices. The circuit is configured with half of the address space implemented as RAM and the other half as PROM. The RAMs used are Advanced Micro Devices Am9128-70 2K x 8 static NMOS RAMs, and the PROMs used are Texas Instruments TBP28S166 2K x 8 TTL PROMs. This memory configuration results in a minimum chip count and provides an even mix between RAM and PROM; however, other RAM/ROM mixes may be used. Note that if PROMs or ROMs are the only external memory required in the system, no additional logic is needed since the interface to most of these devices allows direct connection to the TMS32010.

Using RAM as program memory allows downloading into the address space from slower (possibly EPROM) memory or a host system if required, and also allows for communication between program and data memory spaces using the Table Read (TBLR) and Table Write (TBLW) instructions. If internal program ROM and external program RAM are required, the same external memory configuration may be used with MC/$\overline{\text{MP}}$ set to 1. Note that in this case, external RAM located at addresses coincident with those of the internal program ROM cannot be used without some modification of the address decoding scheme.

Functional Description

The direct memory expansion circuit, shown in Figure 1, consists of the four memory chips, a set of address bus buffers, and logic that controls the address bus buffers and enables the memories' three-state outputs.

The 2K words of PROM are located at addresses 0-7FF, preventing any conflict with I/O ports if present in the system, and RAM is located at addresses 800-FFF. PROM or RAM is selected on the basis of the most significant address line using a single inverter (U8:8,9).

Circuit Operation

Read operations are performed from the PROM/RAM memory space either during opcode or operand fetches or during TBLR instructions. Write operations occur to RAM only during TBLW instructions; write operations to PROM have no effect on the circuit.

The TMS32010 initiates a read operation from either the RAM (U2,3) or the PROM (U4,5) by presenting an address and driving $\overline{\text{MEN}}$ low (see Figure 2). Since the control line to the address latches (buffered $\overline{\text{WE}}$) is high, these latches (U6,7) are transparent, and the address is presented to the memories after a short propagation delay. After the memory-access time delay, data is available internally but is not driven out of the chip since the memories' three-state outputs (controlled by $\overline{\text{RDEN}}$) are not enabled. $\overline{\text{RDEN}}$ is generated so that the memories' outputs are not enabled on a read operation until CLKOUT goes high. This

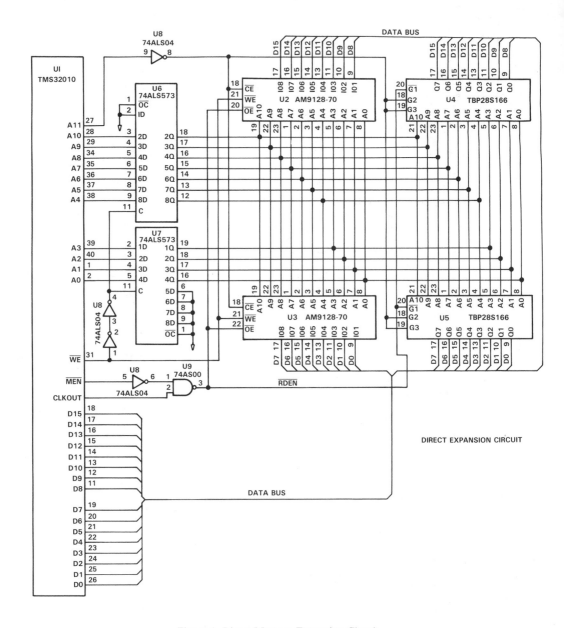

Figure 1. Direct Memory Expansion Circuit

ensures that with fast memories, if $\overline{\text{MEN}}$ occurs early, a data bus conflict will not occur when the read follows a write operation. $\overline{\text{MEN}}$ can be used directly to control the output drivers of the memories if it can be guaranteed that the memory will not drive the data bus immediately following a write operation or if buffers are used on the data bus. Buffers are not used in this design in order to minimize chip count.

After CLKOUT goes high, the memory outputs are enabled. The TMS32010 processor's 50-ns read data setup time to CLKOUT falling is met from CLKOUT rising by an 8-ns propagation delay through the 74AS00 NAND gate

(U9) and the memories' 40-ns maximum output turn-on delay (with a 100-ns high portion of CLKOUT). At the end of the cycle, the TMS32010 requires that read data be held at least until CLKOUT falls or $\overline{\text{MEN}}$ goes high, whichever occurs first. This is guaranteed since $\overline{\text{RDEN}}$ will not go high until at least one of these signals changes state.

Write operations begin in a similar manner to read operations, with the exception that the $\overline{\text{WE}}$ signal is active instead of $\overline{\text{MEN}}$ (see Figure 3). Since the address buffers are controlled by $\overline{\text{WEB}}$ (buffered $\overline{\text{WE}}$), which is high for the first half of the cycle, the buffers are transparent when the address becomes valid. Then, when $\overline{\text{WEB}}$ goes low, the address remains latched until after $\overline{\text{WE}}$ goes high.

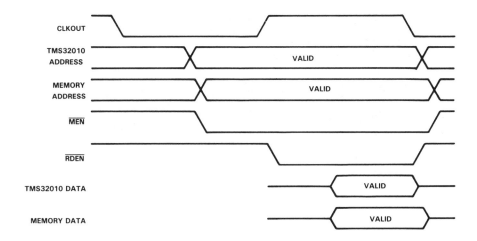

Figure 2. Memory Read Timing

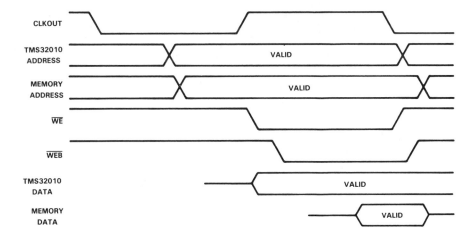

Figure 3. Memory Write Timing

The address bus buffers provide the 5-ns address hold time required by the RAMs following $\overline{WE}$ going high. This hold time is provided by the propagation delay of $\overline{WE}$ resulting from the two 74ALS04s (U8, 1-4) and the buffer propagation delay. While it is not generally good design practice to rely on propagation delays for timing, in this case the technique can be used to eliminate the need for more cumbersome design approaches, such as delay lines, since the minimum propagation delays for these ALS devices are specified. Note that the address bus buffers are not required if memories with 0-ns write-address hold times (such as the INMOS 4K x 4 devices) are used. These devices are not used here because their organization does not suit the desired configuration of this design.

The remaining memory bus timing requirements for write operations are also easily met by this circuit design. The RAMs require a 30-ns data setup time with respect to $\overline{WE}$ rising, of which the TMS32010 provides about 80 ns. The data hold time required with respect to $\overline{WE}$ going high is 5 ns, of which the TMS32010 provides a minimum of 20 ns.

EXTENDED MEMORY INTERFACE

If large program memory expansion is required, bank switching techniques can be employed with the direct expansion scheme to allow greater program memory space, some of which can still be used for small segments of data. These segments of data, however, can only be accessed using TBLR/TBLW instructions. For this reason, the direct expansion scheme is quite inefficient when large amounts of external data storage are required.

To implement large data memory expansion, the extended memory interface can be used. With this approach, memory can be accessed in two cycles once an address has been loaded, making this technique preferable to the direct memory expansion scheme for data storage. Note that the primary savings in cycles required to access the memory result from loading the address only once and having this address increment or decrement with each access. Thus, for the most efficient use of this memory, data should be stored sequentially to avoid having to reload an address for each access. If data is not saved sequentially, four cycles are required for each access, making the direct expansion scheme the preferred approach.

The extended memory interface is more efficient for data storage, but may be used to access instructions even though they cannot be executed directly. Instructions are accessed by using an IN instruction, followed by a TBLW instruction, thereby placing the instruction in program memory. Because the transfer requires a minimum of five cycles, this technique should only be used to store instructions that need to be accessed infrequently. This feature is useful, for example, for implementing downloads from slow memory or a host system.

The extended memory expansion approach may also be used in conjunction with one of the direct memory expansion schemes to expand both program and data memory efficiently.

Design Considerations

A primary consideration of the extended memory expansion design is to implement an efficient interface to large amounts of data memory. The program interface to this memory uses the I/O ports. These ports are accessed in two cycles, whereas three cycles are required to access program memory via the TBLR/TBLW instructions.

This interface is mapped into three port locations:

1. Port 0, which receives the starting address for the memory access
2. Port 1, which decrements the address following each access
3. Port 2, which increments the address following each access.

Functional Description

The extended memory interface circuit, shown in Figure 4, contains the minimum amount of logic required to efficiently communicate with larger amounts of memory at relatively high speeds. Due to the nature of the interface, the devices used for the memory space are not required to be as fast as those used in the direct expansion circuit. The devices used are Synertek SY2128-1 4K x 4 NMOS static RAMs, chosen on the basis of their organization and availability. The RAM organization provides a 4K x 16 memory space using only four chips.

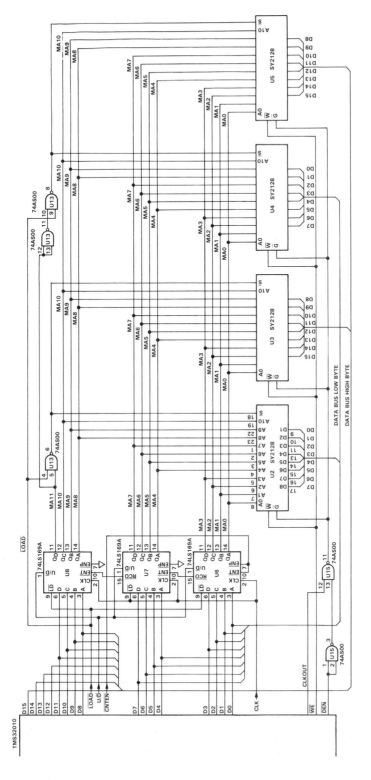

Figure 4. Extended Memory Interface Circuit (page 1 of 2)

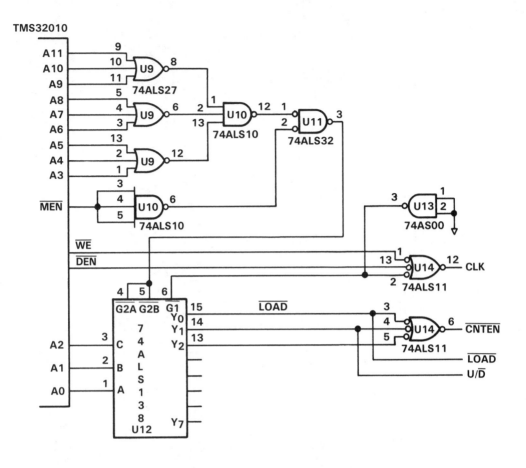

Figure 4. Extended Memory Interface Circuit (page 2 of 2)

The address used to access these RAMs is derived from a 12-bit up/down counter, implemented using three 74LS169A 4-bit counters (U6-8) cascaded together. An address is loaded into the counters, using an OUT instruction to port address 0. Then, with each access to port 1 or 2, this address is decremented or incremented, respectively.

The logic controlling this interface consists of a 74ALS138 decoder (U12), which decodes the three port addresses and some miscellaneous gating that generates strobe and enable signals (U9-11,13,14).

Circuit Operation

The memory in the extended memory interface circuit is accessed using three types of memory cycles:

1. An address load cycle
2. A read cycle to RAM
3. A write cycle to RAM.

Addresses are loaded by writing the desired address to port 0. In order to simplify the design, logic to allow reading of the address counters was not included. Therefore, port 0 should not be read or improper loading of the counters

will occur. Note that although reading port 0 may corrupt the counters' contents, succeeding loads will function properly.

After an address is loaded, each access to either port 1 or 2 decrements or increments, respectively, the memory address after the completion of the cycle. Since the effective address for the next memory cycle becomes valid shortly after the end of the current cycle, the RAMs used can be quite slow. Their speed is limited only by the output enable time, which is generally significantly faster than the address access time. The memories used in this circuit must have an output enable time of no more than 42 ns, but their address access time can be as slow as 150 ns or more. For this reason, less expensive memories can be used.

Figure 5 shows an address load operation. The address presented by the TMS32010 is decoded by the 74ALS138 and some random logic, consisting of AND and OR gates and an inverter, to detect an access to port 0. This decode results in the LOAD and CNTEN signals going active (low). The LOAD signal indicates that this is a load operation, and CNTEN enables the counters. CLK, the clock signal to the counters (in this case, derived from $\overline{WE}$), goes high at the end of the cycle when both the $\overline{LOAD}$ and a $\overline{CNTEN}$ decodes are stable. The rising edge of CLK synchronously clocks the address from the TMS32010 data lines into the counters. Shortly after the rising edge of CLK, the loaded address is available to the memories at the output of the counters.

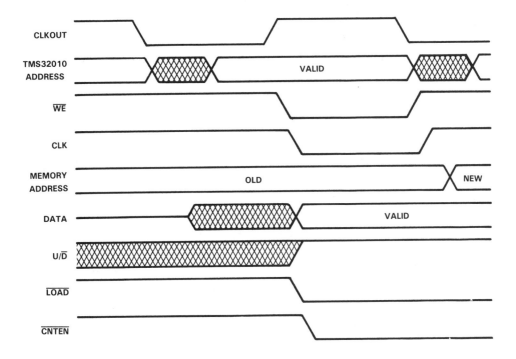

Figure 5. Address Load Timing

In a read operation, as shown in Figure 6, the TMS32010 address is decoded to detect a port address in the same manner as in an address load operation. In this case, however, accessing port addresses 1 and 2 results in the $\overline{\text{CNTEN}}$ signal being active and the $\overline{\text{LOAD}}$ signal inactive. In addition, the U/$\overline{\text{D}}$ signal asserts the correct state of the up/down control input to the counters depending on whether the cycle is an incrementing or decrementing access. In an I/O read cycle from the TMS32010, the $\overline{\text{DEN}}$ signal is active, and the interface uses this signal both to enable the memories'

output buffers and to clock the counters at the end of the buffer access. Since inverted $\overline{\text{DEN}}$ is gated with CLKOUT to enable the memories' output buffers, the buffers will not be enabled until CLKOUT goes high. As in the direct memory interface, this feature is included to avoid any bus conflicts that might occur between the TMS32010 and the memories following a write operation. Note that if the system reads other port addresses, $\overline{\text{DEN}}$ must be further gated to ensure that only the accessed port's output buffers are enabled.

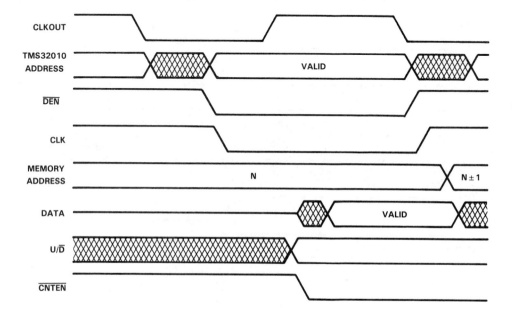

Figure 6. RAM Read Timing

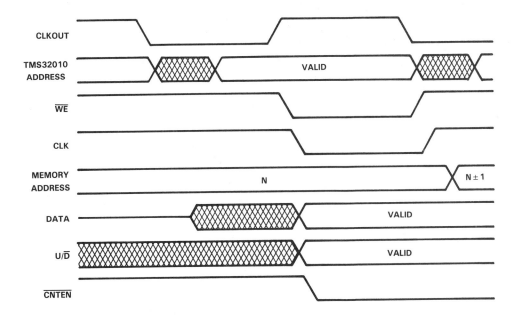

Figure 7. RAM Write Timing

The 35-ns output enable time of the SY2128-1 RAMs (U2-5), added to the delay due to the 74AS00 (U15), acceptably meets the 50-ns data setup time required by the TMS32010. At the end of the cycle, $\overline{\text{DEN}}$ going high causes CLK to go high which either increments or decrements the memory address contained in the counters depending on the state of $U/\overline{D}$. Thus, the following access is made from the next sequential location.

A write cycle, as shown in Figure 7, occurs in much the same manner as a read cycle. The TMS32010 address is decoded to activate $\overline{\text{CNTEN}}$ and produce the correct state of $U/\overline{D}$. At the end of the cycle, when $\overline{\text{WE}}$ goes high, the CLK signal generated from $\overline{\text{WE}}$ strobes the data into the

memories and increments or decrements the address in the counters.

SUMMARY

Two basic low-cost methods for expanding the TMS32010's memory configuration have been described in this application report. The direct memory expansion scheme provides program and small data memory expansion, and the extended memory interface provides large data memory expansion. The design techniques used in these interfaces may be extended to encompass interface of other devices to the TMS32010.

APPENDIX

Pacific Microcircuits Ltd. in British Columbia, Canada, has introduced a peripheral chip to support the TMS32010. The preliminary specification is included in the appendix of this application report to facilitate minimum chip-count design in TMS32010-based systems. In addition to the electrical specification, a schematic for an audio-processor board and an application note for PD32HC01 interrupt handling are provided.

The PD32HC01 is a digital signal processor interface circuit intended for use in voice-band signal processing applications. This CMOS single chip offers an efficient interface between the TMS32010 and external RAM, ROM, and a serial codec (see Figure A-1).

For further information on price, availability, and support, please refer to the list of Pacific Microcircuits Ltd. representatives on the last page of this appendix.

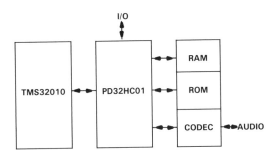

Figure A-1. Voice-Band Signal Processing Interface

PD32HC01
SIGNAL PROCESSOR INTERFACE
for the T.I. TMS32010

Preliminary – September 1985

Features

- Single-chip solution to TMS32010 interfacing
- Serial Codec port
- Serial Data comm. port
- I/O and Interrupt control
- Decoding for external RAM and ROM memory
- I/O expansion interface
- 2400 Hz bit rate generator
- Low-power CMOS technology

Applications

- Voice coders/decoders
- Speech synthesis
- Speech recognition
- Digital telephony
- Data communications
- Digital radio

Description

The PD32HC01 is a DIGITAL SIGNAL PROCESSOR INTERFACE circuit, intended for use in voice band signal processing circuits. **It provides an optimized interface between the Texas Instruments TMS32010 digital signal processor and external RAM, ROM, and Codec.**

Package Availability

- 40 Lead DIL Ceramic (PD32HC01C)
- 40 Lead DIL Plastic (PD32HC01E)
- 44 Lead Surface Mounted Plastic (PD32HC01P)

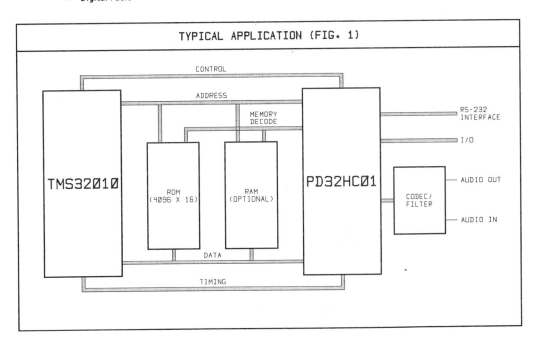

TYPICAL APPLICATION (FIG. 1)

11. Interfacing External Memory to the TMS32010

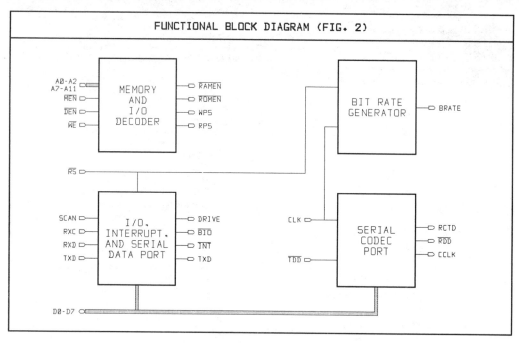

FUNCTIONAL BLOCK DIAGRAM (FIG. 2)

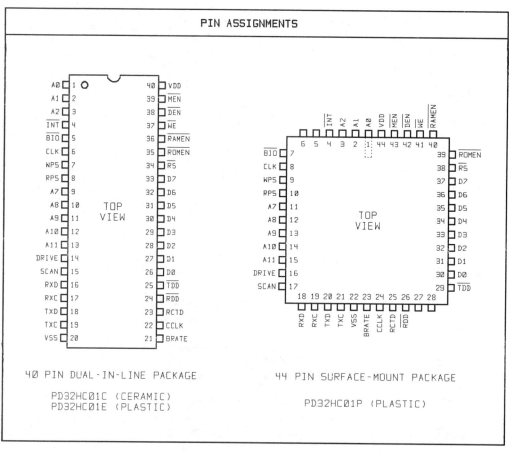

PIN ASSIGNMENTS

40 PIN DUAL-IN-LINE PACKAGE

PD32HC01C (CERAMIC)
PD32HC01E (PLASTIC)

44 PIN SURFACE-MOUNT PACKAGE

PD32HC01P (PLASTIC)

11. Interfacing External Memory to the TMS32010

PD32HC01 Pin Description

Pin	Name	I/O	Description
1-3	A0-A2	Inputs	**Address bus** from processor.
4	$\overline{\text{INT}}$	Output	**Interrupt request** to processor. Responds to RXC, TXC, or Codec A/D interrupts.
5	$\overline{\text{BIO}}$	Output	**Polled output port** bit to processor. Data source to be polled is specified in the Peripheral Status Register.
6	CLK	Input	**4.128 MHz (nominal) clock,** derived from processor clock. Drives the bit rate generator and Codec interface timing.
7,8	WP5, RP5	Output	**Decoded I/O port write and read pulses** for I/O expansion.
9-13	A7-A11	Inputs	**Address bus** from processor.
14	DRIVE	Output	**Output bit** controlled from the Peripheral Status Register.
15	SCAN	Input	**Input bit** selected from the Peripheral Status Register to appear on $\overline{\text{BIO}}$.
16	RXD	Input	**Serial data input.** Must be stable on the rising edge of RXC. Selected from the Peripheral Status Register to appear on $\overline{\text{BIO}}$.
17	RXC	Input	**Serial data receive clock.** Rising edge retimes RXD, and raises an RX clock interrupt.
18	TXD	Output	**Serial data output.** Programmed from the Peripheral Status Register. Edges of TXD are synchronized to the rising edge of TXC.
19	TXC	Input	**Serial data transmit clock.** Rising edge clocks out data onto TXD from the Peripheral Status Register, and raises a TX clock interrupt.
20	V_{SS}	Power	**Negative supply (ground).**
21	BRATE	Output	**2400 Hz square wave** (CLK / 1720-- mask programmable).
22	CCLK	Output	**2.064 MHz (nominal) Codec clock.**
23	RCTD	Output	**Codec framing pulse** for Codec synchronization. Codec A/D interrupt occurs 16 CLK cycles after RCTD goes high.
24	$\overline{\text{RDD}}$	Output	**Serial data output to Codec.** PCM data is shifted out on the rising edges of the first 8 CCLK cycles after the rising edge of RCTD.
25	$\overline{\text{TDD}}$	Input	**Serial data input from Codec.** PCM data is sampled on the first 8 CCLK falling edges after the rising edge of RCTD.
26-33	D0-D7	In/Out	**Data bus** to chip.
34	$\overline{\text{RS}}$	Input	**Master reset** to chip. A low on this input will reset the $\overline{\text{INT}}$ signal, and initalize the bit rate timer. This is a Schmitt trigger input.
35	$\overline{\text{ROMEN}}$	Output	**ROM enable** output. This signal goes low during a valid read from memory locations >000 - >F7F ($\overline{\text{MEN}}$ low).
36	$\overline{\text{RAMEN}}$	Output	**RAM enable** output. This signal goes low during a valid read or write to memory locations >F80 - >FFF ($\overline{\text{MEN}}$ low or $\overline{\text{WE}}$ low).
37	$\overline{\text{WE}}$	Input	**Write enable** to chip. Goes low for I/O or RAM write operations.
38	$\overline{\text{DEN}}$	Input	**Data enable** to chip. Goes low for I/O read operations.
39	$\overline{\text{MEN}}$	Input	**Memory enable** to chip. Goes low for ROM or RAM reads.
40	V_{DD}	Power	**Positive supply (+5 Volts).**

Detailed Description

The PD32HC01 consists of 4 functional blocks: a memory and I/O decoder; I/O, interrupt, and serial data port control; a serial Codec port; and a bit rate generator (see figures. 2 & 3).

Memory and I/O Decoder

The memory and I/O decoder segments the 4K word address space of the TMS32010 into 3 areas: a 3968 word ROM area inclusive of addresses >000 to >F7F; a 128 word RAM area inclusive of addresses >F80 to >FFF; and from addresses >XX0 to >XX7, an I/O expansion port, an I/O, interrupt, and serial data port control; and a serial Codec port.

Memory Decoding

The $\overline{ROMEN}$ signal is used for selecting external program ROM. It goes low during memory read or table read cycles ($\overline{MEN}$ low), and the processor address is less than >F80.

The $\overline{RAMEN}$ signal is used for selecting external data RAM. It goes low during memory read, table read, or table write cycles ($\overline{MEN}$ or $\overline{WE}$ low), and the processor address is above >F7F.

I/O Expansion Port

The RP5 and WP5 signals are used for I/O port expansion. RP5 goes high during an I/O read cycle from port 5 ($\overline{DEN}$ low). WP5 goes high during an I/O write cycle to port 5, or a table write cycle to address >XX5 ($\overline{WE}$ low).

I/O, Interrupt, and Serial Data Port

The Program Status Register (PSR) at port location 6 controls the DRIVE and TXD output signals; the $\overline{INT}$ output operation via the Codec A/D, TX clock, and RX clock interrupt mask bits; and selects inputs to be tested on $\overline{BIO}$ (interrupt flags; the SCAN input; or the retimed RXD input). The bit encoding of the PSR is shown below:

bit 0: RXMSK , RX clock interrupt mask.
bit 1: TXMSK , TX clock interrupt mask.
bit 2: ADMSK , Codec interrupt mask.

Writing 1's to these bits will mask interrupts from the respective sources, and/or clear posted interrupts. Writing 0's will enable interrupts. By testing for the interrupting source on the $\overline{BIO}$ line, interrupt vectoring can be managed (see bits 3,4,5 description).

bits 3,4,5: BIO Source Select . These three bits select one of five input sources (interrupt flags or pin inputs) onto the $\overline{BIO}$ output:

bit 5	bit 4	bit 3	Selected Source
0	0	0	Codec A/D Int. status
0	0	1	TX Clock Int. status
0	1	0	RX Clock Int. status
0	1	1	SCAN bit input
1	X	X	Retimed RXD input

Whenever a posted interrupt is selected, $\overline{BIO}$ will go low. $\overline{BIO}$ will stay high if the selected interrupt is not posted. When the SCAN input or the retimed RXD input is selected, $\overline{BIO}$ follows the polarity of the respective signal.

bit6: TXD , the serial data port transmit data bit. This signal is retimed by the rising edge of the TXC clock, and appears on the TXD output pin.

bit 7: DRIVE , a general purpose output pin.

Serial Codec Port

The Serial Codec Port consists of 8-bit Transmit and Receive data registers, designed to directly interface to Motorola 14400 series PCM Monochips.

The Transmit Register forms incoming serial data on $\overline{TDD}$ into 8-bit parallel PCM samples, while the Receive Register forms 8-bit parallel PCM data samples into serial data on $\overline{RDD}$. The operation of these registers is controlled by the Codec Timing Generator, which also generates the CCLK, the RCTD, and the internal A/D interrupt signals.

Data written to the Receive Register at I/O port location 3 is inverted, and sent MSB first on the $\overline{RDD}$ Pin. Data is shifted out on the 8 rising edges of CCLK following the rising edge of RCTD. To prevent Receive Register underflow, data must be available in the Receive Register within 248 CCLK cycles after an A/D interrupt (nominally 120 usec).

Serial PCM data on the $\overline{TDD}$ pin is inverted, and read into the Transmit Register, MSB first, at I/O port location 3. Data is shifted in on the 8 falling edges of CCLK following the rising edge of RCTD. To prevent Transmit Register overflow, data must be read within 248 CCLK cycles after an A/D interrupt (nominally 120 usec).

Bit Rate Generator

The BRATE output signal is nominally a 2400 Hz square wave, derived from the CLK input divided by 1720 (mask programmable). This signal may be used for bit rate generation, TXC or RXC clocking, or real-time interrupts. BRATE is reset whenever the $\overline{RS}$ signal is low.

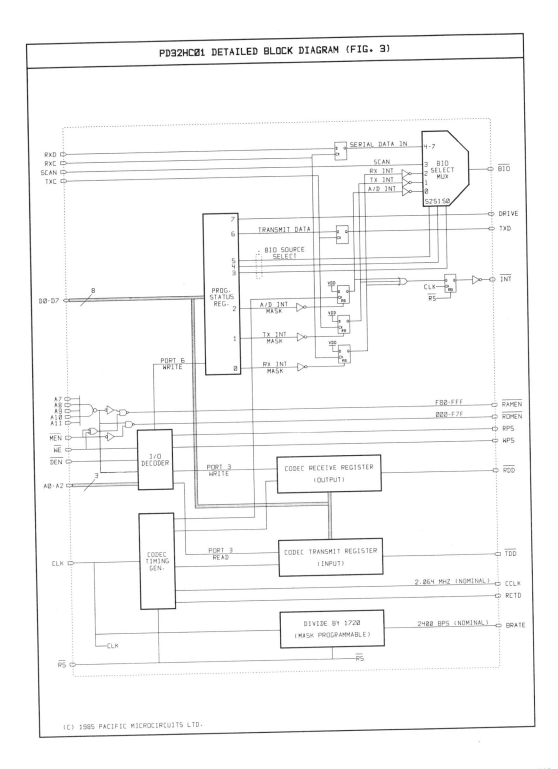

PD32HC01 DETAILED BLOCK DIAGRAM (FIG. 3)

(C) 1985 PACIFIC MICROCIRCUITS LTD.

Address and Input/Output Map [1]

Address/ Port #	R/W [2]	Function
>XX3 Port 3	Read[3]	**Codec Transmit Register.** Valid for 248 CCLK cycles (nominally 120 usec) after an A/D interrupt. This 8-bit register contains the inverted version of the digitized serial PCM signal appearing on $\overline{TDD}$. This register can only be read by using a IN from port 3 instruction.
>XX3/ Port 3	Write	**Codec Receive Register.** Must be valid within 248 CCLK cycles (nominally 120 usec) after an A/D interrupt. Data written to this 8-bit register will be inverted and shifted out on the $\overline{RDD}$ pin. This register can be written using a OUT to port 3 instruction, or by using a TBLW to address >XX3 (address must be less than >F80).
>XX5/ Port 5	Read[3]	**Input Port Expansion.** The RP5 signal will pulse high whenever an IN from port 5 instruction is executed.
>XX5/ Port 5	Write	**Output Port Expansion.** The WP5 signal will pulse high whenever an OUT to port 5, or TBLW to >XX5 instruction is executed (address must be less than >F80).
>XX6/ Port 6	Write	**Program Status Register.** This register is used to: Select I/O and interrupt status bits onto the $\overline{BIO}$ pin; mask and reset interrupts; control the DRIVE pin; and to send data on TXD. This register can be written using an OUT to port 6 instruction, or by using a TBLW to address >XX6 (address must be less than >F80).
>000– >77F	Read	Program ROM space. Notice that the I/O space and the ROM space are mapped to overlapping addresses, but are distinguished by the $\overline{MEN}$ and $\overline{DEN}$ signals. $\overline{MEN}$ will go low for valid instruction reads, while $\overline{DEN}$ will go low for I/O reads.
>780– >7FF	R/W	Data/Program RAM space. For read cycles, $\overline{MEN}$ goes low; for write cycles $\overline{WE}$ goes low.

Notes: 1. When using TBLW to perform Output, many aliases of the I/O port locations exist due to incomplete address decoding. To maintain compatibility with future products, it is recommended that addresses >000 to >007 be used. When using OUT instructions, the TMS32010 always addresses >000 to >007, and the aliases are irrelevant.

2. Some I/O addresses are not used. To prevent data corruption, port locations 0, 1, 2, and 4 should not be written with TBLW or OUT.

3. The I/O read locations can only be accessed with a TMS32010 IN instruction. The actual address locations are shown, however, for applications using other than the TMS32010 processor.

Recommended operating conditions ● V_{SS} = 0 Volts

Parameter	Symbol	Min	Typ	Max	Units	Notes
Supply voltage	V_{DD}	4.75	5.0	5.25	V	
Input high voltage	V_{IH}	2		V_{DD}+0.3	V	
Input low voltage	V_{IL}	-0.3		0.8	V	
Output high current	I_{OH}			3	mA	
Output low current	I_{OL}			3	mA	
Operating Temperature	T_A	0		70	°C	

Electrical characteristics over recommended operating conditions [1]

Parameter	Symbol	Min	Typ [2]	Max	Units	Test Condition
High level output voltage	V_{OH}	3.5	4.5		V	I_{OH} = 1 ma
Low level output voltage	V_{OL}		0.3	0.5	V	I_{OL} = 2 ma
RS hysteresis voltage	V_{HYS}		200		mV	
Off-state leakage current	I_{OZ}		0.5	5	uA	
Input current	I_{IN}		0.5	5	uA	$V_{SS} < V_{IN} < V_{DD}$
Supply current [3]	I_{DD}		2		mA	
Input capacitance	C_I		10		pF	● 1 MHz;
Output capacitance	C_O		10		pF	all other pins 0 V
CLK input frequency	F_{CLK}	0	4.128	6.0	MHz	

Note: 1. See Fig. 4 for DUT test loads.
2. **Typical** specifications are valid at T_A = 25 °C, V_{DD} = 5.0 Volts.
3. I_{DD} Is a function of V_{DD}, clock frequency, and output loading.

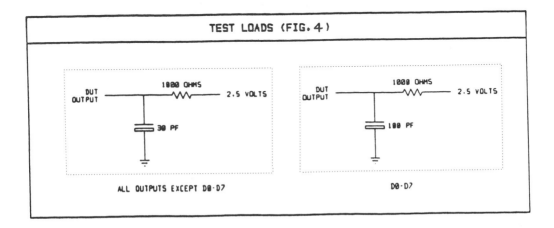

TEST LOADS (FIG. 4)

ALL OUTPUTS EXCEPT D0-D7

D0-D7

Timing Specifications over recommended operating conditions [1]

	Name	Description	Min	Typ[2]	Max	Units
R	(1) t_{ROMHL1}	$\overline{ROMEN}$ select time from addr.		28		nsec
O	(2) t_{ROMLH1}	$\overline{ROMEN}$ deselect time from addr.		30		nsec
M	(3) t_{ROMHL2}	$\overline{ROMEN}$ select time from $\overline{MEN}$		25		nsec
	(4) t_{ROMLH2}	$\overline{ROMEN}$ deselect time from $\overline{MEN}$		18		nsec
	(5) t_{RAMHL1}	$\overline{RAMEN}$ select time from addr.		27		nsec
	(6) t_{RAMLH1}	$\overline{RAMEN}$ deselect time from addr.		15		nsec
R	(7) t_{RAMHL2}	$\overline{RAMEN}$ select time from $\overline{MEN}$		24		nsec
A	(8) t_{RAMLH2}	$\overline{RAMEN}$ deselect time from $\overline{MEN}$		14		nsec
M	(9) t_{RAMHL3}	$\overline{RAMEN}$ select time from addr.		27		nsec
	(10) t_{RAMLH3}	$\overline{RAMEN}$ deselect time from addr.		15		nsec
	(11) t_{RAMHL4}	$\overline{RAMEN}$ select time from $\overline{WE}$		21		nsec
	(12) t_{RAMLH4}	$\overline{RAMEN}$ deselect time from $\overline{WE}$		15		nsec
	(13) t_{DR1}	Data read access time from addr.		33		nsec
R	(14) t_{DRHLD1}	Data read hold time from addr.		60		nsec
E	(15) t_{DR2}	Data read access time from $\overline{DEN}$		29		nsec
G	(16) t_{DRHLD2}	Data read hold time from $\overline{DEN}$		56		nsec
I	(17) t_{RPLH1}	RP5 select time from address		30		nsec
S	(18) t_{RPHL1}	RP5 deselect time from address		32		nsec
T	(19) t_{RPLH2}	RP5 select time from $\overline{DEN}$		26		nsec
E	(20) t_{RPHL2}	RP5 deselect time from $\overline{DEN}$		26		nsec
R	(21) t_{ASUW}	Address set-up time to $\overline{WE}$		4		nsec
	(22) t_{AHLDW}	Address hold time from $\overline{WE}$		-5		nsec
&	(23) t_{DWSU}	Data write set-up time to $\overline{WE}$		-25		nsec
	(24) t_{DWHLD}	Data write hold time from $\overline{WE}$		0		nsec
I	(25) t_{WPLH1}	WP5 select time from address		31		nsec
/	(26) t_{WPHL1}	WP5 deselect time from address		32		nsec
O	(27) t_{WPLH2}	WP5 select time from $\overline{WE}$		27		nsec
	(28) t_{WPHL2}	WP5 deselect time from $\overline{WE}$		27		nsec
S	(29) t_{RXDSU}	RXD set-up time to RXC	20	3		nsec
I	(30) t_{RXDHLD}	RXD hold time from RXC	20	2		nsec
O	(31) t_{TXD}	TXD delay time from TXC		28		nsec
C	(32) t_{RCTD}	RCTD Delay time from CCLK	15	34	100	nsec
O	(33) t_{TDDSU}	$\overline{TDD}$ set-up time to CCLK	50	16		nsec
D	(34) t_{TDDHLD}	$\overline{TDD}$ hold time from CCLK	20	-14		nsec
E	(35) t_{RDD}	$\overline{RDD}$ delay time from CCLK		19	50	nsec
C	(36) t_{ADINT}	$\overline{INT}$ delay time from RCTD		Note 3		

Note: 1. See Fig. 4 for DUT test loads.
2. **Typical** specifications are valid at $T_A = 25\,°C$, $V_{DD} = 5.0$ Volts
3. $t_{ADINT}(max)$ is $17 \times t_{CLK} - 45$ nsec (nominally 4.1 usec).

11. Interfacing External Memory to the TMS32010

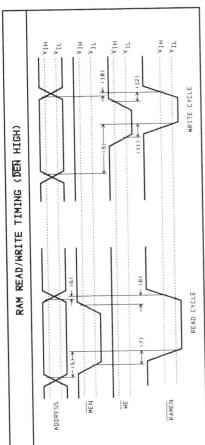

RAM READ/WRITE TIMING (DEN HIGH)

ROM READ TIMING (DEN AND WE HIGH)

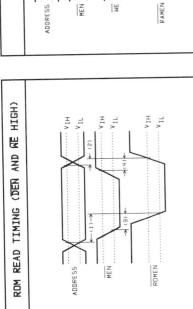

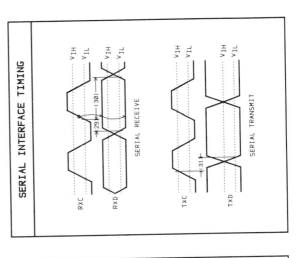

SERIAL INTERFACE TIMING

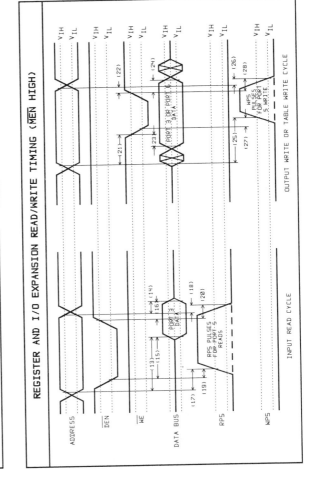

REGISTER AND I/O EXPANSION READ/WRITE TIMING (MEN HIGH)

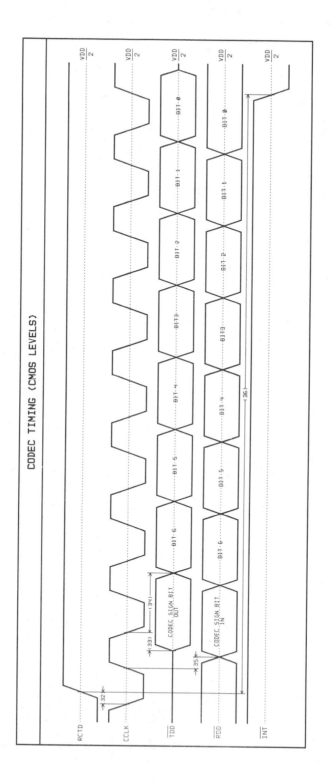

CODEC TIMING (CMOS LEVELS)

To handle interrupts on the PD32HC01, the techniques illustrated below may be used. Not all applications may require the full implementation. Details on TMS32010 interrupts may be obtained from the publication "TMS32010 User's Guide" ,© 1983, Texas Instruments Ltd., Revision B, March 1985. For Mu-law to linear and linear to Mu-law conversion routines, see "Companding Routines for the TMS32010 -- Digital Signal Processing Application Report ", © 1984, Texas Instruments Ltd.

Program Status Register

Set up an 8-bit Program Status Register (PSR) image in data RAM with the following definitions:

Bit	Name	Description
b7	DRIVE	General purpose output bit
b6	TXD	Serial data output
b5	\	3 bit field for selecting BIO input.
b4	┝--	0 = A/D interrupt status, 1 = TX interrupt status,
b3	/	2 = RX interrupt status, 3 = SCAN input, 4 = RXD input.
b2	\	Interrupt mask/acknowledge bits (active high). Used to clear posted interrupts.
b1	┝--	b2 = A/D interrupt mask, b1 = TX interrupt mask, b0 = RX interrupt mask.
b0	/	

The status of the PD32HC01 is defined by writing the contents of the PSR to port 6 with an OUT or a TBLW instruction. In addition to defining the state of the DRIVE and TXD pins, the PSR is used to respond to and acknowledge interrupts from the RX clock, TX clock, and a Codec A/D conversion. The following is an example of TMS32010 software which implements interrupt vectoring with the A/D interrupt at highest priority, and TX and RX interrupts at lower priority.

Interrupt Handling Program

```
*
* CONTEXT SAVE FOR INTERRUPT (INTRP stack and PSR on data page 1)
*
INT   SST   INTRP       Save_machine_state:
      SACH  INTRP+1,0      Save_Status;
      SACL  INTRP+2        Save_ACH;
      MPYK  1              Save_ACL;
      PAC                  Save_T;
      SACL  INTRP+3     End.
*
* VECTORED INTERRUPT ARBITRATION (BIO Mux always initially points to A/D interrupt flag)
*
*                       Vector:
      BIOZ  ADDA          IF A/D_interrupt THEN
*                           AD_interrupt_service
*                         END;
      LAC   PSR,0
      ADD   ONE,3         IF TX_interrupt THEN
      SACL  PSR             TX_interrupt_service
      OUT   PSR,PA6       END;
      BIOZ  TXDAT
```

```
*           ADD     ONE,3           IF RX_Interrupt THEN
            SACL    PSR                 RX_interrupt_service
            OUT     PSR,PA6         END;
            BIOZ    RXDAT
*
* INTERRUPT ERROR RECOVERY PROCEDURE (Should never need execution)
*
            LACK    >C0             IF No_Interrupt THEN
            AND     PSR                 Initialize_PSR;
            SACL    PSR                 Restore_machine_state;
            OUT     PSR,PA6             Return_from_interrupt;
            ZALH    INTRP+1         END;
            ADDS    INTRP+2         END.
            LT      INTRP+3
            LST     INTRP
            EINT
            RET
*
* INTERRUPT SERVICE ROUTINES (A/D, TX DATA, RX DATA)
*
* ADDA:  Clear A/D Interrupt, process, restore machine state, and return.
* TXDAT: Disable TX Interrupt, process, restore machine state (and enable TX interrupt), and return.
* RXDAT: Disable RX Interrupt, process, restore machine state (and enable RX interrupt), and return.
*
*                                   AD_interrupt_service:
ADDA        LAC     ONE,2           Clear_AD_interrupt;  { highest priority interrupts    }
            OR      PSR                                  { should have fast clear to allow}
            SACL    PSR                                  { reposting.                     }
            OUT     PSR,PA6
            SUB     ONE,2
            SACL    PSR
            OUT     PSR,PA6
             .
             .                      (Body of interrupt service routine)
             .
            ZALH    INTRP+1         Restore_machine_State;
            ADDS    INTRP+2         Return_from_interrupt;
            LT      INTRP+3         END.
            LST     INTRP
            EINT
            RET
*                                   TX_interrupt_service:
TXDAT       LACK    >02             Disable_TX_interrupt;
            OR      PSR                                  { lower priority interrupts can  }
            SACL    PSR                                  { be left disabled until interrupt }
            OUT     PSR,PA6                              { servicing is complete.         }
             .
             .                      (Body of interrupt service routine)
             .
            LACK    >C0
            AND     PSR             Initialize_PSR;
            SACL    PSR             Restore_machine_state;
            OUT     PSR,PA6         Return_from_interrupt;
            ZALH    INTRP+1         END.
            ADDS    INTRP+2
            LT      INTRP+3
            LST     INTRP
            EINT
            RET
```

© 1985 Pacific Microcircuits Ltd.
1645 140th Street, White Rock, B.C., Canada V4A 4H1. (604) 536-1886

11. Interfacing External Memory to the TMS32010

```
*                              RX_interrupt_service:
RXDAT  LACK   >21                  Disable_RX_interrupt_select_RXD;
       OR     PSR
       SACL   PSR
       OUT    PSR,PA6
       BIOZ   xxx
        .
        .                       (Body of interrupt service routine)
        .
       LACK   >C0
       AND    PSR              Initialize_PSR;
       SACL   PSR              Restore_machine_state;
       OUT    PSR,PA6          Return_from_interrupt;
       ZALH   INTRP+1       END.
       ADDS   INTRP+2
       LT     INTRP+3
       LST    INTRP
       EINT
       RET
```

A/D Interrupt Handling

When responding to an A/D interrupt, the following TMS32010 code is appropriate:

```
       OUT    SMPL,PA3      (output a sample)
       IN     SMPL,PA3      (input a sample)
        .
        .                    (Remainder of Interrupt Service Routine)
        .
```

Note that if the software is written with a one sample "look-ahead", almost 124 usec (one sample period) can elapse between the interrupt request and execution of the A/D interrupt service routine without loss of data. The above code should be executed early in the interrupt service routine to take maximum advantage of the hardware architecture. This feature is important when several foreground software routines are implemented, or when the A/D interrupt service routine execution time can occasionally exceed 124 usec.

Note: The information contained in this document is for illustrative purposes only. No guarantee as to its suitability for end-use applications is implied.

© 1985 Pacific Microcircuits Ltd.
1645 140th Street, White Rock, B.C., Canada V4A 4H1. (604) 536-1886

Interface Chip Simplifies TMS32010 Based Voice-band Processing

Vernon R. Little, Pacific Microcircuits Ltd., White Rock, B.C., Canada

The TMS32010 is well suited for voice band signal processing. Its powerful instruction set and simple architecture makes cost-effective digital signal processing a reality for modems, vocoders, smart telephones, and a host of other voice and data communications functions.

Pacific Microcircuits Ltd. of White Rock, B.C., Canada has introduced a high-speed, low power CMOS interface chip, the PD32HC01, specifically tailored to interface the TMS32010 to an external Codec, and RAM and ROM memory. With the addition of a few external components, a complete general purpose audio processor can be constructed without any logic 'glue'.

The PD32HC01, besides having high-speed memory decoding and a serial Codec interface, has SCAN and DRIVE input and output pins, a serial data communications port for USRT emulation, a mask-programmable 2400 Hz bit-rate generator or real time clock, multiple source interrupt control, and decoding for expanding I/O off chip. It is available in both 40 pin plastic and ceramic DIP and 44 pin surface-mount plastic packages.

General Purpose Audio Processor

The utility of the PD32HC01 is illustrated in the schematic diagram entitled "GENERAL PURPOSE AUDIO PROCESSOR". This is a complete DSP function capable of implementing a fully functional LPC vocoder, voice band modem, intelligent telephone, or a telecom test set.

In the schematic, the PD32HC01 (U8) ties all of the system resources to the TMS32010 (U7). ROM (U5, U6) is segmented as a 3968 word address space from address >000 to >F7F. An optional RAM (U1-U4) is segmented as a 128 word address space from >F80 to >FFF. This decoding can be arbitrarily modified to increase the RAM space at the expense of ROM (see detail on schematic diagram). This allows 'soft' programming by downloading program instructions into RAM by using a bootstrap loader in ROM. Downloading can occur either on the serial RS-232 interface, or, with additional hardware, the WP5 and RP5 signals on the PD32HC01 can be used to coordinate parallel data transfers with a host processor.

The 8-bit serial Codec interface connects directly to a μ-law or A-law PCM Codec (U12), and is accessible through parallel registers on the PD32HC01. Analog circuitry (U13 etc.), including pre-emphasis and de-emphasis filters, can be connected to the Codec as required. Also shown is a VU-meter circuit for level setting of incoming audio.

Provision for a synchronous RS-232 port is provided. Using the on-chip interrupt control and I/O logic, the TMS32010 can emulate a USRT at 2400 bps. This is very useful for modem and vocoder applications, and it may also be used for program downloading into RAM. With minor modifications in hardware, and appropriate software, asynchronous RS-232 communications can also be emulated at speeds up to 1200 bps. EIA drivers and receivers (U10, U11) are required to interface the RS-232 signals to the PD32HC01.

The SCAN and DRIVE pins are used for miscellaneous control functions. In the example given, SCAN is used to sense the off hook status of the handset, and DRIVE is buffered with a 74HC04 (U9), and used to activate a buzzer or lamp to indicate device operation or error states.

For more information on the PD32HC01 signal processor interface, and other members of the product family, please contact the Product Development g.roup at Pacific Microcircuits Ltd.

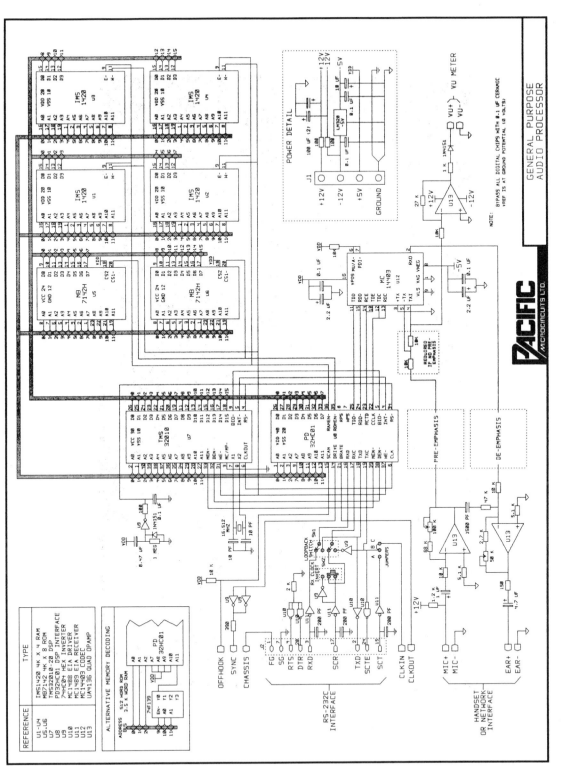

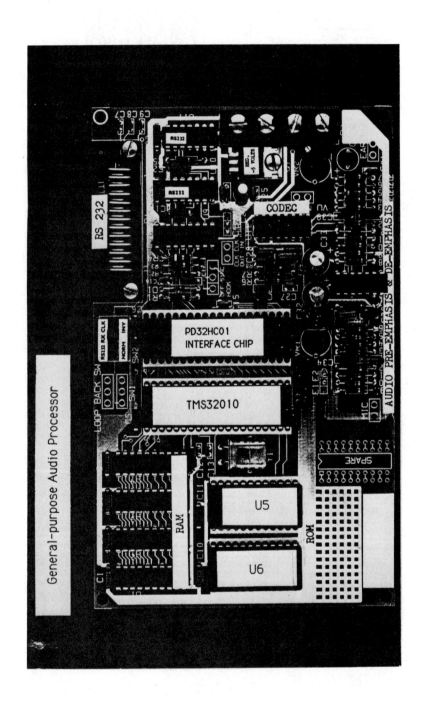

PACIFIC MICROCIRCUITS LTD.

1645 140th Street
White Rock, B.C.
Canada, V4A 4H1
(604) 536-1886

North American Representatives

Astec Components Limited   (416) 669-4022
101 Citation Drive, Unit #7
Concord, Ontario L4K 2S4

Carlisle Technical Sales, Inc.   (617) 890-8800
391 Totten Pond Road
Waltham, Mass. 02154

Electro Sales Associates   (513) 426-5551
1635 Mardon Drive
Dayton, Ohio 54532

Electro Sales Associates   (216) 729-0190
Diplomat Building   729-0191
12575 Chillicothe Rd., Ste. #8
Chesterland Ohio, 44026

Electro Sales Associates   (313) 474-7320
29200 Vassar Road, Suite 505
Livona, Michigan 48152

Electro Sales Associates   (412) 487-3801
3740 Mount Royal Blvd.
Allison Park, Pa. 15101

Electro Sales Associates   (616) 323-2416
9816 Portage Road
Portage, Michigan 49002

L-Tec, Inc.   (312) 773-2900
810 Arlington Heights Rd.
Itasca, Illinois 60143

Mesa Technical Associates   (609) 429-9531
P.O. Box 466
20 Kings Highway West
Haddonfield, N.J. 08033

12. Hardware Interfacing to the TMS32020

Jack Borninski, Jon Bradley, Charles Crowell, and Domingo Garcia
Digital Signal Processing - Semiconductor Group
Texas Instruments

INTRODUCTION

The TMS32020 Digital Signal Processor has the power and flexibility to cost-effectively implement configurations that satisfy a wide range of system requirements. The large address space of the TMS32020 can be filled in those circuits that require external data or program memory. Peripheral devices can be interfaced to the TMS32020 to perform serial communication and analog signal acquisition.

This application report suggests hardware design techniques for interfacing memory devices and peripherals to the TMS32020. Examples of PROM, EPROM, static RAM, and dynamic RAM circuits built around the TMS32020 are demonstrated, with consideration given to the timing requirements of the processor and external devices. A memory-mapped UART (Universal Asynchronous Receiver-Transmitter) interface for communication with a host computer is presented, as well as an interface to a combo-codec (coder-decoder + filter) device for analog signal acquisition.

All circuits shown in this application report have been built and their operation verified at room temperature. Since the logic devices in these circuits have not been optimized as the most cost-effective, the designer may desire to make tradeoffs with respect to speed, cost, performance, and temperature.

TMS32020 CONSIDERATIONS

The TMS32020 has program, data, and I/O address spaces for interfacing with external memory and peripherals. Memory and I/O devices are usually selected by using the TMS32020 $\overline{PS}$, $\overline{DS}$, or $\overline{IS}$ signals (program, data, or I/O select, respectively), combined with the $\overline{STRB}$ (strobe) signal. The signal ($\overline{PS}$, $\overline{DS}$, or $\overline{IS}$) used depends on the memory or I/O space to be addressed. Some of the read-only devices may be selected (enabled) using any of the $\overline{PS}$, $\overline{DS}$, or $\overline{IS}$ signals exclusively, whereas the read-write devices commonly use a $\overline{PS}$, $\overline{DS}$, or $\overline{IS}$ combined with the $\overline{STRB}$ for an enable signal. The $\overline{STRB}$, combined with a select line and the TMS32020 R/$\overline{W}$ (read/write) signal, forms a convenient timing reference to control the write operation of the read/write devices. The read-only devices typically do not require the $\overline{STRB}$ and R/$\overline{W}$ combination.

Memory and I/O devices must respond with the data within a maximum of 90 ns when they are selected with either $\overline{PS}$, $\overline{DS}$, or $\overline{IS}$ during the read cycle. (Refer to the TMS32020 (20-MHz operation) timing diagram in the TMS32020 Data Sheet.) Consequently, the memory or I/O devices used should have correspondingly fast access time. For slower devices, one or more TMS32020 wait states must be inserted for proper operation.

When the $\overline{PS}$, $\overline{DS}$, or $\overline{IS}$ is combined with the $\overline{STRB}$ signal during the read cycle, the memory or I/O devices must respond within 50 ns of the $\overline{STRB}$ signal going low, due to the data-read setup time required by the TMS32020. During the write cycle, the TMS32020 provides the minimum of a 55-ns data-write setup time to the memory or I/O devices (see the timing diagrams in the TMS32020 Data Sheet). Wait states also apply to the configuration using the $\overline{STRB}$ signal.

Tradeoffs with respect to using faster memory and slower decode logic devices, and vice versa, may be made when designing a memory system around the TMS32020.

MEMORIES

Examples of four memory types, PROMs, EPROMs, static RAMs, and dynamic RAMs, are shown in interface to the TMS32020. The selection of which memory device to use in a particular application is determined by speed, cost, and functional requirements. If speed and maximum throughput are desired, the TMS32020 can run with zero wait states and perform memory accesses in a single machine cycle. The TMS32020 can access slower memories by inserting one or more wait states into the memory access operation by using the READY input signal. A circuit using each of the memory devices is described and illustrated in the following subsections of this application report.

PROM

When only fixed program memory is required and speed is a consideration, a PROM device may be chosen for memory interface. A Texas Instruments TBP28S166 PROM (2048 X 8) with a three-state output has been selected as an example interface to the TMS32020. The TBP28S166 has the maximum access time from address of 75 ns, which meets the TMS32020 timing requirements. A basic configuration showing this PROM interfaced to the TMS32020 as program memory is shown in Figure 1.

Another configuration that shows the TBP28S166 interface to the TMS32020 is shown in Figure 2. Here, more memory exists in the system, and the TBP28S166 is decoded and mapped into the program memory address space, starting at >4000.

The timing diagram for the interface of the TBP28S166 to the TMS32020 is shown in Figure 3. No wait states are necessary in both the basic and decoded configurations.

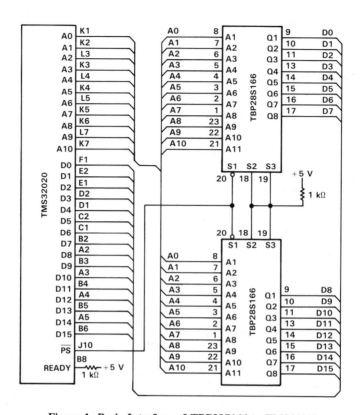

Figure 1. Basic Interface of TBP28S166 to TMS32020

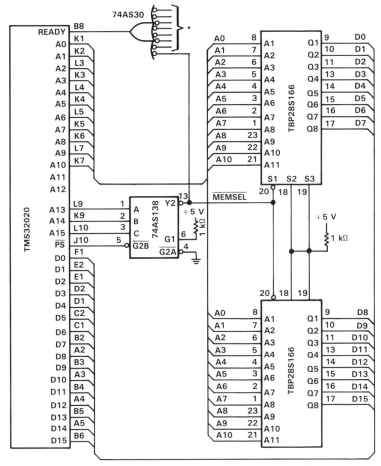

*Connections to other devices in the system. (Inputs not used should be pulled up.)

Figure 2. Decoded Interface of TBP28S166 to TMS32020

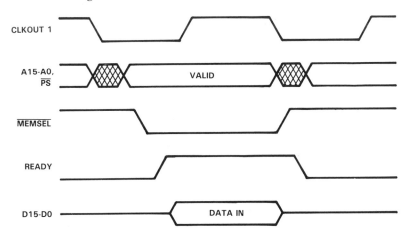

Figure 3. Interface Timing of TBP28S166 to TMS32020

12. Hardware Interfacing to the TMS32020

351

EPROM and Wait-State Generator

During the prototyping and development design stage, an EPROM may be selected as the memory device for interfacing to the TMS32020. The Texas Instruments TMS2764-35 EPROM (8192 X 8) has the access time from address of 350 ns. This does not directly meet the TMS32020 maximum allowable read-data access time of 90 ns, so a series of wait states is used to delay the TMS32020. A wait-state generator circuit for providing up to two 200-ns wait states is shown in Figure 4.

Point A in Figure 4 corresponds to one wait-state input and point B to two wait-state inputs. In the case of the TMS2764-35 EPROM with a 350-ns access time, two 200-ns wait states are needed for proper interfacing. Figure 5 shows

the TMS2764 decoded into the program memory address space 0000-1FFF. Due to the long EPROM turn-off time, the 74ALS244 buffers at the EPROM output prevent data bus conflict.

Figure 6 shows the timing considerations for the interface of the TMS2764 to the TMS32020.

If faster EPROMs are desired, TMS2764-25 (250-ns access time from address) EPROMs can be interfaced to the TMS32020. If the circuit shown in Figure 5 is used, the only change required is to connect the MEMSEL (memory select) signal to input A of the wait-state generator (see Figure 4). The TMS2764-25 operates with one wait state in this configuration.

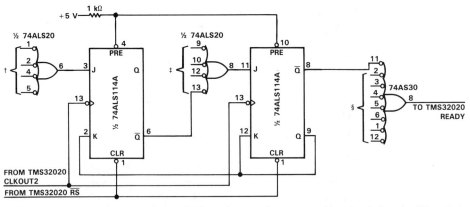

†Connections to other devices in the system that require two wait states. (Inputs not used by other devices should be pulled up.)
‡Connections to other devices in the system that require one wait state. (Inputs not used by other devices should be pulled up.)
§Connections to other devices in the system that require zero wait states. (Inputs not used by other devices should be pulled up.)

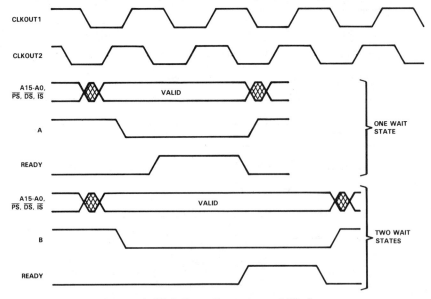

Figure 4. Wait-State Generator and Timing

12. Hardware Interfacing to the TMS32020

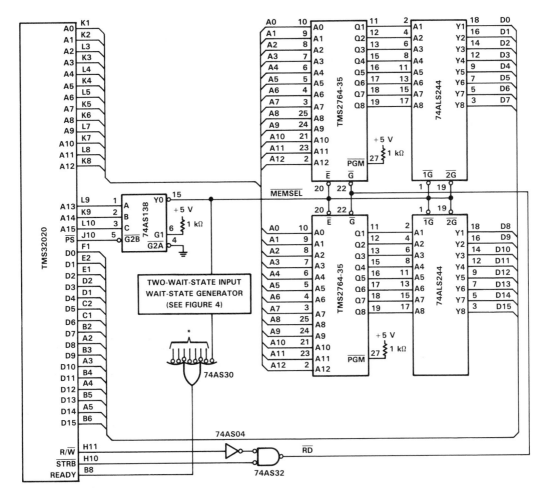

*Connections to other devices in the system. (Inputs not used should be pulled up.)

Figure 5. Decoded Interface of TMS2764 to TMS32020

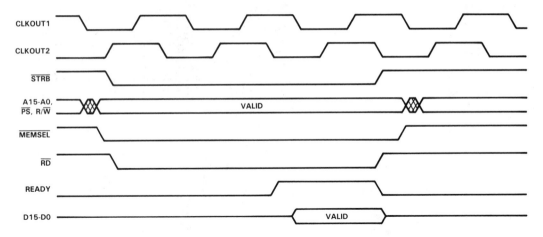

Figure 6. Interface Timing of TMS2764 to TMS32020

Static RAM

Static RAM as program memory is useful for program storage when a remote download capability exists. Static RAM as data memory is used when minimum design overhead and minimum chip count are desired.

Figure 7 shows an example of INMOS IMS1421-50 static RAMs (4096 X 4), interfaced to the TMS32020. These static RAMs are mapped into the program memory address space, starting at >6000.

The read/write timing considerations for the interface of the IMS1421 to the TMS32020 are shown in Figure 8.

During the memory-read cycle, the IMS1421-50 responds with data within 40 ns from being selected. This is in the specified range of the TMS32020, and no wait states are required.

During the memory-write cycle, the falling and rising edges of the $\overline{\text{MEMSEL}}$ signal form the beginning and end of the IMS1421-50 write cycle while the IMS1421 $\overline{\text{W}}$ (write) line is asserted early (early write).

During the early-write cycle, the IMS1421 data-bus drivers never turn on. This helps prevent bus conflicts with the TMS32020. The data supplied to the bus by the TMS32020 meets the timing parameters of the IMS1421-50, and no wait states are required.

The use of slower memories (e.g., IMS1420-45) is possible, provided the designer places external buffers on the data bus between the memory and the TMS32020. Since the internal IMS1420 output buffers do not turn off until the $\overline{\text{W}}$ line goes low, the external buffers (with the appropriate enable logic) prevent data bus conflicts during the write operation.

If CMOS RAMs are required, the IMS1423-35 devices can be used in the circuit shown in Figure 7. These devices operate faster, and their pinouts are identical to that of the IMS1421-50. Slower CMOS RAMs (e.g., IMS1423-45) can also be used; however, the data bus buffers must be added

as outlined previously to prevent data bus buffers must be added as outlined previously to prevent data bus conflicts.

The design of the IMS1421 (NMOS) or IMS1423 (CMOS) interface as data memory is very similar to the one presented in Figure 7, and the data and program memory timing are identical. The only change needed is the use of the $\overline{\text{DS}}$ select line (instead of $\overline{\text{PS}}$) in the decode logic.

Static memories that do not meet the TMS32020 no-wait-state timing requirements can be interfaced, provided the necessary number of wait states is generated and used to drive the TMS32020 READY line.

Dynamic RAM

In systems where large inexpensive memory space is required, the use of dynamic RAM devices with the associated control circuitry may be justified. This section describes an interface between the Texas Instruments TMS4416-15 (16K X 4 dynamic RAM) and the TMS32020. The circuit shown in Figure 9 uses the TMS4500A Dynamic RAM Controller to supply the control functions for the dynamic RAM devices. The TMS4500A provides address generation, timing, access/refresh arbitration, and other functions to control a bank of TMS4416 dynamic RAMs.

The timing diagram for the memory access and refresh cycles in the interface of the TMS4416 to the TMS32020 is shown in Figure 10.

During the regular access cycle, the TMS4500A $\overline{\text{RAS}}$ (row address strobe) and $\overline{\text{CAS}}$ (column address strobe) signals both become active, and one automatic TMS32020 wait state is generated using the $\overline{\text{MSC}}$ (microstate complete) signal (see Figure 9). The generation of at least one TMS32020 wait state is necessary because of the TMS4416-15 access time of 150 ns. The regular (i.e., uninterrupted by refresh) memory access takes two TMS32020 cycles (400 ns at 20-MHz operation).

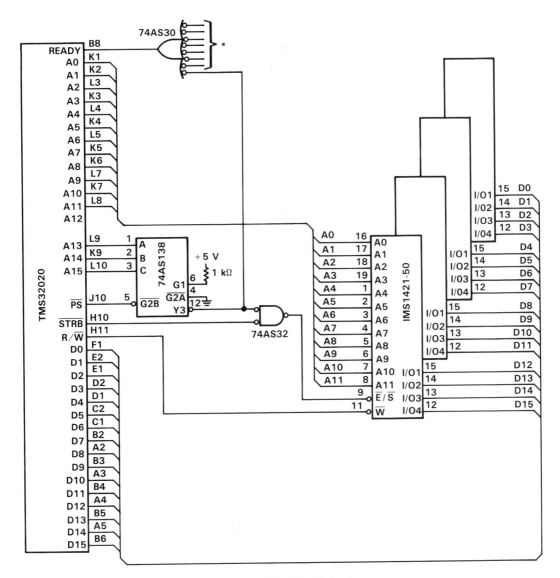

*Connections to other devices in the system. (Inputs not used should be pulled up.)

Figure 7. Interface of IMS1421 to TMS32020

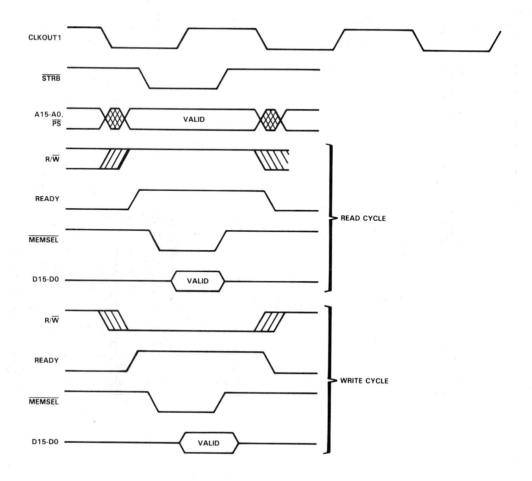

Figure 8. Interface Timing of IMS1421 to TMS32020

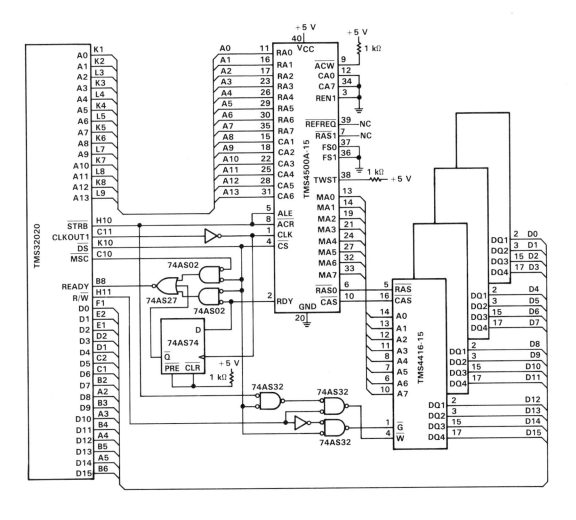

Figure 9. Interface of TMS4416 to TMS32020

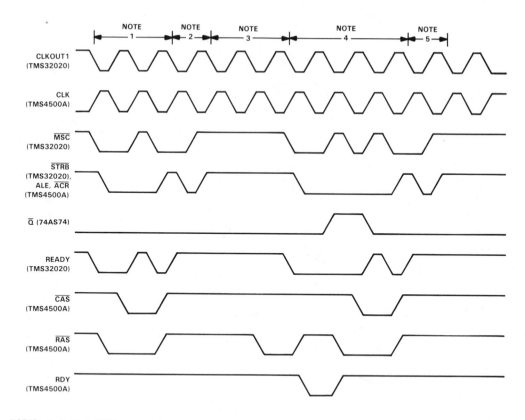

NOTES: 1. Dynamic RAM access cycle
2. Access cycle for memory other than dynamic RAM
3. Internal TMS32020 instruction cycles
4. Dynamic RAM access cycle with refresh
5. Access cycle for memory other than dynamic RAM

Figure 10. Interface Timing of TMS4416 to TMS32020

During the refresh cycle, which may be observed in Figure 10 at the point where $\overline{RAS}$ becomes low and $\overline{CAS}$ stays high, the TMS4500A performs memory access/refresh arbitration. The arbitration must be performed when the TMS32020 tries to access the dynamic RAM memory at the time the TMS4500A begins the refresh operation, as shown in Figure 10. In this case, the TMS32020 must be held for one or two additional wait states, until the refresh operation is completed. The READY signal to the TMS32020 is generated through a combination of three other signals. The $\overline{MSC}$ signal starts driving the TMS32020 READY line, the TMS4500A RDY (ready) signal continues to drive it, and the output of the 74AS74 flip-flop ($\overline{Q}$) extends it to the end of the required period, as shown in Figures 9 and 10. Such a circuit is required since the TMS4500A RDY signal alone is not of the proper duration for the TMS32020. After the memory refresh is performed, a regular access cycle takes place as previously described.

In an access/refresh arbitration case, it takes the TMS32020 three or four clock cycles to access the dynamic RAM, depending on the time relationship of the access and refresh requests. The worst-case RAM memory access timing is 800 ns at 20-MHz operation. This occurs when the requests for access and refresh take place nearly simultaneously. The access/refresh arbitration timing, shown in Figure 10, takes three TMS32020 clock cycles, i.e., 600 ns at 20-MHz operation (not the worst-case condition), since the request for access happens right after the end of the refresh cycle.

If the TMS32020 continuously accesses the dynamic memory, it will be interrupted by a refresh cycle once out of ten thousand memory access cycles, i.e., 99.99 percent of the memory accesses will happen without interruption. This low interruption rate is based on the relationship between the TMS32020 dynamic-memory cycle length (400 ns) and the required TMS4416 refresh period (4 ms).

The transition in the TMS32020 READY line, at the end of the READY signal, is caused by the corresponding behavior of the $\overline{\text{MSC}}$ signal during the TMS32020 operation. The transition does not affect the circuit operation since the READY line is sampled on the rising edge of CLKOUT1, i.e., before the transition takes place.

The interface logic between the TMS32020, TMS4500A, and TMS4416 devices, i.e., the gates and a flip-flop (see Figure 9), can be replaced with one Programmable Array Logic (PAL)[†] integrated circuit for reduced IC count. One such PAL that is well suited to this application is SN74PAL16R4A (see Appendix).

PERIPHERALS

Peripheral devices, such as the UART and combo-codec, can be interfaced to the TMS32020 to perform serial communication and analog signal acquisition. Communication to a host computer can be provided by a memory- or I/O-mapped UART interface. A codec provides analog signal conversion for telecommunications and speech processing.

UART

The UART and TMS32020 configuration may be useful when a general-purpose TMS32020 program development system or upload/download capability from/to the TMS32020-based signal processing system is required.

As an interface example of an asynchronous communication controller to the TMS32020, a General Instruments AY-3-1015D UART is memory-mapped into the TMS32020 data memory space. Figure 11 shows the circuit for interfacing the AY-3-1015D UART to the TMS32020. The transmitter is data memory-mapped into location C000 and the receiver into location E000.

The receiver/transmitter timing diagrams for the interface of the AY-3-1015D to the TMS32020 are shown in Figure 12.

The three-state buffer of the UART receiver has received-data-enable (RDE) timing constraints that preclude its use with the TMS32020 (see the AY-3-1015D specifications sheet). The 74LS244 buffer placed on the UART's receiver port is fast enough to perform the switching function on the TMS32020 data bus.

The one-processor wait state, present during the transmitter operation, allows for positive data strobing into the UART's transmitter data register as required by the AY-3-1015D.

The UART's receiver and transmitter status lines, DAV (data available) and TBMT (transmitter buffer empty), are synchronized with the rest of the system and connected to either the interrupt or $\overline{\text{BIO}}$ (branch on I/O) pin of the TMS32020. This allows either interrupt or polled I/O techniques to be implemented and assures the synchronization of the UART device with the TMS32020 regarding the receiving and transmitting rates. For example, the TMS32020 does not try to read the UART's receiver buffer at a rate greater than the rate at which the characters are coming in or write to the UART's transmitter buffer at a rate greater than the maximum character transmission rate.

The UART can equivalently be mapped into the I/O space of the TMS32020, since the memory and I/O cycles of the TMS32020 are identical. The only change required is the use of the $\overline{\text{IS}}$ (instead of the $\overline{\text{DS}}$) line and of address lines A0-A3 to select an I/O port.

Combo-Codec

In some areas of telecommunications, speech processing, and other applications that require low-cost analog I/O devices, a combo-codec device may be useful. A codec consists of nonlinear A/D and D/A converters with all the associated filters and data-holding registers.

The TMS32020 contains a serial port for communicating to serial devices, such as codecs. The speed and versatility of the TMS32020 allow it to compand (COMpress and exPAND) a PCM (Pulse Code Modulation) data stream, acquired by the codec, through the TMS32020 execution of software conversion routines (see the application report, 'Companding Routines for the TMS32010'). Figure 13 shows an interface example of a Texas Instruments TCM2913 codec to the TMS32020 serial port.

Figure 14 shows the TMS32020 serial port receive and transmit timing considerations in the interface of the TCM2913 to the TMS32020.

In this configuration, the TCM2913 codec functions in the fixed data-rate mode (2.048 MHz) with μ-law operation selected. All timing and synchronization signals are externally generated using independent oscillator and frequency-dividing hardware (an 8-bit counter such as a 74AS867 may be used in place of two 74LS161 4-bit counters to minimize the chip count). Alternatively, the designer may decide to generate the timing signals from the TMS32020 clock by subdividing its frequency.

In some circuits, it may be necessary to wire an op-amp to the analog output of the codec. In such cases or if variable output gain is required, a gain-setting resistor network must be provided as specified in the TCM2913 documentation.

Other linear A/D and D/A converters may be interfaced to the TMS32020 through its parallel ports.

[†]PAL is a trademark of Monolithic Memories, Inc.

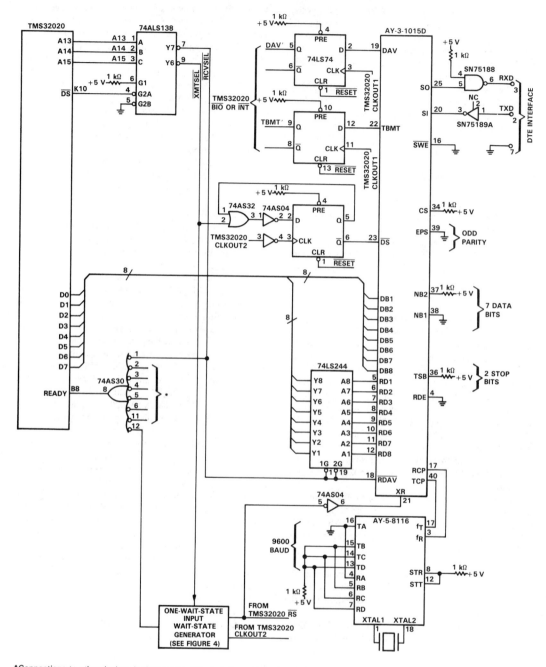

Figure 11. Interface of AY-3-1015D to TMS32020

*Connections to other devices in the system. (Inputs not used by other devices should be pulled up.)

12. Hardware Interfacing to the TMS32020

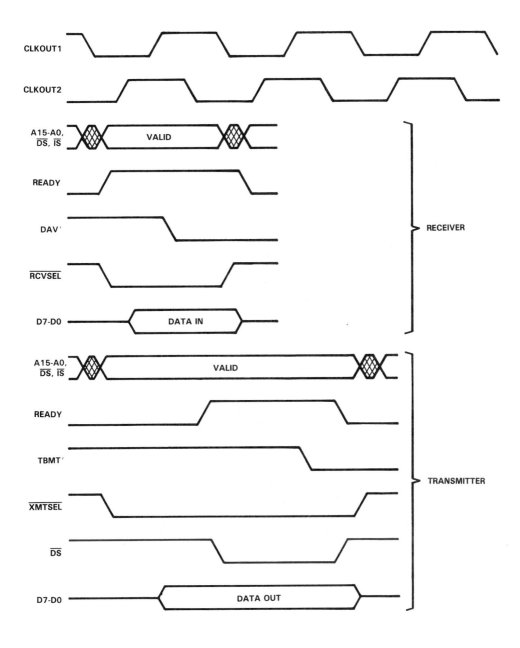

Figure 12. Interface Timing of AY-3-1015D to TMS32020

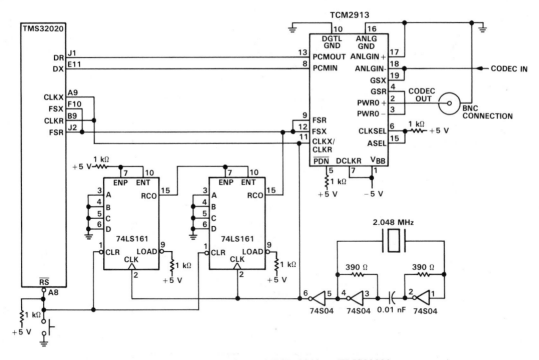

Figure 13. Interface of TCM2913 to TMS32020

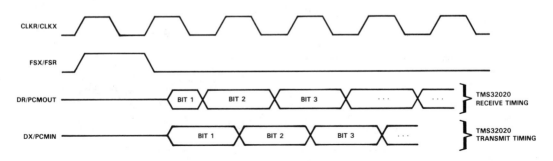

Figure 14. Interface Timing of TCM2913 to TMS32020

SUMMARY

The speed, performance, and flexibility of the TMS32020 allow it to cost-effectively implement configurations that satisfy a wide range of system requirements. This application report has described and demonstrated hardware design techniques for interfacing memory devices and peripherals to the TMS32020.

Examples of PROM, EPROM, static RAM, and dynamic RAM circuits built around the TMS32020 have been provided, with consideration given to the timing requirements of the processor and external devices. Interface examples of UART and codec chips to the TMS32020 are also presented.

Table 1 summarizes the interface requirements of various memory devices to the TMS32020, as described in this report. The table also includes maximum affordable access times for each circuit configuration.

Table 1. Memory Type and TMS32020 Interface Requirements

MEMORY TYPE	PART NUMBER	ACCESS TIME	NUMBER OF WAIT STATES REQUIRED BY THE TMS32020	ACCESS TIME FOR CIRCUIT CONFIGURATION
PROM	TBP28S166	75 ns max (from address)	0	85 ns max (from address)
EPROM	TMS2764-35	350 ns max (from address)	2	371 ns max (from address)
Static RAM	IMS1421-40	30 ns max (from select)	0	39.2 ns max (from select)
Static RAM	IMS1420-55	55 ns max (from select)	0 (buffers reqd)	62 ns max (from select)
Static RAM (CMOS)	IMS1423-35	35 ns max (from select)	0	39.2 ns max (from select)
Static RAM (CMOS)	IMS1423-45	45 ns max (from select)	0 (buffers reqd)	62 ns max (from select)
Dynamic RAM	TMS4416-15	80 ns max (from $\overline{CAS}$)	1	170 ns max (from $\overline{CAS}$)

APPENDIX
PAL[†] DESIGN

PAL Assembler Code

```
PAL16R4                      PAL DESIGN DOCUMENT
0000000-0000                 REV *  5/02/85 BY DANA CROWELL
SHIVA DRAM CONTROLLER

CLK MSC DS RDY CLKOUT1 STROBE RW NC1 NC2 GND
GND /CLK1 /READY /NC3 /NC4 /Q /NC5 /G /W VCC

IF (VCC)
READY      = /MSC*/DS +
             /RDY*/DS +
             Q

IF (VCC)
CLK1       = CLKOUT1

IF (VCC)
W          = /STROBE*/DS*/RW

IF (VCC)
G          = /DS*RW

IF (VCC)
Q          = /RDY


FUNCTION TABLE

CLK MSC DS RDY CLKOUT1 STROBE RW
GND /CLK1 /READY /W /G /Q
```

```
;-------------------------+-------------------+
;                         |                   |
;      C                  |           S       |
;   C  L            A     |   R M     B M     |
; ; C  L          O R R R |   D E     M E     |
; ; L  K        F E E E ; S Y M W I A M       |
; ; K  O      M F A A S ; Y P P A D C C       |
; ; O  U    R E A I D D E ; N A A I L Y Y W   |
; ; U  T  R E A I D D E ; C L L T E C C E  ;  PATH  VECTOR
; ; T  2  W M O O Y Y T ; C L L T E C C E  ;
-------------------------------------------------------------
   C  H H H L H H L L    L L L L H L L L            1
-------------------------------------------------------------
```

†PAL is a trademark of Monolithic Memories, Inc.

12. Hardware Interfacing to the TMS32020

PAL Fuse Map

```
DESCRIPTION:

SHIVA DRAM CONTROLLER

               11 1111 1111 2222 2222 2233
       0123 4567 8901 2345 6789 0123 4567 8901

 0   ---- ---- ---- ---- ---- ---- ---- ----
 1   ---- -X-- ---- ---- -X-- -X-- ---- ----   /STROBE*/DS*/RW
 2   XXXX XXXX XXXX XXXX XXXX XXXX XXXX XXXX
 3   XXXX XXXX XXXX XXXX XXXX XXXX XXXX XXXX
 4   XXXX XXXX XXXX XXXX XXXX XXXX XXXX XXXX
 5   XXXX XXXX XXXX XXXX XXXX XXXX XXXX XXXX
 6   XXXX XXXX XXXX XXXX XXXX XXXX XXXX XXXX
 7   XXXX XXXX XXXX XXXX XXXX XXXX XXXX XXXX

 8   ---- ---- ---- ---- ---- ---- ---- ----
 9   ---- -X-- ---- ---- ---- X--- ---- ----   /DS*RW
10   XXXX XXXX XXXX XXXX XXXX XXXX XXXX XXXX
11   XXXX XXXX XXXX XXXX XXXX XXXX XXXX XXXX
12   XXXX XXXX XXXX XXXX XXXX XXXX XXXX XXXX
13   XXXX XXXX XXXX XXXX XXXX XXXX XXXX XXXX
14   XXXX XXXX XXXX XXXX XXXX XXXX XXXX XXXX
15   XXXX XXXX XXXX XXXX XXXX XXXX XXXX XXXX

16   XXXX XXXX XXXX XXXX XXXX XXXX XXXX XXXX
17   XXXX XXXX XXXX XXXX XXXX XXXX XXXX XXXX
18   XXXX XXXX XXXX XXXX XXXX XXXX XXXX XXXX
19   XXXX XXXX XXXX XXXX XXXX XXXX XXXX XXXX
20   XXXX XXXX XXXX XXXX XXXX XXXX XXXX XXXX
21   XXXX XXXX XXXX XXXX XXXX XXXX XXXX XXXX
22   XXXX XXXX XXXX XXXX XXXX XXXX XXXX XXXX
23   XXXX XXXX XXXX XXXX XXXX XXXX XXXX XXXX

24   ---- ---- -X-- ---- ---- ---- ---- ----   /RDY
25   XXXX XXXX XXXX XXXX XXXX XXXX XXXX XXXX
26   XXXX XXXX XXXX XXXX XXXX XXXX XXXX XXXX
27   XXXX XXXX XXXX XXXX XXXX XXXX XXXX XXXX
28   XXXX XXXX XXXX XXXX XXXX XXXX XXXX XXXX
29   XXXX XXXX XXXX XXXX XXXX XXXX XXXX XXXX
30   XXXX XXXX XXXX XXXX XXXX XXXX XXXX XXXX
31   XXXX XXXX XXXX XXXX XXXX XXXX XXXX XXXX

32   XXXX XXXX XXXX XXXX XXXX XXXX XXXX XXXX
33   XXXX XXXX XXXX XXXX XXXX XXXX XXXX XXXX
34   XXXX XXXX XXXX XXXX XXXX XXXX XXXX XXXX
35   XXXX XXXX XXXX XXXX XXXX XXXX XXXX XXXX
36   XXXX XXXX XXXX XXXX XXXX XXXX XXXX XXXX
37   XXXX XXXX XXXX XXXX XXXX XXXX XXXX XXXX
38   XXXX XXXX XXXX XXXX XXXX XXXX XXXX XXXX
39   XXXX XXXX XXXX XXXX XXXX XXXX XXXX XXXX

40   XXXX XXXX XXXX XXXX XXXX XXXX XXXX XXXX
41   XXXX XXXX XXXX XXXX XXXX XXXX XXXX XXXX
42   XXXX XXXX XXXX XXXX XXXX XXXX XXXX XXXX
43   XXXX XXXX XXXX XXXX XXXX XXXX XXXX XXXX
44   XXXX XXXX XXXX XXXX XXXX XXXX XXXX XXXX
45   XXXX XXXX XXXX XXXX XXXX XXXX XXXX XXXX
46   XXXX XXXX XXXX XXXX XXXX XXXX XXXX XXXX
47   XXXX XXXX XXXX XXXX XXXX XXXX XXXX XXXX

48   ---- ---- ---- ---- ---- ---- ---- ----
49   -X-- -X-- ---- ---- ---- ---- ---- ----   /MSC*/DS
50   ---- -X-- -X-- ---- ---- ---- ---- ----   /RDY*/DS
51   ---- ---- ---- ---X ---- ---- ---- ----   Q
52   XXXX XXXX XXXX XXXX XXXX XXXX XXXX XXXX
53   XXXX XXXX XXXX XXXX XXXX XXXX XXXX XXXX
54   XXXX XXXX XXXX XXXX XXXX XXXX XXXX XXXX
55   XXXX XXXX XXXX XXXX XXXX XXXX XXXX XXXX

56   ---- ---- ---- ---- ---- ---- ---- ----
57   ---- ---- ---- X--- ---- ---- ---- ----   CLKOUT1
58   XXXX XXXX XXXX XXXX XXXX XXXX XXXX XXXX
59   XXXX XXXX XXXX XXXX XXXX XXXX XXXX XXXX
60   XXXX XXXX XXXX XXXX XXXX XXXX XXXX XXXX
61   XXXX XXXX XXXX XXXX XXXX XXXX XXXX XXXX
62   XXXX XXXX XXXX XXXX XXXX XXXX XXXX XXXX
63   XXXX XXXX XXXX XXXX XXXX XXXX XXXX XXXX


LEGEND:   X : FUSE NOT BLOWN (L,N,0)   - : FUSE BLOWN   (H,P,1)

NUMBER OF FUSES BLOWN = 340
```

12. Hardware Interfacing to the TMS32020

PAL Fuse Map Summary

```
                 11  1111  1111  2222  2222  2233
      0123  4567  8901  2345  6789  0123  4567  8901

 0    ----  ----  ----  ----  ----  ----  ----  ----
 1    ----  -X--  ----  ----  -X--  -X--  ----  ----  /STROBE*/DS*/RW

 8    ----  ----  ----  ----  ----  ----  ----  ----
 9    ----  -X--  ----  ----  ----  X---  ----  ----  /DS*RW


24    ----  ----  -X--  ----  ----  ----  ----  ----  /RDY


48    ----  ----  ----  ----  ----  ----  ----  ----
49    -X--  -X--  ----  ----  ----  ----  ----  ----  /MSC*/DS
50    ----  -X--  -X--  ----  ----  ----  ----  ----  /RDY*/DS
51    ----  ----  ----  ---X  ----  ----  ----  ----  Q

56    ----  ----  ----  ----  ----  ----  ----  ----
57    ----  ----  ----  X---  ----  ----  ----  ----  CLKOUT1
```

LEGEND: X : FUSE NOT BLOWN (L,N,0)   - : FUSE BLOWN   (H,P,1)

NUMBER OF FUSES BLOWN = 340

PAL Object Code

```
B B B B B B B B B 3 B B B B B B B B B B B B B B B B B B B B B B B .
3 3 3 3 3 0 3 3 3 3 3 3 3 3 3 3 3 2 3 3 1 2 3 3 3 3 3 3 3 3 3 3 3 .
0 0 0 0 0 0 0 0 0 0 0 0 0 0 0 0 0 0 0 0 0 0 0 0 0 0 0 0 0 0 0 0 0 .
0 0 0 0 0 0 0 0 0 0 0 0 0 0 0 0 0 0 0 0 0 0 0 0 0 0 0 0 0 0 0 0 0 .
0 0 0 0 0 0 0 0 0 0 0 0 0 0 0 0 0 0 0 0 0 0 0 0 0 0 0 0 0 0 0 0 0 .
0 0 0 0 0 0 0 0 0 0 0 0 0 0 0 0 0 0 0 0 0 0 0 0 0 0 0 0 0 0 0 0 0 .
0 0 0 0 0 0 0 0 0 0 0 0 0 0 0 0 0 0 0 0 0 0 0 0 0 0 0 0 0 0 0 0 0 .
0 0 0 0 0 0 0 0 0 0 0 0 0 0 0 0 0 0 0 0 0 0 0 0 0 0 0 0 0 0 0 0 0 .
C C C C C C C C C C C C C C C C C C C C C C C C C C C C C C C C C .
C 8 C C C 8 C C C C C C C 4 C C C C C C C C C C C C C C C C C C C .
```

13. TMS32020 and MC68000 Interface

Charles Crowell
Digital Signal Processing - Semiconductor Group
Texas Instruments

INTRODUCTION

Certain functions in a computer system may be too time consuming for a single processor to perform. A high-speed numeric processor, such as the TMS32020 Digital Signal Processor, may serve as a coprocessor with a slower yet capable host in a computer system. For example, many graphics algorithms must be implemented on a numeric coprocessor to the host so that the host can perform system functions while the coprocessor computes the numeric-intensive algorithms. The TMS32020 is capable of performing numeric functions, such as a multiply-accumulate, in a single cycle (200 ns). Other 16-bit processors, such as the Motorola MC68000, cannot approach the computational speed of the TMS32020, but have other qualities such as 'supervisor mode' and 'user mode' which make them useful as host processors.

This application report shows how the MC68000-10 can be used as a host processor with the TMS32020 serving as a numeric coprocessor to implement the numeric-intensive algorithms often required in computer systems. Applications for such a system include graphic workstations, speech processing, spectrum analysis, and other computational-intensive applications.

The schematic in the appendix has been fully built and tested and has proven functional.

SYSTEM CONFIGURATION

In Figure 1, the basic block diagram for the interface of the MC68000 with the TMS32020 is shown. The MC68000 is interfaced to its own separate program memory (EPROM) and data memory. Although the TMS32020 is interfaced to its own external program memory (PROM), it shares its external data memory with the MC68000. The TMS32020 typically has access to this shared data memory; however, the MC68000 can access this memory by asserting the $\overline{\text{HOLD}}$ line on the TMS32020. In this event, the TMS32020 places all its buses in a high-impedance state and

turns on the buffers between the MC68000 and the shared data memory. This configuration allows the MC68000 to give the TMS32020 instructions and data, and then release the TMS32020 to perform various functions.

HARDWARE CONSIDERATIONS

Acknowledging Hold

After the MC68000 has written to the latch that puts the TMS32020 into the hold mode, the TMS32020 must communicate to the MC68000 that it is ready for the MC68000 to communicate with the shared data memory. The three methods of acknowledging hold to the MC68000 are as follows:

1. The MC68000 waits until it knows the TMS32020 is held.
2. The HOLD Acknowledge ($\overline{\text{HOLDA}}$) signal causes an interrupt to the MC68000.
3. The $\overline{\text{HOLDA}}$ signal writes to a memory-mapped latch.

The first method is implemented in the schematic in the appendix. This method assumes that the MC68000 will allow enough time for the buffers to be turned on before trying to access this memory. For example, the MC68000 could execute several NOP (No OPeration) instructions before attempting to access the shared data memory. Sometimes this method may not be sufficient. For example, if the TMS32020 is in the repeat mode, it does not recognize the $\overline{\text{HOLD}}$ assertion until it has finished the repeat instruction. This could cause a long unpredictable delay before the TMS32020 acknowledges the $\overline{\text{HOLD}}$ interrupt.

The second method of communicating to the MC68000 that the TMS32020 is in the hold mode is to interrupt the MC68000. To implement this, the $\overline{\text{HOLDA}}$ signal can be tied to one of the MC68000 interrupts, thus allowing the MC68000 to access the shared memory as fast as possible. Some method of communicating to the MC68000 as to which device caused the interrupt needs to be considered, since the MC68000 searches all external devices for the originator of the interrupt.

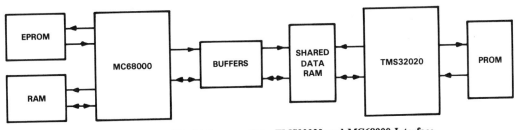

Figure 1. System Block Diagram of the TMS32020 and MC68000 Interface

The third method of communicating to the MC68000 when it can access the shared memory is to allow the HOLDA to be read through a memory-mapped latch. Then, the MC68000 can poll this memory location until it recognizes a change, thus signifying that the TMS32020 has indeed been placed in the hold mode.

Communicating with Shared Memory

Once the TMS32020 has acknowledged the HOLD assertion, the three-state buffers (74LS241) are turned on to allow the MC68000 address bus and R/W line to become valid to the shared memory (IMS1421-40). These buffers are physically enabled by HOLDA, thus assuring that the TMS32020 has three-stated its memory bus. Once the address becomes valid, the transceivers (74LS245) are enabled so that the MC68000 data bus can access the shared memory. These buffers are enabled by the output of the decoder (74ALS138). By doing this, the MC68000 data bus accesses the shared-memory data bus only when MC68000 is trying to access the shared memory. This prevents data bus conflicts when the MC68000 accesses other memory while the TMS32020 is being held. After the communication path is enabled, the MC68000 can read and write to the shared memory. Figure 2 shows the timing when the MC68000 writes to the shared memory. The Enable/Select (E/S) on the shared memory is enabled when the address and the Address Strobe (AS) on the MC68000 become valid. The rising edge of the AS causes E/S to rise, thus writing to the shared memory.

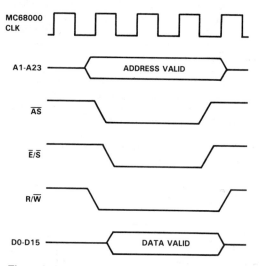

Figure 2. MC68000 Write Cycle to Shared Memory

When the TMS32020 is not in the HOLD mode, it can communicate directly with the shared data memory. The three-state buffers and the transceivers between the MC68000 and the shared data memory are turned off, and the link between the TMS32020 and the shared memory is direct. Figure 3 shows the timing when the TMS32020 writes to the shared memory. The E/S on the IMS1421-40 is enabled by Data Strobe (DS) and Strobe (STRB) becoming valid on the TMS32020. The rising edge of STRB causes E/S to rise, thus writing to the shared memory.

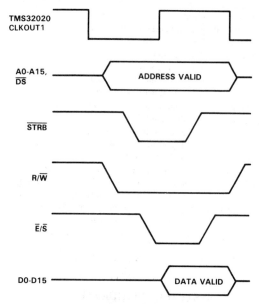

Figure 3. TMS32020 Write Cycle to Shared Memory

LDS and UDS Considerations

The schematic in the appendix is the expansion of the block diagram shown in Figure 1. In this schematic, the Upper Data Strobe (UDS) and Lower Data Strobe (LDS) signals on the MC68000 are not included, i.e., not connected. This method is sufficient if 'word'-specified instructions are the only ones used on the MC68000. Many systems work more efficiently if other length specifications for some of the MC68000 instructions are used. Therefore, a decode scheme, such as in Figure 4, may be implemented. In this scheme, the MC68000 can read or write bytes or words to the Synertek RAMs (SY2128). For example, the MC68000 may write to data bits D0-D7 and not affect the upper data bits by asserting LDS low and leaving UDS at a logic one.

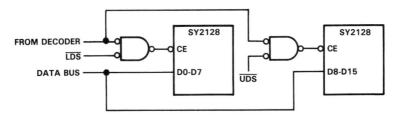

Figure 4. $\overline{\text{LDS}}$ and $\overline{\text{UDS}}$ Scheme for Memory Access

This is an automatic function of the MC68000 if 'byte' lengths are specified on certain instructions.

SUMMARY

The TMS32020 Digital Signal Processor is capable of performing numeric-intensive algorithms faster than other numeric coprocessors used in the past. In addition, the TMS32020 offers a minimal-chip, cost-effective solution to applications requiring a high-performance coprocessor.

This report shows how the TMS32020 can work with the MC68000 to serve as a numeric coprocessor. The interface shown in this report is a generic one and can be used with different host processors. A block diagram of the system configuration is included, as well as hardware considerations. The appendix contains a fully tested schematic of the design presented in this report.

APPENDIX

Schematic of TMS32020 and MC68000 Interface

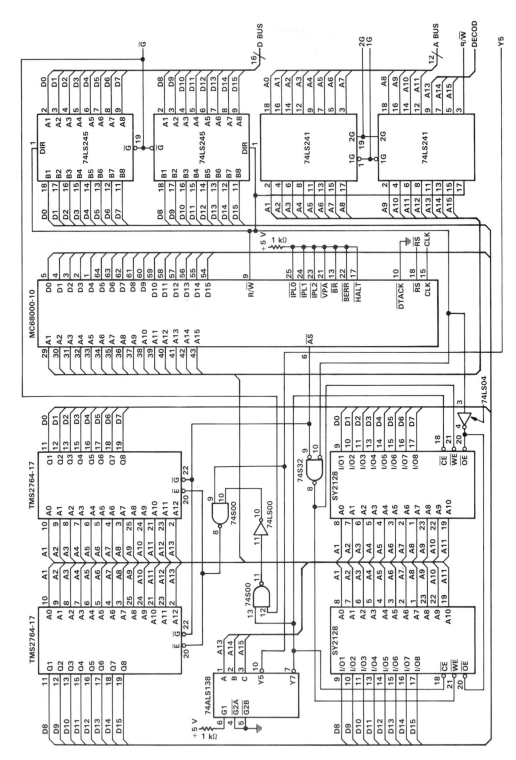

Figure A-1. Schematic of TMS32020 and MC68000 Interface (Sheet 1 of 3)

13. TMS32020 and MC68000 Interface

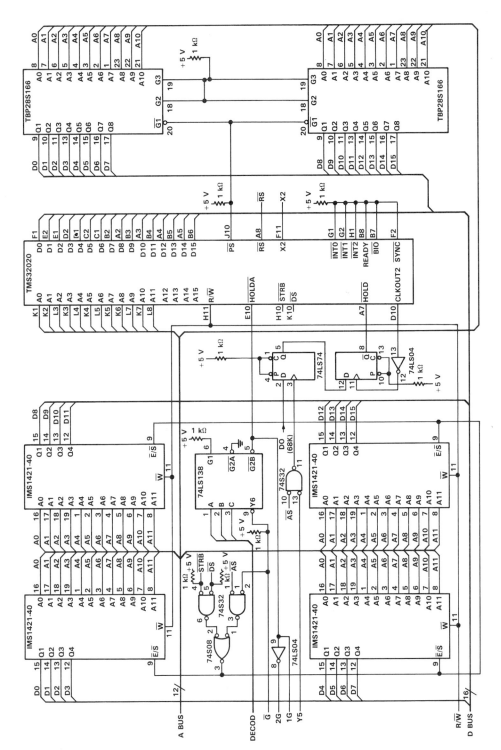

Figure A-1. Schematic of TMS32020 and MC68000 Interface (Sheet 2 of 3)

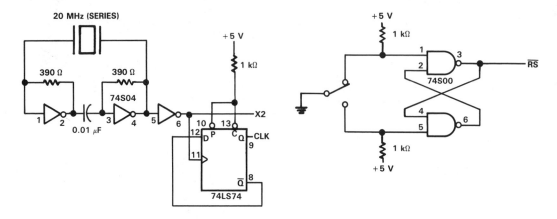

Figure A-1. Schematic of TMS32020 and MC68000 Interface (Sheet 3 of 3)

13. TMS32020 and MC68000 Interface

PART III
DIGITAL SIGNAL PROCESSING APPLICATIONS

In the last few years, many DSP applications have been created using the TMS320 processors. Although it is beyond the scope of this book to cover all of these applications in Part III, some typical examples have been selected in two areas: telecommunications and computer/peripherals.

In the telecommunications area, applications are made possible by taking advantage of the TMS320 processors' high-speed computational power, large on-chip memories, advanced architecture, and comprehensive instruction sets. Because of these and many other features, a high-performance, effective, single-chip solution can be created to solve telecommunications problems. Some applications in the area are echo cancellers, transcoders, DTMF encoders/decoders, and adaptive repeaters.

The first application report in the telecommunications area describes some of the TMS32010 interface circuits frequently used in telecommunications applications. Some of these circuits are interfaces to a codec, a host processor, and external memory devices. A circuit is also included to perform PCM companding functions for time-critical telecommunications applications. The report concludes by suggesting system configurations for some telecommunications applications using these interface circuits. Some of these applications are ADPCM transcoding, hands-free telephone (speakerphone), three-way conferencing, 2/4 wire transformer connection, and DTMF detection. The next report covers both the theory and implementation of a single-chip TMS32020 digital voice echo canceller. The single-chip system can perform a 128-tap or 16-ms echo cancellation for telephone network applications. The echo canceller is implemented in accordance with the CCITT recommendation (G.165). A simulation has been performed to test the echo canceller, and the result exceeds the CCITT requirements. The third report in Part III examines the implementation of a data encryption method using the TMS32010. The encryption algorithm chosen for the report is the Data Encryption Standard (DES). The report details the DES algorithm and its TMS32010 coding. Processor resource requirements are also provided for applying the DES to speech coding at different bit rates. For a 2.4-kbit/s LPC vocoder application, the DES only requires six percent of the TMS32010 CPU loading. The encryption scheme operates on a stream of bits that represent text, computer files, speech, or any other entity in binary form. The result is directly applicable to the design of any secure data/voice communications.

The fourth report in Part III discusses 32-kbit/s Adaptive Differential Pulse Code Modulation (ADPCM) transcoders. A half-duplex ADPCM transcoder, which complies with the CCITT recommendation (G.721), can be achieved with a single TMS32010. If the transcoder is used only for private lines, a full-duplex non-CCITT ADPCM transcoder is more cost-effective and can be designed with a single TMS32010 processor. Both the CCITT and non-CCITT algorithms and code implementations are covered in the report. To date, the CCITT standard for the 16-kbit/s transcoder has not been established. One of the potential candidates for the 16-kbit/s transcoder is the subband coder. A reprinted article on the subject is included, which describes the theory, TMS32010 implementation requirements, and performance figures for the 16-kbit/s subband coder. The complete code for the coder is available for licensing from Atlanta Signal Processors Inc. (ASPI). Interested readers should contact ASPI for further information. Because of the programmability of the digital signal processor, the TMS32010 can also be programmed to handle Dual-Tone MultiFrequency (DTMF) encoding and decoding over telephone lines. For a system already performing digital signal processing functions using the TMS320, this DTMF capability may be obtained at no additional hardware cost. A reprinted article from Electronic Design News, which details the DTMF implementation algorithm and provides TMS32010 program description and code, concludes this portion of Part III.

The computer/peripherals area is a vastly diversified field for DSP applications. It is further divided into three subareas: speech coding/recognition, image/graphics, and digital control. Applications in these areas are usually both computation- and data I/O-intensive. Multiprocessing capability, ease of host interface, and programmability are also required. These requirements have made the TMS320 processors attractive for computer/peripherals applications. In these applications, a general-purpose, TMS320-based digital signal processing system or subsystem is usually built for multitasking functions. For example, such a general-purpose system may be designed to perform a speech compression algorithm for voice store-and-forward applications and to transmit and receive data through the phone line when the system is reprogrammed with a modem algorithm. The TMS320-based system, capable of multitasking, proves to be more cost-/performance-effective than a system consisting of a group of dedicated devices. In addition, the programmability of the TMS320 allows users to update or upgrade at a later date.

The first subarea in computer/peripherals comprises a series of reprinted articles and a technical report from MIT Lincoln Laboratory on speech vocoding and recognition using the TMS32010. Two vocoders at different data rates are discussed: the 2.4-kbit/s pitch-excited Linear Predictive Coding (LPC) and 9.6-kbit/s Adaptive Predictive Coding (APC). Vocoders at these rates are often used in computer voice-mail, voice store-and-forward, and low bit-rate speech communications applications. These two vocoders, combined with the 32-kbit/s ADPCM and the 16-kbit/s subband coder, provide the full spectrum of support by the TMS320 in the area of speech compression. The first article describes the implementation of a full-duplex 2.4-kbit/s LPC vocoder on a single TMS32010 device. Next, the reprinted report from MIT Lincoln Laboratory discusses the design considerations for constructing a compact, low-power, 9.6-kbit/s APC. The report also includes the TMS32010 source codes for the major functional steps in the APC algorithm. The third article discusses speech recognition, usually both computation- and data I/O-intensive. A reprinted article from Electronic Design examines the design of a general-purpose and high-performance speech recognition system using the TMS32010. Algorithms for speech recognition are still being evolved. New techniques for better feature extraction and recognition decision making are being studied. The TMS320 programmable digital signal processors are not only powerful enough to execute recognition algorithms available today, but also flexible enough to be updated with new recognition algorithms at a later date.

The second subarea consists of an image/graphics application where the TMS32020 serves as the graphics engine performing tasks, such as image scaling, translation, and rotation. Circuits are also shown for the interface between the TMS32020 and the Texas Instruments TMS34061 Video System Controller. The TMS32020-TMS34061 graphics system produces a high-speed, high-image-resolution, graphics capability. When integrating such a system into PC or workstations, the TMS320 reduces the host processor burden of performing intensive graphics data computations, thus allowing the host processor to concentrate on other important system and I/O tasks.

The third subarea in computer/peripherals presents an application report on the design of a digital control system and its implementation with the TMS32010. Because of the increased availability and lower cost of suitable digital hardware, microprocessors such as the TMS32010 are increasingly being used to implement algorithms for the control of feedback systems. The report provides an example that uses the design of a digital compensator. Other control applications are possible using the TMS32010, such as computer disk control, laser print-head control, robotic control, automobile-engine system control, flight control, and autopilot systems.

Code for the major DSP routines and applications in Part III is included in the DSP Software Library.

14. Telecommunications Interfacing to the TMS32010

Jeff Robillard
Semiconductor Systems Engineering
Texas Instruments

INTRODUCTION

Signal processing has long been a tool used to improve performance in telecommunications systems. The advent of digital signal processing has given the telecommunications industry a powerful instrument to further enhance performance and reliability. Speech recognition, speech coding, speech synthesis, and high-speed modems are examples of applications now possible because of digital signal processing. Until recent years, real-time digital signal processing was solely the domain of minicomputers. The Texas Instruments TMS320 family offers low-cost VLSI implementations of digital signal processing functions. The TMS32010, the first-generation processor of the TMS320 family, is a high-speed, 16/32-bit digital signal processing microprocessor/microcomputer, ideally suited for telecommunications applications.

The purpose of this application report is to facilitate TMS32010 design by showing various telecommunications applications. The interfaces described in this report can be used in the following applications:

1. Digital modem
2. Echo cancellation
3. Analog repeaters
4. Handsfree telephone (speakerphone)
5. Noise reduction
6. Digital speech interpolation (DSI)
7. Analog-switched network simulator
8. Voice and data encryption
9. Simultaneous voice and data transmission
10. Speech coding and decoding, such as Linear Predictive Coding (LPC)
11. ADPCM transcoding
12. Voice mail
13. DTMF encoding and decoding
14. System identification.

This application report consists of two major sections. The first section details various hardware building blocks required for system interfacing. This provides a simple means of constructing a TMS32010-based system for many telecommunications applications. Included are circuit diagrams and their functional descriptions. These standard telecommunications interfaces provide a realistic medium for implementing digital signal processing algorithms based on the TMS32010. Standard serial and parallel interfaces permit connection to most digital Private Branch Exchange (PBX) backplanes and Time Division Multiplex (TDM) systems. Standard combo-codec filters provide analog interfaces that conform to CCITT recommendations. Description of hardware PCM companding for use with combo-codec filters is useful with time-critical algorithms. Microphone, loudspeaker, and host system interfaces are also included.

The second section of the report examines the following TMS32010 applications based on the building blocks described in the first section:

1. Standalone analog interface
2. Telephony test-set interface
3. PBX backplane interface
4. 2/4-wire transformer interface
5. Three-way conference interface
6. ADPCM interface
7. DTMF detection interface.

HARDWARE INTERFACE BLOCKS AND CIRCUITRY

The hardware circuit described in this section is divided into distinct modules or blocks. This modular approach allows flexibility in accommodating a wide range of telecommunications applications. These blocks can be placed together to fit a specific application. The following hardware blocks are detailed in the next sections of this application report:

1. TMS32010 and support circuit
2. Timing and control circuit
3. PCM linearization circuit
4. PBX/TDM interface circuit
5. Analog interface circuit
6. Host interface circuit.

The system block diagram, shown in Figure 1, illustrates how these blocks fit together around the TMS32010, the focal point of all activities. All blocks are directly connected to the TMS32010 data bus. The TMS32010 accesses all interfaces, except the program memory, using its I/O ports. Each interface is assigned to an I/O port. Port decode logic is used to decode the port address from the TMS32010 I/O space.

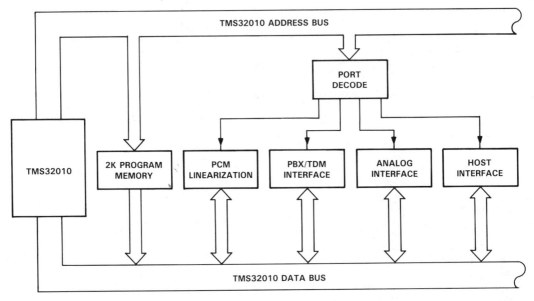

Figure 1. System Block Diagram

TMS32010 and Support Circuit

The TMS32010 serves as the hardware engine that performs the DSP algorithms. It interfaces to its peripherals via its I/O ports. Support circuitry for the TMS32010 includes memory and port decode logic and synchronization flip-flops. Table 1 indicates the TMS32010 port assignments for the application hardware constructed for this application report.

Table 1. TMS32010 Port Assignments

PORT	ASSIGNMENT
0	Analog interface 1
1	Analog interface 2,
	Microphone and loudspeaker interface
2	PBX/TDM interface 1
3	PBX/TDM interface 2
4	μ/A-law to two's-complement linear conversion
5	Two's-complement linear to μ/A-law conversion*
6	Host interface
7	Not used

*To perform conversions, a write to port 4 is required. The converted value is read at port 5.

The TMS32010 can be reset by momentarily activating the reset switch, which provides a 6-ms pulse to the $\overline{RS}$ input to the TMS32010. The device has one hardware interrupt ($\overline{INT}$) and one software interrupt ($\overline{BIO}$). The $\overline{BIO}$ (branch on I/O) input to the TMS32010 is considered the low-priority interrupt and is used for synchronizing to the 8-kHz framing pulse that drives the telecommunications interfaces (i.e., combo-codec filters and PBX/TDM interfaces). The $\overline{INT}$ input is considered the high-priority interrupt and is used for communicating with the host processor via the host interface (see Table 2 for interrupt assignments). The $\overline{INT}$ input becomes active when a command from the host is waiting to be processed. The reset and interrupt signals going to the TMS32010 are all synchronized.

Table 2. TMS32010 Interrupt Assignments

INTERRUPT	ASSIGNMENT
$\overline{RS}$	Reset. Vector to location >000H in program memory. Initialization of all pointers and internal registers recommended.
$\overline{INT}$	Hardware interrupt. Vector to location >002H in program memory. Interrupt service routine for performing host command processing.
$\overline{BIO}$	Software interrupt. Program in wait state when using the BIOZ instruction, waiting for 8-kHz synchronization signal.

Figure 2 shows the TMS32010 and support circuitry. The TMS32010 (U1) has a 16-bit data bus that interfaces to all ports and program memory. All port decoding is accomplished with two SN74ALS138 decoders (U6 and U7), which provide the write and read port strobes, respectively.

14. Telecommunications Interfacing to the TMS32010

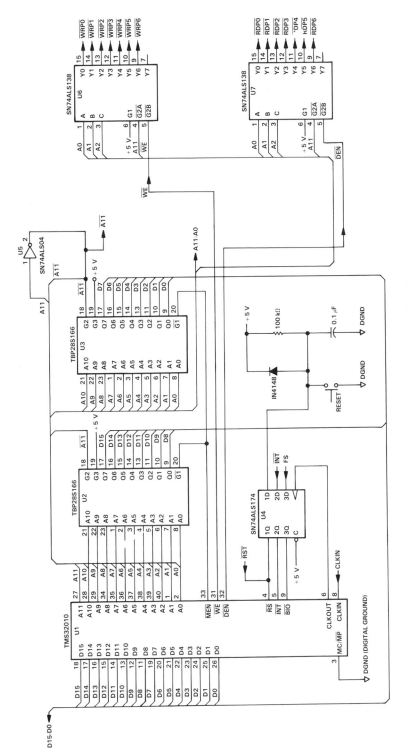

Figure 2. TMS32010 and Support Circuit

The reset and interrupts are synchronized to the TMS32010 with an SN74ALS174 D-type flip-flop (U4). The TMS32010 12-bit address bus is used for addressing program memory and the individual ports. This address bus drives the TBP28S166 program memory PROMs (U2 and U3) and the port decode logic (U6 and U7). The PROMs are mapped into the lower 2K address space (>000H to >7FFH). Address bit A11 is used for decoding the lower and upper 2K program memory spaces.

Timing and Control Circuit

The timing and control circuit supplies all the required clocking, timing, and control signals to operate the TMS32010 and interfaces. A 20.48-MHz clock is provided for the TMS32010. Derived from this clock are all the timing signals to operate the combo-codec filters, such as the 128-kHz clock for shifting data and the 8-kHz framing pulse for synchronization purposes (see Figure 3). External frame synchronization is permitted with an 8-kHz framing pulse input. Note that the internal clock always operates asynchronously with the external system clock.

All timing and control signals are derived from a 20.48-MHz clock source (see Figure 4). The inverters (U12) are configured as an oscillator with a 20.48-MHz crystal and 10-kohm resistor. A decode counter (U8) divides the 20.48-MHz clock by 10 to produce a 2.048-MHz clock signal required by the combo-codec filters. A dual 4-bit binary counter (U9) divides the 2.048-MHz clock by 16 to produce a 128-kHz clock signal used by the combo-codec filters and related support circuitry to shift in/out serial data. U9 also divides the 128-kHz clock by 16 to produce an 8-kHz framing strobe (FS) used in the serial/parallel data conversion process and as a synchronization signal to the TMS32010. U10 and U11 are used to synchronize counters U8 and U9 to an external asynchronous 8-kHz framing pulse (FP). This permits synchronization to an external system, such as a PBX, on a frame-by-frame basis.

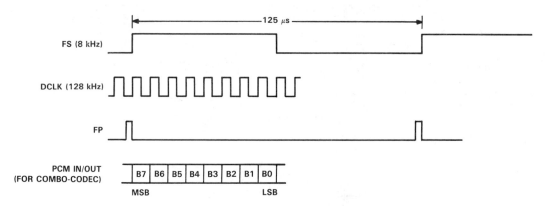

Figure 3. Clocking and Timing Diagram

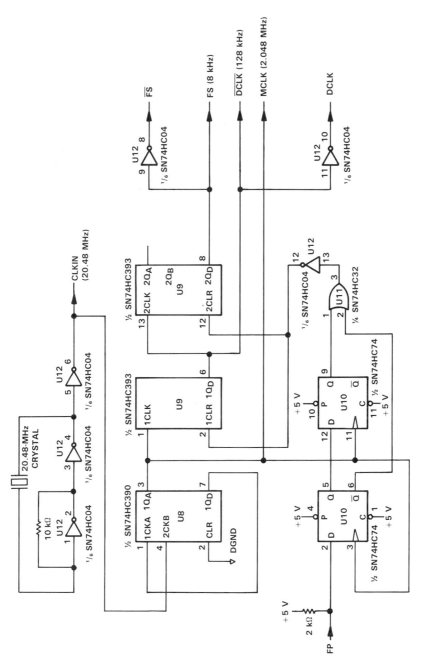

Figure 4. Timing and Control Circuit

PCM Linearization Circuit

In digital telecommunications systems, speech is transmitted in an 8-bit nonlinear PCM code. μ-law and A-law are the two most popular PCM coding schemes. For linear digital signal processing, the PCM code must be converted to two's-complement linear codes for processing by the TMS32010. After processing, the two's-complement linear code must be converted to the nonlinear PCM code. This conversion process can be performed in software on the TMS32010. A TMS32010 source code for the conversion process is provided in the application report, "Companding Routines for the TMS32010."[1] In certain cases, there may not be enough time in an 8-kHz framing period to accomplish this conversion process.

For time-critical applications, hardware linearization is used to provide a faster conversion process. The conversion process is accomplished by using table lookup ROMs. To convert from one code to another, a value is written to port 4, followed by a read from port 4 or 5. Port 4 is read to convert from a nonlinear code value to a two's-complement linear value. Port 5 is read to convert a two's-complement linear value to a nonlinear code value. The TMS32010 performs a port write followed by a port read in 800 ns when driven by a 20-MHz clock source.

The linearization circuit uses a lookup table to perform the conversion process (see Figure 5). U18 converts 14-bit two's-complement linear PCM to 8-bit μ/A-law PCM. U19 converts μ/A-law 8-bit PCM to 14-bit two's-complement linear PCM. Data to be converted is clocked into two SN74ALS574 octal D-type edge-triggered flip-flops (U13 and U14) when the TMS32010 writes to port 4. U13 and U14 contain the least- and most-significant bytes of data, respectively. μ/A-law data is buffered to the data bus via U16. Data is enabled onto the data bus when the TMS32010 reads port 5. Note that only the least-significant byte is active on the data bus. Linear data is buffered to the data bus via U15 and U17, which contain the most- and least-significant bytes of data, repectively. Data is enabled onto the data bus when the TMS32010 reads port 4. The most-significant byte of data is clocked into U15 on the rising edge of the $\overline{\text{MEN}}$ output from the TMS32010.

FORTRAN programs that compute μ-law PCM to 14-bit two's-complement linear PCM and vice versa are provided in the appendix. This computation generates the object file (in the INTEL data record format) for purposes of burning-in PROMs and EPROMs. A lookup ROM can also be programmed to generate patterns for test signal generator and DTMF encoder applications.

PBX/TDM Interface Circuit

The PBX/TDM interface permits connection to serial and parallel digital PBX backplanes and TDM systems. An example of a TDM system is a D1 channel bank that provides T1 carrier digital transmission on two-wire trunks between central offices. Each trunk provides 24 8-bit channels multiplexed in time. Each channel is intended to carry 8-bit companded PCM voice. The application hardware does not support any signalling functions in a T1 carrier connection.

Two PBX/TDM interfaces are provided in the application circuit to accommodate separate connections to receive (RX) and transmit (TX) paths. The hardware, as shown in Figure 6, is configured to interface onto an 8-bit parallel bus or serial PCM highway. The parallel bus interface provides three-state I/O, and the serial interface provides open-collector or three-state I/O. Selection between parallel or serial interface on the application circuit is hardware-selectable via a jumper. Each PBX/TDM interface occupies an I/O port on the TMS32010 I/O port map, as shown in Table 1. These interfaces are assigned to ports 2 and 3. The TMS32010 is synchronized to these ports via an 8-kHz framing pulse on its $\overline{\text{BIO}}$ input.

The two interfaces in this application circuit operate identically; therefore, only one interface, as shown in Figure 6, is described. The serial data input (SERIAL PCM IN) is converted to 8-bit parallel data via shift register U23. All timing and control signals are provided by the external system interface. SYSCLK is the external synchronous clock that drives U23. $\overline{\text{ITSEN}}$ (input time-slot enable) is synchronous with SYSCLK and clocks the 8-bit parallel data from U23 into U20 at an 8-kHz rate. Parallel data P7-P0 is clocked into U21 with the $\overline{\text{ITSEN}}$ signal at an 8-kHz rate. U20 and U21 are the storage registers for serial and parallel PCM data, repectively. Access to either of these registers by the TMS32010 is controlled on the application circuit via a jumper. The jumper routes the read port 2 signal from the TMS32010 to either U20 or U21, thereby selecting the serial or parallel data inputs. The PCM data output from the TMS32010 is clocked into U22 when a write to port 2 occurs. The parallel data in U22 is buffered to the system data bus via U26. U26 is an SN74LS244 line driver that provides high-current drivers to the external system PCM data bus. Data is enabled onto the external system PCM data bus when $\overline{\text{OTSEN}}$ (output time-slot enable) is active low.

For serial interfacing, the $\overline{\text{OTSEN}}$ signal loads the parallel data at U22 into shift register U24. Data is shifted to the PCM highway using the external system clock SYSCLK. The serial input pin (pin 10) is connected to the serial output pin (pin 9) on U24 to permit broadcasting the same PCM data on two or more consecutive time-slots on the PCM highway. $\overline{\text{OTSEN}}$ enables the data onto the PCM highway by controlling buffer U25. U25 drives either a three-state or open collector/drain PCM highway. U18 provides the logic which configures a three-state driver on U25 as an open-collector/drain driver.

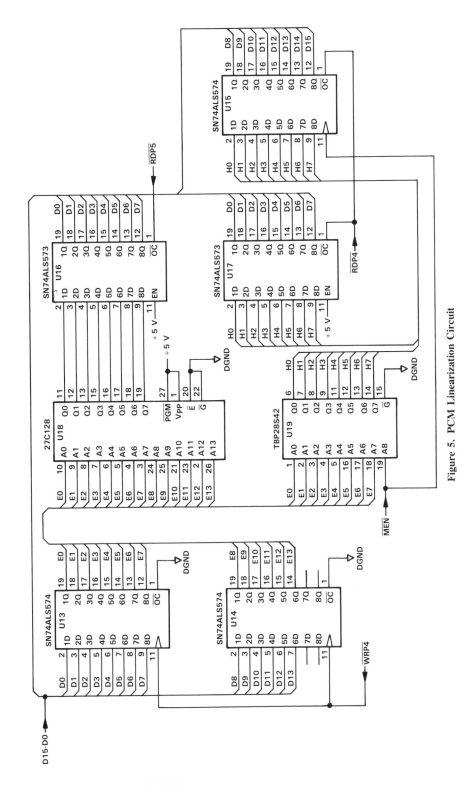

Figure 5. PCM Linearization Circuit

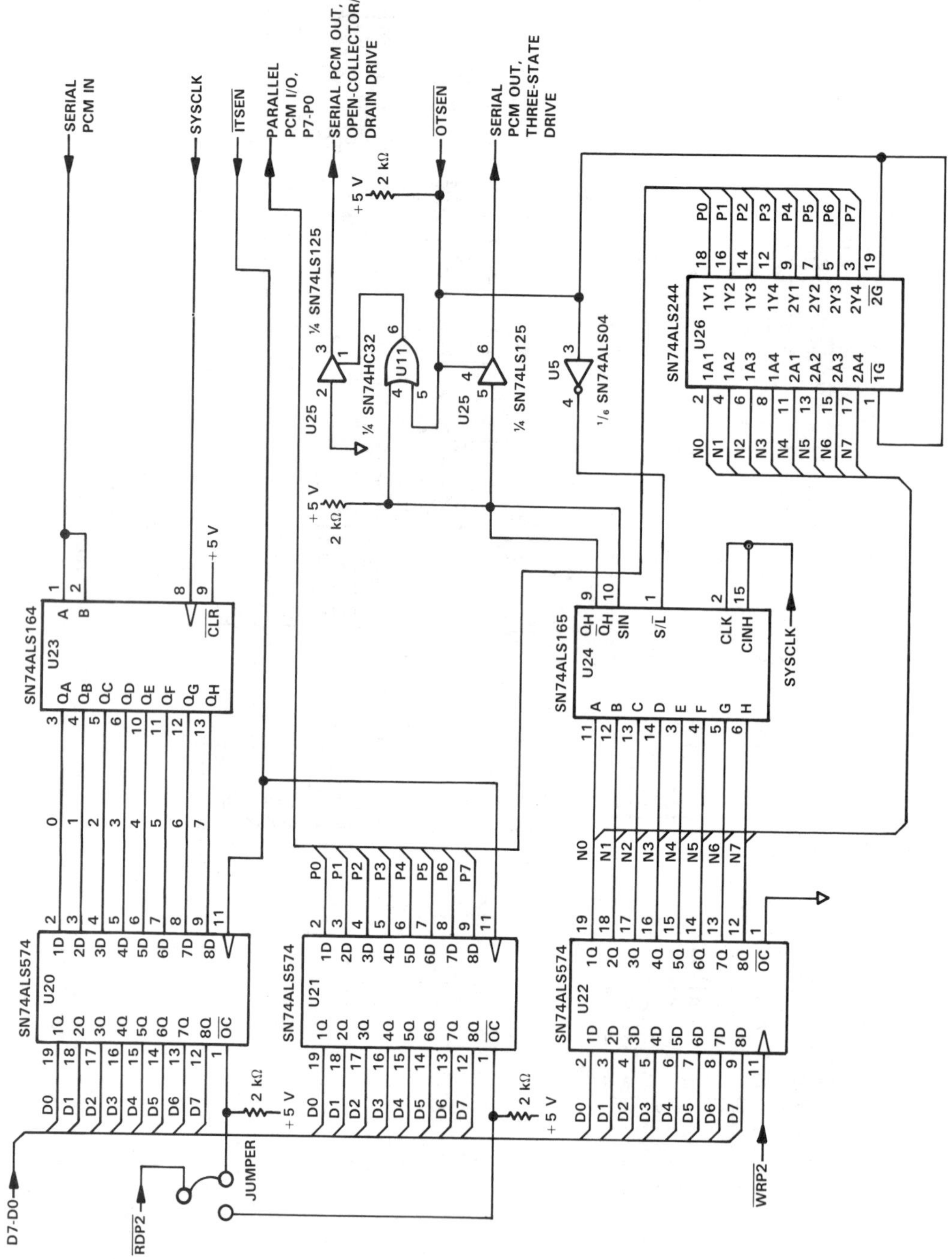

Figure 6. PBX/TDM Interface Circuit

14. Telecommunications Interfacing to the TMS32010

Analog Interface Circuit

Two analog interfaces are provided to access receive (RX) and transmit (TX) analog paths. These interfaces conform to CCITT recommendations through the use of industry-standard combo-codec filters. μ/A-law operation can be hardware-selected via a jumper on the application circuit, as shown in Figure 7. One of the interfaces also contains microphone and loudspeaker amplifiers, which can be used in a handsfree speakerphone application. The gain of both amplifiers is adjustable using potentiometers. The overall gain through the TMS32010 and combo-codec filters is fixed at 3 dB. Each analog interface occupies an I/O port on the TMS32010 I/O port map (see Table 1). These interfaces are assigned to ports 0 and 1. The analog interface containing the microphone and loudspeaker amplifiers is at port 1. The application hardware sets the electret condenser microphone current at 1 mA. The loudspeaker amplifier can drive a standard 8-ohm loudspeaker.

The transmission signal reference points are governed by the combo-codec filters.[2] These reference points for unbalanced line interfacing are summarized as follows:

INPUT: 0 dBmO = 1.064-V RMS 1020-Hz sine wave between $V_{IN}+$ and analog ground; $V_{IN}+$ not connected.

OUTPUT: 0 dBmO = 1.503-V RMS 1020-Hz sine wave between $V_{OUT}+$ or $V_{OUT}-$ and ground.

Referring to Figure 7, the TCM2913 (U27) combo-codec filters[2] contain A/D and D/A converters and filters required for analog interfacing. A jumper is used to select either μ/law or A-law operation on the combo-codec filters by controlling the voltage level of ASEL (pin 15) on U27. A +5 voltage selects μ-law, and a −5 voltage selects A-law. Analog and digital grounds should be kept completely separate to avoid introducing digital switching noise in U27 and ground loops. The analog input to U27 is configured as a unity gain difference amplifier. U32 is the microphone amplifier that is configured as an inverting amplifier. The amplifier gain is controlled by a 1-Mohm potentiometer. The

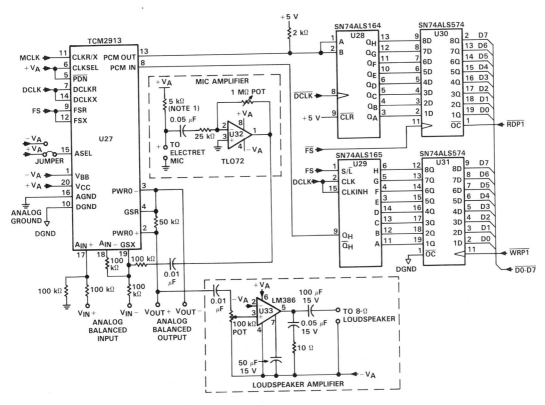

NOTES: 1. Electret MIC current set at 1 mA.
2. Analog supply:
   $+V_A = +5$ V
   $-V_A = -5$ V

Figure 7. Analog Interface Circuit

amplifier input is designed to interface to an electret condenser microphone. The microphone current is set at 1 mA via a 5-kohm resistor connected to the analog +5 V supply. The DC voltage component is removed by capacitively coupling amplifier. The 0.05-μF capacitor sets the highpass filter corner frequency at roughly 120 Hz. U32 is connected to U27, and the input gain at U27 is set at unity. The output transmission level at U27 is set at a 0-dB loss by strapping the PWR and GSR pins together.[2]

The loudspeaker amplifier is capacitively coupled to the U42 output, and fixed with a gain of 26 dB. However, a 100-kohm potentiometer attenuates the input, thereby acting as a volume control. The loudspeaker amplifier can drive an 8-ohm load. PCM data is clocked out of U27 at the PCM output pin (PCM OUT) at a rate of 128 kbit/s. U28 is a shift register that converts the serial data from U27 to 8-bit parallel data clocked into register U30 at an 8-kHz rate. The TMS32010 accesses register U30 when a read to port 1

occurs, which enables 8-bit data onto the least-significant byte of the 16-bit data bus. When data is being sent to the interface, the TMS32010 writes to port 1, which clocks 8-bit data into register U31. This data resides on the least-significant byte of the 16-bit data bus. Data at U31 is loaded into shift register U29 at an 8-kHz rate. This data is shifted out in serial format at a 128-kbit/s rate to the PCM input pin (PCM IN) of U27. The 128-kHz clock signal used for shifting serial data and the 8-kHz loading signal are provided by the system timing and control circuit as DCLK and FS,[2] respectively.

Host Interface Circuit

The TMS32010 can be externally controlled (if desired) by a host processor via the host interface. The host interface contains three registers: an acknowledgement register, ACKR (U34); a command register, COMR (U35); and a status register, SR (U36). Figure 8 shows the host interface

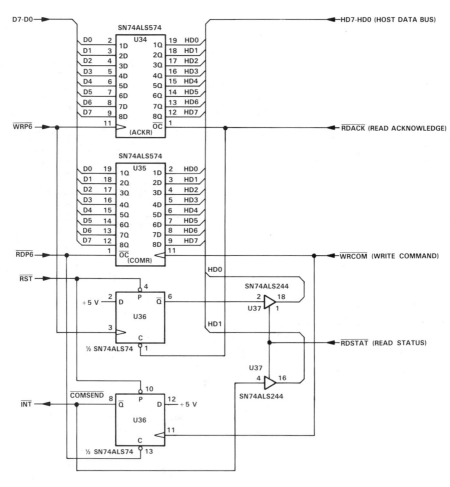

Figure 8. Host Interface Circuit

circuit. The host writes commands to the COMR, which can be read by the TMS32010. The commands are completely defined by the user. The TMS32010 writes to the ACKR, which can be read by the host. The COMR and ACKR are assigned to port 6 in the TMS32010 I/O map. The read and write strobes from the TMS32010 and the host control the bits in the SR that are used for handshaking purposes with the host. Only the host can read the contents of the status register. For PBX applications, the host is located within the PBX.

Handshaking signals are provided to control data transfers across the host interface. The host can access these signals by reading a two-bit SR, which resides on the two least-signficant bits of the host 8-bit data bus. The SR should reside in the host I/O port map or memory map. When reading the SR, the host must mask out the six most-significant bits since they are unknown. When a system reset occurs, the SR is set to 3H (xxxxxx11). When the host writes a command to the COMR, the SR is set to 1H (xxxxxx01), and the TMS32010 is interrupted. The TMS32010 then reads the command value in the COMR. This sets the SR to 3H.

When the TMS32010 completes processing the command, it responds to the host by writing an acknowledgement value to the ACKR, which sets the SR to 2H. The host reads the SR to determine if the TMS32010 has processed the command. If the command has been processed, the host reads the acknowledgement value in the ACKR. This sets the SR to 3H, which indicates that the command transaction has been satisfactorily completed by the TMS32010 and the host. The acknowledgement value can be dummy values or data being retrieved from the TMS32010. If the TMS32010 has not completed processing the command, the SR will read 1H or 3H. If the SR stays at 1H or 3H for an abnormally long period of time, a possible fault or problem is indicated. Table 3 shows the possible SR values and their interpretation.

Table 3. Host Interface Status Register Interpretation

HOST*	SR VALUE	COMSEND (bit 1)	COMACK (bit 0)	DESCRIPTION
A	0H	0	0	These conditions should not occur. Potential problems are indicated.
A	1H	0	1	
A	2H	1	0	
A	3H	1	1	Idle command processing state.
B	0H	0	0	Not allowed. Potential problem is indicated.
B	1H	0	1	The TMS32010 has not yet started processing the host command. Potential problem is indicated.
B	2H	1	0	The TMS32010 has completed processing the host command. Data is waiting for the host.
B	3H	1	1	The TMS32010 has started, but not completed, processing the host command.

* "HOST" refers to the command transaction state. "A" indicates that the command transaction is completed, or that a host command has not been sent. "B" indicates that a host command has been sent and that a command transaction is in progress.

The TMS32010 can process commands from the host in the following suggested manner: The command value written by the host, with an offset added to it, becomes the command vector or subroutine for the TMS32010. This is accomplished in software with a sequence of actions using the TMS32010. The interrupt service routine reads the host command value at port 6 (COMR) and stores it in internal data RAM (using the IN instruction). This value is then stored in the accumulator (using the LAC instruction). Using the AND instruction, the command value is masked to zero the 24 most-significant bits, and the new value is stored in memory. Masking is required because the eight most-significant bits on the TMS32010 data bus are unknown when reading the host interface port. A minimum offset of 4 is added to the masked command value in the accumulator (using the ADDS or ADD instruction). The offset is added to the command value to ensure that the resulting command vector does not overlap the reset and interrupt vector spaces. The next step is to vector to the subroutine via the CALA instruction. The CALA instruction uses the value in the accumulator as the subroutine vector. This forces the TMS32010 to begin executing a subroutine located at the command vector value in program memory space. Shown below is the source code sequence for the TMS32010.

```
ISR  IN     x,6  ; read host command in COMR
     LAC    x,0  ; load host command in ACC
     AND    y    ; mask host command
     ADD    z    ; add an offset to host command
     CALA        ; call subroutine
```

Note that ISR (interrupt service routine) is a label. During the TMS32010 initialization routine, the values >FFH and >4 are stored in locations y and x in data memory, respectively.

The TMS32010 communicates with the host on a byte data-transfer basis. The least-significant byte of the host and TMS32010 data bus is used. When the host issues a command, it writes one byte of data to COMR (U35), as shown in Figure 8. WRCOM (write command) is the write strobe from the host and is active low. Both flip-flops in U36 represent the SR. WRCOM sets a flip-flop in U36, thereby setting COMSEND (command send) low and interrupting the TMS32010. COMSEND is bit 1 in the SR. During the TMS32010's interrupt service routine, the TMS32010 reads host commands at port 6, which enables data from U35 onto the data bus. The port 6 read strobe sets COMSEND high, indicating that the command is being processed. When the TMS32010 responds to the command, it writes one byte into the ACKR U34 at port 6. This sets the second flip-flop in U36, thus forcing COMACK (command acknowledge) low. This indicates command processing is completed. COMACK is bit 0 in the SR. The host reads the ACKR (U34) by enabling data onto its bus using the RDACK (read acknowledge) strobe, which is active low and sets COMACK high. The host can read the SR (U36) with strobe RDSTAT (read status register), which is active low. U36 sets COMACK and COMSEND high when the TMS32010 is reset. A timing diagram illustrating the host transaction is shown in Figure 9.

INTERFACING APPLICATIONS

In this second section of the report, several telecommunications applications are discussed. First, an analog interface hardware application is examined, which is useful in self-contained standalone designs, such as analog repeaters or analog PBX designs. Then, a telephony test-set application is shown. This is useful as a self-contained standalone test tool or as a diagnostic function within a PBX. Finally, digital PBX applications in the following areas are described: PBX backplane interface, 2/4 wire transformer interface, three-way conference interface, ADPCM interface, and DTMF detection interface. It is important to note that the telecommunications designer can combine several applications using one TMS32010 to perform multiple functions.

These applications utilize the interface blocks previously described and the TMS32010 for performing digital signal processing algorithms. However, the scope of this report is not to discuss the DSP algorithms, but rather to demonstrate how these applications can be made possible using the telecommunications interface circuits. Application reports are available, which implement digital signal processing algorithms for telecommunications applications using the TMS320 family.[3,4,5]

Standalone Analog Interface

The standalone analog interface is useful where remote or independent operation is required. Some examples of standalone applications are adaptive repeaters and handsfree telephones (speakerphones), as shown in Figure 10.

The standalone analog interface contains elements from the following hardware blocks previously described:

1. TMS32010 and support circuit,
2. Timing and control circuit, and
3. Analog interface circuit.

Figure 10 illustrates the hardware configuration for a handsfree telephone application. The TMS32010 performs adaptive filtering and voice-switching functions. The analog interface circuit provides two combo-codec filters to interface to the analog telephone line, loudspeaker, and microphone. The combo-codec filters perform filtering of the analog I/O, and A/D and D/A conversion. The serial data streams to and from the combo-codec filters are converted to parallel format that permits interfacing to the TMS32010's parallel I/O. The TMS32010 easily interfaces to combo-codec filters using 8-bit shift registers and octal D-type flip-flop registers. Both combo-codec filters share one set of shift registers at the serial interface. The timing and control circuit provides all the signals required to synchronize and control the combo-codec filters, registers, and TMS32010.

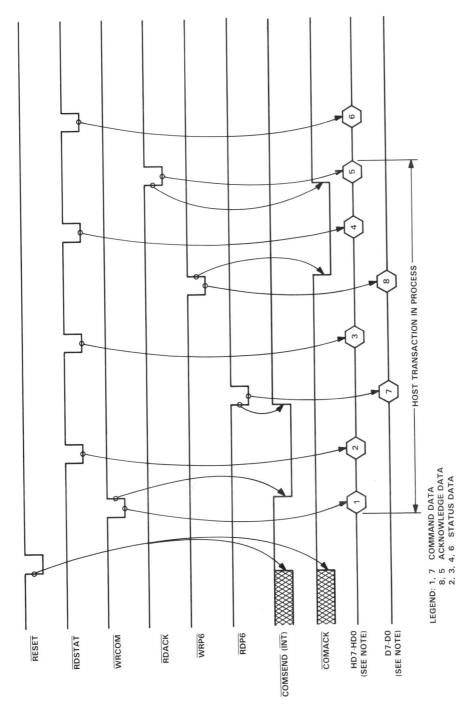

LEGEND: 1, 7 COMMAND DATA
8, 5 ACKNOWLEDGE DATA
2, 3, 4, 6 STATUS DATA

NOTE: Only host transaction data bursts shown.

RESET

RDSTAT

WRCOM

RDACK

WRP6

RDP6

COMSEND (INT)

COMACK

HD7-HD0
(SEE NOTE)

D7-D0
(SEE NOTE)

HOST TRANSACTION IN PROCESS

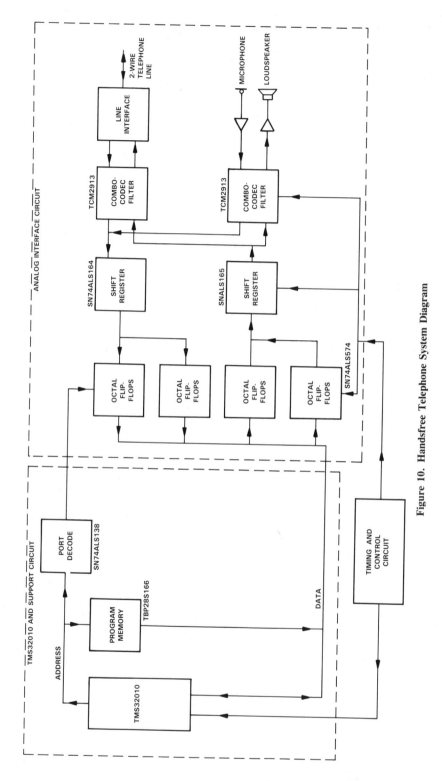

Figure 10. Handsfree Telephone System Diagram

Telephony Test-Set Interface

A telephony test-set is a useful tool for testing analog interfaces in telecommunications. It can also be used in a PBX design where analog line testing diagnostics are required. The application shown in Figure 11 is a standalone system.

A test-set application uses the following hardware blocks:

1. TMS32010 and support circuit,
2. Timing and control circuit,
3. Analog interface circuit,
4. Host interface circuit, and
5. PCM linearization circuit (optional).

Balanced inputs and outputs are required in a test-set application. The TMS32010 can be programmed to measure noise, distortion, signal levels, etc., and serve as a network simulator for analog telephony circuits. The electrical performance of the combo-codec filters is a limiting factor on measurements and simulations that can be performed. Figure 11 shows a possible telephony test-set system configuration. All measurements are under control of a host processor. The host interfaces to a control input, such as a keyboard or human interface I/O, and a display, such as a CRT. Measurement parameters can be modified via the control interface, and the actual measured values displayed. All transfers between the host and application circuit are performed via the host interface.

A possible test-set function may be a sine-wave signal generator. The TMS32010 produces a digital sine wave, which is converted to the analog domain via the combo-codec filters in the analog interface circuit. TMS32010 source code for a precision sine-wave generator is given in the application report, ''Precision Digital Sine-Wave Generation with the TMS32010.''[6] Note that this algorithm produces linear samples that must be converted to μ-law or A-law PCM samples. TMS32010 source code for performing this conversion can be found in the application report, ''Companding Routines for the TMS32010.''[1] Hardware conversion is also possible using the PCM linearization circuit, described in the previous major section of this report.

PBX Backplane Interface

A backplane is an interconnect system with its mechanical, electrical, pin assignments, and interaction protocol rigidly defined. Mechanically, a backplane is a printed circuit board with connectors that reside at the rear of a card-cage, providing electrical interconnection between cards plugged into these connectors. Architecturally, a backplane has a set of protocols that control data transfers between interconnected cards. This application report concerns itself only with some architectural aspects of possible PBX backplanes.

Two types of backplanes are used in most PBX designs: serial and parallel. The serial backplane is the most popular. Interfacing with standard codecs is simple, because they contain serial input and output ports. In some modern PBXs, a parallel backplane is used, which eases hardware interfacing since standard byte-wide logic can be used. In either system,

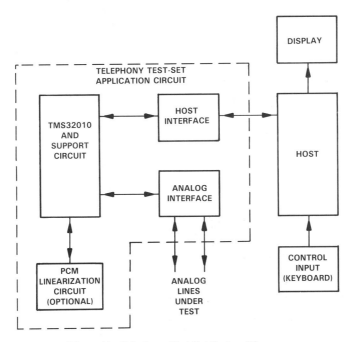

Figure 11. Telephony Test-Set System Diagram

this application circuit can serve as a signal processing module to perform specialized functions, such as DTMF detection, echo cancellation, noise reduction, and conferencing. The host processor determines which PCM channels are assigned to the signal processing module.

A typical PBX backplane contains one or more serial PCM highways or an 8/16-bit PCM bus, system address and data buses, and system control lines (see Figure 12). The system address bus is used by the host processor to access all the peripherals or modules connected to the backplane. The data transfers occur across the data bus between the host processor and modules. The system control lines are typically read and write strobes used by the host processor to control these data transfers. The system clock and 8-kHz framing pulse are used for synchronizing access to the PCM highway or bus. In most systems, time-slot enable control lines are provided to access individual PCM channels. The time-slot enable control lines are provided by either the backplane with a control card using time-slot decode circuitry or card-resident time-slot decode circuitry. The application circuit does not provide any time-slot decoding; however, it provides the flexibility to interface to most systems through the external signals $\overline{\text{ITSEN}}$ and $\overline{\text{OTSEN}}$.

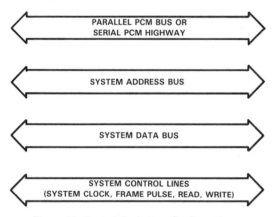

Figure 12. Typical Backplane Configuration

In Figure 13, the framing pulse is used to indicate the start of a frame. The framing pulse, usually one system clock cycle wide, is synchronous with the system clock and synchronizes all peripherals connected to the PCM highway or bus. Many PCM systems are based on a 2.048-MHz system clock, which provides 32 channels on a serial PCM highway or 256 channels on an 8-bit parallel PCM bus.

The backplane application hardware requires the following blocks:

1. TMS32010 and support circuit,
2. Timing and control circuit,
3. Two PBX/TDM interface circuits for use with receive and transmit channels,
4. Host interface circuit, and
5. PCM linearization circuit for critical real-time applications (optional).

Each PBX/TDM interface contains a serial input port, serial open-drain/collector, three-state output ports, and an 8-bit wide three-state parallel output port. These ports directly connect to the PCM bus or highway on the backplane. An external device or circuit is required to provide the time-slot enables $\overline{\text{ITSEN}}$ and $\overline{\text{OTSEN}}$ (see Figure 14). The host determines which channels reside in time-slots on the PCM backplane, and writes information to a time-slot controller residing in the peripheral or module connected to the backplane. The controller then furnishes the input and output time-slot enables $\overline{\text{ITSEN}}$ and $\overline{\text{OTSEN}}$, respectively. $\overline{\text{ITSEN}}$ assigns to the application circuit the PCM data in the corresponding time-slot on the PCM backplane. $\overline{\text{OTSEN}}$ enables PCM data from the application circuit onto the appropriate time-slot on the PCM backplane. The application hardware permits broadcasting PCM data onto several time-slots on the PCM backplane.

2/4-Wire Transformer Interface

The 2/4-wire transformer is useful for interfacing directly to 2-wire trunks or telephone lines. The transformer application requires the following hardware blocks:

1. TMS32010 and support circuit, and
2. Analog interface circuit.

Figure 15 shows how a 2/4-wire transformer is connected to the analog interface circuit. $V_{OUT}+$ drives a load resistor (RL) in series with the transformer. This series arrangement produces a 6-dB loss between $V_{OUT}+$ and the telephone line. The load resistor balances the transmission line. The input is arranged so that the output signal reflected back at the transformer to input $V_{IN}+$ is minimized. This is accomplished by subtracting $V_{OUT}+$ using the $V_{IN}-$ input in a difference amplifier configuration. The return loss will be a function of matching RL to the telephone-line impedance. A mismatch between the load RL and the telephone line will worsen the return loss. The transmission-signal reference points were described in the analog interface circuit section. Information on 2/4-wire analog interfacing that conforms to FCC Rules Part 68 can be found in reference [7].

Three-Way Conference Interface

A popular PBX feature is three-way conferencing. In analog PBXs, conferencing is easily accomplished with op-amps in a summing configuration. The problem is not trivial in digital PBXs. μ/A-law encoded PCM samples cannot be directly summed together with standard adders (i.e., SN74LS283s) because of the nonlinear encoding scheme used. These PCM values must first be converted to linear values, summed, then converted back to μ/A-law PCM. This function is possible with the TMS32010. As mentioned previously, the PCM samples can be linearized in software[1] or in hardware.

For a three-way conference, the application hardware (see Figure 16) contains the following modules:

1. TMS32010 and support circuit,
2. Timing and control circuit,

400

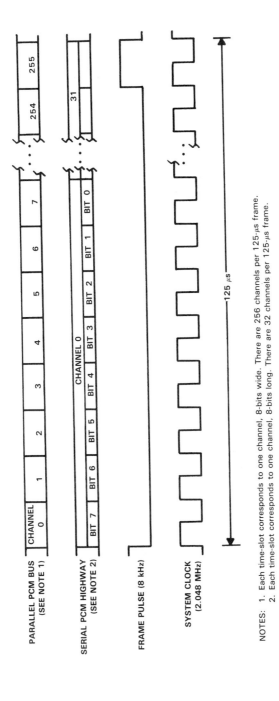

NOTES: 1. Each time-slot corresponds to one channel, 8-bits wide. There are 256 channels per 125-μs frame.
2. Each time-slot corresponds to one channel, 8-bits long. There are 32 channels per 125-μs frame.

Figure 13. Possible Backplane Timing

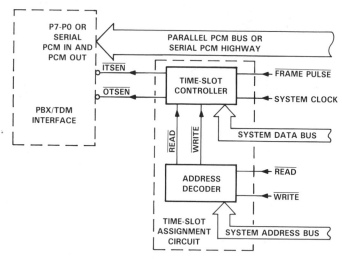

Figure 14. Typical Backplane Interface

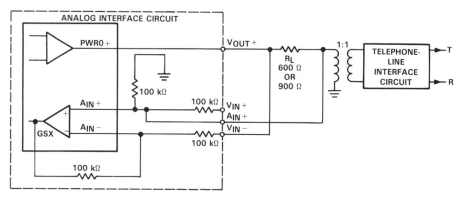

NOTES: 1. R_L balances telephone line.
2. 6-dB loss from V_{OUT} and to telephone line.
3. 6-dB gain from telephone line to GSX.

Figure 15. 2/4-Wire Transformer Connection

3. Three PBX/TDM interface circuits, and

4. PCM linearization circuit (optional).

Three-way conferencing is the simplest method of conferencing; however, some system considerations are required. Return loss and noise performance are most important factors in a conferencing design. Return loss refers to the amount of attenuation of a reflected signal: the larger the return loss, the smaller the reflected signal. Reflections occur where there are 2/4-wire conversions in the network. In long network paths, such as satellite links, these reflections manifest themselves as echoes. (The application report, "Digital Voice Echo Canceller with a TMS32020,"[5] describes the implementation of an echo canceller using the TMS32020, the second-generation digital signal processor of the TMS320 family.) Return loss affects the "singing" margin and echo performance of the conference. "Singing"

refers to an oscillatory state in a closed-loop system that occurs when excessive gain is inserted in the loop. Summing of two or more parties into a conference decreases the return loss, thereby increasing the chances of "singing" and degrading echo performance. In a three-way conference, the simplest method to reduce the effects of return loss is to insert transmission loss into the receive (RX) or transmit (TX) paths.

In the three-way conference node shown in Figure 17, each party is connected only to the other two parties, not to itself. Normally, the return loss is dominated by the party with the worst reflections. In a worst-case condition, if these two parties have similar reflection characteristics, the return loss decreases by 6 dB. To offset this, the transmission levels at summing junctions are decreased by 6 dB, i.e., a 6-dB loss insertion at TX IN or RX OUT. The loss insertion

14. Telecommunications Interfacing to the TMS32010

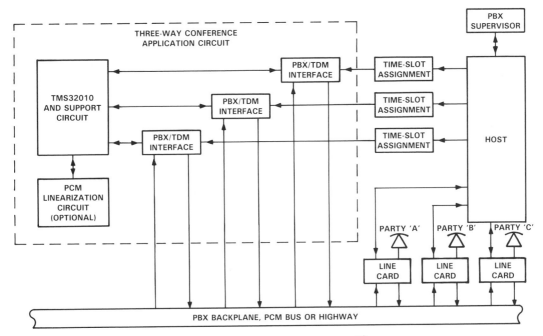

Figure 16. Three-Way Conference System Diagram

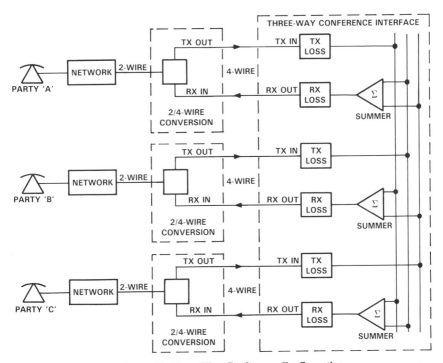

Figure 17. Three-Way Conference Configuration

slightly degrades noise performance. Noise performance refers to the subjective transmission quality of band-limited voice vs. band-limited noise. The noise performance of a conference, where all parties are summed together, always degrades to the party with the worst signal-to-noise ratio. In conferences where a large number of parties are connected together, voice-switching is used to limit the number of talkers summed together at one time, thereby limiting echo and noise performance degradation. Stringing together several three-way conferences to make a larger conference is not recommended, because noise performance will be substantially reduced due to compounded 6-dB insertion losses. The removal of this loss would likely degrade the echo performance of the conference and increase the chances of "singing".

The TMS32010 source code in Figure 18 can be used for a three-way conference. PCM linearization can be performed in hardware or software. The user can insert the appropriate linearization code in the space provided. Memory locations TA, TB, and TC correspond to the TX IN inputs at ports X, Y, and Z, respectively. Memory locations RA, RB, and RC correspond to the RX OUT outputs at port X, Y, and Z, respectively. The PBX supervisor maps the TMS32010 ports to their appropriate time-slots on the backplane. The 6-dB insertion loss is accomplished by a 15-position right-shift on the TX IN samples when loading the accumulator. This results in an effective single-position left-shift when using the upper 16-bits of the accumulator. Note that in the following code, WAIT is a label.

```
WAIT    BIOZ    WAIT    ; wait for 8-kHz interrupt
        IN      TA,X    ; input ports to memory
        IN      TB,Y
        IN      TC,Z
*
* Insert mu-/A-law to two's-complement linear
* conversion code here for ports X, Y, and Z.
        LAC     TA,15   ; sum ports X and Y for
*                       ; port Z
        ADD     TB,15
        SACH    RC,0
        LAC     TB,15   ; sum ports Y and Z for
*                       ; port X
        ADD     TC,15
        SACH    RA,0
        LAC     TA,15   ; sum ports X and Z for
*                       ; port Y
        ADD     TC,15
        SACH    RB,0
*
* Insert two's-complement linear to mu-/A-law
* conversion code here for ports X, Y, and Z.
        OUT     RA,X    ; output memory locations
*                       ; to ports
        OUT     RB,Y
        OUT     RC,Z
        B       WAIT
        END
```

Figure 18. Three-Way Conferencing Source Code

ADPCM Interface

Adaptive Differential Pulse Code Modulation (ADPCM) is a method of increasing voice-band transmission capacity. ADPCM is used in telecommunications systems to transcode 64-kbit/s PCM to 32-kbit/s ADPCM, thereby doubling the capacity of a transmission system. An implementation of a full-duplex ADPCM on a single TMS32010 is described in the application report, "32-kbit/s ADPCM with the TMS32010."[3] Although it does not provide bit-by-bit compatibility with CCITT recommendations, it follows the recommended model. The same report also describes the TMS32010 implementation of a half-duplex 32-kbit/s ADPCM algorithm recommended by CCITT.

The following paragraphs briefly examine two of many possible ADPCM applications: voice mail and transcoding in a PBX design.

Voice-mail in a PBX application is possible by using the ADPCM transcoder and the host processor. The application hardware for a voice-mail function, shown in Figure 19, requires the following modules:

1. TMS32010 and support circuit,
2. PBX/TDM interface circuit, and
3. Host interface circuit.

The ADPCM algorithm is stored on the TMS32010's program memory ROMs. The PBX supervisor instructs the host to initiate the voice-mail function and maps the PBX/TDM interface circuit to the appropriate time-slot on the backplane. The host initiates the transaction by writing a command to the application circuit and waits for ADPCM samples. The TMS32010 reads the μ/A-law PCM sample from the PBX/TDM interface. It then performs the PCM to ADPCM transcoding function. The ADPCM sample is then written to the host interface. The host reads the ADPCM samples at the host interface and stores them in a high-capacity storage device, such as a hard disk or bubble memory. At some point, the PBX supervisor instructs the host to terminate the transaction with the application circuit. When the mail is to be delivered, the PBX supervisor again instructs the host to initiate the transaction with the application circuit, and maps the PBX/TDM interface to the appropriate time-slot on the backplane. The host initiates the transaction by writing a command to the application circuit. The application circuit then waits for ADPCM samples from the host. The host retrieves the ADPCM sample from the storage device and writes it to the host interface. The TMS32010 reads the ADPCM sample, then performs the ADPCM to

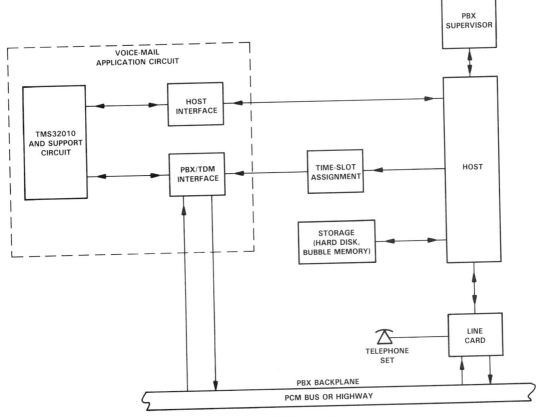

Figure 19. Voice-Mail System Diagram

14. Telecommunications Interfacing to the TMS32010

PCM transcoding function, and writes the PCM sample to the PBX/TDM interface.

Transcoding within a PBX is possible between an 8-bit PCM highway and a 4-bit ADPCM highway. The transcoding application requires the following hardware blocks, as shown in Figure 20:

1. TMS32010 and support circuit,
2. Timing and control circuit, and
3. Two PBX/TDM interface circuits.

The PBX supervisor assigns the PBX/TDM interfaces to their respective time-slots on the PCM and ADPCM highways. The TMS32010 reads the sample from one of the interfaces, performs the transcoding function, then writes the sample to the second interface. The reverse operation is also applied, thereby providing full-duplex operation.

DTMF Detection Interface

In modern PBXs, DTMF detection is used to implement call features. DTMF signalling from the station-set is interpreted by the PBX for various features, such as outside line access, call-forwarding, conferencing, and voice-mail. DTMF detection is an integral part of any PBX design. DTMF detection is a popular signal processing function easily implemented on the TMS32010. TMS32010 source code for the DTMF detector is given in reference [4]. Multiple-channel DTMF detection is possible with a single TMS32010.

The DTMF detection application contains the following modules:

1. TMS32010 and support circuit,
2. Timing and control circuit,
3. Host interface circuit,
4. PBX/TDM interface circuit, and
5. PCM linearization circuit (optional).

The DTMF detection algorithm[4] requires PCM linearization. When the DTMF detection algorithm used is time-critical and software linearization is not possible, the PCM linearization circuit is needed. A system diagram for DTMF detection is shown in Figure 21.

An example of system operation is as follows: The PBX supervisor is informed that a station-set is in an off-hook condition. The PBX supervisor prepares for DTMF detection by mapping the PBX/TDM interface and station-set to the same time-slot, and instructing the host to detect digits and implement features. Note that the station-set is interfaced to the PBX backplane with a line-card. The host writes a command to the host interface instructing the application circuit to begin DTMF detection. The host waits for digits to be returned at the host interface. The TMS32010 reads samples from the PBX/TDM interface and implements the DTMF detection algorithm. When a DTMF digit is detected,

the value is written to the host interface. The host reads the digit and implements the requested feature. The host then writes a command to the host interface, instructing the application circuit to terminate DTMF detection.

SUMMARY

The TMS32010 is well suited for a variety of telecommunications applications, ranging from trunk circuits to PBX systems. The hardware described in this application report allows the designer to apply some new ideas to end-product designs. Various digital PBX applications in the following areas were described: PBX backplane, 2/4-wire transformer connection, telephony test-set, three-way conferencing, ADPCM transcoding, and DTMF detection. Standalone systems applications, such as analog interfaces and testing, were also shown. All the applications can be transposed to the TMS32020, the second-generation digital signal processor.

The TMS32010 offers cost-effective solutions for many telecommunications applications requiring digital signal processing. Further cost reduction is possible using the TMS320M10. This mask version of the TMS32010 provides a 1523-word on-chip program ROM. Texas Instruments provides easy-to-use low-cost tools for circuit and algorithm development. The entire TMS320 family of digital signal processors is extensively supported by a staff of application engineers at the factory and in the field.

REFERENCES

1. *Companding Routines for the TMS32010* (Application Report), Texas Instruments (1984).
2. *Types TCM2913, TCM2914, TCM2916, TCM2917 Combined Single-Chip PCM Codec and Filter* (Data Sheet SCTS012), Texas Instruments (1983).
3. J.B. Reimer, M.L. McMahan, and M. Arjmand, *32-kbit/s ADPCM with the TMS32010* (Application Report), Texas Instruments (1985).
4. P. Mock, "Add DTMF Generation and Decoding to DSP-μP Designs," *EDN* (March 21, 1985).
5. D.G. Messerschmitt, D.J. Hedberg, C.R. Cole, A. Haoui, and P. Winship, *Digital Voice Echo Canceller with a TMS32020* (Application Report), Texas Instruments (1985).
6. *Precision Digital Sine-Wave Generation with the TMS32010* (Application Report), Texas Instruments (1984).
7. G. Dash, "Understand FCC Rules When Designing Telecomm Equipment," *EDN* (May 16, 1985).

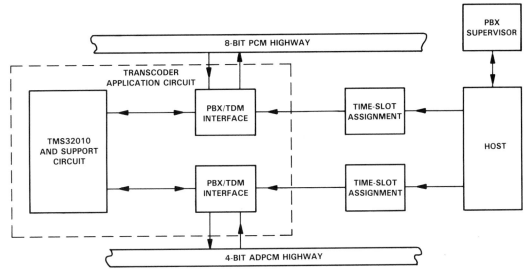

Figure 20. PBX Transcoder System Diagram

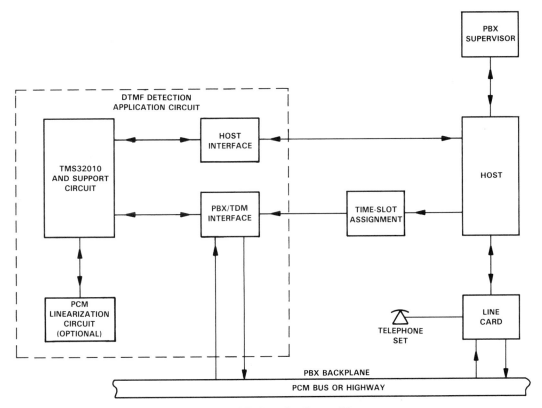

Figure 21. DTMF Detection System Diagram

APPENDIX

PCM Linearization Code

```
D Line# 1       7
1 C**********************************************
2 C
3 C THIS PROGRAM COMPUTES MU-LAW PCM TO 14-BIT TWO'S COMPLEMENT
4 C LINEAR CODES FOR PURPOSES OF PROGRAMMING PROMS AND EPROMS.
5 C
6 C GENERATED IS AN OBJECT FILE IN THE INTEL DATA RECORD FORMAT.
7 C
8 C WRITTEN BY JEFF ROBILLARD, TEXAS INSTRUMENTS INC.
9 C        AUGUST 6, 1985
10 C
11 C**********************************************
12 C
13 C PROGRAM WRITTEN ON THE TEXAS INSTRUMENTS PROFESSIONAL COMPUTER
14 C USING MICROSOFT FORTRAN.
15 C
16 C**********************************************
17 C
18 C MU-LAW LINEARIZATION FORMULA:
19 C S=SEGMENT
20 C Q=QUANTIZATION CHORD
21 C
22 C LIN=(2**S(2*Q+33))-33
23 C
24 C**********************************************
25 C
26 C DEFINE CONSTANTS AND INITIALIZE
27 C
28       INTEGER S,Q,LIN,LINBB
29       DIMENSION LINBB(512)
30 C
31 C 2'S COMPLEMENT LINEARIZATION CALCULATION
32 C
33       DO 100 J=1,128,1
34       JJ=J-1
35       Q=MOD(JJ,16)
36       S=JJ/16
37       LIN=(2**S*(2*Q+33))-33
38       LINBB(128+J)=(65536-LIN)/256
39       LINBB(256+J)=MOD((65536-LIN),256)
40       LINBB(384+J)=MOD(LIN,256)
41  100 CONTINUE
42       LINBB(129)=0
43       LINBB(385)=0
44       CALL HEXFILE (LINBB,'U-LIN.HEX',512)
45 C     STOP LINEARIZATION COMPLETED'
46       END
47
```

Name	Type	Offset	P Class
J	INTEGER*4	2050	
JJ	INTEGER*4	2062	
LIN	INTEGER*4	2074	
LINBB	INTEGER*4	2	
MOD			INTRINSIC
Q	INTEGER*4	2066	
S	INTEGER*4	2070	

```
48 C
49 C INTEL RECORD FORMAT GENERATOR
```

```
D Line# 1       7
50 C
51       SUBROUTINE HEXFILE (DEC,LABL,SIZE)
52       CHARACTER LABL*9,REC,HX
53       INTEGER DEC,SIZE,SUM,ADDR,CH
54       DIMENSION DEC(SIZE),REC(43),HX(16),CH(43)
55       OPEN(1,FILE=LABL,STATUS='NEW')
56       HX(1)='0'
57       HX(2)='1'
58       HX(3)='2'
59       HX(4)='3'
60       HX(5)='4'
61       HX(6)='5'
62       HX(7)='6'
63       HX(8)='7'
64       HX(9)='8'
65       HX(10)='9'
66       HX(11)='A'
67       HX(12)='B'
68       HX(13)='C'
69       HX(14)='D'
70       HX(15)='E'
71       HX(16)='F'
72       CH(2)=1
73       CH(3)=0
74       CH(8)=0
75       CH(9)=0
76       REC(1)=':'
77       I=SIZE/16
78       DO 1000 J=1,I,1
79       ADDR=(J-1)*16
80       CH(4)=ADDR/4096
81       CH(5)=MOD(ADDR,4096)/256
82       CH(6)=MOD(ADDR,256)/16
83       CH(7)=MOD(ADDR,16)
84       DO 2000 L=1,16,1
85       LL=(L-1)*2+10
86       CH(LL)=DEC(ADDR+L)/16
87       CH(LL+1)=MOD(DEC(ADDR+L),16)
88  2000 CONTINUE
89       SUM=0
90       DO 3000 M=2,40,2
91       SUM=SUM+(CH(M)*16)
92  3000 CONTINUE
93       DO 3100 M=3,41,2
94       SUM=SUM+CH(M)
95  3100 CONTINUE
96       SUM=256-MOD(SUM,256)
97       CH(42)=SUM/16
98       CH(43)=MOD(SUM,16)
99       DO 4000 N=2,43,1
100      REC(N)=HX(CH(N)+1)
101 4000 CONTINUE
102      WRITE(1,1300) (REC(K2),K2=1,43,1)
103 1000 CONTINUE
104      WRITE(1,1400)
105      CLOSE(1,1400)
106 1300 FORMAT(43(A1))
107 1400 FORMAT(':000000001FF')
108      RETURN
```

14. Telecommunications Interfacing to the TMS32010

```
D Line#  1      7

   1  C****************************************************************************
   2  C
   3  C   THIS PROGRAM COMPUTES 14-BIT TWO'S COMPLEMENT LINEAR TO MU-LAW PCM
   4  C   CODE FOR PURPOSES OF PROGRAMMING PROMS AND EPROMS.
   5  C
   6  C   GENERATED IS AN OBJECT FILE IN THE INTEL DATA RECORD FORMAT.
   7  C
   8  C   WRITTEN BY JEFF ROBILLARD, TEXAS INSTRUMENTS INC.
   9  C           AUGUST 6, 1985
  10  C
  11  C****************************************************************************
  12  C
  13  C   PROGRAM WRITTEN ON THE TEXAS INSTRUMENTS PROFESSIONAL COMPUTER
  14  C   USING MICROSOFT FORTRAN.
  15  C
  16  C****************************************************************************
  17  C
  18  C   MU-LAW TO LINEAR CONVERSION:
  19  C   S=SEGMENT
  20  C   Q=QUANTIZATION CHORD
  21  C
  22  C   DETERMINE S BY TABLE LOOK-UP
  23  C   Q=INT(INT((LIN+33)/2**S)/2
  24  C
  25  C****************************************************************************
  26  C
  27  $STORAGE:2
  28          INTEGER L,S,CPAND
  29          INTEGER*2 ULAW
  30          DIMENSION ULAW(16384)
  31          DO 100 J=1,32,1
  32          S=0
  33          L=J-1
  34          ULAW(J)=CPAND(L,S)
  35  100     CONTINUE
  36          DO 110 J=33,96,1
  37          S=1
  38          L=J-1
  39          ULAW(J)=CPAND(L,S)
  40  110     CONTINUE
  41          DO 120 J=97,224,1
  42          S=2
  43          L=J-1
  44          ULAW(J)=CPAND(L,S)
  45  120     CONTINUE
  46          DO 130 J=225,480,1
  47          S=3
  48          L=J-1
  49          ULAW(J)=CPAND(L,S)
  50  130     CONTINUE
  51          DO 140 J=481,992,1
  52          S=4
  53          L=J-1
  54          ULAW(J)=CPAND(L,S)
  55  140     CONTINUE
  56          DO 150 J=993,2016,1
  57          S=5
  58          L=J-1
  59          ULAW(J)=CPAND(L,S)
```

```
D Line#  1      7
        109  END

Name    Type        Offset P Class

ADDR    INTEGER*4    2326
CH      INTEGER*4    2138
DEC     INTEGER*4       0   *
HX      CHAR*1       2121
I       INTEGER*4    2310
J       INTEGER*4    2314
K2      INTEGER*4    2358
L       INTEGER*4    2330
LABL    CHAR*9          4   *
LL      INTEGER*4    2342
M       INTEGER*4    2350

N       INTEGER*4    2354                 INTRINSIC
REC     CHAR*1       2078
SIZE    INTEGER*4       8   *
SUM     INTEGER*4    2346


Name    Type        Size   Class

HEXFIL                            SUBROUTINE
MAIN                              PROGRAM

Pass One    No Errors Detected
            109 Source Lines
```

14. Telecommunications Interfacing to the TMS32010

```
D Line# 1    7
1       60  150 CONTINUE
1       61      DO 160 J=2017,4064,1
1       62      S=6
1       63      L=J-1
1       64      ULAW(J)=CPAND(L,S)
1       65  160 CONTINUE
1       66      DO 170 J=4065,8192,1
1       67      S=7
1       68      L=J-1
1       69      ULAW(J)=CPAND(L,S)
1       70  170 CONTINUE
1       71      DO 200 J=1,8192,1
1       72      ULAW(16384-J+1)=128+ULAW(J)
1       73  200 CONTINUE
1       74      CALL HEXFILE (ULAW,'LIN-U.HEX',16384)
1       75      STOP 'U-LAW COMPANDING COMPLETED'
1       76      END

Name    Type        Offset  P Class

CPAND   INTEGER*2                  FUNCTION
J       INTEGER*2   32770
L       INTEGER*2   32778
S       INTEGER*2   32776
ULAW    INTEGER*2   2

77  C
78  C  U-LAW COMPAND FUNCTION
79  C
80      INTEGER FUNCTION CPAND(LIN,SEG)
81      INTEGER LIN,SEG
82      CPAND=SEG*16+(((LIN+33)/(2**SEG))-33)/2
83      RETURN
84      END

Name    Type        Offset  P Class

LIN     INTEGER*2   0 *
SEG     INTEGER*2   4 *

85  C
86  C  INTEL RECORD FORMAT GENERATOR
87  C
88      SUBROUTINE HEXFILE (DEC,LABL,SIZE)
89      CHARACTER LABL*9,REC,HX
90      INTEGER DEC,SIZE,SUM,ADDR,CH
91      DIMENSION DEC(SIZE),REC(43),HX(16),CH(43)
92      OPEN(1,FILE=LABL,STATUS='NEW')
93      HX(1)='0'
94      HX(2)='1'
95      HX(3)='2'
96      HX(4)='3'
97      HX(5)='4'
98      HX(6)='5'
99      HX(7)='6'
100     HX(8)='7'
101     HX(9)='8'
```

```
D Line# 1    7
1       102     HX(10)='9'
1       103     HX(11)='A'
1       104     HX(12)='B'
1       105     HX(13)='C'
1       106     HX(14)='D'
1       107     HX(15)='E'
1       108     HX(16)='F'
1       109     CH(2)=1
1       110     CH(3)=0
1       111     CH(8)=0
1       112     CH(9)=0
1       113     REC(1)=':'
1       114     I=SIZE/16
1       115     DO 1000 J=1,I,1
1       116     ADDR=(J-1)*16
1       117     CH(4)=ADDR/4096
1       118     CH(5)=MOD(ADDR,4096)/256
1       119     CH(6)=MOD(ADDR,256)/16
1       120     CH(7)=MOD(ADDR,16)
1       121     DO 2000 L=1,16,1
2       122     LL=(L-1)*2+10
2       123     CH(LL)=DEC(ADDR+L)/16
2       124     CH(LL+1)=MOD(DEC(ADDR+L),16)
2       125 2000 CONTINUE
1       126     SUM=0
1       127     DO 3000 M=2,40,2
2       128     SUM=SUM+(CH(M)*16)
2       129 3000 CONTINUE
1       130     DO 3100 M=3,41,2
2       131     SUM=SUM+CH(M)
2       132 3100 CONTINUE
1       133     SUM=256-MOD(SUM,256)
1       134     CH(42)=SUM/16
1       135     CH(43)=MOD(SUM,16)
1       136     DO 4000 N=2,43,1
2       137     REC(N)=HX(CH(N)+1)
2       138 4000 CONTINUE
1       139     WRITE(1,1300) (REC(K2),K2=1,43,1)
1       140 1000 CONTINUE
1       141     WRITE(1,1400)
1       142     CLOSE(1)
1       143 1300 FORMAT(43(A1))
1       144 1400 FORMAT(':00000001FF')
1       145     RETURN
1       146     END

Name    Type        Offset  P Class

ADDR    INTEGER*2   32934
CH      INTEGER*2   32840
DEC     INTEGER*2   0 *
HX      CHAR*1      32823
I       INTEGER*2   32926
J       INTEGER*2   32928
K2      INTEGER*2   32950
L       INTEGER*2   32936
LABL    CHAR*9      4 *
LL      INTEGER*2   32942
M       INTEGER*2   32946
```

14. Telecommunications Interfacing to the TMS32010

```
D Line# 1      7                 INTRINSIC
MOD
N       INTEGER*2    32948
REC     CHAR*1       32780
SIZE    INTEGER*2        8 *
SUM     INTEGER*2    32944


Name    Type        Size    Class

CPAND   INTEGER*2            FUNCTION
HEXFIL                       SUBROUTINE
MAIN                         PROGRAM


    Pass One    No Errors Detected
                146 Source Lines
```

15. Digital Voice Echo Canceller with a TMS32020

David Messerschmitt, David Hedberg, Christopher Cole, Amine Haoui, and Peter Winship
Technekron Communications Systems

INTRODUCTION

Echo cancellers using adaptive filtering techniques are now finding widespread practical applications to solve a variety of communications systems problems.[1] These applications are made possible by the recent advances in microelectronics, particularly in the area of Digital Signal Processors (DSPs). Cancelling echoes for long-distance telephone voice communications, full-duplex voiceband data modems, and high-performance "handsfree" audio-conferencing systems (including speakerphones) are a few examples of these applications.

The continuing deployment of all-digital toll switches, satellite-based voice and data networks, and new intercontinental long-haul circuits have been accompanied by more widespread use of all-digital voice echo cancellers in carrier systems.[2] In addition, new low-cost integrated single-channel echo cancellers are expected to see increasing application in smaller systems for audio teleconferencing and low-cost voice/data communications using private satellite earth stations.

Advancements in single-chip programmable digital signal processor technology now make it attractive to implement modular per-channel echo canceller architectures with all the functions required for a single echo canceller

integrated within a single device. A programmable DSP implementation offers the advantages of a short development and test schedule and the flexibility to meet custom product requirements by extending software-based functional building blocks rather than designing new hardware.

This application report describes the implementation of an integrated 128-tap (16-ms span) digital voice echo canceller on the Texas Instruments TMS32020 programmable signal processor. The implementation features a direct interface for standard PCM codecs (e.g., Texas Instruments TCM2913) and meets the requirements of the CCITT (International Telegraph and Telephone Consultive Committee) Recommendation G.165 for echo cancellers.[3] This report presents the requirements for echo cancellation in voice transmission and discusses the generic echo cancellation algorithms. The implementation considerations for a 128-tap echo canceller on the TMS32020 are then described in detail, as well as the software logic and flow for each program module.

A hardware demonstration model of a 128-tap voice echo canceller using the TMS32020 has been constructed and tested. Figure 1 shows a photograph of the echo canceller demonstration system. The main features of this model are described within the report. The appendixes contain complete source code and a schematic for the demonstration system echo canceller module.

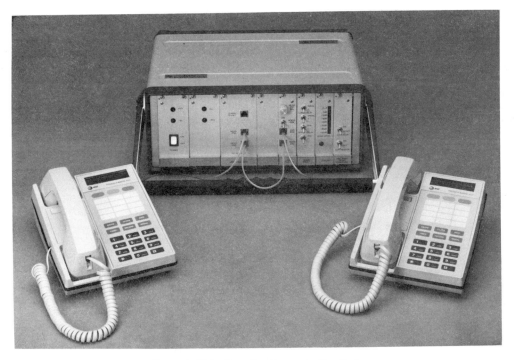

Figure 1. Echo Canceller Demonstration System

ECHO CANCELLATION IN
VOICE TRANSMISSION

Echoes in the Telephone Network

The source of echoes can be understood by considering a simplified connection between two subscribers, S1 and S2, as shown in Figure 2. This connection is typical in that it contains two-wire segments on the ends, a four-wire connection in the center, and a hybrid at each end to convert from two-wire transmission to four-wire transmission. Each two-wire segment consists of the subscriber loop and possibly some portion of the local network. Over this segment, both directions of transmission are carried by the same wire pair, i.e., signals from speakers S1 and S2 are superimposed on this segment. On the four-wire section, the two directions of transmission are segregated. The speech from speaker S1 follows the upper transmission path, as indicated by the arrow, while speech originating from S2 follows the lower path. The segregation of the two signals is necessary where it is desired to insert carrier terminals, amplifiers, or digital switches.

The hybrid is a device that converts two-wire to four-wire transmission. The role of the hybrid on the right-hand side is to direct the signal energy arriving from S1 to the two-wire segment of S2 without allowing it to return to S1 via the lower four-wire transmission path. Because of impedance mismatches (unfortunately occurring in practice), some of this energy will be returned to speaker S1, who then hears a delayed version of his speech. This is the source of "talker echo."

The subjective effect of the talker echo depends on the delay around the loop. For short delays, the talker echo represents an insignificant impairment if the attenuation is reasonable (6 dB or more). This is because the talker echo is indistinguishable from the normal sidetone in the telephone. For satellite connections, the delay in each four-wire path is about 270 ms as a consequence of the high altitude of synchronous satellites. This means that the round-trip echo delay is approximately 540 ms, which makes it very disturbing to the talker, and can in fact make it quite difficult to carry on a conversation. When such is the case, it is essential to find ways of controlling or removing that echo. Since the subjective annoyance of echo increases with delay as well as echo level due to hybrid return energy, the measures for control depend on the circuit length.

For terrestrial circuits under 2,000 miles, the via net loss (VNL) plan,[4] which regulates loss as a function of transmission distance, is used to limit the maximum echo-to-signal ratio. On circuits over this length (e.g., intercontinental circuits), echo suppressors or cancellers are used. An echo suppressor is a voice-operated switch that attempts to open the path from listener to talker whenever the listener is silent. However, echo suppressors perform poorly since echo is not blocked during periods of doubletalk. They impart a choppiness to speech and background noise as the transmission path is opened and closed. Due to recent decreasing trends in DSP costs, digital echo cancellers are now viable as replacements for most of the circuits using echo suppressors.

For satellite circuits with full hop delays of 540 ms, echo suppressors are subjectively inadequate, and cancellers must be employed.

Digital Echo Cancellers in Voice Carrier Systems

The principle of the echo canceller for one direction of transmission is shown in Figure 3. The portion of the four-wire connection near the two-wire interface is shown in this figure, with one direction of voice transmission between ports A and C, and the other direction between ports D and B. All signals shown are sampled data signals that would occur naturally at a digital transmission terminal or digital switch. The far-end talker signal is denoted $y(i)$, the undesired echo $r(i)$, and the near-end talker $x(i)$. The near-end talker is superimposed with the undesired echo on port D. The received signal from far-end talker $y(i)$ is available as a reference signal for the echo canceller and is used by the canceller to generate a replica of the echo called $\hat{r}(i)$. This replica is subtracted from the near-end talker plus echo to yield the transmitted near-end signal $u(i)$ where $u(i) = x(i) + r(i) - \hat{r}(i)$. Ideally, the residual echo error $e(i) = r(i) - \hat{r}(i)$ is very small after echo cancellation.

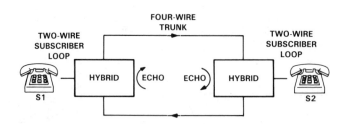

Figure 2. A Simplified Telephone Connection

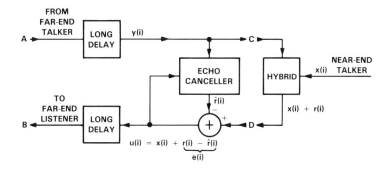

Figure 3. Echo Canceller Configuration

The echo canceller generates the echo replica by applying the reference signal to a transversal filter (tapped-delay line), as shown in Figure 4. If the transfer function of the transversal filter is identical to that of the echo path, the echo replica will be identical to the echo, thus achieving total cancellation. Since the transfer function of the echo path from port C to port D is not normally known in advance, the canceller adapts the coefficients of the transversal filter. To reduce error, the adaptation algorithm infers from the cancellation error $e(i)$ (when no near-end signal is present) the appropriate correction to the transversal filter coefficients.

The number of taps in the transversal filter of Figure 4 is determined by the duration of the impulse response of the echo path from port C to port D. The time span over which this impulse response is significant (i.e., nonzero) is typically 2 to 4 ms. This corresponds to 16 to 32 tap positions with 8-kHz sampling. However, because of the portion of the four-wire circuit between the location of the echo canceller and the hybrid, this response does not begin

at zero, but is delayed. The number of taps N, must be large enough to accommodate that delay. With N = 128, delays of up to 16 ms (or about 1,200 miles of ''tail'' circuit) can be accommodated.

In practice, it is necessary to cancel the echoes in both directions of a trunk. For this purpose, two adaptive cancellers are used, as shown in Figure 5, where one cancels the echo from each end of the connection. The near-end talker for one of the cancellers is the far-end talker for the other. In each case, the near-end talker is the ''closest'' talker, and the far-end talker is the talker generating the echo being cancelled. It is desirable to position these two ''halves'' of the canceller in a split configuration, as shown in Figure 5, where the bulk of the delay in the four-wire portion of the connection is in the middle. The reason is that the number of coefficients required in the echo-cancellation filter is directly related to the delay of the tail circuit between the location of the echo canceller and the hybrid that generates the echo. In the split configuration, the largest delay is not

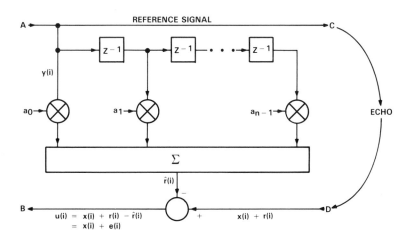

Figure 4. Echo Estimation Using a Transversal Filter

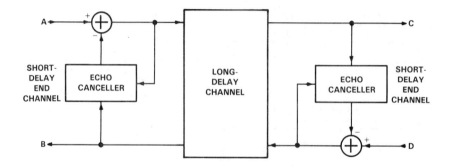

Figure 5. Split-Type Echo Canceller for Two Directions of Transmission

in the echo path of either half of the canceller. Therefore, the number of coefficients is minimized.

The digital voice echo canceller can be applied in a variety of transmission equipment configurations. Some of these are illustrated in Figures 6 through 8.

Figure 6 shows a single-channel echo canceller with a four-wire analog interface. The TMS32020 implementation described in this application report provides for the serial PCM codec interface required for this common configuration.

In digital carrier transmission systems, digital voice channels are usually carried in groups of 24 using the T1 group format.[5] As indicated in Figures 7 and 8, a T1-compatible digital voice echo canceller can be implemented with 24 single-channel echo cancellers connected directly to the serial 1.544-Mbps T1 PCM data streams for the transmit and receive groups.

Figures 9 through 11 show the appropriate architectures for applying digital voice echo cancellers to analog switching and analog transmission channel groups within the telephone network.

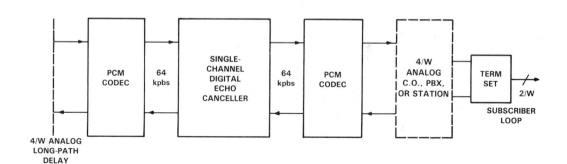

Figure 6. Single-Channel Four-Wire VF Echo Canceller

15. Digital Voice Echo Canceller with a TMS32020

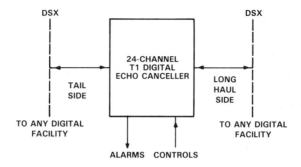

Figure 7. Standalone Digital T1 Echo Canceller

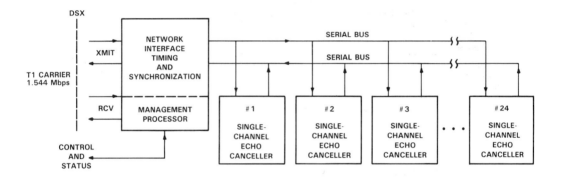

Figure 8. Per-Channel Architecture for a T1 Digital Echo Canceller

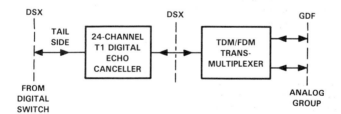

Figure 9. Digital Switch to Analog Facility

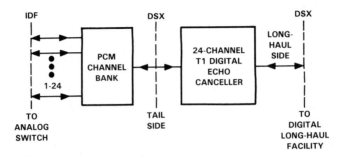

Figure 10. Analog Facility to Digital Facility

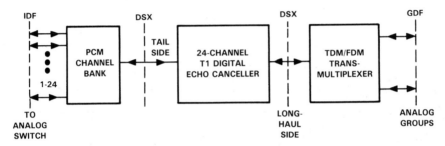

Figure 11. Analog Facility to Analog Facility

ECHO CANCELLATION ALGORITHMS

Generic algorithm requirements for each major signal processing function are discussed in this section. The signal processing flow for a single-channel digital voice echo canceller is shown in the block diagram of Figure 12.

Adaptive Transversal Filter

The reflected echo signal $r(i)$ at time i (see Figure 3) can be written as the convolution of the far-end reference signal $y(i)$ and the discrete representation h_k of the impulse response of the echo path between port C and D.

$$r(i) = \sum_{k=0}^{N-1} h_k \, y(i-k) \tag{1}$$

Linearity and a finite duration N of the echo-path response have been assumed. An echo canceller with N taps adapts the N coefficients a_k of its transversal filter to produce a replica of the echo $\hat{r}(i)$ defined as follows:

$$\hat{r}(i) = \sum_{k=0}^{N-1} a_k \, y(i-k) \tag{2}$$

Clearly, if $a_k = h_k$ for $k=0,\dots,N-1$, then $\hat{r}(i) = r(i)$ for all time i and the echo is cancelled exactly.

Since, in general, the echo-path impulse response h_k is unknown and may vary slowly with time, a closed-loop coefficient adaptation algorithm is required to minimize the average or mean-squared error (MSE) between the echo and its replica. From Figure 3, it can be seen that the near-end error signal $u(i)$ is comprised of the echo-path error $r(i) - \hat{r}(i)$ and the near-end speech signal $x(i)$, which is uncorrelated with the far-end signal $y(i)$. This gives the equation

$$E(u^2(i)) = E(x^2(i)) + E(e^2(i)) \tag{3}$$

where E denotes the expectation operator. The echo term $E(e^2(i))$ will be minimized when the left-hand side of (3) is minimized. If there is no near-end speech ($x(i) = 0$), the minimum is achieved by adjusting the coefficients a_k along the direction of the negative gradient of $E(e^2(i))$ at each step with the update equation

$$a_k(i+1) = a_k(i) - \beta \, \frac{\partial E(e^2(i))}{\partial a_k(i)} \tag{4}$$

where β is the stepsize. Substituting (1) and (2) into (3) gives from (4) the update equation

$$a_k(i+1) = a_k(i) + 2\beta E \left[e(i) \, y(i-k) \right] \tag{5}$$

15. Digital Voice Echo Canceller with a TMS32020

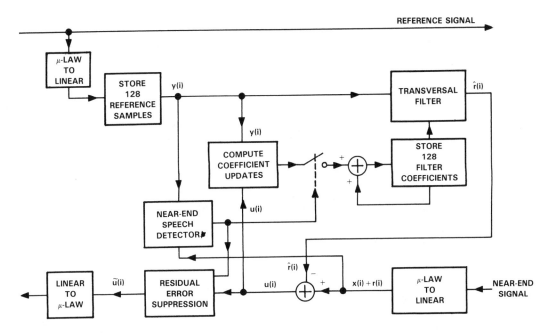

Figure 12. Signal Processing for a Digital Voice Echo Canceller

In practice, the expectation operator in the gradient term $2\beta E[e(i) y(i-k)]$ cannot be computed without a priori knowledge of the reference signal probability distribution. Common practice is to use an unbiased estimate of the gradient, which is based on time-averaged correlation error. Thus, replacing the expectation operator of (5) with a short-time average, gives

$$a_k(i+1) = a_k(i) + 2\beta \frac{1}{M} \sum_{m=0}^{M-1} e(i-m) y(i-m-k) \quad (6)$$

The special case of (6) for $M = 1$ is frequently called the least-mean-squared (LMS) algorithm or the stochastic gradient algorithm. Alternatively, the coefficients may be updated less frequently with a thinning ratio of up to M, as given in

$$a_k(i+M+1) = a_k(i) + 2\beta \sum_{m=0}^{M-1} e(i+M-m) y(i+M-m-k) \quad (7)$$

Computer simulations of this "block update" method show that it performs better than the standard LMS algorithm (i.e., $M = 1$ case) with noise or speech signals.[6] Many cancellers today avoid multiplication for the correlation function in (7), and instead use the signs of $e(i)$ and $y(i-k)$ to compute the coefficient updates. However, this "sign algorithm" approximation results in approximately a 50-percent decrease in convergence rate and an increase in

degradation of residual echo due to interfering near-end speech.

The convergence properties of the algorithm are largely determined by the stepsize parameter β and the power of the far-end signal $y(i)$. In general, making β larger speeds the convergence, while a smaller β reduces the asymptotic cancellation error.

It has been shown that the convergence time constant is inversely proportional to the power of $y(i)$, and that the algorithm will converge very slowly for low-power signals.[7] To remedy that situation, the loop gain is usually normalized by an estimate of that power, i.e.,

$$2\beta = 2\beta(i) = \frac{\beta_1}{P_y(i)} \quad (8)$$

where β_1 is a compromise value of the stepsize constant and $P_y(i)$ is an estimate of the average power of $y(i)$ at time i.

$$P_y(i) = (L_y(i))^2 \quad (9)$$

where $L_y(i)$ is given by

$$L_y(i+1) = (1-\rho) L_y(i) + \rho|y(i)| \quad (10)$$

The estimate $\rho_y(i)$ is used since the calculation of the exact average power is computation-expensive.

Near-End Speech Detector

When both near-end and far-end speakers are talking, the condition is termed "doubletalk." Since the error signal u(i) of Figure 2 contains a component of the near-end talker x(i) in addition to the residual echo-cancellation error, it is necessary to freeze the canceller adaptation during doubletalk in order to avoid divergence. Doubletalk status can be detected by a near-end speech detector operating on the near-end and far-end signals y(i) and s(i), respectively.

A commonly used algorithm by A. A. Geigel[8] consists of declaring near-end speech whenever

$$|s(i)| = |x(i) + r(i)| \geq = \frac{1}{2} \max\{|y(i)|, |y(i-1)|, \ldots, |y(i-N)|\} \tag{11}$$

where N is the number of samples in the echo canceller transversal filter memory. It is necessary to compare s(i) with the recent past of the far-end signal rather than just y(i) because of the unknown delay in the echo path. The factor of one-half is based on the hypothesis that the echo-path loss through a hybrid is at least 6 dB. The algorithm in effect performs an instantaneous power comparison over a time window spanning the echo-path delay range.

A more robust version of this algorithm uses short-term power estimates, $\tilde{y}(i)$ and $\tilde{s}(i)$, for the power estimates of the recent past of the far-end receive signal y(i) and the near-end hybrid signal s(i), respectively. These estimates are computed recursively by the equations

$$\tilde{s}(i+1) = (1-\alpha)\,\tilde{s}(i) + \alpha|s(i)| \tag{12}$$

$$\tilde{y}(i+1) = (1-\alpha)\,\tilde{y}(i) + \alpha|y(i)| \tag{13}$$

where the filter gain $\alpha = 2^{-5}$. For this version of the algorithm, near-end speech is declared whenever

$$\tilde{s}(i) \geq \frac{1}{2} \max(\tilde{y}(i), \tilde{y}(i-1), \ldots, \tilde{y}(i-N)) \tag{14}$$

Since the near-end speech detector algorithm detects short-term power peaks, it is desirable to continue declaring near-end speech for some hangover time after initial detection.

Residual Echo Suppressor

Nonlinearities in the echo path of the telephone circuit and uncorrelated near-end speech limit the amount of achievable suppression in the circuit from 30 to 35 dB. Thus, there is no merit in achieving more than a certain degree of cancellation.

The use of a residual echo suppressor algorithm has been found to be subjectively desirable.[7] During doubletalk, the residual suppressor must be disabled. A common

suppression control algorithm is to detect when the return signal power falls below a threshold based on the receive reference signal power. If the return signal consists only of residual echo and the canceller has properly converged, then the residual echo level will be below the threshold and the transmitted return signal will be set to zero.

The return signal power is estimated by the equation

$$L_u(i+1) = (1-\rho)\,L_u(i) + \rho|u(i)| \tag{15}$$

The reference power estimate $L_y(i)$ is given by (10). Suppression is enabled on the transmitted signal u(i) (i.e., u(i) = 0) whenever $L_u(i)/L_y(i) < 2^{-4}$. This corresponds to a suppression threshold of 24 dB.

IMPLEMENTATION OF A 128-TAP ECHO CANCELLER WITH THE TMS32020

The TMS32020 is ideally suited for the implementation of a single 128-tap digital voice echo canceller channel since it has the capability and features to implement all of the required functions with full precision. This section discusses an implementation approach that meets or exceeds the performance of currently available products and the requirements of the CCITT G.165 recommendations.[3]

Echo Canceller Performance Requirements

Echo cancellers have the following fundamental requirements:
1. Rapid convergence when speech is incident in a new connection
2. Low-returned echo level during singletalking (i.e., echo-return loss enhancement)
3. Slow divergence when there is no signal
4. Rapid return of the echo level to residual if the echo path is interrupted
5. Little divergence during doubletalking

The CCITT recommendation G.165 specifies echo canceller performance requirements with band-limited white-noise (300 – 3400 Hz) test signals at the near-end and far-end input signal ports. The test specifications of G.165 are summarized in Table 1.

Digital voice echo canceller products are typically designed to accommodate circuits with tail delays of 16 ms or more and circuits with echo-return loss levels greater than 3 dB to 6 dB. Typical digital voice echo canceller product specifications are summarized in Table 2.

Table 1. CCITT G.165 Performance Test Specifications

CCITT TEST	DESCRIPTION	PERFORMANCE REQUIREMENT
1. Final echo return loss (ERL) after convergence; singletalk mode	Input noise level: − 10 dbm0 to − 30 dbm0 Circuit ERL: 10 dB Steady-state residual echo level after convergence with no near-end signal	− 40 dbm0
2. Convergence rate; singletalk mode	Input noise level: − 10 dbn 0 Combined echo loss after 500 ms from initialization with cleared register and with near-end signal set to zero at initialization time	≥ 27 dB
3. Leak rate	Degradation of residual echo after 2 minutes from time all signals are removed from fully converged canceller	≤ 10 dB
4. Infinite return loss convergence	Input noise level: − 10 dbm0 to − 30 dbm0 Circuit ERL: 10 dB Returned echo level 500 ms after echo path is interrupted	− 40 dbm0

Table 2. Typical Echo Canceller Product Specifications

PARAMETER	SPECIFICATION
1. Maximum tail circuit length	16, 32, or 48 ms
2. Absolute delay	0.375 ms maximum
3. Minimum echo return loss	6 dB
4. Convergence	24 dB enhancement in 250 ms
5. Residual echo level (− 30 to − 10 dbm0 receive level)	− 40 dbm0 (suppressor disabled) − 65 dbm0 (suppressor enabled)
6. Speech detector threshold	6 dB below receive level
7. Speech detector hangover time	75 ms

Implementation Approach

In the implementation of the generic echo-cancelling algorithms discussed above, the coefficient update process dominates the computational requirement and efficiency of DSP realizations. The DSP efficiency and speed, in turn, determines the maximum number of echo canceller taps that can be achieved with the processor.

The block update approach of (7) with M = 16 was chosen for the TMS32020 implementation because it takes advantage of the efficient multiply and accumulate capabilities of the processor. Using the block update approach, a full-performance 128-tap canceller can be realized with a small margin. During each sample period (125 μs), 8 out of 128 coefficients are updated using correlation of the 16 past error and signal values.

Computer simulation studies were undertaken to verify the performance of the block update algorithm (M = 16) in comparison with the stochastic gradient algorithm (M = 1), taking into account the finite-precision and word-length limitations of the TMS32020. Figures 13 and 14 show the simulation results for three values of the compromize stepsize constant $\beta 1$, defined in (8). The curves represent the average of 600 samples for single convergence runs from a zero initial condition with white-noise input. The block update algorithm performs better than the stochastic gradient algorithm for all three values. For values of $\beta 1$ larger than 2^{-8}, the algorithm can become unstable. Therefore, for both practical and performance reasons, the value $\beta_1 = 2^{-10}$ was chosen for implementation.

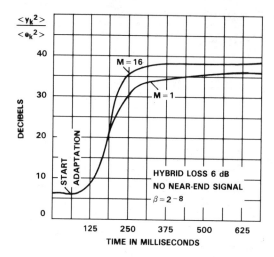

Figure 13
Convergence Performance of the Block Update Algorithm
and Stochastic Gradient Algorithm

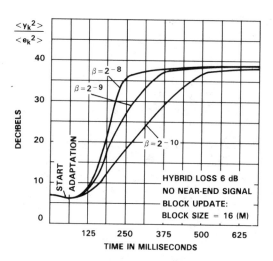

Figure 14
Convergence Performance of the Block Update Algorithm

In the TMS32020 implementation, it is convenient and desirable to normalize both the stepsize and the error variables u(i) by the square root of the power estimate $P_y(i)$, i.e., $L_y(i)$ of (9).

Normalizing u(i) and the stepsize separately enables the product term of (7) to be computed with single precision on the TMS32020 without significant loss of precision or overflow due to varying signal level.

Table 3 gives a description of the program variables together with their names and ranges, and summarizes the number formats chosen for the echo canceller implementation. One of the most important aspects of the implementation approach is the handling of the binary representation of the signal samples, algorithm variables, coefficients, and constant parameters for various stages of the processing. The notation (Q.F) is used to define the representation of either 16-bit numbers or 32-bit accumulator numbers, where F specifies the number of bits which are to the right of the implicit binary point. The assignments of Table 3 ensure that the algorithm can be executed on the TMS32020 with single-precision arithmetic and with no significant loss of precision.

Memory Requirements

The echo canceller algorithm requires the storage of both reference samples and variable coefficients in on-chip data RAM so that the required FIR and block update convolution can be performed efficiently using the RPTK and MACD instructions. Therefore, the coefficients a_k are stored in block B0, which is configured as program memory. The 16 normalized error samples for coefficient updating are also stored in B0. The 128 reference signal samples y(i) are stored in data RAM along with an additional 16 reference samples y(1 − 129), ..., y(i − 143), which are used in the update of coefficients a_{112}, ..., a_{127}. The echo canceller data memory locations are summarized in Table 4.

Software Logic and Flow

A flowchart of the TMS32020 program for a 128 − tap digital voice echo canceller is shown in Figure 15.

In Table 5, the instruction cycle and memory requirements are listed for the various blocks of the program implementation. The blocks are listed in the order of execution.

15. Digital Voice Echo Canceller with a TMS32020

Table 3. Algorithm Number Representation on the TMS32020

VARIABLE	DESCRIPTION	BINARY REPRESENTATION	RANGE
$a_0, a_1,...,a_{127}$	Filter coefficients	(Q.15)	$\left[-1, 1-2^{-15}\right]$
$y(i), y(i-1),...,y(i-143)$	Reference samples	(Q.0)	$\left[-2^{15}, 2^{15}-1\right]$
$s(i)$	Near-end signal	(Q.0)	$\left[-2^{15}, 2^{15}-1\right]$
$r(i)$	Echo estimate	(Q.0)	$\left[-2^{15}, 2^{15}-1\right]$
$L_y(i)$	Average absolute value of $y(k)$	(Q.0)	$\left[0, 2^{15}-1\right]$
$L_y(i)^{-1}$		(Q.15)	$\left[-1, 1-2^{-15}\right]$
$u(i)$	Near-end signal minus echo estimate $s(k) - r(k)$	(Q.0)	$\left[-2^{15}, 2^{15}-1\right]$
$un(i),...,un(i-15)$	Normalized outputs $un(i) = u(i) \times L_y(i)^{-1}$	(Q.15)	$\left[-1, 1-2^{-15}\right]$
$\widetilde{s}(i)$	Short-time average of $2 \times \|s(i)\|$	(Q.0)	$\left[0, 2^{15}-1\right]$
$\widetilde{y}(i)$	Short-time average of $\|y(i)\|$	(Q.0)	$\left[0, 2^{15}-1\right]$

Table 4. Echo Canceller Data Memory Locations

VARIABLE	SYMBOL	LOCATION	REMARK
$a_0,...,a_{127}$	A0,...,A127	Block B0 767,766,....,640	A0 is in higher address
$y(k),...,y(k)-143)$	Y0,...,Y143	Block B1 768,769,....,911	Y128,...,Y143 required for block update
$un(k,...,un(k-15)$	UN0,....,UN15	Block B0 512,...,527	

15. Digital Voice Echo Canceller with a TMS32020

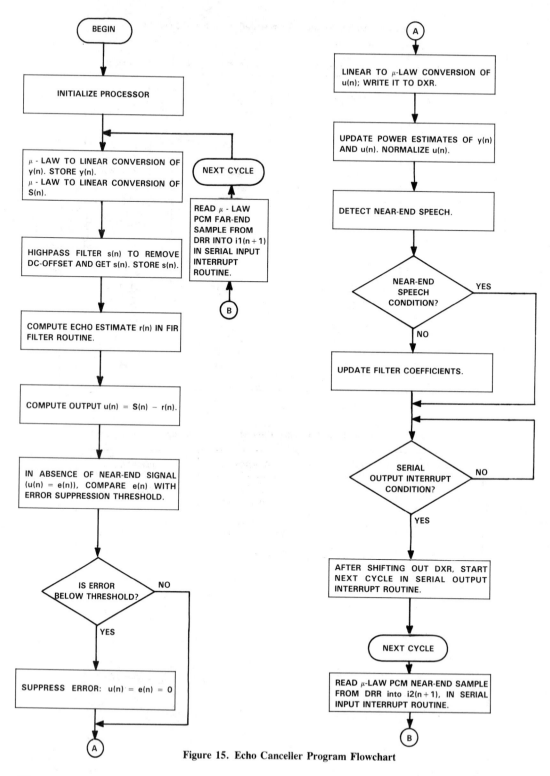

Figure 15. Echo Canceller Program Flowchart

Table 5. Program Module Requirements

STEP	MODULE FUNCTION	CODE LISTING PAGE	DESCRIPTION	CPU CYCLES	PROGRAM MEMORY LOCATIONS	DATA* MEMORY LOCATIONS
1.	Cycle Start Routine	7	μ-law to linear conversions; take absolute value of inputs and high-pan filter s(i).	32	28	11
2.	Echo Estimation Routine	9	FIR convolution of reference samples and filter coefficients to get echo replica r(i).	156	14	258
3.	Compute Output	9	$u(i) = s(i) - r(i)$ Store u)i).	6	6	2
4.	Residual Output Suppression Routine	10	If output power below threshold, set $u(i) = 0$.	12	15	4
5.	Linear to μ-law Compression Routine	11	Convert u(i) to μ-law.	26	35	4
6.	Power Estimation Routine	13	Estimate short-term power of u(i) and y(i).	28	14	6
7.	Output Normalization	14	Comput $u_n(i) = \dfrac{u(i)}{y(i)}$ and clip it.	28	25	19
8.	Near-end Speech Detection	16	Perform maximum test for near-end speech.	54	74	16
9.	Coefficient Increment Update Routine	20	If no near-end speech, compute increments for coefficient group.	183	63	26
10.	Coefficient Update Routine	23	Add increments to coefficient group.	43	43	2
11.	Cycle End Routine	25	Wait for interrupt.	1	3	0
12.	Receive Interrupt Service Routine	25	Save status and read input sample.	2×14	14	3
13.	Transmit Interrupt Service Routine	25	Branch to start.	2	2	0
14.	Interrupt Branches	3		12	6	0
15.	Processor Initialization**	4	Clear memory, initialize status and set parameters.	86 **	86	0
16.	μ-law to Linear Conversion Table*	26		0	256	0
Total				614	676	351

*Locations are entered only for the routine that uses them first.

**Not in main cycle; CPU cycles not counted in total.

15. Digital Voice Echo Canceller with a TMS32020

The program loop is executed once per I/O data sample period of 125 μs. The program loop is interrupt-driven from the output data sample mark of a T1 frame. Depending on the near-end speech detector/hangover status, the coefficient update computation module may be skipped. An input data sample interrupt mark occurs during the program loop at a time dependent on the channel location within the T1 frame. In response to the interrupt, the main program execution is interrupted and saved until the new input samples have been read into memory. At the end of each program loop, the processor waits for the next output sample interrupt.

In the following subsections, the implementation of each major block is described in detail. Each variable used in an equation is referred to by its name in the program enclosed in parentheses.

Cycle Start Routine

The voice echo canceller program has been implemented with either μ-law or A-law conversion routines as a program option.

The μ-law (or A-law) to linear input conversion routine is implemented by table lookup in order to minimize the number of instructions. The 256 14-bit two's-complement number corresponding to the 256 possible 8-bit μ-law numbers are stored in program memory. The 8 bits of the μ-law number specify the relative address of the corresponding linear number in the table, which is added to the first address in the table to form the absolute program memory address for the linear number. The TBLR instruction is then used to move the number from program to data memory.

In the cycle start routine, the μ-law input reference sample is read from memory location DRR2 and converted to its linear representation $y(i)$ (Y0). Its absolute value is also stored in location ABSY0. The near-end input sample is then read and converted to a linear representation $sdc(i)$ (S0DC). The sample $s(i)$ is next put through a highpass filter to remove any residual dc offset. The highpass filter is a first-order filter with a $3-dB$ frequency at 160 Hz. Its output $s(i)$ (S0) is given by

$$s(i+1) = (1-\gamma)\, s(i) + \frac{1}{2}\,(1-\gamma)\,(sdc(i) - sdc(i-1))$$

where $\gamma = 2^{-3}$. $\hspace{3cm}$ (16)

Note that the filter implementation requires double-precision arithmetic, with S0 denoting the MSBs of $s(i)$ and S0LSBS its LSBs.

Echo Estimation

The echo estimate $\hat{r}(i)$ (EEST) is formed by convolving the tap weight coefficients $a_0, \ldots, a_{127}$ (A0, ..., A127) with the 128 most recent reference samples $y(i), \ldots, y(i-127)$ (Y0, ..., Y127).

$$\hat{r}(i) = \sum_{k=0}^{127} a_k\, y(i-k) \hspace{2cm} (17)$$

This operation is most efficiently implemented on the TMS32020 using the RPTK and MACD instruction. The samples $y(i), \ldots, y(i-127)$ are stored in block B1 of data memory while $a_0, \ldots, a_{127}$ are stored in block B0 configured as program memory. Since the MACD instruction also performs a data move,

$$y(i-k+1) \rightarrow y(i-k) \quad \text{for } k = 1, \ldots, 128 \hspace{1cm} (18)$$

no data shifting is required for the computation of the next echo estimate.

The block update routine used for the coefficient adaptation requires the storage of $y(i-128), \ldots, y(i-143)$ (Y128, ..., Y143) in addition to the most recent 128 samples used in the convolution. Since these samples are not used in the convolution, they are updated using the RPTK and DMOV instructions.

$$y(i-k+1) \rightarrow y(i-k) \quad \text{for } k = 129, \ldots, 143 \hspace{1cm} (19)$$

The tap weight coefficients $a_0, \ldots, a_{127}$ are initially set to zero, and are adjusted by the algorithm to converge to the impulse response of the echo path $h_0, \ldots, h_{127}$.

$$a_k(i) \rightarrow h_k \quad \text{for } k = 0, \ldots, 127 \hspace{2cm} (20)$$

The $|h_k| < 1, \forall\, k$, because the power gain of the echo path is smaller than unity. The binary representation for the a_k's was chosen to be of the form (Q.15) with 15 bits after the binary points. This format represents a number between -1 and $(1 - 2^{-15})$. The reference samples and the echo estimate are represented as 16-bit two's-complement integers (no binary point). The 32-bit result of the convolution is therefore of the form (Q.15), and the 16 bits of the echo estimate are the MSB of accumulator low (ACCL) and the 15 LSBs of accumulator high (ACCH). One left shift of the accumulator is required before ACCH is stored in EEST.

Residual Error Suppression

The residual cancellation error is set to zero (or suppressed) whenever the ratio of a long-time average of the absolute value of the output (ABSOUT) to a long-time average of the absolute value of the reference signal (ABSY) is smaller than a fixed threshold. The two long-time averages are updated subsequently in the program as described below. The suppression is, of course, disabled when a near-end speech signal is present (HCNTR > 0). The suppression threshold is set at 1/16 or -24 dB.

Linear to μ-Law (A-Law) Conversion

The linear to μ-law (A-law) conversion routine is an efficient adaptation to the TMS32020 of the conversion routine written for the TMS32010 and described in the application report, "Companding Routines for the TMS32010."[9]

Signal and Output Power Estimation

An estimate of the long-time average of $|u(i)|$ is required by the residual error suppression routine. This estimate $L_u(i)$ (ABSOUT) is obtained by lowpass filtering $|u(i)|$ (ABSU0) using the following infinite impulse response (IIR) filter:

$$L_u(i+1) = (1-\alpha) L_u(i) + \alpha|u(i)| \qquad (21)$$

where $\alpha = 2^{-7}$. In terms of the program variables, the IIR filter is given by

$$ABSOUT = 2^{-16} \left(2^{16} \times ABSOUT - 2^9 \right.$$
$$\left. \times ABSOUT + 2^9 \times ABSU0 \right) \qquad (22)$$

Similarly, the estimate $L_y(i)$ (ABSY) of the long-term average of $y(i)$ (ABSY0) is the output of an IIR filter with the same α, but differs from the above filter by the addition of a cutoff term that prevents the estimate from taking values smaller than a desired level.

$$ABSY = 2^{-16} (2^{16} \times ABSY - 2^9 \times ABSY + 2^9$$
$$\times ABSY0 + 2^9 \times CUTOFF) \qquad (23)$$

This insures that $ABSY \geq CUTOFF$ even if ABSY0 is zero for a long time.

Since $L_y(i)$ is used to normalize the algorithm stepsize, this feature is important in order to prevent excessively large stepsizes when the far-end talker is silent.

The stepsize is normalized according to

$$2\beta(i) = \frac{\beta_1}{L_y^2(i)} \qquad (24)$$

In order to avoid double-precision arithmetic, this normalization is carried out in two stages (as described in the subsection on coefficient adaptation). Each of the stages requires a division by $L_y(i)$. It is more efficient to compute $L_y(i)^{-1}$ (IABSY) and replace the divisions by two multiplications.

Since ABSY is a positive integer, taking its inverse consists simply of repeating the SUBC instruction. IABSY is a positive fractional number of the form (Q.15), taking values between 0 and $1-2^{-15}$.

Output Normalization

The normalized output $u_n(i)$ (UN0) is defined as $\mu(i)/L_y(i)$ and replaces the actual error in the coefficient update routine for finite-precision considerations, described in the subsection on coefficient adaptation. In the absence of near-end speech, $u_n(i)$ is equal to a normalized cancellation error and is used in the coefficient update. In the presence of near-end speech, no coefficient update is carried out, and the normalized outputs are not used.

The block update approach requires the storage of the 16 most recent normalized outputs $u_n(i), \ldots, u_n(i-15)$ (UN0,..., UN15). In a given program cycle, only $u_n(i)$ is computed and stored, while $u_n(i-1),\ldots, u_n(i-15)$ computed in previous program cycles are only updated using the DMOV instruction.

$$u_n(i-k+1) \rightarrow u_n(i-k) \quad \text{for } k = 1,\ldots,14 \qquad (25)$$

In the absence of near-end speech, the normalized output should be a number smaller than one, which is represented as a (Q.15) fraction. To insure that the representation is adequate even in the presence of a near-end signal, the normalized output is clipped at $+1$ or -1, i.e.,

$$\text{if } u_n(i) > 1.0, \text{ then } u_n(i) = 1.0$$
$$\text{if } u_n(i) < -1.0, \text{ then } u_n(i) = -1.0 \qquad (26)$$

Near-End Speech Detection

Near-end speech is declared if

$$\tilde{s}(i) \geq \max \left(\tilde{y}(i),\tilde{y}(i-1),\ldots,\tilde{y}(i-127-h(i)) \right) \qquad (27)$$

where $\tilde{s}(i)$ (ABSS0F) is the output of a lowpass filter with input $2 \times |s(i)|$ (ABSS0). The variable $\tilde{y}(i)$ is a lowpass filtered version of $|y(i)|$, and $h(i)$ (H) a modulo-16 counter. The lowpass filters are IIR filters with short-time constants,

$$s(i+1) = (1-\alpha) s(i) + \alpha \times 2 \times |s(i)| \qquad (28)$$

$$y(i+1) = (1-\alpha) y(i) + \alpha \times |y(i)| \qquad (29)$$

where $\alpha = 2^{-5}$.

The counter $h(i)$ is incremented by one for every input sample. The routines maintain nine partial maxima $m0, m1,\ldots, m8$ ($M_0, M_1, \ldots, M_8$), defined at time $i = 16m + h(i)$ by

$$m_0(i) = \max \left((\tilde{y}(i),\ldots,\tilde{y}(i-h(i)+1) \right)$$
$$m_1(i) = \max \left((\tilde{y}(i-h),\ldots,\tilde{y}(i-h(i)-15) \right) \qquad (30)$$

$$m_8(i) = \max \left(\tilde{y}(i-h-112),\ldots,\tilde{y}(i-h(i) - 127) \right)$$

Figure 16 illustrates how the partial maxima are maintained.

The condition for near-end speech declaration is then equivalent to

$$\tilde{s}(i) \geq \max (m_0,\ldots,m_8) \qquad (31)$$

The partial maxima are updated according to the following recursions:

$$\text{if } h = 0, \text{ then } m_0(i) = \tilde{y}(i+1)$$
$$\text{and } m_j(i) = m_{j-1}(i) \qquad (32)$$
$$\text{where } j = 1,\ldots,8$$

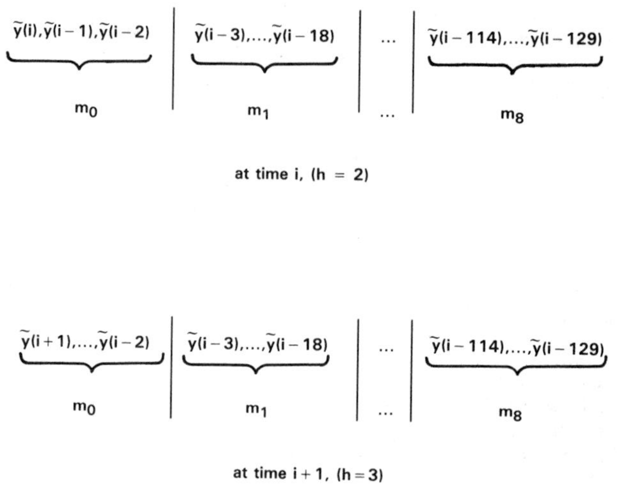

$$\underbrace{\tilde{y}(i), \tilde{y}(i-1), \tilde{y}(i-2)}_{m_0} \Bigg| \underbrace{\tilde{y}(i-3), \ldots, \tilde{y}(i-18)}_{m_1} \Bigg| \cdots \Bigg| \underbrace{\tilde{y}(i-114), \ldots, \tilde{y}(i-129)}_{m_8}$$

$$\cdots$$

at time i, (h = 2)

$$\underbrace{\tilde{y}(i+1), \ldots, \tilde{y}(i-2)}_{m_0} \Bigg| \underbrace{\tilde{y}(i-3), \ldots, \tilde{y}(i-18)}_{m_1} \Bigg| \cdots \Bigg| \underbrace{\tilde{y}(i-114), \ldots, \tilde{y}(i-129)}_{m_8}$$

at time i + 1, (h = 3)

Figure 16. Partial Maxima for Near-End Speech Detection

and

if $0 < h \leq 15$, then $m_0(i+1) = \max(m_0(i), \tilde{y}(i+1))$
and $m_j(i+1) = m_j(i)$
where $j = 1, \ldots, 8$

If near-end speech is declared, a hangover counter (HCNTR) is set equal to a hangover time (HANGT), which was chosen to be 600 samples or 75 ms. If no near-end speech is declared, then the hangover counter is decremented by one, unless it is zero. If the hangover counter is larger than zero, then the coefficient update routine is skipped. Moreover, if the reference signal power estimate $L_y(i)$ is smaller or equal to the cutoff value of -48 dB, then adaptation is also disabled to avoid divergence during long silences of the far-end talker.

Coefficient Adaptation

The 128 coefficients of the transversal filter are divided into 16 groups of 8 coefficients each, as shown in Table 6.

Table 6. The Coefficient Groups

GROUP	COEFFICIENTS
0	$a_0, a_{16}, a_{32}, \ldots, a_{112}$
1	$a_1, a_{17}, a_{33}, \ldots, a_{113}$
.	.
.	.
.	.
15	$a_{15}, a_{31}, a_{47}, \ldots, a_{127}$

The coefficients in only one of the groups are updated in a given program cycle, while the other coefficients are not modified. A modulo-16 counter h(i) (H) points to the index of the group to be updated, and is incremented by one during every program cycle.

The update equation is repeated here for ease of reference.

$$a_k(i+1) = a_k(i) + \frac{\beta_1}{(L_y(i))^2} \sum_{m=0}^{15} e(i-m)\, y(i-k-m) \tag{34}$$

for $k = h, h+16, \ldots, h+112$, where h is the value of the counter and goes from 0 to 15. The error terms $e(i-m)$ $(m = 0, \ldots, 15)$ are the most recent cancellation errors. In this case, the errors are equal to the 15 most recent canceller outputs $u(i), \ldots, u(i-15)$ since the adaptation is carried out only in the absence of a near-end signal.

For finite-precision considerations, the actual implementation of the update equation by the routine is carried out in the following two main steps:

1. Compute eight partial updates:

$$\gamma_k(i) = \sum_{m=k}^{k+15} \frac{u(i-m)}{L_y(i)}\, y(i-k-m) \tag{35}$$

where $k = h, h+16, \ldots, h+112$.

432

The normalized outputs $u_n(i), \ldots, u_n(i-15)$ have already been computed and stored.

2. Update the coefficients:

$$a_k(i+1) = a_k(i) + \left(2^4 \times \left(L_y(i)^{-1} \times 2^G\right) \times \gamma_k(i)\right) 2^{-16}$$

(36)

where G (GAIN) is a program parameter that determines the stepsize of the algorithm and has the value 0, 1, 3, ..., 15.

The partial updates $\gamma_k(i)$ are computed using the MAC instruction in repeat mode. The result is rounded and stored in temporary locations INC0, ..., INC0 + 7 in block B1.

For the second step of the update, $L_y(i)^{-1}$ (IABSY) is first loaded in the T register with a left shift of G (GAIN). It is then multiplied by each of the $\gamma_k(i)$'s. SPM is set to 2 to implement the 2^4 multiplication by shifting the P register four positions to the right before adding it to the accumulator (APAC).

Interrupt Service Routines

At the end of the cycle, the program becomes idle until a receive interrupt occurs followed by a transmit interrupt that sends it back to the beginning of the cycle. The transmit interrupt routine simply enables interrupts and branches back to the start. The receive interrupt must store the status register ST0 and the accumulator, then read the received sample from DRR, zero its eight most significant bits, and store it in DRR1. It restores the accumulator and status register ST0 before returning to the main program.

External Processor Hardware Requirements

Very little external hardware is required to implement a complete single-channel 128-tap echo canceller with the TMS32020. In addition to the processor, only two external 1K x 8 PROMs and some system-dependent interface logic are required. A typical interface circuit for the demonstration system is shown in Appendix A.

The TMS32020 serial I/O ports allow direct interfacing of the echo canceller to a digital T1 carrier data stream.

Three I/O functions must be performed during each T1 frame (125 μs). The far-end and the near-end signals must be read in, and the processed near-end signal must be written out. To perform these functions, a timing circuit must extract the T1 clock and the T1 frame marks for each direction of transmission. The timing circuit uses the frame mark to generate a channel mark that selects the desired channel out of the 24 present in the T1 frame. The channel mark goes to a high level during the clock cycle, immediately preceding the eight serial bits of the desired sample.

The T1 clock, channel mark, and serial data signals are directly input into the TMS32020 serial clock (CLKR), serial input control (FSR), and serial input port (DR), respectively. Because data is read in from two directions of transmission, a triple two-to-one multiplexer (e.g., SN74LS157) is required to select one of the two sets of T1 signals to be input into the TMS32020. During each T1 frame, the multiplexer alternates once between each direction of transmission, under the control of the timing circuit.

Since data is written out in only one direction, the TMS32020 serial output port (DX) is directly tied to the outgoing T1 data line. The serial output clock (CLKX) and the serial output control (FSX) signals are the same as the near-end direction-of-transmission CLKR and FSR signals. If the far-end T1 channel-frame location overlaps the near-end T1 channel location in time, it is necessary to delay each far-end sample external to the TMS32020 to permit it to be read following the sample from the near-end direction. This requires an eight-bit serial shift register and some additional timing circuits.

Description of a Single-Channel Demonstration System

The demonstration system has been constructed in order to verify the TMS32020 implementation. Two photographs and a block diagram of the demonstration system are provided.

Figure 17 is a photograph of the front panel of the demonstration system, and Figure 18 is a closeup photograph of the single-channel echo canceller module.

15. Digital Voice Echo Canceller with a TMS32020

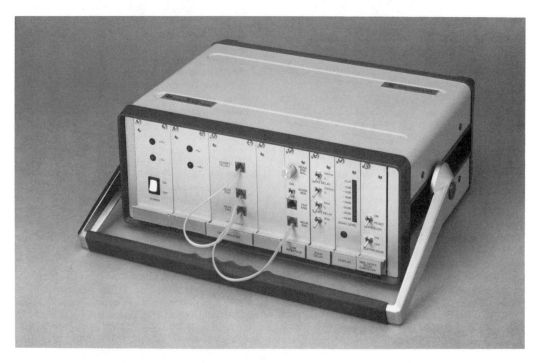

**Figure 17. Front Panel of the Echo Canceller
Demonstration System**

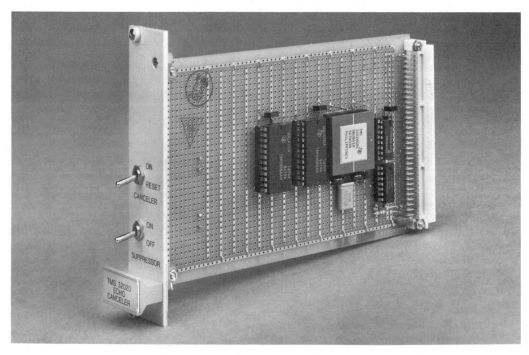

Figure 18. Single-Channel Echo Canceller Module

15. Digital Voice Echo Canceller with a TMS32020

As shown in the block diagram of Figure 19, the demonstration system models two end offices, a delay due to a satellite link, a delay due to a terrestrial link, a typical end-loop line response, and the echo canceller. A phone is connected via a two-wire interface to each end of the path. The two-wire interfaces are converted to four-wire in electronic hybrids. The hybrids also provide the required battery voltage to power the phones. The near-end two-wire line has a series-passive line simulator. The associated hybrid has an adjustable termination to allow a variable amount of hybrid mismatch, and therefore a variable amount of near-end echo response.

At each end, the four-wire analog signal is converted to and from PCM μ-law digital representation by a codec. The PCM signaling is done in a T1 format, with appropriate timing provided by a central timing generator. Variable delay is provided in the near-end and far-end path by digital

memories. The TMS32020 echo canceller is situated in the middle of the path, with signal processing done on the near-end to far-end direction of transmission. The other direction is used as the reference signal. All the TMS32020 signal I/O is performed using the T1 format. A display of the processed signal is used as an indicator of echo suppression in the absence of near-end signal. To aid the testing of the echo canceller, the far-end phone can be switched out and a noise generator switched in as a source of far-end signal.

The performance of the TMS32020 echo canceller was measured for white-noise input, as suggested in the CCITT G.165 recommendation. The measurement results are summarized in Table 7 and show that the TMS32020 echo canceller performance exceeds the CCITT requirements in all the tests described. The subjective performance on speech was also found to be very good in both singletalk and doubletalk modes, with no audible distortion of the signal.

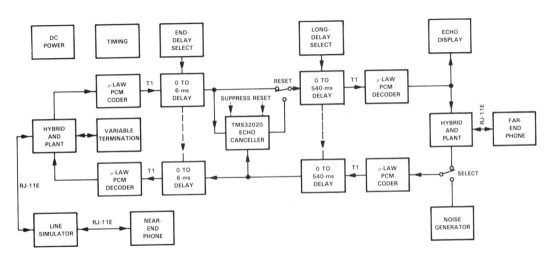

Figure 19. Block Diagram of Echo Canceller Demonstration System

Table 7. TMS32020 Echo Canceller Performance

TEST DESCRIPTION	CCITT G.165 PERFORMANCE REQUIREMENT	TMS32020 ECHO-CANCELLER PERFORMANCE
1. Final echo return loss after convergence; singletalk mode	− 40 dbm0	< − 48 dbm0
2. Convergence rate; singletalk mode	≥ 27 dB	> 38 dB
3. Leak rate	≤ 10 dB	≈ 0 dB
4. Infinite return loss convergence	− 40 dbm0	< − 48 dbm0

CONCLUSION

The development of novel variations of the generic least-mean-squared (LMS) echo cancelling algorithm and the near-end speech and residual suppression control algorithms has resulted in the implementation of a complete 128-tap single-channel echo canceller on a single TMS32020 programmable Digital Signal Processor. The echo canceller performance exceeds all requirements of the CCITT G.165 recommendations and the performance of similar currently available products. The only external hardware required are two program PROMs and a serial data multiplexer. A direct T1-rate serial interface is available to minimize component count in four-wire VF and T1 carrier configurations.

The single-channel TMS32020 echo canceller program provides a high-performance building block for low-cost systems, which can be tailored to a wide variety of system applications. Programmability offers the flexibility to implement custom requirements, such as cascaded sections for longer tail delay range, short-range multichannel versions, or other special-purpose functions.

The echo canceller application illustrates the power and versatility of the TMS32020 single-chip programmable signal processor. Applications of this technology can be expected to benefit many other complex signal processing tasks in communications products, including voiceband data modems, voice codecs, digital subscriber transceivers, and TDM/FDM transmultiplexers.

ACKNOWLEDGEMENTS

Texas Instruments and Teknekron Communications Systems wish to acknowledge the fine work done by the Teknekron project team comprised of Dr. David Messerschmitt, Dr. David Hedberg, Mr. Christopher Cole, Dr. Amine Haoui, and Mr. Peter Winship. Special appreciation is extended to Peter Winship for his excellent work in the design and construction of the prototype demonstration system and to James Hesson for his helpful

suggestions on the interleaved, coefficient update technique.

Further information about the echo canceller applications may be obtained by contacting Texas Instruments or Teknekron Communications Systems, 2121 Allston Way, Berkeley, CA 94704, (415) 548-4100.

Note that Texas Instruments does not warrant or guarantee the applicability of this application report to any particular design or customer use.

REFERENCES

1. M.L. Honig and D.G. Messerschmitt, *Adaptive Filters,* Kluwer Academic Publishers (1984).
2. D.L. Duttweiler and Y.S. Chen, "A Single-Chip VLSI Echo Canceller," *Bell System Technical Journal,* Vol 59, No. 2, 149 (February 1980).
3. "Recommendation G.165, Echo Cancellers," *CCITT,* Geneva (1980).
4. H.R. Huntley, "Transmission Design of Intertoll Telephone Trunks," *Bell System Technical Journal,* Vol 32, No. 5, 1019-36 (September 1953).
5. J.C. Bellamy, *Digital Telephony,* Wiley & Sons (1982).
6. M.J. Gingell, B.G. Hay, and L.D. Humphrey, "A Block Mode Update Echo Canceller Using Custom LSI," *GLOBECOM Conference Record,* Vol 3, 1394-97 (November 1983).
7. D.G. Messerschmitt, "Echo Cancellation in Speech and Data Transmission," *IEEE Journal on Selected Topics in Communications,* SAC-2, No. 2, 283-303 (March 1984).
8. D.L. Duttweiler, "A Twelve-Channel Digital Voice Echo Canceller," *IEEE Transactions on Communications,* COM-26, No. 5, 647-53 (May 1978).
9. L. Pagnucco and C. Ershine, "Companding Routines for the TMS32010/TMS32020," *Digital Signal Processing Applications with the TMS320 Family,* Texas Instruments (1986).

HARDWARE SCHEMATIC OF THE SINGLE-CHANNEL DEMONSTRATION PROCESSOR

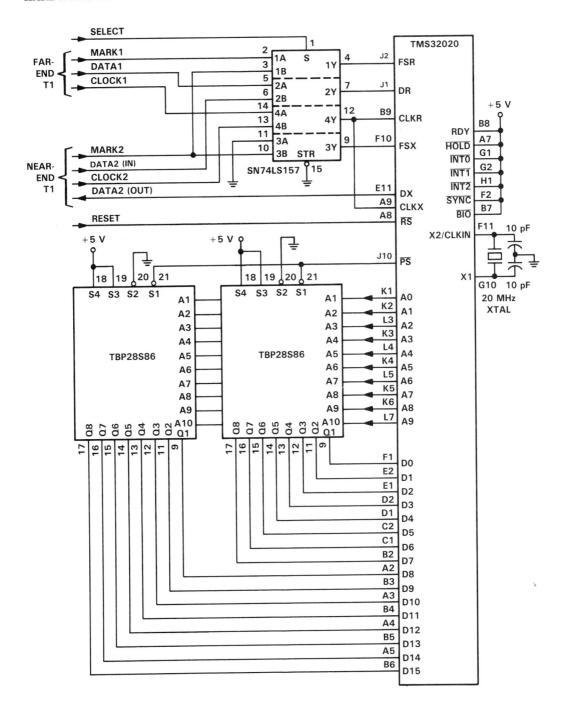

APPENDIX B

SOURCE CODE LISTING

```
0001
0002          *********************************************
0003          *
0004          *        128-TAP ECHO-CANCELLER PROGRAM
0005          *        (C) COPYRIGHT TEXAS INSTRUMENTS INC., 1985
0006          *
0007          *********************************************
0008   0000
0009                  IDT  'EC128'
0010   0000
0011                  *
0012                  *  ALGORITHM CONSTANTS
0013                  *
0014   0000
0015   0003   GAIN    EQU  3       *  COEFF UPDATE GAIN = 2**-10
0016                                *      = 2**(GAIN - 13)
0017
0018   0009   LTAU    EQU  9       *  LPF LONG TAU = 16 MSECS
0019                                *      = 125 USECS * 2**(16-LTAU)
0020   000B   STAU    EQU  11      *  LPF SHORT TAU = 4 MSECS
0021                                *      = 125 USECS * 2**(16-STAU)
0022   000D   HTAU    EQU  13      *  HPF TAU = 1MSEC = 1/2*PI*170HZ
0023                                *      = 125 USECS * 1/(16-HTAU)
0024   0800   THRESO  EQU  >800    *  SUPRESS THRES = 1/16 (-24DB)
0025                                *  2**-THRESO (-6DB * THRESO)
0026   0258   HANGTO  EQU  600     *  HANG OVER TIME = 75 MSECS
0027                                *      = 125 USECS * HANGTO
0028   0001   NER     EQU  1       *  NEAR END SPEECH THRES = -6DB
0029                                *      = -6DB * NER
0030   0021   CUTOF0  EQU  >21     *  UPDATE CUTOFF = -48DB
0031                                *      = -6DB * (13 - LN(CUTOF0))
0032
0033                  *  PAGE 0 DATA MEMORY ALLOCATION
0034
0035   0000
0036   0000   PODM    EQU  0       *  PAGE 0 DATA MEM ADRS
0037
0038   0000   DRR     EQU  0       *  SERIAL PORT DATA RECEIVE REG
0039   0001   DXR     EQU  1       *  SERIAL PORT DATA TRANSMIT REG
0040   0002   TIM     EQU  2       *  TIMER REG (NOT USED)
0041   0003   PRD     EQU  3       *  PERIOD REG (NOT USED)
0042   0004   IMR     EQU  4       *  INTERRUPT MASK REG
0043   0005   GREG    EQU  5       *  MEM ALLOCATION REG (NOT USED)
0044   0000
0045   0060   DRR1    EQU  96      *  STORAGE FOR S(0) BY RINT
0046   0061   DRR2    EQU  97      *  STORAGE FOR Y(0) BY RINT
0047   0062   TST0    EQU  98      *  STORAGE FOR ST0 BY RINT
0048   0063   TACCH   EQU  99      *  STORAGE FOR ACC HIGH BY RINT
0049   0064   TACCL   EQU  100     *  STORAGE FOR ACC LOW BY RINT
0050   0000
0051   0065   BADDR   EQU  101     *  BASE ADR FOR MU-LAW EXPANSION
0052   0066   BIAS2   EQU  102     *  BIAS FOR MU-LAW COMPRESSION
0053   0067   NEG7    EQU  103     *  -7 FOR MU-LAW COMPRESSION
0054   0068   Q       EQU  104     *  MU-LAW COMPRESSION MANTISSA
0055   0069   S1      EQU  105     *  MU-LAW COMPRESSION SIGN
0056   0000
0057   0068   TEMP1   EQU  104     *  TEMPORARY STORAGE LOCATION 1
```

```
0058   0000
0059   0069   TEMP2   EQU  105     *  TEMPORARY STORAGE LOCATION 2
0060   0000
0061   006B   EEST    EQU  107     *  ECHO ESTIMATE
0062   006C   OUTPUT  EQU  108     *  OUTPUT ESTIMATE
0063   006D   THRES   EQU  109     *  RESIDUAL OUTPUT SPRS THRESHOLD
0064   0000
0065   0000
0066   006E   ONE     EQU  110     *  HOLDS 1
0067   0000
0068   006F   ADY142  EQU  111     *  Y142 DATA MEM ADRS
0069
0070   0070   S0      EQU  112     *  NEAR-END SAMPLE MSBS
0071   0071   S0LSBS  EQU  113     *  NEAR-END SAMPLE LSBS
0072   0072   S0DC    EQU  114     *  INPUT NEAR-END SAMPLE (K=0)
0073   0073   S1DC    EQU  115     *  INPUT NEAR-END SAMPLE (K=1)
0074
0075                  *  PAGE 4 DATA MEMORY ALLOCATION
0076
0077   0200   P4DM    EQU  512     *  PAGE 4 DATA MEM ADRS
0078   FF00   P4PM    EQU  65280   *  PAGE 4 PROG MEM ADRS
0079   0000
0080   0000   UN0     EQU  0       *  NORMALIZED OUTPUT (K=0)
0081   000F   UN15    EQU  15      *  NORMALIZED OUTPUT (K=15)
0082   0000
0083                  *  PAGE 5 DATA MEMORY ALLOCATION
0084
0085
0086   0280   P5DM    EQU  640     *  PAGE 5 DATA MEM ADRS
0087   FF80   P5PM    EQU  65408   *  PAGE 5 PROG MEM ADRS
0088
0089   0000   A127    EQU  0       *  FIR FILTER COEFFICIENT (K=127)
0090   007F   A0      EQU  127     *  FIR FILTER COEFFICIENT (K=0)
0091   0000
0092                  *  PAGE 6 DATA MEMORY ALLOCATION
0093
0094
0095
0096   0300   P6DM    EQU  768     *  PAGE 6 DATA MEM ADRS
0097   0000
0098   0000   Y0      EQU  0       *  REFERENCE SAMPLE (K=0)
0099   007F   Y127    EQU  127     *  REFERENCE SAMPLE (K=127)
0100
0101   0000
0102                  *  PAGE 7 DATA MEMORY ALLOCATION
0103
0104
0105   0380   P7DM    EQU  896     *  PAGE 7 DATA MEM ADRS
0106   0000
0107   0000   Y128    EQU  0       *  REFERENCE SAMPLE (K=128)
0108   000F   Y143    EQU  15      *  REFERENCE SAMPLE (K=143)
0109   0000
0110   0010   TEMP3   EQU  16      *  TEMPORARY STORAGE LOCATION 3
0111
0112   0011   CUN0    EQU  17      *  COPY OF UN0 FROM PAGE 4
0113
0114   0000
```

15. Digital Voice Echo Canceller with a TMS32020

```
0170                                    *    PROCESSOR INITIALIZATION ROUTINE
0171
0172
0173
0174
0175  001E
0176  0028
0177  0028
0178  0028 C800    INIT    AORG  40
0179  0029                 LDPK  0        * INITIALIZE ST0 AND ST1
0180  0029 D001            LALK  >2E00         0010 1110 0000 0000  IN BINARY
      002A 2E00
0181  002B 6068            SACL  TEMP1         * DATA FOR ST0
0182  002C
0183  002C 5068            LST   TEMP1         0 -> DP    PG POINTER SET TO 0
0184                                           1 -> INTM  INTERRUPTS DISABLED
0185                                           0 -> OVM   OVERFLOW SATURATION
0186                                           0 -> OV    OVERFLOW REG CLEARED
0187                                           1 -> ARP   AR POINTER SET TO 1
0188  002D
0189  002D D001            LALK  >27F8         0010 0111 1111 1000  IN BINARY
      002E 27F8
0190  002F 6068            SACL  TEMP1         * DATA FOR ST1
0191  0030
0192  0030 5168            LST1  TEMP1         0 -> PM    NO P REG SHIFTING
0193                                           1 -> FSX   FSX IS AN INPUT
0194                                           1 -> FO    DRR, DXR TO 8 BITS
0195                                           1 -> XF    XF PIN SET TO HIGH
0196                                           0 -> SXM   SIGN EXTENSION ON
0197                                           0 -> TC    TC FLAG BIT RESET
0198                                           0 -> CNF   B0 IS DATA MEM
0199                                           1 -> ARB   AND 1 -> ARP
0200  0031
0201
0202                                    *  INITIALIZE PAGE 0
0203
0204  0031
0205  0031 C160            LARK  AR1,96        * LOWEST PAGE 0 LOCATION -> AR1
0206  0032
0207  0032 CA00            ZAC                 * 0 -> ACC
0208  0033
0209  0033 CB1F            RPTK  31
0210  0034
0211  0034 60A0            SACL  *+
0212  0035
0213  0035 D001            LALK  >0030         * ZERO PAGE 0
      0036 0030
0214  0037 6004            SACL  IMR           * ENABLE XINT,RINT
0215  0038
0216  0038 D001            LALK  >FFFF         * DISBALE TINT,INT0,INT1,INT2
      0039 FFFF
0217  003A 6001            SACL  DXR           * MU-LAW FFFF = LINEAR 0
0218  003B
0219  003B 6060            SACL  DRR1
0220  003C
0221  003C 6061            SACL  DRR2
0222  003D
```

```
0115  0012    AONE   EQU  18     * HOLDS 1
0116  0013    SONE   EQU  19     * HOLDS SATURATION 1
0117  0000
0118  0014    ADA0   EQU  20     * A0 DATA MEM ADRS
0119  0015    ADY1   EQU  21     * Y1 DATA MEM ADRS
0120  0016    ADINC0 EQU  22     * INC0 DATA MEM ADRS
0121  0017    ADM7   EQU  23     * M7 DATA MEM ADRS
0122  0018    ADUN14 EQU  24     * UN14 DATA MEM ADRS
0123  0000
0124  0060    H      EQU  96     * MODULO 16 COUNTER
0125  0000
0126  0061    HANGT  EQU  97     * HANG OVER COUNTER RESET VALUE
0127  0062    HCNTR  EQU  98     * HANG OVER COUNTER
0128  0000
0129  0063    ABSS0  EQU  99     * !S0!
0130  0064    ABSS0F EQU  100    * SHORT TAU LPF 2*!S0!
0131  0000
0132  0065    ABSE0  EQU  101    * !OUTPUT!
0133  0066    ABSOUT EQU  102    * LONG TAU LPF !OUTPUT! MSBS
0134  0067    AELSBS EQU  103    * LONG TAU LPF !OUTPUT! LSBS
0135  0000
0136  0068    ABSY0  EQU  104    * !Y0!
0137  0069    ABSY   EQU  105    * LONG TAU LPF !Y0! MSBS
0138  006A    AY1SBS EQU  106    * LONG TAU LPF !Y0! LSBS
0139  006B    1ABSY  EQU  107    * 1/ABSY
0140  006C    CUTOFF EQU  108    * ABSY CUTOFF LVL FOR NO UPDATE
0141  006D    ABSY0F EQU  109    * SHORT TAU LPF !Y0!
0142  0000
0143  006E    M0     EQU  110    * LOCAL MAXIMA (K=0)
0144  0076    M8     EQU  118    * LOCAL MAXIMA (K=8)
0145  0000
0146  0078    INC0   EQU  120    * UPDATE INCREMENT (K=0)
0147  007F    INC7   EQU  127    * UPDATE INCREMENT (K=7)
0148  0000
0149  0000
0150  FF80    A127PM EQU  P5PM+A127   * A127 PROG MEM ADRS
0151  FF00    UN0PM  EQU  P4PM+UN0    * UN0 PROG MEM ADRS
0152  0000
0153  0000
0154                         * INTERRUPT BRANCHES
0155
0156
0157
0158
0159  0000
0160  0000          AORG  0
0161  0000
0162  0000 FF80     B     INIT       * ON HARDWARE RESET GO TO INIT
      0001 0028
0163  0002
0164  001A
0165  001A          AORG  26
0166  001A
0167  001A FF80     B     RXRT       * ON RINT GO TO RXRT
      001B 01C1
0168  001C FF80     B     TXRT       * ON TINT GO TO TXRT
      001D 01CF
```

```
0223  003D D001     LALK  XTBL
      003E 0300
0224  003F 6065     SACL  BADDR
0225  0040 D001     LALK  132
0226  0041 0084
0227  0042 6066     SACL  BIAS2
0228  0043 D001     LALK  -7
0229  0044 FFF9
0230  0045 6067     SACL  NEG7
0231  0046 D001     LALK  THRES0
0232  0047 0800
0233  0048 606D     SACL  THRES
0234  0049 D001     LALK  1
0235  004A 0001
0236  004B 606E     SACL  ONE
0237  004C D001     LALK  P7DM+Y143-1
0238  004D 038E
0239  004E 606F     SACL  ADY142
0240
0241
0242                *       INITIALIZE PAGE 4 AND 5
0243
0244                        • • •
0245  004F D100     LRLK  AR1,512         * LOWEST PAGE 4 ADDRESS -> AR1
      0050 0200
0247  0051 CA00     ZAC                   * 0 -> ACC
0249  0052 CBFF     RPTK  255
0251  0053 60A0     SACL  *+              * ZERO PAGE 4 AND 5
0253
0254                *       INITIALIZE PAGE 6 AND 7
0255
0256                        • • •
0257  0054 CBFF     RPTK  255
0258
0259  0055 60A0     SACL  *+              * ZERO PAGE 6 AND 7
0260
0261  0056 C807     LDPK  7
0262  0057 D001     LALK  1
0263  0058 0001
0264  0059 6012     SACL  AONE
0265  005A D001     LALK  >4FFF
0266  005B 4FFF
0267  005C 6013     SACL  SONE
0268  005D D001     LALK  P5DM+A0
0269
0270  005E 0001
```

```
0271  005E 02FF
      005F 6014     SACL  ADA0
0272  0060 D001     LALK  P6DM+YO+1
0273  0061 0301
0274  0062 6015     SACL  ADY1
0275  0063 D001     LALK  P7DM+1NCO
0276  0064 03F8
0277  0065 6016     SACL  AD1NCO
0278  0066 D001     LALK  P7DM+MB-1
0279  0066 03F5
0280  0067 6017     SACL  ADM7
0281  0068 D001     LALK  P4DM+UN15-1
0282  0069 020E
0283  006A 6018     SACL  ADUN14
0284  006B D001     LALK  HANGT0
0285  006C 0258
0286  006D 6061     SACL  HANGT
0287  006E D001     LALK  >400         * >400 = 1/8 OF MAX ABSY
0288  006F 0400
0289  006F 6069     SACL  ABSY
0290  0070 D001     LALK  >20
0291  0071 0020
0292  0072 606B     SACL  IABSY
0293  0073 D001     LALK  CUTOF0
0294  0075 0021
0295  0076 606C     SACL  CUTOFF
0296  0077 D001     LALK  >400
0297  0078 0400
0298  0079 606D     SACL  ABSYOF
0299  007A 606E     SACL  M0
0300  007B CE00     EINT
0301  007C FF80     B     LOOP
0302  007D 01BE
```

15. Digital Voice Echo Canceller with a TMS32020

```
0306
0307                       ***********************************************
0308             *
0309             *         CYCLE START ROUTINE
0310             *
0311             *         ***********************************************
0312  007F C800  START     LDPK   0
0313  0080
0314
0315             *
0316             *         CONVERT MU-LAW INPUT REFERENCE SAMPLE TO LINEAR   (Y0)
0317             *
0318  0080 4161            ZALS   DRR2          * MU-LAW Y(0) -> ACC
0319  0081
0320  0081 0065            ADD    BADDR         * ADD MU-LAW TABLE BASE ADDRESS
0321  0082
0322  0082 C806            LDPK   6
0323  0083
0324  0083 5800            TBLR   Y0            * LINEAR Y(0) -> Y0
0325  0084
0326
0327             *
0328             *         COMPUTE ABSOLUTE VALUE OF Y0
0329             *
0330  0084 2000            LAC    Y0            * Y0 -> ACC
0331  0085
0332  0085 CE1B            ABS
0333  0086
0334  0086 C807            LDPK   7
0335  0087
0336  0087 6068            SACL   ABSY0         * ;Y0; -> ABSY0 ON PAGE 7
0337  0088
0338  0088 C800            LDPK   0
0339  0089
0340
0341             *
0342             *         CONVERT MU-LAW NEAR END SAMPLE TO LINEAR (S0DC)
0343             *
0344  0089 4160            ZALS   DRR1          * MU-LAW S(0)DC -> ACC
0345  008A
0346  008A 0065            ADD    BADDR         * ADD MU-LAW TABLE BASE ADDRESS
0347  008B
0348  008B 5872            TBLR   S0DC          * LINEAR S(0)DC -> S0DC
0349  008C
0350
0351             *
0352             *         COMPUTE HIGH PASS FILTERED NEAR END SAMPLE (S0)
0353             *
0354  008C 4171            ZALS   SOLSBS        * SOLSBS -> LOW ACC
0355  008D
0356  008D 4870            ADDH   S0            * S0 (MSBS) -> HIGH ACC
0357  008E
0358  008E 1D70            SUB    S0,HTAU       * ACC - S0 * 2**HTAU -> ACC
0359  008F
0360  008F 4872            ADDH   S0DC          * ACC + S0DC * 2**16 -> ACC
0361  0090
0362  0090 1C72            SUB    S0DC,HTAU-1   * ACC - S0DC * 2**HTAU-1 -> ACC
```

```
0363  0091
0364  0091 4473            SUBH   S1DC          * ACC - S1DC * 2**16 -> ACC
0365  0092
0366  0092 0C73            ADD    S1DC,HTAU-1   * ACC + S1DC * 2**HTAU-1 -> ACC
0367  0093
0368  0093 6071            SACL   SOLSBS        * LOW ACC -> SOLSBS
0369  0094
0370  0094 6870            SACH   S0            * HIGH ACC -> S0 (MSBS)
0371  0095
0372  0095 5672            DMOV   S0DC          * S0DC -> S1DC
0373  0096
0374             *
0375             *         COMPUTE ABSOLUTE VALUE OF S0
0376             *
0377  0096 CE1B            ABS
0378  0096
0379  0097 C807            LDPK   7
0380  0097
0381  0098 6863            SACH   ABSS0         * ;S0; -> ABSS0 ON PAGE 7
0382  0098
```

```
EC128    32020 FAMILY MACRO ASSEMBLER   PC 1.0 85.157    14:10:03  11-19-85
                                                                   PAGE 0009

                    *************************************
                    *                                   *
                    *      ECHO ESTIMATION ROUTINE       *
                    *                                   *
                    *************************************
0384
0385
0386
0387
0388
0389 0099 C800    EESTR  LDPK   0
0390 0099
0391 009A                        * MOVE Y128,Y129,....,Y142 TO NEXT HIGHER MEMORY LOCATION
0392
0393
0394
0395 009A 5589           LARP   AR1            * 1 -> AR POINTER
0396 009A
0397 009B 316F           LAR    AR1,ADY142     * ADY142 -> AR1
0398 009B
0399 009C CB0E           RPTK   14             * K = 142,141,....128
0400 009C
0401 009D
0402 009D 5690           DMOV   *-             * Y(K) -> Y(K+1)
0403 009E
0404
0405                *  CONVOLVE REFERENCE SAMPLES WITH FIR COEFFICIENTS
0406
0407 009E
0408 009E 2E6E    FIR    LAC    ONE,14         * ROUND-OFF OFFSET -> ACC
0409 009F
0410 009F A000           MPYK   0              * 0 -> P
0411 00A0
0412 00A0 CE05           CNFP                  * AR1 STILL POINTS AT Y127
0413 00A1
0414 00A1 CB7F           RPTK   127            * K = 127,126,....0
0415 00A2
0416 00A2 5C90           MACD   A127PM,*-      * Y(K) * A(I-K) + ACC -> ACC
0417 00A3 FF80
0418 00A4 CE04           CNFD
0419 00A4
0420 00A5
0421 00A5 CE15           APAC                  * P + ACC -> ACC
0422 00A6
0423 00A6 696B           SACH   EEST,1         * 2 * HIGH ACC -> EEST
0424 00A7
0425
0426                *  COMPUTE THE OUTPUT
0427 00A7
0428 00A7 2070           LAC    S0             * S0 -> ACC
0429 00A8
0430 00A8 106B           SUB    EEST           * ACC - EEST -> ACC
0431 00A9
0432 00A9 606C           SACL   OUTPUT         * ACC -> OUTPUT
0433 00AA
0434 00AA C807           LDPK   7
0435 00AB
0436 00AB CE1B           ABS
0437 00AC
0438 00AC 6065           SACL   ABSE0          * ACC -> ABSE0 ON PAGE 7
```

```
EC128    32020 FAMILY MACRO ASSEMBLER   PC 1.0 85.157    14:10:03  11-19-85
                                                                   PAGE 0010

                    *************************************
                    *                                   *
                    *  RESIDUAL OUTPUT SUPPRESSION ROUTINE *
                    *                                   *
                    *************************************
0440 00AD
0441
0442
0443
0444
0445 00AD
0446 00AD
0447 00AD 3C6B    SPRS   LT     IABSY          * IABSY -> T REG
0448 00AE
0449 00AE 3866           MPY    ABSOUT         * ABSOUT * IABSY -> P REG
0450 00AF
0451 00AF 2062           LAC    HCNTR          * NEAR END SPEECH FLAG -> ACC
0452 00B0
0453 00B0 C800           LDPK   0
0454 00B1
0455 00B1 F180           BGZ    WOUT           * IF N.E. SPEECH NO SPRS
0456 00B2 00BB
0457 00B3
0458 00B3 CE14           PAC                   * P REG -> ACC
0459 00B4
0460 00B4 106D           SUB    THRES          * ACC - THRES -> ACC
0461 00B5
0462 00B5 F180           BGZ    WOUT           * IF THRES EXCEEDED SKIP SPRS
0463 00B6 00BB
0464 00B7 FA80           BIOZ   WOUT           * IF BIO PIN LOW SKIP SPRS
0465 00B8 00BB
0466 00B9
0467 00B9 CA00           ZAC                   * 0 -> ACC
0468 00BA
0469 00BA 606C           SACL   OUTPUT         * ACC -> OUTPUT
     00BB
     00BB 3C6C    WOUT   LT     OUTPUT         * OUTPUT -> T REG   (FOR UNO)
```

15. Digital Voice Echo Canceller with a TMS32020

```
0471
0472                         *************
0473                         *  LINEAR TO MU-LAW COMPRESSION ROUTINE
0474                         *
0475                         *************
0476   00BC 406C    CMPRS ZALH   OUTPUT        * OUTPUT -> ACC
0477   00BD CE18          SFL
0478   00BE CE18          SFL                  * LEFT JUSTIFY ACC
0479
0480
0481
0482   00BF F380          BLZ    NEGCMP        * IF ACC < 0 THEN GO TO NEGCMP
0483   00C0 0000
0484
0485   00C1 4866    POSCMP ADDH  BIAS2
0486
0487   00C2 3167          LAR    AR1,NEG7
0488
0489   00C3 CB06          RPTK   6             * FIND MSB
0490
0491   00C4 CEA2          NORM
0492
0493   00C5 DE04          ANDK   >F000,14      * ZERO 2 MSBS AND ALL LSBS
       00C6 F000
0494
0495   00C7 6868          SACH   Q
0496
0497   00C8 7169          SAR    AR1,S1
0498
0499   00C9 4069          ZALH   S1
0500
0501   00CA CE1B          ABS
0502
0503   00CB 0268          ADD    Q,2
0504
0505   00CC D406          XORK   >FF00,4       * INVERT ALL BITS
       00CD FF00
0506
0507   00CE FF80          B      TXOUT
       00CF 00DE
0508
0509
0510   00D0 CE1B    NEGCMP ABS                 * LEFT JUSTIFIED OUTPUT IN ACC
0511
0512   00D1 4866          ADDH   BIAS2
0513
0514   00D2 3167          LAR    AR1,NEG7
0515
0516   00D3 CB06          RPTK   6             * FIND MSB
0517
0518   00D4 CEA2          NORM
0519
0520   00D5 DE04          ANDK   >F000,14      * ZERO 2 MSBS AND ALL LSBS
       00D6 F000
0521
0522   00D7 6868          SACH   Q
```

```
0523   00D8 7169          SAR    AR1,S1
0524
0525   00D9 4069          ZALH   S1
0526
0527   00DA CE1B          ABS
0528
0529   00DB 0268          ADD    Q,2
0530
0531   00DC D406          XORK   >7F00,4       * INVERT ALL BITS IN Q
       00DD 7F00
0532
0533   00DE 6C01    TXOUT  SACH   DXR,4        * 2**4 * HIGH ACC -> DXR
0534
```

```
                    ********************
                    * POWER ESTIMATION ROUTINE *
                    ********************

0536
0537
0538
0539
0540
0541  00DF
0542  00DF C807   NORM   LDPK   7           * T REG STILL CONTAINS OUTPUT
0543  00E0
0544
0545                     * UPDATE LONG TAU OUTPUT POWER ESTIMATE (ABSOUT)
0546
0547  00E0 4066          ZALH   ABSOUT      * ABSOUT -> HIGH ACC
0548  00E0
0549  00E1 4967          ADDS   AELSBS      * AELSBS -> LOW ACC
0550  00E1
0551  00E2 1966          SUB    ABSOUT,LTAU * ACC - ABSOUT * 2**LTAU -> ACC
0552  00E2
0553  00E3 0965          ADD    ABSE0,LTAU  * ACC + ABSE0 * 2**LTAU -> ACC
0554  00E3
0555                     SACH   ABSOUT      * HIGH ACC -> ABSOUT
0556  00E4 6866
0557  00E4
0558  00E5 6067          SACL   AELSBS      * LOW ACC -> AELSBS
0559  00E5
0560  00E6
0561
0562                     * UPDATE LONG TAU REFERENCE POWER ESTIMATE (ABSY)
0563  00E6 4069          ZALH   ABSY        * ABSY -> HIGH ACC
0564  00E6
0565  00E7 496A          ADDS   AYLSBS      * AYLSBS -> LOW ACC
0566  00E7
0567  00E8
0568  00E8 1969          SUB    ABSY,LTAU   * ACC - ABSY * 2**LTAU -> ACC
0569  00E9
0570  00E9 0968          ADD    ABSY0,LTAU  * ACC + ABSY0 * 2**LTAU -> ACC
0571  00EA
0572  00EA 096C          ADD    CUTOFF,LTAU * ACC + CUTOFF * 2**LTAU -> ACC
0573  00EB
0574  00EB 6869          SACH   ABSY        * HIGH ACC -> ABSY
0575  00EC
0576  00EC 606A          SACL   AYLSBS      * LOW ACC -> AYLSBS
0577  00ED
0578
0579                     * COMPUTE 1/ABSY (DIVIDE 1 BY ABSY)
0580
0581  00ED
0582  00ED 4012          ZALH   AONE
0583  00EE
0584  00EE CB0E          RPTK   14
0585  00EF
0586  00EF 4769          SUBC   ABSY
0587  00F0
0588  00F0 606B          SACL   IABSY
```

```
                    ************************
                    * OUTPUT NORMALIZATION ROUTINE *
                    ************************

0590
0591
0592
0593
0594
0595  00F1
0596
0597
0598                     * MOVE UN0,UN1,....UN14 TO NEXT HIGHER MEMORY LOCATION
0599  00F1
0600  00F1 3118          LAR    AR1,ADUN14  * ADUN14 -> AR1
0601  00F2
0602  00F2 CB0D          RPTK   13          * K=14,13,.....1
0603  00F3
0604  00F3 5690          DMOV   *-          * UN(K) -> UN(K+1)
0605  00F4
0606  00F4 5680          DMOV   *           * UN(0) -> UN(1)
0607  00F5
0608
0609
0610                     * COMPUTE NORMALIZED OUTPUT (UN0)
0611
0612  00F5 386B          MPY    IABSY       * IABSY * T REG(OUTPUT) -> P REG
0613  00F6
0614  00F6 CE14          PAC                * P REG (UN0) -> ACC
0615  00F7
0616
0617
0618                     * SATURATE NORMALIZED OUTPUT (UN0) AT +/- 1.0
0619
0620  00F7 F480          BGEZ   POSUN0      * IF UN0 > 0 THEN GO TO POSUN0
      00F8 0100
0621  00F9
0622  00F9 0013   NEGUN0 ADD    SONE        * ACC + SONE -> ACC
0623  00FA
0624  00FA F480          BGEZ   SMLUN0      * IF -1.0 < UN0 < 0 THEN NO SATR
      00FB 0106
0625  00FC
0626  00FC CA00          ZAC                * 0 -> ACC
0627  00FD
0628  00FD 1013          SUB    SONE        * ACC - SONE -> ACC
0629  00FE
0630  00FE FF80          B      SAVUN0
      00FF 0107
0631
0632  0100 1013   POSUN0 SUB    SONE        * ACC - SONE -> ACC
0633  0101
0634  0101 F280          BLEZ   SMLUN0      * IF 0 < UN0 < 1.0 THEN NO SATR
      0102 0106
0635  0103
0636  0103 2013          LAC    SONE        * SONE -> ACC
0637  0104
0638  0104 FF80          B      SAVUN0
      0105 0107
0639  0106
0640  0106 CE14   SMLUN0 PAC                * P REG (UN0) -> ACC
0641  0107
```

15. Digital Voice Echo Canceller with a TMS32020

```
0646              .............
0647              .............  NEAR-END SPEECH DETECTION ROUTINE
0648              .............
0649
0650
0651 0109        NESP  LDPK  7
0652 0109 C807
0653 010A
0654
0655              . . .
0656              * UPDATE SHORT TAU REFERENCE POWER ESTIMATE (ABSY0F)
0657 010A               ZALH  ABSY0F        * ABSY0F * 2**16 -> ACC
0658 010A 4060D
0659 010B               SUB   ABSY0F,STAU   * ACC - ABSY0F * 2**STAU -> ACC
0660 010B 1B6D
0661 010C               ADD   ABSY0,STAU    * ACC + ABSY0 * 2**STAU -> ACC
0662 010C 0B68
0663 010D               SACH  ABSY0F        * HIGH ACC -> ABSY0F
0664 010D 6860
0665 010E
0666
0667              * UPDATE SHORT TAU NEAR END POWER ESTIMATE (ABSS0F)
0668
0669 010E               ZALH  ABSS0F        * ABSS0F * 2**16 -> ACC
0670 010E 4064
0671 010F               SUB   ABSS0F,STAU   * ACC - ABSS0F * 2**STAU -> ACC
0672 010F 1B64
0673 0110               ADD   ABSS0,STAU+NER * ACC + ABSS0*2**STAU+NER -> ACC
0674 0110 0C63
0675 0111               SACH  ABSS0F        * HIGH ACC -> ABSS0F
0676 0111 6864
0677 0112
0678
0679              * UPDATE MODULO 16 COUNTER (H)
0680
0681 0112               LAC   H             * H -> ACC
0682 0112 2060
0683 0113               ADD   AONE          * ACC + 1 -> ACC
0684 0113 0012
0685 0114               ANDK  >000F         * IF ACC = 16 THEN 0 -> ACC
0686 0114 D004
     0115 000F
0687 0116               SACL  H             * ACC -> H
0688 0116 6060
0689 0117               BGZ   NESP1         * IF H > 0 THEN GO TO NESP1
0690 0117 F180
     0118 011F
0691 0119
0692
0693
0694              * MOVE M0,M1,.....M7 TO NEXT HIGHER MEMORY LOCATION
0695 0119               LAR   AR1,ADM7      * ADM7 -> AR1
0696 0119 3117
0697 011A               RPTK  7             * K=7,6,.....0
0698 011A C807
0699 011B               DMOV  *-            * M(K) -> M(K+1)
0700 011B 5690
```

```
0642 0107 6011   SAVUN0  SACL  CUN0         * ACC -> CUN0
0643 0108
0644 0108 6080           SACL  *            * ACC -> UN(0)
```

15. Digital Voice Echo Canceller with a TMS32020

```
0701 011C
0702 011C 566D          DMOV  ABSY0F    * ABSY0F -> M0
0703 011D FF80          B     NESP3     * ON MEMORY MOVES SKIP DETECTION
0704 011D 0149
0705 011E
0706 011E
0707 011F
                * UPDATE MOST RECENT LOCAL MAXIMA (M0)
0708 011F
0709 011F
0710 011F 206D   NESP1  LAC   ABSY0F    * ABSY0F -> ACC
0711 0120               SUB   M0        * ACC - M0 -> ACC
0712 0120 106E
0713 0121               BLEZ  NESP2     * IF M0 > ABSY0F THEN NO UPDATE
0714 0121 F280
0715 0122 0124
0716 0123               DMOV  ABSY0F    * ABSY0F -> M0
0717 0123 566D
0718 0124
                * COMPARE REFERENCE POWER TO NEAR-END POWER
0719 0124
0720 0124
0721 0124 2064   NESP2  LAC   ABSS0F    * ABSS0F -> ACC
0722 0124 2064
0723 0125               SUB   M0        * ACC - M0 -> ACC
0724 0125 106E
0725 0126               BLEZ  NESP3     * NO N.E. SPEECH IF M0 > ABSS0F
0726 0126 F280
0727 0127 0149
0728 0128
0729 0128 2064          LAC   ABSS0F    * ABSS0F -> ACC
0730 0129               SUB   M0+1      * ACC - M1 -> ACC
0731 0129 106F
0732 012A               BLEZ  NESP3     * NO N.E. SPEECH IF M1 > ABSS0F
0733 012A F280
0734 012B 0149
0735 012C
0736 012C 2064          LAC   ABSS0F    * ABSS0F -> ACC
0737 012D               SUB   M0+2      * ACC - M2 -> ACC
0738 012D 1070
0739 012E               BLEZ  NESP3     * NO N.E. SPEECH IF M2 > ABSS0F
0740 012E F280
0741 012F 0149
0742 0130
0743 0130 2064          LAC   ABSS0F    * ABSS0F -> ACC
0744 0131               SUB   M0+3      * ACC - M3 -> ACC
0745 0131 1071
0746 0132               BLEZ  NESP3     * NO N.E. SPEECH IF M3 > ABSS0F
0747 0132 F280
0748 0133 0149
0749 0134
0750 0134 2064          LAC   ABSS0F    * ABSS0F -> ACC
0751 0135
```

```
0752 0135 1072          SUB   M0+4      * ACC - M4 -> ACC
0753 0136               BLEZ  NESP3     * NO N.E. SPEECH IF M4 > ABSS0F
0754 0136 F280
0755 0137 0149
0756 0138
0757 0138
0758 0138 2064          LAC   ABSS0F    * ABSS0F -> ACC
0759 0139 1073          SUB   M0+5      * ACC - M5 -> ACC
0760 013A               BLEZ  NESP3     * NO N.E. SPEECH IF M5 > ABSS0F
0761 013A F280
0762 013B 0149
0763 013C
0764 013C 2064          LAC   ABSS0F    * ABSS0F -> ACC
0765 013D 1074          SUB   M0+6      * ACC - M6 -> ACC
0766 013D               BLEZ  NESP3     * NO N.E. SPEECH IF M6 > ABSS0F
0767 013E F280
0768 013E 0149
0769 013F
0770 0140
0771 0140 2064          LAC   ABSS0F    * ABSS0F -> ACC
0772 0141 1075          SUB   M0+7      * ACC - M7 -> ACC
0773 0141               BLEZ  NESP3     * NO N.E. SPEECH IF M7 > ABSS0F
0774 0142 F280
0775 0142 0149
0776 0143
0777 0144
0778 0144 2064          LAC   ABSS0F    * ABSS0F -> ACC
0779 0145 1076          SUB   M0+8      * ACC - M8 -> ACC
0780 0145               BLEZ  NESP3     * NO N.E. SPEECH IF M8 > ABSS0F
0781 0146 F280
0782 0146 0149
0783 0147
0784 0148
                * NEAR-END SPEECH DETECTED  SET HANGOVER COUNTER (HCNTR)
0785 0148
0786 0148
0787 0148 5661          DMOV  HANGT     * HANGT -> HCNTR
0788 0148
0789 0149
                * CHECK AND UPDATE HANGOVER COUNTER
0790 0149
0791
0792
0793 0149 2062   NESP3  LAC   HCNTR     * HCNTR -> ACC
0794 0149
0795 014A F680          BZ    NESP4     * IF HCNTR = 0 THEN GO TO NESP4
0796 014B 0150
0797 014C 1012          SUB   AONE      * ACC - 1 -> ACC
0798 014C
0799 014D
0800 014D 6062          SACL  HCNTR     * ACC -> HCNTR
0801 014E
0802 014E FF80          B     LOOP      * GO TO CYCLE END
```

15. Digital Voice Echo Canceller with a TMS32020

```
                  *
                  *
                  *      CHECK IF LTAU REFERENCE POWER ESTIMATE IS BELOW CUTOFF
0803  014F 01BE
0804  0150
0805
0806
0807  0150
0808  0150 2069   NESP4  LAC   ABSY          * ABSY -> ACC
0809  0151
0810  0151 106C          SUB   CUTOFF        * ACC - CUTOFF -> ACC
0811  0152 F280          BLEZ  LOOP          * IF ABSY < CUTOFF THEN LOOP
0812  0153 01BE
0813
```

```
                  *  ......................................................
                  *  ......................................................
                  *  ....    COEFFICIENT INCREMENT UPDATE ROUTINE
                  *  ......................................................
                  *  ......................................................

0814
0815  0154
0816
0817
0818
0819  0154
0820  0154 2015   UPINC  LAC   ADY1          * ADY1 -> ACC   (Y0 IS NOW IN Y1)
0821  0155
0822  0155 0060          ADD   H             * ACC + H -> ACC
0823  0156
0824  0156 6010          SACL  TEMP3
0825  0157
0826  0157 3110          LAR   AR1,TEMP3     * ADY1 + H -> AR1
0827  0158
0828  0158 CE05          CNFP
0829  0159
0830  0159 3C11          LT    CUN0          * UN0 -> T REG
0831  015A
0832  015A
0833  015A 2F12          LAC   AONE,15       * ROUND-OFF OFFSET -> ACC
0834  015B
0835  015B 38A0          MPY   *+            * UN(0) * Y(0+H) -> P REG
0836  015C
0837  015C CB0E          RPTK  14            * K = 1,2,....15
0838  015D
0839  015D 5DA0          MAC   UN0PM+1,*+    * UN(K) * Y(K+H) + ACC -> ACC
      015E FF01
0840  015F
0841  015F 3D11          LTA   CUN0          * P REG + ACC -> ACC   UN0 -> T
0842  0160
0843  0160 6878          SACH  INC0          * HIGH ACC -> INC(0)
0844  0161
0845  0161
0846  0161 2F12          LAC   AONE,15       * ROUND-OFF OFFSET -> ACC
0847  0162
0848  0162 38A0          MPY   *+            * UN(0) * Y(16+H) -> P REG
0849  0163
0850  0163 CB0E          RPTK  14            * K = 1,2,....15
0851  0164 5DA0          MAC   UN0PM+1,*+    * UN(K) * Y(K+16+H) + ACC -> ACC
0852  0164 FF01
0853  0166
0854  0166 3D11          LTA   CUN0          * P REG + ACC -> ACC   UN0 -> T
0855  0167
0856  0167 6879          SACH  INC0+1        * HIGH ACC -> INC(1)
0857  0168
0858  0168
0859  0168 2F12          LAC   AONE,15       * ROUND-OFF OFFSET -> ACC
0860
0861  0169 38A0          MPY   *+            * UN(0) * Y(32+H) -> P REG
0862
0863  016A CB0E          RPTK  14            * K = 1,2,....15
0864  016B
0865  016B 5DA0          MAC   UN0PM+1,*+    * UN(K) * Y(K+32+H) + ACC -> ACC
0866  016C FF01
0867  016D 3D11          LTA   CUN0          * P REG + ACC -> ACC   UN0 -> T
```

15. Digital Voice Echo Canceller with a TMS32020

```
                                                    * HIGH ACC -> INC(2)

0868  016E  687A      SACH    INC0+2
0869  016E
0870  016F
0871  016F
0872  016F  2F12      LAC     AONE.15
0873  0170  38A0      MPY     *+
0874  0170
0875  0171
0876  0171  CB0E      RPTK    14
0877  0172
0878  0172  5DA0      MAC     UN0PM+1,*+
      0173  FF01
0879  0173
0880  0174  3D11      LTA     CUN0
0881  0174
0882  0175  687B      SACH    INC0+3
0883  0175
0884  0176
0885  0176  2F12      LAC     AONE.15
0886  0177
0887  0177  38A0      MPY     *+
0888  0178
0889  0178  CB0E      RPTK    14
0890  0179
0891  0179  5DA0      MAC     UN0PM+1,*+
      017A  FF01
0892  017A
0893  017B  3D11      LTA     CUN0
0894  017B
0895  017C  687C      SACH    INC0+4
0896  017C
0897  017D
0898  017D  2F12      LAC     AONE.15
0899  017E  38A0      MPY     *+
0900  017E
0901  017F
0902  017F  CB0E      RPTK    14
0903  0180
0904  0180  5DA0      MAC     UN0PM+1,*+
      0181  FF01
0905  0182
0906  0182  3D11      LTA     CUN0
0907  0183
0908  0183  687D      SACH    INC0+5
0909  0184
0910  0184
0911  0185  2F12      LAC     AONE.15
0912  0185  38A0      MPY     *+
0913  0186
0914  0186
0915  0186  CB0E      RPTK    14
0916  0187
0917  0187  5DA0      MAC     UN0PM+1,*+
      0188  FF01
0918  0189
0919  0189  3D11      LTA     CUN0
0920  018A
```

```
0921  018A  687E      SACH    INC0+6
0922  018B
0923  018B
0924  018B  2F12      LAC     AONE.15
0925  018C  38A0      MPY     *+
0926  018C
0927  018D
0928  018D  CB0E      RPTK    14
0929  018E
0930  018E  5DA0      MAC     UN0PM+1,*+
      018F  FF01
0931  0190
0932  0190  3D11      LTA     CUN0
0933  0191
0934  0191  687F      SACH    INC0+7
0935  0192
0936  0192
0937  0192  CE04      CNFD
```

15. Digital Voice Echo Canceller with a TMS32020

```
0939
0940            *****************
0941            *
0942            *   COEFFICIENT UPDATE ROUTINE
0943            *
0944 0193       *****************
0945 0193 C010       LARK  AR0,16        * 16 -> AR0   (AR2 INCREMENT)
0947 0194 3116       LAR   AR1,ADINC0    * ADINC0 -> AR1
0949 0195 2014       LAC   ADA0          * ADA0 -> ACC
0951 0196 1060       SUB   H             * ACC - H -> ACC
0953 0197 6010       SACL  TEMP3
0955 0198 3210       LAR   AR2,TEMP3     * ADA0 - H -> AR2
0957 0199 CE0A       SPM   2             * SET 4 BIT LEFT SHIFT OF P REG
0959 019A 236B       LAC   IABSY,GAIN    * IABSY * 2**GAIN -> ACC
0961 019B 6010       SACL  TEMP3         * ACC -> TEMP3
0963 019C 3C10       LT    TEMP3         * TEMP3 -> T REG
0965 019D 38AA       MPY   *+,AR2        * INC(0) * T REG -> P REG
0967 019E 4080       ZALH  *             * A(H) * 2**16 -> ACC
0969 019F CE15       APAC                * P REG + ACC -> ACC
0971 01A0 6BD9       SACH  *0-,0,AR1     * HIGH ACC -> A(H)
0975 01A1 38AA       MPY   *+,AR2        * INC(1) * T REG -> P REG
0977 01A2 4080       ZALH  *             * A(16+H) * 2**16 -> ACC
0979 01A3 CE15       APAC                * P REG + ACC -> ACC
0981 01A4 6BD9       SACH  *0-,0,AR1     * HIGH ACC -> A(16+H)
0984 01A5 38AA       MPY   *+,AR2        * INC(2) * T REG -> P REG
0986 01A6 4080       ZALH  *             * A(32+H) * 2**16 -> ACC
0988 01A7 CE15       APAC                * P REG + ACC -> ACC
0990 01A8 6BD9       SACH  *0-,0,AR1     * HIGH ACC -> A(32+H)
0993 01A9 38AA       MPY   *+,AR2
0995 01AA 4080       ZALH  *
```

```
0996 01AB
0997 01AB CE15       APAC
0999 01AC 6BD9       SACH  *0-,0,AR1
1002 01AD 38AA       MPY   *+,AR2
1004 01AE 4080       ZALH  *
1006 01AF CE15       APAC
1008 01B0 6BD9       SACH  *0-,0,AR1
1011 01B1 38AA       MPY   *+,AR2
1013 01B2 4080       ZALH  *
1015 01B3 CE15       APAC
1017 01B4 6BD9       SACH  *0-,0,AR1
1020 01B5 38AA       MPY   *+,AR2
1022 01B6 4080       ZALH  *
1024 01B7 CE15       APAC
1026 01B8 6BD9       SACH  *0-,0,AR1
1029 01B9 38AA       MPY   *+,AR2
1031 01BA 4080       ZALH  *
1033 01BB CE15       APAC
1035 01BC 6BD9       SACH  *0-,0,AR1
1038 01BD CE08       SPM   0             * SET NO SHIFT OF P REG
```

```
1040                    ****************************
1041                    *      CYCLE END ROUTINE   *
1042                    ****************************
1043
1044
1045   01BE
1046   01BE CEIF   LOOP  IDLE            * WAIT IN LOOP UNITL RINT/XINT
1047   01BF CEIF         IDLE            * EXTRA IDLE FOR TWO RINT
1048   01BF
1049   01C0
1050   01C0 5500         NOP
1051   01C1
1052                    ****************************
1053                    *     RINT SERVICE ROUTINE *
1054                    ****************************
1055
1056
1057   01C1
1058   01C1 7862   RXRT  SST   TST0      * SAVE ST0
1059   01C2
1060   01C2 C800         LDPK  0         * 0 -> PAGE POINTER
1061   01C3
1062   01C3 6863         SACH  TACCH     * SAVE HIGH ACC
1063   01C4
1064   01C4 6064         SACL  TACCL     * SAVE LOW ACC
1065   01C5
1066   01C5 5660         DMOV  DRR1      * DRR1 -> DRR2
1067   01C6
1068   01C6 4100         ZALS  DRR       * DRR -> ACC
1069   01C7
1070   01C7 D004         ANDK  >00FF     * MASK-OFF MSB BYTE
       01C8 00FF
1071
1072   01C9 6060         SACL  DRR1      * ACC -> DRR1
1073   01C9
1074   01CA 4164         ZALS  TACCL     * RESTORE LOW ACC
1075   01CA
1076   01CB 4863         ADDH  TACCH     * RESTORE HIGH ACC
1077   01CB
1078   01CC 5062         LST   TST0      * RESTORE ST0
1079   01CC
1080   01CD CE00         EINT            * INTERRUPTS ENABLED
1081   01CD
1082   01CE CE26         RET             * RETURN TO PROGRAM
1083   01CE
1084   01CF
1085   01CF
1086
1087                    ****************************
1088                    *    XINT SERVICE ROUTINE  *
1089                    ****************************
1090   01CF CE00   TXRT  EINT            * INTERRUPTS ENABLED
1091   01D0
1092   01D0 FF80         B     START     * BRANCH TO PROGRAM START
       01D1 007F
1093
```

```
1095                    *****************************
1096                    *  MU-LAW EXPANSION LOOKUP TABLE *
1097                    *****************************
1098
1099
1100   01D2
1101   0300              AORG  >300
1102   0300
1103   0300      0300 XTBL  EQU   $
1104   0300 0300
1105   0300 E0A1         DATA  >E0A1     * NEGATIVE VALUES FIRST
1106   0301 E1A1         DATA  >E1A1     *  (FF, FE, ETC.)
1107   0302 E2A1         DATA  >E2A1
1108   0303 E3A1         DATA  >E3A1
1109   0304 E4A1         DATA  >E4A1
1110   0305 E5A1         DATA  >E5A1
1111   0306 E6A1         DATA  >E6A1
1112   0307 E7A1         DATA  >E7A1
1113   0308 E8A1         DATA  >E8A1
1114   0309 E9A1         DATA  >E9A1
1115   030A EAA1         DATA  >EAA1
1116   030B EBA1         DATA  >EBA1
1117   030C ECA1         DATA  >ECA1
1118   030D EDA1         DATA  >EDA1
1119   030E EEA1         DATA  >EEA1
1120   030F EFA1         DATA  >EFA1
1121   0310 F061         DATA  >F061
1122   0311 F061         DATA  >F061
1123   0312 F161         DATA  >F161
1124   0313 F1E1         DATA  >F1E1
1125   0314 F261         DATA  >F261
1126   0315 F2E1         DATA  >F2E1
1127   0316 F361         DATA  >F361
1128   0317 F3E1         DATA  >F3E1
1129   0318 F461         DATA  >F461
1130   0319 F4E1         DATA  >F4E1
1131   031A F561         DATA  >F561
1132   031B F5E1         DATA  >F5E1
1133   031C F661         DATA  >F661
1134   031D F6E1         DATA  >F6E1
1135   031E F761         DATA  >F761
1136   031F F7E1         DATA  >F7E1
1137   0320 F841         DATA  >F841
1138   0321 F881         DATA  >F881
1139   0322 F8C1         DATA  >F8C1
1140   0323 F901         DATA  >F901
1141   0324 F941         DATA  >F941
1142   0325 F981         DATA  >F981
1143   0326 F9C1         DATA  >F9C1
1144   0327 FA01         DATA  >FA01
1145   0328 FA41         DATA  >FA41
1146   0329 FA81         DATA  >FA81
1147   032A FAC1         DATA  >FAC1
1148   032B FB01         DATA  >FB01
1149   032C FB41         DATA  >FB41
1150   032D FB81         DATA  >FB81
1151   032E FBC1         DATA  >FBC1
```

15. Digital Voice Echo Canceller with a TMS32020

```
1152  032F  FC01   DATA  >FC01
1153  0330  FC31   DATA  >FC31
1154  0331  FC51   DATA  >FC51
1155  0332  FC71   DATA  >FC71
1156  0333  FC91   DATA  >FC91
1157  0334  FCB1   DATA  >FCB1
1158  0335  FCD1   DATA  >FCD1
1159  0336  FCF1   DATA  >FCF1
1160  0337  FD11   DATA  >FD11
1161  0338  FD31   DATA  >FD31
1162  0339  FD51   DATA  >FD51
1163  033A  FD71   DATA  >FD71
1164  033B  FD91   DATA  >FD91
1165  033C  FDB1   DATA  >FDB1
1166  033D  FDD1   DATA  >FDD1
1167  033E  FDF1   DATA  >FDF1
1168  033F  FE11   DATA  >FE11
1169  0340  FE29   DATA  >FE29
1170  0341  FE39   DATA  >FE39
1171  0342  FE49   DATA  >FE49
1172  0343  FE59   DATA  >FE59
1173  0344  FE69   DATA  >FE69
1174  0345  FE79   DATA  >FE79
1175  0346  FE89   DATA  >FE89
1176  0347  FE99   DATA  >FE99
1177  0348  FEA9   DATA  >FEA9
1178  0349  FEB9   DATA  >FEB9
1179  034A  FEC9   DATA  >FEC9
1180  034B  FED9   DATA  >FED9
1181  034C  FEE9   DATA  >FEE9
1182  034D  FEF9   DATA  >FEF9
1183  034E  FF09   DATA  >FF09
1184  034F  FF19   DATA  >FF19
1185  0350  FF25   DATA  >FF25
1186  0351  FF2D   DATA  >FF2D
1187  0352  FF35   DATA  >FF35
1188  0353  FF3D   DATA  >FF3D
1189  0354  FF45   DATA  >FF45
1190  0355  FF4D   DATA  >FF4D
1191  0356  FF55   DATA  >FF55
1192  0357  FF5D   DATA  >FF5D
1193  0358  FF65   DATA  >FF65
1194  0359  FF6D   DATA  >FF6D
1195  035A  FF75   DATA  >FF75
1196  035B  FF7D   DATA  >FF7D
1197  035C  FF85   DATA  >FF85
1198  035D  FF8D   DATA  >FF8D
1199  035E  FF95   DATA  >FF95
1200  035F  FF9D   DATA  >FF9D
1201  0360  FFA3   DATA  >FFA3
1202  0361  FFA7   DATA  >FFA7
1203  0362  FFAB   DATA  >FFAB
1204  0363  FFAF   DATA  >FFAF
1205  0364  FFB3   DATA  >FFB3
1206  0365  FFB7   DATA  >FFB7
1207  0366  FFBB   DATA  >FFBB
1208  0367  FFBF   DATA  >FFBF
```

```
1209  0368  FFC3   DATA  >FFC3
1210  0369  FFC7   DATA  >FFC7
1211  036A  FFCB   DATA  >FFCB
1212  036B  FFCF   DATA  >FFCF
1213  036C  FFD3   DATA  >FFD3
1214  036D  FFD7   DATA  >FFD7
1215  036E  FFDB   DATA  >FFDB
1216  036F  FFDF   DATA  >FFDF
1217  0370  FFE2   DATA  >FFE2
1218  0371  FFE4   DATA  >FFE4
1219  0372  FFE6   DATA  >FFE6
1220  0373  FFE8   DATA  >FFE8
1221  0374  FFEA   DATA  >FFEA
1222  0375  FFEC   DATA  >FFEC
1223  0376  FFEE   DATA  >FFEE
1224  0377  FFF0   DATA  >FFF0
1225  0378  FFF2   DATA  >FFF2
1226  0379  FFF4   DATA  >FFF4
1227  037A  FFF6   DATA  >FFF6
1228  037B  FFF8   DATA  >FFF8
1229  037C  FFFA   DATA  >FFFA
1230  037D  FFFC   DATA  >FFFC
1231  037E  FFFE   DATA  >FFFE
1232  037F  0000   DATA  >0
1233  0380         * POSITIVE VALUES NEXT
1234  0380  1F5F   DATA  >1F5F   * (POLARITY BIT = 1)
1235  0381  1E5F   DATA  >1E5F
1236  0382  1D5F   DATA  >1D5F
1237  0383  1C5F   DATA  >1C5F
1238  0384  1B5F   DATA  >1B5F
1239  0385  1A5F   DATA  >1A5F
1240  0386  195F   DATA  >195F
1241  0387  185F   DATA  >185F
1242  0388  175F   DATA  >175F
1243  0389  165F   DATA  >165F
1244  038A  155F   DATA  >155F
1245  038B  145F   DATA  >145F
1246  038C  135F   DATA  >135F
1247  038D  125F   DATA  >125F
1248  038E  115F   DATA  >115F
1249  038F  105F   DATA  >105F
1250  0390  0F9F   DATA  >F9F
1251  0391  0F1F   DATA  >F1F
1252  0392  0E9F   DATA  >E9F
1253  0393  0E1F   DATA  >E1F
1254  0394  0D9F   DATA  >D9F
1255  0395  0D1F   DATA  >D1F
1256  0396  0C9F   DATA  >C9F
1257  0397  0C1F   DATA  >C1F
1258  0398  0B9F   DATA  >B9F
1259  0399  0B1F   DATA  >B1F
1260  039A  0A9F   DATA  >A9F
1261  039B  0A1F   DATA  >A1F
1262  039C  099F   DATA  >99F
1263  039D  091F   DATA  >91F
1264  039E  089F   DATA  >89F
1265  039F  081F   DATA  >81F
```

```
1266  03A0  07BF        DATA  >7BF
1267  03A1  077F        DATA  >77F
1268  03A2  073F        DATA  >73F
1269  03A3  06FF        DATA  >6FF
1270  03A4  06BF        DATA  >6BF
1271  03A5  067F        DATA  >67F
1272  03A6  063F        DATA  >63F
1273  03A7  05FF        DATA  >5FF
1274  03A8  05BF        DATA  >5BF
1275  03A9  057F        DATA  >57F
1276  03AA  053F        DATA  >53F
1277  03AB  04FF        DATA  >4FF
1278  03AC  04BF        DATA  >4BF
1279  03AD  047F        DATA  >47F
1280  03AE  043F        DATA  >43F
1281  03AF  03FF        DATA  >3FF
1282  03B0  03CF        DATA  >3CF
1283  03B1  03AF        DATA  >3AF
1284  03B2  038F        DATA  >38F
1285  03B3  036F        DATA  >36F
1286  03B4  034F        DATA  >34F
1287  03B5  032F        DATA  >32F
1288  03B6  030F        DATA  >30F
1289  03B7  02EF        DATA  >2EF
1290  03B8  02CF        DATA  >2CF
1291  03B9  02AF        DATA  >2AF
1292  03BA  028F        DATA  >28F
1293  03BB  026F        DATA  >26F
1294  03BC  024F        DATA  >24F
1295  03BD  022F        DATA  >22F
1296  03BE  020F        DATA  >20F
1297  03BF  01EF        DATA  >1EF
1298  03C0  01D7        DATA  >1D7
1299  03C1  01C7        DATA  >1C7
1300  03C2  01B7        DATA  >1B7
1301  03C3  01A7        DATA  >1A7
1302  03C4  0197        DATA  >197
1303  03C5  0187        DATA  >187
1304  03C6  0177        DATA  >177
1305  03C7  0167        DATA  >167
1306  03C8  0157        DATA  >157
1307  03C9  0147        DATA  >147
1308  03CA  0137        DATA  >137
1309  03CB  0127        DATA  >127
1310  03CC  0117        DATA  >117
1311  03CD  0107        DATA  >107
1312  03CE  00F7        DATA  >F7
1313  03CF  00E7        DATA  >E7
1314  03D0  00DB        DATA  >DB
1315  03D1  00D3        DATA  >D3
1316  03D2  00CB        DATA  >CB
1317  03D3  00C3        DATA  >C3
1318  03D4  00BB        DATA  >BB
1319  03D5  00B3        DATA  >B3
1320  03D6  00AB        DATA  >AB
1321  03D7  00A3        DATA  >A3
1322  03D8  009B        DATA  >9B
```

```
1323  03D9  0093        DATA  >93
1324  03DA  008B        DATA  >8B
1325  03DB  0083        DATA  >83
1326  03DC  007B        DATA  >7B
1327  03DD  0073        DATA  >73
1328  03DE  006B        DATA  >6B
1329  03DF  0063        DATA  >63
1330  03E0  005D        DATA  >5D
1331  03E1  0059        DATA  >59
1332  03E2  0055        DATA  >55
1333  03E3  0051        DATA  >51
1334  03E4  004D        DATA  >4D
1335  03E5  0049        DATA  >49
1336  03E6  0045        DATA  >45
1337  03E7  0041        DATA  >41
1338  03E8  003D        DATA  >3D
1339  03E9  0039        DATA  >39
1340  03EA  0035        DATA  >35
1341  03EB  0031        DATA  >31
1342  03EC  002D        DATA  >2D
1343  03ED  0029        DATA  >29
1344  03EE  0025        DATA  >25
1345  03EF  0021        DATA  >21
1346  03F0  001E        DATA  >1E
1347  03F1  001C        DATA  >1C
1348  03F2  001A        DATA  >1A
1349  03F3  0018        DATA  >18
1350  03F4  0016        DATA  >16
1351  03F5  0014        DATA  >14
1352  03F6  0012        DATA  >12
1353  03F7  0010        DATA  >10
1354  03F8  000E        DATA  >E
1355  03F9  000C        DATA  >C
1356  03FA  000A        DATA  >A
1357  03FB  0008        DATA  >8
1358  03FC  0006        DATA  >6
1359  03FD  0004        DATA  >4
1360  03FE  0002        DATA  >2
1361  03FF  0000        DATA  >0
1362  0400
1363              END

NO ERRORS,  NO WARNINGS
```

15. Digital Voice Echo Canceller with a TMS32020

16. Implementation of the Data Encryption Standard Using the TMS32010

Panos Papamichalis and Jay Reimer
Digital Signal Processing - Semiconductor Group
Texas Instruments

INTRODUCTION

The programmability of the TMS320 family of digital signal processors makes possible the implementation of different signal processing algorithms on the same device, rather than on several different custom chips. As an example, this application report describes the implementation of a data encryption method on the TMS32010. The encryption scheme operates on a stream of bits that represents text, computer files, or anything else presented in binary form. In particular, the encryption method is considered in conjunction with speech coding to achieve secure communication over voice channels. Another example consists of the encryption of data supplied to a transmitting modem and decryption of the corresponding data stream coming out of the receiving modem. The algorithm chosen for the encryption is the Data Encryption Standard (DES) as extablished by the National Bureau of Standards.[1]

The following paragraph has been taken from reference [1] and is brought to the attention of the readers of this report. For more information, please contact the national Bureau of Standards.

> "EXPORT CONTROL: Cryptographic devices and technical data regarding them are subject to Federal Government export controls as specified in Title 22, Code of Federal Regulations, Parts 121 through 128. Cryptographic devices implementing [the DES] and technical data regarding them must comply with these Federal regulations."

THE DATA ENCRYPTION STANDARD (DES)

Encryption/Decryption Algorithm

The Data Encryption Standard (DES), the algorithm of data encryption selected by the United States federal government, is described in detail in the NBS publications.[1] The description is summarized here in order to associate it with the implementation on the TMS32010.

Figure 1 shows a flowchart of the enciphering portion of the algorithm. The stream of bits to be encrypted (the plaintext) is segmented into blocks of 64 bits. Each 64-bit block is submitted to an initial permutation, and then is split into two 32-bit blocks (a left one L_0 and a right one R_0), which are input to the transformation section. The transformation section consists of 16 stages, where the data is scrambled by the use of the encryption key. At each stage i, the inputs are the left block L_{i-1} and right block R_{i-1} of the previous stage, and the outputs are the left block L_i and right block R_i of this stage. The outputs L_i and R_i of each stage are computed from L_{i-1}, R_{i-1}, and a subkey K_i that is generated from the encryption key. Note that the 16th stage is different from the other stages. The output of the last stage is subjected to a permutation, which is the inverse of the initial permutation. The resulting 64 bits are the ciphertext.

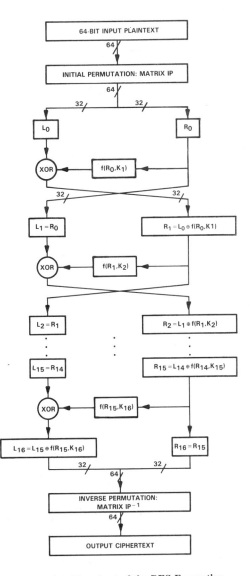

Figure 1. Flowchart of the DES Encryption

The initial permutation and inverse initial permutation matrices IP and IP^{-1} in Figure 1 are shown in Figures 2 and 3. Where there is bit manipulation according to a matrix, the matrix should be read from left to right and row by row. Each entry of the matrix corresponds to a bit in the output block. In other words, the information presented in the matrix is sequential and not two-dimensional. The matrix arrangement is used here only for convenience of presentation.

Each entry of the matrix indicates the order of the bit in the input bit stream, starting from the leftmost bit. For example, during the initial permutation, the leftmost bit after the permutation is the 58th bit from the left in the input stream; the second leftmost bit is the 50th bit, etc. If the matrix has as many entries as the number of bits of the input block, the input and output blocks from the bit manipulation have the same number of bits. If, however, the matrix has more entries (by repeating some of the bits of the input block) or fewer entries (by skipping some input-block bits), the output of the bit manipulation is longer or shorter, respectively, than the input. Such examples, shown later in this report, are matrix E that transforms 32 bits into 48 bits, and matrix PC-2 that transforms 56 bits to 48 bits.

40	8	48	16	56	24	64	32
39	7	47	15	55	23	63	31
38	6	46	14	54	22	62	30
37	5	45	13	53	21	61	29
36	4	44	12	52	20	60	28
35	3	43	11	51	19	59	27
34	2	42	10	50	18	58	26
33	1	41	9	49	17	57	25

Figure 3. Matrix IP-1

The function $f(R_{i-1}, K_i)$ of Figure 1, which combines the right block R_{i-1} of the previous stage and the subkey K_i of the present stage i, is shown in detail in Figure 4. The input to the function is a block of 32 bits, and the output is also a block of 32 bits. Figures 5, 6, and 7 give the matrices E, P, and S1 through S8 of Figure 4. K_i's are subkeys generated from the main key, as described in the next section. The S-boxes, which convert a 6-bit sequence to a 4-bit sequence, are used as follows: the first and last bits of the sequence, taken together, represent a number I between 0 and 3, while the middle 4 bits represent a number J between 0 and 15. The S-boxes are 4 x 16 matrices whose entries take values between 0 and 15. Each of these entries can be represented by a 4-bit number. For each S-box, the I and

58	50	42	34	26	18	10	2
60	52	44	36	28	20	12	4
62	54	46	38	30	22	14	6
64	56	48	40	32	24	16	8
57	49	41	33	25	17	9	1
59	51	43	35	27	19	11	3
61	53	45	37	29	21	13	5
63	55	47	39	31	23	15	7

Figure 2. Matrix IP

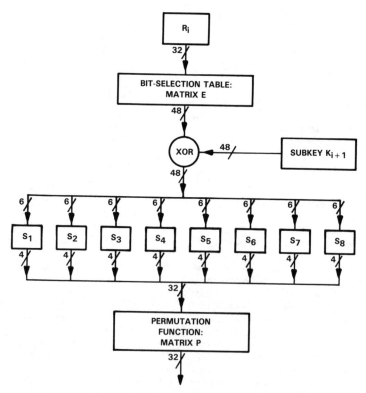

Figure 4. Computation of the Function $f(R_i, K_{i+1})$ of Figure 1

16. Implementation of the Data Encryption Standard Using the TMS32010

J are computed, and the corresponding 4-bit number in the matrix is the output. For example, if the input to S1 box is 001010, then I = 0, J = 5, and the output is the decimal 15, which corresponds to the 4-bit binary number 1111.

32	1	2	3	4	5
4	5	6	7	8	9
8	9	10	11	12	13
12	13	14	15	16	17
16	17	18	19	20	21
20	21	22	23	24	25
24	25	26	27	28	29
28	29	30	31	32	1

Figure 5. Matrix E

16	7	20	21
29	12	28	17
1	15	23	26
5	18	31	10
2	8	24	14
32	27	3	9
19	13	30	6
22	11	4	25

Figure 6. Matrix P

S1

14	4	13	1	2	15	11	8	3	10	6	12	5	9	0	7
0	15	7	4	14	2	13	1	10	6	12	11	9	5	3	8
4	1	14	8	13	6	2	11	15	12	9	7	3	10	5	0
15	12	8	2	4	9	1	7	5	11	3	14	10	0	6	13

S2

15	1	8	14	6	11	3	4	9	7	2	13	12	0	5	10
3	13	4	7	15	2	8	14	12	0	1	10	6	9	11	5
0	14	7	11	10	4	13	1	5	8	12	6	9	3	2	15
13	8	10	1	3	15	4	2	11	6	7	12	0	5	14	9

S3

10	0	9	14	6	3	15	5	1	13	12	7	11	4	2	8
13	7	0	9	3	4	6	10	2	8	5	14	12	11	15	1
13	6	4	9	8	15	3	0	11	1	2	12	5	10	14	7
1	10	13	0	6	9	8	7	4	15	14	3	11	5	2	12

S4

7	13	14	3	0	6	9	10	1	2	8	5	11	12	4	15
13	8	11	5	6	15	0	3	4	7	2	12	1	10	14	9
10	6	9	0	12	11	7	13	15	1	3	14	5	2	8	4
3	15	0	6	10	1	13	8	9	4	5	11	12	7	2	14

S5

2	12	4	1	7	10	11	6	8	5	3	15	13	0	14	9
14	11	2	12	4	7	13	1	5	0	15	10	3	9	8	6
4	2	1	11	10	13	7	8	15	9	12	5	6	3	0	14
11	8	12	7	1	14	2	13	6	15	0	9	10	4	5	3

S6

12	1	10	15	9	2	6	8	0	13	3	4	14	7	5	11
10	15	4	2	7	12	9	5	6	1	13	14	0	11	3	8
9	14	15	5	2	8	12	3	7	0	4	10	1	13	11	6
4	3	2	12	9	5	15	10	11	14	1	7	6	0	8	13

Figure 7. S Matrices

4	11	2	14	15	0	8	13	3	12	9	7	5	10	6	1
13	0	11	7	4	9	1	10	14	3	5	12	2	15	8	6
1	4	11	13	12	3	7	14	10	15	6	8	0	5	9	2
6	11	13	8	1	4	10	7	9	5	0	15	14	2	3	12

S8

13	2	8	4	6	15	11	1	10	9	3	14	5	0	12	7
1	15	13	8	10	3	7	4	12	5	6	11	0	14	9	2
7	11	4	1	9	12	14	2	0	6	10	13	15	3	5	8
2	1	14	7	4	10	8	13	15	12	9	0	3	5	6	11

Figure 7. S Matrices (continued)

For decrypting the ciphertext, the same procedure is used as for encrypting, except that the transformation block in Figure 1 is modified as shown in Figure 8. The left and right 32-bit word locations are exchanged, and the algorithm starts with subkey 16 and ends with subkey 1. The implementation for decryption is similar to encryption. The TMS32010 code (see the appendix) includes a flag to signal if encryption or decryption is performed. In decryption, the subkeys are read backwards, but everything else remains the same.

Generation of the Subkeys

The encryption key is 64 bits long, out of which only 56 bits are used for the encryption. In every group of eight bits, the eighth bit is such that the byte has odd parity. This provides some guarantee of key integrity during transmission. From the 64-bit key, 16 subkeys, each 48-bits long, are generated as shown in Figure 9. The matrices for permuted choices 1 and 2, PC-1 and PC-2, are given in Figures 10 and 11, respectively. Table 1 shows the number of left shifts at each stage. The indicated left shifts are circular left shifts for each half of keys C_i and D_i.

Table 1. Left Shifts for Subkey Generation

Iteration Number	Number of Left Shifts
1	1
2	1
3	2
4	2
5	2
6	2
7	2
8	2
9	1
10	2
11	2
12	2
13	2
14	2
15	2
16	1

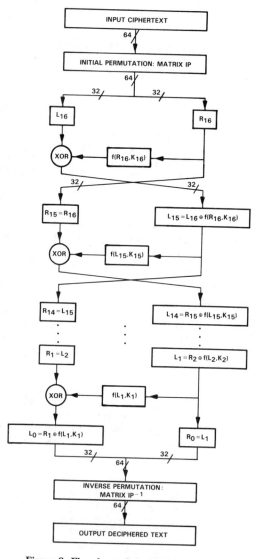

Figure 8. Flowchart of the DES Decryption

16. Implementation of the Data Encryption Standard Using the TMS32010

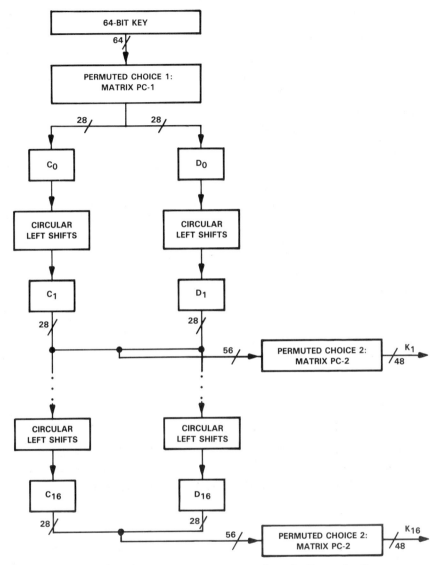

Figure 9. Generation of 16 48-Bit Subkeys from the Encryption Key

57	49	41	33	25	17	9
1	58	50	42	34	26	18
10	2	59	51	43	35	27
19	11	3	60	52	44	36
63	55	47	39	31	23	15
7	62	54	46	38	30	22
14	6	61	53	45	37	29
21	13	5	28	20	12	4

Figure 10. Matrix of PC-1

14	17	11	24	1	5
3	28	15	6	21	10
23	19	12	4	26	8
16	7	27	20	13	2
41	52	31	37	47	55
30	40	51	45	33	48
44	49	39	56	34	53
46	42	50	36	29	32

Figure 11. Matrix PC-2

16. Implementation of the Data Encryption Standard Using the TMS32010

MODES OF OPERATION

The DES algorithm can be implemented in different modes of operation[2], two of which are the Electronic CodeBook (ECB) mode and the Output FeedBack (OFB) mode. Selection of a mode determines the kind of plaintext to serve as input to the algorithm and how the ciphertext is used.

Figure 12 is another representation of Figure 1 and corresponds to the most straightforward application of the algorithm, the Electronic CodeBook (ECB) mode. The plaintext is the actual bit stream to be encrypted, and the ciphertext is what is actually transmitted over the communications medium.

Figure 13 depicts the 64-bit Output FeedBack (OFB) mode. In this case, the 64 bits of the output of the DES algorithm are exclusive-ORed with 64 bits of the actual bit stream to generate the encrypted stream for transmission. The same 64 bits of the DES output are then fed back to the input of the DES as the new input to the algorithm. The whole process is initialized by inputting a predetermined block of 64 bits into the DES algorithm. The receiver has a similar operation. As long as it has the same initial 64-bit block as input, the receiver operates synchronously with the transmitter. Note that in this mode there is no need to implement the decrypting portion of the algorithm, because both the transmitter and the receiver are using only the encrypting portion of the DES. The OFB mode has the advantage over the ECB that it confines any transmission errors to single bits, rather than a block of 64 bits.

TMS32010 IMPLEMENTATION OF THE DES

In addition to implementing the main encryption/decryption algorithm, it is necessary to process the encryption key in order to generate the subkeys used in the DES. The key consists of 64 bits, of which only 56 are active. In every byte of the key, the eighth bit is an odd-parity bit. To avoid any mistakes, the user is asked to supply the 56 bits. The device then generates the correct 64-bit version that can be used for storage and transmission. This function is realized by the routine DESKEY. The whole implementation of the Data Encryption Standard in TMS32010 assembly language can be found in the appendix. Figure 14 is a flowchart illustrating the arrangement of the TMS32010 programs that implement the DES.

The largest part of the code consists of a repetitive construction, which implements the permutation of a block of bits according to a matrix. This construction is discussed here to provide a better understanding when implementing the code.

Assume that a certain step of the algorithm requires the scrambling of 64 bits according to a matrix. The 64 input bits are stored in four words, WI1 to WI4. The scrambled-bit output is also stored in four words, WO1 to WO4. The scrambling is accomplished by constructing the

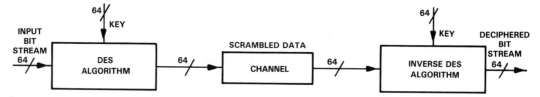

Figure 12. Electronic CodeBook (ECB) Encryption Mode

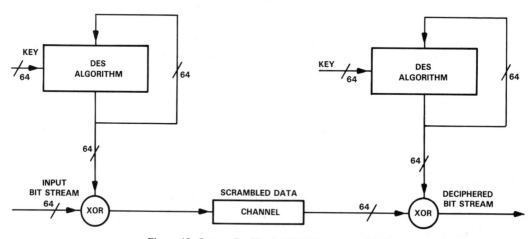

Figure 13. Output FeedBack (OFB) Encryption Mode

16. Implementation of the Data Encryption Standard Using the TMS32010

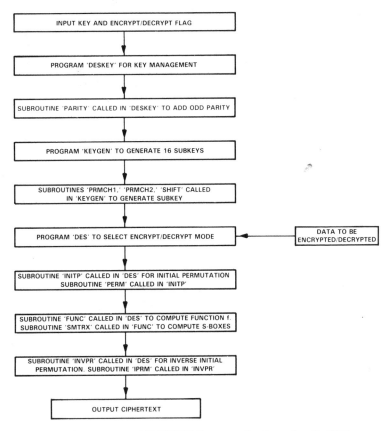

Figure 14. Flowchart of TMS32010 Programs Implementing the DES

WO1 to WO4, one bit at a time from left to right. For example, let the current entry of the matrix be the 35th entry and have the value 53. This means that WO1 and WO2 are already filled (containing $2 \times 16 = 32$ bits), and now the third bit of WO3 is being determined. Since that bit is the 53rd bit of the input stream, it occupies the fifth location in WI4 ($53 = 3 \times 16 + 5$). If M1 contains >8000 (hexadecimal value), the transfer of the bit from the input stream to the scrambled output stream is performed by the following code:

```
LAC     W14,4
AND     M1
ADDH    WO3
SACH    WO3,1
```

In other words, the bit of interest is positioned at the leftmost location of the accumulator's low word, and then masked out by M1. The current contents of WO3 are added to the accumulator's high word, which is then stored back to WO3 shifting by one. In this way, the bit of interest is also picked up.

In order to run the main algorithm, 16 subkeys, each 48-bits long, must be generated, to be used at the 16 stages of the algorithm. The algorithm for the generation of the subkeys involves permutations and shifts of the original key. The method of generating the 16 subkeys is implemented in the routine KEYGEN. For convenience, each subkey is stored in four 16-bit words so that only the top 12 bits are active in each word.

Having generated the subkeys, an incoming stream of bits can be encrypted or decrypted. The body of the DES algorithm is implemented in the routine DES. This routine implements both encryption and decryption, and the appropriate choice is flagged by a bit. In the present implementation, this bit is supplied together with the key.

Table 2 shows the program storage in words and the execution time in instruction cycles and milliseconds, as required by the DES algorithm. The information given for the routine DES is the time required per 64 input data bits. The execution time of the DES algorithm corresponds to a loading of 100 percent of the device for a bit rate of about 42.5 kbit/s. This makes the TMS32010 a faster encryption device than some of the currently available dedicated devices.[9]

Table 2. TMS32010 Memory and Speed for the DES

Function	Program Memory (words)	Execution Time (instruction cycles)	(ms)
Key Management and Subkey Generation	600	4527	0.905
Encryption	1102	7464	1.493
Decryption	(both)	7542	1.508

NOTE: The encryption/decryption execution time is an average for the half-duplex operation and for processing 64 bits, based on a 200-ns instruction cycle. The corresponding actual bit rates are:

        Encryption: 42.86 kbit/s
        Decryption: 42.42 kbit/s

The data memory required for the implementation of the DES includes 64 locations for the subkeys and another 51 locations as scratch registers. If an application running concurrently requires more data memory, the minimum memory requirement is four words, which contain the DES key. The subkeys can be regenerated at the beginning of every encrypting (or decrypting) session (but with a corresponding penalty in execution time).

Note that no attempt of optimization is made in this implementation. Because of tradeoffs between data memory, program memory, and execution speed, the designer must optimize the implementation of the algorithm to fit the needs of the application.

ENCRYPTION OF CODED SPEECH

The TMS32010 Digital Signal Processor has been applied to encoding of speech signals in a variety of approaches. Its architecture and instruction set are particularly convenient for easy implementation and fast execution of speech algorithms, such as those used in vocoders. Several algorithms have been implemented on the TMS32010 to encode speech data at 2.4, 9.6, 16, and 32 kbit/s.[3-8] These speech-encoding algorithms can be readily combined with a data encryption method to provide security both in store-and-forward applications and in telecommunications applications. In this section, a review of the requirements for the speech coding algorithms is presented to show how the DES can be incorporated in a speech coding application. Figure 15 illustrates how the DES may be implemented on the TMS32010 processor for either data or speech encryption.

Table 3 presents the different speech-coding algorithms that have been implemented on the TMS32010. Table 4 summarizes the loading of the TMS32010, when running the DES, for several bit rates corresponding to the bit rates of Table 3. Note that these loadings are for half-duplex operation, i.e., for either encryption or decryption only.

Table 3. Speech-Coding Algorithms Implemented on the TMS32010

Coding Method	Bit Rate (kbit/s)	TMS32010 Loading (percent realtime) Analysis	Synthesis
LPC	2.4	43	44
RELP	9.6	91	79
APC	16	87	35
SBC	16	38	38
ADPCM	32	47	45
ADPCM-CCITT	32	91	95

NOTE: LPC   = Linear Predicitive Coding, 10th-order model
      RELP   = Residual Excited Linear Predictive vocoder
      APC   = Adaptive Predictive Coding
      SBC   = Sub-Band Coding
      ADPCM = Adaptive Differential PCM (CCITT algorithm, not bit-by-bit compatible)
      ADPCM-CCITT = CCITT ADPCM (bit-by-bit compatible)

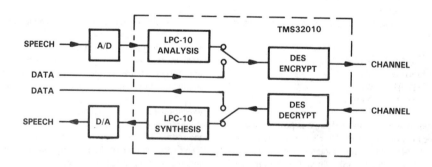

Figure 15. Speech or Data Encryption/Decryption with the TMS32010

16. Implementation of the Data Encryption Standard Using the TMS32010

Table 4. TMS32010 Loading Time for the DES /
(Half-Duplex)

Bit Rate (kbit/s)	TMS32010 Loading (percent realtime)
42.8	100
32	75
16	37
9.6	22
2.4	6

NOTE: The loading (for encryption) is given as a percentage of 42.8 kbit/s. The other bit rates were selected to match the bit rates in Table 3.

A comparison of Tables 3 and 4 shows that the 2.4-kbit/s full-duplex system can be implemented in realtime, including the encryption, in only one TMS32010. However, for the other coding schemes, it is necessary to use a second chip. The speech algorithms used have been selected without consideration for any other processing of the speech signal. It may be possible to restructure the marginal cases in order to implement the combined speech and encryption algorithms with fewer chips. Other speech coding algorithms with lighter computational load may also be considered.

CONCLUSION

The TMS32010 Digital Signal Processor can be used to implement algorithms, such as the DES, which in the past have been implemented on dedicated devices. The 200-ns cycle time of the TMS32010 makes possible a half-duplex rate of 42.5 kbit/s for the DES. This rate is superior to that achieved by some of the available devices. The single-cycle multiplication is used in the computation of the S-boxes and helps increase encryption speed. Applications of the encryption include not only speech, as discussed in this report, but any kind of digital data. For example,

implementation of a V.22 bis modem can be combined with the DES to achieve security of the transmitted information. An important point to be considered is that, having designed the TMS32010 in a system, the data encryption function can be added with minimal system cost because of the programmability of the device. This feature is a natural extension in the functionality of a digital signal processor.

REFERENCES

1. *Data Encryption Standard*, FIPS Publication 46, National Bureau of Standards (January 1977).
2. *DES Modes of Operation*, FIPS Publication 81, National Bureau of Standards (November 1981).
3. A.W. Holk and W.W. Anderson, "A Single-Processor LPC Vocoder,"*Proceedings 1984 IEEE International Conference on Acoustics, Speech and Signal Processing*, 44.13.1-4 (March 1984).
4. W.K. Gass and M.M. Arjmand, "Real-Time 9600 Bits/Sec Speech Coding on the TI Professional Computer," *Proceedings 1984 IEEE International Conference on Acoustics, Speech and Signal Processing*, 27.9.1-4 (March 1984).
5. B.S. Atal and M.R. Schroeder, "Adaptive Predictive Coding of Speech Signals," *Bell System Technology Journal, Vol 49*, 1973-1986 (October 1970).
6. T.P. Barnwell, III, R.W. Schafer, R.M. Mercereau, and D.L. Smith, "A Real Time Speech Subband Coder Using the TMS32010," *Proceedings IEEE Southcon 1984* (1984).
7. "Recommendation G.721: 32 Kbit/s Adaptive Differential Pulse Code Modulation (ADPCM)," *CCITT* (October 1984).
8. J.B. Reimer, M. McMahan, M.M. Arjmand, *32-kbit/s ADPCM with the TMS32010*, Texas Instruments (1985).
9. C.R. Abbruscato, "Data Encryption Equipment," *IEEE Communications Magazine, Vol 22*, 15-21 (September 1984).

Appendix

TMS32010 Assembly Language Programs

The TMS32010 code implementing the DES is not published in this appendix because it falls within the U.S. Department of State Export Control Regulations. If a copy of the code is desired, please contact your local TI representative.

17. 32-kbit/s ADPCM with the TMS32010

Jay Reimer
Digital Signal Processing - Semiconductor Group
Texas Instruments

Mike McMahan
Corporate Engineering Center
Texas Instruments

Masud Arjmand
Central Research Laboratories
Texas Instruments

INTRODUCTION

Digital voice communication is typically transmitted in a 64-kbit/s PCM bit stream. Voice and data communications demand increasing capacities for signal transmission without significant degradation in the quality of the transmitted signal. One of the recommended solutions for accomplishing this task is that of Adaptative Differential Pulse Code Modulation (ADPCM). This solution has been reviewed by CCITT (International Telegraph and Telephone Consultative Committee), and a specific standard* has been recommended. Two solutions, a full-duplex solution and a half-duplex solution, are discussed in this application report. Both follow the model recommended by CCITT for 32-kbit/s ADPCM, although only the half-duplex solution provides a bit-for-bit compatible data stream as required by the recommendation. At 32 kbit/s, the ADPCM solution provides double the channel capacity of the current 64-kbit/s PCM technique. Each solution has been totally incorporated in the internal memory space of the Texas Instruments TMS32010 microprocessor.

This application report presents a brief review of the basic principles of PCM and ADPCM. Hardware requirements, software logic flow, and key features of the TMS32010 microprocessor for the implementation of ADPCM are also given. Source code is provided for the implementation and creation of an ADPCM transmission channel.

DIGITIZATION

Over the past 20 years, the telecommunications industry has changed from totally analog circuits to networks which integrate both analog and digital circuits. Digital signal encoding has the advantages of greater noise immunity, efficient regeneration, easy and effective encryption, and uniformity in transmitting voice and data signals. Increased bandwidth is required to transmit digital signals while maintaining a given analog signal quality at the receiver.

Voice store and forward systems have been changing from totally analog storage media, such as audio tape, to digitized storage which allows random access of stored data, but with the tradeoff of increased storage media requirements.

Signal quality begins with the digitization of the original analog signal. The process of digitization and coding introduces a distortion associated with the quantization of the digitized signal, as shown in Figure 1. This signal distortion or noise is different from the channel noise normally associated with a transmitted signal. After a signal

*Recommendation G.721, 32 kbits/s Adaptive Differential Pulse Code Modulation'', CCITT, 1984.

has been digitized, the signal is much less susceptible to channel noise since the signal can be regenerated as well as amplified along the way, thus reducing the possibility of being corrupted by the transmission system. The overall quality of digital transmission is then limited by the digitization process in an error-free transmission system.

Figures 2 and 3 show general representations of a digital communication channel. The actual transmission (and storage of a digital waveform) uses an analog channel. The outside points of the communications channel are the transmitter and receiver, as shown in Figure 2. These are commonly combined in a single device known as a combo-CODEC (CODing and DECoding device). The codec supplies, on the coding or transmitting side, the necessary filtering to bandlimit the analog signal and avoid signal alias and A/D conversion. On the decoding or receiving side, the codec performs a D/A conversion and then interpolates or smooths the resultant signal.

Figure 3 shows the digitized signal modulated for transmission in the network and then demodulated at the receiving end to retrieve the transmitted digital signal.

PCM

Digitization and coding of the analog signal at the transmitter can be performed in several ways. The complexity of the chosen method is related to availability of encoder memory and to the resultant delay in the encoding process.

When digital signal transmission is implemented, memory and the resultant delay dictate that a simple scheme, such as Pulse Code Modulation (PCM), be implemented. PCM codes each sampled analog value of the input waveform to a unique or discrete value. The digital quantization introduces distortion into the signal waveform, as shown in Figure 1.

A nonuniform quantization scheme may be used to COMPAND (COMpress and exPAND) the signal in the waveform coding and decoding blocks in the system, generating log-PCM. By using larger quantization steps for large amplitude signals and smaller steps for small amplitude signals, efficient use is made of the data bits for digital transmission while maintaining specific signal-to-quantization noise thresholds. With the two current methods of COMPANDING (A-law and μ-law), the signal quality of a 13-bit digitized signal is maintained while transmitting only 8 bits per sample.

While quantizers remove the irrelevancy in a signal, coders remove the redundancy. In PCM encoding, each sample of the input waveform is independent of all previous samples; no encoder memory is required.

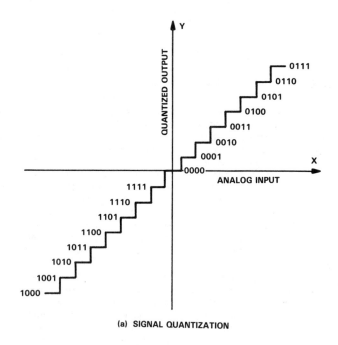

(a) SIGNAL QUANTIZATION

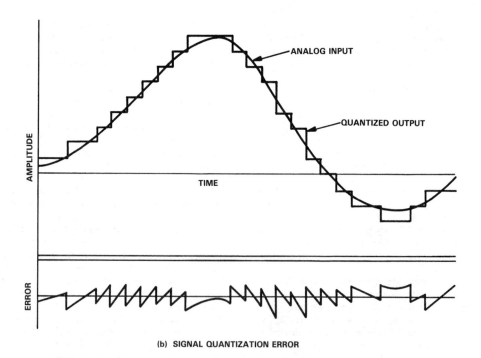

(b) SIGNAL QUANTIZATION ERROR

Figure 1. Quantization Errors in a Digitized Signal

17. 32-kbit/s ADPCM with the TMS32010

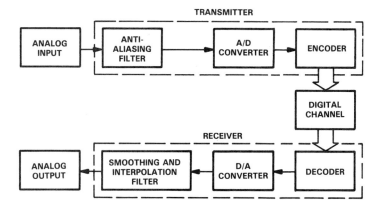

Figure 2. Digital Communication of Waveforms

Figure 3. Digital Channel

ADPCM

Analysis of speech waveforms shows a high sample-to-sample correlation. By taking advantage of this property in speech signals, more efficient coding techniques have been designed to further reduce the transmission bit rate while preserving the overall signal quality.

APCM

Adaptive PCM (APCM) is a method that may be applied to both uniform and nonuniform quantizers. It adapts the stepsize of the coder as the signal changes. This accommodates amplitude variations in a speech signal between one speaker and the next, or even between voiced and unvoiced segments of a continuous signal. The adaptation may be instantaneous, taking place every few samples. Alternatively, it may occur over a longer period of time, taking advantage of more slowly varying features. This is known as syllabic adaptation.

The basic concept for an adaptive feedback system, APCM, is shown in Figure 4. An input signal, $s(k)$, in the transmitter is quantized and coded to an output, $I(k)$. This output is also processed by stepsize adaptation logic to create a signal, $q(k)$, that adapts the stepsize in the quantizer. Correspondingly, in the receiver, the received signal, $I(k)$, is processed by an inverse quantizer (i.e., decoded), producing the reconstructed signal, $s_r(k)$. Like the transmitter, the quantized signal, $I(k)$, is processed by adaptation logic to create a stepsize control signal, $q(k)$, for the inverse quantizer.

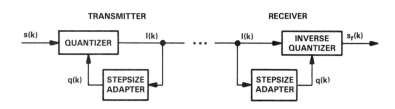

Figure 4. APCM Block Diagram

17. 32-kbit/s ADPCM with the TMS32010

DPCM

The method of using the sample-to-sample redundancies in the signal is known as differential PCM (DPCM). The overall level of high correlation on a sample-by-sample basis indicates that the difference between adjacent samples produces a waveform with a much lower dynamic range. Correspondingly, an even lower variance can be expected between samples in the difference signal. A signal with a smaller dynamic range may be quantized to a specific signal-to-noise ratio with fewer bits.

A differential PCM system, DPCM, is shown in Figure 5. In Figure 5, the signal difference, $d(k)$, is determined using a signal estimate, $s_e(k)$, rather than the actual previous sample. By using a signal estimate, $s_e(k)$, the transmitter uses the same information available to the receiver. Each successive coding actually compensates for the quantization error in the previous coding. In this way, the reconstructed signal, $s_r(k)$, can be prevented from drifting from the input signal, $s(k)$, as a result of an accumulation of quantization errors. The reconstructed signal, $s_r(k)$, is formed by adding the quantized difference signal, $d_q(k)$, to the previous signal estimate, $s_e(k)$. The sum is the input to predictor logic which determines the next signal estimate. A decoding process is used in both the transmitter and receiver to determine the quantized difference signal, $d_q(k)$, from the transmitted signal, $I(k)$.

ADPCM

ADPCM combines the features of both the APCM and DPCM systems. Figure 6 shows the basic blocks combining adaptation and differencing features in an ADPCM system.

Both quantizer adaptation and signal differencing require the storage (in memory) of one or more samples in both the transmitter and receiver. Furthermore, the transmitter must use some method to ensure that the receiver is operating synchronously. This is accomplished by using only the transmitted signal, $I(k)$, to determine stepsize adaptation in the quantizer and inverse quantizer and to predict the next signal estimate. In this way, the blocks in the receiver can be identical to those in the transmitter. Additionally, the specific adaptation techniques are designed to be convergent and thereby help provide quick recovery following transmission errors.

The ADPCM system, as used in digital telephony, is not an original signal coding system, but is actually a transcoder, converting between log-PCM and ADPCM codes. Currently there are a large number of systems using log-PCM for transmission. The ADPCM system incorporates both an adaptive quantizer and an adaptive predictor. The adaptive quantizer contains speed-control and scale-factor adaptation. A measure of the rate-of-change of the difference signal provides a means of determining the speed control. The scale factor adjustments to the difference signal adapt the fit of the quantization levels to minimize the signal-to-noise ratio. With speed control, the system can take advantage of both the instantaneous and syllabic adaptation rates, thereby adapting better to both speech and data signals. In the adaptive predictor, the prediction filter coefficients are updated by a gradient algorithm. Predictor adaptation improves the performance of the predictor for nonstationary signals (e.g., speech).

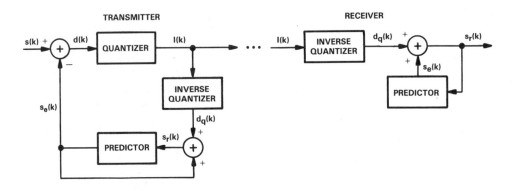

Figure 5. DPCM Block Diagram

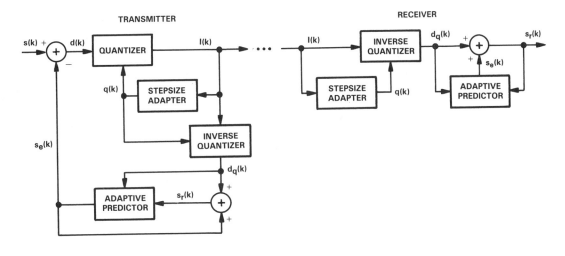

Figure 6. ADPCM Block Diagram

THE ADPCM ALGORITHM

The ADPCM algorithm has a receiver imbedded in the transmitter. This is important since, if the signal feedback used to determine the signal estimate, $s_e(k)$, and consequently the quantized difference signal, $d_q(k)$, is the same as in the decoder, then the compensation for quantization errors can be made with subsequent difference samples. Since the decoder is actually imbedded in the encoder, each of the common blocks for transmitting and receiving is discussed in the following paragraphs.

Figures 7 and 8 show block diagrams of an ADPCM transmitter and receiver as specified by CCITT.

Encoder

The function of the encoder or transmitter, shown in Figure 7, is to receive a 64-kbit/s log-PCM signal and transcode it to a 32-kbit/s ADPCM signal. This is accomplished by converting the log-PCM signal, $s(k)$, to a linear signal, $s_l(k)$, from which an estimate, $s_e(k)$, of the signal is subtracted to obtain a difference signal, $d(k)$. The next step is to adaptively quantize this difference signal, $d(k)$, by first taking the log (base 2), then normalizing by the quantization scale factor, $y(k)$, and finally coding the result, $I(k)$. A more uniform signal-to-noise ratio can be achieved by coding the log of the signal rather than the linear representation. The normalization provides the adaptation to the quantization and is based on past coded samples. Adaptation is controlled bimodally, being comprised of a fast adaptation factor for signals with large amplitude fluctuations (i.e., speech) and a slow adaptation factor for signals which vary more slowly (i.e., data). A speed-control factor, $a_l(k)$, weights the fast and the slow adaptation factors to form a single quantization scale factor, $y(k)$.

The inverse adaptive quantizer uses the same signal, $I(k)$, that has been transmitted to reconstruct a quantized version of the difference, $d_q(k)$, and the same adaptive quantization characteristics as the adaptive quantizer section.

The quantized difference signal, $d_q(k)$, is input to an adaptive predictor which uses this input to compute a signal estimate, $s_e(k)$. The signal estimate, $s_e(k)$, is combined with the difference signal, $d_q(k)$, to determine a reconstructed signal, $s_r(k)$, which is the output in the decoder. This output is then subtracted from the next input sample to complete the feedback loop.

The adaptive predictor makes use of both an all-pole filter and an all-zero filter. The all-pole filter is a second-order filter with constrained adaptive coefficient values designed to match the slowly varying aspects of the speech signal. Since an all-pole predictor is particularly sensitive to errors, the predictor makes use of a sixth-order all-zero filter to offer signal stability even with transmission errors.

Decoder

The function of the decoder or receiver, shown in Figure 8, is to receive a 32-kbit/s ADPCM signal and transcode it to a 64-kbit/s log-PCM signal. To accomplish this, the decoder utilizes many of the elements used by the encoder. The received data, $I(k)$, is processed by an inverse adaptive quantizer, identical to the one in the corresponding encoder, to determine a quantized difference signal, $d_q(k)$. By filtering the difference signal, $d_q(k)$, through the adaptive predictor together with the previously reconstructed signal, $s_r(k)$, a signal estimate, $s_e(k)$, is obtained. The signal estimate, $s_e(k)$, is added to the difference signal, $d_q(k)$, to compute the reconstructed signal, $s_r(k)$. The reconstructed signal, $s_r(k)$, is converted from a linear-PCM to a log-PCM signal, $s_p(k)$, which is then output following a synchronous

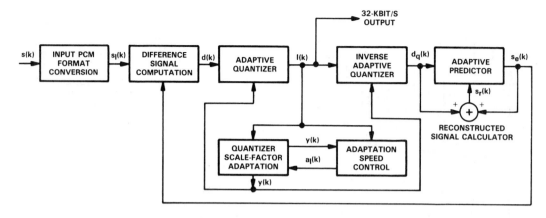

Figure 7. ADPCM Encoder Block Diagram
(Diagram taken from CCITT Recommendation G.721)

coding adjustment. The coding adjustment limits the errors in tandem codings of a signal.

Note that the algorithm design achieves a convergence of the states of the encoder and decoder in spite of transmission errors. This convergence is a part of each of the adaptation computations and is demonstrated equationally in the following sections. The convergence is brought about by the inclusion of $(1-2^{-N})$ terms which provide a finiteness to the memory of the adaptation parameters.

Adaptive Quantization

Adaptive quantization, a multistage process, is used to determine the quantization scale factor and the speed control that controls the rate at which the scale factor is adapted. Quantization is actually a four-bit quantization (a sign bit plus three-bit magnitude), since a four-bit signal is the transmitted output of the ADPCM transcoder. The adaptive quantizer block can be noted in Figure 7.

The difference signal, $d(k)$, an input to the quantization process, is calculated by subtracting the signal estimate, $s_e(k)$, from the linear-PCM signal, $s_l(k)$.

$$d(k) = s_l(k) - s_e(k) \tag{1}$$

This difference signal is normalized by taking the log (base 2) and subtracting from it the quantizer scale factor, $y(k)$.

$$|I(k)| \leftarrow \log_2 |d(k)| - y(k) \tag{2}$$

Table 1 is used to provide the magnitude of the quantization result, $|I(k)|$, from this normalized input. The

sign bit of the ADPCM output value, $I(k)$, is the sign of the difference signal, $d(k)$.

The quantizer scale factor, $y(k)$, is comprised of two parts, and therefore bimodal in nature. The two parts, $y_l(k)$ and $y_u(k)$, are weighted by the speed-control factor, $a_l(k)$. For speech signals, $a_l(k)$ will tend toward a value of one; for voiceband data, $a_l(k)$ will tend toward zero. Refer to both Figures 7 and 8 for the inclusion of the quantizer scale factor and speed-control factor adaptation blocks.

$$y(k) = a_l(k)y_u(k-1) + [1 - a_l(k)] y_l(k-1) \tag{3}$$

where $0 \leq a_l(k) \leq 1$

One of the factors, $y_u(k)$, is considered to be unlocked, since it can adapt quickly to rapidly changing signals (e.g., speech) and has a relatively short-term memory. This factor, $y_u(k)$, is recursively determined from the quantizer factor, $y(k)$, and the discrete function, $W(I)$.

$$y_u(k) = [1 - 2^{-5}] y(k) + 2^{-5}W[I(k)] \tag{4}$$

where $1.06 \leq y_u(k) \leq 10.00$

The factor, $W(I)$, found in Table 2, is a function of I which causes $y_u(k)$ to adapt by larger steps for larger values of I. This gives $y_u(k)$ the freedom to track a signal almost instantaneously. Since $y(k)$ is in the logarithmic domain, $W(I)$ is effectively a multiplier of the scale factor.

476

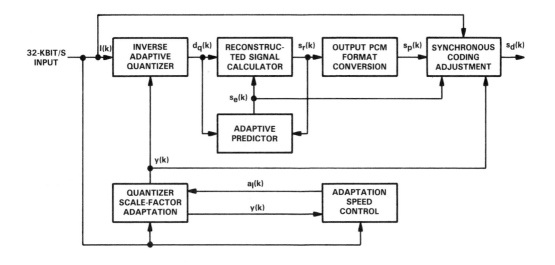

Figure 8. ADPCM Decoder Block Diagram
(Diagram taken from CCITT Recommendation G.721)

Table 1. I/O Characteristics of the Normalized Quantizer

Normalized Quantizer Input Range $\log_2 \|d(k)\| - y(k)$	$\|I(k)\|$	Normalized Quantizer Output $\log_2 \|d_q(k)\| - y(k)$
[3.16, $+\infty$)	7	3.34
[2.78, 3.16)	6	2.95
[2.42, 2.78)	5	2.59
[2.04, 2.42)	4	2.23
[1.58, 2.04)	3	1.81
[0.96, 1.58)	2	1.29
[-0.05, 0.96)	1	0.53
($-\infty$, -0.05)	0	-1.05

The other factor, $y_l(k)$, adapts more slowly and tracks signals which change slowly (e.g., voiceband data). This factor includes a lowpass filtering of the unlocked factor, $y_u(k)$. By including $y_u(k)$ in the manner shown, $y_l(k)$ is implicitly limited to the same range of values as the explicit limit placed on $y_u(k)$. Furthermore, the unity limit of $a_l(k)$ provides the same limit implicitly for $y(k)$ as for $y_l(k)$ and $y_u(k)$.

$$y_l(k) = [1 - 2^{-6}] y_l(k-1) + 2^{-6} y_u(k) \qquad (5)$$

A speed-control factor, $a_l(k)$, adjusts the relative weighting of these two scale factors by making use of the short- and long-term averages, $d_{ms}(k)$ and $d_{ml}(k)$, respectively, of the coded output to determine how rapidly the signal is changing. The combined scale factor, $y(k)$, cannot be larger than either the unlocked, $y_u(k)$, or locked $y_l(k)$, terms. Therefore, $a_l(k)$ is limited to one even if the predicted speed control, $a_p(k)$, is larger than one.

$$a_l(k) = \begin{cases} 1 & \text{,if } a_p(k-1) > 1 \\ a_p(k-1) & \text{,if } a_p(k-1) \le 1 \end{cases} \qquad (6)$$

Note that $a_p(k)$ is implicitly limited to a maximum value of 2, while the speed-control factor used to mix the two scale factors is capped at a value of 1. In determining $a_p(k)$, an additional term of 1/8 is added each time if the difference in the short- and long-term averages becomes too large (i.e., $\| d_{ms}(k) - d_{ml}(k) \| \ge 2^{-3} d_{ml}(k)$) or if there is an idle channel (i.e., $y(k) < 3$). Where neither of these conditions exist, a uniform, slowly varying signal can be assumed, such as occurs in data transmission.

Table 2. Scale-Factor Multipliers

$\|I\|$	7	6	5	4	3	2	1	0
W(I)	69.25	21.25	11.50	6.12	3.12	1.69	0.25	-0.75

$$a_p(k) = \begin{cases} [1 - 2^{-4}] \, a_p(k-1) + 2^{-3}, \text{ if} \\ \quad |d_{ms}(k) - d_{ml}(k)| \geq 2^{-3}d_{ml}(k) \\ [1 - 2^{-4}] \, a_p(k-1) + 2^{-3}, \text{ if } y(k) < 3 \\ [1 - 2^{-4}] \, a_p(k-1), \text{ otherwise} \end{cases} \quad (7)$$

The short-, $d_{ms}(k)$, and long-term, $d_{ml}(k)$, averages of the transmitted ADPCM signal, $I(k)$, are actually determined by averaging a weighted function, $F(I)$, of the transmitted I, shown in Table 3.

$$d_{ms}(k) = [1 - 2^{-5}] \, d_{ms}(k-1) + 2^{-5}F \, [I(k)] \quad (8)$$

$$d_{ml}(k) = [1 - 2^{-7}] \, d_{ml}(k-1) + 2^{-7}F \, [I(k)] \quad (9)$$

The scale-factor and speed-control adaptations are a part of both the encoder and decoder logic. The adaptive quantization block has been specifically included in Figure 7, showing the encoder. For the decoder, the adaptive quantizer is included as part of the synchronization block to aid in the reduction of errors in tandem codings.

Table 3. Rate-of-Change Weighting Function

|I|	7	6	5	4	3	2	1	0
F(I)	7	3	1	1	1	0	0	0

Inverse Adaptive Quantization

Inverse adaptive quantization is a process in which the four-bit ADPCM signal, $I(k)$, is used to determine the normalized log of the difference signal from Table 1. The result is actually a quantized version of the difference signal, $d_q(k)$, determined by adding the scale factor, $y(k)$, to the value specified by Table 1 and calculating, the inverse log (base 2) of this sum.

$$d_q(k) = \log_2^{-1} [\{\log_2 |d_q(k)| - y(k)\} + y(k)] \quad (10)$$

For both the encoder and decoder, this quantized difference signal is the input to the reconstruction signal calculator and the adaptive predictor, as shown in Figures 7 and 8.

Adaptive Prediction

The adaptive predictive filter is a two-pole, six-zero filter used to determine the signal estimate. The combination of both poles and zeroes allows the filter to model more effectively any general input signal. The sixth-order all-zero section helps to stabilize the filter and prevent it from drifting into oscillation. For both the poles and the zeroes, the coefficients, $a_i(k)$ and $b_i(k)$, respectively, are adapted. This adaptation is based upon a gradient algorithm to further adjust the filter model to the input signal. Figures 9 and 10 show the sixth-order and second-order filters, respectively.

The signal estimate, $s_e(k)$, represents the sum of the all-pole filter and the all-zero filter. Since the sum of the all-zero filter is used to aid the determination of the pole coefficients, it is also extracted as a separate sum, $s_{ez}(k)$. The reconstructed signal, the output in the receiver, is the sum determined by the quantized difference signal $d_q(k)$, and the signal estimate, $s_e(k)$.

$$s_e(k) = \sum_{i=1}^{2} a_i(k-1)s_r(k-i) + s_{ez}(k) \quad (11)$$

$$s_{ez}(k) = \sum_{i=1}^{6} b_i(k-1)d_q(k-i) \quad (12)$$

$$s_r(k-i) = s_e(k-i) + d_q(k-i) \quad (13)$$

The adaptation of the pole coefficients, $a_i(k)$, is shown in the equations below. The gradient function is determined from a signal, $p(k)$, that is equivalent to the reconstructed signal minus the contribution of the pole filter output. Stability of the filter is further provided by explicitly limiting the coefficients.

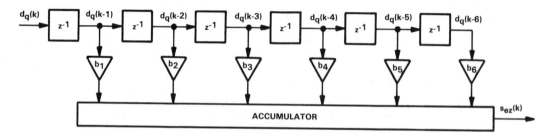

Figure 9. Sixth-Order All-Zero (FIR) Filter

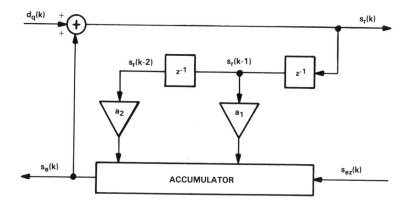

Figure 10. Second-Order IIR Filter

$a_1(k) = [1 - 2^{-8}]\, a_1(k-1)$
$\quad + 3 \cdot 2^{-8} \mathrm{sgn}\ [p(k)]\ \mathrm{sgn}[p(k-1)]$ $\qquad$ (14)

where $|a_1(k)| \leq 1 - 2^{-4} - a_2(k)$

$a_2(k) = [1 - 2^{-7}]\, a_2(k-1)$
$\quad + 2^{-7}\{\mathrm{sgn}\ [p(k)]\ \mathrm{sgn}\ [p(k-2]$
$\quad - f[a_1(k-1)]\ \mathrm{sgn}\ [p(k)]\ \mathrm{sgn}\ [p(k-1]\}$ $\qquad$ (15)

where $|a_2(k)| \leq 0.75$

$p(k) = d_q(k) + s_{ez}(k)$ $\qquad$ (16)

$$f(a_1) = \begin{cases} 4a_1 & , \text{if } |a_1| \leq 1/2 \\ 2\mathrm{sgn}(a_1), & \text{if } |a_1| > 1/2 \end{cases}$$ $\qquad$ (17)

where $\mathrm{sgn}(0) = +1$

For the coefficients, $b_i(k)$, of the sixth-order all-zero filter, the adaptation procedure is similar, but the limit is implicit in the equations to a maximum of ± 2. The gradient function, in this case, is determined by the current difference signal, $d_q(k)$, and corresponding difference signal, $d_q(k-i)$, at the specific filter tap.

$b_i(k) = [1 - 2^{-8}]\, b_i(k-1)$
$\quad + 2^{-7}\mathrm{sgn}\ [d_q(k)]\ \mathrm{sgn}\ [d_q(k-i)]$ $\qquad$ (18)

where $i = 1,2, .. 6$ and $-2 \leq b_i(k) \leq +2$

Signal Conversion

Signal conversion consists of the conversion from an 8-bit log-PCM representation of a signal to a 13-bit linear PCM representation (note Figure 7), or the reverse (note Figure 8). Signal conversions of this type are described in the application report on COMPANDING ROUTINES FOR THE TMS32010. In the encoder, the log-PCM signal, $s(k)$, is expanded to create the linear-PCM value, $s_l(k)$. The decoder, on the other hand, compresses the reconstructed signal, $s_r(k)$, to create the log-PCM signal, $s_p(k)$.

Reconstructed Signal Synchronization

To avoid a cumulative distortion in synchronous tandem codings, an adjustment to the reconstructed signal is specified. The adjustment block, shown in Figure 8, estimates the quantization of the encoder by determining a difference signal and executing the adaptive quantization logic. The quantization result is an estimate of the received value of $I(k)$.

The difference signal, $d_x(k)$, is determined by subtracting the signal estimate, $s_e(k)$, from the linear-PCM signal, $s_{lx}(k)$, which is itself determined by expanding the log-PCM signal, $s_p(k)$.

$d_x(k) = s_{lx}(k) - s_e(k)$ $\qquad$ (19)

The adaptive quantization process produces the estimate of the ADPCM code value, $I_d(k)$. If the estimate implies a difference signal that is lower than the received interval boundary, the log-PCM code is changed to the next most positive value. An estimate implying a difference signal larger than the received interval boundary requires the log-PCM code to be changed to the next most negative value; otherwise, the log-PCM value is left unchanged. The adjusted log-PCM value is denoted as $s_d(k)$ in the following equation to differentiate it from the input value, $s_p(k)$.

$$s_d(k) = \begin{cases} s_p^+(k), \; d_x(k) < \text{lower interval boundary} \\ s_p^-(k), \; d_x(k) \geq \text{upper interval boundary} \\ s_p(k), \; \text{otherwise} \end{cases} \quad (20)$$

where

$s_d(k)$ = output PCM of the decoder

$s_p^+(k)$ = next more positive PCM level (if $s_p(k)$ is the most positive level, then $s_p^+(k) = s_p(k)$)

$s_p^-(k)$ = next more negative PCM level (if $s_p(k)$ is the most negative level, then $s_p^-(k) = s_p(k)$)

FULL-DUPLEX IMPLEMENTATION OF ADPCM ON A TMS32010

The specific implementation of ADPCM presented here involves the use of a single TMS320M10 to accomplish a full-duplex transcoder. The TMS320M10 is a masked ROM, microcomputer version of the TMS32010, which requires no external program memories. A full-duplex transcoder provides transmission in both directions simultaneously. Such a transcoder is depicted in Figure 11. A complete system diagram of a full-duplex communications channel is shown in Figure 12. In comparison to current systems that modulate a 64-kbit/s A-law or μ-law PCM signal on a carrier for transmission, the described system transcodes the 64-kbit/s code to a 32-kbit/s code. This 32-kbit/s code, which requires correspondingly less bandwidth, is modulated on the carrier for transmission.

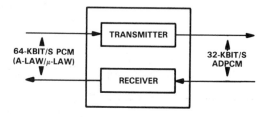

Figure 11. Full-Duplex ADPCM Transcoder

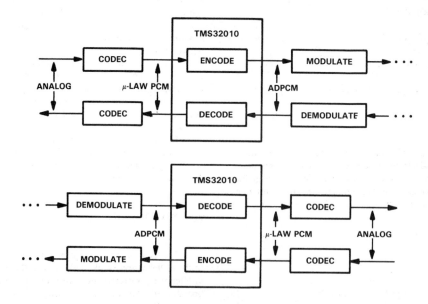

Figure 12. Full-Duplex Telecommunications Channel

17. 32-kbit/s ADPCM with the TMS32010

Hardware Logic and I/O

The hardware required to implement the ADPCM system consists of an addition to an existing circuit. As shown in Figure 13, the TMS32010 addresses the external I/O blocks through its port addressing structure. The lower three address lines, A2-A0, form a port address that can be decoded by port decode logic to provide specific enable lines (e.g., $\overline{\text{WRTEN1}}$ and $\overline{\text{RDEN1}}$) to the various peripheral blocks. The TMS32010 reads and writes the 64-kbit/s data through the codec interface eight bits at a time. The sampling frequency is 8 kHz. For this full-duplex implementation, one sample is written and one sample is read every 125 μs.

Figure 13 also shows the serial interface to the codec that provides the μ-law companded PCM data, although this is not part of the transcoding system itself. The log-PCM signal may already be available (e.g., in existing digital telecom networks) and, as such, may be interfaced to the TMS32010 either directly as parallel data or serially through conversion logic. Parallel codecs are also becoming available to reduce the hardware logic and interface required for those systems which do not already include a codec. The TMS32010 is available at crystal and clock input rates of 20.5 MHz which may be divided down to provide the codec timing and further reduce the logic requirement.

At the other end of the transcoder function, the TMS32010 reads and writes the 32-kbit/s ADPCM data through the ADPCM interface four bits at a time for each 125-μs period. This interface provides four-bit parallel data which may be serialized, if required, for transmission or storage.

Software Logic and Flow

Tables 4 and 5 list the various blocks in the algorithm, directly relating them to Figures 7 and 8 by the signal names given in the description and function. The blocks are listed in the order in which they are executed. Also listed is processor demand or loading which consists of the amount of program memory used to implement the given function and the number of instruction cycles executed in worst case. There are more blocks in the table than are shown in the figures (e.g., the algorithm uses the adaptive predictor at one point to produce the signal estimate, and later returns to update or adapt the predictor coefficients). Each block has been implemented using the equations given in previous sections concerning the ADPCM algorithm. For convenience, the equations implemented in each block are listed in the description section for the block. A more detailed description of the TMS32010 implementation is given in the next section.

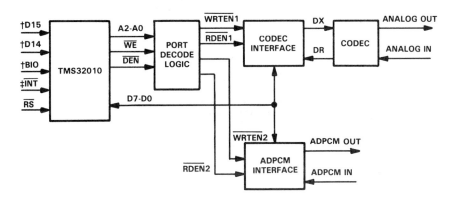

† Half-duplex, CCITT bit-compatible, version only
‡ Full-duplex version only

Figure 13. System Interface of a TMS32010 ADPCM Transcoder

Table 4. Full-Duplex Transmitter

Order	Function	Description	CPU Clocks	Program Memory (Words)
1.	INPUT PCM	Read an 8 bit μ-law PCM sample [s(k)] and linearize it to a 12-bit sample [s_l(k)].	7	0004
2.	COMPUTE SIGNAL ESTIMATE	Calculate the signal estimate [s_e(k)] from the previous data samples [d_q(k)] and reconstructed samples [s_r(k)] through the predictor filter. (12),(11)	30	001E
3.	COMPUTE ADAPTIVE QUANTIZER	Calculate speed control [a_l(k)] and quantizer scale factor [y(k)] from past quantizer output [I(k)]. (6),(3)	33	0021
4.	COMPUTE DIFFERENCE SIGNAL	Calculate the difference signal [d(k)] from the current sample [s_l(k)] and signal estimate [s_e(k)]. (1)	3	0003
5.	COMPUTE QUANTIZED OUTPUT	Calculate the log of the difference signal [d(k)] and adaptively quantize the result to yield the ADPCM output [I(K)]. (2)	46	00AD
6.	OUTPUT ADPCM	Write the ADPCM output [I(k)].	2	0001
7.	COMPUTE RECONSTRUCTED SIGNAL	Calculate the inverse of the adaptively quantized signal [d_q(k)] and the reconstructed signal difference [s_r(k)]. (10),(13)	43	0027
8.	COMPUTE SCALE FACTOR	Calculate the updates for the scale-factor adaptation. (4),(5)	46	002F
9.	COMPUTE SPEED CONTROL	Calculate the update for the speed-control adaptation. (8),(9),(7)	30	001B
10.	COMPUTE PREDICTOR ADAPTATION	Calculate the updates for the adaptive predictor filter coefficients. (18),(16),(17),(14),(15)	102	006B

17. 32-kbit/s ADPCM with the TMS32010

Table 5. Full-Duplex Receiver

Order	Function	Description	CPU Clocks	Program Memory (Words)
1.	INPUT ADPCM	Read the ADPCM input [I(k)].	2	0001
2.	COMPUTE SIGNAL ESTIMATE	Calculate the signal estimate [$s_e(k)$] from the previous data samples [$d_q(k)$] and reconstructed samples [$s_r(k)$] through the predictor filter. (12),(11)	30	001E
3.	COMPUTE ADAPTIVE QUANTIZER	Calculate speed control [$a_l(k)$] and quantizer scale factor [y(k)] from the past quantizer output [I(k)]. (6),(3)	33	0021
4.	COMPUTE QUANTIZED DIFFERENCE	Calculate the inverse of the adaptively quantized signal [$d_q(k)$]. (10)	47	002F
5.	COMPUTE SCALE FACTOR	Calculate the updates for the scale-factor adaptation. (4),(5)	48	002F
6.	COMPUTE SPEED CONTROL	Calculate the update for the speed-control adaptation. (8),(9),(7)	29	001B
7.	COMPUTE RECONSTRUCTED SIGNAL	Calculate the reconstructed signal [$s_r(k)$]. (13)	3	0003
8.	COMPUTE PREDICTOR ADAPTATION	Calculate the updates for the adaptive predictor filter coefficients. (18),(16),(17),(14),(15)	90	006B
9.	COMPUTE LOG-PCM	Convert the reconstructed linear-PCM signal [$s_r(k)$] to a μ-law PCM signal [$s_p(k)$].	39	0074
10.	OUTPUT PCM	Write the μ-law output [s_p)k)].	2	0001
11.	WAIT	Spin until the next interrupt.	—	0006

Implementation and Advantages of TMS32010 Architecture

This implementation is only concerned with μ-law PCM, although A-law PCM may also be used. Additional information on log-PCM companding is found in an application report, COMPANDING ROUTINES FOR THE TMS32010. The implementation is simplified here so that the expansion is a simple table lookup which saves 21 instruction cycles over the algorithmic approach.

The processing of the signal through the predictor filter is similar to the processing discussed in the application report, IMPLEMENTATION OF FIR/IIR FILTERS WITH THE TMS32010. The filter used in this ADPCM algorithm is a combination of a second-order IIR filter and a sixth-order

all-zero or FIR filter. The filters are shown in Figures 9 and 10, respectively, with the system interaction shown in Figure 14.

Several manipulations of data format occur in adapting the predictor coefficients. In updating the coefficients of the all-zero filter (the B_i's), the coefficients that are normally Q14 numbers are loaded with a shift allowing the calculations to be done in a Q29 representation. This greatly simplifies the subtraction of the leakage term and the prediction gain. The leakage term, which occurs here in the predictor coefficient adaptation and also in the speed-control and scale-factor adaptation, controls the rate of change of the parameter away from zero and towards the absolute maximum limits of the particular parameter. The prediction gain also uses

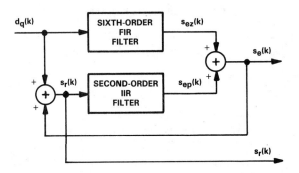

Figure 14. Predictor Filter

an approach whereby the signs are actually stored as a signed Q11 value. In this way, the product is a Q22 value of the correct sign and can be added to the B value, equivalent to a Q29 value times 2^{-7}. As with the filter process itself, the signs of the Dq values are propagated through each filter tap delay with the LTD instruction. An example for one of the B_i values is shown in Figure 15.

A similar process takes place in adapting the prediction coefficients (A_i's) in the second-order filter, although the fixed-point representation of the coefficients is Q26. The remaining requirement is to limit-check the A_i values.

The adaptive quantization section requires that the log (base 2) of the difference signal be taken, the result normalized, and the normalized value quantized. Taking the log (base 2) of a number is accomplished by using the approximation

$$\log_2(1 + x) = x \qquad (21)$$

```
* ; * * * * * * * * * * * * * * * * * * * * * * * * * * * * * * * * * * * * * * * * * * * * * * * * * * *
* ;
* ;   COMPUTE COEFFICIENTS OF THE 6TH-ORDER PREDICTOR
* ;
* ;   Bi(k) = [1 - 2**-8] * Bi(k-1)
* ;           + 2**-7 * SGN[DQ(k)] * SGN[DQ(k-i)]
* ;
* ;         FOR  i = 1 ... 6
* ;         AND Bi IS IMPLICITLY LIMITED TO +/- 2
* ;
* ;         NOTATION:   Bn   -- 16b TC (Q14)
* ;                     SDQn -- +2048 IF DQn POSITIVE (Q11)
* ;                             -2048 IF DQn NEGATIVE (Q11)
* ;
* ; * * * * * * * * * * * * * * * * * * * * * * * * * * * * * * * * * * * * * * * * * * * * * * * * * * *
* ;
GETB6  LT   SDQ6     * (Q11)
       LAC  B6,15    * (Q29)
       SUB  B6,7     * B6 * 2**-8   (Q29)
       MPY  SDQ      * SGN(SDQ)*SGN(SDQ6)*2**-7 (Q29)
       LTD  SDQ5     * (Q11)
       SACH B6,1     * (Q14)
        .
        .
        .
```

Figure 15. Predictor Coefficient Adaptation Code

The characteristic of the result is the bit position of the most significant one digit in the absolute value. The result can be represented as a Q7 value. Finding the most significant digit is most efficiently done by a binary search technique. This technique is discussed in the application report, FLOATING-POINT ARITHMETIC WITH THE TMS32010. Since the exponent is part of the number instead of being stored in a separate register, one of the auxiliary registers is loaded with the exponent value. The auxiliary register stores it in memory and adds it to the mantissa in the accumulator. A short example of this is shown in the excerpted code in Figure 16 where the signal has an assumed exponent value of 9.

Normalization of this log value is simply a subtraction of a scale factor which may be as large in fixed magnitude as the largest logarithmic value represented in Q7 notation. The result of the subtraction may be a negative value. Since the normalized result is to be quantized in a nonuniform manner and one of the quantization levels could contain both positive and negative values, the normalized result is scaled by adding a fixed value of 2048. Nonuniform quantization can be performed by a binary-type search technique. The normalization and quantization are included in the program shown in Figure 16.

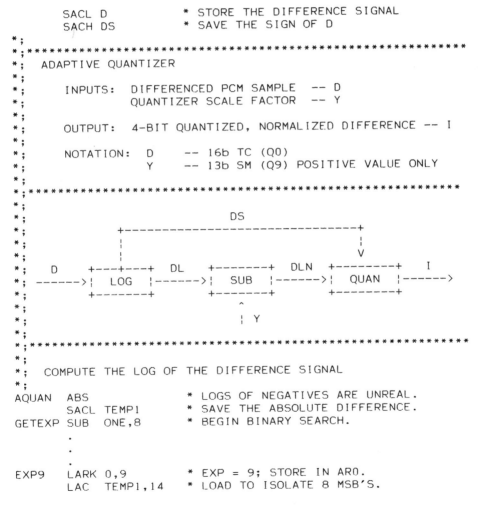

```
          SACL  D              * STORE THE DIFFERENCE SIGNAL
          SACH  DS             * SAVE THE SIGN OF D
* ;
* ;*********************************************************************
* ;
* ;   ADAPTIVE QUANTIZER
* ;
* ;      INPUTS:  DIFFERENCED PCM SAMPLE   -- D
* ;               QUANTIZER SCALE FACTOR   -- Y
* ;
* ;      OUTPUT:  4-BIT QUANTIZED, NORMALIZED DIFFERENCE -- I
* ;
* ;      NOTATION:  D   -- 16b TC (Q0)
* ;                 Y   -- 13b SM (Q9) POSITIVE VALUE ONLY
* ;
* ;*********************************************************************
* ;
* ;                                DS
* ;           +---------------------------------+
* ;           |                                 |
* ;           |                                 V
* ;   D   +---+---+  DL  +-------+  DLN +--------+   I
* ; ----->| LOG   |----->|  SUB  |----->|  QUAN  |----->
* ;       +-------+      +-------+      +--------+
* ;                          ^
* ;                          | Y
* ;
* ;*********************************************************************
* ;
* ;   COMPUTE THE LOG OF THE DIFFERENCE SIGNAL
* ;
AQUAN     ABS              * LOGS OF NEGATIVES ARE UNREAL.
          SACL  TEMP1      * SAVE THE ABSOLUTE DIFFERENCE.
GETEXP    SUB   ONE,8      * BEGIN BINARY SEARCH.
          .
          .
          .
EXP9      LARK  0,9        * EXP = 9; STORE IN AR0.
          LAC   TEMP1,14   * LOAD TO ISOLATE 8 MSB'S.
```

Figure 16. Adaptive Quantization Code

```
          SACH  TEMP1      * SAVE MANTISSA.
          LAC   TEMP1      * RELOAD FOR MANTISSA RECOMBINATION.
          B     GETMAN
          .
          .
          .
GETMAN AND   M127         * MASK TO RETAIN ONLY SEVEN BITS.
       SAR   0,TEMP1      * MOVE EXPONENT TO MEMORY FROM AR0.
       ADD   TEMP1,7      * ADD EXPONENT TO MANTISSA FOR LOG VALUE.
*;
*;   SCALE BY SUBTRACTION
*;
SUBTB  ADD   ONE,11       * ADD AN OFFSET OF 2048.
       SUB   TEMP3        * TEMP3 = Y(K) >> 2
*;
*;   4-BIT QUANTIZER
*;
*; QUANTIZATION TABLE FOR 32KB OUTPUT (OFFSET: 2048)
*;
ITAB1  EQU   2041
ITAB2  EQU   2171
ITAB3  EQU   2250
ITAB4  EQU   2309
ITAB5  EQU   2358
ITAB6  EQU   2404
ITAB7  EQU   2453
*;
QUAN    SUB   K2309       * ITAB4
        BGEZ  CI4TO7
CI0TO3  ADD   K138        * ITAB2          I = 0-3
        BGEZ  CI2TO3
CI0TO1  ADD   K130        * ITAB1          I = 0-1
        BGEZ  IEQ1
IEQ0    LACK  0
        B     GETIM
IEQ1    LACK  1
        B     GETIM
CI2TO3  SUB   K79         * ITAB3          I = 2-3
        BGEZ  IEQ3
IEQ2    LACK  2
        B     GETIM
IEQ3    LACK  3
        B     GETIM
CI4TO7  SUB   K95         * ITAB6          I = 4-7
        BGEZ  CI6TO7
CI5TO6  ADD   K46         * ITAB5          I = 5-6
        BGEZ  IEQ5
IEQ4    LACK  4
        B     GETIM
IEQ5    LACK  5
        B     GETIM
CI6TO7  SUB   K49         * ITAB6          I = 6-7
        BGEZ  IEQ7
```

Figure 16. Adaptive Quantization Code (Continued)

```
IEQ6      LACK 6
          B    GETIM
IEQ7      LACK 7
GETIM     SACL IM          * ACCUM = | I |
          XOR  DS          * ADD SIGN BIT AND FLIP IF NECESSARY.
          AND  M15         * MASK FINAL FOUR-BIT VALUE.
          SACL I           * SAVE ADPCM OUTPUT VALUE.
           .
           .
```

Figure 16. Adaptive Quantization Code (Concluded)

Determining the inverse of the 4-bit quantized ADPCM value involves another technique. The code to complete this task is much shorter in program memory requirement and somewhat faster in execution time. The same type of approximation is involved in determining the antilog as used in taking the log,

$$\log_2^{-1}(x) = 1 + x \tag{22}$$

After separating the exponent from the mantissa in the log representation, the quantized difference signal may be recovered by using the exponent to select a scaling factor. The scaling factor or multiplier is used to shift the mantissa to the proper representation, either right or left. Some of the multipliers may be stored as negative values rather than positive values, using the sign of the result to determine whether the answer is obtained from the high half of the accumulator (effectively a right shift) or from the low half of the accumulator (a left shift). The program for this process is shown in Figure 17.

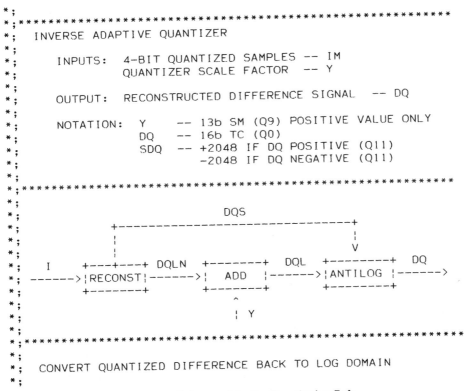

Figure 17. Inverse Adaptive Quantization Code

```
IAQUAN LAC   IM
       ADD   INQTAB      * RECONSTRUCTION TABLE
       TBLR  TEMP1       * READ NORMALIZED VALUE.
*;
*;   ADD NORMALIZING SCALE FACTOR BACK IN
*;
ADDA   LAC   TEMP1
       ADD   TEMP3       * Y >> 2
       AND   M2047
       SACL  TEMP2
*;
*;   CONVERT THE LOG VALUE TO THE LINEAR DOMAIN
*;
*;
ALOG   LAC   TEMP2,9     * EXTRACT EXPONENT.
       SACH  TEMP1       * SAVE EXPONENT VALUE.
       LACK  127
       AND   TEMP2       * MASK FOR LOG MANTISSA ONLY.
       ADD   ONE,7       * 1+x
       SACL  TEMP2       * EXTRACT MANTISSA.
       LT    TEMP2       * PREPARE TO SHIFT.
       LAC   TEMP1
       ADD   SHIFT       * LOOK UP MULTIPLIER.
       TBLR  TEMP3
       MPY   TEMP3       * MULTIPLY MANTISSA BY SHIFT FACTOR.
       PAC
       BLZ   LEFTSF      * NEGATIVE VALUES CORRESPOND TO LEFT SHIFT.
       SACH  DQ,1        * RIGHT SHIFT; SAVE MAGNITUDE OF DQ.
       B     ADDSGN
LEFTSF ABS               * LEFT SHIFT; RESTORE MAGNITUDE.
       SACL  DQ          * SAVE MAGNITUDE OF DQ.
ADDSGN LAC   ONE,11      * ASSUME POSITIVE AND SAVE THE SIGN.
       SACL  SDQ         *   (SIGN IS Q11; REMEMBER FILTER.)
       LAC   I           * CHECK SIGN OF SAMPLE.
       SUB   ONE,3
       BLZ   QSFA        * FINISHED FOR POSITIVE VALUES (I<8).
       ZAC
       SUB   DQ          * COMPUTE TWO'S COMPLEMENT OF THE MAGNITUDE.
       SACL  DQ          * SAVE NEGATIVE DQ VALUE.
       LAC   MINUS,11    * SIGN IS Q11; REMEMBER FILTER.
       SACL  SDQ         * SAVE SIGN.
       .
       .
       .
*;
*;  INVERSE QUANTIZING TABLE
*;
IQTAB  BSS   0
       DATA  65401
       DATA  68
       DATA  165
       DATA  232
       DATA  285
       DATA  332
       DATA  377
       DATA  428
```

Figure 17. Inverse Adaptive Quantization Code (Continued)

```
*;
*; SHIFT MULTIPLIER TABLE
*;
SHFT    BSS   0
        DATA  256
        DATA  512
        DATA  1024
        DATA  2048
        DATA  4096
        DATA  8192
        DATA  16384
        DATA  -1
        DATA  -2
        DATA  -4
        DATA  -8
        DATA  -16
        DATA  -32
        DATA  -64
        DATA  -128
```

Figure 17. Inverse Adaptive Quantization Code (Concluded)

The adaptation of the speed-control and the scale-factor parameters, used to adapt the stepsize in the adaptive quantizer and inverse adaptive quantizer, requires multiple uses of the technique of adjusting the fixed-point representation. The Q point is adjusted for convenience of the table constants which are part of the adaptation process and for saving the output value from the accumulator. Some limit-checking must also take place in calculating the unlocked-scale factor and the speed-control parameter.

In the calculation of the locked-scale factor and its inclusion in the mixing process for determining the overall scale factor used for stepsize quantization, the parameter is maintained with a greater resolution (19 bits of value plus its sign) than can be stored in a single memory. Calculations involving this parameter must then become two stage, both in terms of accumulations and in determining products. The code involving this parameter is listed in Figures 18 and 19.

```
* ;
* ; ***************************************************************
* ;
* ;    QUANTIZER SCALE FACTOR ADAPTATION
* ;
* ;       INPUT:    I : 32KB CODED SAMPLES
* ;
* ;       OUTPUT:  YU,YL : NEXT SAMPLE SCALE FACTOR
* ;
* ;       NOTATION:   Y   -- 13b SM (Q9) POSITIVE VALUE ONLY
* ;                   YU  -- 13b SM (Q9) POSITIVE VALUE ONLY
* ;                   YL  -- 19b SM (Q15) POSITIVE VALUE ONLY
* ;
* ; ***************************************************************
* ;
                .
                .
                .
* ;
* ;    UPDATE SLOW ADAPTATION SCALE FACTOR -- CONSTANT = 1/64
* ;
* ;    YL(k) = (1-2**-6)*YL(k-1) + 2**-6 * YU(k)
* ;
FILTE  LAC   YLH,6        * SHIFT YL LEFT BY 6.
       SACL  TEMP1        * TEMP1 = YLH * 2**6
       LAC   YLL,6
       SACL  TEMP2
       SACH  TEMP3        * TEMP3 ¦ TEMP2 = YLL * 2**6
       LAC   TEMP3        * SUPPRESS SIGN EXTENSION.
       AND   M63
       SACL  TEMP3
       ZALH  TEMP1
       ADDH  TEMP3
       ADDS  TEMP2        * ACCUM = YL * 2**6
       SUBH  YLH
       SUBS  YLL          * ACCUM = YL * 2**6 - YL
       ADD   YU,6         * ACCUM = YL * 2**6 - YL + YU
       SACL  TEMP1
       SACH  TEMP2        * RESULT = YL (SHIFTED LEFT BY 6)
       LAC   TEMP1,10     * SHIFT RESULT RIGHT 6 --> q15
       SACH  TEMP1
       LAC   TEMP1
       AND   M1023        * MASK SIGN EXTENSION.
       ADD   TEMP2,10
       SACL  YLL          * SAVE YLL.
       SACH  YLH
       LACK  7            * MASK UPPER 13 BITS.
       AND   YLH
       SACL  YLH          * SAVE YLH.
         .
         .
         .
```

Figure 18. Quantizer Scale-Factor Adaptation: Locked-Factor Calculation

```
    .
    .
    .
*;
*;  FORM LINEAR COMBINATION OF FAST AND SLOW SCALE FACTORS
*;
*;  Y(k) = (1-AL(k))*YL(k-1) + AL(k)*YU(k-1)
*;
MIX     LAC   YLL,10      * SHIFT YL RIGHT BY 6.
        SACH  TEMP3       * (IE SCALE YL TO MATCH YU SINCE YL
        LAC   TEMP3       *  CONTAINS 6 MORE LSB'S)
        AND   M1023
        ADD   YLH,10
        SACL  TEMP3       * LOW HALF
        LAC   YU
        SUB   TEMP3       * YU-(YLL>>6)
        SACL  TEMP3
        ZALH  YLH
        ADDS  YLL
        LT    AL          * AL IS IN 1.Q6
        MPY   TEMP3
        APAC              * YL + AL*(YU-(YLL>>6))
        SACL  TEMP3
        SACH  TEMP2       * TEMP2 ¦ TEMP3 = Y * 2**6
        LAC   TEMP3,10    * SHIFT RIGHT BY 6.
        SACH  TEMP3
        LAC   TEMP3
        AND   M1023       * MASK SIGN EXTENSION.
        ADD   TEMP2,10
        AND   M8191
        SACL  Y           * SAVE Y.
        LAC   Y,14
        SACH  TEMP3       * SAVE Y >> 2 .
    .
    .
    .
```

Figure 19. Quantizer Scale-Factor Adaptation: Mixing

CCITT IMPLEMENTATION OF ADPCM ON A TMS32010

The implementation of ADPCM that produces a bit-for-bit compatible solution with the CCITT test vectors uses a single TMS320M10 to accomplish a half-duplex transcoder. This solution can provide capability as either a transmitter or a receiver using either A-law or μ-law companding.

Hardware Logic and I/O

The hardware system for this transcoder implementation differs from Figure 13 in that data pins D15 and D14 are used to determine the mode of operation. Table 6 shows the operating mode for the various combined states of the data pin inputs.

Additionally, as has been noted in Figure 13, the interrupt or sample timing is an input to the $\overline{INT}$ pin in

Table 6. Operating Mode Selection

D15*	D14*	Operating Mode
L	L	μ-law transmitter
L	H	μ-law receiver
H	L	A-law transmitter
H	H	A-law receiver

*H = High logic level
L = Low logic level

the full-duplex implementation; here it is an input to the $\overline{BIO}$ pin. Each 125-μs period, the TMS32010 reads a 64-kbit/s sample from the codec and writes a 32-kbit/s sample to the ADPCM interface, or it reads the 4-bit ADPCM sample and writes an 8-bit PCM sample to the codec.

For real-time execution, the TMS32010 requires the use of a 25-MHz clock input.

Software Logic and Flow

Tables 7 and 8 list the various blocks in the algorithm, directly relating them to Figures 7 and 8 by the signal names given in the description and function. No differentiation is made between the transmitter or receiver using A-law or μ-law. The blocks are listed in the order in which they are executed. Also listed is processor demand or loading which consists of the amount of program memory used to implement the given function and the number of instruction cycles executed in worst case. There are more blocks in the tables than are shown in the figures (e.g., the algorithm uses the adaptive predictor at one point to produce the signal estimate, and later returns to update or adapt the predictor coefficients). Each block has been implemented using the equations given in previous sections concerning the ADPCM algorithm. For convenience, the equations implemented in each block are listed in the description section for the block. Additional details of the TMS32010 implementation are given in the next section, especially as they differ from the full-duplex implementation. The appendix contains a complete listing of the code.

Table 7. CCITT Transmitter

Order	Function	Description	CPU Clocks	Program Memory (Words)
1.	INPUT PCM	Read an 8 bit log-PCM sample [s(k)] and linearize it to a 12-bit sample [s_l(k)].	25 μ-law 24 A-law	0024 μ-law 0031 A-law
2.	COMPUTE SIGNAL ESTIMATE	Calculate the signal estimate [s_e(k)] from the previous data samples [d_q(k)] and reconstructed samples [s_r(k)] through the predictor filter. (12), (11)	396	0167
3.	COMPUTE ADAPTIVE QUANTIZER	Calculate speed control [a_l(k)] and quantizer scale factor [y(k)] from past quantizer output [I(k)]. (6), (3)	30	001E
4.	COMPUTE DIFFERENCE SIGNAL	Calculate the difference signal [d(k)] from the current sample [s_l(k)] and signal estimate [s_e(k)]. (1)	3	0003
5.	COMPUTE QUANTIZED OUTPUT	Calculate the log of the difference signal [d(k)] and adaptively quantize the result to yield the ADPCM output [I(k)]. (2)	42	00AD
6.	OUTPUT ADPCM	Write the ADPCM output [I(k)].	6	0005
7.	COMPUTE RECON-STRUCTED SIGNAL	Calculate the inverse of the adaptively quantized signal [d_q(k)] and the reconstructed signal difference [s_r(k)]. (10), (13)	66	00B0
8.	COMPUTE SCALE FACTOR	Calculate the updates for the scale-factor adaptation. (4), (5)	33	0022
9.	COMPUTE SPEED CONTROL	Calculate the update for the speed-control adaptation. (8), (9), (7)	30	001C
10.	COMPUTE PREDICTOR ADAPTATION	Calculate the updates for the adaptive predictor filter coefficients. (18), (16), (17), (14), (15)	111	0074
11.	WAIT	Spin until the next sample is available.	2 +	0004

Table 8. CCITT Receiver

Order	Function	Description	CPU Clocks	Program Memory (Words)
1.	INPUT ADPCM	Read the ADPCM input [I(k)].	10	0009
2.	COMPUTE SIGNAL ESTIMATE	Calculate the signal estimate [$s_e(k)$] from the previous data samples [$d_q(k)$] and reconstructed samples [$s_r(k)$] through the predictor filter. (12), (11)	396	0167
3.	COMPUTE ADAPTIVE QUANTIZER	Calculate speed control [$a_l(k)$] and quantizer scale factor [y(k)] from past quantizer output [I(k)]. (6), (3)	30	001E
4.	COMPUTE QUANTIZED DIFFERENCE AND RECON-STRUCTED SIGNAL	Calculate the inverse of the adaptively quantized signal [$d_q(k)$] and the reconstructed signal [$s_r(k)$]. (10), (13)	66	00B0
5.	COMPUTE SCALE FACTOR	Calculate the updates for the scale-factor adaptation. (4), (5)	33	0022
6.	COMPUTE SPEED CONTROL	Calculate the update for the speed-control adaptation. (8), (9), (7)	30	001C
7.	COMPUTE PREDICTOR ADAPTATION	Calculate the updates for the adaptive predictor filter coefficients. (18), (16), (17), (14), (15)	111	0074
8.	COMPUTE LOG-PCM	Convert the reconstructed linear-PCM signal [$s_r(k)$] to a log-PCM signal [$s_p(k)$].	35 μ-law 33 A-law	0074 μ-law 0072 A-law
9.	SYNCHRON-OUS CODING ADJUSTMENT	Calculate an ADPCM signal from the output [$s_p(k)$] and adjust to create [$s_d(k)$] if it differs from [I(k)].	63	00DA
10.	OUTPUT PCM	Write the log-PCM output [$s_d(k)$].	4	0003
11.	WAIT	Spin until the next interrupt.	2+	0004

Implementation and Advantages of TMS32010 Architecture

Many of the same features are used in the bit-compatible implementation as were discussed in the full-duplex implementation. Some changes are imperative, since performance to the recommended specification requires executing certain calculations in a floating-point representation. These changes or additions require further modifications in order to limit the required amount of program memory to the internal memory space of the TMS32010.

One of the first observed requirements is that the processor must be capable of doing either A-law or μ-law companding and function as either a transmitter or a receiver.

The burden of determining the mode of operation is simplified by selecting one of the four modes from information available at the time of reset, and then executing from one of the four control loops until the next reset. Each loop, therefore, tests the $\overline{\text{BIO}}$ pin to determine when the next input sample is ready, rather than depending on the hardware interrupt.

The requirement of selecting either A-law of μ-law companding also means that a table lookup approach is beyond the program memory capacity. The conversion must be done algorithmically to reduce the amount of memory. Figures 20 and 21 illustrate μ-law companding as it is implemented in this algorithm.

```
XMTMU    IN      SCRACH,ADC
EXPNDU   LAC     SCRACH,8    ; SEEE MMMM 0000 0000
         XOR     KFF00       ; INVERT FROM TRANSMISSION FORMAT
         SACL    TEMP1       ; SAVE VALUE FOR PCM SIGN
         AND     M32767      ; OEEE MMMM 0000 0000
         SACH    TEMP2,4     ; SAVE EXPONENT VALUE
         AND     M4095       ; 0000 MMMM 0000 0000
         ADD     BIAS,7      ; 0001 MMMM 1000 0000
         SACL    SCRACH
         LAC     TEMP1
         SACH    TEMP1       ; SIGN = FFFF OR 0000
         LACK    SBASE
         ADD     TEMP2,1     ; CALCULATE PCM SHIFT ADDRESS
         CALA
         SUB     BIAS,12     ; 0000000X XXXXXXXX XXXX0000 00000000
         SACH    SAMPLE,4
         LAC     SAMPLE      ; 000XXXXX XXXXXXXX
         XOR     TEMP1       ; POS - DO NOTHING : NEG - 1's COMP
         SUB     TEMP1       ; POS - DO NOTHING : NEG - 2's COMP
         SACL    SAMPLE
                 .
                 .
                 .
*;
SBASE    LAC     SCRACH,5    ; 00000000 0000001M MMM10000 00000000
         RET
         LAC     SCRACH,6    ; 00000000 000001MM MM100000 00000000
         RET
         LAC     SCRACH,7    ; 00000000 00001MMM M1000000 00000000
         RET
         LAC     SCRACH,8    ; 00000000 0001MMMM 10000000 00000000
         RET
         LAC     SCRACH,9    ; 00000000 001MMMM1 00000000 00000000
         RET
         LAC     SCRACH,10   ; 00000000 01MMMM10 00000000 00000000
         RET
         LAC     SCRACH,11   ; 00000000 1MMMM100 00000000 00000000
         RET
         LAC     SCRACH,12   ; 00000001 MMMM1000 00000000 00000000
         RET
```

Figure 20. μ-Law Expansion Code

17. 32-kbit/s ADPCM with the TMS32010

```
        LAC    SR              ; GET RECONSTRUCTED SIGNAL
*;
*; COMPRESS--CONVERT TO PCM
*;
CMPRSU  SACH   TEMP4           ; SAVE SIGN OF SR
        ABS
        ADD    BIAS            ; ADD BIAS
        SACL   SCRACH          ; SAVE BIASED PCM VALUE
        SUB    ONE,9           ; EXP = 7 - 4 OR 3 - 0
        BGEZ   SCL427
SCL023  ADD    THREE,7         ; EXP = 3 - 2 OR 1 - 0
        BGEZ   SCL223
SCL021  ADD    ONE,6           ; EXP = 1 OR 0
        BGEZ   SCALE1
SCALE0  LAC    M15,1           ; EXP = 0
        AND    SCRACH          ; MASK FOR MANTISSA
        SACL   SCRACH
        ADD    BIAS
        SACL   SAMPLE          ; BIASED QUANTIZED VALUE
        LAC    SCRACH,15
        LARK   0,0
        B      FINI
SCALE1  LAC    M15,2           ; EXP = 1
        AND    SCRACH          ; MASK FOR MANTISSA
        SACL   SCRACH
        ADD    BIAS,1
        SACL   SAMPLE          ; BIASED QUANTIZED VALUE
        LAC    SCRACH,14
        LARK   0,1
        B      FINI
         .
         .
         .
FINI    SACH   SCRACH          ; SAVE NORMALIZED MANTISSA
        LAC    SCRACH
        SAR    0,TEMP1
        ADD    TEMP1,4         ; ADD EXPONENT
CLNUP   ADD    TEMP4,7
        AND    M255
        SACL   SCRACH          ; 2's COMPLEMENT OF MULAW-PCM
        LAC    SAMPLE          ; REMOVE BIAS FROM QUANTIZED VALUE
        SUB    BIAS
        XOR    TEMP4
        SUB    TEMP4
        SACL   SAMPLE          ; 2's COMPLEMENT OF QUANTIZED SAMPLE
*;
        CALL   AQUAN
*;
        CALL   SYNC
*;
        XOR    M255            ; FLIP BITS FOR TRANSMISSION
        SACL   SCRACH
        OUT    SCRACH,DAC
```

Figure 21. μ-Law Compression Code

The predictor filter implementation is also modified from what has been previously presented. In the CCITT recommendation, the processing of the signal through the predictor filter is performed in a floating-point format. This requirement leads to several modifications. First, all input signals to the filter, $d_q(k)$ and $s_r(k)$, must be converted to a floating-point notation. The conversion to this notation is accomplished by a binary search of the original fixed-point word. As previously mentioned, this technique is explained in some detail in the application report, FLOATING-POINT ARITHMETIC WITH THE TMS32010. Second, the filter coefficients, $a_i(k)$ and $b_i(k)$, must also be floated for each sample so that a floating-point multiply can be executed for each filter tap.

Accumulation of the filter taps is carried out in fixed-point notation. Fixing a floating-point number is equivalent to the scaling presented for taking the anti-log of a number. Some of the floating-point results must be left-shifted, while others need to be right-shifted. The shift is accomplished by use of a scaling factor or multiplier, selected by the exponent sum of the floating-point multiply. Positive multipliers are used to indicate what is effectively a right shift with the result being stored from the high half of the accumulator. Negative multipliers indicate that the result is in the low half of the accumulator and is used for values which have been left shifted.

The process of a single filter tap, not including the code to float the signal and the coefficient, is shown in Figure 22.

```
*;*************************************************************
*;
*;   COMPUTE SEZ -- PARTIAL SIGNAL ESTIMATE
*;
*;   SEZ(k) = B1(k-1)*DQ(k-1) + ... + B6(k-1)*DQ(k-6)
*;
*;        MULTIPLIES ARE DONE IN FLOATING POINT
*;          DQ's ARE STORED IN FLOATING-POINT NOTATION
*;          B's ARE FLOATED EACH PASS
*;
*;        NOTATION: DQnEXP    -- 4 bits + OFFSET
*;                  DQnMAN*8 -- 9 bits
*;                  Bn        -- 16 b TC ; q14
*;                  SEZ       -- 16 b TC ; q0
*;
*;*************************************************************
*;
SIGDIF  LAC     B6,14    ; COMPUTE B6*DQ5.
        CALL    FLOAT    ; RET/W MANTISSA IN TEMP1; EXP IN ACC.
        ADD     DQ5EXP
        SACL    SUM1
        LAR     0,SUM1   ; EXP OF PRODUCT.
        LT      DQ5MAN   ; DQnMAN SCALED BY 2**3.
        LAC     THREE,7  ; PRODUCT FUDGE FACTOR (48*8).
        MPY     TEMP1
        LTA     *,0      ; B6MAN*(DQ5MAN*8)+(48*8)
        AND     KFF80    ; SAVE ONLY 8 MSB'S.
        SACL    TEMP1
        MPY     TEMP1    ; APPLY SHIFT FACTOR.
        PAC
        BLZ     RS1      ; EXP >= 26
        SACH    SUM1,1   ; EXP <  26
CHK1    ZALS    B6       ; CHECK SIGN OF PRODUCT.
        XOR     SDQ6
        AND     K32768
        BZ      POS1
NEG1    ZAC              ; NEGATE IF NECESSARY.
        SUB     SUM1
```

Figure 22. Predictor Filter Execution

17. 32-kbit/s ADPCM with the TMS32010

```
              SACL      SUM1
POS1          LAC       B5,14      ; COMPUTE B5*DQ4.
              .
              .
              .
RS1           ABS                  ; MAKE POSITIVE BEFORE MASK.
              AND       M32767     ; KEEP LOWER 15 BITS.
              SACL      SUM1       ; SAVE RESULT.
              B         CHK1
```

Figure 22. Predictor Filter Execution (Concluded)

SUMMARY

The TMS32010 provides an efficient solution to transcoding a 64-kbit/s PCM signal to a 32-kbit/s bit stream. Transcoding, as described in this application report, is an effective way to maintain the signal quality provided by 7-bit PCM while reducing the data rate.

The basic ADPCM algorithm has been implemented in two slightly different ways. One solution provides CCITT bit-for-bit compatibility. Using this algorithm, a half-duplex transcoder is created that can transcode either A-law or μ-law signals as either a transmitter or a receiver. No external program memory is required for this implementation, although it does require the use of a 25-MHz TMS32010 microprocessor. The second described solution is particularly attractive since it uses a single, 20.5-MHz TMS32010 microprocessor that requires no external program memory to perform a real-time full-duplex (non-CCITT) channel transcoding.

In selecting one of these two solutions, the primary consideration is the network interfacing requirement. For systems that only have analog interfaces to other parts of the network, the full-duplex solution will provide the best choice. On the other hand, a network that may include a digital interface to other ADPCM transcoders will probably require the CCITT bit-compatible solution. Both solutions provide high-quality signal transcoding.

A complete assembled code listing is provided in the appendix of this report and is also available in 1600-BPI VAX/VMS tape format. The software may be purchased by ordering the TMS32010 Software Exchange Library, TMDC3240212-18, from Texas Instruments. For further information, please contact your nearest TI sales representative.

REFERENCES

"Recommendation G.721, 32 kbit/s Adaptive Differential Pulse Code Modulation," *CCITT* (1984).

N.S. Jayant (ed.), *Waveform Quantization and Coding*, IEEE Press (1976).

N.S. Jayant and Peter Noll, *Digital Coding of Waveforms*, Prentice-Hall (1984).

L.R. Rabiner and R.W. Schafer, *Digital Processing of Speech Signals*, Prentice-Hall (1978).

J.C. Bellamy, *Digital Telephony*, John Wiley & Sons (1982).

Bernhard E. Keiser, *Digital Telephony: Speech Digitization*, George Washington University (1981).

Companding Routines for the TMS32010, Texas Instruments Incorporated (1984).

Floating-Point Arithmetic with the TMS32010, Texas Instruments Incorporated (1984).

Implementation of FIR/IIR Filters with the TMS32010, Texas Instruments Incorporated (1984).

TMS32010 User's Guide, Texas Instruments Incorporated (1983).

Appendix
ADPCM Assembly Language Programs

```
0001              COPY   INPUT.ASM
0001              IDT    'CCITT'
0002              OPTION XREF
0003
0004  ;*****************************************************************
0005  ;*
0006  ;*  This is the source module for a half-duplex CCITT
0007  ;*  compatible 32-kbps ADPCM speech system. The transmitter
0008  ;*  assumes that log-PCM data will be available at each
0009  ;*  interrupt (every 125 microseconds) in the lower 8 bits of
0010  ;*  the data bus via I/O port 1 and it supplies ADPCM data on
0011  ;*  the lower 4 bits of the bus via port 2. The receiver does
0012  ;*  the inverse. An interrupt in this case is determined by
0013  ;*  polling the BIO line, with a low signal level signifying
0014  ;*  the presence of a new data sample.
0015  ;*
0016  ;*  The 'R' reset function in the CCITT spec is implemented
0017  ;*  with a hardware reset. At the time of reset, it is
0018  ;*  assumed that the operating mode has been established and
0019  ;*  input via the upper two bits of the data bus. The bit
0020  ;*  condition is read from port 0 so as not to disrupt any
0021  ;*  pending data sample on either of the other two ports.
0022  ;*  Since it is anticipated that the mode pins will be
0023  ;*  selected and maintained in a manner similar to a hardwire
0024  ;*  selection, the actual port from which the mode is read
0025  ;*  is arbitrary.
0026  ;*
0027  ;*****************************************************************
0028
0029  ;*
0030  ;*  System I/O channel assignments
0031  0001 ADC    EQU  1    ; codec input
0032  0001 DAC    EQU  1    ; codec output
0033  0002 CCITT  EQU  2    ; adpcm output/input
0034  0000 CTL    EQU  0    ; control input to select mode
0035                        ;   0000 = mulaw transmitter
0036                        ;   4000 = mulaw receiver
0037                        ;   8000 = alaw transmitter
0038                        ;   C000 = alaw receiver
0039  ;*
0040  ;*****************************************************************
0041  0000           AORG 0
0042  ;*
0043  0000 F900      B    RESET   ; power-up reset
0043  0001 0542
0044
0045  ;*  INTERRUPT HANDLING ROUTINE -- SYSTEM HANDLES CODEC
0046  ;*  SAMPLES ON A SAMPLE BY SAMPLE BASIS.
0047  ;*
0048  ;*
0050  0002 F900 INTRPT B   INTRPT
0050  0003 0002
0051  ;*
```

```
0053
0054  ;*  MU-LAW TRANSMITTER
0055
0056  0004 411B XMTMU IN   SCRACH,ADC  ; input mu-law PCM
0057
0058  ;*  MU-LAW TO LINEAR PCM EXPANSION
0059  ;*
0060  ;*
0061  ;*    INPUT:  MU-LAW PCM SAMPLE  --  S  (SCRACH)
0062  ;*
0063  ;*    OUTPUT: LINEAR PCM SAMPLE  -- SL  (SAMPLE)
0064  ;*
0065  ;*    NOTATION:  S  -- 8b SM (Q4)
0066  ;*               SL -- 14b TC (Q0)
0067  ;*
0068  ;*
0069  ;*               S  +--------+ SL
0070  ;*          ------>; EXPAND ;-------->
0071  ;*               +--------+
0072  ;*
0073  ;*
0074  ;*
0075
0076  0005 281B EXPNDU LAC  SCRACH,8    ; seee mmmm 0000 0000
0077  0006 7875        XOR  KFF00       ; SEEE MMMM 0000 0000
0078  0007 5021        SACL TEMP1       ; save value for PCM sign
0079  0008 7974        AND  M32767      ; 0EEE MMMM 0000 0000
0080  0009 5C22        SACH TEMP2,4     ; save exponent value
0081  000A 7972        AND  M4095       ; 0000 MMMM 0000 0000
0082  000B 074D        ADD  BIAS,7      ; 0001 MMMM 1000 0000
0083  000C 501B        SACL SCRACH
0084  000D 2021        LAC  TEMP1
0085  000E 5821        SACH TEMP1       ; sign = FFFF or 0000
0086  000F 7E24        LACK TEMP2,1     ; calculate PCM shift address
0087  0010 0122        ADD  TEMP2,1     ; calculate PCM shift address
0088  0011 7F8C        CALA
0089  0012 1C4D        SUB  BIAS,12 ; 0000000X XXXXXXX XXXX0000 00000000
0090  0013 5C26        SACH SAMPLE,4 ; 0000000X XXXXXXX XXXX0000 00000000
0091  0014 2026        LAC  SAMPLE
0092  0015 7821        XOR  TEMP1    ; 000X XXXX XXXX XXXX
0093  0016 1021        SUB  TEMP1    ; pos - no change ; neg - 1's compl
0094  0017 5026        SACL SAMPLE   ; pos - no change ; neg - 2's compl
0095
0096  ;*  Now convert PCM value in SAMPLE to ADPCM value in I
0097
0098  0018 F800 GETI  CALL SIGDIF
0099  0019 01B3
0099  001A F800       CALL AQUAN
0099  001B 02AA
0100  001C 5001       SACL 1
0101  001D 4A01       OUT  I,CCITT   ; output ADPCM
0102  001E F800       CALL PRDICT
0102  001F 0355
0103  0020 F600 MULAWX BIOZ XMTMU    ; wait for next sample
0103  0021 0004
0104  0022 F900       B    MULAWX
0104  0023 0020
```

17. 32-kbit/s ADPCM with the TMS32010

```
         *;
SBASE
A0105  0024 251B       LAC   SCRACH,5     ; 00000000 0000001M MMM10000 00000000
A0106  0025 7F8D       RET
A0107  0026 261B       LAC   SCRACH,6     ; 00000000 000001MM MM100000 00000000
A0108  0027 7F8D       RET
A0109  0028 271B       LAC   SCRACH,7     ; 00000000 00001MMM M1000000 00000000
A0110  0029 7F8D       RET
A0111  002A 281B       LAC   SCRACH,8     ; 00000000 0001MMMM 10000000 00000000
A0112  002B 7F8D       RET
A0113  002C 291B       LAC   SCRACH,9     ; 00000000 001MMMM1 00000000 00000000
A0114  002D 7F8D       RET
A0115  002E 2A1B       LAC   SCRACH,10    ; 00000000 01MMMM10 00000000 00000000
A0116  002F 7F8D       RET
A0117  0030 2B1B       LAC   SCRACH,11    ; 00000000 1MMMM100 00000000 00000000
A0118  0031 7F8D       RET
A0119  0032 2C1B       LAC   SCRACH,12    ; 00000001 MMMM1000 00000000 00000000
A0120
A0121  0033 7F8D       RET
```

```
A0123    *;
A0124    *;   A-LAW TRANSMITTER
A0125    *;
A0126  0034 411B   XMTA   IN   SCRACH,ADC   ; input A-law PCM
A0127    *;
A0128    *; ********************************************
A0129    *;   A-LAW TO LINEAR PCM EXPANSION
A0130    *;
A0131    *;   INPUT:  A-LAW PCM SAMPLE   -- S   (SCRACH)
A0132    *;
A0133    *;   OUTPUT: LINEAR PCM SAMPLE  -- SL  (SAMPLE)
A0134    *;
A0135    *;   NOTATION:  S  -- 8b SM (Q4)
A0136    *;              SL -- 14b TC (Q0)
A0137    *;
A0138    *;              +--------+
A0139    *;          S   |        | SL
A0140    *;      --------->| EXPAND |-------->
A0141    *;              |        |
A0142    *;              +--------+
A0143    *;
A0144
A0145  0035 281B   EXPNDA  LAC   SCRACH,8     ; sEEE MMMM 0000 0000
A0146  0036 7848          XOR   K32768       ; SEEE MMMM 0000 0000
A0147  0037 5021          SACL  TEMP1        ; save value for PCM sign
A0148  0038 7974          AND   M32767       ; 0EEE MMMM 0000 0000
A0149  0039 5C22          SACH  TEMP2,4      ; save exponent value
A0150  003A 7972          AND   M4095        ; 0000 MMMM 0000 0000
A0151  003B 501B          SACL  SCRACH
A0152  003C 2021          LAC   TEMP1
A0153  003D 5821          SACH  TEMP1        ; sign = FFFF or 0000
A0154  003E 7E52          LACK  SBASEA       ; calculate PCM shift address
A0155  003F 0222          ADD   TEMP2,2
A0156  0040 7F8C          CALA
A0157  0041 5C26          SACH  SAMPLE,4
A0158  0042 2026          LAC   SAMPLE
A0159  0043 7821          XOR   TEMP1        ; 000X XXXX XXXX XXXX
A0160  0044 1021          SUB   TEMP1        ; pos - no change : neg - 1's compl
A0161  0045 5026          SACL  SAMPLE       ; pos - no change : neg - 2's compl
A0162
A0163    *;
A0164    *;  Now convert PCM value in SAMPLE to ADPCM value in I
A0165    *;
A0166  0046 F800          CALL  SIGD1F
       0047 01B3
A0168  0048 F800          CALL  AQUAN
       0049 02AA
A0169  004A 5001          SACL  TEMP1
A0170  004B 4A01          OUT   I,CCITT      ; output ADPCM
A0171  004C F800          CALL  PRD1CT
       004D 0355
A0172  004E F600   ALAWX  BIOZ  XMTA
       004F 0034
A0173  0050 F900          B     ALAWX        ; wait for next sample
       0051 004E
A0174    *;
```

17. 32-kbit/s ADPCM with the TMS32010

```
A0175 0052 261B  SBASEA LAC  SCRACH,6   ;; 00000000 000000MM MM000000 00000000
A0176 0053 0D4C         ADD  ONE,13     ;; 00000000 000000MM MM100000 00000000
A0177 0054 7F8D         RET
A0178 0055 7F80         NOP
A0179 0056 261B         LAC  SCRACH,6   ;; 00000000 000001MM MM100000 00000000
A0180 0057 0671         ADD  BIASA,6    ;; 00000000 000001MM MM100000 00000000
A0181 0058 7F8D         RET
A0182 0059 7F80         NOP
A0183 005A 271B         LAC  SCRACH,7   ;; 00000000 00000MMM M0000000 00000000
A0184 005B 0771         ADD  BIASA,7    ;; 00000000 00001MMM M1000000 00000000
A0185 005C 7F8D         RET
A0186 005D 7F80         NOP
A0187 005E 281B         LAC  SCRACH,8   ;; 00000000 0000MMMM M0000000 00000000
A0188 005F 0871         ADD  BIASA,8    ;; 00000000 0001MMMM M0000000 00000000
A0189 0060 7F8D         RET
A0190 0061 7F80         NOP
A0191 0062 291B         LAC  SCRACH,9   ;; 00000000 000MMMMM M0000000 00000000
A0192 0063 0971         ADD  BIASA,9    ;; 00000000 001MMMMM M0000000 00000000
A0193 0064 7F8D         RET
A0194 0065 7F80         NOP
A0195 0066 2A1B         LAC  SCRACH,10  ;; 00000000 00MMMMMM M0000000 00000000
A0196 0067 0A71         ADD  BIASA,10   ;; 00000000 01MMMMMM M0000000 00000000
A0197 0068 7F8D         RET
A0198 0069 7F80         NOP
A0199 006A 2B1B         LAC  SCRACH,11  ;; 00000000 0MMMMMMM 00000000 00000000
A0200 006B 0B71         ADD  BIASA,11   ;; 00000000 1MMMMMMM 00000000 00000000
A0201 006C 7F8D         RET
A0202 006D 7F80         NOP
A0203 006E 2C1B         LAC  SCRACH,12  ;; 00000000 MMMMMMM0 00000000 00000000
A0204 006F 0C71         ADD  BIASA,12   ;; 00000001 MMMMMMM0 00000000 00000000
A0205 0070 7F8D         RET
```

```
A0207                  *;
A0208                  *;  MU-LAW RECEIVER
A0209                  *;
A0210 0071 4201  RCVMU IN   I,CCITT       ; input ADPCM
A0211                  *;                 ; determine magnitude of ADPCM
A0212 0072 2001        LAC  I
A0213 0073 5002        SACL IM
A0214 0074 134C        SUB  ONE,3
A0215 0075 FA00        BLZ  DO32KU
      0076 007A
A0216 0077 2002        LAC  IM
A0217 0078 786D        XOR  M15
A0218 0079 5002        SACL IM
A0219                  *;
A0220                  *;  compute pcm output
A0221                  *;
A0222 007A F800  DO32KU CALL SIGDIF
      007B 01B3
A0223 007C F800        CALL PRDICT
      007D 0355
A0224  ***************************************************************
A0225                  *;
A0226                  *;  LINEAR TO U-LAW PCM COMPRESSION/U-LAW TO LINEAR EXPANSION
A0227                  *;
A0228                  *;  INPUT:  LINEAR PCM SAMPLE  -- SR
A0229                  *;
A0230                  *;  OUTPUT: A-LAW PCM SAMPLE   -- SP  (SCRACH)
A0231                  *;          LINEAR PCM SAMPLE  -- SLX (SAMPLE)
A0232                  *;
A0233                  *;  NOTATION:  SR  -- 16b TC (Q0)
A0234                  *;             SP  -- 8b SM (Q4)
A0235                  *;             SLX -- 14b TC (Q0)
A0236                  *;
A0237                  *;
A0238                  *;        SR  +-----------+
A0239                  *;     ----------->: COMPRESS :---------* SP    +-----------+ SLX
A0240                  *;            +-----------+           *------>; EXPAND ;------->
A0241                  *;                                    +-----------+
A0242                  *;                                           +-----------+ SP
A0243                  *;                                           +----------------->
A0244                  *;                                           +
A0245                  *;
A0246                  *;
A0247 007E 2013        LAC  SR      ; get reconstructed signal
A0248                  *;
A0249                  *;  compress--convert to pcm
A0250                  *;
A0251 007F 5824  CMPRSU SACH TEMP4   ; save sign of SR
A0252 0080 7F88         ABS
A0253 0081 004D         ADD  BIAS    ; add bias
A0254 0082 501B         SACL SCRACH  ; save biased PCM value
A0255 0083 194C         SUB  ONE,9   ; exp = 7 - 4 or 3 - 0
A0256 0084 FD00         BGEZ SCL427
      0085 00B3
A0257 0086 077D  SCL023 ADD  THREE,7 ; exp = 3 - 2 or 1 - 0
A0258 0087 FD00         BGEZ SCL223
      0088 009E
```

```
CCITT         32010 FAMILY MACRO ASSEMBLER      PC2.1 84.107      16:36:03  03-20-85
                                                                           PAGE 0007

A0259  0089 064C  SCL021  ADD   ONE,6       ; exp = 1 or 0
A0260  008A FD00          BGEZ  SCALE1
       008B 0095
A0261  008C 2160  SCALE0  LAC   M15,1       ; exp = 0
A0262  008D 791B          AND   SCRACH      ; mask for mantissa
A0263  008E 501B          SACL  SCRACH
A0264  008F 004D          ADD   BIAS
A0265  0090 5026          SACL  SAMPLE
A0266  0091 2F1B          LAC   SCRACH,15   ; biased quantized value
A0267  0092 7000          LARK  0,0
A0268  0093 F900          B     FINI
       0094 00E6
A0269  0095 2260  SCALE1  LAC   M15,2       ; exp = 1
A0270  0096 791B          AND   SCRACH      ; mask for mantissa
A0271  0097 501B          SACL  SCRACH
A0272  0098 014D          ADD   BIAS,1
A0273  0099 5026          SACL  SAMPLE
A0274  009A 2E1B          LAC   SCRACH,14   ; biased quantized value
A0275  009B 7001          LARK  0,1
A0276  009C F900          B     FINI
       009D 00E6
A0277  009E 174C  SCL223  SUB   ONE,7       ; exp = 3 or 2
A0278  009F FD00          BGEZ  SCALE3
       00A0 00AA
A0279  00A1 236D  SCALE2  LAC   M15,3       ; exp = 2
A0280  00A2 791B          AND   SCRACH      ; mask for mantissa
A0281  00A3 501B          SACL  SCRACH
A0282  00A4 024D          ADD   BIAS,2
A0283  00A5 5026          SACL  SAMPLE
A0284  00A6 2D1B          LAC   SCRACH,13   ; biased quantized value
A0285  00A7 7002          LARK  0,2
A0286  00A8 F900          B     FINI
       00A9 00E6
A0287  00AA 2460  SCALE3  LAC   M15,4       ; exp = 3
A0288  00AB 791B          AND   SCRACH      ; mask for mantissa
A0289  00AC 501B          SACL  SCRACH
A0290  00AD 034D          ADD   BIAS,3
A0291  00AE 5026          SACL  SAMPLE
A0292  00AF 2C1B          LAC   SCRACH,12   ; biased quantized value
A0293  00B0 7003          LARK  0,3
A0294  00B1 F900          B     FINI
       00B2 00E6
A0295  00B3 197D  SCL427  SUB   THREE,9     ; exp = 7 - 6 or 5 - 4
A0296  00B4 FD00          BGEZ  SCL627
       00B5 00CB
A0297  00B6 0A4C  SCL425  ADD   ONE,10      ; exp = 5 or 4
A0298  00B7 FD00          BGEZ  SCALE5
       00B8 00C2
A0299  00B9 256D  SCALE4  LAC   M15,5       ; exp = 4
A0300  00BA 791B          AND   SCRACH      ; mask for mantissa
A0301  00BB 501B          SACL  SCRACH
A0302  00BC 044D          ADD   BIAS,4
A0303  00BD 5026          SACL  SAMPLE
A0304  00BE 2B1B          LAC   SCRACH,11   ; biased quantized value
A0305  00BF 7004          LARK  0,4
A0306  00C0 F900          B     FINI
       00C1 00E6
```

```
CCITT         32010 FAMILY MACRO ASSEMBLER      PC2.1 84.107      16:36:03  03-20-85
                                                                           PAGE 0008

A0307  00C2 2660  SCALE5  LAC   M15,6       ; exp = 5
A0308  00C3 791B          AND   SCRACH      ; mask for mantissa
A0309  00C4 501B          SACL  SCRACH
A0310  00C5 054D          ADD   BIAS,5
A0311  00C6 5026          SACL  SAMPLE
A0312  00C7 2A1B          LAC   SCRACH,10   ; biased quantized value
A0313  00C8 7005          LARK  0,5
A0314  00C9 F900          B     FINI
       00CA 00E6
A0315  00CB 1B4C  SCL627  SUB   ONE,11      ; exp = 7 or 6
A0316  00CC FD00          BGEZ  SCALE7
       00CD 00D7
A0317  00CE 2760  SCALE6  LAC   M15,7       ; exp = 6
A0318  00CF 791B          AND   SCRACH      ; mask for mantissa
A0319  00D0 501B          SACL  SCRACH
A0320  00D1 064D          ADD   BIAS,6
A0321  00D2 5026          SACL  SAMPLE
A0322  00D3 291B          LAC   SCRACH,9    ; biased quantized value
A0323  00D4 7006          LARK  0,6
A0324  00D5 F900          B     FINI
       00D6 00E6
A0325  00D7 1C4C  SCALE7  SUB   ONE,12      ; exp = 7
A0326  00D8 FA00          BLZ   NORMAL      ; mag > 8191 ?
       00D9 00DE
A0327  00DA 2767  SATCH   LAC   K63,7       ; save max biased quantized value
A0328  00DB 7E7F          LACK  127         ; set maximum mulaw magnitude
A0329  00DC F900          B     CLNUP
       00DD 00EA
A0330  00DE 286D  NORMAL  LAC   M15,8       ; mask for mantissa
A0331  00DF 791B          AND   SCRACH
A0332  00E0 501B          SACL  SCRACH
A0333  00E1 074D          ADD   BIAS,7
A0334  00E2 5026          SACL  SAMPLE
A0335  00E3 7007          LARK  0,7
A0336  00E4 281B  FINI    LAC   SCRACH,8    ; biased quantized value
A0337  00E5 581B          SACH  SCRACH
A0338  00E6 201B          LAC   SCRACH
A0339  00E7 3021          SAR   0,TEMP1     ; save normalized mantissa
A0340  00E8 0421          ADD   TEMP1,4     ; add exponent
A0341  00E9 0724          ADD   TEMP4,7
A0342  00EA 7947  CLNUP   AND   M255
A0343  00EB 501B          SACL  SCRACH      ; signed magnitude of mulaw-PCM
A0344  00EC 2026          LAC   SAMPLE
A0345  00ED 104D          SUB   BIAS        ; remove bias from quantized value
A0346  00EE 7824          XOR   TEMP4
A0347  00EF 1024          SUB   TEMP4
A0348  00F0 5026          SACL  SAMPLE      ; 2's complement of quantized sample
A0349  00F1 F800          CALL  AQUAN
       00F2 02AA
A0350  00F3 F800          CALL  SYNC
       00F4 0188
A0351  00F5 7847          XOR   M255        ; flip bits for transmission
A0352  00F6 501B          SACL  SCRACH
```

```
A0357  00F8 491B   MULAWR OUT   SCRACH,DAC   ; output mu-law PCM
A0358  00F9 F600          BIOZ  RCVMU        ; wait for next sample
       00FA 0071
A0359  00FB F900          B     MULAWR
       00FC 00F9
```

```
A0361          *;
A0362          *;  A-LAW RECEIVER
A0363          *;
A0364  00FD 4201  RCVA   IN    1,CCITT      ; input ADPCM
A0365          *;                           ; determine magnitude of ADPCM
A0366  00FE 2001         LAC   I
A0367  00FF 5002         SACL  IM
A0368  0100 134C         SUB   ONE,3
A0369  0101 FA00         BLZ   DO32KA
       0102 0106
A0370  0103 2002         LAC   IM
A0371  0104 786D         XOR   M15
A0372  0105 5002         SACL  IM
A0373          *;
A0374          *;  compute pcm output
A0375          *;
A0376  0106 F800  DO32KA CALL  SIGDIF
       0107 01B3
A0377  0108 F800         CALL  PRDICT
       0109 0355

A0378          *;***********************************************************
A0379          *;  LINEAR TO A-LAW PCM COMPRESSION/A-LAW TO LINEAR EXPANSION
A0380          *;
A0381          *;     INPUT:  LINEAR PCM SAMPLE  -- SR
A0382          *;
A0383          *;
A0384          *;     OUTPUT: A-LAW PCM SAMPLE  -- SP  (SCRACH)
A0385          *;             LINEAR PCM SAMPLE  -- SLX (SAMPLE)
A0386          *;
A0387          *;     NOTATION:  SR  -- 16b TC (Q0)
A0388          *;                SP  -- 8b SM (Q4)
A0389          *;                SLX -- 14b TC (Q0)
A0390          *;
A0391          *;
A0392          *;                       SP  +----------+
A0393          *;     SR  +----------+     +-->: EXPAND :-----> SLX
A0394          *;    ------>: COMPRESS :----•   +----------+
A0395          *;           +----------+              SP
A0396          *;                            +------------------> SP
A0397          *;
A0398          *;
A0399          *;
A0400          *;
A0401  010A 2013         LAC   SR           ; get reconstructed signal
A0402          *;  compress--convert to pcm
A0403          *;
A0404          *;
A0405  010B 5824  CMPRSA SACH  TEMP4        ; save sign of SR
A0406  010C 7F88         ABS
A0407  010D 0024         ADD   TEMP4        ; add 1 for negative vals
A0408  010E 501B         SACL  SCRACH       ; save PCM value
A0409  010F 194C         SUB   ONE,9        ; exp = 7 - 4 or 3 - 0
A0410  0110 F000         BGEZ  SCL4T7
       0111 013F
A0411  0112 077D  SCL0T3 ADD   THREE,7      ; exp = 3 - 2 or 1 - 0
A0412  0113 F000         BGEZ  SCL2T3
       0114 012A
```

17. 32-kbit/s ADPCM with the TMS32010

```
A0413 0115 064C  SCLOT1  ADD  ONE,6       ; exp = 1 or 0
A0414 0116 FD00          BGEZ SCAL1A
      0117 0121
A0415 0118 2260  SCAL0A  LAC  M15,2       ; exp = 0
A0416 0119 791B          AND  SCRACH      ; mask for mantissa
A0417 011A 501B          SACL SCRACH
A0418 011B 014C          ADD  ONE,1
A0419 011C 5026          SACL SAMPLE
A0420 011D 2E1B          LAC  SCRACH,14
A0421 011E 7000          LARK 0,0         ; quantized value
A0422 011F F900          B    FINISH
      0120 0172
A0423 0121 2260  SCAL1A  LAC  M15,2       ; exp = 1
A0424 0122 791B          AND  SCRACH      ; mask for mantissa
A0425 0123 501B          SACL SCRACH
A0426 0124 014D          ADD  BIAS,1
A0427 0125 5026          SACL SAMPLE
A0428 0126 2E1B          LAC  SCRACH,14   ; quantized value
A0429 0127 7001          LARK 0,1
A0430 0128 F900          B    FINISH
      0129 0172
A0431 012A 174C  SCL2T3  SUB  ONE,7       ; exp = 3 or 2
A0432 012B FD00          BGEZ SCAL3A
      012C 0136
A0433 012D 2360  SCAL2A  LAC  M15,3       ; exp = 2
A0434 012E 791B          AND  SCRACH      ; mask for mantissa
A0435 012F 501B          SACL SCRACH
A0436 0130 024D          ADD  BIAS,2
A0437 0131 5026          SACL SAMPLE
A0438 0132 2D1B          LAC  SCRACH,13   ; quantized value
A0439 0133 7002          LARK 0,2
A0440 0134 F900          B    FINISH
      0135 0172
A0441 0136 2460  SCAL3A  LAC  M15,4       ; exp = 3
A0442 0137 791B          AND  SCRACH      ; mask for mantissa
A0443 0138 501B          SACL SCRACH
A0444 0139 034D          ADD  BIAS,3
A0445 013A 5026          SACL SAMPLE,12
A0446 013B 2C1B          LAC  SCRACH,12   ; quantized value
A0447 013C 7003          LARK 0,3
A0448 013D F900          B    FINISH
      013E 0172
A0449 013F 1970  SCL4T7  SUB  THREE,9     ; exp = 7 - 6 or 5 - 4
A0450 0140 FD00          BGEZ SCL6T7
      0141 0157
A0451 0142 0A4C  SCL4T5  ADD  ONE,10      ; exp = 5 or 4
A0452 0143 FD00          BGEZ SCAL5A
      0144 014E
A0453 0145 2560  SCAL4A  LAC  M15,5       ; exp = 4
A0454 0146 791B          AND  SCRACH      ; mask for mantissa
A0455 0147 501B          SACL SCRACH
A0456 0148 044D          ADD  BIAS,4
A0457 0149 5026          SACL SAMPLE
A0458 014A 2B1B          LAC  SCRACH,11   ; quantized value
A0459 014B 7004          LARK 0,4
A0460 014C F900          B    FINISH
      014D 0172
```

```
A0461 014E 266D  SCAL5A  LAC  M15,6       ; exp = 5
A0462 014F 791B          AND  SCRACH      ; mask for mantissa
A0463 0150 501B          SACL SCRACH
A0464 0151 054D          ADD  BIAS,5
A0465 0152 5026          SACL SAMPLE,10
A0466 0153 2A1B          LAC  SCRACH,10
A0467 0154 7005          LARK 0,5         ; quantized value
A0468 0155 F900          B    FINISH
      0156 0172
A0469 0157 1B4C  SCL6T7  SUB  ONE,11
A0470 0158 FD00          BGEZ SCAL7A
      0159 0163
A0471 015A 276D  SCAL6A  LAC  M15,7       ; exp = 6
A0472 015B 791B          AND  SCRACH      ; mask for mantissa
A0473 015C 501B          SACL SCRACH
A0474 015D 064D          ADD  BIAS,6
A0475 015E 5026          SACL SAMPLE,9
A0476 015F 291B          LAC  SCRACH,9    ; quantized value
A0477 0160 7006          LARK 0,6
A0478 0161 F900          B    FINISH
      0162 0172
A0479 0163 1C4C  SCAL7A  SUB  ONE,12      ; exp = 7
A0480 0164 FA00          BLZ  NORMLA      ; mag > 8191 ?
      0165 016B
A0481 0166 2767  SATCHA  LAC  K63,7       ; save maximum quantized value
A0482 0167 5026          SACL SAMPLE
A0483 0168 7E7F          LACK 127         ; save maximum alaw magnitude
A0484 0169 F900          B    CLNUPA
      016A 0176
A0485 016B 286D  NORMLA  LAC  M15,8       ; mask for mantissa
A0486 016C 791B          AND  SCRACH
A0487 016D 501B          SACL SCRACH
A0488 016E 074D          ADD  BIAS,7
A0489 016F 5026          SACL SCRACH,8    ; quantized value
A0490 0170 281B          LAC  SCRACH
A0491 0171 7007          LARK 0,7
A0492 0172 581B  FINISH  SACH SCRACH      ; save normalized mantissa
A0493 0173 201B          LAC  SCRACH
A0494 0174 3021          SAR  0,TEMP1
A0495 0175 0421          ADD  TEMP1,4     ; add exponent
A0496 0176 0724  CLNUPA  ADD  TEMP4,7
A0497 0177 7947          AND  M255
A0498 0178 501B          SACL SCRACH      ; signed magnitude of alaw-PCM
A0499 0179 2026          LAC  SAMPLE
A0500 017A 7824          XOR  TEMP4
A0501 017B 1024          SUB  TEMP4
A0502 017C 5026          SACL SAMPLE
A0503 017D F800          CALL AQUAN       ; 2's complement of quantized sample
A0504 017E 02AA
A0505              *;
A0506 017F F800          CALL SYNC
      0180 0188
A0507              *;
A0508 0181 787F          XOR  M0080       ; flip bits for transmission
A0509 0182 501B          SACL SCRACH
A0510 0183 491B          OUT  SCRACH,DAC  ; output A-law PCM
```

17. 32-kbit/s ADPCM with the TMS32010

```
A0511 0184 F600  ALAWR   BIOZ  RCVA         ; wait for next sample
      0185 00FD
      0186 F900
A0512 0187 0184          B     ALAWR
```

```
A0514         *****************************************************
A0515         *  SYNCHRONOUS CODING ADJUSTMENT
A0516         *
A0517         *  INPUT:  LOG PCM SAMPLE       -- SP  (SCRACH)
A0518         *          RECEIVED ADPCM       -- I   (TEMP1)
A0519         *          REGENERATED ADPCM    -- ID  (TEMP1)
A0520         *
A0521         *  OUTPUT: ADJUSTED LOG PCM     -- SD  (SCRACH)
A0522         *
A0523         *  NOTATION:   I   -- 4b SM (Q0)
A0524         *              ID  -- 4b SM (Q0)
A0525         *              SP  -- 8b SM (Q4)
A0526         *              SD  -- 8b SM (Q4)
A0527         *
A0528         *                  +-----+
A0529         *                  |     |
A0530         *             ---->| ID  |           SD
A0531         *                  |     |-------->
A0532         *             ---->| SYNC|
A0533         *             SP   |     |
A0534         *             ---->|     |
A0535         *                  +-----+
A0536         *
A0537         *
A0538         *
A0539         *
A0540         *
A0541 0188 7838  SYNC    XOR   EIGHT
A0542 0189 5021          SACL  TEMP1        ; flip the polarity bit in ID
A0543 018A 234C          LAC   ONE,3
A0544 018B 7801          XOR   I
A0545 018C 1021          SUB   TEMP1        ; flip the polarity bit in I
A0546 018D FA00          BLZ   IDGT1M       ; ID > IM ... -
      018E 01A0
A0547 018F FF00          BZ    IDEQ1M       ; ID = IM
      0190 01B1
A0548 0191 201B  IDLT1M  LAC   SCRACH       ; ID < IM ... +
A0549 0192 106F          SUB   M127
A0550 0193 FC00          BGZ   SUBONE
      0194 019B
A0551 0195 FF00          BZ    MAXPOS
      0196 0199
A0552 0197 074C          ADD   ONE,7        ; SD = SP + 1 : 0 <= SP < 127
A0553 0198 7FBD          RET
A0554 0199 7E7F  MAXPOS  LACK  127          ; SD = 127 : SP = 127
A0555 019A 7FBD          RET
A0556 019B 104C  SUBONE  SUB   ONE          ; SD = 0 : SP = 128
A0557 019C FF00          BZ    ANOMLE
      019D 019F
A0558 019E 006F          ADD   M127         ; SD = SP + 1 : 255 >= SP > 128
A0559 019F 7FBD  ANOMLE  RET
A0560 01A0 201B  IDGT1M  LAC   SCRACH
A0561 01A1 174C          SUB   ONE,7
A0562 01A2 FD00          BGEZ  ADDONE
      01A3 01A8
A0563 01A4 006F          ADD   M127         ; SD = SP - 1 : 0 < SP <= 127
A0564 01A5 FA00          BLZ   ANOMLY
```

17. 32-kbit/s ADPCM with the TMS32010

```
A0565 01A6 01AF        RET
A0566 01A7 7FBD  ADDONE SUB   M127
A0567 01A8 106F        BGEZ  MAXNEG
      01A9 FD00
      01AA 01AD
A0568 01AB 084C        ADD   ONE,8    ; SD = SP - 1  ; 255 > SP >= 128
A0569 01AC 7FBD        RET
A0570 01AD 7EFF  MAXNEG LACK  255      ; SD = 255
A0571 01AE 7FBD        RET
A0572 01AF 7E80  ANOMLY LACK  128      ; SD = 128
A0573 01B0 7FBD        RET
A0574 01B1 201B  IDEQ1M LAC   SCRACH   ; SD = 0
A0575 01B2 7FBD        RET              ; SD = SP
```

```
0002                     COPY   SIGD1F.ASM
B0001  *****************************************************
B0002  *
B0003  *  SIGD1F
B0004  *
B0005  *  Implements the following modules (per CCITT spec):
B0006  *
B0007  *  DELAY D -- delay of DQ and SR derivatives
B0008  *  FMULT   -- Bn * DQn, An * SRn
B0009  *  DELAY A -- (implicit in use of last frames data)
B0010  *  ACCUM   -- Accumulate partial products for SEZ, SE
B0011  *
B0012  *  LIMA    -- compute AL(k)
B0013  *  MIX     -- compute Y(k)
B0014  *
B0015  *****************************************************
B0016  *  compute SEZ-- partial signal estimate
B0017  *
B0018  *  SEZ(k) = B1(k-1)*DQ(k-1) + ... + B6(k-1)*DQ(k-6)
B0019  *
B0020  *  Multiplies are done in floating pt
B0021  *  DQ's are stored in f.p. notation
B0022  *  B's are floated each pass
B0023  *
B0024  *****************************************************
B0025  *  FLOATING POINT MULTIPLY (FMULT)
B0026  *
B0027  *  INPUT:  QUANTIZED DIFFERENCE    -- DQn (DQnEXP/DQnMAN)
B0028  *          PREDICTOR COEFFICIENTS  -- Bn
B0029  *
B0030  *  OUTPUT: FILTER TAP OUTPUTS      -- WBn (SUMn)
B0031  *
B0032  *  NOTATION: DQnEXP   -- 4b + offset
B0033  *            DQnMAN*8 -- 9b magnitude
B0034  *            Bn       -- 16b TC (Q14)
B0035  *            SUMn     -- 16b TC (Q1)
B0036  *
B0037  *
B0038  *
B0039  *
B0040  *
B0041  *
B0042  *
B0043  *         -1      -1      -1      -1      -1      -1
B0044  * DQ      N       N       N       N       N       N
B0045  *  >---o---->--o---->--o---->--o---->--o---->--o----o---> SEZ
B0046  *
B0047  *      √B1(k)  √B2(k)  √B3(k)  √B4(k)  √B5(k)  √B6(k)
B0048  *
B0049  *
B0050  *      WB1     WB2     WB3     WB4     WB5     WB6     SEZ
B0051  *
B0052  *
B0053  01B3 2E0F  SIGD1F LAC   B6,14    ; compute B6*DQ5
B0054  01B4 F800         CALL  FLOAT    ; ret/w mantissa in TEMP1; exp in ac
       01B5 04DE
B0055  01B6 001A         ADD   DQ5EXP
```

17. 32-kbit/s ADPCM with the TMS32010

```
B0056  01B7  5022        SACL  SUM1
B0057  01B8  3822        LAR   0,SUM1       ; exp of product offset by table add
B0058  01B9  6A60        LT    DQ5MAN       ; scaled up by 2**3
B0059  01BA  277D        LAC   THREE,7      ; multiply fudge factor
B0060  01BB  6D21        MPY   TEMP1
B0061  01BC  6C80        LTA   *,0          ; mult mant, add 48, fetch shift fac
B0062  01BD  796E        AND   KFF80
B0063  01BE  5021        SACL  TEMP1
B0064  01BF  6D21        MPY   TEMP1        ; apply shift factor = f(exp)
B0065  01C0  7F8E        PAC
B0066  01C1  FA00        BLZ   RS1
       01C2  04B6
B0067  01C3  5922        SACH  SUM1,1       ; exp >= 26
B0068  01C4  660F  CHK1  ZALS  B6
B0069  01C5  785A        XOR   SDQ6         ; check sign of product
B0070  01C6  7948        AND   K32768
B0071  01C7  FF00        BZ    POS1
       01C8  01CC
B0072  01C9  7F89  NEG1  ZAC                ; negate if necessary
B0073  01CA  1022        SUB   SUM1
B0074  01CB  5022        SACL  SUM1
B0075  01CC  2E0E  POS1  LAC   B5,14        ; compute B5*DQ4
B0076  01CD  F800        CALL  FLOAT        ; ret/w mantissa in TEMP1; exp in ac
       01CE  04DE
B0077  01CF  0019        ADD   DQ4EXP       ; exp of product offset by table add
B0078  01D0  6919        DMOV  DQ4EXP       ; scaled up by 2**3
B0079  01D1  5023        SACL  SUM2         ; multiply fudge factor
B0080  01D2  3823        LAR   0,SUM2
B0081  01D3  6B5F        LTD   DQ4MAN
B0082  01D4  277D        LAC   THREE,7
B0083  01D5  6D21        MPY   TEMP1
B0084  01D6  6C80        LTA   *,0          ; mult mant, add 48, fetch shift fac
B0085  01D7  796E        AND   KFF80
B0086  01D8  5021        SACL  TEMP1
B0087  01D9  6D21        MPY   TEMP1        ; apply shift factor = f(exp)
B0088  01DA  7F8E        PAC
B0089  01DB  FA00        BLZ   RS2
       01DC  04BB
B0090  01DD  5923        SACH  SUM2,1       ; exp >= 26
B0091  01DE  660E  CHK2  ZALS  B5
B0092  01DF  7859        XOR   SDQ5         ; check sign of product
B0093  01E0  7948        AND   K32768
B0094  01E1  FF00        BZ    POS2
       01E2  01E6
B0095  01E3  7F89  NEG2  ZAC                ; negate if necessary
B0096  01E4  1023        SUB   SUM2
B0097  01E5  5023        SACL  SUM2
B0098  01E6  2E0D  POS2  LAC   B4,14        ; compute B4*DQ3
B0099  01E7  F800        CALL  FLOAT        ; ret/w mantissa in TEMP1; exp in ac
       01E8  04DE
B0100  01E9  0018        ADD   DQ3EXP       ; exp of product offset by table add
B0101  01EA  6918        DMOV  DQ3EXP       ; scaled up by 2**3
B0102  01EB  5025        SACL  SUM3         ; multiply fudge factor
B0103  01EC  3825        LAR   0,SUM3
B0104  01ED  6B5E        LTD   DQ3MAN
B0105  01EE  277D        LAC   THREE,7
B0106  01EF  6D21        MPY   TEMP1
```

```
B0107  01F0  6C80        LTA   *,0
B0108  01F1  796E        AND   KFF80        ; mult mant, add 48, fetch shift fac
B0109  01F2  5021        SACL  TEMP1
B0110  01F3  6D21        MPY   TEMP1        ; apply shift factor = f(exp)
B0111  01F4  7F8E        PAC
B0112  01F5  FA00        BLZ   RS3
       01F6  04C0
B0113  01F7  5925        SACH  SUM3,1       ; exp >= 26
B0114  01F8  660D  CHK3  ZALS  B4
B0115  01F9  7858        XOR   SDQ4         ; check sign of product
B0116  01FA  7948        AND   K32768
B0117  01FB  FF00        BZ    POS3
       01FC  0200
B0118  01FD  7F89  NEG3  ZAC                ; negate if necessary
B0119  01FE  1025        SUB   SUM3
B0120  01FF  5025        SACL  SUM3
B0121  0200  2E0C  POS3  LAC   B3,14        ; compute B3*DQ2
B0122  0201  F800        CALL  FLOAT        ; ret/w mantissa in TEMP1; exp in ac
       0202  04DE
B0123  0203  0017        ADD   DQ2EXP       ; exp of product offset by table add
B0124  0204  6917        DMOV  DQ2EXP       ; scaled up by 2**3
B0125  0205  501E        SACL  SUM4         ; multiply fudge factor
B0126  0206  381E        LAR   0,SUM4
B0127  0207  6B5D        LTD   DQ2MAN
B0128  0208  277D        LAC   THREE,7
B0129  0209  6D21        MPY   TEMP1
B0130  020A  6C80        LTA   *,0          ; mult mant, add 48, fetch shift fac
B0131  020B  796E        AND   KFF80
B0132  020C  5021        SACL  TEMP1
B0133  020D  6D21        MPY   TEMP1        ; apply shift factor = f(exp)
B0134  020E  7F8E        PAC
B0135  020F  FA00        BLZ   RS4
       0210  04C5
B0136  0211  591E        SACH  SUM4,1       ; exp >= 26
B0137  0212  660C  CHK4  ZALS  B3
B0138  0213  7857        XOR   SDQ3         ; check sign of product
B0139  0214  7948        AND   K32768
B0140  0215  FF00        BZ    POS4
       0216  021A
B0141  0217  7F89  NEG4  ZAC                ; negate if necessary
B0142  0218  101E        SUB   SUM4
B0143  0219  501E        SACL  SUM4
B0144  021A  2E0B  POS4  LAC   B2,14        ; compute B2*DQ1
B0145  021B  F800        CALL  FLOAT        ; ret/w mantissa in TEMP1; exp in ac
       021C  04DE
B0146  021D  0016        ADD   DQ1EXP       ; exp of product offset by table add
B0147  021E  6916        DMOV  DQ1EXP       ; scaled up by 2**3
B0148  021F  501F        SACL  SUM5         ; multiply fudge factor
B0149  0220  381F        LAR   0,SUM5
B0150  0221  6B5C        LTD   DQ1MAN
B0151  0222  277D        LAC   THREE,7
B0152  0223  6D21        MPY   TEMP1
B0153  0224  6C80        LTA   *,0          ; mult mant, add 48, fetch shift fac
B0154  0225  796E        AND   KFF80
B0155  0226  5021        SACL  TEMP1
B0156  0227  6D21        MPY   TEMP1        ; apply shift factor = f(exp)
B0157  0228  7F8E        PAC
```

```
B0158  0229 FA00         BLZ    RS5       ; exp >= 26
       022A 04CA
B0159  022B 591F  CHK5   SACH   SUM5,1    ; exp < 26
B0160  022C 660B         ZALS   B2        ; check sign of product
B0161  022D 7856         XOR    SDQ2
B0162  022E 7948         AND    K32768
B0163  022F FF00         BZ     POS5
       0230 0234
B0164  0231 7F89  NEG5   ZAC              ; negate if necessary
B0165  0232 101F         SUB    SUM5
B0166  0233 501F         SACL   SUM5
B0167  0234 2E0A  POS5   LAC    B1,14     ; compute B1*DQ
B0168  0235 F800         CALL   FLOAT
       0236 04DE
B0169  0237 0015         ADD    DQEXP
B0170  0238 6915         DMOV   DQEXP
B0171  0239 5020         SACL   SUM6
B0172  023A 3820         LAR    0,SUM6    ; exp of product offset by table addr
B0173  023B 6B5B         LTD    DQMAN     ; scaled up by 2**3
B0174  023C 277D         LAC    THREE,7   ; multiply fudge factor
B0175  023D 6D21         MPY    TEMP1
B0176  023E 6C80         LTA    *,0       ; mult mant, add 48, fetch shift factor
B0177  023F 796E         AND    KFF80
B0178  0240 5021         SACL   TEMP1
B0179  0241 6D21         MPY    TEMP1     ; apply shift factor = f(exp)
B0180  0242 7F8E         PAC
B0181  0243 FA00         BLZ    RS6       ; exp >= 26
       0244 04CF
B0182  0245 5920  CHK6   SACH   SUM6,1    ; exp < 26
B0183  0246 660A         ZALS   B1        ; check sign of product
B0184  0247 7855         XOR    SDQ1
B0185  0248 7948         AND    K32768
B0186  0249 FF00         BZ     POS6
       024A 024E
B0187  024B 7F89  NEG6   ZAC              ; negate if necessary
B0188  024C 1020         SUB    SUM6
B0189  024D 5020         SACL   SUM6
B0190  024E        POS6   EQU    $
B0191        ; ************************************************
B0192        ;
B0193        ; compute SE -- signal estimate
B0194        ;
B0195        ; SE = A1(k-1)*SR(k-1) + A2(k-1)*SR(k-1) + SEZ(k)
B0196        ;
B0197        ; Multiplies are DONE in floating pt
B0198        ; SR's are stored in f.p. notation
B0199        ; A's are floated each pass
B0200        ;
B0201        ; ************************************************
B0202        ;* FLOATING POINT MULTIPLY (FMULT)
B0203        ;*
B0204        ;* INPUT:  RECONSTRUCTED SIGNAL  -- SRn (SRnEXP/SRnMAN)
B0205        ;*         PREDICTOR COEFFICIENTS -- An
B0206        ;*
B0207        ;* OUTPUT: FILTER TAP OUTPUTS    -- WAn (SUMn+6)
B0208        ;*
B0209        ;*
```

```
         ;* NOTATION: SRnEXP   -- 4b + offset
         ;*           SRnMAN*8 -- 9b magnitude
         ;*           An       -- 16b TC (Q14)
         ;*           SUMn+6   -- 16b TC (Q1)
         ;*
         ;*                   SR          -1        -1
         ;*                 o-->--z--->--z--->--o
         ;*                                vA1(k)  vA2(k)
         ;*            o---->---<----<---<---o
         ;*            SE     WA1      WA2        SEZ
```

```
B0210
B0211
...
B0229
B0230  024E 2E12  GETSE  LAC    A2,14     ; compute A2*SR1
B0231  024F F800         CALL   FLOAT
       0250 04DE
B0232  0251 001D         ADD    SR1EXP
B0233  0252 5027         SACL   SUM7
B0234  0253 3827         LAR    0,SUM7    ; exp of product offset by table addr
B0235  0254 6A53         LT     SR1MAN    ; scaled up by 2**3
B0236  0255 277D         LAC    THREE,7   ; multiply fudge factor
B0237  0256 6D21         MPY    TEMP1
B0238  0257 6C80         LTA    *,0       ; mult mant, add 48, fetch shift factor
B0239  0258 796E         AND    KFF80
B0240  0259 5021         SACL   TEMP1
B0241  025A 6D21         MPY    TEMP1     ; apply shift factor = f(exp)
B0242  025B 7F8E         PAC
B0243  025C FA00         BLZ    RS11      ; exp >= 26
       025D 04D4
B0244  025E 5927  CHK11  SACH   SUM7,1    ; exp < 26
B0245  025F 6612         ZALS   A2        ; check sign of product
B0246  0260 7814         XOR    SR1
B0247  0261 7948         AND    K32768
B0248  0262 FF00         BZ     POS11
       0263 0267
B0249  0264 7F89  NEG11  ZAC              ; negate if necessary
B0250  0265 1027         SUB    SUM7
B0251  0266 5027         SACL   SUM7
B0252  0267 2E11  POS11  LAC    A1,14     ; compute A1*SR
B0253  0268 F800         CALL   FLOAT
       0269 04DE
B0254  026A 001C         ADD    SREXP     ; exp of product offset by table addr
B0255  026B 691C         DMOV   SREXP     ; scaled up by 2**3
B0256  026C 5028         SACL   SUM8      ; multiply fudge factor
B0257  026D 3828         LAR    0,SUM8
B0258  026E 6852         LTD    SRMAN
B0259  026F 277D         LAC    THREE,7
B0260  0270 6D21         MPY    TEMP1
B0261  0271 6C80         LTA    *,0       ; mult mant, add 48, fetch shift factor
B0262  0272 796E         AND    KFF80
```

17. 32-kbit/s ADPCM with the TMS32010

```
CCITT      32010 FAMILY MACRO ASSEMBLER    PC2.1 84.107    16:36:03 03-20-85
                                                                  PAGE 0021

B0263  0273  5021          SACL   TEMP1
B0264  0274  6D21          MPY    TEMP1
B0265  0275  7F8E          PAC
B0266  0276  FA00          BLZ    RS21        ; apply shift factor = f(exp)
       0277  04D9
B0267  0278  5928   CHK21  SACH   SUM8,15     ; exp >= 26
B0268  0279  6913          DMOV   SR          ; exp < 26
B0269  027A  6613          ZALS   SR          ; check sign of product
B0270  027B  7811          XOR    A1
B0271  027C  7948          AND    K32768
       027D  FF00
B0272  027E  0282          BZ     POS21
B0273  027F  7F89   NEG21  ZAC                ; negate if necessary
B0274  0280  1028          SUB    SUM8
B0275  0281  5028          SACL   SUM8
B0276  0282  0282   POS21  EQU    $

B0277  ;****
B0278  ;*  ACCUMULATE FILTER TAP OUTPUTS (ACCUM)
B0279  ;*
B0280  ;*  INPUT: FILTER TAP OUTPUTS -- WAn & WBn (SUMm)
B0281  ;*
B0282  ;*  OUTPUT: PARTIAL SUM OF ZEROES FILTER -- SEZ
B0283  ;*          SIGNAL ESTIMATE              -- SE
B0284  ;*
B0285  ;*  NOTATION: SUMm -- 16b TC (Q1) [sign extended]
B0286  ;*            SEZ  -- 15b TC (Q0) [sign extended]
B0287  ;*            SE   -- 15b TC (Q0) [sign extended]
B0288  ;*
B0289  ;****
B0290  ;*

B0292  0282  2F20          LAC    SUM6,15     ; accumulate products
B0293  0283  0F1F          ADD    SUM5,15
B0294  0284  0F1E          ADD    SUM4,15
B0295  0285  0F25          ADD    SUM3,15
B0296  0286  0F23          ADD    SUM2,15
B0297  0287  0F22          ADD    SUM1,15
B0298  0288  5904          SACH   SEZ,1
B0299  0289  0F27          ADD    SUM7,15
B0300  028A  0F28          ADD    SUM8,15
B0301  028B  5903          SACH   SE,1
B0302  028C  2F03          LAC    SE,15
B0303  028D  5803          SACH   SE

B0304  ;****
B0305  ;*  limit speed control parameter: AL <= 1.0
B0306  ;*
B0307  ;*   AL = 1     if APP > 1
B0308  ;*   AL = APP   if APP <= 1
B0309  ;*
B0310  ;*   INPUT: UNLIMITED SPEED CONTROL -- AP (APP)
B0311  ;*
B0312  ;*   OUTPUT: LIMITED SPEED CONTROL -- AL
B0313  ;*
B0314  ;*   NOTATION: APP -- unsigned 10b (Q8)
B0315  ;*             AL  -- unsigned 7b (Q6)
B0316  ;*
B0317  ;****
```

```
CCITT      32010 FAMILY MACRO ASSEMBLER    PC2.1 84.107    16:36:03 03-20-85
                                                                  PAGE 0022

B0318
B0319
B0320  028E  2C4C   LIMA   LAC    ONE,12
B0321  028F  5006          SACL   AL
B0322  0290  2005          LAC    APP
B0323  0291  18AC          SUB    ONE,8
B0324  0292  FD00          BGEZ   MIX         ; check if APP >=1
       0293  0297
B0325  0294  2405          LAC    APP,4       ;APP >= 1
B0326  0295  7970          AND    MFFC0
B0327  0296  5006          SACL   AL          ;APP < 1

B0329  ;****
B0330  ;*  MIX
B0331  ;*  Form linear combination of fast and slow scale factors
B0332  ;*
B0333  ;*  Y(k) = (1-AL(k))*YL(k-1) + AL(k)*YU(k-1)
B0334  ;*
B0335  ;*  INPUT: SLOW QUANTIZER SCALE FACTOR -- YL (YLL/YLH)
B0336  ;*         FAST QUANTIZER SCALE FACTOR -- YU
B0337  ;*         LIMITED SPEED CONTROL       -- AL
B0338  ;*
B0339  ;*  OUTPUT: QUANTIZER SCALE FACTOR          -- Y
B0340  ;*          RESCALED QUANTIZER SCALE FACTOR -- YOVER4
B0341  ;*
B0342  ;*  NOTATION: YL     -- 19b unsigned (Q15)
B0343  ;*               stored as:
B0344  ;*                   low 15b -- YLL
B0345  ;*                   hi  4b  -- YLH
B0346  ;*            YU     -- 13b unsigned (Q9)
B0347  ;*            AL     -- 7b unsigned (Q6)
B0348  ;*            Y      -- 13b unsigned (Q9)
B0349  ;*            YOVER4 -- 11b unsigned (Q7)
B0350  ;*
B0351  ;****
B0352  ;*

B0353  0297  2A4A   MIX    LAC    YLL,10      ; shift yl right by 6
B0354  0298  5823          SACH   YLH,9
B0355  0299  2949          LAC    YLH,9
B0356  029A  0023          ADD    TEMP3
B0357  029B  5023          SACL   TEMP3
B0358  029C  204E          LAC    YU
B0359  029D  1023          SUB    TEMP3       ; YL>>6
B0360  029E  5021          SACL   TEMP1
B0361  029F  6A06          LT     AL          ; YU-(YL>>6)
B0362  02A0  6D21          MPY    TEMP1
B0363  02A1  7F8E          PAC                ; AL*(YU-(YL>>6))
B0364  02A2  FD00          BGEZ   NONNEG      ; negative truncation
       02A3  02A5
B0365  02A4  0072          ADD    M4095
B0366  02A5  0C23   NONNEG ADD    TEMP3,12
B0367  02A6  5C09          SACH   Y,4
B0368  02A7  2E09          LAC    Y,14
B0369  02A8  5829          SACH   YOVER4      ; compute and save y>>2
B0370  02A9  7F80          RET                ; ret from SIGDIF
```

```
CCITT

0003            COPY    AQUAN.ASM
C0001
C0002   *; ********************************************************
C0003   *; DIFFERENCE SIGNAL COMPUTATION
C0004   *; ********************************************************
C0005   *;
C0006   *; INPUT:  LINEAR PCM SAMPLE -- SL (SAMPLE)
C0007   *;         SIGNAL ESTIMATE   -- SE
C0008   *;
C0009   *; OUTPUT: DIFFERENCE SIGNAL -- D (accumulator)
C0010   *;
C0011   *; NOTATION:  SL -- 14b TC (Q0) [sign extended]
C0012   *;            SE -- 15b TC (Q0) [sign extended]
C0013   *;            D  -- 16b TC (Q0)
C0014   *;
C0015   *;                  +----+
C0016   *;          SL ---->|    |       D
C0017   *;                  | SUBTA |----->
C0018   *;          SE ---->|    |
C0019   *;                  +----+
C0020   *;
C0021   *;
C0022   *; *******************************************************
C0023   *;
C0024   02AA 2026   AQUAN  LAC   SAMPLE   ; compute difference sig
C0025   02AB 1003          SUB   SE
C0026   *;
C0027   *; *******************************************************
C0028   *; ADAPTIVE QUANTIZER
C0029   *; *******************************************************
C0030   *;
C0031   *; Implements the following modules (per CCITT spec):
C0032   *;
C0033   *;  LOG   -- computes log of difference signal
C0034   *;  SUBTB -- scales log by subtracting Y
C0035   *;  QUAN  -- computes 4b output
C0036   *;
C0037   *; *******************************************************
C0038   *; INPUT:  DIFFERENCED PCM SAMPLE -- D (accumulator)
C0039   *;         QUANTIZER SCALE FACTOR -- Y (YOVER4)
C0040   *;
C0041   *; OUTPUT: ADPCM OUTPUT SAMPLE    -- I
C0042   *;
C0043   *; NOTATION:  D      -- 16b TC (Q0)
C0044   *;            YOVER4 -- 11b SM (Q7)  POSITIVE VALUE ONLY
C0045   *;            I      -- 4b SM (Q0)
C0046   *;
C0047   *;                           DS
C0048   *;               +-----+  DL    +-----+  DLN   +-----+   I
C0049   *;         D     |     |------->|     |------->|     |--->
C0050   *;       ------->| LOG |        |SUBTB|        |QUAN |
C0051   *;               |     |  +---->|     |        |     |   ^
C0052   *;               +-----+  |     +-----+        +-----+
C0053   *;                        |
C0054   *;                        | QUAN
C0055   *;                        |
C0056   *;
```

```
CCITT

C0057   *;
C0058   *;                          ; Y
C0059   *;
C0060   *; *******************************************************
C0061   *; First get log of difference signal -- express
C0062   *; as unsigned 11b number (4b exp/7b mantissa)
C0063   *;
C0064   *; First order log approximation: log2 (1+x) = x.
C0065   *;
C0066   02AC 5824        SACH   TEMP4    ; -1 if neg; 0 if positive (DS)
C0067   02AD 7F88        ABS
C0068   02AE 5021        SACL   TEMP1
C0069   02AF 184C        SUB    ONE,8    ; binary search to get exponent
C0070   02B0 F000        BGEZ   C8TO14
        02B1 02E7
C0071   02B2 0460 GETEXP ADD    M15,4    ; TEMP1-16   ; exp = 0-7
C0072   02B3 F000        BGEZ   C4TO7
        02B4 02CE
C0073   02B5 027D C0TO7  ADD    THREE,2  ; TEMP1-4    ; exp = 0-3
C0074   02B6 F000        BGEZ   C2TO3
        02B7 02C3
C0075   02B8 014C C0TO3  ADD    ONE,1    ; TEMP1-2    ; exp = 0-1
C0076   02B9 F000        BGEZ   C0TO1
        02BA 02BF
C0077   02BB 7000 EXP0   LARK   0,0      ; exp = 0
C0078   02BC 2721        LAC    TEMP1,7
C0079   02BD F900        B      GETMAN   ; save exponent and get mantissa
        02BE 0321
C0080   02BF 7001 EXP1   LARK   0,1      ; exp = 1
C0081   02C0 2621        LAC    TEMP1,6
C0082   02C1 F900        B      GETMAN
        02C2 0321
C0083   02C3 124C C2TO3  SUB    ONE,2    ; TEMP1-8    ; exp = 2-3
C0084   02C4 F000        BGEZ   EXP3
        02C5 02CA
C0085   02C6 7002 EXP2   LARK   0,2      ; exp = 2
C0086   02C7 2521        LAC    TEMP1,5
C0087   02C8 F900        B      GETMAN
        02C9 0321
C0088   02CA 7003 EXP3   LARK   0,3      ; exp = 3
C0089   02CB 2421        LAC    TEMP1,4
C0090   02CC F900        B      GETMAN
        02CD 0321
C0091   02CE 147D C4TO7  SUB    THREE,4  ; TEMP1-64   ; exp = 4-7
C0092   02CF F000        BGEZ   C6TO7
        02D0 02DC
C0093   02D1 054C C4TO5  ADD    ONE,5    ; TEMP1-32   ; exp = 4-5
C0094   02D2 F000        BGEZ   EXP5
        02D3 02D8
C0095   02D4 7004 EXP4   LARK   0,4      ; exp = 4
C0096   02D5 2321        LAC    TEMP1,3
C0097   02D6 F900        B      GETMAN
        02D7 0321
C0098   02D8 7005 EXP5   LARK   0,5      ; exp = 5
C0099   02D9 2221        LAC    TEMP1,2
C0100   02DA F900        B      GETMAN
        02DB 0321
```

```
C0101 02DC 164C   C6TO7  SUB  ONE,6        ; TEMP1-128
C0102 02DD FD00          BGEZ EXP7           exp = 6-7
      02DE 02E3
C0103 02DF 7006   EXP6   LARK 0,6          ; exp = 6
C0104 02E0 2121          LAC  TEMP1,1
C0105 02E1 F900          B    GETMAN
      02E2 0321
C0106 02E3 7007   EXP7   LARK 0,7          ; exp = 7
C0107 02E4 2021          LAC  TEMP1
C0108 02E5 F900          B    GETMAN
      02E6 0321
C0109 02E7 1B6D   C8TO14 SUB  M15,8        ; TEMP1-4096
C0110 02E8 FD00          BGEZ CATOB          exp = 8-14
      02E9 030B
C0111 02EA 0A7D   C8TO11 ADD  THREE,10     ;TEMP1-1024
C0112 02EB FD00          BGEZ CATOB          exp = 8-11
      02EC 02FC
C0113 02ED 094C   C8TO9  ADD  ONE,9        ; TEMP1-512
C0114 02EE FD00          BGEZ EXP9           exp = 8-9
      02EF 02F6
C0115 02F0 7008   EXP8   LARK 0,8          ; exp = 8
C0116 02F1 2F21          LAC  TEMP1,15
C0117 02F2 5821          SACH TEMP1
C0118 02F3 2021          LAC  TEMP1
C0119 02F4 F900          B    GETMAN
      02F5 0321
C0120 02F6 7009   EXP9   LARK 0,9          ; exp = 9
C0121 02F7 2E21          LAC  TEMP1,14
C0122 02F8 5821          SACH TEMP1
C0123 02F9 2021          LAC  TEMP1
C0124 02FA F900          B    GETMAN
      02FB 0321
C0125 02FC 1A4C   CATOB  SUB  ONE,10       ; TEMP1-2048
C0126 02FD FD00          BGEZ EXP11          exp = 10-11
      02FE 0305
C0127 02FF 700A   EXP10  LARK 0,10         ; exp = 10
C0128 0300 2D21          LAC  TEMP1,13
C0129 0301 5821          SACH TEMP1
C0130 0302 2021          LAC  TEMP1
C0131 0303 F900          B    GETMAN
      0304 0321
C0132 0305 700B   EXP11  LARK 0,11         ; exp = 11
C0133 0306 2C21          LAC  TEMP1,12
C0134 0307 5821          SACH TEMP1
C0135 0308 2021          LAC  TEMP1
C0136 0309 F900          B    GETMAN
      030A 0321
C0137 030B 1C70   CCTOE  SUB  THREE,12     ; TEMP1-16384
C0138 030C FD00          BGEZ EXP14          exp = 12-14
      030D 031D
C0139 030E 0D4C   CCTOD  ADD  ONE,13       ; TEMP1-8192
C0140 030F FD00          BGEZ EXP13          exp = 13-14
      0310 0317
C0141 0311 700C   EXP12  LARK 0,12         ; exp = 12
C0142 0312 2B21          LAC  TEMP1,11
C0143 0313 5821          SACH TEMP1
C0144 0314 2021          LAC  TEMP1
```

```
C0145 0315 F900          B    GETMAN
      0316 0321
C0146 0317 700D   EXP13  LARK 0,13         ; exp = 13
C0147 0318 2A21          LAC  TEMP1,10
C0148 0319 5821          SACH TEMP1
C0149 031A 2021          LAC  TEMP1
C0150 031B F900          B    GETMAN
      031C 0321
C0151 031D 700E   EXP14  LARK 0,14         ; exp = 14
C0152 031E 2921          LAC  TEMP1,9
C0153 031F 5821          SACH TEMP1
C0154 0320 2021   GETMAN LAC  TEMP1
C0155 0321 796F          AND  M127
C0156 0322 3021          SAR  0,TEMP1
C0157 0323 0721          ADD  TEMP1,7       ; DL  4e...7m (sign=SGN(D))
C0158
C0159 *
C0160 *
C0161 *  scale LOG D by subtraction (Y>>2 is in YOVER4)
C0162 *
C0163
C0164 *
C0165 0324 0B4C   SUBTB  ADD  ONE,11       ; offset by 2K
C0166 0325 1029          SUB  YOVER4
C0167
C0168 *
C0169 *  16 LEVEL quantizer
C0170 *
C0171 *  Table values defined in CCITT spec p67
C0172 *  Implemented table is offset by 2048
C0173 *
C0174 *
C0175 07F9   ITAB1  EQU  2041                ; bottom of level 1
C0176 087B   ITAB2  EQU  2171                ; bottom of level 2
C0177 08CA   ITAB3  EQU  2250                ; bottom of level 3
C0178 0905   ITAB4  EQU  2309                ; bottom of level 4
C0179 0936   ITAB5  EQU  2358                ; bottom of level 5
C0180 0964   ITAB6  EQU  2404                ; bottom of level 6
C0181 0995   ITAB7  EQU  2453                ; bottom of level 7
C0182 *
C0183 0326 107C   QUAN   SUB  K2309          ; TEMP2-2309
C0184 0327 FD00          BGEZ C14TO7
      0328 033E
C0185 0329 007B   C10TO3 ADD  K138           ; TEMP2-2171    1 = 0-3
C0186 032A FD00          BGEZ C12TO3
      032B 0335
C0187 032C 007A   C10TO1 ADD  K130           ; TEMP2-2041    1 = 0-1
C0188 032D FD00          BGEZ IEQ1
      032E 0332
C0189 032F 7E00   IEQ0   LACK 0              ; 0
      0330 F900          B    GETIM
      0331 0351
C0190 0330 F900
      0331 0351
C0191 0332 7E01   IEQ1   LACK 1              ; 1
C0192 0333 F900          B    GETIM
      0334 0351
C0193 0335 1078   C12TO3 SUB  K79            ; TEMP2-2250    1 = 2-3
C0194 0336 FD00          BGEZ IEQ3
```

17. 32-kbit/s ADPCM with the TMS32010

```
       0337 033B  IEQ2   LACK 2
C0195  0338 7E02         B    GETIM
C0196  0339 F900
       033A 0351
C0197  033B 7E03  IEQ3   LACK 3
C0198  033C F900         B    GETIM
       033D 0351
C0199  033E 1079  C14TO7 SUB  K95     ; TEMP2-2404   I = 4-7
C0200  033F F000         BGEZ C16TO7
       0340 034A
C0201  0341 0064  C15TO6 ADD  K46     ; TEMP2-2358   I = 5-6
C0202  0342 FD00         BGEZ IEQ5
       0343 0347
C0203  0344 7E04  IEQ4   LACK 4
C0204  0345 F900         B    GETIM
       0346 0351
C0205  0347 7E05  IEQ5   LACK 5
C0206  0348 F900         B    GETIM
       0349 0351
C0207  034A 1065  C16TO7 SUB  K49     ; TEMP2-2453   I = 6-7
C0208  034B FD00         BGEZ IEQ7
       034C 0350
C0209  034D 7E06  IEQ6   LACK 6
C0210  034E F900         B    GETIM
       034F 0351
C0211  0350 7E07  IEQ7   LACK 7
C0212  0351 5002  GETIM  SACL IM      ; accumulator = !I!
C0213  0352 7824         XOR  TEMP4   ; add sign bit and flip if necessary
C0214  0353 7960         AND  M15     ; mask for final four-bit value
C0215  0354 7F8D  QDONE  RET          ; return from AQUAN
```

```
               COPY   PRDICT.ASM

D0004    *;  *************************************************
D0001    *;
D0002    *;  ADAPTATION/PREDICTION
D0003    *;
D0005    *;  Implements the following modules per CCITT spec:
D0006    *;
D0007    *;  Inverse Adaptive Quantizer
D0008    *;    RECONST   -- reconstructs D from I
D0009    *;    ADDA      -- adds back scale factor
D0010    *;    ANTILOG   -- log to lin conversion to get DQ
D0011    *;    FLOAT A   -- float DQ
D0012    *;  Scale Factor Adaptation
D0013    *;    FUNCTW    -- map I to log scale factor
D0014    *;    FILTD     -- update fast scale factor
D0015    *;    LIMB      -- limit scale factor
D0016    *;    FILTE     -- update slow scale factor
D0017    *;  Adaptation Speed Control
D0018    *;    FUNCTF    -- map I to F function
D0019    *;    FILTA     -- update short term ave of F
D0020    *;    FILTB     -- update long term ave of F
D0021    *;    SUBTC     -- determ speed control update
D0022    *;              technique
D0023    *;    FILTC     -- update speed control
D0024    *;  Adaptive Predictor
D0025    *;    ADDB      -- compute reconstructed signal
D0026    *;    FLOAT A   -- float SR
D0027    *;    ADDC      -- compute sign of PK
D0028    *;    UPA2      -- update A2 coeff of 2nd order pred
D0029    *;    LIMC      -- limit A2
D0030    *;    UPA1      -- update A1 coeff of 2nd order pred
D0031    *;    LIMD      -- limit A1
D0032    *;    UPB       -- update coeffs of 6th order pred
D0033    *;    XOR       -- compute sign of DQ*DQn
D0034    *;
D0035    *;  NOTE: DELAY A/B/C implicit in timing of MIX/LIMA
D0036    *;        and computation of SEZ/SE
D0037    *;
D0038    *;  *************************************************
D0039    *;
D0040    *;  First convert quantized difference back to log domain.
D0041    *;  This is done by table look-up. Also use ADPCM magnitude
D0042    *;  to look-up the scale-factor multipliers WI and rate-of-
D0043    *;  change weighting function FI.
D0044    *;
D0045  0355 2002  PRDICT LAC  IM
D0046  0356 006A         ADD  INQTAB   ; reconst table
D0047  0357 6721         TBLR TEMP1    ; DQLN
D0048  0358 034C         ADD  ONE,3
D0049  0359 6769         TBLR WI       ; WI table address and offset
D0050  035A 034C         ADD  ONE,3    ;   lookup WI
D0051  035B 6768         TBLR FI       ; FI table address and offset
D0052    *;                                lookup FI
D0053    *;
D0054    *;  INVERSE ADAPTIVE QUANTIZER
D0055    *;
D0056    *;  INPUT:  ADPCM INPUT SAMPLE     -- I (IM->TEMP1)
```

```
;*:                QUANTIZER SCALE FACTOR -- Y (YOVER4)
;*:
;*:     OUTPUT: QUANTIZED DIFFERENCE SIGNAL -- DQ
;*:
;*:     NOTATION:  I           -- 4b SM (Q0)
;*:                IM          -- 3b magnitude (Q0)
;*:                DQLN(TEMP1) -- 12b TC (Q7) [sign extended]
;*:                YOVER4      -- 11b SM (Q7) POSITIVE VALUE ONLY
;*:                DQ          -- 15b TC (Q0) [sign extended]
;*:                DQMAN*8     -- 9b magnitude
;*:                DQEXP       -- 4b magnitude
;*:
;*:                                    DQS
;*:        +----------+      +------+          +---------+        DQ
;*: I      |          | DQLN |      | DQL      |         |  -->
;*: ---->  | RECONST  |------| ADDA |--------->| ANTILOG |----->
;*:        +----------+      +------+          +---------+
;*:                            ^                    ---> V
;*:                            |
;*:                          : Y
;*:
;             ; add back scale factor
D0085 035C 2521  ADDA    LAC   TEMP1,5
D0086 035D 0529          ADD   YOVER4,5
;
;             ; now covert to linear domain
D0090 035E 0C4C  ALOG    ADD   ONE,12        ; inc exponent for floated value
D0091 035F 5C15          SACH  DQEXP,4       ; save exponent + sign ext
D0092 0360 7972          AND   M4095         ; isolate mantissa * 2**5
D0093 0361 0C4C          ADD   ONE,12        ; Alog x = 1 + x
D0094 0362 505B          SACL  DQMAN         ; DQMAN = 0001 XXXX XXX0 0000
D0095 0363 2015          LAC   SHIFT
D0096 0364 0066          ADD   TEMP3         ; add table ptr
D0097 0365 6723          TBLR  TEMP3
D0098 0366 6A23          LT    TEMP3         ; get multiplier
D0099 0367 044C          ADD   ONE,4         ; offset to mask table
D0100 0368 6723          TBLR  TEMP3         ; mask for dqman
D0101 0369 6D5B          MPY   TEMP3         ; adjust mantissa
D0102 036A 7F8E          PAC
D0103 036B 5C10          SACH  DQ,4
D0104 036C 2B4C  ADDSGN  LAC   ONE,11        ; +2048 represents +sign
D0105 036D 5054          SACL  SDQ
D0106 036E 2001          LAC   I             ; check sign
D0107 036F 134C          SUB   ONE,3
D0108 0370 FA00          BLZ   FLTDQ
           0371 0377
D0109 0372 7F89          ZAC
D0110 0373 1010          SUB   DQ            ; I carried negative sign
D0111 0374 5010          SACL  DQ
D0112 0375 2B4B          LAC   MINUS,11      ; -2048 represents -sign
```

```
D0113 0376 5054          SACL  SDQ
;
;****************************************************************************
; FLOAT DQ -- convert 2's comp number to floating
;
;   INPUT:  DQ
;
;   OUTPUT: 4b exponent in DQEXP (saved from log value)
;           6b mantissa*8 in DQMAN (adjusted from log)
;           sign preserved in DQ
;
D0126 0377 255B  FLTDQ   LAC   DQMAN,5 ; 00000000 00000001X XXXXX00 00000000
D0127 0378 5C5B          SACH  DQMAN,4 ; DQMAN = 2**3
D0128 0379 235B          LAC   DQMAN,3 ; DQMAN * 2**3
D0129 037A 7923          AND   TEMP3
D0130 037B 505B          SACL  DQMAN
;
;****************************************************************************
; QUANTIZER SCALE FACTOR ADAPTATION
;****************************************************************************
;
;   INPUT: ADPCM SAMPLE -- I
;
;   OUTPUT: FAST QUANTIZER SCALE FACTOR -- YU (YLL/YLH)
;           SLOW QUANTIZER SCALE FACTOR -- YL (YLL/YLH)
;
;   NOTATION: I   -- 4b SM (Q0)
;             YU  -- 13b unsigned (Q9)
;             YL  -- 19b unsigned (Q15)
;                  stored as:
;                       low 15b -- YLL
;                       hi   4b -- YLH
;
;   WI +-------+YUT +------+YUP +------+
;   -->: FILTD :--->: LIMB :-+->: DELAYB :-----------+     YU
;      +-------+    +------+  |  +------+             :---->
;                            |                       ^
;       +---+                |  +------+YLP +------+YL:
;       :   :                +->: FILTE :--->: DELAYC :--+
;     : Y -->                   +------+    +------+   :
;       +---+                                          v
;
; Update fast adaptation scale factor
;
;  YU(k) = (1-2**-5)*Y(k) + (2**-5)*W(I(k))
;
;   INPUT:  QUANTIZER SCALE FACTOR -- Y
;           SCALE FACTOR MULTIPLIER -- WI
;
;   OUTPUT: FAST QUANTIZER SCALE FACTOR -- YU
;
;   NOTATION: WI -- 12b TC (Q4) [sign extended]
```

```
D0170                              *;                Y  -- 13b unsigned (Q9)
D0171                              *;                YU -- 13b unsigned (Q9)
D0172                              *;
D0173                              *;***********************************
D0174
D0175  037C  2C09   FILTD   LAC   Y,12          ; Y      (Q21)
D0176  037D  1709           SUB   Y,7           ; Y/32   (Q21)
D0177  037E  0069           ADD   WI,12         ; WI/32  (Q21)
D0178  037F  5C4E           SACH  YU,4          ; YU     (Q9)
D0179
D0180          *; limit quant scale factor 1.06 <= YU <= 10.0
D0181
D0182  0380  1C6B   LIMB    SUB   K544,12       ; check lo threshold
D0183  0381  FD00           BGEZ  CHKHI
       0382  0386
D0184  0383  2068           LAC   K544
       0384  F900           B     STRLIM        ; go store limited value
D0185  0385  038A
D0186  0386  1C61   CHKHI   SUB   K4576,12      ; check hi threshold
D0187  0387  FB00           BLEZ  FILTE         ; within limits--continue
       0388  038B
D0188  0389  206C           LAC   K5120
D0189  038A  504E   STRLIM  SACL  YU
D0190
D0191   *; Update slow adaptation scale factor
D0192   *;
D0193   *; YL(k) = (1-2**-6)*YL(k-1) + 2**-6 * YU(k)
D0194   *;
D0195   *; INPUT:  SLOW QUANTIZER SCALE FACTOR -- YL (YLL/YLH)
D0196   *;         FAST QUANTIZER SCALE FACTOR -- YU
D0197   *;
D0198   *; OUTPUT: SLOW QUANTIZER SCALE FACTOR -- YL (YLL/YLH)
D0199   *;
D0200   *; NOTATION: YU -- 13b unsigned (Q9)
D0201   *;           YL -- 19b unsigned (Q15)
D0202   *;                   stored as:
D0203   *;                     low 15b -- YLL
D0204   *;                     hi   4b -- YLH
D0205   *;
D0206
D0207
D0208
D0209  038B  2649   FILTE   LAC   YLH,6         ; shift yl left by 6
D0210  038C  5021           SACL  TEMP1
D0211  038D  2F21           LAC   TEMP1,15      ; YL      (Q21)
D0212  038E  064A           ADD   YLL,6
D0213  038F  1F49           SUB   YLH,15        ; YL/64   (Q21)
D0214  0390  104A           SUB   YLL
D0215  0391  064E           ADD   YU,6          ; YU/64   (Q21)
D0216  0392  5921           SACH  TEMP1,1
D0217  0393  7974           AND   M32767
D0218  0394  5022           SACL  TEMP2         ; result = yl (shifted left by 6)
D0219  0395  2A22           LAC   TEMP2,10      ; shift result right 6 --> 4.Q15
D0220  0396  5822           SACH  TEMP2
D0221  0397  2921           LAC   TEMP1,9       ; YL      (Q15)
D0222  0398  0022           ADD   TEMP2
D0223  0399  5949           SACH  YLH,1
```

```
D0224  039A  7974           AND   M32767
D0225  039B  504A           SACL  YLL
D0226
D0227   *;**********************************************
D0228   *; ADAPTATION SPEED CONTROL
D0229   *;
D0230   *; INPUT:  ADPCM SAMPLE -- I
D0231   *;
D0232   *; OUTPUT: UNLIMITED SPEED CONTROL -- AP (APP)
D0233   *;
D0234   *; NOTATION: I   -- 4b SM  (Q0)
D0235   *;           APP -- 10b unsigned (Q8)
D0236   *;
D0237   *;                                      Y
D0238   *;                                      v
D0239   *;      +--------+DMSP+--+   AX +--------+   +APP +--------+ AP
D0240   *;  FI  |        |----+-->| SUBTC |---->| FILTC |---->| DELAYA |--+-->
D0241   *; +-->| FILTA  |--+      +--------+   +--------+   +--------+
D0242   *;      +--------+  |              ^
D0243   *;                  v
D0244   *;  DMS: +--------+
D0245   *;       +-| DELAYA |
D0246   *;          +--------+
D0247   *;                  v
D0248   *;       +--------+ DMLP
D0249   *;  +-->| FILTB  |
D0250   *;       +--------+
D0251   *;  DML: +--------+
D0252   *;       +-| DELAYA |
D0253   *;          +--------+
D0254
D0255   *; update short term average of FI
D0256   *;
D0257   *; DMS(k) = (1-2**-5)*DMS(k-1) + 2**-5 * FI(k)
D0258   *;
D0259   *; INPUT:  SHORT TERM AVERAGE -- DMS
D0260   *;         FI   -- RATE-OF-CHANGE FUNCTION -- FI
D0261   *;
D0262   *; OUTPUT: SHORT TERM AVERAGE -- DMS
D0263   *;
D0264   *; NOTATION: DMS -- 12b unsigned (Q9)
D0265   *;           FI  -- 7b unsigned  (Q4)
D0266   *;
D0267
D0268
D0269
D0270
D0271
D0272  039C  2F68   FILTA   LAC   FI,15         ; FI/32   (Q24)
D0273  039D  0F07           ADD   DMS,15        ; DMS     (Q24)
D0274  039E  1A07           SUB   DMS,10        ; DMS/32  (Q24)
D0275  039F  5907           SACH  DMS,1
D0276
D0277   *; update long term average of FI
D0278   *;
D0279   *; DML(k) = (1-2**-7)*DML(k-1) + 2**-7 * FI(k)
D0280   *;
```

17. 32-kbit/s ADPCM with the TMS32010

```
D0281       *;
D0282       *;    INPUT:   LONG TERM AVERAGE  -- DML
D0283       *;             RATE-OF-CHANGE FUNCTION -- F1
D0284       *;
D0285       *;    OUTPUT:  LONG TERM AVERAGE -- DML
D0286       *;
D0287       *;    NOTATION:  DML -- 14b unsigned (Q11)
D0288       *;               F1  -- 7b unsigned (Q4)
D0289       *;
D0290       *;
D0291       *;
D0292  03A0 2F68    FILTB  LAC   F1,15        ; F1/128   (Q26)
D0293  03A1 0F08           ADD   DML,15       ; DML      (Q26)
D0294  03A2 1808           SUB   DML,8        ; DML/128  (Q26)
D0295  03A3 5908           SACH  DML,1
D0296       *;
D0297       *;-----------------------------------------------------
D0298       *; Compute mag of diff of short and long term functions of
D0299       *; quantizer output sequence and perform threshold
D0300       *; comparison to compute speed control parameter--low-pass
D0301       *; result.
D0302       *;
D0303       *; APP(k) = (1-2**-4)*APP(k-1) + 2**-3 , if Y < 3 or
D0304       *;                              if |DMS-DML| > 2**-3 * DML
D0305       *;
D0306       *; else
D0307       *;
D0308       *; APP(k) = (1-2**-4)*APP(k-1)
D0309       *;
D0310       *;    INPUT:   SHORT TERM AVERAGE  -- DMS
D0311       *;             LONG TERM AVERAGE   -- DML
D0312       *;             UNLIMITED SPEED CONTROL -- APP
D0313       *;             QUANTIZER SCALE FACTOR -- Y
D0314       *;
D0315       *;    OUTPUT:  UNLIMITED SPEED CONTROL -- APP
D0316       *;
D0317       *;    NOTATION:  APP -- 10b unsigned (Q8)
D0318       *;               Y   -- 13b unsigned (Q9)
D0319       *;               DMS -- 12b unsigned (Q9)
D0320       *;               DML -- 14b unsigned (Q11)
D0321       *;
D0322       *;
D0323  03A4 6505    FILTC  ZALH  APP          ; APP        (Q24)
D0324  03A5 1C05           SUB   APP,12       ; APP/16     (Q24)
D0325  03A6 5805           SACH  APP          ; (1-2**-4)*APP  (Q8)
D0326  03A7 2009           LAC   Y
D0327  03A8 197D           SUB   THREE,9      ; 3          (Q9)
D0328  03A9 FA00           BLZ   ADD18
       03AA 03B3
D0329  03AB 2D08           LAC   DML,13       ; DML/8      (Q27)
D0330  03AC 5823           SACH  TEMP3        ; DML/8      (Q11)
D0331  03AD 2207           LAC   DMS,2        ; DMS        (Q11)
D0332  03AE 1008           SUB   DML          ; DMS-DML    (Q11)
D0333  03AF 7F88           ABS
D0334  03B0 1023           SUB   TEMP3        ; |DMS-DML|-DML/8
D0335  03B1 FA00           BLZ   APRED
       03B2 03B6
```

```
D0336  03B3 2005    ADD18  LAC   APP          ; APP    (Q8)
D0337  03B4 054C           ADD   ONE,5        ; + 1/8  (Q8)
D0338  03B5 5005           SACL  APP
D0339       *
D0340       *;******************************************************
D0341       *; ADAPTIVE PREDICTOR
D0342       *;******************************************************
D0343       *
D0344       *
D0345  03B6         APRED  EQU   $
D0346       *
D0347       *; compute coeff of 6th order predictor
D0348       *
D0349       *; Bi(k) = (1-2**-8)*Bi(k-1) + 2**-7*SGN[DQ(k)]*SGN[DQ(k-i)]]
D0350       *;    for i=1,..6
D0351       *; and Bi is implicitly limited to +/- 2
D0352       *
D0353       *
D0354       *;    NOTATION:     Bn -- 16b TC (Q14)
D0355       *;                  SDQn -- +2048 if sign positive
D0356       *;                          -2048 if sign negative
D0357       *
D0358       *
D0359       *
D0360  03B6 6A5A    GETB6  LT    SDQ6
D0361  03B7 280F           LAC   B6,8         ; B6 * 2**-8 TRUNCATED
D0362  03B8 5821           SACH  TEMP1,15
D0363  03B9 2F0F           LAC   B6,15        ; Q29
D0364  03BA 1F21           SUB   TEMP1,15
D0365  03BB 6D54           MPY   SDQ          ; SGN(SDQ)*SGN(SDQ6) * 2**-7   (Q29)
D0366  03BC 6859           LTD   SDQ5
D0367  03BD 590F           SACH  B6,1         ; Q14
D0368  03BE 280E    GETB5  LAC   B5,8         ; B5 * 2**-8 TRUNCATED
D0369  03BF 5821           SACH  TEMP1,15
D0370  03C0 2F0E           LAC   B5,15        ; Q29
D0371  03C1 1F21           SUB   TEMP1,15
D0372  03C2 6D54           MPY   SDQ
D0373  03C3 6858           LTD   SDQ4
D0374  03C4 590E           SACH  B5,1         ; Q14
D0375  03C5 280D    GETB4  LAC   B4,8         ; B4 * 2**-8 TRUNCATED
D0376  03C6 5821           SACH  TEMP1,15
D0377  03C7 2F0D           LAC   B4,15        ; Q29
D0378  03C8 1F21           SUB   TEMP1,15
D0379  03C9 6D54           MPY   SDQ
D0380  03CA 6857           LTD   SDQ3
D0381  03CB 590D           SACH  B4,1         ; Q14
D0382  03CC 280C    GETB3  LAC   B3,8         ; B3 * 2**-8 TRUNCATED
D0383  03CD 5821           SACH  TEMP1
D0384  03CE 2F0C           LAC   B3,15        ; Q29
D0385  03CF 1F21           SUB   TEMP1,15
D0386  03D0 6D54           MPY   SDQ
D0387  03D1 6856           LTD   SDQ2
D0388  03D2 590C           SACH  B3,1         ; Q14
D0389  03D3 280B    GETB2  LAC   B2,8         ; B2 * 2**-8 TRUNCATED
D0390  03D4 5821           SACH  TEMP1
D0391  03D5 2F0B           LAC   B2,15        ; Q29
D0392  03D6 1F21           SUB   TEMP1,15
```

17. 32-kbit/s ADPCM with the TMS32010

```
D0393  03D6 6054   MPY  SDQ
D0394  03D7 6B55   LTD  SDQ1
D0395  03D9 590B   SACH B2,1         ; Q14
D0396  03DA 280A   LAC  B1,8         ; B1.* 2**-8 TRUNCATED
D0397  03DB 5821   SACH TEMP1
D0398  03DC 2F0A   LAC  B1,15
D0399  03DD 1F21   SUB  TEMP1,15
D0400  03DE 6054   LTD  SDQ
D0401  03DF 6B54   MPY  SDQ
D0402  03E0 590A   SACH B1,1         ; Q14
D0403
D0404  *;*****************************************************
D0405  *;  To update coefficients of 2nd order predictor,
D0406  *;  First get sign of sum of SEZ and DQ
D0407  *;
D0408  *;  NOTATION: if SEZ+DQ >= 0 then PK0 =  512
D0409  *;            else                PK0 = -512
D0410  *;
D0411  *;*****************************************************
D0412
D0413  03E1 6950   ADDC   DMOV PK1           ; PK1==>PK2
D0414  03E2 694F          DMOV PK0           ; PK0==>PK1
D0415  03E3 2004          LAC  SEZ
D0416  03E4 0110          ADD  DQ,1
D0417  03E5 5821          SACH TEMP1         ; FFFF or 0000
D0418  03E6 2A21          LAC  TEMP1,10      ; FC00 or 0000
D0419  03E7 094C          ADD  ONE,9         ; FE00 or 0200  ; -512 or +512
D0420  03E8 504F          SACL PK0
D0421  03E9 6AAF   SUMGT0 LT   PK0
D0422
D0423  *;
D0424  *;  now calculate 1/2 * f[A1(k-1)]
D0425  *;
D0426  *;          = 2*A1      if |A1| <= 1/2
D0427  *;          = SGN(A1)   if |A1| >  1/2
D0428  *;
D0429  *;*****************************************************
D0430
D0431  03EA 2111   GETF   LAC  A1,1          ; 2*A1
D0432  03EB 5023          SACL TEMP3
D0433  03EC FA00          BLZ  GETF2
       03ED 03F4
D0434  03EE 1E4C   GETF1  SUB  ONE,14        ; is |A1| < 1/2
       03EF FA00
D0435  03F0 03FA          BLZ  GETA1
D0436  03F1 2062          LAC  K16382        ; approx 1
       03F2 F900
D0437  03F3 03F9          B    DONEF
D0438  03F4 7F88   GETF2  ABS                ; is |A1| < 1/2
D0439  03F5 1E4C          SUB  ONE,14
       03F6 FA00
D0440  03F7 03FA          BLZ  GETA1
D0441  03F8 2063          LAC  M16382        ; approx -1
D0442  03F9 5023   DONEF  SACL TEMP3
D0443
D0444  *;
D0445  *;  Compute A1 coeff of 2nd order predictor
```

```
D0446  *;*****************************************************
D0447  *;  A1(k) = (1-2**-8)*A1(k-1)
D0448  *;                  + (3*2**-8)*SGN[p(k)]*SGN[p(k-1)]
D0449  *;
D0450  *;  NOTATION: A1  -- 16b TC (Q14)
D0451  *;            PKn -- +512 if SGN[p(k)] =  1
D0452  *;                   -512 if SGN[p(k)] = -1
D0453  *;
D0454  *;*****************************************************
D0455
D0456  03FA 2811   GETA1 LAC  A1,8          ; A1*2**-8 TRUNCATED
D0457  03FB 5822         SACH TEMP2
D0458  03FC 2C11         LAC  A1,12          ; Q26
D0459  03FD 1C22         SUB  TEMP2,12
D0460  03FE 6050         MPY  PK1
D0461  03FF 7F8F         APAC                ; SGN[p(k-1)]*SGN[p(k)]
D0462  0400 7F8F         APAC
D0463  0401 7F8F         APAC                ; +3*SGN[p(k-1)]*SGN[p(k)]
D0464  0402 5C11         SACH A1,4           ; store as Q14
D0465  0403 7F8E         PAC                 ; save sign
D0466
D0467  *;*****************************************************
D0468  *;  Compute A2 coeff of 2nd order predictor
D0469  *;
D0470  *;  A2(k) = (1-2**-7)*A2(k-1)
D0471  *;          + (2**-7)*{SGN[p(k)]*SGN[p(k-2)]
D0472  *;          - f[A1(k-1)]*SGN[p(k)]*SGN[p(k-1)]}
D0473  *;
D0474  *;  NOTATION: A2        -- 16b TC (Q14)
D0475  *;            F(),TEMP3 -- 16b TC (Q14)
D0476  *;            PKn.      -- +512 if SGN[p(k)] =  1
D0477  *;                      -- -512 if SGN[p(k)] = -1
D0478  *;
D0479  *;*****************************************************
D0480
D0481  0404 F000   GETA2 BGEZ SUBF          ; if sign + --> subtract F
       0405 0409
D0482  0406 7F89         ZAC                 ; else negate F and subtract
D0483  0407 1023         SUB  TEMP3
D0484  0408 5023         SACL TEMP3
D0485
D0486  0409 2912   SUBF  LAC  A2,9          ; A2*2**-7 TRUNCATED
D0487  040A 581E         SACH SUM4
D0488  040B 6051         MPY  PK2
D0489  040C 7F8E         PAC
D0490  040D 7F8E         APAC                ; SGN[p(k-2)]*SGN[p(k)]
D0491  040E 1623         SUB  TEMP3,6        ; 2*2**-8*above     (Q26)
D0492  040F 5C23         SACH TEMP3,4        ; Q14
D0493  0410 2012         LAC  A2
D0494  0411 101E         SUB  SUM4
D0495  0412 0023         ADD  TEMP3          ; leak factor
D0496  0413 5012         SACL A2             ; Q14
D0497
D0498  0414 5821         SACH TEMP1          ; save sign to make +/- .75
D0499
D0500  *;  limit A2 to +/- .75 and prevent overflow
D0501  *;
```

17. 32-kbit/s ADPCM with the TMS32010

```
D0502 0415 7F88  LIMC  ABS
D0503 0416 1C7D        SUB   THREE,12
D0504 0417 F800        BLEZ  LIMD          ; :value; must be < .75
      0418 041D
D0505 0419 2C7D        LAC   THREE,12      ; .75  (Q14)
D0506 041A 7821        XOR   TEMP1         ; 1's complement if negative
D0507 041B 1021        SUB   TEMP1         ; 2's complement if negative
D0508 041C 5012        SACL  A2            ; Q14
D0509
D0510 *; limit A1(k) to +/- [1-2**-4 - A2(k)]
D0511
D0512 041D 2A6D  LIMD  LAC   M15,10
D0513 041E 1012        SUB   A2            ; 1-2**-4    (Q14)
D0514 041F 5021        SACL  TEMP1
D0515 0420 2011        LAC   A1
D0516 0421 5824        SACH  TEMP4         ; 1-2**-4-A2P (Q14)
D0517 0422 7F88        ABS
D0518 0423 1021        SUB   TEMP1         ; save sign to make +/- LIMIT
D0519 0424 F800        BLEZ  FLTSR
      0425 042A
D0520 0426 2021  AILIM LAC   TEMP1         ; ABS value of LIMIT
D0521 0427 7824        XOR   TEMP4         ; 1's complement if negative
D0522 0428 1024        SUB   TEMP4         ; 2's complement if negative
D0523 0429 5011        SACL  A1            ; Q14
D0524
D0525 ;***********************************************
D0526 ;*  COMPUTE RECONSTRUCTED SIGNAL
D0527 ;*
D0528 ;*  INPUT:  QUANTIZED DIFFERENCE SIGNAL -- DQ
D0529 ;*          SIGNAL ESTIMATE            -- SE
D0530 ;*
D0531 ;*  OUTPUT: RECONSTRUCTED SIGNAL -- SR
D0532 ;*
D0533 ;*  NOTATION:  DQ -- 15b TC (Q0) [sign extended]
D0534 ;*             SE -- 15b TC (Q0) [sign extended]
D0535 ;*             SR -- 16b TC (Q0)
D0536 ;*
D0537 ;*  FLOAT SR -- convert 2's comp number to floating
D0538 ;*
D0539 ;*  INPUT:  accumulator
D0540 ;*
D0541 ;*  OUTPUT: --4b exponent left in SREXP
D0542 ;*          --6b mantissa*8 left in SRMAN
D0543 ;*          --sign preserved in SR
D0544
D0545
D0546
D0547 ;***********************************************
D0548 042A 2010  FLTSR LAC   DQ            ; compute reconstructed signal
D0549 042B 0003        ADD   SE
D0550 042C 5013        SACL  SR            ; convert to floating point notation
D0551 042D 7F88        ABS
D0552 042E 5052        SACL  SRMAN
D0553 042F 174C        SUB   ONE,7         ; binary search to get exponent
D0554 0430 F000        BGEZ  D8TOF
      0431 0469
D0555 0432 0360  DOTO7 ADD   M15,3         ; TEMP1-8 -- exp = 0-7
```

```
D0556 0433 F000        BGEZ  D4TO7
      0434 044A
D0557 0435 017D        ADD   THREE,1       ; TEMP1-2 -- exp = 0-3
D0558 0436 F000        BGEZ  D2TO3
      0437 043D
D0559 0438 2052  DOTO1 LAC   SRMAN         ; exp = 0-1
D0560 0439 501C        SACL  SREXP
D0561 043A 284C        LAC   ONE,8
D0562 043B 5052        SACL  SRMAN
D0563 043C 7F8D        RET
D0564 043D 114C  D2TO3 SUB   ONE,1         ; TEMP1-4 -- exp = 2-3
D0565 043E F000        BGEZ  EXX3
      043F 0445
D0566 0440 2752  EXX2  LAC   SRMAN,7       ; exp=2
D0567 0441 5052        SACL  SRMAN
D0568 0442 7E02        LACK  2
D0569 0443 501C        SACL  SREXP
D0570 0444 7F8D        RET
D0571 0445 2652  EXX3  LAC   SRMAN,6       ; exp=3
D0572 0446 5052        SACL  SRMAN
D0573 0447 7E03        LACK  3
D0574 0448 501C        SACL  SREXP
D0575 0449 7F8D        RET
D0576 044A 137D  D4TO7 SUB   THREE,3       ; TEMP1-32 -- exp = 4-7
D0577 044B F000        BGEZ  D6TO7
      044C 045A
D0578 044D 044C  D4TO5 ADD   ONE,4         ; TEMP1-16 -- exp = 4-5
D0579 044E F000        BGEZ  EXX5
      044F 0455
D0580 0450 2552  EXX4  LAC   SRMAN,5       ; exp=4
D0581 0451 5052        SACL  SRMAN
D0582 0452 7E04        LACK  4
D0583 0453 501C        SACL  SREXP
D0584 0454 7F8D        RET
D0585 0455 2452  EXX5  LAC   SRMAN,4       ; exp=5
D0586 0456 5052        SACL  SRMAN
D0587 0457 7E05        LACK  5
D0588 0458 501C        SACL  SREXP
D0589 0459 7F8D        RET
D0590 045A 154C  D6TO7 SUB   ONE,5         ; TEMP1-64 -- exp = 6-7
D0591 045B F000        BGEZ  EXX7
      045C 0462
D0592 045D 2352  EXX6  LAC   SRMAN,3       ; exp=6
D0593 045E 5052        SACL  SRMAN
D0594 045F 7E06        LACK  6
D0595 0460 501C        SACL  SREXP
D0596 0461 7F8D        RET
D0597 0462 2F52  EXX7  LAC   SRMAN,15
D0598 0463 5852        SACH  SRMAN,3
D0599 0464 2352        LAC   SRMAN,3
D0600 0465 5052        SACL  SRMAN
D0601 0466 7E07        LACK  7
D0602 0467 501C        SACL  SREXP
D0603 0468 7F8D        RET
D0604 0469 1177  D8TOF SUB   K960,1        ; TEMP1-2048 -- exp = 8-15
D0605 046A F000        BGEZ  DCTOF
      046B 0491
```

17. 32-kbit/s ADPCM with the TMS32010

```
D0606 046C 097D        D8TOB   ADD    THREE,9   ; TEMP1-512 -- exp = 8-11
D0607 046D F000                BGEZ   DATOB
      046E 0480
D0608 046F 084C        D8TO9   ADD    ONE,8     ; TEMP1-256 -- exp = 8-9
D0609 0470 F000                BGEZ   EXX9
      0471 0479
D0610 0472 2E52        EXX8    LAC    SRMAN,14  ; exp=8
D0611 0473 5852                SACH   SRMAN
D0612 0474 2352                LAC    SRMAN,3
D0613 0475 5052                SACL   SRMAN
D0614 0476 7E08                LACK   8
D0615 0477 501C                SACL   SREXP
D0616 0478 7F8D                RET
D0617 0479 2D52        EXX9    LAC    SRMAN,13  ; exp=9
D0618 047A 5852                SACH   SRMAN
D0619 047B 2352                LAC    SRMAN,3
D0620 047C 5052                SACL   SRMAN
D0621 047D 7E09                LACK   9
D0622 047E 501C                SACL   SREXP
D0623 047F 7F8D                RET
D0624 0480 194C        DATOB   SUB    ONE,9     ; TEMP1-1024 -- exp=10-11
D0625 0481 F000                BGEZ   EXX11
      0482 048A
D0626 0483 2C52        EXX10   LAC    SRMAN,12  ; exp=10
D0627 0484 5852                SACH   SRMAN
D0628 0485 2352                LAC    SRMAN,3
D0629 0486 5052                SACL   SRMAN
D0630 0487 7E0A                LACK   10
D0631 0488 501C                SACL   SREXP
D0632 0489 7F8D                RET
D0633 048A 2B52        EXX11   LAC    SRMAN,11  ; exp=11
D0634 048B 5852                SACH   SRMAN
D0635 048C 2352                LAC    SRMAN,3
D0636 048D 5052                SACL   SRMAN
D0637 048E 7E0B                LACK   11
D0638 048F 501C                SACL   SREXP
D0639 0490 7F8D                RET
D0640 0491 1B7D        DCTOF   SUB    THREE,11  ; TEMP1-8192 -- exp=12-15
D0641 0492 F000                BGEZ   DETOF
      0493 04A5
D0642 0494 0C4C        DCTOD   ADD    ONE,12    ; TEMP1-4096 -- exp=12-13
D0643 0495 F000                BGEZ   EXX13
      0496 049E
D0644 0497 2A52        EXX12   LAC    SRMAN,10  ; exp=12
D0645 0498 5852                SACH   SRMAN
D0646 0499 2352                LAC    SRMAN,3
D0647 049A 5052                SACL   SRMAN
D0648 049B 7E0C                LACK   12
D0649 049C 501C                SACL   SREXP
D0650 049D 7F8D                RET
D0651 049E 2952        EXX13   LAC    SRMAN,9   ; exp=13
D0652 049F 5852                SACH   SRMAN
D0653 04A0 2352                LAC    SRMAN,3
D0654 04A1 5052                SACL   SRMAN
D0655 04A2 7E0D                LACK   13
D0656 04A3 501C                SACL   SREXP
D0657 04A4 7F8D                RET
```

```
D0658 04A5 1D4C        DETOF   SUB    ONE,13    ; TEMP1-16384 -- exp=14-15
D0659 04A6 F000                BGEZ   EXX15
      04A7 04AF
D0660 04A8 2852        EXX14   LAC    SRMAN,8   ; exp=14
D0661 04A9 5852                SACH   SRMAN
D0662 04AA 2352                LAC    SRMAN,3
D0663 04AB 5052                SACL   SRMAN
D0664 04AC 7E0E                LACK   14
D0665 04AD 501C                SACL   SREXP
D0666 04AE 7F8D                RET
D0667 04AF 2752        EXX15   LAC    SRMAN,7   ; exp=15
D0668 04B0 5852                SACH   SRMAN
D0669 04B1 2352                LAC    SRMAN,3
D0670 04B2 5052                SACL   SRMAN
D0671 04B3 7E0F                LACK   15
D0672 04B4 501C                SACL   SREXP
D0673 04B5 7F8D                RET
```

17. 32-kbit/s ADPCM with the TMS32010

```
                0005              COPY  UTILITY.ASM
                E0001     *;
                E0002     *;  code to do left shifts for SEZ/SE calculations
                E0003     *;
E0004  0486 7F88  RS1      ABS
E0005  0487 7974           AND   M32767   ; make positive before mask
E0006  0488 5022           SACL  SUM1     ; keep lower 15 bits
E0007  0489 F900           B     CHK1     ; save result
       048A 01C4                          ; return
E0008  048B 7F88  RS2      ABS
E0009  048C 7974           AND   M32767
E0010  048D 5023           SACL  SUM2
E0011  048E F900           B     CHK2
       048F 01DE
E0012  04C0 7F88  RS3      ABS
E0013  04C1 7974           AND   M32767
E0014  04C2 5025           SACL  SUM3
E0015  04C3 F900           B     CHK3
       04C4 01F8
E0016  04C5 7F88  RS4      ABS
E0017  04C6 7974           AND   M32767
E0018  04C7 501E           SACL  SUM4
E0019  04C8 F900           B     CHK4
       04C9 0212
E0020  04CA 7F88  RS5      ABS
E0021  04CB 7974           AND   M32767
E0022  04CC 501F           SACL  SUM5
E0023  04CD F900           B     CHK5
       04CE 022C
E0024  04CF 7F88  RS6      ABS
E0025  04D0 7974           AND   M32767
E0026  04D1 5020           SACL  SUM6
E0027  04D2 F900           B     CHK6
       04D3 0246
E0028  04D4 7F88  RS11     ABS
E0029  04D5 7974           AND   M32767
E0030  04D6 5027           SACL  SUM7
E0031  04D7 F900           B     CHK11
       04D8 025F
E0032  04D9 7F88  RS21     ABS
E0033  04DA 7974           AND   M32767
E0034  04DB 5028           SACL  SUM8
E0035  04DC F900           B     CHK21
       04DD 0279
```

```
E0037     *;
E0038     *;
E0039     *;  FLOAT SUBROUTINE--convert 2's comp number to floating
E0040     *;        INPUT: accumulator
E0041     *;        OUTPUT:
E0042     *;           --4b exponent left in accum
E0043     *;           --6b mantissa left in TEMP1
E0044     *;           --sign preserved in original number
E0045     *;
E0046     *;
E0047  002A      FLTSFT  EQU   42      ; address of shift multipliers
E0048     *;
E0049     *;
E0050  04DE 5821  FLOAT   SACH  TEMP1
E0051  04DF 2021          LAC   TEMP1
E0052  04E0 7F88          ABS
E0053  04E1 5021          SACL  TEMP1
E0054  04E2 164C          SUB   ONE,6   ; binary search to get expONEnt
E0055  04E3 F000          BGEZ  E7TOD
       04E4 0511
E0056  04E5 0076  EOTO7   ADD   K56     ; TEMP1-8 -- exp = 0-6
E0057  04E6 F000          BGEZ  E4TO6
       04E7 0501
E0058  04E8 017D  EOTO3   ADD   THREE,1 ; TEMP1-2 -- exp = 0-3
E0059  04E9 F000          BGEZ  E2TO3
       04EA 04F6
E0060  04EB 2021  EOTO1   LAC   TEMP1   ; exp = 0-1
E0061  04EC FE00          BNZ   E1
       04ED F000
E0062  04EE 254C  E0      LAC   ONE,5   ; exp=0
E0063  04EF 5021          SACL  TEMP1
E0064  04F0 7E2A          LACK  FLTSFT+0
E0065  04F1 7F8D          RET
E0066  04F2 2521  E1      LAC   TEMP1,5 ; exp=1
E0067  04F3 5021          SACL  TEMP1
E0068  04F4 7E2B          LACK  FLTSFT+1
E0069  04F5 7F8D          RET
E0070  04F6 114C  E2TO3   SUB   ONE,1   ; TEMP1-4 -- exp = 2-3
E0071  04F7 F000          BGEZ  E3
       04F8 04FD
E0072  04F9 2421  E2      LAC   TEMP1,4 ; exp=2
E0073  04FA 5021          SACL  TEMP1
E0074  04FB 7E2C          LACK  FLTSFT+2
E0075  04FC 7F8D          RET
E0076  04FD 2321  E3      LAC   TEMP1,3 ; exp=3
E0077  04FE 5021          SACL  TEMP1
E0078  04FF 7E2D          LACK  FLTSFT+3
E0079  0500 7F8D          RET
E0080  0501 134C  E4TO6   SUB   ONE,3   ; TEMP1-16 -- exp = 4-6
E0081  0502 F000          BGEZ  E5TO6
       0503 0508
E0082  0504 2221  E4      LAC   TEMP1,2 ; exp=4
E0083  0505 5021          SACL  TEMP1
E0084  0506 7E2E          LACK  FLTSFT+4
E0085  0507 7F8D          RET
E0086  0508 144C  E5TO6   SUB   ONE,4   ; TEMP1-32 -- exp = 5-6
E0087  0509 FD00          BGEZ
```

17. 32-kbit/s ADPCM with the TMS32010

```
E0088  050A 050F        LAC   TEMP1,1   ; exp=5
E0089  050B 2121        SACL  TEMP1
E0090  050C 5021  E5    LACK  FLTSFT+5
E0091  050D 7E2F        RET
E0092  050E 7F8D
E0093  050F 7E30  E6    LACK  FLTSFT+6   ; exp=6
E0094  0510 7F8D        RET
E0095  0511 1077  E7TOD SUB   K960   ; TEMP1-1024 -- exp = 7-13
E0096  0512 F000        BGEZ  EBTOD
       0513 052D
E0097  0514 087D  E7TOA ADD   THREE,8 ; TEMP1-256  -- exp = 7-10
       0515 F000        BGEZ  E9TOA
E0098  0516 0522
       0517 074C  E7TOB ADD   ONE,7  ; TEMP1-128  -- exp = 7-8
       0518 F000        BGEZ  E8
E0099  0519 051E
E0100  051A 2F21  E7    LAC   TEMP1,15  ; exp=7
E0101  051B 5821        SACH  TEMP1
E0102  051C 7E31        LACK  FLTSFT+7
E0103  051D 7F8D        RET
E0104  051E 2E21  E8    LAC   TEMP1,14  ; exp=8
E0105  051F 5821        SACH  TEMP1
E0106  0520 7E32        LACK  FLTSFT+8
E0107  0521 7F8D        RET
E0108  0522 184C  E9TOA SUB   ONE,8  ; TEMP1-512  -- exp = 9-10
       0523 F000        BGEZ  E10
E0109  0524 0529
E0110  0525 2021  E9    LAC   TEMP1,13  ; exp=9
E0111  0526 5821        SACH  TEMP1
E0112  0527 7E33        LACK  FLTSFT+9
E0113  0528 7F8D        RET
E0114  0529 2C21  E10   LAC   TEMP1,12  ; exp=10
E0115  052A 5821        SACH  TEMP1
E0116  052B 7E34        LACK  FLTSFT+10
E0117  052C 7F8D        RET
E0118  052D 1A7D  EBTOD SUB   THREE,10  ; TEMP1-2048 -- exp=11-13
E0119  052E F000        BGEZ  EBTOC
E0120  052F 053B
E0121  0530 0B4C  EBTOC SUB   ONE,11  ; TEMP1-4096 -- exp=11-12
       0531 F000        BGEZ  E12
E0122  0532 0537
E0123  0533 2B21  E11   LAC   TEMP1,11  ; exp=11
E0124  0534 5821        SACH  TEMP1
E0125  0535 7E35        LACK  FLTSFT+11
E0126  0536 7F8D        RET
E0127  0537 2A21  E12   LAC   TEMP1,10  ; exp=12
E0128  0538 5821        SACH  TEMP1
E0129  0539 7E36        LACK  FLTSFT+12
E0130  053A 7F8D        RET
E0131  053B 1C4C  EDTOE SUB   ONE,12  ; TEMP1-8192 -- exp=13
       053C F000        BGEZ  E0
       053D 04EE
E0132  053E 2921  E13   LAC   TEMP1,9  ; exp=13
E0133  053F 5821        SACH  TEMP1
E0134  0540 7E37        LACK  FLTSFT+13
E0135  0541 7F8D        RET
```

```
0006                   ;*****************************
F0001                  ;* SYSTEM INITIALIZATION
F0002                  ;*****************************
F0003  004A    NOCONS  EQU   74
F0004  0036    PTCONS  EQU   54
F0005                  ;*
F0006                  ;*
F0007  0542 7F81 RESET DINT         ; Disable interrupts
F0008  0543 6E00       LDPK  0      ; Initialize data page
F0009                  ;*
F0010  0544 7035 SETPAC LARK 0,53
F0011  0545 6880        LARP  0
F0012  0546 7F89        ZAC
F0013  0547 5080        SACL  *,0,0
F0014  0548 F400 ZRAMA  BANZ  ZRAMA   ; Zero iram
       0549 0547
F0015                  ;*
F0016  054A 7E01        LACK  1
F0017  054B 504C        SACL  ONE
F0018  054C 6A4C        LT    ONE
F0019  054D 859B        MPYK  CONS
F0020  054E 7F8E        PAC
F0021  054F 7136        LARK  1,PTCONS
F0022  0550 7049        LARK  0,NOCONS-1
F0023  0551 6881 NXCONS LARP  1
F0024  0552 67A0        TBLR  *+,0
F0025  0553 004C        ADD   ONE
F0026  0554 F400        BANZ  NXCONS
       0555 0551
F0027                  ;*
F0028  0556 4021        IN    TEMP1,CTL
F0029  0557 2021        LAC   TEMP1
F0030  0558 FA00 MULAW  BLZ   ALAW
       0559 055E
F0031  055A FF00        BZ    MULAWX
       055B 0020
F0032  055C F900        B     MULAWR
       055D 00F9
F0033  055E 7974 ALAW   AND   M32767  ; ROM ADDR
F0034  055F FF00        BZ    ALAWX   ; RAM ADDR
       0560 004E
F0035  0561 F900        B     ALAWR
       0562 0184
F0036                  ;*
```

17. 32-kbit/s ADPCM with the TMS32010

```
0007                              COPY    ROMRAM.ASM
G0001            *;
G0002            *;**************************************
G0003            *;   ROM
G0004            *;**************************************
G0005  0563      ROMLOC  BSS     0
G0006            *; SHIFT MULT TABLE
G0007  0563      SHFT    BSS     0
G0008  0563 0000         DATA    0
G0009  0564 0001         DATA    1
G0010  0565 0002         DATA    2
G0011  0566 0004         DATA    4        ; log dq to flt dq exponent adjustment
G0012  0567 0008         DATA    8
G0013  0568 0010         DATA    16
G0014  0569 0020         DATA    32
G0015  056A 0040         DATA    64
G0016  056B 0080         DATA    128
G0017  056C 0100         DATA    256
G0018  056D 0200         DATA    512
G0019  056E 0400         DATA    1024
G0020  056F 0800         DATA    2048
G0021  0570 1000         DATA    4096
G0022  0571 2000         DATA    8192
G0023  0572 4000         DATA    16384
G0024            *; DQMAN MASK TABLE
G0025  0573 FE00         DATA    65024
G0026  0574 FF00         DATA    65280
G0027  0575 FF80         DATA    65408
G0028  0576 FFC0         DATA    65472
G0029  0577 FFE0         DATA    65504
G0030  0578 FFF0         DATA    65520
G0031  0579 FFF8         DATA    65528
G0032  057A FFF8         DATA    65528
G0033  057B FFF8         DATA    65528
G0034  057C FFF8         DATA    65528
G0035  057D FFF8         DATA    65528
G0036  057E FFF8         DATA    65528
G0037  057F FFF8         DATA    65528
G0038  0580 FFF8         DATA    65528
G0039  0581 FFF8         DATA    65528
G0040  0582 FFF8         DATA    65528
G0041            *; INVERSE QUANTIZING TABLE
G0042  0583      IQTAB   BSS     0
G0043  0583 FF79         DATA    65401
G0044  0584 0044         DATA    68
G0045  0585 00A5         DATA    165
G0046  0586 00E8         DATA    232
G0047  0587 011D         DATA    285
G0048  0588 014C         DATA    332
G0049  0589 0179         DATA    377
G0050  058A 01AC         DATA    428
G0051            *; WI TABLE
G0052  058B      WTABLE  BSS     0
G0053  058B FFF4         DATA    65524
G0054  058C 0004         DATA    4
G0055  058D 001B         DATA    27
G0056  058E 0032         DATA    50
```

```
0057
G0057  058F 0062         DATA    98
G0058  0590 00B8         DATA    184
G0059  0591 0154         DATA    340
G0060  0592 0454         DATA    1108
G0061            *; FI TABLE
G0062  0593      FITABL  BSS     0
G0063  0593 0000         DATA    0
G0064  0594 0000         DATA    0
G0065  0595 0000         DATA    0
G0066  0596 0000         DATA    0
G0067  0597 0010         DATA    16
G0068  0598 0010         DATA    16
G0069  0599 0030         DATA    48
G0070  059A 0070         DATA    112
G0071
G0072            *;
G0073            *; misc constants to be initialized
G0074
G0075  059B      CONS    BSS     0
G0076  059B 0002         DATA    2       ; 12th entry of shift table (1st 11 all 0)
G0077  059C 0004         DATA    4
G0078  059D 0008         DATA    8
G0079  059E 0010         DATA    16
G0080  059F 0020         DATA    32
G0081  05A0 0040         DATA    64
G0082  05A1 0080         DATA    128
G0083  05A2 0100         DATA    256
G0084  05A3 0200         DATA    512
G0085  05A4 0400         DATA    1024
G0086  05A5 0800         DATA    2048
G0087  05A6 1000         DATA    4096
G0088  05A7 2000         DATA    8192
G0089  05A8 4000         DATA    16384
G0090  05A9 FFFF         DATA    -1
G0091  05AA FFFE         DATA    -2
G0092  05AB FFFC         DATA    -4
G0093  05AC 00FF         DATA    255     ;M255
G0094  05AD 8000         DATA    32768   ;K32768
G0095  05AE 0001         DATA    1       ;YLH
G0096  05AF 0800         DATA    2048    ;YLL
G0097  05B0 FFFF         DATA    -1
G0098  05B1 0001         DATA    1       ; BIAS
G0099  05B2 0021         DATA    33      ;YU
G0100  05B3 0220         DATA    544
G0101  05B4 0200         DATA    512     ;PK0
G0102  05B5 0200         DATA    512     ;PK1
G0103  05B6 0200         DATA    512     ;PK2
G0104  05B7 0100         DATA    256     ;SRMAN
G0105  05B8 0100         DATA    256     ;SR1MAN
G0106  05B9 0800         DATA    2048    ;SDQ
G0107  05BA 0800         DATA    2048    ;SDQ1
G0108  05BB 0800         DATA    2048    ;SDQ2
G0109  05BC 0800         DATA    2048    ;SDQ3
G0110  05BD 0800         DATA    2048    ;SDQ4
G0111  05BE 0800         DATA    2048    ;SDQ5
G0112  05BF 0800         DATA    2048    ;SDQ6
G0113  05C0 0100         DATA    256     ;DQMAN
```

17. 32-kbit/s ADPCM with the TMS32010

```
G0114 05C1 0100      DATA  256       ;DQ1MAN
G0115 05C2 0100      DATA  256       ;DQ2MAN
G0116 05C3 0100      DATA  256       ;DQ3MAN
G0117 05C4 0100      DATA  256       ;DQ4MAN
G0118 05C5 0100      DATA  256       ;DQ5MAN
G0119 05C6 11E0      DATA  4576      ;K4576
G0120 05C7 3FFE      DATA  16382     ;K16382
G0121 05C8 C002      DATA  -16382    ;M16382
G0122 05C9 002E      DATA  46        ;K46
G0123 05CA 0031      DATA  49        ;K49
G0124 05CB 0563      DATA  SHFT      ;SHIFT
G0125 05CC 003F      DATA  63        ;K63
G0126 05CD 0000      DATA  0         ;F1
G0127 05CE 0000      DATA  0         ;WI
G0128 05CF 0583      DATA  IQTAB     ;INQTAB
G0129 05D0 0220      DATA  544       ;K544
G0130 05D1 1400      DATA  5120      ;K5120
G0131 05D2 000F      DATA  15        ;M15
G0132 05D3 FF80      DATA  -128      ;KFF80
G0133 05D4 007F      DATA  127       ;K127
G0134 05D5 FFC0      DATA  -64       ;MFFC0
G0135 05D6 1080      DATA  4224      ; BIAS*2**7
G0136 05D7 0FFF      DATA  4095      ;M4095
G0137 05D8 0000      DATA  0         ;spare
G0138 05D9 7FFF      DATA  32767     ;M32767
G0139 05DA FF00      DATA  -256      ;KFF00
G0140 05DB 0038      DATA  56        ;K56
G0141 05DC 03C0      DATA  960       ;K960
G0142 05DD 004F      DATA  79        ;K79
G0143 05DE 005F      DATA  95        ;K95
G0144 05DF 0082      DATA  130       ;K130
G0145 05E0 008A      DATA  138       ;K138
G0146 05E1 0905      DATA  2309      ;K2309
G0147 05E2 0003      DATA  3         ;THREE
G0148 05E3 0000      DATA  0         ;spare
G0149 05E4 0080      DATA  128       ;M0080
```

```
G0151                *;**********
G0152                *;   RAM
G0153                *;**********
G0154 05E5           RAMLOC BSS   0
G0155 0000                  DORG  0
G0156                *;
G0157                *;
G0158 0000 0000             RAM Location # 000
                            DATA  0      ; spare
G0159                *;
G0160                *;
G0161 0001 0000      I      RAM Location # 001
                            DATA  0      ; 32Kb output
G0162                *;
G0163                *;
G0164 0002 0000      IM     RAM Location # 002
                            DATA  0      ; 8-level version of I
G0165                *;
G0166                *;
G0167 0003 0000      SE     RAM Location # 003
                            DATA  0      ; signal estimate
G0168                *;
G0169                *;
G0170 0004 0000      SEZ    RAM Location # 004
                            DATA  0      ; partial signal estimate
G0171                *;
G0172                *;
G0173 0005 0000      APP    RAM Location # 005
                            DATA  0      ; unlimited speed control parm
G0174                *;
G0175                *;
G0176 0006 0000      AL     RAM Location # 006
                            DATA  0      ; limited speed control parm
G0177                *;
G0178                *;
G0179 0007 0000      DMS    RAM Location # 007
                            DATA  0      ; short term average of F
G0180                *;
G0181                *;
G0182 0008 0000      DML    RAM Location # 008
                            DATA  0      ; long term average of F
G0183                *;
G0184                *;
G0185 0009 0000      Y      RAM Location # 009
                            DATA  0      ; quantizer scale factor
G0186                *;
G0187                *;
G0188 000A 0000      B1     RAM Location # 010
                            DATA  0      ; 6th order predictor coefficient
G0189                *;
G0190                *;
G0191 000B 0000      B2     RAM Location # 011
                            DATA  0      ; 6th order predictor coefficient
G0192                *;
G0193 000C 0000      B3     RAM Location # 012
                            DATA  0      ; 6th order predictor coefficient
G0194                *;
G0195                *;
G0196                *;
G0197 000D 0000      B4     RAM Location # 013
                            DATA  0      ; 6th order predictor coefficient
G0198                *;
G0199                *;
G0200 000E 0000      B5     RAM Location # 014
                            DATA  0      ; 6th order predictor coefficient
G0201                *;
G0202                *;
G0203 000F 0000      B6     RAM Location # 015
                            DATA  0      ; 6th order predictor coefficient
G0204                *;
G0205                *;
G0206 0010 0000      DQ     RAM Location # 016
                            DATA  0      ; quantized diff signal
G0207                *;
```

17. 32-kbit/s ADPCM with the TMS32010

```
G0208
G0209  0011 0000    *;      RAM Location # 017
G0210              A1      DATA    0    ; coefficients of 2nd order predictor
G0211              *;
G0212  0012 0000            RAM Location # 018
G0213              A2      DATA    0    ; coefficients of 2nd order predictor
G0214              *;
G0215  0013 0000            RAM Location # 019
G0216              SR      DATA    0    ; reconstructed signal frame k
G0217              *;
G0218  0014 0000            RAM Location # 020
G0219              SR1     DATA    0    ; reconstructed signal frame k
G0220              *;
G0221  0015 0000            RAM Location # 021
G0222              DQEXP   DATA    0    ; exponent of DQ
G0223              *;
G0224  0016 0000            RAM Location # 022
G0225              DQ1EXP  DATA    0    ; exponent of DQ1
G0226              *;
G0227  0017 0000            RAM Location # 023
G0228              DQ2EXP  DATA    0    ; exp of DQ2
G0229              *;
G0230  0018 0000            RAM Location # 024
G0231              DQ3EXP  DATA    0    ; exp of DQ3
G0232              *;
G0233  0019 0000            RAM Location # 025
G0234              DQ4EXP  DATA    0    ; exp of DQ4
G0235              *;
G0236  001A 0000            RAM Location # 026
G0237              DQ5EXP  DATA    0    ; exp of DQ5
G0238              *;
G0239  001B 0000            RAM Location # 027
G0240              SCRACH  DATA    0    ; scrach variable
G0241              *;
G0242  001C 0000            RAM Location # 028
G0243              SREXP   DATA    0    ; exp of SR
G0244              *;
G0245  001D 0000            RAM Location # 029
G0246              SR1EXP  DATA    0    ; exp of SR1
G0247              *;
G0248  001E 0000            RAM Location # 030
G0249              SUM4    DATA    0    ; temp
G0250              *;
G0251  001F 0000            RAM Location # 031
G0252              SUM5    DATA    0    ; temp
G0253              *;
G0254  0020 0000            RAM Location # 032
G0255              SUM6    DATA    0    ; temp
G0256              *;
G0257  0021 0000            RAM Location # 033
G0258              TEMP1   DATA    0    ; temp
G0259              *;
G0260  0022                 RAM Location # 034
G0261  0022 0000    SUM1    BSS     0    ; temp
                    TEMP2   DATA    0    ; temp
G0262              *;
G0263  0023                 RAM Location # 035
G0264              SUM2    BSS     0    ; temp
```

```
G0265  0023 0000    TEMP3   DATA    0    ; temp
G0266              *;
G0267                       RAM Location # 036
G0268  0024 0000    TEMP4   DATA    0    ; temp
G0269              *;
G0270                       RAM Location # 037
G0271  0025 0000    SUM3    DATA    0    ; temp
G0272              *;
G0273                       RAM Location # 038
G0274  0026 0000    SAMPLE  DATA    0    ; Linear sample
G0275              *;
G0276                       RAM Location # 039
G0277  0027 0000    SUM7    DATA    0    ; temp storage of SR1*A1 tap
G0278              *;
G0279                       RAM Location # 040
G0280  0028 0000    SUM8    DATA    0    ; temp storage of SR2*A2 tap
G0281              *;
G0282                       RAM Location # 041
G0283  0029 0000    YOVER4  DATA    0    ; Y>>2
G0284              *;
G0285                       RAM Location # 042
G0286  002A 0000            DATA    0    ; first location of shift table
G0287              *;
G0288                       RAM Location # 043
G0289  002B 0000            DATA    0
G0290              *;
G0291                       RAM Location # 044
G0292  002C 0000            DATA    0
G0293              *;
G0294                       RAM Location # 045
G0295  002D 0000            DATA    0
G0296              *;
G0297                       RAM Location # 046
G0298  002E 0000            DATA    0
G0299              *;
G0300                       RAM Location # 047
G0301  002F 0000            DATA    0
G0302              *;
G0303                       RAM Location # 048
G0304  0030 0000            DATA    0
G0305              *;
G0306                       RAM Location # 049
G0307  0031 0000            DATA    0
G0308              *;
G0309                       RAM Location # 050
G0310  0032 0000            DATA    0
G0311              *;
G0312                       RAM Location # 051
G0313  0033 0000            DATA    0
G0314              *;
G0315                       RAM Location # 052
G0316  0034 0000            DATA    0
G0317              *;
G0318                       RAM Location # 053
G0319  0035 0000            DATA    0
G0320              *;
G0321                       RAM Location # 054
```

17. 32-kbit/s ADPCM with the TMS32010

```
G0322 0036 0000              DATA    0
G0323                       *;
G0324                       *;      RAM Location # 055
G0325 0037 0000              DATA    0
G0326                       *;
G0327                       *;      RAM Location # 056
G0328 0038 0000 EIGHT        DATA    0
G0329                       *;
G0330                       *;      RAM Location # 057
G0331 0039 0000              DATA    0
G0332                       *;
G0333                       *;      RAM Location # 058
G0334 003A 0000              DATA    0
G0335                       *;
G0336                       *;      RAM Location # 059
G0337 003B 0000              DATA    0
G0338                       *;
G0339                       *;      RAM Location # 060
G0340 003C 0000              DATA    0
G0341                       *;
G0342                       *;      RAM Location # 061
G0343 003D 0000              DATA    0
G0344                       *;
G0345                       *;      RAM Location # 062
G0346 003E 0000              DATA    0
G0347                       *;
G0348                       *;      RAM Location # 063
G0349 003F 0000              DATA    0
G0350                       *;
G0351                       *;      RAM Location # 064
G0352 0040 0000              DATA    0
G0353                       *;
G0354                       *;      RAM Location # 065
G0355 0041 0000              DATA    0
G0356                       *;
G0357                       *;      RAM Location # 066
G0358 0042 0000              DATA    0
G0359                       *;
G0360                       *;      RAM Location # 067
G0361 0043 0000              DATA    0
G0362                       *;
G0363                       *;      RAM Location # 068
G0364 0044 0000              DATA    0
G0365                       *;
G0366                       *;      RAM Location # 069
G0367 0045 0000              DATA    0
G0368                       *;
G0369                       *;      RAM Location # 070
G0370 0046 0000              DATA    0           ; last loc of table (42-70)
G0371                       *;
G0372                       *;      RAM Location # 071
G0373 0047 0000 M255         DATA    0
G0374                       *;
G0375                       *;      RAM Location # 072
G0376 0048 0000 K32768       DATA    0           ; sign bit
G0377                       *;
G0378                       *;      RAM Location # 073
```

```
G0379 0049 0000 YLH          DATA    0           ; fast quant scale factor (hi word)
G0380                       *;
G0381                       *;      RAM Location # 074
G0382 004A 0000 YLL          DATA    0           ; slow quant scale factor (lo word)
G0383                       *;
G0384                       *;      RAM Location # 075
G0385 004B 0000 MINUS        DATA    0           ; -1
G0386                       *;
G0387                       *;      RAM Location # 076
G0388 004C 0000 ONE          DATA    0           ; 1
G0389                       *;
G0390                       *;      RAM Location # 077
G0391 004D 0000 BIAS         DATA    0           ; constant for mulaw conversions
G0392                       *;
G0393                       *;      RAM Location # 078
G0394 004E 0000 YU           DATA    0           ; fast quant scale factor
G0395                       *;
G0396                       *;      RAM Location # 079
G0397 004F 0000 PK0          DATA    0           ; sign of p(k)
G0398                       *;
G0399                       *;      RAM Location # 080
G0400 0050 0000 PK1          DATA    0           ; sign of p(k-1)
G0401                       *;
G0402                       *;      RAM Location # 081
G0403 0051 0000 PK2          DATA    0           ; sign of p(k-2)
G0404                       *;
G0405                       *;      RAM Location # 082
G0406 0052 0000 SRMAN        DATA    0           ; mantissa of SR
G0407                       *;
G0408                       *;      RAM Location # 083
G0409 0053 0000 SR1MAN       DATA    0           ; mantissa of SR1
G0410                       *;
G0411                       *;      RAM Location # 084
G0412 0054 0000 SDQ          DATA    0           ; sign DQ(k)
G0413                       *;
G0414                       *;      RAM Location # 085
G0415 0055 0000 SDQ1         DATA    0           ; sign DQ(k-1)
G0416                       *;
G0417                       *;      RAM Location # 086
G0418 0056 0000 SDQ2         DATA    0           ; sign DQ(k-2)
G0419                       *;
G0420                       *;      RAM Location # 087
G0421 0057 0000 SDQ3         DATA    0           ; sign DQ(k-3)
G0422                       *;
G0423                       *;      RAM Location # 088
G0424 0058 0000 SDQ4         DATA    0           ; sign DQ(k-4)
G0425                       *;
G0426                       *;      RAM Location # 089
G0427 0059 0000 SDQ5         DATA    0           ; sign DQ(k-5)
G0428                       *;
G0429                       *;      RAM Location # 090
G0430 005A 0000 SDQ6         DATA    0           ; sign DQ(k-6)
G0431                       *;
G0432                       *;      RAM Location # 091
G0433 005B 0000 DQMAN        DATA    0           ; mantissa of DQ
G0434                       *;
G0435                       *;      RAM Location # 092
```

17. 32-kbit/s ADPCM with the TMS32010

```
G0436 005C 0000 DQ1MAN DATA 0          ; mantissa of DQ1
G0437
G0438          *; RAM Location # 093
G0439 005D 0000 DQ2MAN DATA 0          ; mantissa of DQ2
G0440
G0441          *; RAM Location # 094
G0442 005E 0000 DQ3MAN DATA 0          ; mantissa of DQ3
G0443
G0444          *; RAM Location # 095
G0445 005F 0000 DQ4MAN DATA 0          ; mantissa of DQ4
G0446
G0447          *; RAM Location # 096
G0448 0060 0000 DQ5MAN DATA 0          ; mantissa of DQ5
G0449
G0450          *; RAM Location # 097
G0451 0061 0000 K4576  DATA 4576       ; 4576
G0452
G0453          *; RAM Location # 098
G0454 0062 0000 K16382 DATA 16382      ; +16382
G0455
G0456          *; RAM Location # 099
G0457 0063 0000 M16382 DATA 16382      ; -16382
G0458
G0459          *; RAM Location # 100
G0460 0064 0000 K46    DATA 46         ; 46
G0461
G0462          *; RAM Location # 101
G0463 0065 0000 K49    DATA 49         ; 49
G0464
G0465          *; RAM Location # 102
G0466 0066 0000 SHIFT  DATA 0          ; SHIFT table address
G0467
G0468          *; RAM Location # 103
G0469 0067 0000 K63    DATA 63         ; 63
G0470
G0471          *; RAM Location # 104
G0472 0068 0000 FI     DATA 0          ; FI value
G0473
G0474          *; RAM Location # 105
G0475 0069 0000 WI     DATA 0          ; WI value
G0476
G0477          *; RAM Location # 106
G0478 006A 0000 INQTAB DATA 0          ; Inverse quan table address
G0479
G0480          *; RAM Location # 107
G0481 006B 0000 K544   DATA 544        ; 544
G0482
G0483          *; RAM Location # 108
G0484 006C 0000 K5120  DATA 5120       ; 5120
G0485
G0486          *; RAM Location # 109
G0487 006D 0000 M15    DATA 15         ; 15
G0488
G0489          *; RAM Location # 110
G0490 006E 0000 KFF80  DATA >FF80      ; >FF80
G0491
G0492          *; RAM Location # 111
```

```
G0493 006F 0000 M127   DATA 0          ; 127
G0494
G0495          *; RAM Location # 112
G0496 0070 0000 MFFC0  DATA 0          ; -64
G0497
G0498          *; RAM Location # 113
G0499 0071 0000 BIASA  DATA 0          ; 33*128
G0500
G0501          *; RAM Location # 114
G0502 0072 0000 M4095  DATA 0          ; 4095
G0503
G0504          *; RAM Location # 115
G0505 0073 0000        DATA 0          ; spare
G0506
G0507          *; RAM Location # 116
G0508 0074 0000 M32767 DATA 0          ; 32767
G0509
G0510          *; RAM Location # 117
G0511 0075 0000 KFF00  DATA 0          ; >FF00
G0512
G0513          *; RAM Location # 118
G0514 0076 0000 K56    DATA 0          ; 56
G0515
G0516          *; RAM Location # 119
G0517 0077 0000 K960   DATA 0          ; 960
G0518
G0519          *; RAM Location # 120
G0520 0078 0000 K79    DATA 0          ; constants used for quantizing table
G0521
G0522          *; RAM Location # 121
G0523 0079 0000 K95    DATA 0          ; constants used for quantizing table
G0524
G0525          *; RAM Location # 122
G0526 007A 0000 K130   DATA 0          ; constants used for quantizing table
G0527
G0528          *; RAM Location # 123
G0529 007B 0000 K138   DATA 0          ; constants used for quantizing table
G0530
G0531          *; RAM Location # 124
G0532 007C 0000 K2309  DATA 0          ; constants used for quantizing table
G0533
G0534          *; RAM Location # 125
G0535 007D 0000 THREE  DATA 0          ; 3
G0536
G0537          *; RAM Location # 126
G0538 007E 0000        DATA 0          ; spare
G0539
G0540          *; RAM Location # 127
G0541 007F 0000 M0080  DATA 0          ; alaw mask
G0542
NO ERRORS, NO WARNINGS
```

17. 32-kbit/s ADPCM with the TMS32010

CCITT LABEL	VALUE	DEFN	REFERENCES
A1	0011	G0209	B0252 B0270 D0431 D0456 D0458 D0464 D0515 D0523
A1L1M	0426	D0520	D0520
A2	0012	G0212	B0230 B0245 D0486 D0493 D0496 D0508 D0513
ADC	0001	A0031	A0056 A0126
ADD18	03B3	D0336	D0328
ADDA	035C	D0085	
ADDC	03E1	D0413	
ADDDONE	01A8	A0566	A0562
ADDSGN	036C	G0176	
AL	0006	F0033	
ALAW	055E	A0511	B0321 B0327 B0361
ALAWR	004E	A0172	F0030
ALAWX	035E	D0090	A0512 F0035
ALOG	019F	A0559	A0173 F0034
ANOMLE	01AF	A0572	
ANOMLY	0005	G0173	A0557
APP	03B6	D0345	A0564
APRED	02AA	C0024	B0325 D0323 D0324 D0325 D0336 D0338
AQUAN	000A	G0188	D0335
B1	000B	G0191	A0099 A0168 A0351 A0504
B2	000C	G0194	B0167 B0183 D0396 D0398 D0402
B3	000D	G0197	B0144 B0160 D0389 D0391 D0395
B4	000E	G0200	B0121 B0137 D0382 D0384 D0388
B5	000F	G0203	B0098 B0114 D0375 D0377 D0381
B6	004D	G0391	B0075 B0091 D0368 D0370 D0374
B1AS	0071	G0499	B0053 B0068 D0361 D0363 D0367
B1ASA	02B8	C0075	A0082 A0089 A0253 A0264 A0272 A0282 A0290 A0302 A0310
COTO1	02B5	C0073	A0320 A0334 A0346 A0426 A0436 A0444 A0456 A0464 A0474
COTO3	02B2	C0071	A0488
COTO7	02C3	C0083	A0180 A0184 A0188 A0192 A0196 A0200 A0204
C2TO3	0201	C0093	
C4TO5	02CE	C0091	C0074
C4TO7	02DC	C0101	
C6TO7	02EA	C0111	C0072
C8TO11	02E7	C0113	C0092
C8TO14	02FC	C0125	
C8TO9	02ED	A0033	C0070
CATOB	0002	C0137	
CCITT	030E	B0068	C0112
CCTOD	01C4	B0245	A0101 A0170 A0210 A0364
CCTOE	025F	B0091	C0110
CHK1	01DE	B0114	E0007
CHK2	01F8	B0137	E0011
CHK21	0212	B0160	E0015
CHK3	022C	B0183	E0019
CHK4	0246	D0186	E0023
CHK5	0386	C0187	E0027
CHK6	032C	C0185	D0183
CHKHI	0329	C0193	
C10TO1	0335	C0199	C0186
C10TO3	033E	C0201	C0184
C12TO3	0341		
C14TO7			
C15TO6			

CCITT LABEL	VALUE	DEFN	REFERENCES
C16TO7	034A	C0207	C0200
CLNUP	0EA	A0342	A0330
CLNUPA	0176	A0496	A0484
CMPRSA	010B	A0405	
CMPRSU	007F	A0251	
CONS	059B	G0075	
CTL	0000	A0034	F0019
DOTO1	0598	D0559	F0028
DOTO3	0435	D0557	
DOTO7	0432	D0555	
D2TO3	0438	D0564	D0558
D4TO5	0559	D0578	
D4TO7	0557	D0576	D0556
D6TO7	0555	D0590	D0577
D8TO9	043D	D0608	
DBTOB	0440	D0606	
DBTOF	0576	D0604	D0554 A0357 A0510
DAC	044A	A0032	D0607
DATOB	045A	D0624	
DCTOD	046F	D0642	D0605
DETOF	046C	D0640	D0641
DML	0469	D0658	
DMS	0001	G0182	D0293 D0294 D0295 D0329 D0332
DO32KA	0480	G0179	D0273 D0274 D0275 D0331
DO32KU	0494	A0376	A0369
DONEC	0491	A0222	A0215
DONEF	04A5	D0558	
DQ	0008	D0442	D0437
DQ1EXP	0007	G0206	D0103 D0110 D0111 D0416 D0548
DQ1MAN	0106	G0224	B0146 B0147
DQ2EXP	007A	B0150	B0124
DQ2MAN	041C	B0123	B0101
DQ3EXP	03F9	B0127	B0078
DQ3MAN	0010	B0104	
DQ4EXP	0016	B0100	B0077 B0078
DQ4MAN	005C	G0230	B0081
DQ5EXP	0017	G0442	B0055
DQ5MAN	005D	G0233	B0058
DQEXP	0018	G0445	B0169 B0170 D0091 D0095
DQMAN	005E	G0236	B0173 D0094 D0101 D0126 D0127 D0128 D0130
E0	0019	G0448	E0131
E0TO1	005F	G0221	
E0TO3	001A	G0433	E0061
E0TO7	0060	E0062	E0109
E1	0015	E0060	
E10	005B	E0058	E0121
E11	04EE	E0056	
E12	04E8	E0066	
E13	04E5	E0114	E0059
E2	04F2	E0122	E0071
E2TO3	0529	E0126	
E3	0533	E0132	E0057
E4	053E	E0072	
E4TO6	04F9	E0070	E0081
E5	04F6	E0076	
E5TO6	04FD	E0082	
	0504	E0080	
	0501	E0088	
	050B	E0086	
	0508		

CCITT LABEL	VALUE	DEFN	REFERENCES
E6	050F	E0092	E0087
E7	051A	E0100	
E7TO8	0517	E0098	
E7TOA	0514	E0096	
E7TOD	0511	E0094	E0055
E8	051E	E0104	E0099
E9	0525	E0110	
E9TOA	0522	E0108	E0097
EBTOC	0530	E0118	
EBTOD	052D	E0120	E0095
EDTOE	053B	E0130	E0119
EIGHT	0038	G0328	A0541
EXP0	02BB	C0077	
EXP1	02BF	C0080	C0076
EXP10	0305	C0127	
EXP11	0311	C0132	C0126
EXP12	0311	C0141	
EXP13	0317	C0146	C0140
EXP14	031D	C0151	C0138
EXP2	02CA	C0085	
EXP3	02CA	C0088	C0084
EXP4	02D4	C0095	
EXP5	02DB	C0098	C0094
EXP6	02DF	C0103	
EXP7	02E3	C0106	C0102
EXP8	02F0	C0115	
EXP9	02F6	C0120	C0114
EXPNDA	0035	A0146	
EXPNDU	0005	A0076	
EXX01	0439	D0560	
EXX10	0483	D0626	
EXX11	048A	D0633	D0625
EXX12	0497	D0644	
EXX13	049E	D0651	D0643
EXX14	04A8	D0660	
EXX15	04AF	D0667	D0659
EXX2	0440	D0566	
EXX3	0445	D0571	D0565
EXX4	0450	D0580	
EXX5	0455	D0585	D0579
EXX6	045D	D0592	
EXX7	0462	D0597	D0591
EXX8	0472	D0610	
EXX9	0479	D0617	D0609
F1	0068	G0472	D0051 D0272 D0292
FILTA	039C	D0272	
FILTB	03A0	D0292	
FILTC	03A4	D0323	
FILTD	037C	D0175	
FILTE	038B	D0209	
FINI	00E6	A0338	D0187
FINISH	0172	A0492	A0268 A0276 A0286 A0294 A0306 A0314 A0324
FITABL	0593	G0062	A0422 A0430 A0440 A0448 A0460 A0468 A0478
FLOAT	040E	G0050	B0054 B0076 B0099 B0122 B0145 B0168 B0231 B0253
FLTDQ	0377	D0126	D0108 E0068 E0074 E0078 E0084 E0090 E0092 E0102 E0106
FLTSFT	002A	E0047	E0064 E0112 E0116 E0124 E0128 E0134

CCITT LABEL	VALUE	DEFN	REFERENCES
FLTSR	042A	D0548	D0519
GETA1	03FA	D0456	D0435 D0440
GETA2	0404	D0481	
GETB1	030A	D0396	
GETB2	03D3	D0389	
GETB3	03CC	D0382	
GETB4	03C5	D0375	
GETB5	03BE	D0368	
GETB6	03B6	D0360	
GETEXP	02AF	C0069	
GETF	03EA	D0431	
GETF1	03EE	D0434	
GETF2	03F4	D0438	D0433
GETI	0018	A0098	
GETIM	0351	C0212	C0190 C0192 C0196 C0198 C0204 C0206 C0210
GETMAN	0321	C0155	C0079 C0082 C0087 C0090 C0097 C0100 C0105 C0108 C0119
GETSE	024E	B0230	C0124 C0131 C0136 C0145 C0150
I	0001	G0161	A0100 A0101 A0169 A0170 A0210 A0212 A0364 A0366 A0544
IDEQ1M	01B1	A0574	D0106
IDGT1M	01A0	A0560	A0547
IDLT1M	0191	A0548	A0546
IEQ0	032F	C0189	C0188
IEQ1	0332	C0191	
IEQ2	0338	C0195	C0194
IEQ3	033B	C0197	
IEQ4	0344	C0203	C0202
IEQ5	0347	C0205	
IEQ6	034D	C0209	C0208
IEQ7	0350	C0211	
IM	0002	G0164	A0213 A0216 A0218 A0367 A0370 A0372 C0212 D0045
INQTAB	006A	A0478	D0046
INTRPT	0002	A0050	A0050
IQTAB	0583	G0042	G0128
ITAB1	07F9	C0175	C0187
ITAB2	087B	C0176	C0185
ITAB3	08CA	C0177	G0436
ITAB4	0905	C0178	C0183
ITAB5	0936	C0179	A0147
ITAB6	0964	C0180	
ITAB7	0995	C0181	
K130	007A	G0526	B0070 B0093 B0116 B0139 B0162 B0185 B0247 B0271
K138	007B	G0529	
K16382	0062	G0454	
K2309	007C	G0532	
K32768	0048	A0376	
K4576	0061	D0186	D0186
K46	0064	G0460	C0201
K49	0065	G0463	C0207
K5120	006C	G0484	D0184
K544	006B	G0481	D0182 D0184
K56	0076	G0514	E0056
K63	0067	G0469	A0327 A0481
K79	0078	G0520	C0193
K95	0079	G0523	C0199
K960	0077	G0517	D0604 E0094

17. 32-kbit/s ADPCM with the TMS32010

CCITT LABEL	VALUE	DEFN	REFERENCES
KFF00	0075	G0511	A0077
KFF80	006E	G0490	B0062 B0085 B0108 B0131 B0154 B0177 B0239 B0262
LIMA	028E	B0320	
LIMB	0380	D0182	
LIMC	0415	D0502	D0504
LIMD	041D	D0512	
M0080	007F	G0541	A0508
M127	006F	G0493	A0549 A0558 A0563 A0566 C0155
M15	006D	G0487	A0217 A0261 A0269 A0279 A0287 A0299 A0307 A0317 A0331
			A0371 A0415 A0423 A0433 A0441 A0453 A0461 A0471 A0485
			C0071 C0109 C0214 D0512 D0555
			D0441
M16382	0063	G0457	A0343 A0355 A0497
M255	0047	G0373	A0079 A0149 D0217 D0224 E0005 E0009 E0013 E0017 E0021
M32767	0074	G0508	A0081 A0151 D0092
			E0025 E0029 E0033 F0033
M4095	0072	G0502	A0343 A0355 A0497 D0217 D0224 E0005 E0009 E0013 E0017 E0021
			A0079 A0149 E0029 E0033 F0033 B0365 D0092
			E0025 E0029 F0033
MAXNEG	01AD	A0570	A0551
MAXPOS	0199	A0554	A0551
MFFC0	0070	G0496	B0326
MINUS	004B	G0385	B0353
MIX	0297	F0031	B0324
MULAW	055A	A0358	A0359 F0032
MULAWR	00F9	A0103	A0104 F0031
MULAWX	0020	B0072	
NEG1	01C9	B0249	
NEG2	0264	B0095	
NEG11	01E3	B0273	
NEG21	027F	B0118	
NEG3	0200	B0141	
NEG4	0217	B0164	
NEG5	0231	B0187	
NEG6	024B	B0187	
NOCONS	004A	F0004	F0022
NONNEG	02A5	B0366	B0364
NORMAL	00DF	A0331	A0326
NORMLA	016B	A0485	A0480
NXCONS	0551	G0023	F0026
ONE	004C	G0388	A0176 A0214 A0255 A0259 A0277 A0297 A0315 A0325 A0368
			A0409 A0413 A0418 A0431 A0451 A0469 A0479 A0543 A0552
			A0556 A0561 A0568 B0320 C0069 C0075 C0083 C0093
			C0101 C0113 C0125 C0139 C0165 D0048 D0050 D0090 D0093
			D0099 D0104 D0107 D0337 D0419 D0434 D0439 D0553 D0561
			D0564 D0578 D0590 D0608 D0624 D0642 D0658 E0054 E0062
			E0070 E0080 E0086 E0098 E0108 E0120 E0130 F0017 F0018
			F0025
PK0	004F	G0397	D0414 D0420 D0421
PK1	0050	G0400	D0413 D0460
PK2	0051	G0403	D0488
POS1	01CC	B0075	B0071
POS11	0267	B0252	B0248
POS2	01E6	B0098	B0094
POS21	0282	B0276	B0272
POS3	0200	B0121	B0117
POS4	021A	B0144	B0140
POS5	0234	B0167	B0163
POS6	024E	B0190	B0186
PRO1CT	0355	D0045	A0102 A0171 A0223 A0377

CCITT LABEL	VALUE	DEFN	REFERENCES
PTCONS	0036	F0005	F0021
QDONE	0354	C0215	
QUAN	0326	C0183	
RAMLOC	05E5	G0154	
RCVA	00FD	A0364	A0511
RCVMU	0071	A0210	A0358
RESET	0542	F0007	A0043
ROMLOC	0563	G0005	
RS1	04B6	E0004	B0066 B0243
RS11	04D4	E0028	B0089
RS2	04BB	E0008	B0266
RS21	04D9	E0032	B0112
RS3	04C0	E0012	B0135
RS4	04C5	E0016	B0158
RS5	04CA	E0020	B0181
RS6	04CF	E0024	
SAMPLE	0026	G0274	A0090 A0091 A0094 A0158 A0159 A0162 A0265 A0273 A0283
			A0291 A0303 A0311 A0321 A0328 A0335 A0345 A0349 A0419
			A0427 A0437 A0445 A0457 A0465 A0475 A0482 A0489 A0499
			A0502 C0024
SATCH	00DA	A0327	
SATCHA	0166	A0481	
SBASE	0024	A0106	A0086
SBASEA	0052	A0175	A0155
SCAL0A	0118	A0415	
SCAL1A	0121	A0423	A0414
SCAL2A	012D	A0433	
SCAL3A	0136	A0441	A0432
SCAL4A	0145	A0453	
SCAL5A	014E	A0461	A0452
SCAL6A	015A	A0471	
SCAL7A	0163	A0479	A0470
SCALE0	008C	A0261	
SCALE1	0095	A0269	A0260
SCALE2	00A1	A0279	A0278
SCALE3	00AA	A0287	
SCALE4	0089	A0299	A0298
SCALE5	00C2	A0307	
SCALE6	00CE	A0317	A0316
SCALE7	00D7	A0325	
SCL021	0089	A0259	
SCL023	0086	A0257	A0258
SCL0T1	0115	A0413	A0412
SCL0T3	0112	A0411	
SCL2T3	009E	A0277	A0256
SCL425	012A	A0431	
SCL427	00B6	A0297	
SCL4T5	00B3	A0295	A0410
SCL4T7	0142	A0451	A0296
SCL627	013F	A0449	A0450
SCL6T7	00CB	A0315	
	0157	A0469	

17. 32-kbit/s ADPCM with the TMS32010

CCITT 32010 FAMILY MACRO ASSEMBLER PC2.1 84.107

CCITT LABEL	VALUE	DEFN	REFERENCES
TEMP1	0021	G0257	A0093 A0148 A0153 A0154 A0160 A0161 A0340 A0341 A0494
			A0495 A0542 A0545 A0152 B0060 B0063 B0064 B0083 B0086 B0087
			B0106 B0109 B0110 B0129 B0132 B0133 B0152 B0155 B0156
			B0175 B0178 B0179 B0237 B0240 B0241 B0260 B0263 B0264
			B0360 B0362 C0068 C0078 C0081 C0086 C0089 C0096 C0099
			C0104 C0107 C0116 C0117 C0118 C0121 C0122 C0123 C0128
			C0129 C0130 C0133 C0134 C0135 C0142 C0143 C0144 C0147
			C0148 C0149 C0152 C0153 C0154 C0156 C0157 D0047 D0085
			D0210 D0211 D0216 D0221 D0362 D0364 D0369 D0371 D0376
			D0378 D0383 D0385 D0390 D0392 D0397 D0399 D0417 D0418
			D0498 D0506 D0507 D0514 D0518 D0520 E0050 E0051 E0053
			E0060 E0063 E0066 E0067 E0072 E0073 E0076 E0077 E0082
			E0083 E0088 E0089 E0101 E0104 E0105 E0110 E0111 E0111
			E0114 E0115 E0122 E0123 E0126 E0127 E0132 E0133 F0028
			F0029
TEMP2	0022	G0261	A0080 A0087 A0150 A0156 D0218 D0219 D0220 D0222 D0457
			D0459
TEMP3	0023	G0265	B0354 B0356 B0357 B0359 B0366 D0098 D0100 D0129
			D0330 D0334 D0432 D0442 D0483 D0484 D0491 D0492 D0495
TEMP4	0024	G0268	A0251 A0342 A0347 A0348 A0405 A0407 A0496 A0500 A0501
			C0066 C0213 D0516 D0521 D0522
THREE	007D	G0535	A0257 A0295 A0411 A0449 B0059 B0082 B0105 B0128 B0151
			B0174 B0236 B0259 C0073 C0091 C0111 C0137 C0327 D0503
			D0505 D0557 D0576 D0606 D0640 E0058 E0096 E0118
WI	0069	G0475	D0049 D0177
WTABLE	058B	G0052	
XMTA	0034	A0056	A0172
XMTMU	0004	A0126	A0103
Y	0009	G0185	B0367 B0368 D0175 D0176 D0326
YLH	0049	G0379	B0355 B0209 D0213 D0223
YLL	004A	G0382	B0353 D0212 D0214 D0225
YOVER4	0029	G0283	B0369 C0166 D0086
YU	004E	G0394	D0358 D0178 D0189 D0215
ZRAMA	0547	F0013	F0014

CCITT 32010 FAMILY MACRO ASSEMBLER PC2.1 84.107

CCITT LABEL	VALUE	DEFN	REFERENCES
SCRACH	001B	G0239	A0056 A0076 A0083 A0106 A0108 A0110 A0112 A0114 A0116
			A0118 A0120 A0126 A0146 A0152 A0175 A0179 A0183 A0187
			A0191 A0195 A0199 A0203 A0254 A0262 A0263 A0266 A0270
			A0271 A0274 A0280 A0281 A0284 A0288 A0289 A0292 A0300
			A0301 A0304 A0308 A0309 A0312 A0318 A0319 A0322 A0332
			A0333 A0336 A0338 A0339 A0344 A0356 A0357 A0408 A0416
			A0343 A0420 A0424 A0428 A0434 A0435 A0438 A0442
			A0443 A0446 A0454 A0455 A0458 A0462 A0463 A0466 A0472
			A0473 A0476 A0486 A0487 A0490 A0492 A0493 A0498 A0509
			A0510 A0548 A0560 A0574
			D0105 D0113 D0365 D0372 D0379 D0386 D0393 D0400 D0401
SDQ	0054	G0412	B0184 D0394
SDQ1	0055	G0415	B0161 D0387
SDQ2	0056	G0418	B0138 D0380
SDQ3	0057	G0421	B0115 D0373
SDQ4	0058	G0424	B0092 D0366
SDQ5	0059	G0427	B0069 D0360
SDQ6	005A	G0430	B0301 B0302 B0303 C0025 D0549
SE	0003	G0167	
SETPAC	0544	F0010	B0298 D0415
SEZ	0004	G0170	G0124
SHFT	0563	G0007	D0096
SHIFT	0066	G0466	A0098 A0167 A0222 A0376
SIGD1F	01B3	G0053	A0247 A0401 B0268 B0269 B0550
SR	0013	G0215	B0235
			B0246
			B0232
SR1	0014	G0218	
SRIEXP	001D	G0245	
SRIMAN	0053	G0409	
SREXP	001C	G0242	
SRMAN	0052	G0406	B0254 B0255 D0560 D0569 D0574 D0583 D0588 D0595 D0602
			D0615 D0622 D0631 D0638 D0649 D0656 D0665 D0672
			D0258 D0552 D0559 D0562 D0566 D0567 D0571 D0572 D0580
			D0581 D0585 D0586 D0592 D0593 D0597 D0598 D0599 D0600
			D0611 D0612 D0613 D0617 D0618 D0619 D0620 D0626
			D0628 D0629 D0633 D0634 D0635 D0636 D0644 D0645
			D0647 D0651 D0652 D0653 D0654 D0660 D0661 D0662
			D0646 D0667 D0668 D0669 D0670
			D0663
STRLIM	038A	D0189	B0056 B0057 B0067 B0073 B0074 B0297 E0006
SUBF	0409	D0486	B0079 B0080 B0090 B0096 B0097 B0296 E0010
SUBONE	019B	A0556	B0102 B0103 B0113 B0119 B0120 B0295 E0014
SUBTB	0324	C0165	B0125 B0126 B0136 B0142 B0143 B0294 D0487 D0494 E0018
SUM1	0022	G0260	B0148 B0149 B0159 B0165 B0166 B0293 E0022
SUM2	0023	G0264	B0171 B0172 B0182 B0188 B0189 B0292 E0026
SUM3	0025	G0271	E0030
SUM4	001E	G0248	E0034
SUM5	001F	G0251	
SUM6	0020	G0254	
SUM7	0027	G0277	B0233 B0234 B0244 B0250 B0251 B0299
SUM8	0028	G0280	B0256 B0257 B0267 B0274 B0275 B0300
SUMGT0	03E9	D0421	
SYNC	0188	A0541	A0353 A0506
		D0185	
		D0481	
		A0550	

17. 32-kbit/s ADPCM with the TMS32010

18. A Real Time Speech Subband Coder Using the TMS32010

T. Barnwell, R. Schafer, R. Mersereau, and D. Smith
Atlanta Signal Processors Incorporated

PREFACE

A full-duplex 16-kbit/s subband coder can be implemented using a single TMS32010 (without external program memory) with a 1K-word off-chip RAM data buffer. The compression algorithm achieves toll-quality speech with a DRT (Diagnostic Rhyme Test) score of 94 percent. Because of the compactness of the subband coding, it only requires 80 percent of the TMS32010 CPU utilization, thus allowing the processor to perform other tasks in addition to speech coding. A simplified block diagram of the coder is shown in the following figure:

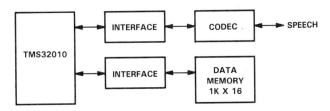

Simplified Block Diagram of the 16-kbit/s Subband Coder

The subband coder can be used in many applications, such as telecommunications and computers. In a transcoder design for digital telephony, the subband coder is valuable because of the following features: relative insensitivity to transmission errors, small bandwidth requirement, toll-quality speech, and simple, economical hardware configuration. In a voice-mail system, speech data rate directly corresponds to the storage required in the memory. The 16-kbit/s subband coder provides an efficient tradeoff between speech quality and data rate. In addition, the use of a single TMS32010 coder allows a cost-effective system implementation.

The following reprinted article details the theory and realization of the subband coder. The TMS32010 source code and documentation of the 16-kbit/s coder is available for licensing from Atlanta Signal Processors Incorporated. For more information, contact:

Atlanta Signal Processors Incorporated
770 Spring Street, Suite 208
Atlanta, Georgia 30308
(404) 892-7265

INTRODUCTION

Over the past several years, the *Subband Coder* has taken its place as one of the highest quality and most cost effective approaches to the coding of speech at medium bit rates. Table 1 illustrates the relationship of the subband coder to a number of other popular approaches to speech coding. Fundamentally, of course, any coding operation must distort the coded signal in some way. The goal of a good coding technique is to constrain the distortion so as to have minimum impact on the overall system performance. In most speech communications systems and voice response systems, this translates into minimizing the perceived differences between the coded and uncoded speech signals.

Modern speech coding systems typically use three classes of features in order to minimize the perceived coding distortion: the characteristics of the auditory system; the characteristics of the vocal tract; and the characteristics of language and individual talkers. Of these, only the first two can be effectively used in medium bit rate systems such as subband coders. Typically, three separate characteristics of the auditory system can be used to advantage. These are (1) the aural noise masking effect, (2) the frequency variant sensitivity of auditory perception (critical bands), and (3) the relative phase insensitivity of the ear (within critical bands). Likewise, a vocal tract model can either be used in a long term statistical sense, or in a short-time stationary time-varying model. If the short-time stationary model is used, then it, in turn, may be applied in three ways: (1) by using syllabic energy variations; (2) by using an explicit slowly time-varying vocal tract model; or (3) by using the pseudo-periodicity (pitch) in voiced sounds.

The simplest common speech coding technique, linear Pulse Code Modulation (PCM), makes essentially no assumptions (except as to dynamic range) about the characteristics of the signal being coded or the eventual use of the coded signal. As a result, linear PCM systems require the greatest bit rate to generate toll quality speech. Such systems have the advantage, however, that since they show no preference toward any particular class of signals, they may be used to code signals which are neither spoken nor heard (such as data signals). Companded Pulse Code Modulation systems, such as the 64 Kbps mu-law companded system used in telephone switching networks, make direct use of the ear's noise masking characteristics in its most basic form. In the resulting systems, the noise is correlated with the signal, which is good, but noise also spreads throughout the frequency range with no regard to the presence of signal energy, which is bad. Companded PCM systems are extremely simple and inexpensive to implement, but the rate at which they must operate to achieve toll quality speech is still relatively high.

Systems such as Differential Pulse Code Modulation (DPCM) and Delta Modulation (DM) make direct use of a long-term stationary statistical model for speech production. Adaptive Differential Pulse Code Modulation (ADPCM) and Adaptive Delta Modulation (ADM) also make use of the slowly varying nature of the short-time energy, causing the noise to be heavily correlated with the speech signal, and causing a dramatic drop in the idle channel noise. Various forms of both the Adaptive Transform Coder (ATC) and the Adaptive Predictive Coder (sometimes called the residual excited vocoder) make use of all of the auditory and vocal tract features, and such systems are capable of generating excellent quality speech at medium to low bit rates. However, such systems are also generally both very complex and quite sensitive to both background noise and transmission errors. Pitch-excited vocoders in general, and LPC vocoders in particular, are capable of operating at low to very low bit rates, but they generally can never achieve toll quality and do not perform well on either noisy speech or non-speech signals.

With this perspective, it is easy to see the advantages and disadvantages of the subband coder as compared to other speech coding systems. Since the subband coder uses only the characteristics of the auditory system in reducing the perceived distortion, it is capable of coding both speech and non-speech signals with good perceived fidelity. Similarly, subband coders can handle multiple simultaneous voices with no increase in degradation. In addition, since subband coders do not include the use of a time-varying vocal tract model, they are far less complex and far less sensitive to transmission bit errors than APC, ATC, or pitch-excited vocoders. This makes the subband coder a very good choice for real communications and telephony systems where the background noise is not controlled and where high transmission error rates might occasionally be expected. Likewise, the low complexity of subband coder realizations make them well matched to implementations on such signal processing micro-computers as the Texas Instruments TMS320 and the NEC 7720. The obvious applications for subband coders for speech include mobile radio, mobile telephony, and voice switching networks.

SUBBAND CODER THEORY

The concept of subband coding for speech was first introduced by Crochiere et al. [1-2]. In a subband coding system, the speech signal is divided into bands by a filter bank (Fig. 1), and then the outputs of the individual bands are first decimated, typically at their nominal Nyquist rate, and then coded for transmission. At the receiver, the coded signals from the individual bands are first decoded, and then padded with zeros and passed through a set of interpolation filters before being summed to form an estimate of the input signal. Such systems can be thought of as exhibiting two classes of distortion: analysis-reconstruction distortion, which is defined as the distortion which would be present if the subband signals were not coded; and coding distortion, which is the distortion due directly to the coding operations themselves. The analysis-reconstruction system alone can exhibit three separate types of distortion: interband aliasing distortion; frequency domain magnitude distortion (which will be called "frequency distortion"); and phase distortion. In the original subband coding systems, infinite impulse response (IIR) filters were used, and the resulting coding systems exhibited all four types of distortion: interband aliasing; frequency distortion; phase distortion; and coding distortion. In later work, Esteban et al. [3] introduced the concept of quadrature mirror filters, which can be used to realize two-band analysis-reconstruction systems which exhibit no interband aliasing. Further, Esteban showed that if equal length linear phase filters are used for the band splitting, then the overall analysis-reconstruction transfer function is a linear phase. Hence, quadrature mirror filters allow for analysis-reconstruction systems which exhibit only frequency distortion.

Based on these results, Esteban et al. [4] and Crochiere et al. [5-6] developed subband speech coding systems based on octave band tree structures of quadrature mirror filters. The octave band structure has the advantage that it can simultaneously approximate the critical band structure of the ear while also utilizing quadrature mirror filters. Such systems exhibit an overall system response which is linear phase, and as such they can be used to code data signals as well as speech signals. The filters used by Crochiere et al. [5] were designed by Johnston [7] using an iterative approach which sought to minimize the frequency distortion in the analysis-reconstruction system. Another characteristic of these later subband coders is that the band-splitting filters were realized using a two-band polyphase filter bank [8] to improve the computational efficiency. In very recent work, Smith and Barnwell [9-10] have presented a technique for designing tree-structured analysis-reconstruction filter banks which allow for exact reconstruction. Such systems, which allow for the design of filter banks of arbitrary quality, are very attractive since they exhibit only coding distortions. However, they sometimes require slightly more computation since they cannot always use the polyphase structure directly in implementing the filter banks.

The octave band subband coders used in speech coding applications are typically formed as a tree structure of analysis-reconstruction pairs based on quadrature mirror or exact reconstruction filters. Fig. 2(a) shows the block diagram for such a two-band system. For the lower channel, the input signal, x(n), is first passed through a half-band low-pass filter with impulse response $h_0(n)$, and then decimated at a 2-to-1 rate. For the upper channel, the signal is passed through a half-band high-pass filter with the impulse response $h_1(n)$, followed by another 2-to-1 decimator. The output of these two channels, $y_0(n)$ and $y_1(n)$, are the two signals which would be coded in a two-band subband coding system. For the reconstruction, these two signals are padded with zeros at a 1-to-2 rate, passed through the two interpolation

filters with impulse responses $g_0(n)$ and $g_1(n)$, respectively, and the outputs are summed to form the estimation of the original input, $\hat{x}(n)$.

Let us define the transfer functions of the two analysis filters as

$$H_0(z) = \sum_{n=-\infty}^{\infty} h_0(n)z^{-n} \qquad (1a)$$

$$H_1(z) = \sum_{n=-\infty}^{\infty} h_1(n)z^{-n} \qquad (1b)$$

and the Fourier transform of these two transfer functions by

$$H_0[\omega] = H_0(e^{j\omega}) = \sum_{n=-\infty}^{\infty} h_0(n)e^{-j\omega} \qquad (2a)$$

$$H_1[\omega] = H_1(e^{j\omega}) = \sum_{n=-\infty}^{\infty} h_1(n)e^{-j\omega} \qquad (2b)$$

Then, for quadrature mirror filters, the high-pass filter transfer function can be defined in terms of the low-pass filter by

$$H_1(z) = H_0(-z) \qquad (3)$$

or equivalently by

$$H_1[\omega] = H_0[\omega - \pi] \qquad (4)$$

In comparison, the Smith-Barnwell exact reconstruction filters are related by

$$H_1(z) = H_0(-1/z) \qquad (5)$$

or equivalently by

$$H_1[\omega] = H_0[-\omega - \pi] \qquad (6)$$

For quadrature mirror filters, this leads to the relationship

$$h_1(n) = (-1)^n h_0(n) \qquad (7)$$

and for the Smith-Barnwell filters, the relationship becomes

$$h_1(n) = (-1)^n h_0(-n) \qquad (8)$$

Fig. 2(b) gives the equivalent polyphase filter structure [8] for the two-band analysis-reconstruction system of Fig. 2(a). For the quadrature mirror filters, the transfer function for the low-pass filter can be expressed as

$$H_0(z) = P_0(z^2) + P_1(z^2)z^{-1} \qquad (9)$$

and the polyphase filters can be defined as

$$p_0(n) = h_0(2n) \qquad (10a)$$

$$p_1(n) = h_0(2n+1) \qquad (10b)$$

This is the well known polyphase result that the impulse response of the polyphase filters for a two-band system is formed by taking every second sample of the impulse response of the prototype low-pass filter. The polyphase filter structure reduces the required number of multiplies by a factor of two over the filter structure of Fig. 2(a). The polyphase filter structure can also be used with Smith-Barnwell exact reconstruction filters under the condition that

$$h_0(n) = h_0(-n) \qquad (11)$$

or, equivalently, $h_0(n)$ must be a zero (or equivalently, a linear) phase filter. Although this condition can sometimes be met for Smith-Barnwell filters, the filter design problem for this special case is not well understood, and no realizations of this type have yet been demonstrated [9-10]. For those cases which are well understood (including the ones discussed below), this condition is not met and polyphase realizations are not possible.

In the traditional quadrature mirror reconstruction, the reconstruction filters can be defined in terms of their polyphase filters as

$$q_0(n) = p_1(n) \qquad (12a)$$

$$q_1(n) = p_0(n) \qquad (12b)$$

From Fig. 2(c), it is obvious that this leads to an overall system transfer function given by

$$\hat{X}(z) = C(z)X(z) = P_0(z^2)P_1(z^2)X(z) \qquad (13)$$

What this shows is that for the case of quadrature mirror filters, the entire analysis-reconstruction system behaves like a linear filter.

For the Smith-Barnwell exact reconstruction filters, the equivalent characterization of the reconstruction system is given by

$$G_0(z) = H_0(1/z) = H_1(-z) \qquad (14a)$$

$$G_1(z) = H_1(1/z) = H_0(-z) \qquad (14b)$$

This, of course, leads to exact reconstruction, or equivalently, a linear system $C(z)$ which is distortionless. In particular, if the analysis and reconstruction filters chosen are both length L FIR filters, then $C(z)$ is an ideal delay given by

$$C(z) = z^{-2L} \qquad (15)$$

This condition is also approximately true for the Johnston quadrature mirror filters. The primary point here in terms of the realization of subband coding algorithms is that, even under the best of conditions, a two-band analysis-reconstruction system still behaves as an ideal delay. This implies the need for additional storage for delay memory in tree structured realizations.

TREE STRUCTURED SUBBAND CODERS

Two-band subband coders based on the two-band analysis-reconstruction systems of the type we have been discussing are of relatively little interest for speech coding themselves, since they offer little gain over full-band ADPCM. However, such two-band systems can be combined in tree structures to produce effective coding structures for speech. This tree-structuring procedure has the advantage that it can take direct advantage of the properties of half-band quadrature mirror filters and Smith-Barnwell filters in filter structures more appropriate to the critical band nature of aural perception.

The basic structure of an octave band subband coder for speech is illustrated in Fig. 3 where the band-splitters and band-mergers are typically implemented using the polyphase forms shown in Fig. 4. A key issue in the octave band structure is the compensation filters for bands which have not been split. It has been shown [11] that for both the quadrature mirror filters and for Smith-Barnwell filters, the compensation filters can be adequately realized using ideal delays. The basic problem is that in a tree structured environment, these delays "stack", and the required delay memory size can become quite large. In particular, if there are N band-splitters in an octave band subband coder, then the compensation delay required by the qth band is given by

$$C_q(z) = z^{-d(q)} \qquad (16)$$

where $d(q)$ is given by

$$d(q) = \sum_{p=1}^{N-q} L(p+q)2^p \qquad (17)$$

and where $L(p)$ is the length of the FIR analysis and reconstruction filters in the pth band-splitting operation. This delay storage requirement can be quite large if a large number of octave bands is used. This translates into a total delay storage requirement for the entire system of

$$D = \sum_{p=1}^{N} L(p)2^p \qquad (18)$$

TMS320 REALIZATIONS

In designing subband coder implementations for a fast signal processing microprocessor such as the TMS320, it is important to determine exactly which of the microcomputer resources (data memory, program memory, or real-time) constrains the form of the implementations. Since the TMS320 has a relatively large program address space (4K sixteen bit words) and since the subband coder algorithms are quite compactly programmed, it is clear that program address space is not an issue of concern. However, this is certainly not true of data memory, which is in rather short supply (144 words total), and real-time might also be an issue. A program for a subband coder can be functionally divided into five parts: input-output operations; analysis and reconstruction filter implementations; delay compensation implementation; channel quantization implementation; and control.

The main loop in a subband coder program must operate on frames of input samples where the frame length is such as to result in an integer number of data points being coded by the deepest branch coder in the tree structure. In order to compute the number of cycles per sample required by the subband coder main loop, it is hence correct to compute the number of cycles required by the entire main loop and then to divide by the frame length. To compute the actual number of samples required by the entire realization, the number of cycles per sample which are required by the interrupt based I/O buffering routines must also be included.

The minimum number of sample points which is required for a data frame in order to meet the above constraint is given by

$$M = 2^N \qquad (19)$$

where, as before, N is the number of band-splitters in the coder structure. In an interrupt driven implementation, at least M data locations must be available for I/O buffering. In addition, the total number of data locations necessary to support the filter and delay structure is given by

$$K = \sum_{p=1}^{N} L(p)2^p \qquad (20)$$

This includes both the delay storage needed in the filter implementations and the delay storage needed for the compensation delays. In addition to this storage, our current implementations use two storage locations per APCM coder, two storage locations for the interrupt routine, and four temporary storage locations for various utility operations. Hence, the total number of storage locations required can be approximated as

$$NN = M + K + 2N + 6 \qquad (21)$$

Table 2 gives a tabulation of NN versus N for a "typical" FIR filter structure given by $L(1) = 32$, $L(2) = L(3) = L(4) = 16$, and $L(5) = 8$ for both full duplex and half duplex realizations. Since the TMS320 has only 144 data storage locations, clearly all implementations which use more than 144 locations must be realized through the addition of a read and write capable off-chip random access memory. This, of course, adds considerably to the cost of the implementation. In addition, this "external program" RAM must be accessed using the Table Read and Table Write TMS320 operations, which require three cycles per access and which also require the use of the arithmetic accumulator. Hence, this memory can only be accessed at a comparatively low rate. Table 2 shows that a three band (two band-splitters) system is the largest subband coder which can be realized in full duplex without external program RAM, and a four band system (three band-splitters) is the largest subband coder which can be realized in half duplex without external program RAM. It is clear from this analysis that data memory is a major constraining factor in subband coder realizations.

The real-time constraints in any subband coder realization, of course, are highly dependent on the sampling rate at which the coder is implemented. For toll quality speech, a sampling rate of eight kilohertz is appropriate, whereas for lower bit rate systems, sampling rates as low as 6.4 kilohertz might well be used. Since access to external program RAM, if used, must be performed by means of Table Reads and Table Writes, the programs must run much slower if more than 144 data storage locations are required. For the purposes of this paper, we will first compute the approximate number of cycles per sample point required for all of the operations in the subband coder excluding the data exchanges needed to support external program RAM. This will show approximately how much real-time remains for the implementation of the external program RAM data transfers. From this analysis and an analysis of the data transfer rates required by the Table Read and the Table Write operations, it will be obvious that essentially any reasonable subband coder for speech can be implemented in real-time on a single TMS320 equipped with adequate external program RAM.

For a subband coder environment, the total number of cycles per sample point necessary to realize the arithmetic operations of the analysis and reconstruction filters is simply computed as

$$MM = \sum_{p=1}^{N} L(p)2^{1-p} \qquad (22)$$

current realization requires 24 cycles, while the control overhead per sample requires less than 50 cycles per sample per band-splitter. In addition, the APCM coders require approximately 60 cycles for both the coder and the decoder (120 cycles total for full duplex operation), while the delay functions require less than 10 cycles per sample. Hence, the approximate number of cycles per sample can be written as

$$LL = MM + 120N + 24 + 60 \sum_{p=1}^{N} 2^{1-p} \tag{23}$$

Table 3 gives a tabulation for the approximate percent of real-time required as a function of N excluding the time required to access external program RAM for a sampling rate of eight kilohertz. As would be expected, the real-time requirement is nowhere near linear with N, and there is still over 25% of real-time available even at N = 5. Now let us consider the additional requirement imposed by the slow access of the external program RAM. The first point to note is that the vast majority of the required delay memory is not used in the filter computations, but is used to implement the compensation delays. The number of memory locations required by the filters is given by

$$MA = \sum_{p=1}^{N} 2L(p) \tag{24}$$

Since, for the example shown, MA has a maximum value of 176, 84 for the receiver and 84 for the transmitter, then at most two data context switches must be computed for a full duplex realization and no data context switches must be performed for half duplex realizations. Since the data delay operations can all be performed using only one data memory location and paired Table Read and Table Write operations and since only block accesses are required, then no additional data memory is required and the total number of memory locations accessed per frame is given by

$$MB = \sum_{p=1}^{N} 2^{p-1} \tag{25}$$

Since these operations must be done once per frame, and they can be performed in six cycles per data point, the total number of additional cycles is given by

$$NA = [2MA + 2MB]6/2^N \tag{26}$$

As can be seen in Table 4, all of the example subband coders can still be implemented in real-time.

CONCLUSIONS

Based on the above analyses and the associated realizations, it is clear that a TMS320 is well matched to the subband coder implementation problem for voice applications. For the subband coder realization, it was found that the delay compensation memory requirements dominated the realization, and reasonable implementations always required external program RAM. However, when this is made available, virtually any subband coder of interest can be implemented using the same hardware environment.

REFERENCES

1. R. E Crochiere, S. A. Webber, and J. L. Flanagan, "Digital coding of speech in sub-bands," *Bell Syst. Tech. J.*, vol. 55, pp. 1069-1085, Oct. 1976.
2. R. E. Crochiere, "On the design of sub-band coders for low-bit-rate speech communications," *Bell Syst. Tech. J.*, vol. 56, pp. 747-770, May-June, 1977.
3. A. Croisier, D. Esteban, and G. Galand, "Perfect channel splitting by use of interpolation/decimation/tree decomposition techniques," presented at the 1976 Int. Conf. Inform. Sci. Syst., Patras, Greece, 1976.
4. D. Esteban and C. Galand, "Applications of quadrature mirror filters to split band or voice coding schemes," *Proc. 1977 Int. Conf. Acoust., Speech, Signal Processing*, Hartford, CT, May 1977, pp. 191-195.
5. A. J. Barabell and R. E. Crochiere, "Sub-band coder design incorporating quadrature mirror filters and pitch detection," *Proc. 1979 Int. Conf. Acoust., Speech, Signal Processing*, Washington, D.C. Apr. 1979.
6. R. E. Crochiere, "A novel approach for implementing pitch prediction in sub-band coders," *Proc. 1979 Int. Conf. Acoust., Speech, Signal Processing*, Washington, D.C. Apr. 1979.
7. J. D. Johnston, "A filter family designed for use in quadrature mirror filter banks," *Proc. 1980 Int. Conf. Acoust., Speech, Signal Processing*, Denver, CO, Apr. 1980.
8. M. G. Bellanger, G. Bonnerst, and M. Coudreuse, "Digital filtering by polyphase network: Application to sample-rate alteration and filter banks," *IEEE Trans. Acoust., Speech, Signal Processing*, vol. ASSP-24, pp. 109-114, Apr. 1976.
9. M. J. T. Smith and T. P. Barnwell III, "Exact reconstruction techniques for tree-structured subband coders," accepted for publication, *IEEE Trans. Acoust., Speech, Signal Processing.*
10. M. J. T. Smith and T. P. Barnwell III, "A procedure for designing exact reconstruction filter banks for tree-structures," *Proc. 1984 Int. Conf. Acoust., Speech, Signal Processing*, San Diego, CA, Mar. 1984.
11. T.P. Barnwell III, "Subband coder design incorporating recursive quadrature filters and optimum ADPCM coders," *IEEE Trans. Acoust., Speech, Signal Processing*, vol. ASSP-30, pp. 751-765, Oct. 1982.

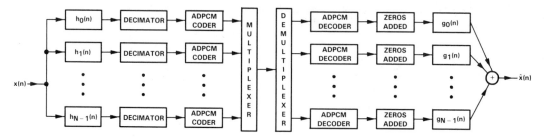

Fig. 1. N-band analysis-reconstruction system for speech coding

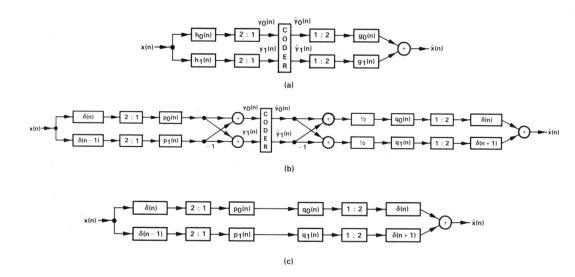

(a)

(b)

(c)

Fig. 2. (a) Two-band analysis-reconstruction band splitter
(b) Equivalent polyphase structure for a two-band analysis-reconstruction band-splitter
(c) Equivalent polyphase structure for analyzing the reconstruction

TABLE 1

TABLE 1
Summary of Speech Coding Techniques

Coding Technique	Effective Bit Rates	Aural Noise Masking	Aural Frequency	Aural Phase	Syllabic Energy	Vocal Tract Model	Short Time Stationarity	Pitch
Linear PCM	120 Kbps- 80 Kbps	No	No	No	No	No	No	No
Companded PCM	100 Kbps- 50 Kbps	Yes	No	No	No	No	No	No
Delta Modulation	80 Kbps- 50 Kbps	Yes	No	No	No	Yes	No	No
DPCM	80 Kbps- 40 Kbps	Yes	No	No	No	Yes	No	No
ADM	40 Kbps- 16 Kbps	Yes	No	No	Yes	Yes	No	Maybe
ADPCM	40 Kbps- 16 Kbps	Yes	No	No	Yes	Yes	No	Maybe
Subband Coder	32 Kbps- 10 Kbps	Yes	Yes	Maybe	Yes	No	No	Maybe
ATC	32 Kbps- 7 Kbps	Yes	Yes	Maybe	Yes	Yes	Yes	Yes
APC	32 Kbps- 4 Kbps	Yes	No	No	Yes	Yes	Yes	Maybe
LPC Vocoder	2.4 Kbps- .6 Kbps	Yes	No	Yes	Yes	Yes	Yes	Yes

TABLE 2
Storage Requirements For Typical Subband Coders

Number of Band-Splits	Locations (Full Duplex)	Locations (Half Duplex)	I/O Buffer (M)	Delay Memory (K)
1	76	52	2	64
2	144	77	4	128
3	278	144	8	256
4	536	275	16	512
5	818	364	32	768

TABLE 3
Total Cycles of TMS320 and Percent Real-Time Excluding External Program RAM Access Time at Eight Kilohertz Sampling Rate

Number of Bands	Cycles Per Samples	Percent Real-Time
1	268	43
2	374	59
3	427	68
4	454	72
5	467	74

TABLE 4
Total Cycles of TMS320 and Percent Real-Time Including External Program RAM Access Time at Eight Kilohertz Sampling Rate

Number of Bands	Cycles Per Samples	Percent Real-Time
1	268	43
2	374	59
3	617	98
4	584	93
5	540	87

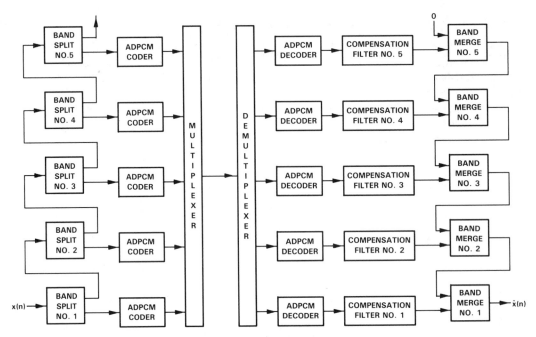

Fig. 3. Octave band subband coding system

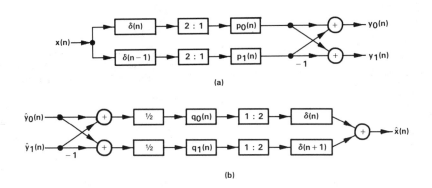

(a)

(b)

Fig. 4. (a) Band-splitting module for octave band subband coder
(b) Band merging module for octave band subband coder

19. Add DTMF Generation and Decoding to DSP-μP Designs

Pat Mock

Semiconductor Systems Engineering

Texas Instruments

Add DTMF generation and decoding to DSP-μP designs

In a computer system that employs a digital-signal-processing μP and that's equipped for phone-line communications, the DSP μP can generate and decode DTMF dialing signals as well as handle typical DSP functions. Therefore, the system can both dial out to establish communications links and accept Touchtone inputs for remote control of its functions.

Patrick Mock, *Texas Instruments*

A digital-signal-processing (DSP) μP can handle Touchtone (DTMF) dialing and decoding over telephone lines in addition to its customary signal-processing chores. As a consequence, if a computer system already has a DSP μP and A/D and D/A converters in place, then the system can decode DTMF signals, and any Touchtone telephone can serve as a data-entry terminal or a remote-control console. The only cost for these DTMF enhancements is additional program space in the μP's ROM.

This article outlines a DTMF generating scheme and describes in detail the implementation of DTMF decoding in a specific DSP μP, the TMS32010. Although the DTMF decoder functions as intended, it fails to meet AT&T specs exactly because it's designed to detect DTMF tones in the presence of speech and because it suits computer applications like voice-mail and electronic-mail systems, which are not pure telephone applications. DTMF tone decoders that *do* meet AT&T specs usually stop decoding tones if they detect speech. With a more exacting program, the TMS32010 could meet AT&T specs to the letter. One of the goals of this project, however, was to make the DTMF code as compact as possible to allow the DSP μP to do other jobs. Some performance was sacrificed as a consequence.

Tone generation is easy

A DTMF tone generator (**Ref 1**) can consist of a pair of programmable, second-order harmonic oscillators (**Fig 1**). The sample-generation rate of the oscillators determines the total harmonic distortion of the output. The higher the sampling rate, the more nearly exact the signal will be. In all cases, you must choose a sample-generation rate greater than approximately 7k samples/sec to achieve an acceptable signal. (**Fig 2** explains the DTMF tone-coding scheme.)

Because the telephone company's official digitizing rate is 8k samples/sec, most generating circuits run at this rate. According to the Nyquist criterion, which specifies that the sampling rate must be at least twice

*If a computer can decode DTMF signals,
then any Touchtone telephone can serve as
a data-entry terminal.*

the frequency of the highest-frequency signal being sampled, 8k samples/sec is more than adequate for generating any valid pair of tones using the TMS32010; the highest frequency involved is 1633 Hz. Because of a limitation in the system used to develop the chip's tone-generating and -decoding programs, the decoding-program version listed in **Fig 3** runs at 9766 samples/sec, and all testing was done using this version. However, **Table 1** presents coefficients for running at 8k samples/sec; **Fig 4**'s listing shows the portion of the code that must be amended for 8k-sample/sec operation.

Fig 5 shows the flow chart for the DTMF tone-generating algorithm. (The DTMF tone-generating routine described in **Ref 1** takes up 160 words in the program ROM.) The first step of the algorithm initializes the processor and the interfaces and performs all other required initialization. The next step retrieves the digit that's to be dialed (0 through 9, A through D, or "#" or "*") from a specified location in memory. The digit serves as a pointer within a table that contains the values required to initialize the resonators.

Because this design uses two oscillators for eight possible frequencies rather than eight oscillators, to provide the correct frequencies you must load the

Text continues on pg 212

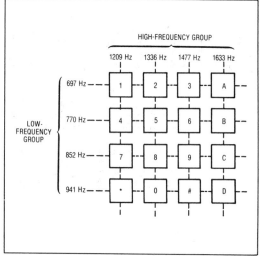

Fig 2—Pressing a button on a Touchtone telephone's 4×4 keypad *generates DTMF signaling tones in pairs. For example, pressing "6" generates a 770-Hz tone from the low-frequency group and a 1477-Hz tone from the high-frequency group. Note that the keypad has four keys (A through D) that are not normally seen on most phones. They're available with some special instruments.*

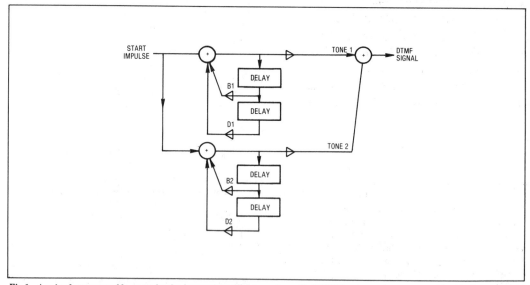

Fig 1—A pair of programmable, second-order harmonic oscillators *make up the DTMF tone generator represented by this directed graph. The delay boxes temporarily hold samples for one iteration. The delayed samples are multiplied by coefficients B and D and summed to generate a tone sample. The tones, in turn, are summed and sent to a D/A converter.*

19. Add DTMF Generation and Decoding to DSP-μP Designs

```
v: 1.00      DTMF TONE DECODER                    TMS320 Assembler vers 1.3

     0                    *
     1                    *                         DTMF TONE DECODER
     2                    *                    c Copyright Texas Instruments, 1984
     3                    *                         by Patrick C. Mock
     4                    *                    Northeast Syste…s Engineering
     5                    *
     6                    *
     7                            TITL    ′ DTMF TONE DECODER ′
     8                            IDT     ′v: 1.00′
     9                    *
    10   0000   f900      B       START               Go To The Beginning
         0001   0017
    11                    *
    12                    *
    13          0007      CS8     EQU     7           Define Variables
    14          0008      CS9     EQU     8
    15          000c      CS13    EQU     12
    16          000f      CS16    EQU     15
    17          0010      CLCK    EQU     16
    18          0011      MODE    EQU     17
    19          0010      ROWMX   EQU     16          10 Contains decoded row
    20          0011      COLMX   EQU     17          11    "         "      column
    21          0012      POSMAX  EQU     18
    22          0013      NEGMAX  EQU     19
    23          0014      ONE     EQU     20
    24          0015      LAST    EQU     21          16   Contains last decode
    25          0016      LAST2   EQU     22
    26          0017      COUNT   EQU     23
    27          0018      RC      EQU     24
    28          0019      CC      EQU     25
    29          001a      ROWMAX  EQU     26
    30          001b      COLMAX  EQU     27
    31          001c      DAT11   EQU     28
    32          0021      DAT23   EQU     33
    33          0022      DAT14   EQU     34
    34          0024      DAT15   EQU     36
    35          0028      DAT17   EQU     40
    36          0029      DAT27   EQU     41
    37          002a      DAT18   EQU     42
    38          002b      DAT28   EQU     43
    39          002d      DAT29   EQU     45
    40          0035      DAT213  EQU     53
    41          003b      DAT216  EQU     59
    42          003c      DATIN   EQU     60
    43                    *
    44                    *   BEGIN DATA TABLES
    45                    *
    46   0002   738b              DATA    29579       Real Coeff N=226
    47   0003   704e              DATA    28750
    48   0004   6cb8              DATA    27832
    49   0005   68cb              DATA    26827
    50   0006   5b23              DATA    23331
    51   0007   5355              DATA    21333
    52   0008   4af3              DATA    19187
    53   0009   3ef8              DATA    16120
    54                    *
    55   000a   4eff              DATA    20223       Real Coeff N=222
    56   000b   462b              DATA    17963       2nd harmonic
    57   000c   39a0              DATA    14752
    58   000d   2c58              DATA    11352
    59   000e   01d0              DATA      464
    60   000f   ec28              DATA    -5080
    61   0010   d712              DATA    -10478
    62   0011   c000              DATA    -16384
    63                    *
    64   0012   0200              DATA    512         CLCK = Sample Frequency
    65   0013   000a              DATA    >000A       MODE
    66                    *
    67   0014   7fff              DATA    >7FFF       POSMAX = Mask for data in
    68   0015   8000              DATA    >8000       NEGMAX = Mask for data out
    69                    *                           Program continues on pg 208
```

Fig 3—This tone-decoding program for the TMS32010 runs at 9766 samples/sec. However, the official digitizing rate specified by the phone company is 8k samples/sec; *Fig 4* shows the section of code that adapts this program to 8k-sample/sec operation.

```
70                      *
71    0016   0001   TABLE    DATA   1              ONE
72                      *    Start of Program     ********************
73                      *
74    0017   7f8b   START    SOVM
75    0018   6e00            LDPK   0
76    0019   6880            LARP   0              <Break>
77    001a   7014            LARK   0,ONE
78    001b   7e16            LACK   TABLE
79    001c   6788   NEXT     TBLR   *              Initialize Coefficients
80    001d   1014            SUB    ONE
81    001e   f400            BANZ   NEXT
      001f   001c
82    0020   4811            OUT    MODE,0         Set AIB Mode
83    0021   4910            OUT    CLCK,1         Set AIB Clock
84                      *
85                      *
86                      * Load not recognized symbol
87                      *
88    0022   7eff   NOT      LACK   >FF
89    0023   6915            DMOV   LAST
90    0024   5015            SACL   LAST
91                      *
92    0025   7f89   AGAIN    ZAC                   Zero DFT Loop Variables
93    0026   701f            LARK   0,31
94    0027   711c            LARK   1,DAT11
95    0028   f500            BV     ZERO
      0029   002a
96    002a   6881   ZERO     LARP   1
97    002b   50a0            SACL   *+,0,0
98    002c   f400            BANZ   ZERO
      002d   002a
99                      *
100                     *    Take data and calculate DFT loop
101                     *
102   002e   7ee2            LACK   226            SET DFT LOOP VARIABLE
103                     *
104   002f   5017   LOOP     SACL   COUNT
105   0030   700f            LARK   0,CS16
106   0031   713b            LARK   1,DAT216
107   0032   1214            SUB    ONE,2
108   0033   fc00            BGZ    WAIT
      0034   0037
109   0035   7007            LARK   0,CS8
110   0036   712b            LARK   1,DAT28
111                     *
112   0037   f600   WAIT     BIOZ   CALC           Wait for A/D
      0038   003b
113   0039   f900            B      WAIT
      003a   0037
114                     *
115   003b   423c   CALC     IN     DATIN,2
116   003c   2012            LAC    POSMAX
117   003d   783c            XOR    DATIN          Convert data to 320 format
118   003e   503c            SACL   DATIN
119                     *
120                     *  BEGIN DFT LOOPS
121                     *
122                     *
123   003f   6a81   FRPT     LT     *,1
124   0040   2c3c            LAC    DATIN,12       X(n)
125   0041   6298            SUBH   *-             X(n)-Y(n-2)
126   0042   6d88            MPY    *              cos(8*C)*Y(n-1)
127   0043   6b88            LTD    *              Y(n-1)->Y(n-2) and
128   0044   7f8f            APAC
129   0045   7f8f            APAC
130   0046   7f8f            APAC                  X(n)+2cos(8*C)*Y(n-1)-Y(n-2)
131   0047   5890            SACH   *-,0,0         --> Y(n-1)
132   0048   f500            BV     CHECK
      0049   0050
133   004a   f400            BANZ   FRPT
      004b   003f
134                     *
135   004c   2017            LAC    COUNT
136   004d   1014            SUB    ONE
```

```
137   004e   fe00           BNZ     LOOP
      004f   002f
138                 *
139                 *      Calculate Energy at each frequency
140                 *
141                 *
142   0050   7007   CHECK   LARK    0,CS8
143   0051   712b           LARK    1,DAT28
144                 *
145   0052   f800   MAGLP   CALL    ENERGY
      0053   00eb
146   0054   5990           SACH    *-,1,0
147   0055   f400           BANZ    MAGLP
      0056   0052
148                 *
149                 *   Compare Energies And Determine Decode Value
150                 *
151   0057   7e03           LACK    3
152   0058   5010           SACL    ROWMX
153   0059   5011           SACL    COLMX
154                 *
155                 * Find Row Peak
156                 *
157   005a   7102   ROWS    LARK    1,2
158   005b   7021           LARK    0,DAT23
159   005c   2022           LAC     DAT14
160   005d   501a           SACL    ROWMAX
161   005e   6880   ROWL    LARP    0
162   005f   6898           MAR     *-
163   0060   201a           LAC     ROWMAX
164   0061   1088           SUB     *
165   0062   fd00           BGEZ    ROWBR
      0063   0067
166   0064   3110           SAR     1,ROWMX
167   0065   2088           LAC     *
168   0066   501a           SACL    ROWMAX
169   0067   6891   ROWBR   MAR     *-,1
170   0068   f400           BANZ    ROWL
      0069   005e
171                 *
172                 * Find Column Peak
173                 *
174   006a   7102   COLUMN  LARK    1,2
175   006b   7029           LARK    0,DAT27
176   006c   202a           LAC     DAT18
177   006d   501b           SACL    COLMAX
178   006e   6880   COLL    LARP    0
179   006f   6898           MAR     *-
180   0070   201b           LAC     COLMAX
181   0071   1088           SUB     *
182   0072   fd00           BGEZ    COLBR
      0073   0077
183   0074   3111           SAR     1,COLMX
184   0075   2088           LAC     *
185   0076   501b           SACL    COLMAX
186   0077   6891   COLBR   MAR     *-,1
187   0078   f400           BANZ    COLL
      0079   006e
188                 *
189                 * Check For Valid Signal Strength
190                 *
191   007a   201b           LAC     COLMAX
192   007b   101a           SUB     ROWMAX
193   007c   fd00           BGEZ    COLBIG
      007d   008a
194   007e   201b   ROWBIG  LAC     COLMAX      Reverse Twist
195   007f   1414           SUB     ONE,4
196   0080   fa00           BLZ     NOT
      0081   0022
197   0082   6a1b           LT      COLMAX
198                 *
199   0083   800c           MPYK    12          Ideal 8db = 6
200                 *
201   0084   7f8e           PAC
202   0085   101a           SUB     ROWMAX
```

```
203   0086   fa00            BLZ    NOT
      0087   0022
204   0088   f900            B      VROW
      0089   0094
205   008a   201a    COLBIG  LAC    ROWMAX          Twist
206   008b   1414            SUB    ONE,4
207   008c   fa00            BLZ    NOT
      008d   0022
208   008e   6a1a            LT     ROWMAX
209                  *
210   008f   8003            MPYK   3               Ideal 4db = 3
211                  *
212   0090   7f8e            PAC
213   0091   101b            SUB    COLMAX
214   0092   fa00            BLZ    NOT
      0093   0022
215                  *
216                  * Check for valid row tone
217                  *
218   0094   6a1a    VROW    LT     ROWMAX
219                  *
220   0095   82ab            MPYK   683             683 = 1/6 = -8dB
221                  *
222   0096   7003            LARK   0,3
223   0097   711c            LARK   1,DAT11
224   0098   2014            LAC    ONE
225   0099   5017            SACL   COUNT
226   009a   6881    RVL     LARP   1
227   009b   2ca8            LAC    *+,12
228   009c   7f90            SPAC
229   009d   68a0            MAR    *+,0
230   009e   fb00            BLEZ   RCNT
      009f   00a3
231   00a0   2017            LAC    COUNT
232   00a1   1014            SUB    ONE
233   00a2   5017            SACL   COUNT
234   00a3   f400    RCNT    BANZ   RVL
      00a4   009a
235   00a5   2017            LAC    COUNT
236   00a6   fe00            BNZ    NOT
      00a7   0022
237                  *
238                  * Check for valid column tone
239                  *
240   00a8   6a1b    VCOL    LT     COLMAX
241                  *
242   00a9   82ab            MPYK   683             683 = 1/6 = -8dB
243                  *
244   00aa   7003            LARK   0,3
245   00ab   7124            LARK   1,DAT15
246   00ac   2014            LAC    ONE
247   00ad   5017            SACL   COUNT
248   00ae   6881    CVL     LARP   1
249   00af   2ca8            LAC    *+,12
250   00b0   7f90            SPAC
251   00b1   68a0            MAR    *+,0
252   00b2   fb00            BLEZ   CCNT
      00b3   00b7
253   00b4   2017            LAC    COUNT
254   00b5   1014            SUB    ONE
255   00b6   5017            SACL   COUNT
256   00b7   f400    CCNT    BANZ   CVL
      00b8   00ae
257   00b9   2017            LAC    COUNT
258   00ba   fe00            BNZ    NOT
      00bb   0022
259                  *
260                  * Check 2ND Harmonic Energy Levels
261                  *
262   00bc   7e2d            LACK   DAT29           Calculate address of
263   00bd   0110            ADD    ROWMX,1         row data locations
264   00be   5017            SACL   COUNT
265   00bf   3917            LAR    1,COUNT
266   00c0   7e08            LACK   CS9
267   00c1   0010            ADD    ROWMX
```

```
268    00c2   5017          SACL    COUNT
269    00c3   3817          LAR     0,COUNT
270    00c4   f800          CALL    ENERGY      Calculate energy level
       00c5   00eb
271    00c6   6a1a          LT      ROWMAX
272                  *
273    00c7   8fff          MPYK    4095        ROWMAX/8
274                  *
275    00c8   7f90          SPAC
276    00c9   7f90          SPAC                ROWMAX/4 > 2nd Har
277    00ca   fd00          BGEZ    NOT
       00cb   0022
278                  *
279    00cc   7e35          LACK    DAT213      Calculate address of
280    00cd   0111          ADD     COLMX,1     col data locations
281    00ce   5017          SACL    COUNT
282    00cf   3917          LAR     1,COUNT
283    00d0   7e0c          LACK    CS13
284    00d1   0011          ADD     COLMX
285    00d2   5017          SACL    COUNT
286    00d3   3817          LAR     0,COUNT
287    00d4   6880          LARP    0
288    00d5   f800          CALL    ENERGY      Calculate energy level
       00d6   00eb
289    00d7   6a1b          LT      COLMAX        TEST CODE
290                  *
291    00d8   8800          MPYK    2048          "      "
292                  *
293    00d9   7f90          SPAC                  "      "
294    00da   fd00          BGEZ    NOT           -12dB = 1/16
       00db   0022
295                  *
296                  *   Load recognized number and check that it is new
297                  *
298    00dc   6916          DMOV    LAST2
299    00dd   6915          DMOV    LAST
300    00de   2210          LAC     ROWMX,2
301    00df   0011          ADD     COLMX
302    00e0   5015          SACL    LAST
303    00e1   1017          SUB     COUNT       Return if same number
304    00e2   ff00          BZ      NOT
       00e3   0022
305    00e4   2015          LAC     LAST
306    00e5   1016          SUB     LAST2
307    00e6   fe00          BNZ     AGAIN       2 Passes to recognize
       00e7   0025
308                  *
309    00e8   4a15          OUT     LAST,2
310    00e9   f900          B       AGAIN       <break>
       00ea   0025
311                  *
312                  *   Energy Calculation Subroutine
313                  *
314    00eb   2f13   ENERGY LAC     NEGMAX,15   NEGMAX = >8000
315    00ec   0f81          ADD     *,15,1
316    00ed   5817          SACH    COUNT       -1/2 + CSn/2
317    00ee   6a98          LT      *-
318    00ef   6d17          MPY     COUNT
319    00f0   7f8e          PAC
320    00f1   5917          SACH    COUNT,1     D2(CSn-1)/2
321    00f2   6aa8          LT      *+
322    00f3   6d17          MPY     COUNT
323    00f4   7f8e          PAC
324    00f5   5917          SACH    COUNT,1     D1*D2(CSn-1)/2
325    00f6   2f98          LAC     *-,15
326    00f7   1f88          SUB     *,15
327    00f8   7f88          ABS
328    00f9   5888          SACH    *           abs(D2-D1)/2
329    00fa   6a88          LT      *
330    00fb   6d88          MPY     *
331    00fc   7f8e          PAC                 ((D2-D1)/2)^2
332    00fd   1f17          SUB     COUNT,15    ((D2-D1)^2)/4-D1*D2(CSn-1)/2
333    00fe   7f8d          RET
334                  *
335                         END
```

DTMF decoding doesn't necessarily require elaborate DSP routines; the routine presented here leaves room for several other DSP routines in the DSP μP.

oscillators' coefficients (B1, D1, B2, and D2 in **Fig 1**) prior to the start of signaling. (The use of eight oscillators would decrease execution time but would require four times as much memory as two oscillators.) After initializing the resonators, the program loops repeatedly through the resonator code and generates samples of the appropriate high- and low-frequency tones. Then the program sums the pairs of tone samples. The DSP μP then feeds this sum to an external D/A converter, and the resulting analog output is the DTMF signal.

Frequency specs aren't the only ones DTMF tones have to meet; duration specs apply also. According to AT&T specs, 10 digits/sec (or 100 msec/digit) is the maximum data rate for Touchtone signals. AT&T specifications state that within its allotted 100-msec interval, a tone must be present for at least 45 msec and no more than 55 msec. During the remainder of the 100-msec interval, the tone generator must be quiet to allow the receiver's DTMF decoder to settle. Therefore, a counter makes sure that the generated tone's duration meets the minimum time—approximately 45 msec—to minimize computing time. After the tone's on for a sufficiently long time, the D/A converter is zeroed and maintained at the zero-output level so that the total on time and off time equals 100 msec.

Although DTMF tone-decoding schemes require con-

Glossary

Center frequency offset—the offset of the center of the recognition bandwidth from the nominal DTMF frequencies.

DFT—discrete Fourier transform.

DTMF—dual-tone multifrequency signaling system used by the telephone company for dialing.

Guard time—the duration of the shortest DTMF tone a detector will recognize.

IIR—infinite impulse response (a type of digital filter).

Log-Linear—transformation of logarithmically compressed data from a codec back to linear form.

Recognition bandwidth—the percent change from the nominal frequencies that a detector will tolerate.

Reverse twist—the condition that exists when a DTMF signal's row amplitude is greater than the column amplitude.

Standard twist—the condition that exists when a DTMF signal's column amplitude is greater than the row amplitude.

Talk-off—a measure of the detector's ability to ignore speech signals that look like DTMF signals.

Twist—the difference, in decibels, between the loudest row tone's amplitude and the loudest column tone's amplitude.

siderably more code than do generation schemes, the decoding program in **Fig 3** takes less than twice as much code as the simpler generating program. Furthermore, both programs are much smaller than the total capacity of the DSP μP. In this case, rather than being called as a result of a keystroke (as the generating algorithm is), the decoding algorithm continually processes signal samples and so must be interleaved with other DSP functions. The algorithm must run continually because, after all, it doesn't know whether or not DTMF tones are present until after it processes the input.

The discrete Fourier transform (DFT) algorithm employed in the program listing is known as Goertzel's algorithm (**Ref 2**). This algorithm is compact and needs only one real coefficient per frequency to determine magnitude (**Fig 6**); although extracting magnitude and phase requires complex coefficients and hence more complex programming, you can decode DTMF signals

---

TABLE 1—RECOMMENDED DFT LENGTHS AT AN 8-kHz SAMPLING RATE

1ST HARMONIC
N = 205 DUR = 25.6 mSEC = (18 20 22 24 31 34 38 42)
2ND HARMONIC
N = 201 DUR = 25.1 mSEC = (35 39 43 47 61 67 74 82)

COEFFICIENTS	N = 205 1ST HARMONIC	N = 201 2ND HARMONIC
697	27906	15036
770	26802	11287
852	25597	7363
941	24295	3323
1209	19057	−10805
1336	16529	−16384
1477	12945	−22153
1633	9166	−27440

TABLE 2—DFT PROGRAM SPECIFICATIONS

DFT SIZE:	FIRST HARMONIC	N = 226
		K = (16, 18, 20, 22, 28, 31, 34, 38)
	SECOND HARMONIC	N = 222
		K = (32, 35, 39, 43, 55, 61, 67, 74)

PROGRAM WORDS	=	255 WORDS
DATA MEMORY WORDS	=	60 WORDS
SAMPLING FREQUENCY	=	9766 SAMPLES/SEC
SAMPLING INTERVAL	=	102.4 μSEC
DFT LOOP TIME	=	45 μSEC
TOTAL DFT TIME	=	23.2 mSEC
TIME REQUIRED BY THE DECISION LOGIC	=	150 μSEC (ONE SAMPLE MISSED BETWEEN DFTs)

19. Add DTMF Generation and Decoding to DSP-μP Designs

```
*
*BEGIN DATA TABLES
*                                     Real Coeff N=205
           DATA      27906              697
           DATA      26802              770
           DATA      25597              851
           DATA      24295              941
           DATA      19057             1209
           DATA      16527             1336
           DATA      12945             1477
           DATA       9166             1633
*    2nd harmonic                     Real Coeff N=201
           DATA      15036             1394
           DATA      11287             1540
           DATA       7363             1702
           DATA       3323             1882
           DATA     -10805             2418
           DATA     -16384             2672
           DATA     -22153             2954
           DATA     -27440             3266
*
           DATA      419              CLCK = Sample Frequency
           DATA      >000A            MODE
*
           DATA      >7FFF            POSMAX = Mask for data in
           DATA      >8000            NEGMAX = Mask for data out
*
*
TABLE      DATA      1                ONE
*    Start of Program     **********************
*
START      SOVM
           LDPK      0
           LARP      0                <Break>
           LARK      0,ONE
           LACK      TABLE
NEXT       TBLR      *                Initialize Coefficients
           SUB       ONE
           BANZ      NEXT
           OUT       MODE,0           Set AIB Mode
           OUT       CLCK,1           Set AIB Clock
*
*
* Load not recognized symbol
*
NOT        LACK      >FF
           DMOV      LAST
           SACL      LAST
*
AGAIN      ZAC                        Zero DFT Loop Variables
           LARK      0,31
           LARK      1,DAT11
           BV        ZERO
ZERO       LARP      1
           SACL      *+,0,0
           BANZ      ZERO
*
*    Take data and calculate DFT loop
*
           LACK      205              SET DFT LOOP VARIABLE
*
LOOP       SACL      COUNT
           LARK      0,CS16
           LARK      1,DAT216


           SUB       ONE,2
           BGZ       WAIT
           LARK      0,CS8
           LARK      1,DAT28
```

Fig 4—This amendment of Fig 3's listing adapts the tone-generating routine to 8k-sample/sec operation. It's a substitute routine for lines 43 to 122.

DTMF decoders that meet AT&T specs usually stop decoding tones if they detect the presence of speech.

---

simply by extracting the magnitude of a tone's frequency components and ignoring their phase. In addition, instead of waiting for a complete sample set to begin processing, Goertzel's algorithm processes each sample as it arrives.

Goertzel's algorithm takes the form of a series of second-order IIR (infinite-impulse-response) filters. Notice that, in **Fig 6,** you can divide the directed graph into two parts: a left-hand part that includes the two feedback elements (boxes marked "delay") and a right-hand portion leading to the output that has no feedback elements. For DTMF decoding, you are interested only in the last iteration $(N-1)$ of the algorithm. Consequently, because the right-hand branches don't involve feedback, there's no need to execute these branches of

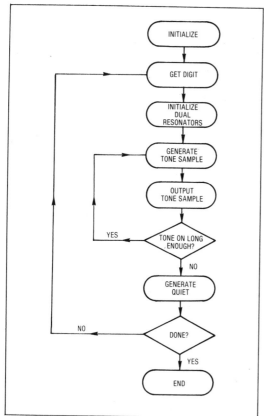

Fig 5—The directed graph in Fig 1 translates into a program that follows this flowchart. Note the step that produces a quiet period.

INITIALIZE

GET DIGIT

INITIALIZE DUAL RESONATORS

GENERATE TONE SAMPLE

OUTPUT TONE SAMPLE

TONE ON LONG ENOUGH? — YES

NO

GENERATE QUIET

DONE? — NO

YES

END

TABLE 3—TONE DECODER TEST RESULTS

BANDWIDTH TEST RESULTS:

TONE	%HIGH	%LOW	TONE	%HIGH	%LOW
697	2.5	3.5	1209	2.4	3.0
770	3.7	2.3	1336	2.3	2.5
852	3.9	1.7	1477	1.3	2.9
941	3.3	1.7	1633	2.4	1.6

SPECIFICATIONS REQUIRE: MIN = 1.5% AND MAX = 3.5%

AMPLITUDE RATIO TEST RESULTS

	TWIST	REVERSE TWIST
SPECIFICATIONS	>4.0 dB	>8.0 dB
DIGIT 1	5.3	8.4
DIGIT 5	5.7	9.0
DIGIT 9	8.3	9.7
DIGIT 16	5.4	9.5

DYNAMIC RANGE: 25 dB (SPECIFICATION 25 dB)

NOISE TEST: PASSES AT –24 dBV

TALK-OFF IMMUNITY: ONE FALSE RECOGNITION PER 1000 CALLS (SPEC 1500)

the algorithm until the last iteration of the algorithm.

What's not obvious from the directed graph of the algorithm is that the left-hand constant, $2cos(2\pi k/N)$, is the same as the right-hand constant, W^k_N, for calculating the magnitude of DTMF signals. W^k_N is a complex number, and the left-hand constant is not. However, the program calculates the magnitude *squared* of the output of the algorithm. Squaring a complex number always yields a real number, and in this case, squaring the right-hand constant yields a real number that's the same as the left-hand constant. Therefore, Goertzel's algorithm, adapted to DTMF decoding, not only executes quickly because it has few steps, but it also takes up little memory space because it uses few constants.

Given a time-ordered sample set of size N, processing each sample means you'll do N iterations of the algorithm. If k is the frequency you're solving the transform for, then the values k and N determine the coefficients of each IIR filter. The values of k and N and the sampling rate also determine how accurately the transform discriminates between in-band and out-of-band frequencies.

Specifically, k is a discrete integer corresponding to the frequency you're solving for. It's defined as

$$k=N\times frequency/(sample\ rate).$$

(Note that you must round off the frequency of interest to an integer.) W^k_N is a frequently encountered constant in digital signal processing. It's defined as

$$W^k_N=exp(-2\pi jk/N).$$

Because the sampling rate and the frequencies you're

19. Add DTMF Generation and Decoding to DSP-μP Designs

The decoding algorithm processes signal samples continuously, so the DSP μP must interleave the decoding with other DSP functions.

extracting are fixed (by the phone company), the sample-set length (N) is the only parameter you can vary. In order to obtain the best performance from the transform, the length of the sample set must be optimized with respect to two conflicting criteria: The sample set must be small enough so that the decoder can accumulate a complete set in an interval that's short enough to keep up with the DTMF digit-transmission rate; conversely, the sample set must be long enough so that the transform discriminates between in-band and out-of-band signals. An exhaustive and inelegant computerized search of all possible combinations of k and N resulted in the sample lengths listed in **Tables 1** and **2** and used in the **Fig 3** and **4** program listings.

Companding not accounted for

This design assumes that the analog input is linearly encoded. This is often not the case, because many phone signals are compressed logarithmically by a codec. In such cases, you must first perform a log-to-linear expansion before submitting the samples to the DFT algorithm. (**Ref 3** describes how to do companding with the TMS32010 if your system doesn't incorporate companding hardware.)

Fig 7 shows how samples are fed, in effect, into a

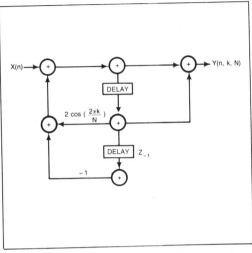

Fig 6—A simplified form of Goertzel's algorithm, *represented by this directed graph, decodes DTMF tones. A program that implements it can save processing steps by performing the calculations illustrated by the right-hand portion of this graph for just the last iteration. Furthermore, for DTMF-decoding purposes, this compact algorithm requires only one constant per frequency because both the right-hand and left-hand constants have the same value.*

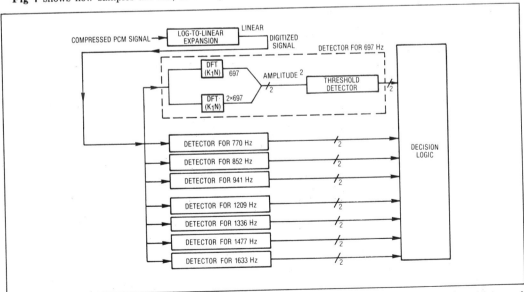

Fig 7—The tone decoder employs 16 of the transforms *shown in* **Fig 6.** *They extract each of the eight Touchtone frequencies and their second harmonics.*

19. Add DTMF Generation and Decoding to DSP-μP Designs

Goertzel's algorithm, a discrete Fourier transform algorithm, operates as a series of second-order IIR filters.

parallel array of 16 DFT algorithms. There is one DFT for each of the eight frequencies and each of the second harmonics of the eight frequencies. You need the second harmonics as well as the fundamentals to discriminate between speech and DTMF tones. Of course, they execute serially because they are sections of code—not physical devices.

As **Fig** 8's flowchart shows, after initializing the processor, the program feeds the first sample to the IIR filters. After all the samples have been processed, each filter's current value is squared. This operation yields the magnitude of the strength of the signal at each of the eight DTMF frequencies and each second harmonic of the DTMF frequencies.

The program next compares the data against several thresholds. It performs four principal checks. First, after finding the strongest signals in the high- and low-frequency groups, it simply determines whether any valid DTMF tones are present at all. If the strongest signals are not above a minimal value (2^{-12} in **Fig** 3's program), the program does no more processing and begins collecting another sample set.

Second, if valid DTMF tones are present, the program checks the strongest signals in the low and high (row and column) groups for twist—the ratio of the row-tone amplitude to the column-tone amplitude. This ratio must be between certain values for the DTMF tones to be valid. (Because of the frequency response of telephone systems, the high tones are attenuated. Consequently, the phone company doesn't expect the high and low groups to have exactly the same amplitude at the receiver, even though they were transmitted at the same strength.)

Third, the program compares the amplitude of the strongest signal in each group to the amplitudes of the rest of the tones in its group. Again, the strongest tone must stand out from the other tones in its group by a certain ratio.

Finally, the program checks to see that the strongest signals are above one threshold while their corresponding second harmonics are below another threshold. Checking for strong harmonics insures that the DSP system won't confuse speech for DTMF signals. (Speech has significant even-order harmonics; DTMF signals don't.)

Fig 8—The tone-decoder program does far more than simply detect the presence of DTMF signals; it also performs an elaborate series of checks to ensure that the tones are within specifications and that a valid tone is new data that must be acted on.

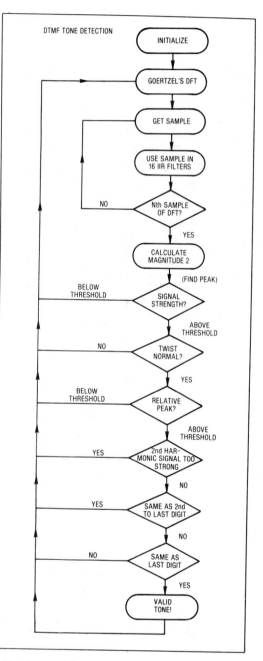

The DTMF-decoding program checks the signal pair to establish tone validity, and then it determines whether the pair constitutes new information.

If the DTMF signal pair passes all these comparisons, then it's a valid tone pair that corresponds to a digit. Just because it's valid, however, doesn't mean that the corresponding digit is necessarily new information. The remaining two steps of the program compare the current digit to the two most recently derived digits. First, the program checks to see if the current digit is the same as the second-to-last digit. If they match, then the program assumes the tone hasn't changed lately. If they differ, it performs one final check to see if the current digit matches the last digit received. If these are the same, then the DTMF tone has changed recently and remained stable for two iterations. This means you finally have a valid new digit. If they don't match, it means the tone has changed since the last sample was acquired but hasn't remained stable long enough. Consequently, the program loops back without signaling that a new digit has arrived. If the new tone is really valid and stable, the next iteration of the algorithm will recognize the digit as valid because the new current digit will now match the previously received digit.

There are two reasons for checking three successive digits at each pass. First, the check eliminates the need to generate hits every time a tone is present; acknowledging it only once is enough. As long as the tone is present, it can be ignored until it changes. Second, comparing digits improves noise characteristics and speech immunity.

The implementation of the decoder algorithm follows the specification listed in **Table 2**. The TMS32010's Harvard architecture separates data and program memory. The data memory is on chip. The program keeps the tables required by the decoding algorithm in on-chip data memory. These tables take up more than half the available data-memory locations. Depending on your application, you might have to store the tables in program memory and move the tables onto the chip every time the decoding algorithm runs. This will free the on-chip memory for other uses, but it will obviously increase the decoding algorithm's execution time.

Checking the decoder's performance

Evaluating the performance of a DTMF decoder is more difficult than evaluating the performance of a DTMF generator. You can check the generator very simply with a spectrum analyzer. To test the decoder, on the other hand, you have to determine not only that it will decode valid tones, but that it will both reject invalid signals and operate properly in the presence of noise. Testing the decoder using AT&T's published test method is an all-day affair and requires a specific instrumentation suite. Prerecorded tapes of various test tones speed things considerably. For example, Mitel's (San Diego, CA) $90 CM7291 test cassette tape cuts the evaluation time of DTMF tone receivers to less than 90 minutes, according to the company (**Ref 4**).

The TMS32010's DTMF decoder was tested against the Mitel test tape. The test results given in **Table 3** indicate that the receiver can detect all tones. And the receiver bandwidths conformed almost exactly to all AT&T specs. There were only three tones for which the decoder was slightly off. In two instances, results were 0.2% too large and, in one instance, 0.2% too small. The other AT&T specs were met perfectly, including the twist's dynamic range at 25 dB, the guard time at 20 msec, and the white-noise test at 24 dBV. **EDN**

References

1. Clark, N V, "DTMF Encoder Demonstration," Texas Instruments internal publication, February 1984.

2. Oppenheim, A V, and Schafer, R W, *Digital Signal Processing*, Prentice-Hall, Englewood Cliffs, NJ, 1975 (see Section 6.1).

3. *Companding Routines for the TMS32010*, Application Note SPRA001, Texas Instruments, Dallas, TX.

4. *Tone Receiver Test Cassette #CM7291*, Mitel Technical Data Manual, Mitel Semiconductor, 2321 Morena Blvd, Suite M, San Diego CA 92110. Phone (619) 276-3421.

5. *Touch-Tone/RTM calling—Requirements for Central Office*, AT&T Compatibility Bulletin No 105, August 8, 1975.

20. A Single-Processor LPC Vocoder

Andrew Holck
Digital Signal Processing - Semiconductor Group
Texas Instruments

Wallace Anderson
Central Research Laboratories
Texas Instruments

Article reprinted with permission from ICASSP '84 Proceedings, IEEE International Conference on Acoustics, Speech, and Signal Processing, San Diego, CA, March 19-21, 1984; copyright IEEE, 1984.

A SINGLE-PROCESSOR LPC VOCODER

Andrew W. Holck
Texas Instruments Incorporated
Semiconductor Group
9901 S. Wilcrest, M/S 6437
Houston, Texas 77099

Wallace W. Anderson
Texas Instruments Incorporated
Corporate Engineering Center
M/S 247
Dallas, Texas 75265

ABSTRACT

This paper presents the design of a full-duplex 2400-bit/sec vocoder which implements an LPC-10 algorithm in real time. A single commercially available digital signal processor IC, the TMS32010, is used to perform the digital processing. The TMS32010 code for this vocoder has been run on a software simulator and has been found to be within the real-time constraints of the vocoder design.

INTRODUCTION

Tenth-order Linear Predictive Coding (LPC) algorithms have been used successfully in digital speech data compression systems to achieve low bit rates. A main limitation to the use of these systems is the cost of implementation. These costs have been steadily declining as advances in semiconductor technology are made. Recently, programmable signal processing devices have been introduced which incorporate high-speed multiply/accumulate hardware with RAM, ROM, and program control onto a single IC. Compact, low-cost vocoder implementations based upon such signal processing devices have been demonstrated.

One of the challenges of designing a system based upon these devices is how to efficiently structure the algorithms which are to be implemented. Traditional LPC algorithms perform processing on a frame-by-frame basis which requires the buffering of at least one frame of input speech data. The RAM required to perform this buffering is larger than the internal RAM of currently available programmable signal processing ICs (<512 bytes).

This restriction has been averted in several ways. Many LPC algorithms can be restructured so that the processing is done in a sample-by-sample or "streamed" manner instead of frame-by-frame[1]. This eliminates the need to buffer a frame of speech. The processing required for an LPC vocoder can also be distributed between several processors. The LPC algorithm can be split into the operations of autocorrelation analysis, pitch detection, and synthesis. With a distributed processing approach, a separate processor can be used to implement each of these functions[2].

An alternative approach is to provide additional external RAM. The TMS32010 digital signal processor has the capability of accessing external RAM. With added hardware, a relatively large external RAM space can be accessed efficiently with a modest overhead penalty. This feature makes it possible to implement a full-duplex 2400-bit/sec frame-oriented LPC-10 vocoder using a single TMS32010 to perform the digital processing.

VOCODER HARDWARE

The vocoder block diagram is shown in Figure 1. Analog speech input is digitized by the analog conversion system, coded into a 2400-bit/sec data stream, and transmitted to an external system through the coded data I/O block. Simultaneously, 2400-bit/sec coded speech data is received from an external system through the coded data I/O block. This data is used to synthesize speech samples which are then converted to a synthesized analog speech signal by the analog conversion system. The extended RAM system provides a larger RAM space than that available internally on the TMS32010.

TMS32010 Digital Signal Processor

The TMS32010 is a general-purpose digital signal processor IC which is used in this vocoder to perform the number crunching required to implement the LPC-10 algorithm. The vocoder design requires a processor with fast execution time, processing power, and general-purpose features for stand-alone operation. Fast execution time is achieved through a 200-ns instruction cycle time and a pipelined architecture which enables most instructions to execute in a single cycle. A two's complement 16×16-bit single-cycle hardware multiplier with 32-bit result and a 32-bit wide accumulator provide processing power. The multiplier and the accumulator enable the TMS32010 to perform a multiply-accumulate operation in 400 ns. General-purpose features which allow the vocoder to be implemented with a single processor include: full-speed execution from external program memory, ability to access external parallel I/O ports, both interrupt ($\overline{INT}$) and polling input ($\overline{BIO}$) pins, and a full set of bit manipulation and branch instructions. This combination of features, along with internal ROM ($1.5K \times 16$) and internal RAM (144×16) in a single 40-pin package, enables this vocoder design to be much more compact than a similar bit-slice design.

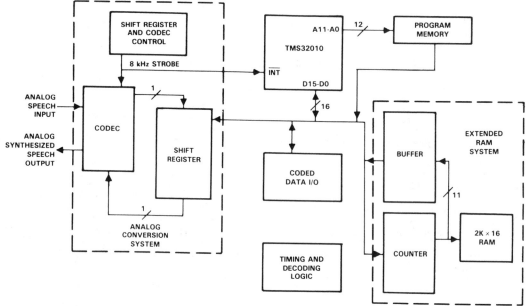

FIGURE 1 – VOCODER BLOCK DIAGRAM

Analog Conversion System

Codec/filter combochips are commercially available which are tailored for the A/D and D/A conversion of telephone quality signals. One of these chips is used in the analog conversion system to perform full-duplex A/D and D/A conversion, input antialiasing filtering, and lowpass filtering of the output. An oscillator is used to provide the clock for the codec's switched-capacitor filters. The oscillator also generates an 8-kHz strobe which controls the sampling rate of the codec. This strobe is also the interrupt signal for the TMS32010. A shift register is used to perform serial-to-parallel conversion of the codec A/D data and parallel-to-serial conversion of the data sent to the codec D/A by the TMS32010. The TMS32010 uses one of its eight parallel I/O ports to access the analog conversion system. Nine devices are required to implement this block.

Program Memory

The vocoder requires about 3K words (16-bit words) of the TMS32010's 4K-word program memory address space. Program memory can be implemented in two ways. All 3K words can be implemented off-chip, or the 1.5K words of internal masked ROM can be used and the remaining 1.5K words implemented off-chip.

Extended RAM System

The TMS32010 accesses the extended RAM system using two of the eight parallel I/O ports. One of the ports is used to ac-

cess the address counter for the RAM and another port is used to access the RAM. The counter increments each time the 2K-word RAM is accessed. The counter hardware can be thought of as an external autoincrementing index register. This feature enables the external RAM to be accessed with a minimum of overhead. The TMS32010 accesses one of the I/O ports in two instruction cycles (400 ns). A buffer is needed to read the output of the counter so that the interrupt routine can restore the counter's contents. Seven devices are required to implement this block.

Coded Data I/O

One of the TMS32010's eight I/O ports is used to access the coded data I/O block. This block handles the transfers of coded speech data to and from the external system. The implementation of this block will depend on the characteristics of the external system.

VOCODER SOFTWARE

The software for the system consists of straightforward LPC routines, organized to take advantage of the TMS32010's architecture. The software is based on the 2400-bit/sec analysis/synthesis modules used in the Speech Command System for the Texas Instruments Professional Computer, which are used only in half-duplex mode. The switch to full-duplex mode was made by interleaving analysis and synthesis modules on a frame basis, using the external RAM to buffer the sampled data.

20. A Single-Processor LPC Vocoder

The analog speech data is processed through a standard codec, operating at an 8-kHz sample rate. The 8-kHz strobe which sets the sampling rate for the codec is also connected to the interrupt pin of the TMS32010. For each interrupt, the TMS32010 reads in an A/D sample and writes out a D/A sample. The input samples are decoded from the codec μ-law representation to linear representation through a lookup table stored in the TMS32010 program memory, and are then buffered in the external RAM. Synthesized speech samples in linear format are read from the external RAM, coded to μ-law representation through a tree-structured algorithm, and output to the codec.

The analysis is performed by standard autocorrelation analysis, using a 30-ms Hamming window with a 20-ms frame interval. The autocorrelation is done with blocks of 20 samples. For each autocorrelation coefficient, 10 terms are multiplied and accumulated in the 32-bit wide accumulator before performing the processing for the next coefficient. The reflection coefficients are calculated from the autocorrelation coefficients by the LeRoux-Guegen algorithm[3]. Pitch tracking is done with the Gold-Rabiner pitch tracker[4]. The energy, pitch, and 10 reflection coefficients are then quantized to 48 bits per frame, providing the basic 2400-bit/sec data rate. The lattice structure is used for synthesis, with the parameters being interpolated only at the start of pitch periods.

With the TMS32010 instruction cycle of 200 nanoseconds, there are 625 instruction cycles per sample period, for a total of 100,000 instruction cycles per 20-ms frame period. Out of this total, 87% of the processing capacity is used by the full-duplex processing, exclusive of the coded data I/O. The breakdown of the processing is shown in Figure 2. An interesting result is that the processing time required for analysis (37%) is less than the processing time required for synthesis (39%). This is due to the efficiency of the TMS32010 in performing the autocorrelation function in the spectral analysis.

Spectral Analysis	17%
Pitch Tracking	20%
Synthesis	39%
I/O Overhead	11%
Total	87%

FIGURE 2 – PROCESSOR UTILIZATION (% OF REAL TIME USED)

These timing numbers were derived using an internal simulator on a VAX computer, using previously collected digitized speech files. The simulator models the entire system, including the external RAM and the codec companding, through FORTRAN subroutines linked to the basic TMS32010 simulator. The TMS32010 simulator itself is written in VAX Macro, providing detailed simulation of each instruction, and allowing break and trace points for following the program flow and timing.

The code for the full-duplex vocoder, exclusive of the code required for the coded data I/O, occupies about 70% of the TMS32010's 4K-word (16-bit words) address space. The breakdown of program memory usage is shown in Figure 3. The internal RAM of 144 words is not sufficient for all of the

full-duplex processing. It is used as 128 words for the current process and 16 words associated with the interrupt processing. The 128 words are swapped between internal and external RAM as the process is swapped between analysis and synthesis, so that external RAM is holding the memory state of the non-current process. The breakdown of the external RAM usage is shown in Figure 4.

Initialization and Interrupt Processing	448
Analysis	1106
Synthesis	440
Tables	854
Total	2848

FIGURE 3 – PROGRAM MEMORY USAGE (16-BIT WORDS)

A/D and D/A Sample Buffers	960
Coded Speech Data Buffers	256
Internal RAM Buffers	384
Hamming Window Constants	240
Pitch Tracker Variables	40
Total	1880

FIGURE 4 – EXTERNAL RAM USAGE (16-BIT WORDS)

SUMMARY AND COMMENTS

A compact LPC-10 vocoder can be implemented based upon this design. Approximately 20 commercially available ICs are required to implement this design, excluding the coded data I/O block. Processor loading, ROM usage, and external RAM usage are within system constraints with sufficient margin to handle the interface to an external system. The vocoder hardware is not limited to LPC-10 analysis and synthesis. Other speech or signal processing algorithms can be implemented with similar hardware structures. Since the TMS32010 can execute at full speed from external program memory, one possibility is to implement the program memory with RAM. This feature would give a host system the ability to use the same hardware for different signal processing functions by downloading different programs to the program RAM.

A significant portion of the vocoder hardware is the interface logic for the codec. Future developments are anticipated which will reduce the hardware required for the analog conversion system, provide more flexible A/D and D/A conversion, and reduce the I/O overhead for the A/D and D/A samples[5]. Another significant portion of the vocoder hardware is the extended RAM system. The ability of the TMS32010 to efficiently access external RAM is one of the main features which allows a single TMS32010 to perform the full-duplex LPC-10 processing. Advances in semiconductor technology will allow larger internal RAM spaces, but there will still be the need to efficiently access external RAM. Features could be incorporated into future digital signal processors which would allow efficient access of external RAM with a minimum of external hardware. These features, along with increased internal RAM spaces and processing speed, will allow more and more compact implementations of complex signal processing algorithms.

REFERENCES

[1] Kaltenmeier, A., "Implementation of Various LPC Algorithms Using Commercial Digital Signal Processors," IEEE INT. CONF. ASSP, April 1983, 487-490.

[2] Feldman, J.A., Hofstetter, E.M., and Malpass, M.L., "A Compact, Flexible LPC Vocoder Based on a Commercial Signal Processing Microcomputer," IEEE TRANS. ON ASSP, Vol. ASSP-31, February 1983, 252-257.

[3] LeRoux, J., and Gueguen, C., "A Fixed Point Computation of Partial Correlation Coefficients in Linear Prediction," IEEE TRANS. ON ASSP, June 1977, 257-259.

[4] Gold, B., and Rabiner, L.R., "Parallel Processing Techniques for Estimating Pitch Periods of Speech in the Time Domain," JOURNAL ACOUST. SOC. AMER., Vol. 46, No. 2, August 1969, 442-448.

[5] Robertson, S.J., Feger, W.E., and Hester, R.K., "Analog Interface Chips for Audio Band Digital Signal Processing," IEEE INT. CONF. ASSP, March 1984.

20. A Single-Processor LPC Vocoder

21. The Design of an Adaptive Predictive Coder Using a Single-Chip Digital Signal Processor

M. Randolph
Linclon Laboratory
Massachusetts Institute of Technology

Technical report reprinted with permission from Lincoln Laboratory, Massachusetts Institute of Technology, Lexington, MA, January 11, 1985.

PREFACE

Speech processing is one of the first applications of digital signal processing. With the advancement of commercially available digital signal processors, speech processing algorithms can now be implemented in compact, low-power systems. These systems are now widely used in both commercial and military applications. Some of the applications are voice-mail, voice store-and-forward, secure communications, voice response for secure/warning systems, and speaker-dependent/-independent recognition.

Speech algorithms are typically complex and large in program size. For the algorithm to be executed in real time, the speech processor must finish computing all necessary steps before the next frame of data is received. A speech processing system also carries a heavy I/O load intensity. The Texas Instruments TMS320 processors' fast instruction cycle and parallelism in the architecture allow many speech processing algorithms to be performed in real time. The processors' large address range (4K words of program for the TMS32010 and 64K words for the TMS32020) meet large memory requirements of many complex speech algorithms. The TMS320 can also access external I/O devices efficiently, thus eliminating the need of buffering large arrays of speech data. Finally, the TMS320 exhibits some common characteristics of general-purpose microprocessors, such as a rich instruction set and hardware/software interrupts, which make them very easy to use. Because of these features, the TMS320 digital signal processors have been widely adapted as the speech processor for many speech processing systems. Some of these systems are: IBM's Voice Communications Option,[1] Votan's Plug-in Card for IBM PC,[2] TI-Speech System,[3] and DECtalk Text-to-Speech System.[4]

The TMS320 can be used to implement speech algorithms at a variety of data rates. In the Telecommunications section of the book, two speech algorithms, 32-kbit/s ADPCM and 16-kbit/s subband coding, have been included. In the Computers and Peripherals section, this report from MIT was chosen to give the designer an appreciation of the requirements of a processor for speech processing applications. In the report, a speech algorithm at 9.6-kbit/s using Adaptive Predictive Coding (APC) was thoroughly studied. The TMS32010 was selected as the most suitable device for implementing a compact APC speech system. The design considerations of critical loop, control logic, and memory allocation are performed and included. A preliminary hardware design consisting of a TMS32010, 2K words of program memory, and 2K words of data memory is included. TMS32010 codes for performing major steps of the APC algorithm are listed in the appendices of the report. These include Average Magnitude Difference Function (AMDF) pitch estimation, pitch prediction coefficient calculation, combined-first residual and autocorrelation computation, Leroux-Gueguen recursion for LPC Parcor coefficients computation, predictive quantizer loop, receiver loop, and I/O service routine.

Although the report does not give a cookbook solution to the implementation of APC, the material included provides enough ingredients for readers to construct an APC system and a wide range of vocoder derivatives. For example, the hardware configuration and a subset of the TMS32010 code in the appendices can be used as references for creating a TMS32010-based pitch-excited LPC vocoder at 2400 bps. Some of the useful routines for the 2400 bps LPC are AMDF pitch estimation, autocorrelation computation, Leroux-Gueguen recursion for LPC Parcor coefficients computation, receiver loop, and I/O service routine.

The references given below may be helpful to the reader of this report.

REFERENCES

1. IBM Product Announcement: *IBM Personal Computer Voice Communications Option* (October 15, 1985).
2. Votan Product Brochure: *VPC 2000 Plug-in Card for IBM PC*.
3. Texas Instruments Product Brochure: *TI-Speech System*.
4. Digital Equipment Corporation Product Brochure: *DECtalk Text-to-Speech System*.

MASSACHUSETTS INSTITUTE OF TECHNOLOGY
LINCOLN LABORATORY
THE DESIGN OF AN ADAPTIVE PREDICTIVE CODER USING A SINGLE -
CHIP DIGITAL SIGNAL PROCESSOR
M.A. RANDOLPH Group 24
TECHNICAL REPORT 11 JANUARY 1985
Approved for public release; distribution unlimited
LEXINGTON MASSACHUSETTS

ABSTRACT

A speech coding processor architecture design study has been performed in which the Texas Instruments TMS32010 has been selected from among three commercially available digital signal processing integrated circuits and evaluated in an implementation study of real-time Adaptive Predictive Coding (APC). The TMS32010 has been compared with the AT&T Bell Laboratories DSP I and Nippon Electric Co. μPD7720 and was found to be most suitable for a single chip implementation of APC. A preliminary system design based on the TMS32010 has been performed, and several of the hardware and software design issues are discussed. Particular attention was paid to the design of an external memory controller which permits rapid sequential access of external RAM. As a result, it has been determined that a compact hardware implementation of the APC algorithm is feasible based on the TMS32010.

1. INTRODUCTION

Recently, several digital signal processing integrated circuits (DSPs) have become commercially available. These devices possess significant computational capability and permit a variety of speech processing algorithms to be implemented in compact, low power systems. In this report we summarize the results of a processor design study in which the Texas Instruments (TI) TMS32010, the Nippon Electric Co. (NEC) μPD7720, and the AT&T Bell Laboratories DSP I have been evaluated for the task of implementing real-time Adaptive Predictive Coding (APC). We have surveyed the architectural features of these three DSPs and have compared and contrasted their expected performance in implementing real-time APC.

Digital signal processors are typically benchmarked using some of the more common signal processing algorithms such as digital filters and FFTs. They are usually compared solely on the basis of the execution times of these computations. Unfortunately, we have found that in evaluating DSPs for real-time speech coding applications, these typical signal processing benchmarks do not provide us with complete information. For this reason, we have chosen to use an actual speech coding algorithm, real-time APC, as a benchmark. The decision to use APC was based on a number of factors. First, it is an algorithm of moderate to high complexity that requires a processor with considerable numerical processing capability. In addition, it requires a processor which can access an extensive amount of memory.

It therefore provides a reasonable indication of the processing power of a particular digital signal processor. Secondly, it fits within the category of medium- to low-bit rate speech coding algorithms which we are currently interested in implementing at Lincoln Laboratory. We postulate that a DSP's ability to implement APC is reasonable assurance that comparable algorithms could also be implemented on that DSP.

This report shall be organized as follows. In section 2, we describe pertinent aspects of the APC algorithm as they relate to the algorithm's implementation. In Section 3, we briefly review and compare the architectural features of the three digital signal processors that we have considered. In Section 4, we summarize a software/hardware implementation of the APC algorithm based on the TMS32010.

2. ADAPTIVE PREDICTIVE CODING

In the present section, we briefly outline the fundamentals of the APC algorithm. This discussion is intended to serve as a means of introducing our own terminology and notation. We have chosen not to develop the theory upon which the algorithm is based. For a more complete treatment of APC in terms of its theoretical aspects, we refer the reader to a report by Viswanathan, et al., [16], which is a comprehensive review of the theory and also describes several variations and improvements that have been made upon the APC algorithm. For the purpose of this study, we have considered the APC algorithm in its most basic form. This particular version of the APC algorithm is similar in structure to the original proposed by Atal and Schroeder [2]. A block diagram is shown in Fig. 1. Figures 1 (a) and (b) are the APC analyzer and synthesizer, respectively. In the analyzer, two predictors are employed for removing presumed redundancy in the input speech signal and are arranged in a feedback loop surrounding a one-bit quantizer. The predictor A(z) is a spectral predictor and is intended to remove the redundancy in the speech signal which is due to its quasi-stationary spectral properties. The spectral predictor has the polynomial transfer function

$$A(z) = \sum_{i=1}^{P} a_i z^{-1} \qquad (1)$$

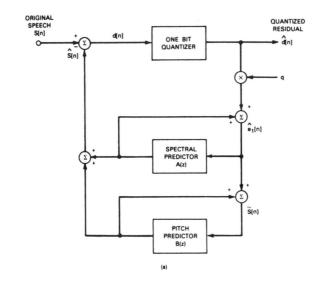

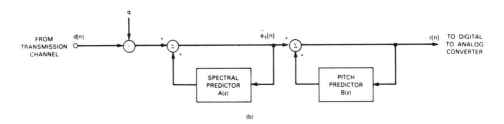

Figure 1. Block Diagrams of APC (a) Analyzer and (b) Synthesizer

where the coefficients a_i are the spectral predictor coefficients and the parameter P is the prediction order. The spectral predictor coefficients are obtained from linear prediction methods which will be outlined below. The prediction order, P, is usually a specification of the particular implementation and is typically equal to 4. The second predictor, B(z), is the pitch predictor. It removes redundancy in the speech signal that is due to the quasi-periodicity of voiced sounds. The pitch predictor has the transfer function

$$B(z) = \alpha z^{-T} \qquad (2)$$

where α is the pitch prediction coefficient and T is the estimated pitch period.

In the block diagram of Fig. 1, we show pitch prediction being followed by spectral prediction. At sample n, the predicted speech signal, s[n], is subtracted from the incoming speech signal, s[n], and the resulting residual, d[n], is quantized, coded and transmitted to the receiver. The quantized residual is also fed back within the analyzer loop.

In the receiver, the spectral prediction signal is computed first and is added to the received residual before the pitch predic tion signal. The pitch prediction signal is then added in, and the resultant synthesized speech signal is passed to a digital to analog converter for the reconstruction of the analog speech signal.

The one-bit quantized residual fed back within the analyzer and the decoded residual d̃[n] in the receiver are both unit variance signals and are scaled by a multiplicative factor q. The factor q is an estimate of the standard deviation of d[n] and forces the quantized residual signals in both the analyzer and the receiver to be equal to the original residual, d[n], with the addition of quantization noise.

The methods used to compute the APC side parameters a, T, ai and q are understood by viewing the prediction operations in the time domain. Removal of the pitch redundancy in the input speech signal can be written as the difference equation

$$e_1[n] = s[n] - \alpha s[n-T] \qquad (3)$$

21. An Adaptive Predictive Coder Using a Single Chip DSP

in which the signal $e_1[n]$ is referred to as the first residual. If the signal $s[n]$ were exactly periodic, and if T were computed without error, the coefficient a would equal one, and the first residual would be identically equal to zero. However, since speech is never exactly periodic and since the pitch estimation process employed in APC does produce errors, the first residual can never be identically zero in any practical sense. Thus, once the pitch period has been determined, a is estimated in a manner which minimizes the mean square energy in the first residual. This results in the following expression for a, the normalized correlation coefficient

$$\alpha = \frac{\sum\limits_{n=0}^{N-1} s[n]s[n-T]}{\sum\limits_{n=0}^{N-1} s[n-T]s[n-T]} \qquad (4)$$

Computing the spectral predictor is actually the linear prediction analysis problem. In this context, instead of performing the analysis on the speech signal, linear predictive analysis is performed on the first residual, $e_1[n]$. Linear prediction methods have been examined thoroughly in the literature and several solutions to the resulting normal equations have been proposed (see for example [10], [11], and [12] for solution methods). In the implementation of APC given in this report, we have used the autocorrelation method of solving the linear prediction normal equations. Standard algorithms, such as Levinson's recursion and the LeRoux-Gueguen recursion, which allow the predictor coefficients to be obtained from the first P+1 autocorrelation values, have both been considered.

The parameter q, as we have mentioned above, is an estimate of the standard deviation of the residual $d[n]$ and is used to scale the standard deviation of the quantized residual $\hat{d}[n]$ in the analyzer and $\tilde{d}[n]$ in the receiver to the same level as $\tilde{d}[n]$. Thus, when the quantized residual is used in the feedback loop in the analyzer, the reconstructed first residual $\hat{e}_1[n]$ and the reconstructed speech $\hat{s}[n]$ become equal to $\tilde{e}_1[n]$ and $r[n]$, the first residual and speech signals in the receiver in the absence of transmission errors. Note also that all of these signals will have been degraded by the identical quantization noise introduced at the analyzer.

The pitch period, T, can be estimated using any number of methods in APC. However, practical considerations permit only simple methods to be employed and pitch errors will typically be introduced. This is not a major disadvantage, however, since pitch errors do not severely degrade the quality of the resynthesized speech in APC. In most APC implementations, pitch period estimates are obtained using either autocorrelation analysis or the average magnitude difference function (AMDF) [14]. In the auto correlation method, the autocorrelation function is computed for each frame of speech. The distance in lags between its peaks is taken as the pitch estimate. In the AMDF method, the method

most often employed in real-time APC, the average magnitude difference function is substituted for the autocorrelation function and is computed in a manner very similar to the autocorrelation function for each frame of speech. The distance between nulls in the AMDF is taken as the pitch estimate. In this study we have used the AMDF method and have employed the following technique. To limit the number of computations, only certain values of T, Ti, corresponding to fundamental frequency values in the range from 50 to 400 Hz are used. These values of Ti are precomputed and stored in a table. The AMDF, $D[Ti]$, rewritten as

$$D[T_i] = \sum_{n=0}^{N-1} |s[n] - s[n-T_i]| \qquad (5)$$

is computed for each value of Ti. The value of Ti which gives the minimum AMDF value is taken as the pitch estimate. Another measure often taken to minimize the number of computations is to compute the AMDF summation for every fourth sample of $s[n]$ and $s[n-T]$, skipping three samples in between. We have also taken this step to minimize the computation time.

2.1 Implementation Issues and Discussion

In a real-time implementation of APC, the side parameters T, a, ai, and q are usually computed as background tasks while in coming speech is pre-emphasized and buffered by foreground I/O handling routines. This method of arranging the computations into background and foreground routines immediately imposes several requirements on the DSP that is used to implement the algorithm. The most obvious requirement is speed. In order to process speech data in real time, the DSP must be capable of executing a large number of machine instructions within the duration of a speech analysis frame. Another important issue is the size of the memory that the DSP either possesses on chip or can access externally. The memory requirements are large because computing much of the APC algorithm in the background demands that the speech data be double buffered. Multi-tasking foreground and background routines also requires a DSP capable of controlling a relatively large number of external I/O sources through the use of an interrupt mechanism. In the following section we shall see how these computational and control requirements of real-time APC translate into DSP architectural requirements.

In this section we have summarized some basic aspects of the APC algorithm. For the purpose of brevity, we have chosen not to describe several of the measures taken which improve APC's performance. For example, much of the research in Adaptive Predictive Coding has been involved with improving the performance of the predictive quantizer loop. These efforts include modifying the spectral predictor filter so that it shapes the quantization noise in ways which better match the masking properties of the auditory system [3]. Other efforts have included the development of various segmental quantization techniques that require that the quantization gain term, q, be computed several times per frame (on the order of 10) instead of just once as we have described [16]. The effect here, also, is to better control the properties of the quantization noise. Although including these techniques would strongly affect the execution of APC in any of the processors evaluated in this study, we felt that including them in the evaluation process would not provide further insight in assessing the relative performances of the DSPs.

3. DSP SELECTION

Before the existence of digital signal processor integrated circuits, DSPs employed in real-time signal processing architectures served primarily as number-crunching peripherals. In these systems, conventional microprocessors (e.g., the Intel 8085 and the Motorola 68000) served as central processing units that controlled the flow of data throughout the system and in and out of these peripherals. Although these implementations have been relatively small in size, smaller configurations have become possible by integrating more system control functions inside the DSP itself.

For the remainder of this section we will review the architectures of the AT&T Bell Laboratories DSP I, the Nippon Electric μPD7720, and the Texas Instruments TMS32010. We shall focus on each DSP's on-chip memory size and on the feasibility of supplementing this internal memory with external RAM. Secondly, we shall focus on each DSP's system control capabilities and evaluate its ability to manipulate the various data paths in a system.

3.1 AT&T Bell Laboratories DSP I

The digital signal processing integrated circuit initially considered in our study was the Bell Laboratories DSP I. The DSP has been successfully employed in other moderate complexity mid-rate coders at Bell Laboratories, such as Sub-band Coding [6] and ADPCM [4]. It therefore became a candidate for implementing real-time APC.

A complete description of the Bell Labs DSP architecture can be found in [5]. It features:

- a 1024 x 16 bit on-chip ROM for program and coefficient storage,
- a 128 x 20 bit on-chip data RAM,
- an extensive set of memory address registers and a separate address arithmetic unit,
- an Arithmetic and Logic Unit which features a 16 x 20 bit multiplier and a 40-bit accumulator.

The Bell Labs DSP architecture also features a great deal of parallelism which permits its relatively slow 800 nsec machine cycle time to be effectively reduced to 200 nsec by a 4-stage instruction pipelining mechanism. However, instruction pipelining is not always possible in many signal processing operations and is generally effective only in the type of computations required of digital filters and correlators

(i.e, register transfers, multiplies and adds). Therefore, the Bell Laboratories DSP's 800 nsec cycle time is prohibitively slow for the implementation of many signal processing algorithms such as autocorrelation coefficient calculations and the other computations required in APC.

Aside from its relatively slow machine cycle time, we perceive that the major problem involved in using the Bell Laboratories DSP for real-time APC implementation is the necessity for transferring speech and other data between external memory and the limited on-chip RAM. The use of RAM is essential because the 128 word internal RAM that the DSP provides is inadequate for storing the large speech buffers required for background computation of the APC side parameters. The DSP allows external memory to be substituted for the internal 1K program/coefficient ROM through a reconfiguration of the device and by using the multiplexed address/data bus which is brought off-chip. Unfortunately, external RAM cannot be substituted for the on-chip ROM because the DSP has no external memory write capability that directly utilizes this data bus.

The architecture of the DSP supports serial I/O with external devices through the use of asynchronous serial interface lines which could be used for transferring data to an off-chip memory. However, there are two problems associated with an approach which would utilize the serial data ports for memory I/O. The first problem is that the serial lines must be multiplexed between the memory and normal I/O devices, such as codecs and modems that would ordinarily communicate with the DSP. This problem could possibly be fixed by using external hardware that would arbitrate among these sources of data. The second, more critical problem, is data throughput. The following sample calculation determines the amount of time that would be required for serial I/O in computing an AMDF pitch estimate and illustrates the nature of the data throughput problem. At the maximum input clock rate, the DSP requires 400 nsec/bit to bring data on-chip. Therefore, reading a 16-bit word from memory would require 6.4 usec. For each point in the AMDF summation, two speech samples, s[n] and s[n-T], must be read from external memory. Assuming that a frame consists of 160 samples and that 3 samples are skipped between summations, computing the AMDF for a single value of T would require 2x40x6.4 sec or 0.51 msec. Computing the AMDF for 60 values of T would require that 30.7 msec be spent in memory I/O alone. This is in excess of common speech analysis frame durations and, thus, demonstrates that using the Bell Labs DSP to implement APC in this fashion is infeasible.

3.2 The Nippon Electric Co. μPD7720 Signal Processing Interface

The NEC μPD7720 Signal Processing Interface (SPI) has been used previously at Lincoln Laboratory in a compact linear predictive vocoder implementation [7]. The success experienced with the NEC μPD7720 in this project prompted us to consider the NEC μPD7720 as a candidate for implementing real-time APC. The NEC μPD7720 architecture is described in [13] and features:

- a 512 x 23-bit program ROM,
- a 510 x 13-bit data coefficient ROM,
- a 128 x 16-bit data RAM,
- a 250 nsec 16 x 16-bit parallel multiplier which gives a 31-bit result.

Although the NEC μPD7720 has several characteristics in common with the Bell Labs DSP, significant advantages are apparent when the two DSPs are compared. These advantages include an enhanced I/O structure, interrupt service capabilities, and a 4-level stack which provides for up to 4-level nesting of subroutines. The I/O structure contains several features which enable the NEC μPD7720 to be easily interfaced with a conventional microprocessor/controller, including an 8/16-bit parallel I/O data port which attaches directly to a system data bus. The NEC μPD7720 allows this data port to be configured for DMA mode data transfers between it and other system devices. This DMA capability ideally permits speech and other data to be loaded into the NEC μPD7720 in 128 word blocks for parameter computation (see discussion below).

A typical system architecture employing several NEC μPD7720 SPIs is shown in Fig. 2. An architecture similar to this was implemented in the Lincoln Laboratory compact LPC vocoder, and we assert that an APC implementation based on the NEC μPD7720 SPI would also resemble the architecture shown in Fig. 2. A conventional microprocessor is employed as a system controller. In the figure, we have indicated that an Intel 8085 could serve in this function; however, a number of other commercially available microprocessors could be used as well. The primary purpose of the system controller is to manipulate the data paths among the multiple NEC μPD7720 SPIs in the system.

Although we have shown an indefinite number of SPIs being deployed in the system shown in Fig. 2, we can assume that at least two (most likely three) SPIs will be needed to implement APC. As was true of the Bell Labs DSP, the SPI possesses only 128 words of internal RAM which is insufficient for the data buffering requirements of APC. However, since the data throughput of the NEC μPD7720 SPI is considerably faster, it is possible to spread the memory requirements among the several SPIs in the system and have these devices pass data among themselves under the direction of a system wide controller. Another alternative is to deploy a separate external RAM with a seperate DMA controller. Both of these approaches have been included in the architecture shown in Fig. 2. The DMA controller in the system architecture is used to handle block (and stream) data transfers between the SPIs and the external RAM, and the control microcomputer is used to control the data flow between the SPIs.

In addition to the large amount of memory required, the computational requirements of real-time APC also make it necessary that several SPIs be used. We base this assumption on the distribution of the computational load in the Lincoln Laboratory LPC vocoder design [7]. The real-time LPC implementation, although requiring very little memory for data buffering, requires three SPIs. The LPC and APC algorithms are of comparable complexity and three NEC μPD7720 SPIs, or possibly more, will most likely be necessary for APC.

When discussing the drawbacks of the Bell Labs DSP, we pointed out that most of the processing time required for computing the APC critical loops is spent accessing external memory. AMDF pitch estimation was the particular example cited. The Bell Labs DSP was ruled out because its I/O

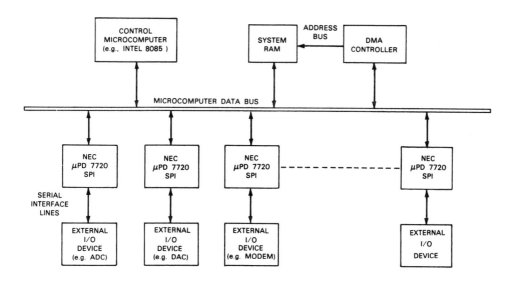

Figure 2. A Typical System Architecture Employing the NEC μPD7720

structure was not conducive to extensive use of external memory. A similar situation exists with the NEC μPD7720 SPI; however, it is less severe. The NEC μPD7720 possesses an equal amount of internal RAM as the Bell Labs DSP, and factors alluded to previously dictate that speech and other data be stored in external memory. An additional factor which makes the use of the NEC μPD7720 internal RAM undesirable is the internal memory pointer system that the NEC μPD7720 provides. It was discovered when programming the NEC μPD7720 for LPC [9] that a significant amount of overhead was devoted to manipulating these internal memory pointers. This programming inconvenience makes stream processing of data in the APC algorithm preferable over the use of block processing methods, which would buffer data needed for parameter computation in internal RAM.

Assuming data is to be processed in a stream fashion, we were able to approximate the execution times of some of the critical loops of the APC algorithm. Assuming external control of the system data paths, as shown in Fig. 2, approximately 4 μsec per 16-bit word are required to exchange data with external memory [8]. If one adds these numbers up, the approximated execution time of the AMDF pitch estimation algorithm is in excess of an analysis frame duration. However, the execution times of the remaining APC critical loops are each shorter than the assumed frame duration of 20 msec, thereby making a NEC μPD7720 based implementation of real-time APC feasible if an alternative pitch estimation algorithm to the AMDF method is used. These results are particularly encouraging, since it has already been demonstrated in the Lincoln Laboratory compact LPC vocoder that the more sophisticated, but less memory intensive, Gold pitch estimation algorithm can be programmed to run in the NEC μPD7720 in real time.

From these observations it seems that a real-time implementation of APC based on the NEC μPD7720 SPI is feasible. The architecture of such a system would most likely resemble the one shown in Fig. 2. Further determination of the specific hardware and software complexity, such as the exact number of NEC μPD7720 SPIs that would be required, has not been undertaken and, of course, would be the next step. Instead, our attention has been directed towards determining the feasibility of the Texas Instruments TMS32010.

3.3 Texas Instruments TMS32010

The Texas Instruments TMS32010 effectively combines the high numerical processing power of the NEC μPD7720 SPI with the control, data manipulation, and storage capabilities previously found only in general purpose microprocessors. A full description of the Texas Instruments TMS32010 architecture is given in [15]. We have highlighted some of its features here. They include:

- a 1500 x 16-bit internal ROM, an external ROM/RAM memory address space of up to 4K words,
- a 144 x 16-bit internal RAM, eight 16-bit parallel I/O ports,

- a 200 nsec 16 x 16-bit parallel multiplier with a 32-bit ALU/accumulator,
- a 200 nsec machine cycle time.

The most important advantage that the TMS32010 offers over the Bell Laboratories DSP and the NEC μPD7720 SPI is its 12 bits of external memory address space. A 16-bit word can be accessed from external ROM/RAM in two to three machine cycle instructions. We have found that based on this relatively short memory access time, the sum of the execution times of all of the critical loops of the APC algorithm is less than a 20 msec frame duration (see Section 4.2 below). Therefore, the APC algorithm could most likely run in real-time in a single chip, TMS32010-based system.

The TMS32010 provides two modes of accessing off-chip memory. In mode I, the TMS32010 generates the necessary memory addresses internally, and data is transferred over the 16-bit data bus. In mode II, the TMS32010 transfers data over its parallel I/O ports. Although both modes of external memory access physically involve the data bus, the differences between these two modes are important. Mode I memory transfers are effected by executing a three machine cycle TBLR or TBLW instruction in the TMS32010. When executing these instructions, the contents of the CPU accumulator are taken and used directly as the memory address. In mode II memory transfers, a two machine cycle IN or OUT instruction is executed. For these instructions the least significant three bits of the address bus contain a port address which can be decoded externally to select one of eight devices that are to send or receive data from the TMS32010 over the data bus. When the I/O ports are used for memory access, ROM/RAM addresses must be provided externally. The trade-off between using these two modes of memory I/O is one made between hardware and software efficiencies. Although the TBLR and TBLW instructions nominally require three machine cycles to execute, extra machine cycles are typically required for saving and restoring the contents of the accumulator which are involved in the ongoing computation. As a result, instead of three machine cycles being required for memory I/O, the amount of time often turns out to be on the order of seven to eight machine cycles. On the other hand, memory I/O involving the data ports is guaranteed to require no more than two machine cycles. The disadvantage in the case of mode II is that external hardware is needed for generating the required memory addresses.

3.4 Discussion and Summary

With ample external support hardware, any of these DSP integrated circuits could be used to implement real-time APC. However, given that the implementation should require a minimal amount of hardware, the TMS32010 seems to be the best alternative among the three. Using the AMDF pitch estimation computation as a comparison task, the Bell Labs DSP would require the most extensive amount of external support hardware followed by the NEC μPD7720 requiring an external memory controller and a control microcomputer, and, lastly, the TMS32010 requiring just an external memory controller.

In this section we have made comparisons of these DSPs based primarily on their memory accessing capabilities. Another important distinguishing feature which allows these DSPs to be compared is their capability for providing foreground/background multi-tasking of computations. In this respect, we require a DSP to have the ability to handle interrupts. As far as satisfying this particular requirement, the Bell Labs DSP does not support interrupts while the NEC μPD7720 and the TMS32010 do.

Based on the issues discussed in this section the TMS32010 is the processor of choice among the three DSPs that we have evaluated for the task of implementing APC. Although complete evaluations based on speech coding algorithms other than APC have not been carried out, it is reasonable to assert that the TMS32010 would be most appropriate for a variety of other moderate complexity speech coding algorithms as well.

4. APC PROCESSOR DESIGN

For the remainder of this report, we shall summarize the results of this APC processor design study by briefly describing one possible hardware/software implementation of real-time Adaptive Predictive Coding using the Texas Instruments TMS32010. Because the basic APC structure described in Section 2 will support a variety of speech sampling and data transmission rates, we begin this summary of the algorithm's implementation by describing, in closer detail, the specifications of the APC algorithm that we propose to implement. In Section 4.2 we give results of a critical loop timing study, and we describe a control strategy for fitting together the various software components of the APC algorithm. We have concluded from this programming exercise, in which the critical loops of the APC algorithm were coded in the actual TMS32010 instruction set, that the most important consideration in designing the software for APC is the fashion in which data is stored in memory. In Section 4.3 we describe one possible memory allocation scheme. We close our discussion of the APC implementation by describing the hardware requirements. We have concluded from performing a preliminary hardware design that the majority of the hardware design effort must be directed towards providing a high speed interface with memory external to the TMS32010, and to developing a method of communicating with several external input/output devices under interrupt control. The details of this preliminary hardware design are summarized in Section 4.4.

4.1 Algorithm Specifications

In Section 2 we outlined the structure of the APC algorithm. The structure described in Section 2 will support a range of data transmission and speech sampling rates. The version of the APC algorithm chosen for this study is

intended to operate at a transmission rate of 9.6 Kbps and at a sampling rate of 8000 samples/sec. The average frame duration is in tended to be 20 msec which corresponds to an analysis frame size of 160[1] samples. The transmission frame size for this sampling rate is 192 bits which are allocated among quantized residual and side parameters as shown in Table I.

Table 1. Bit Allocation per Frame

QUANTITY		BITS/ FRAME
d[n]	Residual	157
T	Pitch	6
α	Pitch Predictor Coefficient	4
q	Quantizer Level	5
k_1		5
k_2	Reflection Coefficients	5
k_3		5
k_4		5
	TOTAL	192

4.2 Critical Loop Timing and Control Strategy

The first step actually taken in the evaluation of the TMS32010 was a critical loop timing study in which the various portions of the APC algorithm were coded in the TMS32010 instruction set and approximate execution times of the code were calculated. This critical loop timing study was useful in obtaining two types of information. First, it has given us some benchmark timing figures which could be used as an objective measure for comparing the TMS32010 against the other digital signal processing integrated circuits. Secondly, this timing data provided information which was later used in decisions affecting the hardware design.

Two versions of several of the critical loops were coded. In version I of the code, TBLR and TBLW instructions were used to transfer data to and from off-chip memory. In version II, the data port I/O instructions, IN and OUT, were used to transfer data to external RAM. We have summarized the execution times of the software units in Table II. All of these execution times are based on a 200 nsec machine cycle time. Listings of the code written for these critical loops appear in the appendices. From examining the

---

[1]Note that the frame duration is measured with respect to the transmission and receiver modem clocks which are asynchronous to the sampling rate clock. Therefore, the analysis frame may deviate slightly from the 160 sample normal size.

Table 2. Summary of Critical Loop Execution Times

OPERATION	VERSION I		VERSION II	
	EXECUTION* TIME (msec)	% REAL TIME	EXECUTION* TIME (msec)	% REAL TIME
AMDF Pitch Estimation	10.80-11.00	54.0-55.0	6.60	33.0
ALPHA Calculation	.92	4.6	.62	3.1
1st Residual Calculation & LPC Autocorrelation Analysis	3.18	15.9	2.92	14.6
Reflection Coefficient Calculations	.16	.8	.16	.8
Predictive Quantizer	2.84	14.2	1.40-2.16	7.0-10.8
Receiver Loop	2.00	10.0	1.60	8.0
ADC-DAC I/O	1.09-1.28	5.4-6.4	1.09-1.28	5.4-6.4
Transmit Modem I/O Handler**	1.40	7.0	1.4	7.0
Receive Modem I/O Handler**	1.40	7.0	1.4	7.0
Total	23.79-24.18	119.0-120.9	17.19-18.14	86.0-90.7

*Execution time per frame
**These are foreground routines. Execution times were calcultaed by multiplying the per sample execution times by the 160 sample frame size.

execution times of the critical loops in Table II, it is apparent that the code which incorporates the TBLR and TBLW mode of external memory access could not execute within a 20 msec frame duration.

Thus, if only one TMS32010 were used, we would not expect the algorithm to execute in real time. One therefore would have to partition the APC algorithm among two or more TMS32010 DSPs. On the other hand, if the I/O ports are used for transferring data to external memory in conjunction with external memory address generators, it seems possible that a single TMS32010 would be all that is needed.

During our study we briefly examined the trade-offs between implementing a single TMS32010-based APC processor versus one which incorporates two TMS32010 DSPs with version I of the APC software partitioned. Although a dual TMS32010-based processor possesses potentially more processing power, it is difficult to make effective use of it due to interprocessor communication overhead. In designing a dual TMS32010 architecture, the first task is to find a reasonable partition of the APC software between the two TMS32010 DSPs. The most straightforward partition, a direct split between the analyzer and synthesizer, would result in an unbalanced distribution of the

computational load. The analyzer requires a significantly greater proportion of the computational resources. In fact, given the execution times of the analyzer loops, it is improbable that a single TMS32010 would be able to execute all of the analyzer routines within the 20 msec frame duration. Therefore, a more uniform partitioning of the APC algorithm, in terms of computational requirements, is needed. This alternative has a more subtle drawback in terms of data communication overhead. Although the APC analyzer software can be segmented into several autonomous units, practically all of these units process the same speech data. If these units are contained in separate TMS32010s, then either entire speech buffers would have to be passed among DSPs or each of the DSPs would have to access identical copies of the same speech data. The first option entails a significant amount of processing time being devoted to I/O among the processors. The second option would require that either memory be shared or data be copied to both TMS32010 processors. Both of these memory management schemes are unduly complex.

After recognizing the difficulties involved in using a dual TMS32010 system, we decided not to pursue this effort and, instead, adopted the single-chip design which uses the I/O ports for transferring data to external memory. For the remainder of this section and the next, we describe a software control strategy and external memory allocation for this one chip design. We use the term, *software control strategy*, to refer to the method used to combine software units. Our philosophy in adopting a software control strategy in this APC implementation has been to relegate as much of the computation to background tasks as possible. This allows the foreground routines, which are executed upon interrupt from the external I/O sources, to be simple I/O handlers that merely control the pointers required for buffering the data. In Table II, foreground routines are identified with two asterisks.

The obvious disadvantage of computing the APC routines as background routines is the overall increased demand for memory. However, as we shall illustrate in the following section, the memory requirements of this APC implementation fit safely within the confines of commercially available RAMs.

4.3 Memory Allocation

In Fig. 3 we show how memory is allocated among the APC software units. A total of 2048 words of RAM are required and have been divided among eight 256-word pages. The 256-word page size is used to accommodate the use of 8-bit external address generators which are used as memory pointers as described in the next section. Computing the analyzer routines as background tasks normally requires that the incoming speech be double buffered. Actually, triple buffering is used. The extra buffer, stored on a separate page, is provided for storing the previous pitch period of speech that is necessary in computing the pitch period estimate, T, the pitch predictor coefficient, α, and the first residual signal, $e_1[n]$. During these calculations, the speech sample $s[n-T]$ is needed and, depending on the value of T, could reside on the previous pitch period page. The two buffers of input speech used in these background computations, the current processing frame and the previous pitch period, are arranged contiguously on pages 0 and 1 so that the same 8-bit memory pointer, with the addition of a 9th bit used for page crossing, can be used to access these two pages of data as a single 512-word block. We have thus eliminated the overhead in software involved in page crossing. The 9th bit of the pointer to speech sample $s[n-T]$ is set by the hardware when it reaches the end of page 0. A separate pointer to the speech sample $s[n]$ is initialized to the bottom of page 1 and never crosses the 0/1 or the 1/2 page boundaries.

The reconstructed speech signals in the analyzer and synthesizer (i.e., the state space of the recursive pitch prediction filters) are stored in pages 2 and 3 which are implemented in hardware as circular buffers. The 256-word buffer size is more than adequate for storing the state space which has a maximum length of 160 points.

Since we have decided to compute the predictive quantizer and receiver loops as background routines, the single bit/sample residual data must also be double buffered, as do the APC side parameters. However, since this residual data stream is serial, we can reduce its storage requirements

MEMORY PAGE	FUNCTION
0	PREVIOUS PITCH PERIOD OF SPEECH DATA
1	CURRENT SPEECH PROCESSING FRAME
2	RECONSTRUCTED SPEECH FOR ANALYZER PREDICTIVE QUANTIZER LOOP
3	RECONSTRUCTED SPEECH FOR SYNTHESIZER LOOP
4	SPEECH INPUT BUFFER FROM ADC
5	SYNTHESIZER OUTPUT BUFFER NO. 1
6	SYNTHESIZER OUTPUT BUFFER NO.2
7	DOUBLE BUFFERED RESIDUAL OUTPUT (Analyzer)
	DOUBLE BUFFERED RESIDUAL INPUT (Synthesizer)
	DOUBLE BUFFERED APC SIDE PARAMETERS (Analyzer)
	DOUBLE BUFFERED APC SIDE PARAMETERS (Synthesizer)

Figure 3. Ram Allocation for APC Implementation

by packing it into 16-bit words. This data is stored on page 7, along with the APC side parameters.

We have eliminated the need in the analyzer for buffering the first residual signal by combining the autocorrelation computation with the calculation of the first residual signal. Through the use of a first-in first-out buffer maintained in internal RAM (see code in the appendix) for storing only $P+1$ first residual values, we are able to compute the first residual autocorrelation values that are necessary for computing the LPC spectral predictor coefficients directly from the speech signal.

ROM is needed for storing program instructions and data constants. Although we have not completely specified the amount of RCM which will be needed, we have assumed that no more than 2048 words of ROM will be required. The hardware implementation of both ROM and RAM will be discussed in the following section.

4.4 Hardware Design

There are two principal features of the APC processor hardware. The first is a high speed external memory interface circuit. This circuitry provides two separate memory address generators which are operated under programmed control of the TMS32010. The second major feature is an interface between the TMS32010 and four external I/O devices (the analog to digital and digital to analog converters, the transmit modem, and the receive modem). This interface allows the external devices to communicate with the TMS32010 CPU on an interrupt basis. For the remainder of this section, we outline our approach for designing the APC processor hardware.

A functional block diagram of an APC processor architecture is shown in Fig. 4. In this figure, we have labeled the external memory controller circuit and the external I/O interface portions of the architecture explicitly. The architecture permits access to external memory from the TMS32010 under the two modes described in the previous sections, using either the memory address bus in conjunction with the TBLR/W instructions or the port address bus for faster access. The memory address bus is used primarily for fetching program instructions and constants from RCM via mode I. Under mode II, the data stored in locations 0-1023 of RAM are designed to be accessible using the address generation logic which is contained in the external memory controller circuitry. In order to retain maximum flexibility, we have made all 2K of RAM accessible to the TMS32010 under both modes by multiplexing the address bits input to the RAM devices.

The data ports can be thought of as eight physical ports which are directly tied to the TMS32010. In actuality, data transfers involving the ports will utilize the data bus as well. A 3-bit port address (PAC-3) is decoded and is used to select one of eight devices which is to send or receive data from the TMS32010 over the bus. In the proposed system shown in Fig. 4, three of the eight ports are used to interface the TMS32010 to external I/O devices. The I/O devices requiring separate ports are the analog to digital converter (ADC), the digital to analog converter (DAC), and the parallel to serial converter (PSC) which subsequently connects to a serial transmit modem. We have been able to input data from the serial receive modem without the use of an explicit I/O port. An additional port is used to interface the TMS32010 with external interrupt control logic. The other four ports are used to interface the TMS32010 with external RAM.

4.4.1 Hardware Design External Memory Controller

Figure 5 is a more detailed schematic of the External Memory Controller circuit. In this schematic we have shown explicitly the memory I/O control signals which are generated by the TMS32010 CPU, the logic used for decoding these

signals, and the port/memory address bus. For flexibility in the software design, external memory can be used as either a 4K word block, consisting of both ROM and RAM to be accessed under mode I type memory transfers (i.e., by executing TBLR and TBLW instructions), or as separate ROM and RAM each consisting of 2K words. In mode II, the first 1K words of RAM are accessed via the data I/O ports in conjuction with the external address generation logic which shall be described below. These two modes of memory access are distinquished in the hardware by decoding the TMS32010 control signals MEN, DEN, and WE, along with the memory address bit A_{11}.

For the most part, mode I memory transfers are used primarily for fetching program instructions and data constants from ROM. However, we also must access pages 4 through 7 of RAM under mode I. For mode I, the 4K block of memory has been partitioned into a lower 2K section of ROM and an upper 2K section of RAM. A simple 2-level hardware decoding of address bit A11 distinguishes a read from ROM from a read from RAM. For a read operation from ROM, 7 of RAM under mode I. For mode I, the 4K block of memory has been partitioned into a lower 2K section of ROM and an upper 2K section of RAM. A simple 2-level hardware decoding of address bit A_{11} distinguishes a read from ROM from a read from RAM. For a read operation from ROM, the TMS32010 signal MEN will become active low, along with address bit A11. These two signals cause the ROM-ENABLE signal to become active low, which is directly tied to the chip-select (CS) inputs of the ROM devices. If a read from RAM is to take place, MEN will again be active low, but A_{11} will be high since RAM is contained in the upper 2K section of the memory address space. The address bit A_{11} is inverted and combined with MEN to generate chip select signals for the RAM devices (see Fig.5.).

Modes I and II memory transfers are distinguished by decoding the address bit, A_{11}, along with the TMS32010 control signals WE and DEN. These signals are both active low and are combined with $A11$ to generate a PORT-ENABLE signal (also active low) which enable a 3-to-8 line decoding of the port/address but which is assumed to contain a valid 3-bit port address. The decoder circuit signals to one of the eight devices tied to its output lines to communicate with the TMS32010 CPU over the data bus. Two separate DMA controllers are being employed as external memory address generators. In the schematic in Fig. 5, Advanced Micro Devices (AMD) Am2940s [1] are used. These particular devices have been chosen primarily because of their speed. The machine cycle time of the TMS32010 is nominally 200 nsec, and a reasonable, but fast, access time for commercially available RAMs is 50 nsec. According to the specifications given for the Am2940, its propagation delay, combined with the delays of the other combinational logic in the external memory interface, provide adequate time for data being accessed from RAM to settle on the data bus before the end of the TMS32010 memory read cycle. Similar time constraints are met for the memory write cycle.

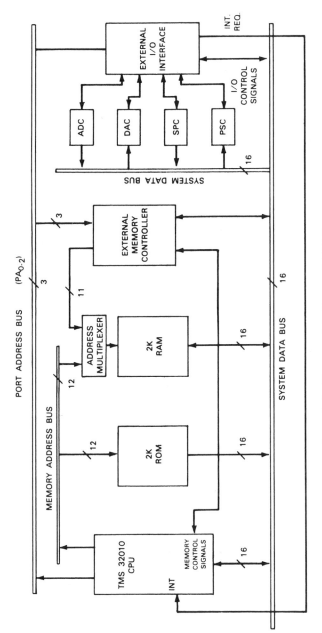

Figure 4. Architecture of the Proposed APC Processor

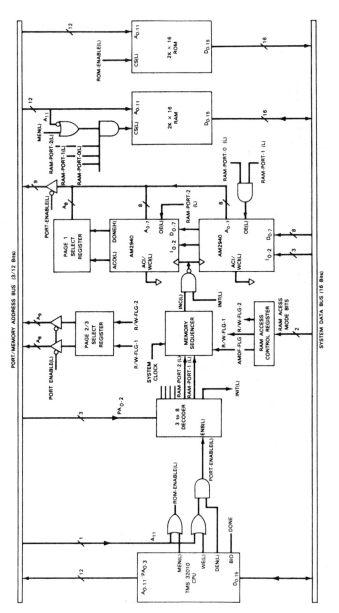

Figure 5. Schematic of the External Memory Interface Circuit

21. An Adaptive Predictive Coder Using a Single Chip DSP

The Am2940s are programmable and receive instructions from the TMS32010 over the data bus via the port I/O mechanism. One of the eight data ports is dedicated entirely to providing initialization and other instructions to the Am2940s. When this port is selected, the INIT signal becomes active low (see Fig. 5) and the Am2940s receive instructions over the data bus. The format of the data instructions which are given to the Am2940s is described below.

Although the memory requirements of the APC algorithm are extensive, an advantage that the algorithm provides is that memory access is primarily sequential within a page. In other words, speech samples and other data that are used within the same software routine will generally reside on the same page and will be arranged sequentially within that page. This way, after the DMA controllers have been programmed at the beginning of a routine, there is little interaction between the TMS32010 CPU and the address generators during the remainder of the routine's execution. In addition, most of the computationally intensive signal processing routines involve sequential data fetches from two memory locations. The Am2940 address generators will increment their present addresses after an INC signal is generated by the Memory Sequencer circuit shown in Fig. 5. The Memory Sequencer is a relatively simple Finite State Machine (FSM) which ensures that the memory pointers to RAM are incremented by the proper amount after each data transfer. Accessing the I/O ports using either RAM-PORT-1 or RAM-PORT-2 causes both addresses to increment while an access via RAM-PORT-0 causes no incrementation. Thus, a typical autocorrelation computation would require first reading [n] through RAM-PORT-0 and the reading s[n-i] through RAM-PORT-1 or 2. Since address incrementation

occurs following the second read, subsequent data fetches will access the proper data. All of the routines which access external memory using the I/O ports have the memory pointers incremented once after each transfer, the exception being the routine which computes the AMDF pitch estimate. In this routine, three speech samples are skipped between each point in the AMDF summation (see Eq. 5); and the memory pointers must, therefore, be incremented three times.

The state transition diagram for the memory sequencer FSM is shown in Fig. 6. Shown in the figure are three inputs to the FSM, RAM-PORT-1. RAM-PORT-2 and AMDF-FLG, in addition to the system clock. The output of the FSM is an INC control signal which causes both of the Am2940 address generators to increment their memory pointers. The memory sequencer FSM steps through its sequence while the CPU is processing the data it has just read, or, in the case of a memory write, while the CPU is preparing to output another word to RAM. We have managed to save a considerable amount of time by having these operations carried out concurrently. The memory sequencer FSM normally sits in an idle state. After the CPU has finished its read or write cycle, signalled by either the RAM-PORT-1 or the RAM-PORT-2 signals becoming inactive, the FSM will enter the INCREMENT #1 state. During the transition it generates the INC control signal which is tied to both Am2940s. If the AMDF pitch estimation is not being performed, the FSM will return to the IDLE state. If an AMDF pitch estimate is being computed, the FSM will continue to its INCREMENT #2 and INCREMENT #3 states, incrementing the Am2940s three more times in the process.

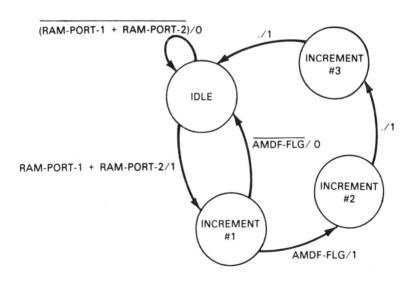

Figure 6. Memory Sequencer State Transition Diagram

The external memory interface circuit is programmed by issuing a sequence of 16-bit instruction words over one of the I/O ports. Each instruction word is broken into several fields which are labelled in Fig. 7(a). The least significant eight bits contain the initialization data for the Am2940s (e.g., initial addresses, etc.). Bits 8 through 10 and 11 through 13 contain the Am2940 instructions, and bits 14 and 15 contain the RAM access mode bits which are described in Fig. 7(b). A list of the Am2940 program instructions is given in [1]. The appropriate setting of the RAM access mode bits indicates to the external memory controller which memory page is to be accessed and the number of times the memory pointers are to be incremented during each memory I/O cycle. The access mode bits will control the setting of the R/W-FLG-1, R/W-FLG-2 and the AMDF-FLG signals, which are output from the RAM Access Control Register and are input to another register which selects pages 2 and 3 of RAM, and are also input to the memory sequencer FSM. A table is provided in Fig. 7(b) that summarizes the settings of the RAM access mode bits.

4.4.2 Hardware Design-I/O Interface Circuit

The second major task in the hardware design is to interface the four external I/O devices with the TMS32010. These external devices are to communicate with the TMS32010 on an interrupt basis. Since the TMS32010 has only one interrupt line, the control signals output from these I/O devices must be multiplexed. Our approach has been to design an interrupt control register which will uniquely indicate the presence of an interrupting device by one of its bits being set. The register is read by the TMS32010 any time after the interrupt has occurred.

The I/O interface circuit is shown in Fig. 8. Buffer registers are provided between the TMS32010 and the analog to digital converter (ADC) and the digital to analog converter (DAC). These registers allow the data transferred between the TMS32010 and these devices to be double buffered, eliminating the need for the TMS32010 to be directly involved in any type of handshaking procedure. A parallel to serial converter (PSC) is provided between the transmit modem and the TMS32010. The parallel to serial converter

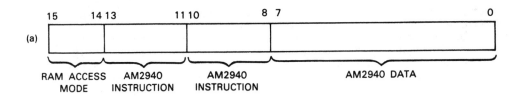

| | ACCESS MODE | | MEMORY PAGE |
	BIT 1	BIT 0	AND FUNCTION
(b)	0	0	PAGES 0 & 1 AMDF MODE ACCESS (R)
	0	1	PAGES 0 & 1 NON AMDF MODE (R)
	1	0	PAGE 2 (R/W)
	1	1	PAGE 3 (R/W)

Figure 7 (a). External Memory Interface Program Instruction Word
(b). RAM Access Mode Bits

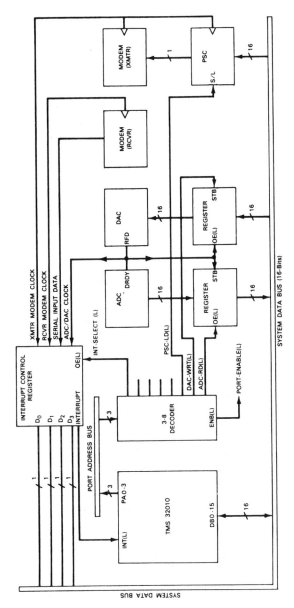

Figure 8. External I/O Device Interface Circuit

changes the parallel data output from TMS32010 into a serial bit stream appropriate for the transmit modem. It also serves as a data buffer as well.

The interrupt control register (ICR) is shown in Fig. 9. It multiplexes three externally generated I/O control signals on to the single interrupt line of the TMS32010 through the use of a single logic gate. These three control signals are the ADC/DAC sampling clock, the transmit modem clock, and the receive modem clock. The ICR is actually a 4-bit register, with the fourth bit being used to store the serial input data from receive modem. The ICR is implemented as a set of four D-latches. A typical interrupt service scenario would proceed as follows. When one of the external devices interrupts the TMS32010, the interrupt signal is passed on directly to the TMS32010 when the corresponding bit is set in the ICR. A bit set in the ICR prohibits the interrupting device from reinterrupting the TMS32010 until the bit is reset through an acknowledgment procedure as follows. The bit is reset by the TMS32010 after the contents of the ICR are read over one of the I/O ports and a word is written back with the corresponding bit set to 1.

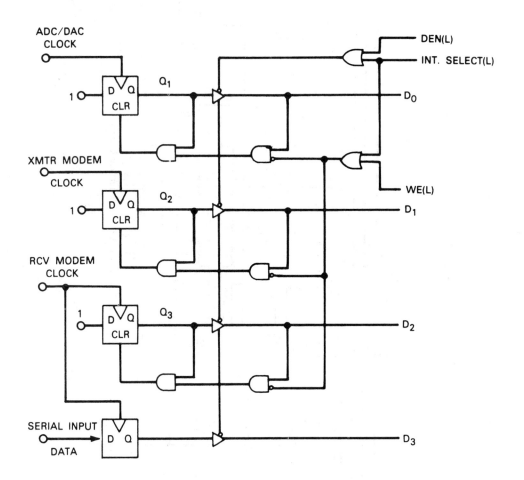

Figure 9. Interrupt Control Register

5. SUMMARY

In this report we have given the results of an APC processor design study. The system that we have proposed is based on the Texas Instruments TMS32010. We began by outlining the basic features of the algorithm and pointing out those aspects of the algorithm which make its implementation challenging. The major problem associated with implementing APC in real time is memory. We acknowledged the necessity of computing the APC side parameters as background tasks which inherently requires an extensive amount of RAM.

As a part of this study, we examined two other DSPs besides the TMS32010: the AT&T Bell Laboratories DSP I and the Nippon Electric µPD7720 Signal Processing Interface. We have compared these DSPs against the TMS32010. The Texas Instruments TMS32010 was determined to be most suitable for a real time implementation of APC because of its speed, its ability to perform the numerically intensive signal processing operations required by APC, and its relatively sophisticated control features which enable it to handle the memory addressing and I/O requirements of APC.

The objective of this study has been two-fold. We wanted to learn the relative strengths and weaknesses of the three DSP ICs, and we wanted to determine whether a compact implementation of APC and other moderate bit rate speech coders of comparable complexity were feasible using the TMS32010. Towards these ends a critical loop timing study was performed which allowed us to obtain some benchmark timing figures which were used to characterize the TMS32010. These timing figures also helped us in making some hardware decisions and allowed us to determine the approximate hardware requirements.

There are several steps which can follow from this work. The most logical step would be for the architecture outlined in Section 4 of this report to be constructed. Although a preliminary hardware design was given in this report, many of the details still need to be more fully developed.

ACKNOWLEDGMENT

The author wishes to acknowledge the assistance of E. Hofstetter, E. Singer, and J. Feldman throughout this investigation. They have provided helpful and insightful comments and criticism and, in the case of Hofstetter and Singer, have carefully edited earlier drafts of this report.

REFERENCES

1. Advance Micro Devices, *"Bipolar Microprocessor Logic and Interface,"* 1983 Data Book.
2. B.S. Atal and M.R. Schroeder, *"Adaptive Predictive Coding of Speech Signals,"* Bell System Technical Journal 55:1973-1985 (1975).
3. B.S. Atal and M.R. Schroeder, *"Predictive Coding of Speech Signals and Subjective Error Criteria,"* in Proceedings of the International Conference on Acoustics, Speech, and Signal Processing, IEEE, Tulsa (1978).
4. J.R. Boddie, J.D. Johnson, C.A. McGonegal, J.W. Upton, P.A. Berkley, R.E. Crochiere and J.L. Flanagan, *"Adaptive Differential Pulse-Code Modulation Coding,"* Bell System Technical Journal 60: No. 7 1547-1561 (September 1981).
5. J.R. Boddie, G.T. Daryanani, I.I. Eldumiati, R.N. Gadenz, J.S. Thompson, and S.M. Walters, *"Digital Signal Processor: Architecture and Performance,"* Bell System Technical Journal 60: No. 7, 1449-1462 (September 1981).
6. R.E. Crochiere, *"Sub-Band Coding,"* Bell System Technical Journal No. 60: 1633-1653 (September 1981).
7. J.A. Feldman, *"LPC Chip Set User's Guide,"* Project Report, PSST-1, Lincoln Laboratory, M.I.T. (November 1982).
8. J.A. Feldman, Personal Communication.
9. E.M. Hofstetter, Personal Communication.
10. J. LeRoux and C. Gueguen, *"A Fixed Point Computation of Partial Correlation Coefficients in Linear Prediction,"* in Proceedings of the International Conference on Acoustics, Speech, and Signal Processing, IEEE (1977).
11. J. Makhoul, *"Linear Prediction: A Tutorial Review,"* Proceedings of the IEEE 63 (12) (April 1975).
12. J.D. Markel and A.H. Gray, *"Linear Prediction of Speech"* (Springer-Verlag 1976).
13. Nippon Electric Company, NEC 7720 Data Manual.
14. M.J. Ross, H.L. Shaffer, A Cohen, R. Freudberg and H.J. Manley, *"Average Magnitude Difference Function Pitch Extractor,"* IEEE Transactions on Acoustics, Speech, and Signal Processing ASSP-22(5) (October 1974).
15. Texas Instruments, *TMS320 Data Manual.*
16. R. Viswanathan, W. Russell, and A.W.F. Huggins, *"Design and Real-Time Implementation of a Robust APC Coder for Speech Transmission over 16Kbps Noisy Channels,"* Bolt Beranek and Newman Inc., BBN Rep. 4565 (December 1980).

```
* Computes a pitch estimate from one out of 60 values which are stored in a
* table in internal RAM.  Assumes speech data to be stored in external RAM
* accessed using TBLR

INIT    ZALS    BIG-NUM-L,0
        ADDH    BIG-NUM-H,0
        SACL    MIN-AMDF-L,0      MIN-AMDF maintains min AMDF
        SACH    MIN-AMDF-H,0        value, init it to something large
        LACK    #60
        SACL    NUM-PITCHS,0      Num-Pitchs is loop counter

        LARK    AR0,#P-TBL-ADDR   AR0 points to pitch table

LOOP-1  LACK    #S-ADDRR          Initialize pointers to speech
        SACL    S-PTR, 0            Data s[n] and s[n-T]

        SUB     *,AR0,0           Compute pointer to s[n-T] by
        SACL    ST-PTR,0            subtracting away current pitch
        ZAC
        SACL    AMDF-L,0          Initialize Accumulated AMDF
        SACH    AMDF-H,0            value
        LARK    AR0, #NUM-SAMPLES  AR0 is loop counter

LOOP-2  ZALS    S-PTR, 0          Load ACC w/ pointer to s[n]
        TBLR    S                 Read s[n]
        ADD     THREE, 0          Update pointer skipping
        SACL    S-PTR, 0          Three speech samples
        ZALS    ST-PTR, 0         Do same w/ s[n-T]
        TBLR    ST
        ADD     THREE, 0
        SACL    ST-PTR, 0

        ZALS    S, 0              Compute /s[n] - s[n-T]/
        SUB     ST,0
        ABS
        ADD     AMDF-L,0          Update AMDF value
        ADD     AMDF-H,15
        SACL    AMDF-L,0
        SACH    AMDF-H,0
        MAR     *-,AR1            If not finished w/ current
        BNAZ    LOOP-2              pitch value, loop

        ZALS    MIN-AMDF-L,0      Compare current AMDF value with
        ADDH    MIN-AMDF-H,0        previous minimum
        SUB     AMDF-L,0
        SUB     AMDF-H, 15        If current is larger, loop
        BLZ     SAME                to next pitch value
```

```
        ZALS    AMDF-L, 0         If current AMDF is smaller
        ADDH    AMDF-H,0           make it the new minimum
        SACL    MIN-AMDF-L,0
        SACH    MIN-AMDF-H,0
        ZALS    T,0               Save the present pitch value
        SACL    *,AR0

SAME    MAR     *+,AR0
        ZALS    NUM-PITCHS,0
        SUB     ONE,0
        SACL    NOM-PITCHS,0      Loop to next pitch value
        BGEZ    LOOP-1
```

AMDF PITCH ESTIMATION - VERSION II

```
*  Compute a pitch estimate from one of 60 values stored in a table in internal
*  RAM.  Assumes speech data to be stored in external RAM which is accessed via
*  the data ports using the IN instruction.  External memory interface is initial-
*  ized by a separate initialization routine.

INIT    ZALS   BIG-NUM-L,0          Initialize minimum AMDF value to some
        ADDH   BIG-NUM-N,0             big number
        SACL   MIN-AMDF-L,0
        SACH   MIN-AMDF-H,0
        LACK   #60                  Initialize counter for number
        SACL   NUM-PITCHS,0            of pitch values
        LARK   AR0,#P-TBL-ADDR      Initialize AR0 to point to pitch table
        ZAC                         Clear RAM access control word
        SACL   ACCESS-CTR-WD,0        to indicate AMDF mode in
        CALL   EXT-MEM-INIT           initialization of external memory
                                      interface

LOOP-1  LACK   #ADDR-S              Init external pointers to s[n] and s[n-T]
        SUB    *,0,AR0              Issue Reinitialization instructions to
        ADD    REIN-INS,11            AM2940's
        ADD    LOAD-INS,8
        SACL   INSTR,0
        OUT    INSTR, INTRFC-PORT   Output Instructions over Data
                                    Port

LOOP-2  IN     S, RAM-PORT-1        Read s[n] and s[n-T] from external RAM
        IN     ST, RAM-PORT-2
        ZALS   S,0                  Compute /s[n]-s[n-T] /
        SUB    ST,0
        ABS
        ADD    AMDF-L,0             Update AMDF Value
        ADD    AMDF-H,15
        SACL   AMDF-L,0
        SACH   AMDF-H,0
        BIOZ   LOOP-2               Use hardware to detect end of loop

        ZALS   MIN-AMDF-L,0         Compare current AMDF value w/
        ADDH   MIN-AMDF-H,0           previous minimum
        SUB    AMDF-L,0
        SUB    AMDF-H,15
        BLZ    SAME
        ZALS   AMDF-L,0             If smaller, update minimum
        ADDH   AMDF-H,0
        SACL   MIN-AMDF-L,0
        SACH   MIN-AMDF-H,0
        ZALS   *,AR0                Save present pitch value
        SACL   T,0
```

```
SAME    MAR    *+,AR0
        ZALS   NUM-PITCHS,0
        SUB    ONE,0
        SACL   NUM-PITCH,0
        BGEZ   LOOP-1               Loop if not finished
```

PITCH PREDICTION COEFFICIENT – VERSION I

```
* Compute the pitch prediction coefficient.  Assumes speech is stored in external
* RAM access using TBLR
*
INIT  ZAC
      SACH  NUM-L,0            Initialize numerator and denominator
      SACL  NUM-H,0
      SACL  DEN-L,0
      SACH  DEN-H,0
      LACK  #S-ADDR
      SACL  S-PTR,0            Initialize pointers to s[n] & s[n-T]
      SUB   T,0
      SACL  ST-PTR,0
      LARK  AR0,#N             Initialize loop counter AR0

LOOP  ZALS  S-PTR
      TBLR  S                  Read s[n]
      ADD   ONE, 0
      SACL  S-PTR, 0
      ZALS  ST-PTR, 0
      TBLR  ST                 Read s[n-T]
      ADD   ONE
      SACL  ST-PTR
      ZALS  NUM-L, 0
      ADDH  NUM-H, 0
      LT    ST
      MPY   S
      APAC
      SACL  NUM-L, 0           Update numerator
      SACH  NUM-H, 0
      ZALS  DEN-L, 0
      ADDH  DEN-H, 0
      MPY   ST
      APAC
      SACL  DEN-L, 0           Update denominator
      SACH  DEN-H, 0
      MAR   *, AR0
      BNAZ  LOOP
      CALL  DIVIDE             Perform divide in a subroutine
      ZACH  QUOTIENT
      SACH  ALPHA              Store result
```

PITCH PREDICTION COEFFICIENT – VERSION II

```
* Compute pitch predictor coefficient α.  Assume speech is stored in external RAM
* accessed via the I/O ports using the IN instruction.
*
INIT  ZAC
      SACL  NUM-L,0            Initialize numerator & denominator to
      SACH  NUM-H,0            zero
      SACL  DEN-L,0
      SACH  DEN-H,0
      LACK  #1
      SACL  ACCESS-CTR-WD,0    Initialize external memory interface
      CALL  EXT-MEM-INIT       for Mode 1 type memory interface

LOOP  IN    S                  Input s[n] & s[n-T]
      IN    ST
      ZALS  NUM-L,0
      ADDH  NUM-H,0
      LT    S
      MPY   S
      APAC
      SACL  NUM-L              Update numerator
      SACH  NUM-H
      ZALS  DEN-L
      ADDH  DEN-H
      MPY   ST
      APAC
      SACL  DEN-L,0            Update denominator
      SACH  DEN-H,0
      BIOZ  LOOP               Detect end of loop in hardware
      CALL  DIVIDE
      ZALH  QUOTIENT, 0        Compute result and store it
      SACH  ALPHA,0
```

21. An Adaptive Predictive Coder Using a Single Chip DSP

APPENDIX V

COMBINED 1ST RESIDUAL AND AUTOCORRELATION COMPUTATION

```
*   Computes 5 autocorrelation values along with the 1st residual direct
*   from the speech signal.  Assumes that speech is stored in external
*   RAM and that it is accessed using TBLR

INIT    LACK    #S-ADDR         Initialize pointers to speech
        SACL    S-PTR, Ø          data
        SUB     T, Ø
        SACL    ST-PTR, Ø
        LACK    #N               Initialize main loop counter
        SACL    COUNT, Ø
        LARK    ARØ, #R-ADDR
        LARK    AR1, #ORDER

LOOP-4  SACL    *+, ARØ          Initialize autocorrelation values
        MAR     *-, AR1             to zero
        BNAC    Loop-4

LOOP-1  ZALS    S-PTR, Ø         Read s[n]
        TBLR    S
        SUB     ONE, Ø
        SACL    S-PTR, Ø
        LT      ALPHA
        ZALS    ST-PTR, Ø
        TBLR    S                Read s[n-T]
        SUB     ONE, Ø
        SACL    ST-PTR, Ø
        ZALS    S
        MPY     S                Compute el[n] = s[n]- s[n-T]
        SPAC
        LARK    ARØ, #e-ADDR     Place el [n] value on a First-in
        SACL    *, Ø, ARØ          first-out, stack which retains
        LT      *, ARØ             five most previous residuals values
        LACK    #ORDER
        SACL    COR-COUNT, Ø     Set up loop to compute correlation
        LARK    AR1, # -ADDR        values

LOOP-2  ZALS    *+, AR1, Ø       Update, RØ through R4
        ADDH    *-, AR1, Ø          values
        MPY     *+, AR1, Ø
        APAC
        SACL    *+, AR1, Ø
        SACH    *+, AR1, Ø
        ZALS    COR-COUNT, Ø
        SUB     ONE
```

APPENDIX V (continued)

```
        SACL    COR-COUNT, Ø
        BNEZ    LOOP-2
        LARK    ARØ, #64-ADDR    Reorder values in the residual FIFO
        LARK    AR1, #ORDER

Loop-3  DMDV    *-, ARØ, Ø
        MAR     *-, AR1
        ZALS    LOOP-3
        SUB     ONE, Ø           Update loop counter
        SACL    COUNT, Ø
        BNEZ    LOOP-1           REDO LOOP
```

21. An Adaptive Predictive Coder Using a Single Chip DSP

APPENDIX VI

LEROUX-GUEGUEN RECURSION FOR COMPUTING LPC PARCOR COEFFICIENTS

```
        ***
LPC-INIT    LACK    #4                  Set up pointers to transfer
            SACL    INIT-COUNTER,0      autocorrelation values to init recursion
            LARK    AR0, #ADDR-R0       Point to R[0]
            LARK    AR1, #ADDR-E0       Point to e'[0]
INIT-LP-1   ZALS    *+,AR0              Auto correlations are double word
            SACH    *+,AR1              (use upper word only)
            ZALS    *+,AR0              (skip a word)
            SUB     ONE,0
            SACL    INIT-COUNTER
            BNEZ    INIT-LP-1
            LACK    #3                  Load the other three autocor. values
            SACL    INIT-COUNTER
            LARK    AR0, #ADDR-R1       Aux registers pointed to data
            LARK    AR1, #ADDR-E-1
INIT-LP-2   ZALS    *+,AR0
            MAR     *+,AR1,0            skip word for double worded
            ZALS    *+,AR0              autocorrelation
            SACL    INIT-COUNTER,0
            BNEZ    INIT-LP-2
INIT-PTRS   LARK    AR0, #ADDR E-ARRAY      Init pointer to e(n)[-r+n+2]
            SAR     AR0, E-ARRAY-START
            LARK    AR0, #ADDR-k0
            SAR     AR0, K-PTR
            LACK    #7                  Init number of iterations
            SACL    NUM-ITERATIONS, 0
            LARK    AR0, #ADDR-EN0      Init pointer to e(n)[0]
            SAR     AR0, EN0-PTR
            LARK    AR0, #ADDR-EN1      Init pointer to e(n)[n+1]
            SAR     AR0, EN1-PTR
LG-REC      LAR     AR0, EN1-PTR
            ZALS    *+,AR0
            SACL    NUMERATOR
            SAR     AR0, ENT-PTR
            LAR     AR0, FN0-PTR
            ZALS    *+,AR0
            SACL    DENOMINATOR
            CALA    DIVIDE


            ZALS    QUOTIENT
            LAR     AR0,K-PTR
            SACL    *,AR0,0
            LT      *+,AR0

            SAR     AR0, K-PTR          Init pointers for from data
            LARK    AR0, #ADDR-E-END
            LAR     AR1, E-ARRAY-START
            SAR     AR1, CROSS-PTR      Init ptr to put cway data
            LARK    AR1, #ADDR-SCR-END
            SAR     AR1, TO-PTR
            ZALS    NUM-ITERATIONS
            BEZ     DONE
            SACL    LG-COUNTER
LG LOOP     ZALS    *+,AR0,0            e(n)[i]    Accumulator
            LAR     ARI, CROSS-PTR
            MPY     *+, ARI             k[n]e(n)[n+1-1]
            SPAC
            SAR     ARI, TO-PTR
            LAR     AR1, ø
            SACL    AR1, TO-PTR
            ZALS    LG-COUNTER
            SUB     ONE
            SACL    LG-COUNTER,0
            BNEZ    LG-LOOP
            ZALS    NUM-ITERATIONS
            SACL    LG-COUNTER, 0
            LARK    AR0, #ADDR-E-END
TERLP       LARK    *+, AR1, 0
            BALS    *+, AR1, 0
            ZALS    LG-COUNTER
            SLB     ONE
            SACL    LG-COUNTER, ø
            BNEZ    TFR-LP
            ZALS    NUM-ITERATIONS
            SUB     ONE, ø
            SACL    NUM-ITERATIONS, ø
            LAR     AR0, E-ARRAY-START
            MAR     *+, AR0
            SAR     AR0, E-ARRAY-START
            B       LG-REC
```

APPENDIX VII

PREDICTIVE QUANTIZER LOOP – VERSION I

```
*   Assumes speech data as well as the pitch predictor state space
*   is stored in external RAM. The pitch predictor state space is kept
*   in a circular buffer which is accessed using TBLR and TBLW
*   instructions

INIT      ZALS   N, 0                   Initialize Loop Counter
          SACL   COUNTER, 0
          ZALS   #S-ADDR, 0             Initialize Pointer to incoming speech
          SACL   S-PTR

LOOP-1    LARK   AR0, #A-ADDR           AR0 points to predictor coefficients
          LARK   AR1, # E1-ADDR         AR1 points to spectral filter
                                         state space
          ZAC
          LT     *+, AR1                Compute spectral predictor
          MPY    *+, AR0

LOOP-2    LTD    *+, AR1
          MPY    *+, AR0
          BANZ   LOOP-2
          APAC
          SACH   Y, 0                   Y contains Spectral Prediction
          ZAC                           COMPUTE PITCH PREDICTION
          LT     ALPHA
          MPY    RT
          APAC
          SACH   X, 0                   X CONTAINS PITCH PREDICTION
          ZALS   S-PTR, 0
          TBLR
          ADD    ONE, 0
          SACL   S-PTR, 0
          ZALS   S                      Subtract two predictions from
          SUB    X, 0                   input speech
          SUB    Y, 0
          BGEZ   DIFF-POS               Quantize residual, if neg.
          ZAC                           transmit zero
          SACL   D, 0                   Scale variance of quantized
          LT     MINUS-1                residual
          MPYK
          APAC
          B      UPDATE

DIFF-POS  LACK   ONE                    If residual is positive, transmit
          SACL   D, 0                   one
          ZALS   Q, 0

UPDATE    ADD    Y, 0                   Compute spectral prediction filter
          SACH   X, 0                   state variable
          SACH   R, 0                   Compute Pitch Predictor State
                                         variable and store it

          ZALS   R-OUT-PTR, 0           Compute pointer into circular buffer
          ADD    ONE, 0                 for storing data
          SACL   R-OUT-PTR, 0
          SUB    R-BOTTOM, 0
          BLZ    OUTPUT-R               If at end of buffer, point
          LACK   #ADDR-RBUF             back to beginning
          SACL   R-OUT-PTR, 0

OUTPUT-R  ZALS   R-OUT-PTR, 0
          TBLW   R

          ZALS   R-IN-PTR, 0            Compute pointer into circular
          ADD    ONE, 0                 buffer for retrieving data
          SACL   R-IN-PTR, 0
          SUB    R-BOTTOM, 0
          BLZ    INPUT-R
          LACK   #ADDR-RBUF
          SACL   R-IN-PTR, 0

INPUT-R   ZALS   R-IN-PTR, 0
          TBLR   RT

          ZALS   COUNTER, 0
          SUB    ONE
          SACH   COUNTER, 0
          BNEZ   LOOP-1
```

21. An Adaptive Predictive Coder Using a Single Chip DSP

APPENDIX VIII

PREDICTIVE QUANTIZER LOOP - VERSION II

```
*   Assumes input speech and reconstructed speech data to be stored
*   in external RAM. The input speech is accessed using TBLR. The
*   reconstructed speech is accessed using the I/O ports and the exter-
*   nal memory interface. The serial quantized residual signal is packed
*   into 16-bit words.

INIT    ZALS    N,0
        SACL    COUNTER,0           Initialize loop counter
        LACK    #S-ADDR,0
        SACL    S-PTR,0             Initialize Pointer to input speech
        LACK    #D-ADDR                 data
        SACL    D-OUT-PTR,0
        ZALS    THREE
        SACL    ACCESS-CTR-WD,0     Initialize external memory
        CALL    EXT-MEM-INIT            interface for read/write mode
        LACK    #16
        SACL    COUNTER,0
        ZAC     BYTE
        SACL    RESIDUAL-BYTE,0

Loop-1  LARK    AR0,#A-ADDR
        LARK    AR1,#E1-ADDR
        ZAC
        LT      *-,AR1              Compute spectral prediction
        MPY     *-,AR0

Loop-2  LTD     *-,AR1
        MPY     *-,AR0
        BANZ    LOOP-2
        APAC
        SACH    Y,0                 Compute pitch prediction
        ZAC
        LT      ALPHA
        MPY     RT
        APAC
        SACH    X,0
        ZALS    S-PTR,0
        TBLR    S-PTR               Read input speech sample
        ADD     ONE,0
        SACL    S-PTR
        ZALS    S
        SUB     X,0
        SUB     Y,0                 Compute residual and quantize
        BGEZ    DIFF-POS
```

```
        ZAC     D,0
        SACL    Q
        LT      MINUS 1
        MPYK
        APAC    UPDATE
        B
DIFF-POS LACK   #ONE                If residual is positive transmit one
        SACL    D,0
        ZALS    Q,0
UPDATE  ADD     Y,0                 Combine quantized first residual
        SACH    E1,0
        ADD     X,0                 Compute reconstructed speech and
        OUT     R,RAM-PORT-1            store it
        IN      RT,RAM-PORT-2
        ZALS    BYTE-COUNTER,0
        SUB     ONE,0
        SACL    BYTE-COUNTER
        BLEZ    NEW-OUT-BYTE
        ZALS    RESIDUAL-BYTE,1
        ADD     D,0
        B       GO-ON
NEW     ZALS    D-OUT-PTR,0
        TBLW    RESIDUAL-BYTE       Pack residual bit stream
        ADD     ONE,0
        SACL    D-OUT-PTR,0
        ZALS    D,0
        SACL    RESIDUAL-BYTE
        LACK    #16
        SACL    BYTE-COUNTER,0
GO-ON   ZALS    COUNTER,0
        SUB     ONE,0
        SACL    COUNTER,0
        BGEZ    LOOP-1
```

If quantized residual is negative transmit zero

RECEIVER LOOP

```
* Assumes residual input is packed in 16-bit words in external RAM.
* It is accessed using TBLR. Synthesized speech is stored in a circular
* buffer in external RAM. This buffer is accessed via the external I/O
* Ports.

INIT   LACK  #N                      Initialize a loop counter
       SACL  COUNTER,Ø
       LACK  #D-IN-ADDR              Initialize pointer to input residual
       SACL  D-IN-PTR,Ø
       ZALS  FOUR                    Initialize external memory interface
       SACL  ACCESS-CTR-WD,Ø            for Read/Write mode
       CALL  EXT-MEM-INIT
       ZAC                           Initialize counter for parallel
       SACL  BYTE-COUNTER               to serial conversion of input residual

LOOP-1 ZALS  BYTE-COUNTER            Obtain quantized input residual
       BNEZ  SHIFT                      from 16-bit residual word
       ZALS  D-IN-PTR
       TBLR  RESIDUAL-BYTE,Ø
       ADD   ONE,Ø
       SACL  D-IN-PTR
       LACK  #16
       SACL  BYTE-COUNTER,Ø

SHIFT  ZALS  RESIDUAL-BYTE,1         Current residual value is high order
       SACH  RESIDUAL-BYTE,Ø            bit of residual-byte
       ZALS  P,Ø
       SUB   ONE,Ø
       SACL  BYTE-COUNTER,Ø

       LT    Q                       Scale variance of residual
       MPY   D
       ZAC
       APAC
       ADD   Y,Ø                     Add in spectral prediction
       SACH  E,Ø                        put result on spectral pred.
       ADD   X,Ø                        filter state space
       SACH  S-HAT,Ø                  Store reconstructed speech
       OUT   S-HAT, RAM-PORT-1          in circular buffer
       LARK  ARØ,#A-ADDR             Compute next spectral prediction
       LARK  AR1, #E-ADDR               value
       LT    *-,AR1
       MPY   *-,ARØ
       ZAC
```

```
       LTD   *-,ARI
       MPY   *-,ARØ
       BNAC  LOOP-2
       SACH  Y,Ø                     Compute next pitch prediction

       IN    ST-HAT
       LT    ALPHA
       ZAC
       APAC  ST-HAT
       SACH  X,Ø
       ZALS  COUNTER,Ø
       SUB   ONE,Ø
       SACL  COUNTER,Ø
       BNEZ  LOOP-1
```

21. An Adaptive Predictive Coder Using a Single Chip DSP

```
                    ADDA A/D-D/A SERVICE ROUTINE INVOKED BY

                         INTERRUPT FROM A/D CLOCK

*  A/D portion
*
*

AD      IN      SN, ADC         Input speech from ADC Register
        ZALS    SN, Ø           Pre-emphasis
        LT      OLDSN
        MPYK    PRE-FAC
        SPAC
        SACL    TEMP,Ø          Store preemphasized speech temp.
        ZALS    S-IN-PTR,Ø      Load pointer to input speech buffer
        TBLW    TEMP            Write out preemphasized speech in buffer
        ADD     ONE,Ø               increment pointer
        SACL    S-IN-PTR,Ø
        ZALS    SN,Ø            delay s[n]
        SACL    OLDSN

*  D/A Portion
*
*
        ZALS    S-OUT-PTR-1,Ø   Retrieve pntr to output speech buffer
        TBLR    YN              Real in processed speech sample
        ADD     ONE,Ø           Increment pointer
        SACL    S-OUT-PTR-1,Ø   Re-store pointer
        ZALS    YN              Do De-emphasis
        LT      OLD SHATN
        MPYK    PRE-FAC
        APAC
        SACL    OLD SHATN,Ø     Delay output speech sample
        OUT     OLD-SHATN,DAC   Output speech sample

*  Check for end of buffer.  If the end, switch speech buffers.
*  This is done by switching pointers.

        ZALS    S-PTR-1,Ø
        SUB     S-OUT-END,Ø     Check for end of Data
        BGZ     DONE
        ZALS    S-OUT-PTR-1,Ø   Toggle bit 8, switching from page
        XOR     H'100               5 to 6 (and vice versa)
        SACL    S-OUT-PTR-1,Ø
        ZALS    S-OUT-PTR-2,Ø
        XOR     H'100
        SACL    S-AT-PTR-2

DONE    RET                     Return from interrupt
```

22. Firmware-Programmable μC Aids Speech Recognition

Tom Schalk and Mike McMahan,
Central Research Laboratories
Texas Instruments

Article reprinted with permission from Electronic Design, Vol. 30,
No. 15, July 22, 1982; copyright Hayden Publishing Co., Inc., 1982.

Recognizing speech patterns is a complex process. However, a firmware-programmable microcomputer with special algorithms forms the foundation for a simple but accurate speech-recognition system.

Firmware-programmable µC aids speech recognition

This article is the third in a series on signal processing with a fast single-chip microcomputer. The series began with an interview with Kevin McDonough, programmable-products design manager at TI, in the May 27 issue (p. 42). The first article, in the same issue (p. 129), described how to implement a vocoder with the chip; the second, in the June 24 issue (p. 109), applied the chip to implementing high-speed modems digitally.

Microprocessor-based VLSI systems can easily provide effective speech analysis and synthesis, but speech recognition is still a very formidable task. A general-purpose, high-performance speech-recognition system requires a computer system of substantial power. Some low-cost, low-performance systems have been built, but they have proved too specialized and too inflexible to adapt to any reasonably wide range of applications (see "Key Features of Speech Recognizers"). The programmability and algorithms of a digital signal processor, however, offer speech recognition the cost-effective flexibility that this developing field needs.

Currently, speech-recognition algorithms are rather primitive. But with a signal processor like the TMS320 microcomputer, only firmware changes will be required as recognition technology advances. System redefinition, which would necessitate new integrated-circuit designs, will not be necessary.

Basically, speech recognition involves four steps: feature extraction, pattern-similarity measurement, time registration, and a decision strategy. An additional task, called enrollment, creates speaker-dependent reference patterns, which key the recognition to a certain speaker.

Feature extraction transforms speech signals into time-varying parameters (or features) that not only can be recognized by the system but also can reduce the amount of data in normal speech to something more manageable. Of all the many available feature-extraction methods, linear predictive coding (LPC) is usually the most effective.

A 320 programmed as a feature extractor based on a tenth-order LPC model (which runs at an 8-kHz sample rate) operates on a 30-ms Hamming window with a frame period of 20 ms, which is short enough to capture most dynamic-speech events adequately. However, extracting the compressed-speech features from the input signal and encoding them into LPC-10 parameters are not simple. Nevertheless, a modified Le Roux-Gueguen algorithm can be readily implemented in just part of a 320 to take care of this speech-recognition step.

The next step—computing the similarity, or dis-

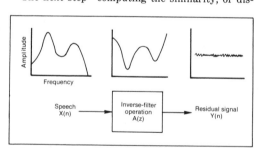

1. Input speech utterances, after conversion into LPC-10 form, are compared for similarity with stored templates of speech patterns via a digital inverse-filter operation. Evaluation of the residual output signal from the filter determines whether a match (recognition) has been obtained.

Tom Schalk, Member, Technical Staff
Michael McMahan, Member, Technical Staff
Texas Instruments Inc.
13510 N. Central Expwy., Dallas, Texas 75266

tance, measured between the extracted speech parameters and stored reference patterns—is basically a frame-by-frame comparison of speech data with reference data, or templates. This operation can be looked upon as passing the input signal for the current frame through an all-zero inverse filter that represents the reference data. When the reference data match the input data—the spectral valleys in the inverse match the spectral peaks in the input signal—a low-energy residual output will be passed through (Fig. 1). If the inverse filter matches the input data perfectly, the residual-error energy will be normalized at a value of 1. Typical normalized values of error, to be within recognizable limits, must be less than 1.2 in speaker-dependent systems within a 1.05 to 1.8 range.

More precisely, however, the parameters that measure distance between input samples and stored-template reference patterns are computed by autocorrelation and residual-energy algorithms (Fig. 2). Autocorrelation is a fairly simple software procedure, since it consists mostly of a long sequence of register loadings and multiplication operations. The residual is thus easy to compute: multiplying respective terms of the two sequences—the inner (dot) product of the autocorrelation functions of the input speech parameters and the impulse response of an inverse filter—and summing them.

The dynamic-programming function then scans for minimum-distance matches of the spoken-word parameters among all the stored templates to find suitable word-recognition candidates. The can-didates are passed to a high-level decision-logic operation, which applies a threshold-comparison test and a next-closest-error threshold test to determine which word or words are to be selected.

Compensating for timing length

One important aspect of the dynamic-programming operation is a procedure that compensates for variations in the length and timing of the input utterances, but which does not change the meaning of the inputs—performing a so-called time warping procedure of the input parameters. A straightforward approach would attempt to derive the reference-data and the speech-signal parameters from frames of the same period (20 ms in the 320's LPC-10), and make a comparison for every frame.

However, for speech recognition, comparing at every second input frame is adequate and leads to two important advantages: the amount of reference data that must be stored and the number of dynamic programming computations that must be performed are cut in half, and the dynamic-programming computation is simplified by eliminating a memory element that would be required when comparing every frame (Fig. 3).

On the other hand, although the time-warping approach can accept speech inputs with a varying time factor of 2 to 1, the system prefers that utterances be spoken at a more or less constant average rate for greatest accuracy. In addition, the time-warping routine need not start and finish on specific speech-input frames. Although the process-

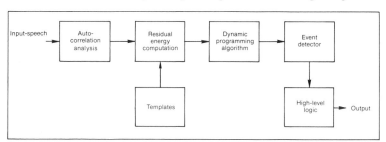

2. The algorithm flow diagram represents the sequence of operations required to implement the inverse-filter comparison operations—a very complex procedure.

Advances in the capabilities and performance of speech-recognition equipment center on three major features:
- Connected-speech vs single-word recognition.
- Speaker-independence.
- Vocabulary size.

Most current speech recognizers depend on a small period of silence between words—typically 200 ms—to determine word end points. Such speech recognizers are discrete, single-word types. Connected-speech systems, on the other hand, can recognize words without explicit knowledge of the end-points. Such systems are more complex and expensive, but they do perform far better than discrete word systems and can accommodate much wider timing variations of the words.

Often, a relatively simple speaker-dependent recognition system can offer reasonable performance for a limited range of speakers other than the one for which the system is tuned. But to handle a wider range of speakers, the speech recognizer would have to contain enough reference patterns for each vocabulary word to be representative for the expected speaker population.

Accordingly, speaker-independent recognition systems should be characterized by population-performance statistics—able to provide less than a certain percentage of the population. The better the system, the smaller the error and the larger the population. Despite the theoretical superiority that speaker-independent systems should offer, of the currently available systems, the speaker-dependent types perform better than the speaker-independent systems.

For now, vocabulary size is limited mainly by a system's ability to keep up with the rate of incoming speech data. The system's computer processing time is spent mostly in making speech-input to reference-data comparisons. The time this process consumes is linearly proportional to the size of the vocabulary. In addition, the error rate usually increases with the size of the vocabulary.

For that matter, the similarity of various words to one another in the composition of the vocabulary generates errors. Thus, the larger the vocabulary, the greater the potential for this type of error. Even a two-word vocabulary might be trouble, if the words happen to be similar sounding like "seen" and "seem." In general, better performance is obtained with vocabularies consisting of long multisyllabic words.

ing time then increases substantially, the unconstrained end points offer two major benefits: system reliability improves greatly, and words embedded in adjacent or overlapping utterances can still be recognized.

For 98%-accurate recognition or better with fixed end points, the reliability of the end-point determination must be at least 98% accurate. This expectation is unrealistic, since end-point information is not normally specific to words as normally spoken. However, with an unconstrained end-point dynamic-programming algorithm, the system can recognize vocabulary words even when not discretely separated in time.

But because of wide variations among different speakers, today's speech-recognition devices must be tuned, or enrolled, to accept speech from only one (or a few) users. (Recognizing only a specific individual may even be a desired attribute.)

Matching speaker to machine

Enrollment creates a set of feature vectors for each vocabulary word in the system's repertoire. The vectors, which are included in the similarity measurement as part of the recognition process, define the spectral shape of the reference pattern for individual vocabulary words.

Before enrollment, the initial end points of the reference patterns are established by energy measures. Actual start and stop times include some leeway—one frame, typically—to allow more reliable end-point detection. This data becomes part of the stored speech autocorrelation coefficients for each frame that determine the inverse filters.

After the initial template end points are formed, samples from the speaker update, the original templates. For each training input, the utterance is compared with the stored reference data for that word. If the utterance is recognized—if the detected error is under a certain threshold—the template will be updated by averaging the appropriate input-speech autocorrelation coefficients with those previously stored as the reference. If the utterance is not recognized, then it must be reprompted, as many times as necessary.

Training applies both to the individual- and multiple-word utterances of the overall vocabulary. Successive repetition of the training procedure leads to significant performance improvement, and the amount of updating, or substitution, decreases substantially with each additional training session. After five training inputs, for example, substitutions will decrease to about one-third that of the first training pass.

To implement all those speech-recognition and training algorithms, more memory than the 320 possesses is needed. In addition, analog I/O is needed and interface circuits must be provided to a host computer (Fig. 4). A standard peripheral speech-processor (PSP) board can be used as such a speech recognizer, at least to demonstrate the system's feasibility. All the signal processing on the board can be provided by a 320, whose features include a 200-ns instruction cycle time, a 16-by-16-bit multiplier-accumulator (400 ns), 1500 words of program ROM, 144 words of data RAM, and 16-bit parallel I/O.

Speech-recognition hardware on a board

A speech-recognition system implemented with a PSP board can recognize both isolated words and connected speech. Its maximum capacity of 32 seconds of vocabulary represents about 40 utterances. Any one vocabulary item can take a maximum of 3 seconds, and a single utterance can contain up to 21 connected items.

The 320 communicates with the rest of the system via its 16-bit bidirectional parallel data bus in either of two modes: polled I/O or a four-level priority interrupt. Eight external status conditions multiplex into the 320, and these can be polled to control the chip's program flow. For example, the status of the input-data FIFO buffer in the analog I/O can be tested to determine whether all pending data has been used. Or, the 320's interrupt input can receive a high-priority command from the external con-

troller or a notification of a FIFO-full or empty condition from the analog I/O subsystem.

The next major PSP function is supplied by its analog I/O subsystem, which provides completely independent input and output channels. Each contains a data converter—an analog-to-digital and a digital-to-analog unit, respectively, with programmable sample rates of 6.25 to 12.5 kHz—a low-pass filter (programmable 3.125 to 6.25 kHz), and an eight-word FIFO buffer. (FIFO-full and FIFO-empty status bits notify the 320 processor when the analog channels require servicing.)

Finally, a 16-by-16-kword RAM subsystem handles the memory requirements that overrun the 320's internal memory. (By incorporating 1.5 kbytes of ROM, a 320 can perform as a true microcomputer by itself. However, it can also execute programs stored in an external memory. This feature is useful in general for program development or for supporting applications that require more memory than is available in the on-chip memory.)

This additional memory is necessary in the PSP application to store the vocabulary templates. Since the 320 cannot directly address such a large external memory, an external address register that can be automatically incremented or decremented following each read or write operation is also provided on the board.

The PSP operates as a memory-mapped peripheral

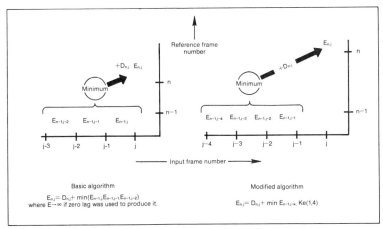

Basic algorithm

$$E_{n,j} = D_{n,j} + \min(E_{n-1,j}, E_{n-1,j-1}, E_{n-1,j-2})$$
where $E \rightarrow \infty$ if zero lag was used to produce it.

Modified algorithm

$$E_{n,j} = D_{n,j} + \min E_{n-1,j-k}, Ke(1,4)$$

3. The dynamic-programming algorithm is simplified in the modified version because it is carried out for every other input frame, instead of for every input frame. For both algorithms, $D_{n,j}$ is the distance measure between reference frame n and input frame j; $E_{n,j}$ is the minimum subsequence distance up through reference frame n, given that input frame j corresponds to reference frame n.

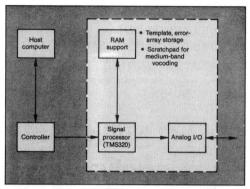

4. To make even a minimal working speech-recognition system, the TMS320 needs the support of more memory and an analog I/O circuit. Mounted on a standard circuit board with the 320, such support must be coordinated by an external host computer and controller for speech recognition. However, the board, or peripheral speech processor, is not limited to speech recognition. A generalized configuration, it can be used for many other signal-processing applications.

of the host computer via an external controller. Four registers support controller-to-PSP communications, as follows,

■ An 8-bit command register, when loaded from the external speech-recognition-system controller, generates an interrupt to the 320 microcomputer.

■ A 16-bit input-data register passes data from the controller to the PSP.

■ An 8-bit status register, an exact analogy of the command register, forces an interrupt of the external controller when the 320 commands it.

■ A 16-bit output-data register passes data from the PSP back to the controller.

The interface between the controller and the PSP generally supports just the relatively low-bandwidth communications, as required by the command and control signals. However, the interface can support data transfers at rates from 10^5 to 1.5×10^6 words per second, depending on the complexity of the interrupt software.☐

Bibliography

1. Gibbs, J., "Windowing Boosts Performance of Dynamic Signal Analyzers," *EDN*, Aug. 5, 1981, p. 109.

2. Leroux, J., and G. Gueguen, "A Fixed Point Computation of Partial Correlation Coefficients," *IEEE Transactions on ASSP*, June, 1977, p. 257.

3. Wiggins, R. H. and G.R. Doddington, "Speech Recognition Spurred by Speech-Synthesis Success," ELECTRONIC DESIGN, July 9, 1981, p. 107.

22. Firmware-Programmable μC Aids Speech Recognition

23. A Graphics Implementation Using the TMS32020 and TMS34061

Jay Reimer and Charles Crowell
Digital Signal Processing - Semiconductor Group
Texas Instruments

INTRODUCTION

Graphics systems are now commonplace with the proliferation of personal and professional computers, particularly in the engineering and business workplace. Workstations and other graphics display stations, which in the past were too expensive to be widely available, are now cost-effective in a larger variety of applications. As the technology has advanced into more areas, the increasing requirement for graphics resolutions has demanded proportionally greater computational and data throughput capabilities. Since graphics is at least two-dimensional, a doubling of the display resolution in two dimensions (X and Y) requires a quadrupling of memory, throughput, and computational resources.

A graphics system, in general, consists of a display monitor, a display controller, a list memory, and an image or main control processor that modifies the list memory. Each basic element in the system represents an item of choice, which impacts the overall performance of the system.

Increasing graphics resolution requires advances in display and display memory control technologies and in processors that can provide the increasing throughput and computational demands of image generation and manipulation. The TMS32020 Digital Signal Processor (DSP)[1-3] is capable of meeting these demands. Its ability to perform 16 x 16-bit multiplies yielding a 32-bit result in a single instruction cycle is significant for graphics, which includes matrix operations for translation, rotation, and scaling.

In this application report, an example of a graphics display subsystem based on a raster-scan display is described. The system consists of the TMS32020 functioning as the image control processor, dynamic video memories, and a display/memory controller. A proposed system configuration is presented and discussed from both hardware and software viewpoints, including some of the necessary tradeoffs and some available options. This report also includes a working schematic and the software used to test the design.

OVERVIEW

System designers for graphics display systems have been faced with classic tradeoffs of cost versus quality. For higher-end systems where crisp images of many line segments are necessary, the decision has frequently led to vector-scan display systems. This has been especially true in such areas as computer-aided design (CAD) systems, which must support high-quality, wire-frame-based images and dynamic operations. At the other end of the spectrum, raster-scan graphics systems have been able to take advantage of the advances and popularity of television-type display systems. This has reduced the cost to affordable levels, but has had a pronounced effect on the display quality.

A raster graphics system is basically a pipeline, such that a bottleneck anywhere in the pipe drastically reduces performance and inhibits other hardware in the system from operating at peak efficiency. Upgrades with these systems have gone from text display drivers to bit-mapped graphics systems with limited input/setup access. The present state of bit-mapped graphics is limited by memory access times and processing capacity.

In the past, computer systems utilizing a raster graphics display might have relied on the host processor in the system to update the bit-mapped pixel display. The bit-mapped display represents the screen image by a series of memory bits read and displayed on the monitor by the raster-scan display system. Since memory access was limited, two display processors were often used: one to drive the display from a bit-mapped memory and the other to drive the display from a text-code memory. By allowing the text to be displayed by a separate processor, the host could avoid the bottleneck of memory access by simply supplying the ASCII code of the characters to be displayed rather than the actual bit pattern of the character itself. This improved the total throughput on a bit-rate basis, but limited the flexibility in the display image. Additionally, the display was limited in the extent of bit-mapped graphic interaction due to the long setup access time required.

The new generation of video controllers,[4-6] together with the multiported video memories,[7-9] permits a unified text and graphics display system with unlimited access to the display memories. Figure 1 shows a system-level block diagram of a unified text and graphics display system. The memory access changes the location of the bottleneck in the system away from the video display system itself and back to the host processor or central processing unit (CPU) supplying the information to be displayed. Since the memory access limitation has been eliminated, the system is now limited more by the CPU. Dynamic modification of high-quality, wire-frame-based displays and other applications are now feasible with a raster-scan display system. This dynamic modification involves updating the pixels in the bit-mapped display memory using seed fills and multiple line draws. The computational load required for object rotation, scaling, translation, and other numeric-intensive calculations to determine the orientation, size, and position of the object is also involved. Even with coprocessors, some graphics applications seem slow to the user since the coprocessor cannot compute the numeric-intensive algorithms fast enough to simulate a realistic response. Therefore, a need exists for faster processors, such as the TMS32020, if realistic responses are desired.

The TMS32020 Digital Signal Processor (DSP) is capable of multiplying a 16-bit number by another 16-bit number and producing the 32-bit result in a single instruction cycle of 200 ns. This feature makes the TMS32020 proficient at graphics operations, such as translation, scaling, and

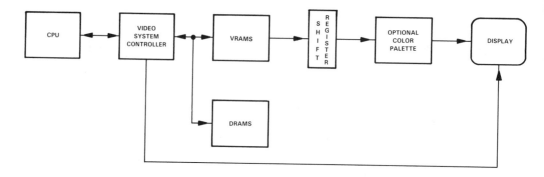

Figure 1. Unified Text and Graphics Display System

rotation. Furthermore, a TMS32020-based graphics system has a high I/O throughput and is easy to interface to any host processor. Additional features, such as a large address reach (common with more generic microprocessors), separate program and data memory spaces (64K words each), multiprocessor interface capabilities, an architecture designed with auxiliary registers to aid with indirect addressing, and a powerful, easy-to-program instruction set, make the TMS32020 an attractive master processor for a graphics display subsystem. Figure 2 shows a block diagram of a graphics display subsystem, which utilizes a TMS32020 as the graphics "engine".

This application report provides an example of a graphics display subsystem using the TMS32020 (DSP), the TMS34061 Video System Controller (VSC), and TMS4161 dynamic video memories (VRAMs). The proposed system serves as the basis for discussions on both hardware and

software and insights into several tradeoffs that a designer must consider. By building the hardware and programming some basic graphics software functions, as shown in the appendices, the design has been taken from a mere paper design through the first several steps toward a practical application. Design flexibility of the TMS32020 is thereby demonstrated as well as the efficiency of the TMS32020 in some of the numeric-intensive tasks required for graphics applications.

The benefits of the proposed system are inherent in the design itself. Basic features of the TMS32020, TMS34061, and TMS4161s simplify the design since the devices are easily interfaced with one another and with other parts of a system. After the hardware is designed, it remains flexible because a large portion of the design is user-definable due to the programmability of the TMS32020 and the TMS34061. Application reports,[10-12] including discussions on

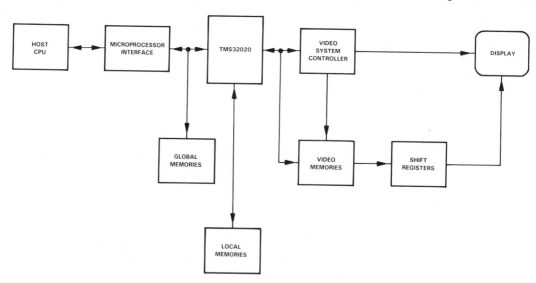

Figure 2. TMS32020-Based Graphics Display Subsystem

programming the matrix operations[10] required for graphics applications for the TMS32020, are useful references.

SYSTEM CONFIGURATION

A graphics system implemented using the TMS32020 consists of a host processor, the TMS32020 as the graphics processor, the TMS34061 Video System Controller (VSC), the multiported video RAMs (VRAMs), and various random logic. The host processor interface may be designed in various ways to accommodate the host itself and the integration of various processing units in the system. A wide variety of host processors used in graphics systems may actually be used. Since the potential is largely user preference, it is not discussed further in this application report. More information and an example of a TMS32020 host interface can be found in the application report, "TMS32020 and MC68000 Interface."[11]

The graphics system discussed here uses the TMS34061 VSC and TMS4161 VRAMs to drive a display monitor with 720 x 300 display-bit resolution. In this system, the VSC supplies all of the necessary $\overline{RAS}$ and $\overline{CAS}$ signals to the VRAMs and provides the necessary synchronization and blanking signals to the display monitor. The VSC expands the addressing range of the TMS32020 to the VRAMs' 20-bit address, thus allowing the TMS32020 to access the large address space required for graphics applications. The TMS4161 VRAMs are 150-ns dynamic RAMs with dual-port access, designed specifically for graphics applications.

From the standpoint of the TMS32020, the system has been designed around a VSC Evaluation Module (EVM) board. This board is characterized by an eight-bit data bus and a memory array that can support display resolutions up to 1024 x 512 pixels with 4 bits per pixel.

With this design, the TMS32020 can provide an efficient implementation of the numeric- and pixel-processing capabilities required. The VSC supplies the necessary memory and display control for the system. Figure 3 shows a block diagram of a graphics system configuration. The detailed schematic of the circuit actually built is found in Appendix A.

TMS32020 Digital Signal Processor

The TMS32020 Digital Signal Processor is a high-speed processor well designed for computation-intensive applications such as graphics. Its 200-ns instruction cycle time, modified Harvard architecture, hardware multiplier, multiprocessor control features, and powerful instruction set make it particularly efficient in many applications. Some of the additional features of the TMS32020 (see Figure 4) are a 32-bit accumulator, parallel shifters, on-chip data RAM, and an on-chip timer.

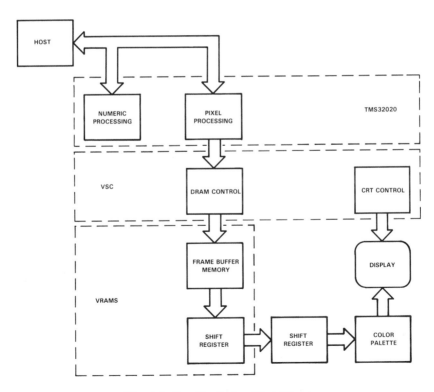

Figure 3. Graphics System Block Diagram

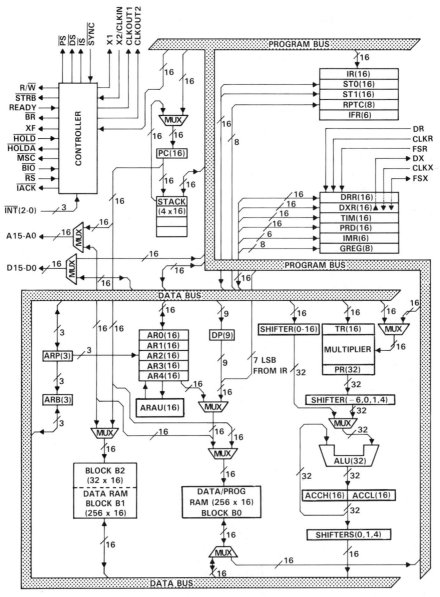

LEGEND:

ACCH	= Accumulator high	**DRR**	= Serial port data receive register	**IR**	= Instruction register
ACCL	= Accumulator low	**DXR**	= Serial port data transmit register	**PR**	= Product register
ARAU	= Auxiliary register arithmetic unit	**GREG**	= Global memory allocation register	**PRD**	= Period register for timer
ARB	= Auxiliary register pointer buffer	**IFR**	= Interrupt flag register	**TIM**	= Timer
ARP	= Auxiliary register pointer	**IMR**	= Interrupt mask register	**TR**	= Temporary register
DP	= Data memory page pointer	**RPTC**	= Repeat instruction counter	**ST0, ST1**	= Status registers

Figure 4. Block Diagram of the TMS32020 Digital Signal Processor

23. A Graphics Implementation Using the TMS32020 and TMS34061

TMS34061 Video System Controller

The TMS34061 Video System Controller, shown in Figure 5, is a high-performance device that controls the video display and the dynamic memories of a bit-mapped graphics system. The VSC relieves the host of memory control, memory refresh, and VRAM shift-register reload functions. It is easily programmed via 18 registers to accommodate a wide variety of display sizes. The VSC also allows the host to access the memory either directly or indirectly through X-Y address registers. When the indirect access method is used, the VSC further assists the host by allowing on-the-fly address adjustments (increment, decrement, clear, and no change). By using the VSC in conjunction with the multiport VRAMs, memory access time by the host is virtually unlimited (better than 94 percent).

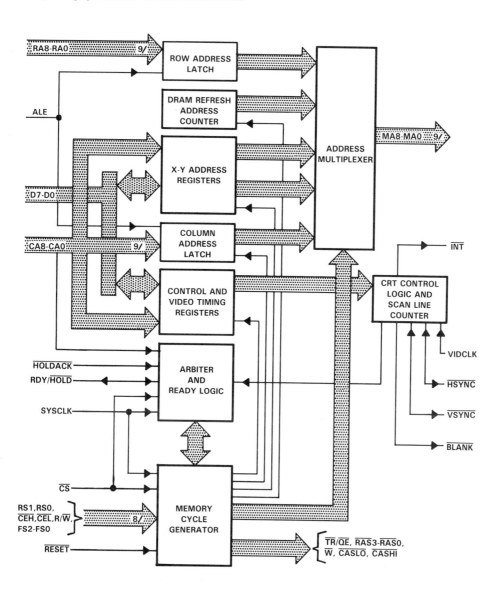

Figure 5. Block Diagram of the TMS34061 Video System Controller

TMS4161 Dynamic Video RAM

The TMS4161 multiport video RAM is a 64K x 1 memory array with a 256-bit shift register. The shift register is divided into four cascaded 64-bit shift register segments, allowing data access in a serial mode to be tapped at four different lengths. A functional block diagram of a TMS4161 is shown in Figure 6. Data may be shifted into or out of the internal shift register while normal random accesses are taking place with the memory array. In a graphics system, the data is moved by block transfer from the memory array to the shift register and shifted out at the dot/chip rate.

Program and Data Memory Requirements

The TMS32020 provides independent 64K-word address spaces for both program and data memory. The program space may be divided among all memory types as the system designer deems necessary. This enables allocation or utilization of memory resources both from a cost and a size perspective since the memory may be RAMs (both static and dynamic), ROMs, PROMs, EPROMs, or EEROMs of any and/or differing speeds. Slower memories are typically less expensive and can prove to be a significant cost advantage for portions of the system infrequently accessed. Where time is critical, faster memories are required to enhance the system performance. In the design described in this report, the program memory consists of 4K words of bipolar PROM (TBP28S166).

This program memory contains the initialization code, basic display control or function algorithms, a command parsing routine, and a table of commands and required data.

By designing the software to be structured around a table-driven approach, the system can easily be modified to allow the host to supply and update the command and data table contents. Through the use of shared memory or other multiprocessing communication designs, a host processor can be conveniently interfaced to the TMS32020. As an example, the system could consist of initialization and interface procedures and some basic functional routines programmed into PROM and additional program memory space implemented with RAM. The RAM area allows the design to expand in capability by having the host download routines to the TMS32020. The routines can then be executed to enhance the system by building on the basic functions.

The circuit presented utilizes only the data memory internal to the TMS32020 (544 words). For systems designed to maintain large blocks of data, such as the display information for character formation in text displays or tables of display coordinates of a large number of dynamically changing objects, it is necessary to also include RAM and/or ROM in the external data memory space. This memory, along with the internal data memory of the TMS32020, the video system controller, and video memories, must be mapped into the 64K-word data memory space.

Considerations for memory selection and interface requirements have been discussed in significant detail in an application report, ''Hardware Interfacing to the TMS32020.''[12] This report provides a comprehensive discussion on interfacing the TMS32020 to memory devices (PROMs, EPROMs, static RAMs, and dynamic RAMs) and peripherals. It also supplies circuit and timing diagrams for each of the interfaces.

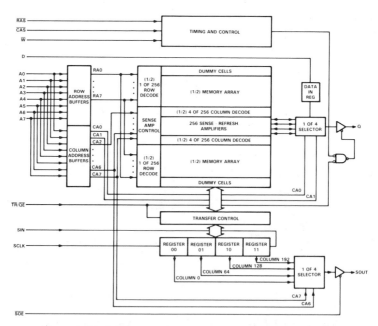

Figure 6. Block Diagram of the TMS4161 Multiport Video RAM

TMS32020 and TMS34061 Interface

Access to the VSC (and the VRAMs) is gained through a data memory-mapped design. The VSC contains a number of internal registers and mode-function selections in addition to video memory access to the video memories, thus providing a homogeneous access methodology for the TMS32020 software design. The VSC is mapped into the lower 32K words of the data memory space, as shown in Figure 7. By paging the VSC at pages 8 through 255, the software differentiates between it and the internal memory of the TMS32020. In addition to the address bus lines and the bidirectional data bus, the TMS32020 supplies the VSC with enable and read/write signals while the VSC in turn supplies a ready or hold signal to the TMS32020.

Figure 8 shows the signal and bus interface between the TMS32020 and the VSC. The design maps the VSC into the lower 32K-word address range of the data memory space by combining the $\overline{\text{DS}}$ signal and A15 address line to gate the $\overline{\text{STRB}}$ signal that allows access to the VSC. This single enable signal is the input to three individual enable inputs (ALE, $\overline{\text{CEH}}$, $\overline{\text{CEL}}$), providing implementation of all modes of VSC access. Since the VRAMs have a 150-ns access time, $\overline{\text{MSC}}$ (from the TMS32020) is used to provide an automatic one-wait state and allow time for the VSC to activate its RDY/HLD line to generate additional wait states to the TMS32020.

On the data bus, the lower eight bits are isolated from the VSC by using an SN74LS245 transceiver to avoid bus conflicts with the VRAMs. The three-state outputs of this buffer are enabled whenever $\overline{\text{STRB}}$ is gated to the VSC. The R/$\overline{\text{W}}$ signal from the TMS32020 is used to determine the direction that the transceiver drives the bus. In this configuration, the upper eight bits of the TMS32020 data bus are not used whenever access is made to the VSC or VRAMs. However, the eight MSBs of the data bus are supplied the low output of an SN74LS244 buffer in such cases. This buffer is normally in a high-impedance state and supplies the low-signal output only when the VSC or VRAMs are being read. The benefit here is that the software of the TMS32020 does not have to mask the upper eight bits whenever the VSC or VRAMs are read. Note that the circuit being presented in this report does not support host-direct access to the VRAMs.

The algorithms that are coded for this design use the X-Y indirect addressing capabilities of the VSC to access each bit location in the VRAMs. Actually, for each address enable to the VRAMs, two pixels (four bits each) are read or written. This approach makes full use of the eight-bit data bus already utilized for VSC register access. However, whenever a single pixel is to be modified, a pixel pair must be read, the selected pixel modified, and then the pixel pair written back to memory. This produces a significant overhead for line-drawing algorithms since a read-modify-write must be performed each time a pixel is changed, rather than a simple write. A possible alternative is to access the video memories on the basis of a single pixel bus. In the case of a four-bit bus, the throughput of a line-drawing algorithm may be nearly doubled since the pixel access is reduced to a simple write. On the other hand, for cases where the

Address	Contents
>0000	TMS32020 INTERNAL REGISTERS
>0005	
>0006	TMS32020 RESERVED
>005F	
>0060	TMS32020 – BLOCK B2 (DATA PAGE 0)
>007F	
>0080	TMS32020 – RESERVED (DATA PAGES 1-3)
>01FF	
>0200	TMS32020 – BLOCK B0 (DATA PAGES 4 AND 5)
>02FF	
>0300	TMS32020 – BLOCK B1 (DATA PAGES 6 AND 7)
>03FF	
>0400	VSC – REGISTERS (DATA PAGES 8-15)
>07FF	
>0800	VSC – X-Y INDIRECT (DATA PAGES 16-31)
>0FFF	
>1000	VSC – REGISTERS (DATA PAGES 32-47)
>17FF	
>1800	VSC – HOST DIRECT (DATA PAGES 48-63)
>1FFF	
>2000	VSC – SHIFT REGISTER (DATA PAGES 64-79)
>27FF	
>2800	VSC – SHIFT REGISTER (DATA PAGES 80-95)
>2FFF	
>3000	VSC – RESERVED (DATA PAGES 96-127)
>3FFF	
>4000	AVAILABLE (EXTERNALLY OVERLAPPING VSC ACCESS)
>7FFF	

Figure 7. TMS32020 Data Memory Map

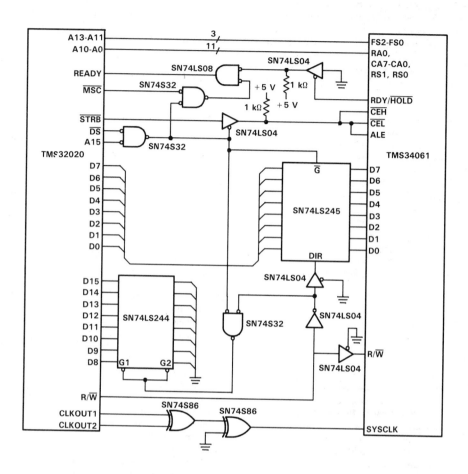

Figure 8. TMS32020 and TMS34061 Interface

algorithms test the pixel value, such as in seed fills, it is advantageous to have the data bus of maximum width (16 bits) and actually read four pixels at a time. This minimizes the number of read accesses per potential test condition, thereby maximizing throughput. It is also beneficial to use a 16-bit data bus for block transfers of data into (or out of) the VRAMs, as in the case when bit-mapped text is to be displayed. For a 16-bit bus, its appropriate use can nearly double the throughput since the number of decisions per pixel group access is maximized.

The consideration involved in choosing among all these cases is the requirement of additional logic to configure the data bus to various widths. The 16-bit data bus, as described

for the bit-mapped text, also benefits from the TMS32020 being able to perform host-direct accesses to the VRAMs. If a programmable-width (4-, 8-, or 16-bit) bus is necessary for speed considerations in the design, then additional logic is required to configure the bus to each of the desired widths.

The VSC needs a 10-MHz system clock, which may be created from the 20-MHz input clock to the TMS32020 in one of two ways. In the current diagram shown in Figure 8, the 10-MHz clock is produced by an exclusive-OR of the two 5-MHz clock output signals (CLKOUT1 and CLKOUT2) of the TMS32020. An alternate approach consists of using a flip-flop to convert the original 20-MHz signal down to the desired 10-MHz clock.

23. A Graphics Implementation Using the TMS32020 and TMS34061

TMS34061 and TMS4161 Interface

The interface between the VRAMs and the VSC is straightforward. Eight address output lines (MA7-MA0) on the VSC are tied to the address lines (A7-A0) of each of the VRAMs. $\overline{RAS0}$-$\overline{RAS3}$ and $\overline{CASHI}$ are control signal outputs of the VSC used to provide the necessary $\overline{RAS}$ and $\overline{CAS}$ signal inputs to each of the VRAMs, thus properly enabling access to the pixels being addressed. $\overline{CASHI}$ is tied commonly to all the $\overline{CAS}$ inputs of the VRAMs, while $\overline{RAS0}$-$\overline{RAS3}$ split the VRAMs into four blocks. The other control lines, $\overline{TR/QE}$ and $\overline{W}$, are tied commonly to all VRAMs, providing signals that enable shift register transfer cycles and write cycles, respectively. Since the VRAMs are each a 64K x 1-bit memory array, an address access of any single pixel (represented by four bits) actually causes four of the VRAMs to be enabled simultaneously.

The shift-register output of the VRAMs is used to provide the RGBI (red, green, blue, and intensity) signals to the display monitor. The overall display memory array in this design is a 1024 x 512-pixel array. Note that the active display area of the total array is a 720 x 300-pixel array, as shown in Figure 9.

Output from each of the VRAMs is shifted in parallel to an eight-bit external shift register for each of the RGBI signals (see Figure 10). More detailed information of the VSC and VRAM interface may be found by consulting the TMS34061 Evaluation Module User's Guide.[13]

Color Palette Option

An option to be considered for a graphics display system is the addition of a color palette. The TMS34070 color palette[14] is designed to generate up to 4096 colors on a single display. It provides a 16-register lookup, expanding the color selection to 4 bits each of RGB in three 4-bit Digital-to-Analog Converters (DAC). The analog output of each DAC is capable of directly driving standard 75-ohm monitor cables. The 16 14-bit registers of the color lookup table are loaded directly from the video memory, eliminating the need for a separate processor interface through which the registers can be loaded. A mode selection allows the color registers to be loaded (1) prior to the start of each individual scan line, (2) prior to the start of each frame, or (3) only on explicit command.

Figure 11 shows the internal architecture of a TMS34070 video palette, and Figure 12 illustrates a typical video memory system that includes the palette.

SOFTWARE DESIGN

In addition to the selection of the hardware to be used, consideration must be given to the design of the software system. When using a raster-scan display system, it is necessary to determine which functions must be included, which would be useful and efficient if included, and which are infrequently used and able to be created from other basic functions by the user. Some functions, such as clearing the screen and drawing lines, are quite obviously necessary.

The TMS32020, together with the VSC, can provide efficient implementations of incremental algorithms utilized in pixel processing. This is possible because of the single-cycle instruction execution of most TMS32020 instructions and the X-Y indirect addressing capabilities of the VSC. In addition, matrix multiplication and other object manipulations are efficiently implemented on the TMS32020 by using the single-cycle hardware multiplier and block processing capabilities. Many other features may need to be evaluated in designing an interactive graphics system (i.e., x-y locators/cursors, output primitives, coordinate

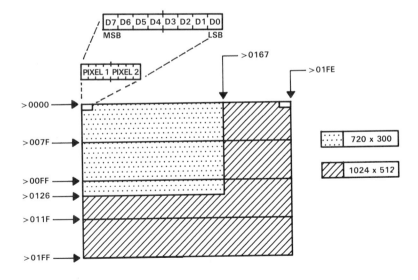

Figure 9. Display Memory Pixel Array

transformations, windows and clipping, zooming, panning, and library functions).[15,16]

Software to run on a TMS32020 for a graphics system incorporating a VSC and VRAMs mapped into the data memory space must be designed with certain considerations. The TMS32020 is designed to access data memory in a direct

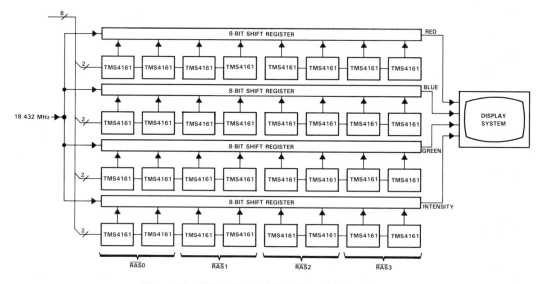

Figure 10. 1024 x 512-Pixel Four-Plane Display Memory

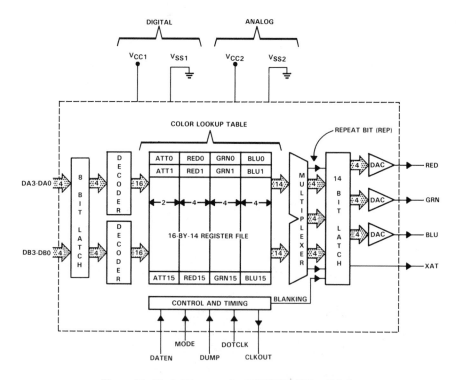

Figure 11. Block Diagram of a TMS34070 Video Palette

23. A Graphics Implementation Using the TMS32020 and TMS34061

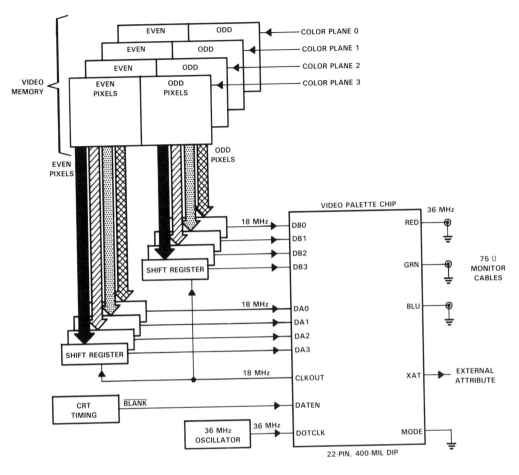

COLOR PLANE 0
COLOR PLANE 1
COLOR PLANE 2
COLOR PLANE 3

EVEN
ODD

VIDEO
MEMORY

EVEN
PIXELS

ODD
PIXELS

ODD
PIXELS

EVEN
PIXELS

VIDEO PALETTE CHIP

36 MHz

18 MHz   DB0   RED
          DB1
          DB2
SHIFT REGISTER   DB3   GRN

75 Ω
MONITOR
CABLES

18 MHz   DA0   BLU
          DA1
          DA2
          DA3

SHIFT REGISTER

18 MHz   CLKOUT   XAT   EXTERNAL
                           ATTRIBUTE

CRT
TIMING   BLANK   DATEN

36 MHz   36 MHz   DOTCLK   MODE
OSCILLATOR

22-PIN, 400-MIL DIP

Figure 12. Video Memory System with a Video Palette Enhancement

memory accessing mode that limits accesses to pages of memory (128 words in length) or in an indirect memory accessing mode through the use of the auxiliary registers. Therefore, accessing a VSC or VRAMs along with the internal data memory space of the TMS32020 requires the simultaneous use of both addressing techniques. (If only direct addressing is used, frequent use of the LDPK instruction is required, resulting in a less efficient coding of the algorithms.) Because of the demand to use both addressing techniques, it is necessary to make the clearest, most efficient use of the auxiliary registers. The VSC and VRAMs are best accessed through a layered software approach with algorithms of basic functionality being called and utilized by routines with a higher level of functionality.

In the following sections, a description of the software design is given. Various considerations with regard to the necessity and level of each functional routine are discussed.

Accessing the VSC

Access to the VSC (and to the VRAMs) is provided through the data memory address space of the TMS32020. The VSC (and consequently the VRAMs) is enabled whenever an external memory access is made with A15 in a low state. The VSC is function-selected by the TMS32020 using a three-bit function-select code input on FS2-FS0. Table 1 gives the function-select codes and their descriptions. These inputs are driven by address lines A13-A11 of the TMS32020. Whenever A11, A12, and A13 are all zero, the A10 address line should be set to a one to assure that an external address is generated.

From Table 1, it can be determined that the registers may be accessed at TMS32020 addresses starting at >0400 or at >1000. X-Y indirect access occupies the address range starting at >0800. The host-direct access, although not fully implemented in this circuit, begins with addresses at >1800.

Table 1. Function-Select Decoding

TMS32020 STARTING DATA MEMORY ADDRESS	A13 FS2	A12 FS1	A11 FS0	SELECTED FUNCTION CYCLE
>0400	0	0	0	Register access
>0800	0	0	1	X-Y indirect access
>1000	0	1	0	Register access
>1800	0	1	1	Host-direct access
>2000	1	0	0	Shift-register access (SR to memory)
>2800	1	0	1	Shift-register access (memory to SR)
>3000	1	1	0	Reserved
>3800	1	1	1	Reserved (put VSC in test mode)

Shift-register accesses are located at either >2000 or >2800 depending on whether data is being moved from the shift register to memory or from memory to the shift register.

Descriptions of the VSC internal registers corresponding to specific TMS32020 data memory addresses are shown in Table 2. Note that in the accesses to the VSC internal registers, the useful addresses are actually eight address units apart; i.e., eight consecutive addresses actually access the same internal register.

Table 2. VSC Register Address Map

TMS32020 DATA MEMORY ADDRESS	TMS34061 REGISTER DESCRIPTION
0400	Horizontal End Sync - Low Byte
0408	Horizontal End Sync - High Byte
0410	Horizontal End Blank - Low Byte
0418	Horizontal End Blank - High Byte
0420	Horizontal Start Blank - Low Byte
0428	Horizontal Start Blank - High Byte
0430	Horizontal Total - Low Byte
0438	Horizontal Total - High Byte
0440	Vertical End Sync - Low Byte
0448	Vertical End Sync - High Byte
0450	Vertical End Blank - Low Byte
0458	Vertical End Blank - High Byte
0460	Vertical Start Blank - Low Byte
0468	Vertical Start Blank - High Byte
0470	Vertical Total - Low Byte
0478	Vertical Total - High Byte
0480	Display Address Update - Low Byte
0488	Display Address Update - High Byte

TMS32020 DATA MEMORY ADDRESS	TMS34061 REGISTER DESCRIPTION
0490	Display Start - Low Byte
0498	Display Start - High Byte
04A0	Vertical Interrupt - Low Byte
04A8	Vertical Interrupt - High Byte
04B0	Control Register 1 - Low Byte
04B8	Control Register 1 - High Byte
04C0	Control Register 2 - Low Byte
04C8	Control Register 2 - High Byte
04D0	Status Register - Low Byte
04D8	Status Register - High Byte
04E0	X-Y Offset Register - Low Byte
04E8	X-Y Offset Register - High Byte
04F0	X-Y Address Register - Low Byte
04F8	X-Y Address Register - High Byte
0500	Display Address Register - Low Byte
0508	Display Address Register - High Byte
0510	Vertical Count Register - Low Byte
0518	Vertical Count Register - High Byte

Information is provided about the X-Y indirect address access in Table 3. Note that in the accesses to the X-Y indirect mode, the useful addresses are actually eight address units apart; i.e., eight consecutive addresses actually perform the same X-Y indirect function. For more general information about accessing the VSC, refer to the TMS34061 User's Guide.

Table 3. Address Map of X-Y Adjustment Codes

TMS32020 DATA MEMORY ADDRESS	X-Y ADJUSTMENT FUNCTION	
	X	Y
0800	No adjustment	
0808	Increment X	
0810	Decrement X	
0818	Clear X	
0820		Increment Y
0828	Increment X	Increment Y
0830	Decrement X	Increment Y
0838	Clear X	Increment Y
0840		Decrement Y
0848	Increment X	Decrement Y
0850	Decrement X	Decrement Y
0858	Clear X	Decrement Y
0860		Clear Y
0868	Increment X	Clear Y
0870	Decrement X	Clear Y
0878	Clear X	Clear Y

Initializing the VSC Registers

Whenever a reset signal is applied to the VSC, the internal registers of the VSC assume specific values. These registers must be initialized to values matching the actual hardware configuration before any memory accesses to the VRAMs take place. The code to initialize these values is contained in the appendix in the section beginning with the label VSCIN.

The first four registers define the timing for each horizontal line. Note that the number of time units between the "End Blank" signal at the beginning of the line and the "Start Blank" signal at the end of the line is 90. The reason for this is that each shift clock signal applied to the VRAMs accesses data from eight VRAMs, shifting data in parallel to an external shift register. This provides 720 pixels per line in total (90 time units x 8 pixel data bits/time unit = 720 pixels). The next four registers define the vertical timing control, with the difference in the two blanking times resulting in 300 lines on the display. The next two registers set up the starting address of the display in the VRAMs and

define the address increment to be applied whenever the internal shift registers must be reloaded. The remaining registers are set to provide feedback to the host, basic function control, and the X-Y description for host-indirect access to the VRAMs.

The X-Y Offset Register is set to split the X-Y Address Register into nine bits of Y and seven bits of X, and to supply an additional two bits of X-Y address extension as the two LSBs of X. This splits the eighteen bits of X-Y address evenly, allowing X and Y addressing from 0 to 511. The X addressing is further expanded by accessing two pixels at a time (eight-bit data bus), resulting in effective X addressing from 0 to 1023.

Clearing the Display Screen

Certain operations are absolutely vital in the design of a graphics system. One of these vital operations or functions is the ability to clear the entire display to a single color. Several features of the VSC and the TMS32020 are used to optimize the implementation of this function. The flow diagram shown in Figure 13 describes the process required to clear the screen to any given color.

To optimize the task of clearing the screen, the VSC provides an active strobe on $\overline{RAS0}$, $\overline{RAS1}$, $\overline{RAS2}$, and $\overline{RAS3}$ simultaneously. The end result is the expansion of the data bus to effectively four times its width. In this system, the same pixel data can be written to eight pixels (instead of two) for each write operation from the TMS32020.

In an efficient implementation of the screen-clearing algorithm, all the pixels are written in a row that is not being displayed. Because of the use of the internal shift register in the VRAMs, writing a row of VRAM data that is consistent with the shift-register size requires 256 writes. Since eight pixels are being written with each write, it requires only 256 writes for a full row of data to be written to all the memories. The TMS32020 RPTK instruction aids in the program execution of this operation by providing a single instruction execution loop, as shown below.

```
RPTK    255
SACL    *+
```

After a row of off-display memory is filled, the TMS32020 must wait for the vertical blanking period to occur. When the vertical blanking period begins, a second feature of the VSC and VRAMs is utilized. The VSC is modified to inhibit display-update cycles, and the contents of the off-screen memory that were filled in previously are transferred to the shift registers of the VRAMs. Using the RPT instruction, the shift-register contents are moved to every row of the display memory. Because of the organization of the VRAMs in representing the display, the contents of the shift register are equivalent to two rows of data on the display. The number of writes required to fill display memory are equal to one-half the actual number of rows in the display. Utilization of the auxiliary registers in the TMS32020 and the capability to modify the contents of

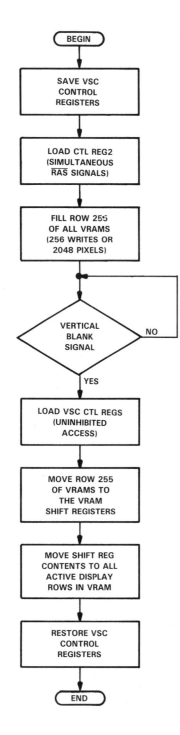

Figure 13. Clear Screen Flow Diagram

an auxiliary register by the contents of the base auxiliary register are critical to the efficient implementation of this operation.

The routine labeled CLEAR and contained in the code listed in the appendix provides an example of this implementation.

Bresenham's Line-Drawing Algorithm

A figure or object is constructed of lines that must be drawn on a pixel-by-pixel basis in any bit-mapped, raster-scan graphics system. The lines used must produce an equally satisfying image with consideration given for actual display resolution when the object is drawn in any orientation on the screen; i.e., when a line of a given length is drawn between two points, the line must be as smooth as possible and without holes or gaps. It is also important that the lines be drawn in an efficient manner with the minimum number of calculations possible since the basic line-drawing algorithm is used frequently in creating and modifying images. An attractive and popular choice for vector generation in a raster-scan display system is Bresenham's algorithm. The algorithm requires only integer arithmetic, thus the reason for its popularity. A discussion of the theory of Bresenham's algorithm is found in reference [17]. The specific TMS32020 implementation is described in the following paragraphs and in the flow diagram shown in Figure 14. The code for this implementation is provided in the appendix beginning with the label LINE.

Figure 15 shows the approach to a line drawing in a two-dimensional Cartesian-coordinate system. Figure 15(a) shows the coordinate system divided into octants. Given two points A and B (see Figure 15(b)), it is convenient to translate the points so that one of them (e.g., A) lies at the coordinate system's origin (see Figure 15(c)). It can then be seen that the line from A to B must lie in one of the octants. (Here it is assumed that the lines shown to divide the octants actually belong to one or the other of the two octants.) Using Bresenham's algorithm, a line may be drawn from A to B in any of the octants. Unfortunately, the coding for each of the octants is unique, and a significant amount of program space is required. A technique used to reduce the program memory requirement by half is to perform the translation of the points in such a way that the point not at the origin always has a positive y-coordinate value (see Figure 15(d)). In this way, it is necessary only to provide code for four of the eight octants.

Each of the four octants begins at the point that has been translated to the origin and utilizes the X-Y indirect-addressing function of the VSC to access memory and modify the address register. The modification of the address for each octant is one of the four combinations of incrementing and decrementing X and Y. On a pixel-by-pixel basis, an increment (or decrement) in the x coordinate does not always result in an increment (or decrement) in the X address. Table 4 shows the selection of X-Y adjustment codes loaded into the auxiliary registers to facilitate access to the video memories.

23. A Graphics Implementation Using the TMS32020 and TMS34061

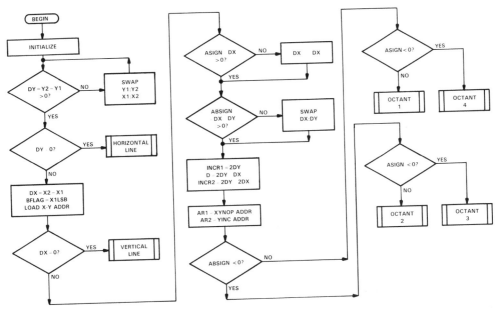

(a) LINE-DRAWING SETUP PROCEDURE

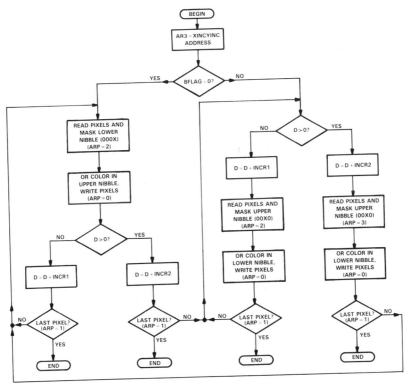

(b) OCTANT-2 LINE-DRAWING PROCEDURE

Figure 14. Flow Diagram for a Line-Drawing Algorithm

23. A Graphics Implementation Using the TMS32020 and TMS34061

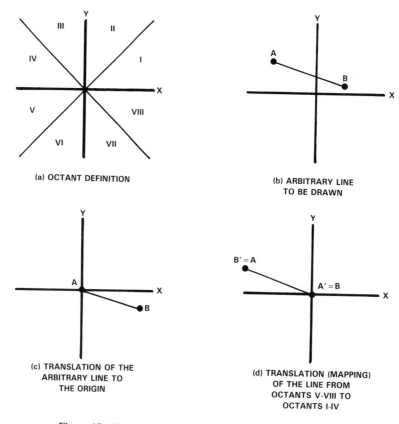

(a) OCTANT DEFINITION

(b) ARBITRARY LINE
TO BE DRAWN

(c) TRANSLATION OF THE
ARBITRARY LINE TO
THE ORIGIN

(d) TRANSLATION (MAPPING)
OF THE LINE FROM
OCTANTS V-VIII TO
OCTANTS I-IV

Figure 15. Line Drawing in a Cartesian-Coordinate System

Table 4. X-Y Adjustment Code Utilization in Line Drawing

OCTANT	AUXILIARY REGISTERS			
	AR1	AR2	AR3	AR4
#1	XNOP YNOP	XNOP YINC	XINC YINC	XINC YNOP
#2	XNOP YNOP	XNOP YINC	XINC YINC	-----
#3	XNOP YNOP	XNOP YINC	XDEC YINC	-----
#4	XNOP YNOP	XNOP YINC	XDEC YINC	XDEC YNOP

23. A Graphics Implementation Using the TMS32020 and TMS34061

The algorithm is designed using the following approach for all octants. The distance function that determines the location of the next point is tested before writing to the present point. Based on whether the distance function is positive or negative, a branch is taken. Then, the corresponding increment constant is added to the difference function to modify it in preparation for the decision at the next point. The pixel pair with the desired pixel is read from memory and masked using the access through auxiliary register 1 (AR1). After the new color value has been added, the pixel pair is written back to video memory using an auxiliary register (AR2, AR3, or AR4), which provides the appropriate address adjustment for the next pixel modification. Auxiliary register 0 (AR0) contains the number of pixels that must be modified to connect the two designated points. AR0 is decremented, with appropriate branch control to the next difference function test, until all pixels between A and B are modified.

Extensive use is made of the auxiliary registers to allow interleaved access of the video memories and internal memory of the TMS32020. This implementation minimizes execution time, a primary concern, as well as reduces the program memory requirement. In addition, tests for horizontal and vertical lines are also made to improve the rate at which those lines are drawn. Table 5 shows the performance per pixel for line drawing in the present implementation.

Table 5. TMS32020 Performance per Pixel for Line Drawing

FUNCTION	CYCLES	TIME AT 200 NS/CYCLE
Horizontal line	5	1.0 μs
Vertical line	13	2.6 μs
Diagonal line	18	3.6 μs

Further optimization can be accomplished by designing the hardware to establish a single pixel-width data bus. By implementing this modification, it is no longer necessary to perform read-modify-writes with the video memory. The reduction to write-only for line drawing nearly doubles the throughput of the TMS32020 for this function. A second benefit of a single pixel-width bus implementation is the compression of the algorithm to the equivalent of the code required for a single octant, thus further reducing the amount of program memory required.

Rotating a Wire-Frame Cube

The usefulness of a line-drawing function can easily be seen in the construction of such simple objects as a wire-frame cube. The importance of the efficiency of the function is observed when the task is expanded to include the rotation of that object. The discussion here encompasses the definition of a three-dimensional object and the rotation of that object in its coordinate space. It further includes the viewpoint perspective of that object and its perspective projection onto the two-dimensional display space.

An object can be defined by its vertices and a list of the connections of those vertices. In the case of a cube, the definition is relatively simple; eight vertices are connected to form 12 edges. To draw the cube on the screen, the vertices in the three-dimensional space of the cube must first be translated, rotated, and scaled with respect to an observer's viewpoint, and then projected onto the two-dimensional display space. The net result of these operations is a second list of points describing the cube in a two-dimensional projection. After the second list of cube vertices is generated, a list that links those points appropriately can be used to supply the line-drawing algorithm with pairs of points, thus generating the image of a wire-frame cube on the screen.

In order to effect a rotation of the wire-frame cube on the display, it is necessary to rotate the vertices of the cube in its originally defined three-dimensional space, apply the viewpoint and projection, and redraw the object on the screen. Two additional elements must be considered. First, in order not to leave multiple images of the object on the screen, it is necessary not only to draw the object in its modified position but also to erase the previous image of the object. Second, the erasure and modified drawing must occur within a period of time that has an upper bound of the refresh cycle of the display (typically 1/60 second). This assures that the viewer always sees a complete object on the display. An alternative, which can allow the time to be longer for display modification and still permit continual observation of fully present objects, is to expand the video memory. By expanding the video memory, a complete screen can be created in memory, which is not being actively displayed. During the vertical blanking period following completion of the image modification, the starting display address can be modified to show that portion of memory, while the next cycle of modification can begin with the memory previously being displayed. In other words, the video memory can be divided into two display windows, and displayed and modified in a "ping-pong" approach.

A code implementation of the first approach discussed for the rotation of a wire-frame cube is included in the appendix beginning with the label CUBER. This routine calls or uses five other subroutines: CLEAR and LINE (mentioned previously), and ROTZ, MTRX4, and PRJCTN (to be discussed later). The flow diagram shown in Figure 16 describes the process that is used to perform the wire-frame cube rotation.

The basic process of cube rotation begins by loading the definition table of the cube and the matrices into the designated areas of memory. If it is required that the viewpoint matrix and projection matrix be dynamically defined, the matrix definition may occur at this point. After this overhead is completed, the overall rotation loop is entered.

Inside the main rotation loop (label CTLOOP), the first task is to build the modified rotation matrix. Two alternatives exist to continuously rotate an object. One of the alternatives is to rotate the present coordinate values for all vertices using a fixed-rotation matrix, and update the coordinates with the modified (rotated) values. This method allows a degradation

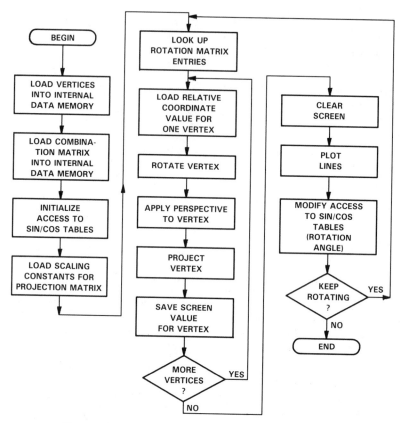

Figure 16. Flow Diagram for a Wire-Frame Cube Rotation

in the object definition since the modified results are recursive, requiring truncation or rounding, and eventually propagate an observable error. The other alternative and the one chosen here is to maintain the original coordinate values and modify the rotation matrix. The benefit of this approach is that the modification can be integer, beginning with zero and increasing to a maximum value at which point a reset to zero occurs. This is expected since the rotation is a circular function. After the rotation matrix has been determined, the matrix manipulation of each of the vertices can begin. Following all the matrix calculations, the screen coordinates of each vertex are in a table. Using straightline code as a list implementation, the screen is cleared, and coordinates are supplied to the line-drawing routine in 12 pairs. When

all 12 lines are drawn, forming the cube, the angle of rotation is updated for the next iteration, and the main loop is repeated.

An inner loop is used to perform the matrix manipulation of each of the vertices. Within this loop, the coordinates of a point are loaded, the three matrix subroutines for rotation, viewpoint perspective, and projection are called in succession, and the results are stored in a table of screen coordinates. This loop is executed once for each vertex point that is used to define the object (eight in the case of a cube). Table 6 summarizes the performance of the TMS32020 in cube rotation and in matrix manipulation that converts the object coordinates to screen coordinates on a per-vertex basis.

Table 6. TMS32020 Performance per Vertex for Cube Rotation

FUNCTION	CYCLES	TIME AT 200 NS/CYCLE
Move vertex from cube definition table	6	1.2 μs
Rotation matrix - rotate point	20	4.0 μs
Combination matrix - viewpoint perspective	35	7.0 μs
Projection matrix - project 3-D to 2-D	49	9.8 μs
Update screen location table	4	0.8 μs
Loop overhead	3	0.6 μs
TOTAL	117	23.4 μs
Total computation for eight vertices	936	187.2 μs

Some assumptions were made in implementing each of the matrix subroutines. These assumptions relating to rotation, viewpoint perspective, and projection are sufficiently general for most cases but are worthy of an explanation. The overall matrix transformation for each point is defined by the following matrix equation:

$$[X_p\ Y_p\ Z_p] = [X\ Y\ Z\ 1] \bullet \begin{bmatrix} R_{11} & R_{12} & R_{13} & 0 \\ R_{21} & R_{22} & R_{33} & 0 \\ R_{31} & R_{32} & R_{33} & 0 \\ 0 & 0 & 0 & 1 \end{bmatrix}$$

$$\bullet \begin{bmatrix} C_{11} & C_{12} & C_{13} & 0 \\ C_{21} & C_{22} & C_{33} & 0 \\ C_{31} & C_{32} & C_{33} & 0 \\ C_{41} & C_{42} & C_{43} & 1 \end{bmatrix} \bullet \begin{bmatrix} P_x & 0 & 0 & 0 \\ 0 & P_y & 0 & 0 \\ 0 & 0 & P_z & 0 \\ T_x & T_y & T_z & 1 \end{bmatrix}$$

$$(1)$$

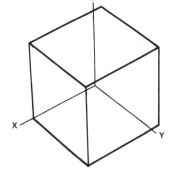

Figure 17. "Real-World" Cube Definition

Rotation

In rotating each vertex of the object, it is assumed that the object is simply rotating about some axis through the body. The actual cube definition, used as the example in the code presented in the appendix, begins with the cube centered at the origin of a Cartesian-coordinate system with the cube faces being orthogonally centered on the x, y, and z axes. This orientation is shown in Figure 17. By using this orientation and rotating the cube about the z axis, the cube is observed to spin on one of its faces. If it is desired to have the cube spin on one of its corners, then the original orientation may be altered to be like the one shown in Figure 18. The key to the assumption is not the axis about which the object is rotating, but rather the fact that the object is simply rotating. If the desire is to observe the object in a more complicated motion, i.e., by adding precession and nutation and/or revolution, then the rotation matrix, which describes the object's motion, necessarily increases in complexity.

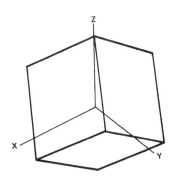

Figure 18. Alternate "Real-World" Cube Definition

23. A Graphics Implementation Using the TMS32020 and TMS34061

With the assumptions as described, the rotation matrix, which begins as a 4 x 4 matrix using homogeneous coordinates, can actually be reduced to a simple 2 x 2 matrix for the necessary calculations. This is possible since the z-coordinate value of each point does not change with rotation and the motion is a simple rotation, involving no additional translation in x or y.

$$[X_r \ Y_r \ Z_r] = [X_o \ Y_o \ Z_o \ 1] \bullet \begin{bmatrix} R_{11} & R_{12} & R_{13} & 0 \\ R_{21} & R_{22} & R_{33} & 0 \\ R_{31} & R_{32} & R_{33} & 0 \\ 0 & 0 & 0 & 1 \end{bmatrix}$$

(2)

$$[X_r \ Y_r] = [X_o \ Y_o] \bullet \begin{bmatrix} R_{11} & R_{12} \\ R_{21} & R_{22} \end{bmatrix}$$

(3)

One additional condition is assumed for this example. The cube being rotated is a "unit" cube in its coordinate system, with no side longer than one. The orthogonal orientation of the cube results in all coordinate values being ± 0.5 for all coordinates (x,y,z). This condition is required so that the coordinate values are always defined as a Q15 number (i.e., 15 bits following an assumed binary point), thus simplifying the calculational load for the TMS32020 code implementation.

Viewpoint Perspective

Placing the cube in a perspective position is a basic, yet involved, operation that must be performed before the object can be projected onto the screen. In most cases, if a stationary viewpoint is used, a combination matrix can be developed prior to its real-time use and stored as a single matrix. The combination matrix is actually the product of seven individual matrices for the specific case of the cube rotation code in the appendix.

A matrix supplying the viewpoint perspective of an object includes translation, rotation, and scaling. Its general form is therefore a complete 4 x 4 matrix, although for purposes of implementation, the calculations reduce to a 4 x 3 matrix since the known zeroes and ones can be eliminated.

$$[X_c \ Y_c \ Z_c \ 1] = [X_r \ Y_r \ Z_r \ 1] \bullet \begin{bmatrix} C_{11} & C_{12} & C_{13} & 0 \\ C_{21} & C_{22} & C_{33} & 0 \\ C_{31} & C_{32} & C_{33} & 0 \\ C_{41} & C_{42} & C_{43} & 1 \end{bmatrix}$$

(4)

To determine the composition of the viewpoint matrix, the following elements must be considered: First, a point of observation in the "real-world" coordinate system, outside the object, must be selected. The coordinates of this point actually become the origin, and the object is subject to a translation equivalent to this from the origin. Second, the X-Y address coordinates of a point on the display screen are like those shown in Figure 19. Since a right-hand coordinate system is the normal expectation, the z axis in the screen coordinate system points into the screen. The observer is thus looking parallel, down the z axis at the point of observation rather than at the object. This means that the next three simple matrices used to build the combination matrix turn the line of sight from parallel with the z axis to pointing at the center of the object. Third, it is generally desirable to observe positive z-coordinate values as if they were projecting from the screen. The next matrix accomplishes this task, thus converting from a right-hand to a left-hand coordinate representation, as shown in Figure 20. In the same way, positive y values are normally observed as going up rather than down. Another similar matrix is utilized, which converts back to a right-hand coordinate system (see Figure 21). Fourth and finally, the last element going into the combination matrix is a matrix that provides a scale normalization defining the size of the object relative to the viewing position of the actual observer. In other words, this matrix incorporates the relative scaling of the object with respect to the assumed distance of the actual observer from the display screen.

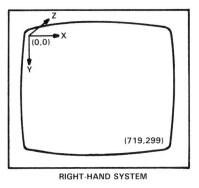

RIGHT-HAND SYSTEM

Figure 19. X-Y Address Coordinates of a Display System

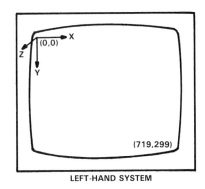

LEFT-HAND SYSTEM

Figure 20. Z-Axis Conversion to the Normal Observation

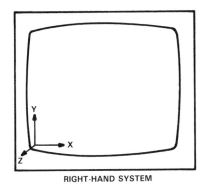

RIGHT-HAND SYSTEM

Figure 21. Y-Axis Conversion to the Normal Observation

Projection

Thus far, discussion has been presented with regard to orienting an object with respect to "normal" screen observation, and all of the matrices have defined and manipulated the object in three dimensions. It is necessary to convert the coordinates to only two dimensions, corresponding to the X-Y addressing of the video memory, in order to "draw" the object on the screen. The conversion must incorporate the z-coordinate value into both the x and y coordinates, producing the X and Y memory addresses.

Figure 22 shows the method used to incorporate z into x and y in the viewer's coordinate system. The fundamental process is to modify the value of x and y, and correspondingly the position of the point on the screen relative to the depth or projection of the z-coordinate value. In other words, in a screen-projected coordinate system, a line positioned close to the observer is seen as a longer line than one positioned farther away, assuming that both have the same relative length in the "real-world" coordinate system.

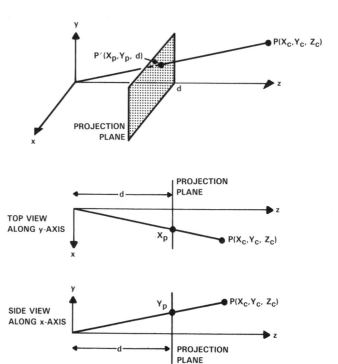

Figure 22. Z-Coordinate Projection onto the X-Y Screen

The matrix to apply this projection is a diagonal matrix plus translation. The diagonal elements also apply a scaling, thus allowing the object to be sized to fill a designated area of the screen. This sizing factor is equivalent to using a "zoom." The translation elements allow the object to be positioned with its center at any point on the screen.

$$[X_p \; Y_p \; Z_p \; 1] = [X_c \; Y_c \; Z_c \; 1]$$

$$\bullet \begin{bmatrix} VSX' & 0 & 0 & 0 \\ 0 & VSY' & 0 & 0 \\ 0 & 0 & VSZ' & 0 \\ VCX & VCY & VCZ & 1 \end{bmatrix} \quad (5)$$

$$VSX' = VSX/Z_c \qquad VSY' = VSY/Z_c \qquad (6)$$

An analysis of the calculations for this matrix shows that it requires the equivalent computations of a 2 x 2 matrix, in addition to the division necessary to calculate $1/Z_c$. This is due to the fact that the matrix is diagonal. The parameters, VSX' and VSY', are dynamically generated for the projection matrix, with the stored parameters for the matrix being the X and Y screen scaling (VSX, VSY) and centering (VCX, VCY) parameters.

Filled-Ellipse Drawing

An additional graphics function for filled-elipse drawing is worthy of discussion although it represents somewhat of a special case. This function is technically composed of two separate functions: curve drawing and filling regular objects. While the task of filling as presented here is specialized, it contains some useful insight into utilizing the power of the VSC. A curve-drawing algorithm, as an extension to the line-drawing algorithm previously discussed, has an even wider scope of potential.

Curve drawing, in general, provides a useful function in many graphics applications. In some display systems, even drawing what appears to be a circle requires a general ellipse-drawing algorithm. This is true for the display monitor used in the project described in this report. An efficient algorithm for drawing ellipses is the midpoint algorithm.[18] This algorithm represents a general approach that can be used to build incremental algorithms for drawing curves other than circles or ellipses.

The basis of an incremental curve drawing is a minimization of the linear error of the drawn curve in reference to the true curve. This is similar to the basis and approach of the line-drawing algorithm. A second basis for the approach must be its efficiency and accuracy in implementation.

23. A Graphics Implementation Using the TMS32020 and TMS34061

The linear error can be minimized to a maximum of one-half the distance between vertically- or horizontally-adjacent pixels if an eight-way-stepping technique is used or minimized to a maximum of one-half the distance between diagonally adjacent pixels if a four-way-stepping technique is used. A comparison of the pixel-stepping directions is shown in Figure 23. Since the task here is to draw a filled ellipse, rather than the ellipse boundary, the four-way-stepping technique is utilized.

Although the distance function used to determine the linear error in a line-drawing algorithm is linear, the distance function in a curve-drawing algorithm is quadratic. The consequence of a quadratic distance function is that the calculations of the function must be carried out with greater than 16-bit precision. The TMS32020 code implementation for determining the curve boundary of an arbitrary ellipse is found in the appendix beginning with the label FELIPS. Figure 24 shows a flow diagram of the filled-ellipse drawing routine.

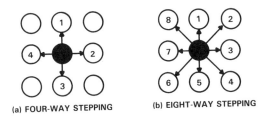

(a) FOUR-WAY STEPPING   (b) EIGHT-WAY STEPPING

Figure 23. Alternative Pixel-Stepping Directions

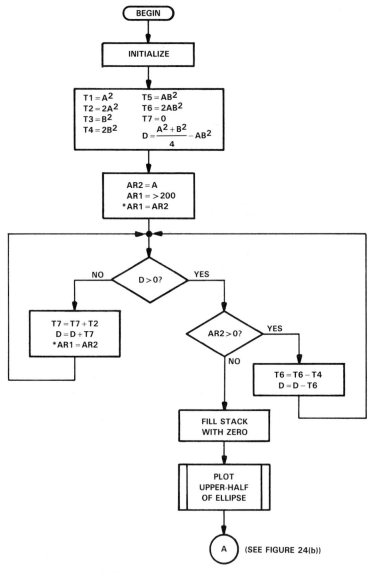

(a) FILLED-ELLIPSE SETUP PROCEDURE

Figure 24. Filled-Ellipse Drawing Flow Diagram

23. A Graphics Implementation Using the TMS32020 and TMS34061

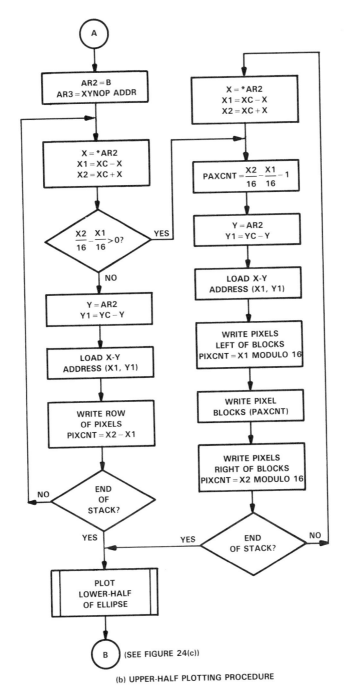

Figure 24. Filled-Ellipse Drawing Flow Diagram
(continued)

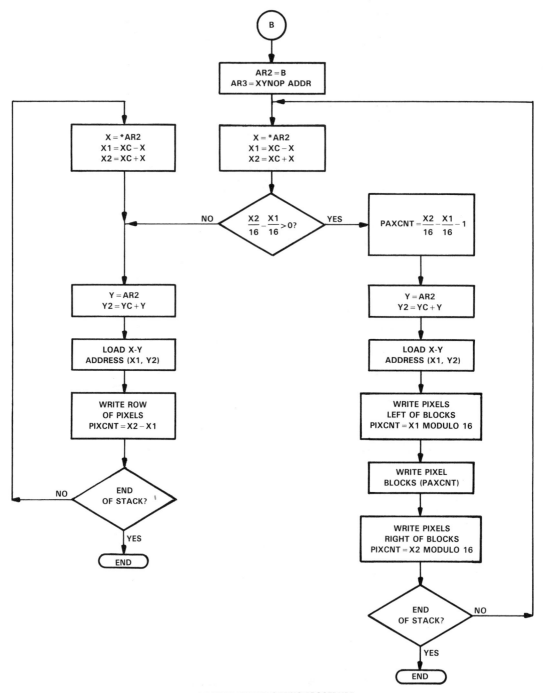

(c) LOWER-HALF PLOTTING PROCEDURE

Figure 24. Filled-Ellipse Drawing Flow Diagram
(concluded)

23. A Graphics Implementation Using the TMS32020 and TMS34061

In the filled-ellipse routine, after the parameter initialization has taken place, the x-coordinate value (in the memory-space coordinate system) is saved on a stack for each y-coordinate value in a single quadrant arc. An example of this approach is shown in Figure 25. The values stored in the memory are the x-coordinate values relative to the center of the ellipse being located at the origin.

To draw the filled ellipse on the screen, four loops are used instead of one or two to allow $\overline{RAS3}$, $\overline{RAS2}$, $\overline{RAS1}$, and $\overline{RAS0}$ to be utilized simultaneously wherever possible, thereby enhancing the performance. This technique is possible since the approach to drawing a filled ellipse is to draw a series of horizontal lines.

Starting at the top of the ellipse, the horizontal lines may be of the short variety and include no groups of eight pixels addressed as a single block. Going down the ellipse toward its center, the lines get longer and begin to include groups of eight pixels addressed as a single block. (This is only guaranteed to be true if the ellipse is specified to have a horizontal axis value of eight or larger.) Figure 26 depicts the difference between a "short" line and a "long" line. When the loop drawing the short lines is terminated by detecting the presence of eight pixel blocks, the second loop

is entered. This loop breaks the line up into a left segment part of one to eight pixels, a center segment part of a number of eight pixel blocks, and a right segment part of one to eight pixels. When the stack of x-coordinate values has been exhausted, the center of the ellipse is reached and the top half of the ellipse is drawn.

Since the stack is a memory implementation, processing begins by accessing the stack in the reverse direction. The series of wide lines that include the eight pixel blocks are drawn until it is determined that a line contains no such blocks. When this condition is observed, the loop is terminated. The last loop, which draws the short lines, is executed until the stack is exhausted, and the drawing of the ellipse is completed.

Several specific features are noted when using this coding of a filled-ellipse algorithm on the TMS32020. First, the symmetry of the ellipse is utilized to reduce the loop calculations to determine the boundary points. Second, in the loops to draw the short lines, the coding is similar to the special-case horizontal-line coding in the standard line-drawing algorithm. Third, the multi$\overline{RAS}$ capability of the VSC and the RPT instruction of the TMS32020 enhance the throughput on the longer lines.

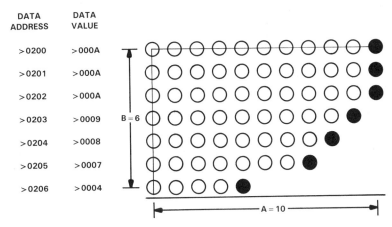

Figure 25. Filled-Ellipse Plotting Stack

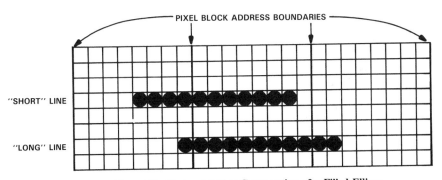

Figure 26. Horizontal-Line Construction of a Filled Ellipse

SUMMARY

A proposed graphics system is presented in this report, which consists of an image control processor (the TMS32020), a display controller (the VSC), video RAMs, and program memory for the TMS32020. A display monitor is incorporated to view the graphics memory. Although a specific host processor and interface to the TMS32020 is not included in this specific project, a typical system may include a host. The TMS32020 is easily interfaced to any of a wide range of host processors. On the other hand, a designer may choose to incorporate the host functions into the TMS32020, in which case the present hardware can be expanded by adding additional program memories and a communications interface.

The TMS32020, while designed for optimizing digital signal processing tasks, has a large number of features commonly found on many general-purpose microprocessors. By combining these features with a high-speed architecture, the TMS32020 provides significant advantages over those same microprocessors. The high-speed architecture is based on separate program and data memory spaces that allow full overlap of instruction fetch and execution, a 16 x 16-bit hardware multiplier with a 32-bit accumulator capable of performing multiply-accumulates in a single instruction cycle of 200 ns, auxiliary registers with a separate arithmetic unit to provide efficient indirect addressing and loop control, and a powerful instruction set that includes multifunction and repeat instructions.

The system presented is efficient in graphics image generation because of several important features. First, by using the VSC and VRAMs, a large video display address space can be addressed with essentially no additional support logic, and video memory accesses are virtually unlimited. Second, the TMS32020 is easily interfaced to the VSC and VRAMs, again with essentially no additional logic. Third, the TMS32020 provides an instruction set that executes largely in a single instruction cycle, thus benefiting even highly optimized graphics algorithms, such as Bresenham's line-drawing algorithm. Finally, the hardware multiplier in the TMS32020 minimizes execution time for matrix multiplications utilized in graphics for rotation, translation, and scaling.

The designer is thus presented with a comprehensive solution to the problem of designing a graphics display system. A large number of product applications, such as matrix printers, laser printers, PC-based engineering workstations, graphics display terminals, CAD systems for logic design, and interactive application simulators, are all benefited with a TMS32020-based design. The powerful TMS32020 processor provides speed comparable to that of custom-logic designs and ease of development that matches conventional processors. The programmability of the TMS32020, as well as the VSC, permits shorter design cycles, simpler maintenance, and easier upgrade capability. The inherent flexibilities of a TMS32020-based system allow the design of a system powerful today and expandable tomorrow.

REFERENCES

1. *Texas Instruments Digital Signal Processor: TMS32020 Product Description*, Texas Instruments Incorporated (1985).
2. *TMS32020 Digital Signal Processor Data Sheet*, Texas Instruments Incorporated (1985).
3. *TMS32020 User's Guide*, Texas Instruments Incorporated (1985).
4. *Texas Instruments Graphics: Video System Controller Product Description*, Texas Instruments Incorporated (1985).
5. *TMS34061 Video System Controller Data Sheet*, Texas Instruments Incorporated (1985).
6. *TMS34061 User's Guide*, Texas Instruments Incorporated (1985).
7. *TMS4161 65,536-Bit Multiport Video RAM Data Sheet*, Texas Instruments Incorporated (1985).
8. G.B. Clark, *Topological Structure of the TMS4161 Multiport Video RAM* (Application Report), Texas Instruments (1984).
9. R. Pinkham et al., "Video RAM Excels at Fast Graphics," *Electronic Design* (August 18, 1983).
10. C.D. Crowell, *Matrix Multiplication with the TMS32010 and TMS32020* (Application Report), Texas Instruments Incorporated (1985).
11. C.D. Crowell, *TMS32020 and MC68000 Interface* (Application Report), Texas Instruments Incorporated (1985).
12. J. Bradley et al., *Hardware Interfacing to the TMS32020* (Application Report), Texas Instruments Incorporated (1985).
13. *TMS34061 Evaluation Module User's Guide*, Texas Instruments Incorporated (1985).
14. *TMS34070 User's Guide*, Texas Instruments Incorporated (1985).
15. J.D. Foley and A. Van Dam, *Fundamentals of Interactive Computer Graphics*, Addison-Wesley (1982).
16. W.M. Newman and R.F. Sproull, *Principles of Interactive Computer Graphics*, McGraw-Hill (1979).
17. J.E. Bresenham, "Algorithm for Computer Control of a Digital Plotter," *IBM Systems Journal*, Vol 4, No. 1, 24-30 (1965).
18. J.R. Van Aken, "An Efficient Ellipse-Drawing Algorithm," *IEEE Computer Graphics and Applications*, Vol 4, No. 9, 24-35 (1984).

23. A Graphics Implementation Using the TMS32020 and TMS34061

APPENDIX A

TMS32020 AND TMS34061 GRAPHICS SYSTEM SCHEMATIC

Parts List for the Graphics System Schematic

U1	TMS32020	Digital Signal Processor
U2	TMS34061	Video System Controller
U3-U6	TBP28S166	2K x 8 Programmable ROM
U7-U38	TMS4161	64K x 1 Multiport Video RAM
U39	SN74S138	3-to-8 Line Decoders/Demultiplexers
U40	SN74LS244	Octal Buffers/Line Drivers/Line Receivers
U41	SN74LS125	Quad Bus Buffer Gates with Three-State Outputs
U42,U48	SN74LS04	Hexadecimal Inverters
U43	SN74S00	Quad 2-Input Positive-NAND Gates
U44	SN74S32	Quad 2-Input Positive-OR Gates
U45	SN74S86	Quad 2-Input Exclusive-OR Gates
U46	SN74LS245	Octal Bus Tranceivers
U50	SN74AS74	Dual D-Type Positive-Edge-Triggered Flip-Flops
U52,U55	SN74S113	Dual J-K Negative-Edge-Triggered Flip-Flops
U53	SN74LS241	Octal Buffers/Line Drivers/Line Receivers
U54	SN74AS02	Quad 2-Input Positive-NOR Gates
U56	SN74AS27	Triple 3-Input Positive-NOR Gates
U57	SN74AS08	Quad 2-Input Positive-NAND Gates
U58-U61	SN74F299	8-Bit Bidirectional Shift/Store Registers
X1		20.000 MHz Oscillator
X2		18.432 MHz Oscillator
U64		33 Ohm 8-Resistor Pack
R1,R2		390 Ohm Resistors
R3-R8		1K Ohm Resistors
R9-R19		4.7K Ohm Resistors
C1-C4		39 μF Bypass Capacitors
C5		0.1 μF Capacitor
C6-C13		0.1 μF Bypass Capacitors
C14-C63		0.001 μF Bypass Capacitors
S1		SPST Momentary Contact Switch

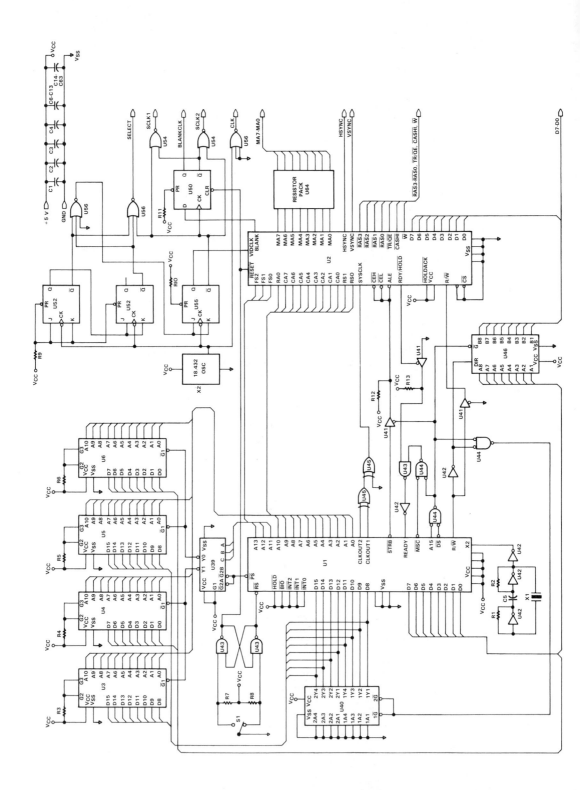

23. A Graphics Implementation Using the TMS32020 and TMS34061

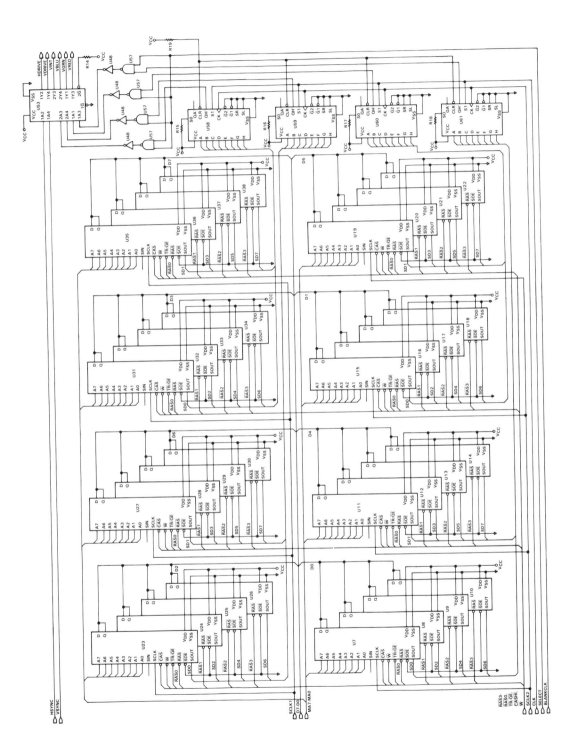

APPENDIX B

TMS32020 GRAPHICS SOFTWARE ALGORITHMS

```
0001            COPY    PRELUDE.ASM
A0002           IDT     'GRAPHIC'
A0003           OPTION  XREF
A0004  *
A0005  ********************************************************************
A0006  *  This is the source module for the TMS32020 for a graphics
A0007  *  demonstration.  The system is built using a VSC EVM board
A0008  *  containing the TMS34061 (VSC) and 2 Mbit of TMS4161 (VRAM)
A0009  *  configured in a 1024 x 512 pixel array with 4 bit pixels.
A0010  *  The VSC displays and maintains the contents of the VRAMs.
A0011  *  The TMS32020 is responsible for creating (plotting) the
A0012  *  graphic images which are displayed.  Access to the VRAMs
A0013  *  by the TMS32020 is provided through the VSCs X-Y indirect
A0014  *  addressing access.  Memory control by the VSC is
A0015  *  programmable through 18 accessible registers to tailor the
A0016  *  available memory and memory access to the selected
A0017  *  monitor.  The memory on the VSC EVM has been configured
A0018  *  for use with a TIPC monitor which has a 720 x 300 pixel
A0019  *  display.
A0020  *
A0021  ********************************************************************

A0022  0000                  AORG  >0000
A0023
A0024  0000 FF80      RESET  B     INIT     * branch to processor initialization
       0001 0022
A0025
A0026  0002 FF80      INT0   B     DUMMY    * no interrupts are utilized
       0003 0020
A0027  *
A0028  0004 FF80      INT1   B     DUMMY
       0005 0020
A0029  *
A0030  0006 FF80      INT2   B     DUMMY
       0007 0020
A0031  *
A0032  0018                  AORG  >0018
A0033  *
A0034  0018 FF80      TINT   B     DUMMY
       0019 0020
A0035  *
A0036  001A FF80      RINT   B     DUMMY
       001B 0020
A0037  *
A0038  001C FF80      XINT   B     DUMMY
       001D 0020
A0039  *
A0040  001E FF80      TRAP   B     DUMMY
       001F 0020
A0041  *
A0042  0020 FF80      DUMMY  B     DUMMY
       0021 0020
0002                  COPY   INIT.ASM
```

```
B0002  *
B0003  *  This routine is executed each time the RS- pin is pulled
B0004  *  active low and released.  All internal registers
B0005  *  (memory-mapped, auxiliary, status, data-memory) of the
B0006  *  TMS32020 are initialized.  All control registers and video
B0007  *  data-memory of the TMS34061 are initialized .
B0008  *
B0009  ********************************************************************
B0010
B0011  0022           INIT   EQU   $
B0012  0022 CE01             DINT               * disable all interrupts
B0013  0023 C800             LDPK  0            * set-up to access Block B2
B0014  0024 CE02             ROVM               * reset overflow mode
B0015  0025 5588             LARP  ARO          * set auxiliary pointer to reg 0
B0016  0026 CE08             SPM   0            * set P reg shift mode to no shift
B0017  0027 CE20             RTXM               * set external serial framing
B0018  0028 CE0D             SXF                * set external flag pin
B0019  0029 CE07             SSXM               * set ALU sign-extension mode
B0020  002A CE04             CNFD               * configure Block B0 as data memory
B0021  002B D001             LALK  >FFFF
       002C FFFF
B0022  002D 6003             SACL  PRD          * initialize period register
B0023  002E CA00             ZAC
B0024  002F 6004             SACL  IMR          * disable each interrupt
B0025  0030 6005             SACL  GREG         * declare all memory local
B0026  0031 CA01             LACK  1
B0027  0032 6062             SACL  ONE
B0028
B0029  0033           VSCIN  EQU   $
B0030  0033 C008             LARK  ARO,8
B0031  0034 5589             LARP  AR1
B0032  0035 0100             LRLK  AR1,VSCREG
       0036 0400
B0033  0037 CA0A             LACK  >000A
B0034  0038 60E0             SACL  *0+
B0035  0039 68E0             SACH  *0+          * load Horizontal End Sync
B0036  003A CA17             LACK  >0017
B0037  003B 60E0             SACL  *0+
B0038  003C 68E0             SACH  *0+          * load Horizontal End Blank
B0039  003D CA71             LACK  >0071
B0040  003E 60E0             SACL  *0+
B0041  003F 68E0             SACH  *0+          * load Horizontal Start Blank
B0042  0040 CA76             LACK  >0076
B0043  0041 60E0             SACL  *0+
B0044  0042 68E0             SACH  *0+          * load Horizontal Total
B0045  0043 CA02             LACK  >0002
B0046  0044 60E0             SACL  *0+
B0047  0045 68E0             SACH  *0+          * load Vertical End Sync
B0048  0046 CA13             LACK  >0013
B0049  0047 60E0             SACL  *0+
B0050  0048 68E0             SACH  *0+          * load Vertical End Blank
B0051  0049 D001             LALK  >103F
       004A 103F
B0052  004B 60E0             SACL  *0+
B0053  004C 6CE0             SACH  *0+,4        * load Vertical Start Blank
B0054  004D D001             LALK  >1042
       004E 1042                                * load Vertical Total
```

23. A Graphics Implementation Using the TMS32020 and TMS34061

```
B0055 004F 60E0    SACL   *0+
B0056 0050 6CE0    SACH   *0+,4
B0057 0051 CA02    LACK   >0002
B0058 0052 60E0    SACL   *0+
B0059 0053 68E0    SACH   *0+              * load Display Update Register
B0060 0054 CA00    LACK   >0000
B0061 0055 60E0    SACL   *0+
B0062 0056 68E0    SACH   *0+              * load Display Start Register
B0063 0057 CA00    LACK   >0000
B0064 0058 60E0    SACL   *0+
B0065 0059 68E0    SACH   *0+              * load Vertical Interrupt
B0066 005A CA10    LACK   >0010
B0067 005B 68E0    SACH   *0+
B0068 005C 60E0    SACL   *0+              * load Control Register 1
B0069 005D CA20    LACK   >0020
B0070 005E 68E0    SACH   *0+
B0071 005F 60E0    SACL   *0+              * load Control Register 2
B0072 0060 CA00    LACK   >0000
B0073 0061 60E0    SACL   *0+
B0074 0062 68E0    SACH   *0+              * load Status Register
B0075 0063 CA20    LACK   >0020
B0076 0064 60E0    SACL   *0+
B0077 0065 68E0    SACH   *0+              * load X-Y Offset Register
B0078 0066 CA00    LACK   >0000
B0079 0067 60E0    SACL   *0+
B0080 0068 68E0    SACH   *0+              * load X-Y Address Register
B0081
                   *
B0082 0069 CA00    LACK   >0000
B0083 006A FE80    CALL   CLEAR            * initialize VRAMs
      006B 04DC
                   *
B0084
B0085 006C CA00    ZAC
B0086 006D 5588    LARP   AR0
B0087 006E D000    LRLK   AR0,>0200
      006F 0200
B0088 0070 CBFF    RPTK   255             * initialize Block B0
B0089 0071 60A0    SACL   *+
B0090 0072 CBFF    RPTK   255             * initialize Block B1
B0091 0073 60A0    SACL   *+
B0092 0074 C063    LARK   AR0,>63         * initialize Block B2
B0093 0075 CB1C    RPTK   28
B0094 0076 60A0    SACL   *+
B0095
                   *
B0096 0077 CA01    LACK   1
B0097 0078 C806    LDPK   6
B0098 0079 602E    SACL   ONE6
B0099 007A 0001    LALK   ROMDAT          * load pixel processing constants
      007B 086D
B0100 007C 5837    TBLR   XMAX            * maximum X address (TIPC = 719)
B0101 007D 002E    ADD    ONE6
B0102 007E 5839    TBLR   YMAX            * maximum Y address (TIPC = 299)
B0103 007F 002E    ADD    ONE6
B0104 0080 583F    TBLR   AMAX            * ellipse demo constants
B0105 0081 002E    ADD    ONE6
B0106 0082 5840    TBLR   AMIN
B0107 0083 002E    ADD    ONE6
B0108 0084 5841    TBLR   BMAX
```

```
B0109 0085 002E    ADD    ONE6
B0110 0086 5842    TBLR   BMIN
B0111 0087 CA03    LACK   3
B0112 0088 6043    SACL   THREE6
B0113 0089 CA05    LACK   5
B0114 008A 6044    SACL   FIVE6
B0115 008B CA09    LACK   9
B0116 008C 6045    SACL   NINE6
B0117 008D FF80    B      PARSER
      008E 008F

0003               COPY   PARSER.ASM
```

23. A Graphics Implementation Using the TMS32020 and TMS34061

```
C0002       *********************************************************
C0003       *
C0004       * The PARSER routine controls the demonstration of the
C0005       * graphics display system by reading the control data table
C0006       * and routing execution to the appropriate function. A
C0007       * synchronization word, 32020, must occur at the beginning
C0008       * of the command table and between each command. The parser
C0009       * uses this synchronization word to recover from errors in
C0010       * entry of data in the command table.  Following the
C0011       * synchronization word, the command word is read and used to
C0012       * select one of the appropriate functional routines.  In the
C0013       * event that the command word is not within the command
C0014       * sequence, searching continues until the next
C0015       * synchronization word is found and the process repeats.
C0016       * When a valid command word is identified, execution is
C0017       * transfered to the appropriate routine which must read all
C0018       * data from the list appropriate to that routine.
C0019       *
C0020       *********************************************************
C0021 008F          PARSER EQU   $
C0022 008F C800            LDPK  0              * set-up to access Block B2
C0023 0090 0001            LALK  BEGIN          * top of command table
      0091 0B9B
C0024 0092 606C     PLOOP  SACL  CMDADR         * save current table address
C0025 0093 C800            LDPK  0              * set-up to access Block B2
C0026 0094 D000            LRLK  AR0,32020      * load synchronization word
      0095 7D14
C0027 0096 5589            LARP  AR1
C0028 0097 206C            LAC   CMDADR         * get current table address
C0029 0098 5869     SYNC   TBLR  SYNCH          * read word; expect a sync value
C0030 0099 0062            ADD   ONE
C0031 009A 3169            LAR   AR1,SYNCH
C0032 009B 55D0            MAR   *0-
C0033 009C FB90            BANZ  SYNC           * test for sync value
      009D 0098
C0034       *
C0035 009E 586B            TBLR  CMND           * read word; expect a valid command
C0036 009F 0062            ADD   ONE
C0037 00A0 606C            SACL  CMDADR         * save current table address
C0038 00A1 CA17            LACK  CMDLEN
C0039 00A2 106B            SUB   CMND
C0040 00A3 F280            BLEZ  PLOOP          * verify command is valid
      00A4 0093
C0041       *
C0042 00A5 D001            LALK  CMDTBL
      00A6 00AB
C0043 00A7 016B            ADD   CMND,1
C0044 00A8 CE24            CALA                 * go execute selected command
C0045 00A9 FF80            B     PLOOP
      00AA 0093
C0046       *
```

```
C0048       *
C0049 00AB FF80     CMDTBL B     INIT    * 0000
      00AC 0022
C0050 00AD FF80            B     DELAY   * 0001
      00AE 000C
C0051 00AF FF80            B     CLRSCR  * 0002
      00B0 00F7
C0052 00B1 FF80            B     PIXELC  * 0003
      00B2 0100
C0053 00B3 FF80            B     LINEC   * 0004
      00B4 010F
C0054 00B5 FF80            B     DUMONE  * 0005
      00B6 00DB
C0055 00B7 FF80            B     POLYC   * 0006
      00B8 0120
C0056 00B9 FF80            B     DUMONE  * 0007
      00BA 00DB
C0057 00BB FF80            B     DUMONE  * 0008
      00BC 00DB
C0058 00BD FF80            B     FELPSC  * 0009
      00BE 0158
C0059 00BF FF80            B     SPIN    * 000A
      00C0 0169
C0060 00C1 FF80            B     ELPSET  * 000B
      00C2 01DD
C0061 00C3 FF80            B     WELPSC  * 000C
      00C4 0212
C0062 00C5 FF80            B     DUMONE  * 000D
      00C6 00DB
C0063 00C7 FF80            B     DUMONE  * 000E
      00C8 00DB
C0064 00C9 FF80            B     STOP    * 000F
      00CA 0009
C0065 00CB FF80            B     DEMO0   * 0010
      00CC 0280
C0066 00CD FF80            B     DEMO1   * 0011
      00CE 0281
C0067 00CF FF80            B     DEMO2   * 0012
      00D0 02FC
C0068 00D1 FF80            B     DEMO3   * 0013
      00D2 0343
C0069 00D3 FF80            B     DEMO4   * 0014
      00D4 03BF
C0070 00D5 FF80            B     DEMO5   * 0015
      00D6 0450
C0071 00D7 FF80            B     CUBER   * 0016
      00D8 08FC
C0072 0017        CMDLEN EQU   $-CMDTBL/2
C0073       *
```

23. A Graphics Implementation Using the TMS32020 and TMS34061

```
C0126 0108 5804          TBLR  COLOR     * read word; expect a color value
C0127 0109 C800          LDPK  0
C0128 010A 0062          ADD   ONE
C0129 010B 606C          SACL  CMDADR    * save current table address
C0130 010C FE80          CALL  LINE      * plot the point
      010D 0524
C0131 010E CE26          RET
C0132
C0133                   * LINE DRAWING CONTROL
C0134                   *
C0135 010F 206C  LINEC   LAC   CMDADR    * get current table address
C0136 0110 C806          LDPK  6
C0137 0111 5800          TBLR  X1        * read word; expect x1 coordinate
C0138 0112 002E          ADD   ONE6
C0139 0113 5801          TBLR  Y1        * read word; expect y1 coordinate
C0140 0114 002E          ADD   ONE6
C0141 0115 5802          TBLR  X2        * read word; expect x2 coordinate
C0142 0116 002E          ADD   ONE6
C0143 0117 5803          TBLR  Y2        * read word; expect y2 coordinate
C0144 0118 002E          ADD   ONE6
C0145 0119 5804          TBLR  COLOR     * read word; expect a color value
C0146 011A C800          LDPK  0
C0147 011B 0062          ADD   ONE
C0148 011C 606C          SACL  CMDADR    * save current table address
C0149 011D FE80          CALL  LINE      * plot the line
      011E 0524
C0150 011F CE26          RET
C0151
C0152                   * POLYGON DRAWING CONTROL
C0153                   *
C0154 0120 206C  POLYC   LAC   CMDADR    * get current table address
C0155 0121 C806          LDPK  6
C0156 0122 5830          TBLR  NPOINT    * read word; expect # polygon sides
C0157 0123 002E          ADD   ONE6
C0158 0124 5804          TBLR  COLOR     * read word; expect a color
C0159 0125 002E          ADD   ONE6
C0160 0126 3030          LAR   AR0,NPOINT
C0161 0127 5588          LARP  AR0
C0162 0128 FB80          BANZ  POLY1
      0129 012C
C0163 012A FF80          B     POLYD
      012B 0154
C0164
C0165 012C 7030  POLY1   SAR   AR0,NPOINT
C0166 012D 5822          TBLR  XA        * read word; expect x1 value
C0167 012E 5824          TBLR  XB
C0168 012F 002E          ADD   ONE6
C0169 0130 5823          TBLR  YA        * read word; expect y1 value
C0170 0131 5825          TBLR  YB
C0171 0132 C800          LDPK  0
C0172 0133 0062          ADD   ONE
C0173 0134 606C          SACL  CMDADR    * save current table address
C0174 0135 FF80          B     POLYN
      0136 0145
C0175
C0176 0137 7030  POLYL   SAR   AR0,NPOINT
C0177 0138 C800          LDPK  0
```

```
C0075
C0076 0009 FF80  STOP    B     STOP
      000A 0009
C0077 000B CE26  DUMONE  RET
C0078
C0079
C0080                   * TIME DELAY CONTROL
C0081 000C 206C  DELAY   LAC   CMDADR    * get current table address
C0082 000D 586A          TBLR  TIMDLY    * read word; expect a time value
C0083 000E 0062          ADD   ONE
C0084 000F 606C          SACL  CMDADR    * save current table address
C0085 00E0 306A  DELAY1  LAR   AR0,TIMDLY
C0086 00E1 C168          LARK  AR1,TMP1
C0087 00E2 5588          LARP  AR0
C0088 00E3 FF80          B     DECDLY
      00E4 00F4
C0089 00E5 C808  WAITI   LDPK  >0008     * VSC register page
C0090 00E6 CA01          LACK  1
C0091 00E7 2060          LAC   VSBLL     * read Vertical Start Blank
C0092 00E8 0868          ADD   VSBLH,8
C0093 00E9 6080          SACL  *
C0094 00EA C809          LDPK  >0009     * VSC register page
C0095 00EB 6020          SACL  VINTL
C0096 00EC 2888          LAC   *,8,0
C0097 00ED 6828          SACH  VINTH     * load Vertical Interrupt
C0098 00EE 2050          LAC   STATL
C0099 00EF 2050  WAITD   LAC   STATL
C0100 00F0 D004          LAC   ANDK      * wait for vertical interrupt
      00F1 0001
C0101 00F2 F680          BZ    WAITD
      00F3 00EF
C0102 00F4 FB99  DECDLY  BANZ  WAITI,*-,1  * decrement delay counter
      00F5 00E5
C0103 00F6 CE26          RET
C0104
C0105                   * CLEAR SCREEN CONTROL
C0106                   *
C0107 00F7 206C  CLRSCR  LAC   CMDADR    * get current table address
C0108 00F8 5868          TBLR  TMP1
C0109 00F9 0062          ADD   ONE
C0110 00FA 606C          SACL  CMDADR    * save current table address
C0111 00FB CAFF          LACK  >00FF
C0112 00FC 4E68          AND   TMP1
C0113 00FD FE80          CALL  CLEAR     * initialize the VRAMs
      00FE 04DC
C0114 00FF CE26          RET
C0115
C0116                   * WRITE PIXEL CONTROL
C0117                   *
C0118 0100 206C  PIXELC  LAC   CMDADR    * get current table address
C0119 0101 C806          LDPK  6
C0120 0102 5800          TBLR  X1        * read word; expect x coordinate
C0121 0103 5802          TBLR  X2
C0122 0104 002E          ADD   ONE6
C0123 0105 5801          TBLR  Y1        * read word; expect y coordinate
C0124 0106 5803          TBLR  Y2
C0125 0107 002E          ADD   ONE6
```

23. A Graphics Implementation Using the TMS32020 and TMS34061

```
C0178 0139 206C        LAC   CMDADR    * load current table address
C0179 013A C806        LDPK  6
C0180 013B 5802        TBLR  X2        * read word; expect next x value
C0181 013C 5824        TBLR  XB
C0182 013D 002E        ADD   ONE6
C0183 013E 5803        TBLR  Y2        * read word; expect next y value
C0184 013F 5825        TBLR  YB
C0185 0140 C800        LDPK  0
C0186 0141 0062        ADD   ONE
C0187 0142 606C        SACL  CMDADR    * save current table address
C0188 0143 FE80        CALL  LINE      * plot the line
      0144 0524
C0189
*  POLYN
C0190 0145 C806 POLYN  LDPK  6
C0191 0146 2024        LAC   XB
C0192 0147 6000        SACL  X1        * x(n)   = x(n+1)
C0193 0148 2025        LAC   YB
C0194 0149 6001        SACL  Y1        * y(n)   = y(n+1)
C0195 014A 3030        LAR   AR0,NPOINT
C0196 014B FB90        BANZ  POLYL     * last point ?
      014C 0137
C0197 014D 2022        LAC   XA        * x(n+1) = x(1)
C0198 014E 6002        SACL  X2
C0199 014F 2023        LAC   YA        * y(n+1) = y(1)
C0200 0150 6003        SACL  Y2
C0201 0151 FE80        CALL  LINE      * plot the line
      0152 0524
C0202 0153 CE26        RET
C0203
*  POLYD
C0204 0154 C800 POLYD  LDPK  0
C0205 0155 0062        ADD   ONE
C0206 0156 606C        SACL  CMDADR    * save current table address
C0207 0157 CE26        RET
C0208
*  FILLED ELLIPSE DRAWING CONTROL
C0209
C0210
C0211 0158 206C FELPSC LAC   CMDADR    * get current table address
C0212 0159 C806        LDPK  6
C0213 015A 5810        TBLR  XC        * read word; expect center x value
C0214 015B 002E        ADD   ONE6
C0215 015C 5811        TBLR  YC        * read word; expect center y value
C0216 015D 002E        ADD   ONE6
C0217 015E 5812        TBLR  A         * read word; expect horizontal axis
C0218 015F 002E        ADD   ONE6
C0219 0160 5813        TBLR  B         * read word; expect vertical axis
C0220 0161 002E        ADD   ONE6
C0221 0162 5804        TBLR  COLOR     * read word; expect a color
C0222 0163 C800        LDPK  0
C0223 0164 0062        ADD   ONE
C0224 0165 606C        SACL  CMDADR    * save current table address
C0225 0166 FE80        CALL  FELIPS    * draw a filled ellipse
      0167 06BB
C0226 0168 CE26        RET
C0227
*  SPINNING FILLED ELLIPSE DRAWING CONTROL
C0228
C0229
C0230 0169 206C SPIN   LAC   CMDADR    * get current table address
```

```
C0231 016A C806        LDPK  6
C0232 016B 5822        TBLR  XA        * read word; expect center x value
C0233 016C 002E        ADD   ONE6
C0234 016D 5823        TBLR  YA        * read word; expect center y value
C0235 016E 002E        ADD   ONE6
C0236 016F 5824        TBLR  XB        * read word; expect horizontal axis
C0237 0170 002E        ADD   ONE6
C0238 0171 5832        TBLR  SAVEA
C0239 0172 5825        TBLR  YB        * read word; expect vertical axis
C0240 0173 002E        ADD   ONE6
C0241 0174 5831        TBLR  SAVCLR    * read word; expect the colors
C0242 0175 002E        ADD   ONE6
C0243 0176 5834        TBLR  SPEED     * read word; expect rate of change
C0244 0177 C800        LDPK  0
C0245 0178 0062        ADD   ONE
C0246 0179 606C        SACL  CMDADR    * save current table address
C0247 017A 2162        LAC   ONE,1
C0248 017B 606A        SACL  TIMDLY    * load time delay constant
C0249 017C C806        LDPK  6
C0250 017D 2831        LAC   SAVCLR,8
C0251 017E 6833        SACH  CLRSAV    * seperate colors : back side
C0252 017F CA0F        LACK  >000F
C0253 0180 4E31        AND   SAVCLR
C0254 0181 6031        SACL  SAVCLR                        : front side
C0255 0182 2032        LAC   SAVEA
C0256 0183 CB0F        RPTK  15
C0257 0184 4734        SUBC  SPEED     * # cycles = a / delta
C0258 0185 6032        SACL  SAVEA
C0259
C0260 0186 3032        LAR   AR0,SAVEA
C0261 0187 FF80        B     SPIN1E
      0188 01A0
C0262
*  SPIN0
C0263 0189 702F        SAR   AR0,SAVAR0
C0264 018A 2022        LAC   XA        * load center x value
C0265 018B 6010        SACL  XC
C0266 018C 2023        LAC   YA        * load center y value
C0267 018D 6011        SACL  YC
C0268 018E 2024        LAC   XB        * load horizontal axis value
C0269 018F 6012        SACL  A
C0270 0190 1034        SUB   SPEED     * a = a - delta
C0271 0191 6024        SACL  XB
C0272 0192 2025        LAC   YB        * load vertical axis value
C0273 0193 6013        SACL  B
C0274 0194 2031        LAC   SAVCLR
C0275 0195 6004        SACL  COLOR     * load color
C0276 0196 FE80        CALL  FELIPS    * draw a filled ellipse
      0197 06BB
*  SPIN1
C0277 0198 C800        LDPK  0
C0278 0199 FE80        CALL  DELAY1    * wait a while
      019A 00E0
C0279 019B C806        LDPK  6
C0280 019C CA00        ZAC
C0281 019D FE80        CALL  CLEAR     * clear the screen
      019E 04DC
C0282 019F 302F        LAR   AR0,SAVAR0
C0283 01A0 5588 SPIN1E LARP  AR0
```

23. A Graphics Implementation Using the TMS32020 and TMS34061

```
C0284 01A1 FB90        BANZ  SPIN1
      01A2 0189
C0285                  *
C0286 01A3 2022        LAC   XA
C0287 01A4 6000        SACL  X1         * X1 = XC
C0288 01A5 6002        SACL  X2         * X2 = XC
C0289 01A6 2023        LAC   YA
C0290 01A7 1025        SUB   YB
C0291 01A8 6001        SACL  Y1         * Y1 = YC - B
C0292 01A9 0125        ADD   YB,1
C0293 01AA 6003        SACL  Y2         * Y2 = YC + B
C0294 01AB 2031        LAC   SAVCLR
C0295 01AC 6004        SACL  COLOR      * load color
C0296 01AD FE80        CALL  LINE       * plot the line
      01AE 0524
C0297 01AF C800        LDPK  0
C0298 01B0 FE80        CALL  DELAY1     * wait a while
      01B1 00E0
C0299 01B2 C806        LDPK  6
C0300 01B3 CA00        ZAC
C0301 01B4 FE80        CALL  CLEAR      * clear the screen
      01B5 04DC
C0302                  *
C0303 01B6 3032        LAR   AR0,SAVEA
C0304 01B7 FF80        B     SPIN2E
      01B8 01D0
C0305                  *
C0306 01B9 702F  SPIN2 SAR   AR0,SAVAR0
C0307 01BA 2022        LAC   XA
C0308 01BB 6010        SACL  XC         * load center x value
C0309 01BC 2023        LAC   YA
C0310 01BD 6011        SACL  YC         * load center y value
C0311 01BE 2024        LAC   XB
C0312 01BF 6012        SACL  XB         * load horizontal axis value
C0313 01C0 0034        ADD   A          * a = a + delta
C0314 01C1 6024        SACL  SPEED
C0315 01C2 2025        LAC   XB
C0316 01C3 6013        SACL  B          * load vertical axis value
C0317 01C4 2033        LAC   CLRSAV
C0318 01C5 6004        SACL  COLOR      * load color
C0319 01C6 FE80        CALL  FELIPS     * draw a filled ellipse
      01C7 068B
C0320 01C8 C800        LDPK  0
C0321 01C9 FE80        CALL  DELAY1     * wait a while
      01CA 00E0
C0322 01CB C806        LDPK  6
C0323 01CC CA00        ZAC
C0324 01CD FE80        CALL  CLEAR      * clear the screen
      01CE 04DC
C0325 01CF 302F  SPIN2E2 LAR  AR0,SAVAR0
C0326 01D0 5588  SPIN2E  LARP AR0
C0327 01D1 FB90        BANZ  SPIN2
      01D2 01B9
C0328                  *
C0329 01D3 4031        ZALH  SAVCLR     * swap colors : front and back
C0330 01D4 4933        ADDS  CLRSAV
C0331 01D5 6833        SACH  CLRSAV
```

```
C0332 01D6 6031        SACL  SAVCLR
C0333 01D7 2025        LAC   YB
C0334 01D8 112E        SUB   ONE6,1     * b = b - 2
C0335 01D9 6025        SACL  YB
C0336 01DA F180        BGZ   SPIN0
      01DB 0186
C0337 01DC CE26        RET
C0338                  *
C0339                  *   FILLED ELLIPSE SET DRAWING CONTROL
C0340                  *
C0341 01DD 206C ELPSET LAC   CMDADR     * get current table address
C0342 01DE C806        LDPK  6
C0343 01DF 5822        TBLR  XA         * read word; expect center x value
C0344 01E0 002E        ADD   ONE6
C0345 01E1 5823        TBLR  YA         * read word; expect center y value
C0346 01E2 002E        ADD   ONE6
C0347 01E3 5824        TBLR  XB         * read word; expect horizontal axis
C0348 01E4 002E        ADD   ONE6
C0349 01E5 5825        TBLR  YB         * read word; expect vertical axis
C0350 01E6 002E        ADD   ONE6
C0351 01E7 5831        TBLR  SAVCLR     * read word; expect a color
C0352 01E8 002E        ADD   ONE6
C0353 01E9 5835        TBLR  DA         * read word; expect horizontal delt
C0354 01EA 002E        ADD   ONE6
C0355 01EB 5836        TBLR  DB         * read word; expect vertical delta
C0356 01EC 002E        ADD   ONE6
C0357 01ED 5832        TBLR  SAVEA      * read word; expect # of ellipses
C0358 01EE C800        LDPK  0
C0359 01EF 0062        ADD   ONE
C0360 01F0 606C        SACL  CMDADR     * save current table address
C0361 01F1 C806        LDPK  6
C0362                  *
C0363 01F2 3032 SETE0  LAR   AR0,SAVEA
      01F3 FF80        B     SETE1E
      01F4 020E
C0364                  *
C0365                  *
C0366 01F5 702F SETE1  SAR   AR0,SAVAR0
C0367 01F6 2022        LAC   XA
C0368 01F7 6010        SACL  XC         * load center x value
C0369 01F8 2023        LAC   YA
C0370 01F9 6011        SACL  YC         * load center y value
C0371 01FA 2024        LAC   XB
C0372 01FB 6012        SACL  A          * load horizontal axis value
C0373 01FC 0035        ADD   DA         * a = a + da
C0374 01FD 6024        SACL  XB
C0375 01FE 2025        LAC   YB
C0376 01FF 6013        SACL  B          * load vertical axis value
C0377 0200 0036        ADD   DB         * b = b + db
C0378 0201 6025        SACL  YB
C0379 0202 2031        LAC   SAVCLR
C0380 0203 6004        SACL  COLOR      * load a color
C0381 0204 002E        ADD   ONE6
C0382 0205 6031        SACL  SAVCLR
C0383 0206 132E        SUB   ONE6,3     * color = (color + 1) & 7
C0384 0207 F380        BLZ   SETE2
      0208 020B
C0385 0209 202E        LAC   ONE6
```

```
C0386 020A 6031        SACL  SAVCLR      * if color = 0 then color = 1
C0387 020B FE80 SETE2  CALL  FELIPS      * draw a filled ellipse
      020C 06BB
C0388 020D 302F        LAR   AR0,SAVAR0
C0389 020E 5588 SETE1E LARP  AR0
C0390 020F F990        BANZ  SETE1
      0210 01F5
C0391 0211 CE26        RET

C0392
C0393              FILLED ELLIPSE DEMONSTRATION CONTROL
C0394
C0395 0212 206C WELPSC LAC   CMDADR      * get current table address
C0396 0213 C806        LDPK  6
C0397 0214 5822        TBLR  XA
C0398 0215 002E        ADD   ONE6
C0399 0216 5823        TBLR  YA          * read word; expect center x value
C0400 0217 002E        ADD   ONE6
C0401 0218 5824        TBLR  XB          * read word; expect center y value
C0402 0219 002E        ADD   ONE6
C0403 021A 5825        TBLR  YB          * read word; expect horizontal axis
C0404 021B 002E        ADD   ONE6
C0405 021C 5831        TBLR  SAVCLR      * read word; expect vertical axis
C0406 021D 002E        ADD   ONE6
C0407 021E 5835        TBLR  DA          * read word; expect a color
C0408 021F 002E        ADD   ONE6
C0409 0220 5836        TBLR  DB          * read word; expect horizontal delt
C0410 0221 002E        ADD   ONE6
C0411 0222 5805        TBLR  DX          * read word; expect vertical delta
C0412 0223 002E        ADD   ONE6
C0413 0224 5806        TBLR  DY          * read word; expect center x delta
C0414 0225 002E        ADD   ONE6
C0415 0226 5832        TBLR  SAVEA       * read word; expect center y delta
C0416 0227 C800        LDPK  0
C0417 0228 0062        ADD   ONE         * read word; expect # of ellipses
C0418 0229 606C        SACL  CMDADR      * save current table address
C0419 022A C806        LDPK  6
C0420 022B 2037        LAC   XMAX
C0421 022C 1024        SUB   XB
C0422 022D 603B        SACL  XMAXA       * xcmax = xmax - a
C0423 022E 2038        LAC   XMIN
C0424 022F 0024        ADD   XB
C0425 0230 603C        SACL  XMINA       * xcmin = xmin + a
C0426 0231 2039        LAC   YMAX
C0427 0232 1025        SUB   YB
C0428 0233 603D        SACL  YMAXB       * ycmax = ymax - b
C0429 0234 203A        LAC   YMIN
C0430 0235 0025        ADD   YB
C0431 0236 603E        SACL  YMINB       * ycmin = ymin + b
C0432
C0433 0237 3032 WELP0  LAR   AR0,SAVEA
C0434 0238 FF80        B     WELP1E
      0239 027C
C0435
C0436 023A 702F WELP1  SAR   AR0,SAVAR0
C0437 023B 2024        LAC   XB
C0438 023C 6012        SACL  A
C0439 023D 0035        ADD   DA          * load horizontal axis value
```

```
C0440 023E 6024        SACL  XB          * a = a + da
C0441 023F 103F        SUB   AMAX
C0442 0240 F180        BGZ   WELP2       * a > amax ?
      0241 0246
C0443 0242 003F        ADD   AMAX
C0444 0243 1040        SUB   AMIN
C0445 0244 F180        BGZ   WELP3       * a < amin ?
      0245 0249
C0446 0246 2035 WELP2  LAC   DA
C0447 0247 CE23        NEG
C0448 0248 6035        SACL  DA          * da = -da
C0449 0249 2025 WELP3  LAC   YB
C0450 024A 6013        SACL  B           * load vertical axis value
C0451 024B 0036        ADD   DB
C0452 024C 6025        SACL  YB          * b = b + db
C0453 024D 1041        SUB   BMAX
C0454 024E F180        BGZ   WELP4       * b > bmax ?
      024F 0254
C0455 0250 0041        ADD   BMAX
C0456 0251 1042        SUB   BMIN
C0457 0252 F180        BGZ   WELP5       * b < bmin ?
      0253 0257
C0458 0254 2036 WELP4  LAC   DB
C0459 0255 CE23        NEG
C0460 0256 6036        SACL  DB          * db = -db
C0461 0257 2022 WELP5  LAC   XA
C0462 0258 6010        SACL  XC          * load center x value
C0463 0259 0005        ADD   DX
C0464 025A 6022        SACL  XA          * xc = xc + dx
C0465 025B 103B        SUB   XMAXA
C0466 025C F180        BGZ   WELP6       * xc > xcmax ?
      025D 0262
C0467 025E 003B        ADD   XMAXA
C0468 025F 103C        SUB   XMINA
C0469 0260 F180        BGZ   WELP7       * xc < xcmin ?
      0261 0265
C0470 0262 2005 WELP6  LAC   DX
C0471 0263 CE23        NEG
C0472 0264 6005        SACL  DX          * dx = -dx
C0473 0265 2023 WELP7  LAC   YA
C0474 0266 6011        SACL  YC          * load center y value
C0475 0267 0006        ADD   DY
C0476 0268 6023        SACL  YA          * yc = yc + dy
C0477 0269 103D        SUB   YMAXB
C0478 026A F180        BGZ   WELP8       * yc > ycmax ?
      026B 0270
C0479 026C 0030        ADD   YMAXB
C0480 026D 103E        SUB   YMINB
C0481 026E F180        BGZ   WELP9       * yc < ycmin ?
      026F 0273
C0482 0270 2006 WELP8  LAC   DY
C0483 0271 CE23        NEG
C0484 0272 6006        SACL  DY          * dy = -dy
C0485 0273 2031 WELP9  LAC   SAVCLR
C0486 0274 6004        SACL  COLOR
C0487 0275 002E        ADD   ONE6
C0488 0276 D004        ANDK  >0007       * load a color
      0277 0007
```

23. A Graphics Implementation Using the TMS32020 and TMS34061

```
D0002
D0003  0280 CE26  DEMO0  RET
D0004                    *
D0005                    * LINE DRAWING DEMONSTRATION #1
D0006                    *
D0007  0281 C806  DEMO1  LDPK  6
D0008  0282 CAF0         LACK  240
D0009  0283 6022         SACL  XA        * x1 = 240
D0010  0284 CA64         LACK  100
D0011  0285 6023         SACL  YA        * y2 = 100
D0012  0286 CA00         ZAC
D0013  0287 6024         SACL  XB        * x2 = 0
D0014  0288 6025         SACL  YB        * y2 = 0
D0015  0289 CA01         LACK  1
D0016  028A 6031         SACL  SAVCLR    * color = 1
D0017                    *
D0018  028B C0EE  D1L1   LARK  AR0,238
D0019  028C 702F         SAR   AR0,SAVAR0
D0020  028D 2022         LAC   XA
D0021  028E 6000         SACL  X1        * load x1 coordinate value
D0022  028F 002E         ADD   ONE6
D0023  0290 6022         SACL  XA        * x1 = x1 + 1
D0024  0291 2024         LAC   XB
D0025  0292 6002         SACL  X2        * load x2 coordinate value
D0026  0293 0043         ADD   THREE6
D0027  0294 6024         SACL  XB        * x2 = x2 + 3
D0028  0295 2023         LAC   YA
D0029  0296 6001         SACL  Y1        * load y1 coordinate value
D0030  0297 2025         LAC   YB
D0031  0298 6003         SACL  Y2        * load y2 coordinate value
D0032  0299 2031         LAC   SAVCLR
D0033  029A 6004         SACL  COLOR     * load a color
D0034  029B 002E         ADD   ONE6
D0035  029C 6031         SACL  SAVCLR    * color = (color + 1) & 7
D0036  029D 132E         SUB   ONE6,3
D0037  029E F380         BLZ   D1L1A
D0038  029F 02A2
D0039  02A0 202E         LAC   ONE6      * if color = 0 then color = 1
D0040  02A1 6031         SACL  SAVCLR
       02A2 FE80  D1L1A  CALL  LINE      * plot the line
       02A3 0524
D0041  02A4 302F         LAR   AR0,SAVAR0
D0042  02A5 55B8         LARP  AR0
D0043  02A6 FB90         BANZ  D1L1
       02A7 028C
D0044                    *
D0045  02A8 C062  D1L2   LARK  AR0,98
D0046  02A9 702F         SAR   AR0,SAVAR0
D0047  02AA 2022         LAC   XA
D0048  02AB 6000         SACL  X1        * load x1 coordinate value
D0049  02AC 2024         LAC   XB
D0050  02AD 6002         SACL  X2        * load x2 coordinate value
D0051  02AE 2023         LAC   YA
D0052  02AF 6001         SACL  Y1        * load y1 coordinate value
D0053  02B0 002E         ADD   ONE6
D0054  02B1 6023         SACL  YA        * y1 = y1 + 1
D0055  02B2 2025         LAC   YB
```

```
       0277 0007         SACL  SAVCLR    * color = (color + 1) & 7
C0489  0278 6031         CALL  FELIPS    * draw a filled ellipse
C0490  0279 FE80
       027A 06BB
C0491  027B 302F  WELP1E LAR   AR0,SAVAR0
C0492  027C 55B8         LARP  AR0
C0493  027D FB90         BANZ  WELP1
       027E 023A
C0494  027F CE26         RET
C0495                    *
0004                     COPY  LDEMOS.ASM
```

23. A Graphics Implementation Using the TMS32020 and TMS34061

```
D0056 02B3 6003          SACL  Y2          * load y2 coordinate value
D0057 02B4 0043          ADD   THREE6
D0058 02B5 6025          SACL  YB          * y2 = y2 + 3
D0059 02B6 2031          LAC   SAVCLR
D0060 02B7 6004          SACL  COLOR       * load a color
D0061 02B8 102E          SUB   ONE6
D0062 02B9 F180          BGZ   D1L2A       * color = (color - 1) & 7
      02BA 02BC
D0063 02BB CA07          LACK  7
D0064 02BC 6031   D1L2A  SACL  SAVCLR      * if color = 0 then color = 7
D0065 02BD FE80          CALL  LINE        * plot the line
      02BE 0524
D0066 02BF 302F          LAR   AR0,SAVAR0
D0067 02C0 5588          LARP  AR0
D0068 02C1 FB90          BANZ  D1L2
      02C2 02A9
D0069                    *
D0070 02C3 C0EE          LARK  AR0,238
D0071 02C4 702F   D1L3   SAR   AR0,SAVAR0
D0072 02C5 2022          LAC   XA
D0073 02C6 6000          SACL  X1          * load x1 coordinate value
D0074 02C7 102E          SUB   ONE6
D0075 02C8 6022          SACL  XA          * x1 = x1 - 1
D0076 02C9 2024          LAC   XB
D0077 02CA 6002          SACL  X2          * load x2 coordinate value
D0078 02CB 1043          SUB   THREE6
D0079 02CC 6024          SACL  XB          * x2 = x2 - 3
D0080 02CD 2023          LAC   YA
D0081 02CE 6001          SACL  Y1          * load y1 coordinate value
D0082 02CF 2025          LAC   YB
D0083 02D0 6003          SACL  Y2          * load y2 coordinate value
D0084 02D1 2031          LAC   SAVCLR
D0085 02D2 6004          SACL  COLOR       * load a color
D0086 02D3 002E          ADD   ONE6
D0087 02D4 6031          SACL  SAVCLR
D0088 02D5 132E          SUB   ONE6,3
D0089 02D6 F380          BLZ   D1L3A       * color = (color + 1) & 7
      02D7 02DA
D0090 02D8 202E          LAC   ONE6
D0091 02D9 6031          SACL  SAVCLR      * if color = 0 then color = 1
D0092 02DA FE80   D1L3A  CALL  LINE        * plot the line
      02DB 0524
D0093 02DC 302F          LAR   AR0,SAVAR0
D0094 02DD 5588          LARP  AR0
D0095 02DE FB90          BANZ  D1L3
      02DF 02C4
D0096                    *
D0097 02E0 C062          LARK  AR0,98
D0098 02E1 702F   D1L4   SAR   AR0,SAVAR0
D0099 02E2 2022          LAC   XA
D0100 02E3 6000          SACL  X1          * load x1 coordinate value
D0101 02E4 2024          LAC   XB
D0102 02E5 6002          SACL  X2          * load x2 coordinate value
D0103 02E6 2023          LAC   YA
D0104 02E7 6001          SACL  Y1          * load y1 coordinate value
D0105 02E8 102E          SUB   ONE6
D0106 02E9 6023          SACL  YA          * y1 = y1 - 1
```

```
D0107 02EA 2025          LAC   YB
D0108 02EB 6003          SACL  Y2          * load y2 coordinate value
D0109 02EC 1043          SUB   THREE6
D0110 02ED 6025          SACL  YB          * y2 = y2 - 3
D0111 02EE 2031          LAC   SAVCLR
D0112 02EF 6004          SACL  COLOR       * load a color
D0113 02F0 102E          SUB   ONE6
D0114 02F1 F180          BGZ   D1L4A       * color = (color - 1) & 7
      02F2 02F4
D0115 02F3 CA07          LACK  7
D0116 02F4 6031   D1L4A  SACL  SAVCLR      * if color = 0 then color = 7
D0117 02F5 FE80          CALL  LINE        * plot the line
      02F6 0524
D0118 02F7 302F          LAR   AR0,SAVAR0
D0119 02F8 5588          LARP  AR0
D0120 02F9 FB90          BANZ  D1L4
      02FA 02E1
D0121 02FB CE26          RET
D0122                    *
D0123                    *
D0124                    * LINE DRAWING DEMONSTRATION #2
D0125                    *
D0126 02FC C806   DEMO2  LDPK  6
D0127 02FD CA00          ZAC
D0128 02FE 6022          SACL  XA          * x1 = 0
D0129 02FF 6023          SACL  YA          * y1 = 0
D0130 0300 2037          LAC   XMAX
D0131 0301 002E          ADD   ONE6
D0132 0302 6024          SACL  XB          * x2 = xmax + 1
D0133 0303 2039          LAC   YMAX
D0134 0304 002E          ADD   ONE6
D0135 0305 6025          SACL  YB          * y2 = ymax + 1
D0136 0306 202E          LAC   ONE6
D0137 0307 6031          SACL  SAVCLR      * color = 1
D0138 0308 C04F          LARK  AR0,79
D0139 0309 702F   D2L1   SAR   AR0,SAVAR0
D0140 030A 2022          LAC   XA
D0141 030B 6000          SACL  X1          * load x1 coordinate value
D0142 030C 0045          ADD   NINE6
D0143 030D 6022          SACL  XA          * x1 = x1 + 9
D0144 030E 2024          LAC   XB
D0145 030F 6002          SACL  X2          * load x2 coordinate value
D0146 0310 1045          SUB   NINE6
D0147 0311 6024          SACL  XB          * x2 = x2 - 9
D0148 0312 2023          LAC   YA
D0149 0313 6001          SACL  Y1          * load y1 coordinate value
D0150 0314 2025          LAC   YB
D0151 0315 6003          SACL  Y2          * load y2 coordinate value
D0152 0316 2031          LAC   SAVCLR
D0153 0317 6004          SACL  COLOR       * load a color
D0154 0318 002E          ADD   ONE6
D0155 0319 6031          SACL  SAVCLR
D0156 031A 132E          SUB   ONE6,3
D0157 031B F380          BLZ   D2L1A       * color = (color + 1) & 7
      031C 031F
D0158 031D 202E          LAC   ONE6
D0159 031E 6031          SACL  SAVCLR      * if color = 0 then color = 1
```

23. A Graphics Implementation Using the TMS32020 and TMS34061

```
GRAPHIC    32020 FAMILY MACRO ASSEMBLER  PC 1.0 85.157   15:17:30  12-05-85
TMS32020 - TMS34061 DEMONSTRATION                                   PAGE 0019

D00160 031F FE80  D2L1A  CALL  LINE         * plot the line
       0320 0524
D00161 0321 302F         LAR   AR0,SAVAR0
D00162 0322 5588         LARP  AR0
D00163 0323 FB90         BANZ  D2L1
       0324 0309
D00164
D00165 0325 C03B  D2L2   LARK  AR0,59
D00166 0326 702F         SAR   AR0,SAVAR0
D00167 0327 2022         LAC   XA
D00168 0328 6000         SACL  X1           * load x1 coordinate value
D00169 0329 2024         LAC   XB
D00170 032A 6002         SACL  X2           * load x2 coordinate value
D00171 032B 2023         LAC   YA
D00172 032C 6001         SACL  Y1           * load y1 coordinate value
D00173 032D 0044         ADD   FIVE6
D00174 032E 6023         SACL  YA           * y1 = y1 + 5
D00175 032F 2025         LAC   YB
D00176 0330 6003         SACL  Y2           * load y2 coordinate value
D00177 0331 1044         SUB   FIVE6
D00178 0332 6025         SACL  YB           * y2 = y2 - 5
D00179 0333 2031         LAC   COLOR
D00180 0334 6004         SACL  SAVCLR       * load a color
D00181 0335 002E         ADD   ONE6
D00182 0336 6031         SACL  COLOR        * color = (color + 1) & 7
D00183 0337 132E         SUB   ONE6,3
D00184 0338 F380         BLZ   D2L2A        * if color = 0 then color = 1
       0339 033C
D00185 033A 202E         LAC   ONE6
D00186 033B 6031         SACL  COLOR
D00187 033C FE80  D2L2A  CALL  LINE         * plot the line
       033D 0524
D00188 033E 302F         LAR   AR0,SAVAR0
D00189 033F 5588         LARP  AR0
D00190 0340 FB90         BANZ  D2L2
       0341 0326
D00191 0342 CE26         RET
D00192                   *
D00193                   * LINE DRAWING DEMONSTRATION #3
D00194                   *
D00195 0343 C806  DEMO3  LDPK  6
D00196 0344 D001         LALK  360
       0345 0168
D00197 0346 6022         SACL  XA           * x1 = 360
D00198 0347 6024         SACL  XB           * x2 = 360
D00199 0348 CA00         ZAC
D00200 0349 6023         SACL  YA           * y1 = 0
D00201 034A CA96         LACK  150
D00202 034B 6025         SACL  YB           * y2 = 150
D00203 034C CA01         LACK  1
D00204 034D 6031         SACL  COLOR        * color = 1
D00205
D00206 034E C01D  D3L1   LARK  AR0,29
D00207 034F 702F         SAR   AR0,SAVAR0
D00208 0350 2022         LAC   XA
D00209 0351 6000         SACL  X1           * load x1 coordinate value
D00210 0352 2024         LAC   XB
```

```
GRAPHIC    32020 FAMILY MACRO ASSEMBLER  PC 1.0 85.157   15:17:30  12-05-85
TMS32020 - TMS34061 DEMONSTRATION                                   PAGE 0020

D00211 0353 6002         SACL  X2           * load x2 coordinate value
D00212 0354 0243         ADD   THREE6,2     * x2 = x2 + 12
D00213 0355 6024         SACL  XB
D00214 0356 2023         LAC   YA
D00215 0357 6001         SACL  Y1           * load y1 coordinate value
D00216 0358 0044         ADD   FIVE6
D00217 0359 6023         SACL  YA           * y1 = y1 + 5
D00218 035A 2025         LAC   YB
D00219 035B 6003         SACL  Y2           * load y2 coordinate value
D00220 035C 2031         LAC   COLOR
D00221 035D 6004         SACL  SAVCLR       * load a color
D00222 035E 002E         ADD   ONE6
D00223 035F 6031         SACL  COLOR        * color = (color + 1) & 7
D00224 0360 132E         SUB   ONE6,3
D00225 0361 F380         BLZ   D3L1A        * if color = 0 then color = 1
       0362 0365
D00226 0363 202E         LAC   ONE6
D00227 0364 6031         SACL  COLOR
D00228 0365 FE80  D3L1A  CALL  LINE         * plot the line
       0366 0524
D00229 0367 302F         LAR   AR0,SAVAR0
D00230 0368 5588         LARP  AR0
D00231 0369 FB90         BANZ  D3L1
       036A 034F
D00232
D00233 036B C01D  D3L2   LARK  AR0,29
D00234 036C 702F         SAR   AR0,SAVAR0
D00235 036D 2022         LAC   XA
D00236 036E 6000         SACL  X1           * load x1 coordinate value
D00237 036F 2024         LAC   XB
D00238 0370 6002         SACL  X2           * load x2 coordinate value
D00239 0371 1243         SUB   THREE6,2     * x2 = x2 - 12
D00240 0372 6024         SACL  XB
D00241 0373 2023         LAC   YA
D00242 0374 6001         SACL  Y1           * load y1 coordinate value
D00243 0375 0044         ADD   FIVE6
D00244 0376 6023         SACL  YA           * y1 = y1 + 5
D00245 0377 2025         LAC   YB
D00246 0378 6003         SACL  Y2           * load y2 coordinate value
D00247 0379 2031         LAC   COLOR
D00248 037A 6004         SACL  SAVCLR       * load a color
D00249 037B 102E         SUB   ONE6         * color = (color - 1) & 7
D00250 037C F180         BGZ   D3L2A        * if color = 0 then color = 7
       037D 037F
D00251 037E CA07         LACK  7
D00252 037F 6031         SACL  COLOR
D00253 0380 FE80  D3L2A  CALL  LINE         * plot the line
       0381 0524
D00254 0382 302F         LAR   AR0,SAVAR0
D00255 0383 5588         LARP  AR0
D00256 0384 FB90         BANZ  D3L2
       0385 036C
D00257
D00258 0386 C01D  D3L3   LARK  AR0,29
D00259 0387 702F         SAR   AR0,SAVAR0
D00260 0388 2022         LAC   XA
D00261 0389 6000         SACL  X1           * load x1 coordinate value
```

```
00262  038A 2024         LAC    XB           * load x2 coordinate value
00263  038B 6002         SACL   X2
00264  038C 1243         SUB    THREE6,2     * x2 = x2 - 12
00265  038D 6024         SACL   XB
00266  038E 2023         LAC    YA           * load y1 coordinate value
00267  038F 6001         SACL   Y1
00268  0390 1044         SUB    FIVE6        * y1 = y1 - 5
00269  0391 6023         SACL   YA
00270  0392 2025         LAC    YB           * load y2 coordinate value
00271  0393 6003         SACL   Y2
00272  0394 2031         LAC    COLOR        * load a color
00273  0395 6004         SACL   ONE6
00274  0396 002E         ADD    SAVCLR
00275  0397 6031         SACL   COLOR        * color = (color + 1) & 7
00276  0398 132E         SUB    ONE6,3
00277  0399 F380         BLZ    D3L3A
       039A 039D
00278  039B 202E         LAC    ONE6
00279  039C 6031         SACL   SAVCLR       * if colot = 0 then color = 1
00280  039D FE80         CALL   LINE         * plot the line
       039E 0524
00281  039F 302F  D3L3A  LAR    AR0,SAVAR0
00282  03A0 5588         LARP   AR0
00283  03A1 FB90         BANZ   D3L3
       03A2 0387
00284                 *
00285  03A3 C01D  D3L4   LARK   AR0,29
00286  03A4 702F         SAR    AR0,SAVAR0
00287  03A5 2022         LAC    XA           * load x1 coordinate value
00288  03A6 6000         SACL   X1
00289  03A7 2024         LAC    XB           * load x2 coordinate value
00290  03A8 6002         SACL   X2
00291  03A9 0243         ADD    THREE6,2     * x2 = x2 + 12
00292  03AA 6024         SACL   XB
00293  03AB 2023         LAC    YA           * load y1 coordinate value
00294  03AC 6001         SACL   Y1
00295  03AD 1044         SUB    FIVE6        * y1 = y1 - 5
00296  03AE 6023         SACL   YA
00297  03AF 2025         LAC    YB           * load y2 coordinate value
00298  03B0 6003         SACL   Y2
00299  03B1 2031         LAC    COLOR        * load a color
00300  03B2 6004         SACL   ONE6
00301  03B3 102E         SUB    ONE6         * color = (color - 1) & 7
00302  03B4 F180         BGZ    D3L4A
       03B5 03B7
00303  03B6 CA07         LACK   7            * if color = 0 then color = 7
00304  03B7 6031  D3L4A  SACL   SAVCLR
00305  03B8 FE80         CALL   LINE         * plot the line
       03B9 0524
00306  03BA 302F         LAR    AR0,SAVAR0
00307  03BB 5588         LARP   AR0
00308  03BC FB90         BANZ   D3L4
       03BD 03A4
00309  03BE CE26         RET
00310
00311                 *  LINE DRAWING DEMONSTRATION #4
00312                 *
```

```
00313  03BF C806  DEMO4  LDPK   6
00314  03C0 CA00         ZAC
00315  03C1 6022         SACL   XA           * x1 = 0
00316  03C2 6024         SACL   XB           * x2 = 0
00317  03C3 6025         SACL   YB           * y2 = 0
00318  03C4 CA96         LACK   150
00319  03C5 6023         SACL   YA           * y1 = 150
00320  03C6 CA01         LACK   1
00321  03C7 6031         SACL   SAVCLR       * color = 1
00322  03C8 C01E         LARK   AR0,30
00323  03C9 702F  D4L1   SAR    AR0,SAVAR0
00324  03CA 2022         LAC    XA
00325  03CB 6000         SACL   X1           * load x1 coordinate value
00326  03CC 2024         LAC    XB
00327  03CD 6002         SACL   X2           * load x2 coordinate value
00328  03CE 0243         ADD    THREE6,2     * x2 = x2 + 12
00329  03CF 6024         SACL   XB
00330  03D0 2023         LAC    YA           * load y1 coordinate value
00331  03D1 6001         SACL   Y1
00332  03D2 1044         SUB    FIVE6        * y1 = y1 - 5
00333  03D3 6023         SACL   YA
00334  03D4 2025         LAC    YB           * load y2 coordinate value
00335  03D5 6003         SACL   Y2
00336  03D6 2031         LAC    COLOR        * load a color
00337  03D7 6004         SACL   ONE6
00338  03D8 002E         ADD    SAVCLR
00339  03D9 6031         SACL   SAVCLR       * color = (color + 1) & 7
00340  03DA 132E         SUB    ONE6,3
00341  03DB F380         BLZ    D4L1A
       03DC 03DF
       03DD 202E         LAC    ONE6
00342  03DE 6031         SACL   SAVCLR       * if color = 0 then color = 1
00343  03DF FE80  D4L1A  CALL   LINE         * plot the line
00344  03E0 FE80
       03E0 0524
00345  03E1 302F         LAR    AR0,SAVAR0
00346  03E2 5588         LARP   AR0
00347  03E3 FB90         BANZ   D4L1
       03E4 03C9
00348                 *
00349  03E5 2037         LAC    XMAX
00350  03E6 6022         SACL   XA           * x1 = xmax
00351  03E7 D001         LALK   359
       03E8 0167
00352  03E9 6024         SACL   XB           * x2 = 359
00353  03EA CA00         ZAC
00354  03EB 6023         SACL   YA           * y1 = 0
00355  03EC 6025         SACL   YB           * y2 = 0
00356  03ED C01E         LARK   AR0,30
00357  03EE 702F  D4L2   SAR    AR0,SAVAR0
00358  03EF 2022         LAC    XA
00359  03F0 6000         SACL   X1           * load x1 coordinate value
00360  03F1 2024         LAC    XB
00361  03F2 6002         SACL   X2           * load x2 coordinate value
00362  03F3 0243         ADD    THREE6,2     * x2 = x2 + 12
00363  03F4 6024         SACL   XB
00364  03F5 2023         LAC    YA           * load y1 coordinate value
00365  03F6 6001         SACL   Y1
```

23. A Graphics Implementation Using the TMS32020 and TMS34061

```
D0366 03F7 0044         ADD   FIVE6        * y1 = y1 + 5
D0367 03F8 6023         SACL  YA
D0368 03F9 2025         LAC   YB
D0369 03FA 6003         SACL  Y2           * load y2 coordinate value
D0370 03FB 2031         LAC   SAVCLR
D0371 03FC 6004         SACL  COLOR        * load a color
D0372 03FD 102E         SUB   ONE6         * color = (color - 1) & 7
D0373 03FE F180         BGZ   D4L2A
      03FF 0401
D0374 0400 CA07         LACK  7            * if color = 0 then color = 7
D0375 0401 6031         SACL  SAVCLR
D0376 0402 FE80         CALL  LINE         * plot the line
      0403 0524
D0377 0404 302F  D4L2A  LAR   AR0,SAVAR0
D0378 0405 5588         LARP  AR0
D0379 0406 FB90         BANZ  D4L2A
      0407 03EE
                 *
D0380
D0381 0408 2039         LAC   YMAX         * y2 = ymax
D0382 0409 6025         SACL  YB
D0383 040A CA95         LACK  149          * y1 = 149
D0384 040B 6023         SACL  YA
D0385 040C 2037         LAC   XMAX
D0386 040D 6022         SACL  XA           * x1 = xmax
D0387 040E 6024         SACL  XB           * x2 = xmax
D0388 040F C01E         LARK  AR0,30
D0389 0410 702F         SAR   AR0,SAVAR0
      0411 2022
D0390
D0391 0412 6000  D4L3   LAC   XA
D0392 0413 2024         SACL  X1           * load x1 coordinate value
D0393 0414 6002         LAC   XB
D0394 0415 1243         SACL  X2
D0395 0416 6024         SUB   THREE6,2     * x2 = x2 - 12
D0396 0417 2023         SACL  XB
D0397 0418 6001         LAC   YA
D0398 0419 0044         SACL  Y1           * load y1 coordinate value
D0399 041A 6023         ADD   FIVE6        * y1 = y1 + 5
D0400 041B 2025         SACL  YA
D0401 041C 6003         LAC   YB
D0402 041D 2031         SACL  Y2           * load y2 coordinate value
D0403 041E 6004         LAC   SAVCLR
D0404 041F 102E         SACL  COLOR        * load a color
D0405 0420 6031         ADD   ONE6,3       * color = (color + 1) & 7
D0406 0421 132E         SUB   ONE6,3
D0407 0422 F380         BLZ   D4L3A
      0423 0426
D0408 0424 202E         LAC   ONE6
D0409 0425 6031         SACL  SAVCLR       * if color = 0 then color = 1
D0410 0426 FE80         CALL  LINE         * plot the line
      0427 0524
D0411 0428 302F  D4L3A  LAR   AR0,SAVAR0
D0412 0429 5588         LARP  AR0
D0413 042A FB90         BANZ  D4L3
      042B 0410
                 *
D0414
D0415 042C CA00         ZAC                * x1 = 0
D0416 042D 6022         SACL  XA
```

```
D0417 042E D001         LALK  360
      042F 0168
D0418 0430 6024         SACL  XB           * x2 = 360
D0419 0431 2039         LAC   YMAX
D0420 0432 6023         SACL  YA           * y1 = ymax
D0421 0433 6025         SACL  YB           * y2 = ymax
D0422 0434 C01E         LARK  AR0,30
D0423 0435 702F         SAR   AR0,SAVAR0
D4L4
D0424 0436 2022         LAC   XA
D0425 0437 6000         SACL  X1           * load x1 coordinate value
D0426 0438 2024         LAC   XB
D0427 0439 6002         SACL  X2           * load x2 coordinate value
D0428 043A 1243         SUB   THREE6,2     * x2 = x2 - 12
D0429 043B 6024         SACL  XB
D0430 043C 2023         LAC   YA
D0431 043D 6001         SACL  Y1           * load y1 coordinate value
D0432 043E 1044         SUB   FIVE6        * y1 = y1 - 5
D0433 043F 6023         SACL  YA
D0434 0440 2025         LAC   YB
D0435 0441 6003         SACL  Y2           * load y2 coordinate value
D0436 0442 2031         LAC   SAVCLR
D0437 0443 6004         SACL  COLOR        * load a color
D0438 0444 102E         SUB   ONE6         * color = (color - 1) & 7
D0439 0445 F180         BGZ   D4L4A
      0446 0448
D0440 0447 CA07         LACK  7            * if color = 0 then color = 7
D0441 0448 6031         SACL  SAVCLR
D0442 0449 FE80         CALL  LINE         * plot the line
      044A 0524
D0443 044B 302F  D4L4A  LAR   AR0,SAVAR0
D0444 044C 5588         LARP  AR0
D0445 044D FB90         BANZ  D4L4
      044E 0435
D0446 044F CE26         RET
                 *
D0447                        *
D0448                        LINE DRAWING DEMONSTRATION #5
D0449                        *
D0450 0450 C806  DEMO5  LDPK  6
D0451 0451 CA00         ZAC
D0452 0452 6022         SACL  XA           * x1 = 0
D0453 0453 6023         SACL  YA           * y1 = 0
D0454 0454 D001         LALK  360
      0455 0168
D0455 0456 6024         SACL  XB           * x2 = 360
D0456 0457 CA96         LACK  150
D0457 0458 6025         SACL  YB           * y2 = 150
D0458 0459 CA01         LACK  1
D0459 045A 6031         SACL  SAVCLR       * color = 1
D0460                        *
D0461 045B C01D  D5L1   LARK  AR0,29
D0462 045C 702F         SAR   AR0,SAVAR0
D0463 045D 2022         LAC   XA
D0464 045E 6000         SACL  X1           * load x1 coordinate value
D0465 045F 0243         ADD   THREE6,2     * x1 = x1 + 12
D0466 0460 6022         SACL  XA
D0467 0461 2024         LAC   XB
D0468 0462 6002         SACL  X2           * load x2 coordinate value
```

```
00469 0463 0243          ADD   THREE6,2
00470 0464 6024          SACL  XB        * x2 = x2 + 12
00471 0465 2023          LAC   YA
00472 0466 6001          SACL  Y1        * load y1 coordinate value
00473 0467 0044          ADD   FIVE6
00474 0468 6023          SACL  YA        * y1 = y1 + 5
00475 0469 2025          LAC   YB
00476 046A 6003          SACL  Y2        * load y2 coordinate value
00477 046B 1044          SUB   FIVE6
00478 046C 6025          SACL  YB        * y2 = y2 - 5
00479 046D 2031          LAC   COLOR
00480 046E 6004          SACL  SAVCLR    * load a color
00481 046F 002E          ADD   ONE6
00482 0470 6031          SACL  SAVCLR    * color = (color + 1) & 7
00483 0471 132E          SUB   ONE6,3
00484 0472 F380          BLZ   D5L1A
      0473 0476
00485 0474 202E          LAC   ONE6      * if color = 0 then color = 1
00486 0475 6031          SACL  SAVCLR
00487 0476 FE80    D5L1A CALL  LINE      * plot the line
      0477 0524
00488 0478 302F          LAR   AR0,SAVAR0
00489 0479 5588          LARP  AR0
00490 047A FB90          BANZ  D5L1
      047B 045C
00491
00492 047C C01D    D5L2  LARK  AR0,29
00493 047D 702F          SAR   AR0,SAVAR0
00494 047E 2022          LAC   XA
00495 047F 6000          SACL  X1        * load x1 coordinate value
00496 0480 0243          ADD   THREE6,2
00497 0481 6022          SACL  XA        * x1 = x1 + 12
00498 0482 2024          LAC   XB
00499 0483 6002          SACL  X2        * load x2 coordinate value
00500 0484 1243          SUB   THREE6,2
00501 0485 6024          SACL  XB        * x2 = x2 - 12
00502 0486 2023          LAC   YA
00503 0487 6001          SACL  Y1        * load y1 coordinate value
00504 0488 0044          ADD   FIVE6
00505 0489 6023          SACL  YA        * y1 = y1 + 5
00506 048A 2025          LAC   YB
00507 048B 6003          SACL  Y2        * load y2 coordinate value
00508 048C 0044          ADD   FIVE6
00509 048D 6025          SACL  YB        * y2 = y2 + 5
00510 048E 2031          LAC   COLOR
00511 048F 6004          SACL  SAVCLR    * load a color
00512 0490 102E          SUB   ONE6
00513 0491 F180          BGZ   D5L2A     * color = (color - 1) & 7
      0492 0494
00514 0493 CA07          LACK  7         * if color = 0 then color = 7
00515 0494 6031    D5L2A SACL  SAVCLR
00516 0495 FE80          CALL  LINE      * plot the line
      0496 0524
00517 0497 302F          LAR   AR0,SAVAR0
00518 0498 5588          LARP  AR0
00519 0499 FB90          BANZ  D5L2
      049A 047D
```

```
00520
00521 049B C01D    D5L3  LARK  AR0,29
00522 049C 702F          SAR   AR0,SAVAR0
00523 049D 2022          LAC   XA
00524 049E 6000          SACL  X1        * load x1 coordinate value
00525 049F 1243          SUB   THREE6,2
00526 04A0 6022          SACL  XA        * x1 = x1 - 12
00527 04A1 2024          LAC   XB
00528 04A2 6002          SACL  X2        * load x2 coordinate value
00529 04A3 1243          SUB   THREE6,2
00530 04A4 6024          SACL  XB        * x2 = x2 - 12
00531 04A5 2023          LAC   YA
00532 04A6 6001          SACL  Y1        * load y1 coordinate value
00533 04A7 1044          SUB   FIVE6
00534 04A8 6023          SACL  YA        * y1 = y1 - 5
00535 04A9 2025          LAC   YB
00536 04AA 6003          SACL  Y2        * load y2 coordinate value
00537 04AB 0044          ADD   FIVE6
00538 04AC 6025          SACL  YB        * y2 = y2 - 5
00539 04AD 2031          LAC   COLOR
00540 04AE 002E          ADD   ONE6      * load a color
00541 04AF 6031          SACL  COLOR     * color = (color + 1) & 7
00542 04B0 6004          SACL  SAVCLR
00543 04B1 132E          SUB   ONE6,3
00544 04B2 F380          BLZ   D5L3A
      04B3 04B6
00545 04B4 202E          LAC   ONE6      * if color = 0 then color = 1
00546 04B5 6031          SACL  SAVCLR
00547 04B6 FE80    D5L3A CALL  LINE      * plot the line
      04B7 0524
00548 04B8 302F          LAR   AR0,SAVAR0
00549 04B9 5588          LARP  AR0
00550 04BA FB90          BANZ  D5L3
      04BB 049C
00551
00552 04BC C01D    D5L4  LARK  AR0,29
00553 04BD 702F          SAR   AR0,SAVAR0
00554 04BE 2022          LAC   XA
00555 04BF 6000          SACL  X1        * load x1 coordinate value
00556 04C0 1243          SUB   THREE6,2
00557 04C1 6022          SACL  XA        * x1 = x1 - 12
00558 04C2 2024          LAC   XB
00559 04C3 6002          SACL  X2        * load x2 coordinate value
00560 04C4 0243          ADD   THREE6,2
00561 04C5 6024          SACL  XB        * x2 = x2 + 12
00562 04C6 2023          LAC   YA
00563 04C7 6001          SACL  Y1        * load y1 coordinate value
00564 04C8 1044          SUB   FIVE6
00565 04C9 6023          SACL  YA        * y1 = y1 - 5
00566 04CA 2025          LAC   YB
00567 04CB 6003          SACL  Y2        * load y2 coordinate value
00568 04CC 1044          SUB   FIVE6
00569 04CD 6025          SACL  YB        * y2 = y2 - 5
00570 04CE 2031          LAC   COLOR
00571 04CF 6004          SACL  ONE6      * load a color
00572 04D0 102E          SUB   ONE6      * color = (color - 1) & 7
00573 04D1 F180          BGZ   D5L4A
```

23. A Graphics Implementation Using the TMS32020 and TMS34061

```
       0402 04D4        LACK    7        * if color = 0 then color = 7
D0574  04D3 CA07        SACL    SAVCLR
D0575  04D4 6031        CALL    LINE              * plot the line
D0576  04D5 FE80
       04D6 0524   D5L4A
D0577  04D7 302F        LAR     AR0,SAVAR0
D0578  04D8 5588        LARP    AR0
D0579  04D9 FB90        BANZ    D5L4
       04DA 04BD
D0580  04DB CE26        RET
       0005             COPY    CLEAR.ASM
```

```
E0002  ****************
E0003  *
E0004  * ROUTINE:  CLEAR
E0005  *
E0006  *  This routine is called with the display color to be used
E0007  *  to clear the display screen in the least significant byte
E0008  *  of the accumulator.  The following approach is taken:
E0009  *
E0010  *  1.  The present control register contents are read and
E0011  *      saved.
E0012  *
E0013  *  2.  Control Register 2 is modified to allow eight pixels
E0014  *      to be written with each indirect access through the X-Y
E0015  *      register.
E0016  *
E0017  *  3.  The last row of the display memory (off screen) is
E0018  *      filled with the display information.  Each write access
E0019  *      fills eight pixels.  Each row of display memory
E0020  *      corresponds to two lines of CRT data.
E0021  *
E0022  *  4.  The Vertical Start Blank register is read to determine
E0023  *      the size of the active display and to update the vertical
E0024  *      Interrupt register to generate a vertical interrupt at the
E0025  *      next vertical sync.
E0026  *
E0027  *  5.  The Status Register is read and tested until the
E0028  *      vertical interrupt occurs.
E0029  *
E0030  *  6.  Screen refresh and the display are disabled.
E0031  *
E0032  *  7.  The VRAM shift registers are loaded from the last row
E0033  *      of VRAM memory.
E0034  *
E0035  *  8.  The rows in VRAM memory corresponding to the active
E0036  *      display area are loaded with the data in the VRAM shift
E0037  *      registers.
E0038  *
E0039  *  9.  Screen refresh and the display are re-enabled for
E0040  *      normal operation.
E0041  *
E0042  ****************
E0043   04DC            CLEAR   EQU     $
E0044   04DC 7860               SST     ST0        * save current data page pointer
E0045   04DD C809               LDPK    >0009      * VSC register page
E0046   04DE C263               LARK    AR2,FILLC  * internal mem on DP=0
E0047   04DF 558A               LARP    AR2
E0048   04E0 60A0               SACL    *+         * save fill/clear color
E0049   04E1 2030               LAC     CTLR1L     * save Control Register 1
E0050   04E2 60A0               SACL    *+
E0051   04E3 2038               LAC     CTLR1H
E0052   04E4 60A0               SACL    *+
E0053   04E5 2040               LAC     CTLR2L     * save Control Register 2
E0054   04E6 60A0               SACL    *+
E0055   04E7 2048               LAC     CTLR2H
E0056   04E8 60A0               SACL    *+
E0057   04E9 CABC               LACK    >00BC      * X-Y mode RAS override
E0058   04EA 6040               SACL    CTLR2L     * load Control Register 2
```

23. A Graphics Implementation Using the TMS32020 and TMS34061

```
E0059 04EB CAFF         LACK  >00FF        * row 255 of VRAM
E0060 04EC 6870         SACH  XYADRL       * load X-Y Address Register
E0061 04ED 6078         SACL  XYADRH
E0062 04EE C263         LARK  AR2,FILLC
E0063 04EF 2090         LAC   *-
E0064 04F0 C810         LDPK  >0010        * X-Y indirect control page
E0065 04F1 CBFF         RPTK  255          * write 8 pixels 256 times: 2 lines
E0066 04F2 6008         SACL  XINC
E0067 04F3 C808         LDPK  >0008        * VSC register page
E0068 04F4 2060         LAC   VSBLL        * read Vertical Start Blank
E0069 04F5 D004         ANDK  >00FF
      04F6 00FF
E0070 04F7 0868         ADD   VSBLH,8
E0071 04F8 00A0         ADD   *+           * increment to one line below last
E0072 04F9 6080         SACL  *
E0073 04FA C809         LDPK  >0009        * VSC register page
E0074 04FB 6020         SACL  VINTL        * load Vertical Interrupt
E0075 04FC 2880         LAC   *,8
E0076 04FD 6828         SACH  VINTH
E0077 04FE 2050         LAC   STATL
E0078 04FF 2050         LAC   STATL
E0079 0500 D004         ANDK  >0001
      0501 0001
                 WAITI
E0080 0502 F680         BZ    WAITI        * wait for vertical interrupt
      0503 04FF
E0081 0504 2030         LAC   CTLR1L
E0082 0505 D005         ORK   >0020        * turn screen refresh off
      0506 0020
E0083 0507 6030         SACL  CTLR1L
E0084 0508 2048         LAC   CTLR2H
E0085 0509 D004         ANDK  >00DF        * blank display
      050A 00DF
E0086 050B 6048         SACL  CTLR2H
E0087 050C 2080         LAC   *            * load display line count
E0088 050D CE19         SFR                * divide by 2: update 2 lines/write
E0089 050E 6089         SACL  *,0,1        * save for repeat count
E0090 050F D100         LRLK  AR1,>2BFC
      0510 2BFC
E0091 0511 608A         SACL  *,0,2        * load shift register from memory
E0092 0512 C004         LARK  AR0,4
E0093 0513 D100         LRLK  AR1,>2000
      0514 2000
E0094 0515 4BA9         RPT   *+,1         * load loop count for screen clear
E0095 0516 60E0         SACL  *0+          * load memory from shift register
E0096 0517 60E0         SACL  *0+
E0097 0518 558A         LARP  AR2
E0098 0519 20A0         LAC   *+
E0099 051A 6030         SACL  CTLR1L       * load Control Register 1
E0100 051B 20A0         LAC   *+
E0101 051C 6038         SACL  CTLR1H
E0102 051D 20A0         LAC   *+
E0103 051E 6040         SACL  CTLR2L       * load Control Register 2
E0104 051F 20A0         LAC   *+
E0105 0520 6048         SACL  CTLR2H
E0106 0521 C800         LDPK  0
E0107 0522 5060         LST   STS0         * restore data page pointer
E0108 0523 CE26         RET
```

```
0006             COPY    LINES.ASM
```

23. A Graphics Implementation Using the TMS32020 and TMS34061

```
F0002   ************************************************
F0003   * ROUTINE: LINE
F0004   *
F0005   * This routine is called with the following information
F0006   * located in the 5 consecutive words on page 6 labeled as
F0007   * X1, Y1, X2, Y2, COLOR. The following approach is taken to
F0008   * draw a line of the color specified from one of the
F0009   * specified points to the other following Bresenham's Line
F0010   * Algorithm.
F0011   *
F0012   *    octant           \ #2 /
F0013   *   definition      \  #3 | #1 /
F0014   *                  #4 \   |   / #1
F0015   *                       \ | /
F0016   *                  #5 ---X--- #8
F0017   *                       / | \
F0018   *                  #6 /   |   \ #7
F0019   *                    /  #6 | #7 \
F0020   *
F0021   *
F0022   * 1. Determine the values for dx and dy, swapping the
F0023   *    points if necessary to assure that dy is positive.
F0024   *
F0025   *       dx = x2 - x1
F0026   *       dy = y2 - y1
F0027   *
F0028   * This maps octants 5 through 8 onto 1 through 4
F0029   * respectively and reduces the code correspondingly. Each
F0030   * octant is coded seperately to simplify the decisions in
F0031   * selecting between pixels.
F0032   *
F0033   *                d   d
F0034   *                 \ /
F0035   *            c     c       OCTANTS 1 & 2
F0036   *             \   / \
F0037   *              \ /   \
F0038   *          0----0     0----X
F0039   *          a    b     a    b
F0040   *
F0041   * 2. Since the data bus is eight bits wide and addresses
F0042   *    two pixels at a time, it is necessary to determine whether
F0043   *    the first pixel of the line is the left or right pixel of
F0044   *    the pair (BFLAG). For each of the four octants there are
F0045   *    at least two cases depending on the location left/right of
F0046   *    the pixel that is to be modified.
F0047   *
F0048   *   OCTANT 1              OCTANT 2
F0049   *
F0050   * +-----+-----+        +-----+-----+
F0051   * | O O2| O O |        | O2O2| O O | O O2| O2O |
F0052   * +-----+-----+        +-----+-----+
F0053   * | O O2| O O |        | O O | O O2| O O2| O2O |
F0054   * +-----+-----+        +-----+-----+
F0055   * | X O2| O2O |        | X O | O2O |     | O X |
F0056   * +-----+-----+        +-----+-----+
F0057   *
F0058   *   OCTANT 3              OCTANT 4
```

```
F0059   *  +-----+-----+        +-----+-----+
F0060   *  | O X | O O2|        | O2X | O O2| O2O |
F0061   *  +-----+-----+        +-----+-----+
F0062   *  | O O2| O O |        | O O2| O O |
F0063   *  +-----+-----+        +-----+-----+
F0064   *
F0065   * 3. Octants 1 and 2 are seperated from octants 3 and 4 by
F0066   *    determining whether dx is positive or negative (ASIGN).
F0067   *
F0068   * 4. Octant 1 can be seperated from octant 2 by determining
F0069   *    whether the absolute value of dx is greater than dy
F0070   *    (octant 1) or less than dy (octant 2). Correspondingly
F0071   *    octant 3 is seperated from octant 4 by determining whether
F0072   *    the absolute value of dy is less than dx (octant 3) or
F0073   *    greater than dy (octant 4). (ABSIGN)
F0074   *
F0075   * 5. Within the coding for each octant, processing
F0076   *    oscillates through two basic blocks (unique for each
F0077   *    octant). These blocks assume the pixel to be modified is
F0078   *    the left or right pixel of a pixel pair and determine
F0079   *    where the next pixel in the line is located.
F0080   *
F0081   * 6. Two special cases of horizontal and vertical lines are
F0082   *    isolated to speed up the processing for those cases.
F0083   ************************************************
F0084   0524          LINE  EQU  $
F0085   0524 7860           SST  STS0          * save current data page pointer
F0086   0525 C806           LDPK 6
F0087   0526 C008           LARK AR0,8         * set-up access to VSC registers
F0088   0527 D100           LRLK AR1,VSCREG+>0080+XYOFFL
        0528 04E0
F0089   0529 5589           LARP AR1
F0090   052A 2004           LAC  COLOR
F0091   052B 600A           SACL COLORL
F0092   052C 640B           SACL COLORH,4
F0093   052D 0404           ADD  COLOR,4
F0094   052E 600C           SACL COLOR2
F0095   052F 2003   DELTA   LAC  Y2            * dy = y2 - y1
F0096   0530 1001           SUB  Y1
F0097   0531 F480           BGEZ POSY          * test for positive dy
        0532 0530
F0098   0533 4001           ZALH Y1            * negative dy : swap y2 and y1
F0099   0534 4903           ADDS Y2
F0100   0535 6803           SACH Y2
F0101   0536 6001           SACL Y1
F0102   0537 4000           ZALH X1                        swap x2 and x1
F0103   0538 4902           ADDS X2
F0104   0539 6802           SACH X2
F0105   053A 6000           SACL X1
F0106   053B FF80           B    DELTA
        053C 052F
F0107   053D 6006   POSY    SACL DY            * positive dy
F0108   053E F680           BZ   HLINE
        053F 066A
F0109   0540 2002           LAC  X2            * dx = x2 - x1
F0110   0541 1000           SUB  X1
F0111   0542 6005           SACL DX
```

```
F0112 0543 CA20    LACK  >0020      * load X-Y Offset Register (lo byte
F0113 0544 60E0    SACL  *0+
F0114 0545 2F00    LAC   X1,15
F0115 0546 600D    SACL  BFLAG      * bit flag for even/odd bit (0/1)
F0116 0547 DF04    ANDK  >0006,15   * mask for next 2 lsb's of x1
      0548 0006
F0117 0549 68E0    SACH  *0+
F0118 054A 2000    LAC   X1,13      * load X-Y Offset Register (hi byte
F0119 054B 6800    SACH  X1         * seperate 7 msb's of x1
F0120 054C 2701    LAC   Y1,7
F0121 054D 0000    ADD   X1         * add lsb of y1 to x1
F0122 054E 60E0    SACL  *0+
F0123 054F 2F01    LAC   Y1,15      * load low byte of X-Y Address Reg
F0124 0550 6880    SACH  *0+        * seperate 8 msb's of y1
F0125 0551 2005    LAC   DX         * load high byte of X-Y Address Reg
F0126 0552 680E    SACH  ASIGN
F0127 0553 F680    BZ    VLINE      * sign of dx (+/-) (0/-1)
      0554 06A2
F0128 0555 F180    BGZ   POSDX
      0556 0558

F0129 0557 CE23  POSDX  NEG
F0130 0558 6005         SACL  DX
F0131 0559 1006         SUB   DY         * |dx| - dy or a - b
F0132 055A 680F         SACH  ABSIGN     * sign of |dx| - dy (+/-) (0/-1)
F0133 055B F480         BGEZ  OTEST
      055C 0561

F0134 055D 4005         ZALH  DX
F0135 055E 4906         ADDS  DY
F0136 055F 6806         SACH  DY         * |dx| - dy : swap dx & dy
F0137 0560 6005         SACL  DX         * or swap a and b
F0138 0561 2106  OTEST  LAC   DY,1
F0139 0562 6008         SACL  INCR1      * incr1 = 2b > 0
F0140 0563 1005         SUB   DX         * dx corresponds to b
F0141 0564 6007         SACL  D          * d corresponds to a
F0142 0565 1005         SUB   DX         * d = 2b - a
F0143 0566 6009         SACL  INCR2      * incr2 = 2b - 2a < 0
F0144 0567 3005         LAR   AR0,DX     * AR0 = # of pixels in line = a
F0145 0568 D100         LRLK  AR1,XYIND+XYNOP
      0569 0800

F0146 056A D200         LRLK  AR2,XYIND+YINC
      056B 0820
F0147 056C 200F         LAC   ABSIGN
F0148 056D F380         BLZ   O2OR3
      056E 05BC

F0149 056F 200E         LAC   ASIGN
F0150 0570 F380         BLZ   OCT4
      0571 061D

F0151            * * *   DRAW A LINE IN OCTANT #1
F0152
F0153

F0154 0572 D300  OCT1   LRLK  AR3,XYIND+X1Y1
      0573 0828
F0155 0574 D400         LRLK  AR4,XYIND+X1NC
      0575 0808
F0156 0576 200D         LAC   BFLAG
F0157 0577 F380         BLZ   OCTILO
      0578 0592
```

```
                    +----+----+
                    | 0 0?| 0 0 |
                    +----+----+
                    | X 0?| 0 0 |
                    +----+----+

F0158            * * *   OCTANT 1 (LEFT PIXEL)
F0159
F0160
F0161
F0162
F0163
F0164
F0165 0579 2007  OCT1HI  LAC   D
F0166 057A F480          BGEZ  O1INC2
      057B 0588
F0167 057C 0008          ADD   INCR1          * d = d + incr1
F0168 057D 5588          MAR   *,0
F0169 057E F89B          BANZ  OCT1HI,*-,3     * write left pixel only
      057F 05A9
F0170 0580 5589          MAR   *,1
F0171 0581 6007          SACL  D
F0172 0582 CA0F          LACK  >000F
F0173 0583 4E80          AND   COLORH
F0174 0584 4D0B          OR    COLORH
F0175 0585 6080          SACL  B
F0176 0586 FF80          B     DONE
      0587 0667

F0177 0588 0009  O1INC2  ADD   INCR2          * d = d + incr2
F0178 0589 6007          SACL  D
F0179 058A CA0F          LACK  >000F
F0180 058B 4E8A          AND   COLORH
F0181 058C 400B          OR    COLORH
F0182 058D 6080          SACL  OCT1LO,*-,1     * write left pixel only
F0183 058E 6899          SACL  .... 
      058F 0592
F0184 0590 FF80          B     DONE
      0591 0592

F0185
F0186
F0187            * * *   OCTANT 1 (RIGHT PIXEL)
F0188
F0189
F0190
F0191
F0192 0592 2007  OCT1LO  LAC   D
F0193 0593 F380          BLZ   O1INC1
      0594 059F
F0194 0595 0009          ADD   INCR2          * d = d + incr2
F0195 0596 6007          SACL  D
F0196 0597 CAF0          LACK  >00F0
F0197 0598 4E8B          AND   COLORL
F0198 0599 4D0A          OR    COLORL
F0199 059A 6088          SACL  OCT1HI,*-,1     * write right pixel only
F0200 059B FB99          BANZ  OCT1HI
      059C 0579
F0201 059D FF80          B     DONE
      059E 0667
F0202 059F 0008  O1INC1  ADD   INCR1          * d = d + incr1
F0203 05A0 6007          SACL  D
F0204 05A1 CAF0          LACK  >00F0
F0205 05A2 4E8C          AND   COLORL
F0206 05A3 4D0A          OR    COLORL          * write right pixel only
```

```
+----+----+
| 0 0 | 020 |
+----+----+
| 0 X | 020 |
+----+----+
```

23. A Graphics Implementation Using the TMS32020 and TMS34061

```
F0207  05A4 6088          SACL  *,0,0
F0208  05A5 FB99          BANZ  OCT1HI,*-,1
       05A6 0579
F0209  05A7 FF80          B     DONE
       05A8 0667
F0210  05A9 F480   OCT1HL  BGEZ  O1X1Y1
       05AA 0584
F0211  05AB 0008          ADD   INCR1          * d = d + incr1
F0212  05AC 6007          SACL  D              * write both pixels of the pair
F0213  05AD 200C          LAC   COLOR2
F0214  05AE 558C          MAR   *,4
F0215  05AF 6088          SACL  *,0,0
F0216  05B0 FB99          BANZ  OCT1HI,*-,1
       05B1 0579
F0217  05B2 FF80          B     DONE
       05B3 0667
F0218  05B4 0009   O1X1Y1  ADD   INCR2          * d = d + incr2
F0219  05B5 6007          SACL  D              * write both pixels of the pair
F0220  05B6 200C          LAC   COLOR2
F0221  05B7 6088          SACL  *,0,0
F0222  05B8 FB99          BANZ  OCT1HI,*-,1
       05B9 0579
F0223  05BA FF80          B     DONE
       05BB 0667
F0224  05BC 200E   O2OR3   LAC   ASIGN
F0225  05BD F380          BLZ   OCT3
       05BE 05EE

F0227                      * DRAW A LINE IN OCTANT #2
F0228
F0229
F0230  05BF D300   OCT2    LRLK  AR3,XY1N0+XIY1
       05C0 0828
F0231  05C1 200D          LAC   BFLAG
F0232  05C2 F380          BLZ   OCT2LO
       05C3 0507

F0234
F0235
F0236                      * OCTANT 2 (LEFT PIXEL)

              +----+----+
              : O2O2: O O :
              +----+----+
              : X O : O O :
              +----+----+

F0239
F0240  05C4 CAOF   OCT2HI  LACK  >000F
F0241  05C5 4E8A          AND   *,2            * write left pixel only
F0242  05C6 400B          OR    COLORH
F0243  05C7 6088          SACL  *,0,0
F0244  05C8 2007          LAC   D
F0245  05C9 F480          BGEZ  O2INC2
       05CA 05D1
F0246  05CB 0008   O2INC1  ADD   INCR1          * d = d + incr1
F0247  05CC 6007          SACL  D
F0248  05CD FB99          BANZ  OCT2HI,*-,1
       05CE 05C4
F0249  05CF FF80          B     DONE
       05D0 0667
F0250  05D1 0009   O2INC2  ADD   INCR2
```

```
F0251  05D2 6007          SACL  D              * d = d + incr2
F0252  05D3 FB99          BANZ  OCT2LO,*-,1
       05D4 0507
F0253  05D5 FF80          B     DONE
       05D6 0667
F0254
F0255
F0256                      * OCTANT 2 (RIGHT PIXEL)

              +----+----+
              : O O2: O2O :
              +----+----+
              : O X : O O :
              +----+----+

F0260
F0261  05D7 2007   OCT2LO  LAC   D
F0262  05D8 F480          BGEZ  O2X1Y1
       05D9 05E4
F0263  05DA 0008   O2X0Y1  ADD   INCR1          * d = d + incr1
F0264  05DB 6007          SACL  D              * write right pixel only
F0265  05DC CAF0          LACK  >00F0
F0266  05DD 4E8A          AND   *,2
F0267  05DE 400A          OR    COLORL
F0268  05DF 6088          SACL  *,0,0
F0269  05E0 FB99          BANZ  OCT2LO,*-,1
       05E1 0507
F0270  05E2 FF80          B     DONE
       05E3 0667
F0271  05E4 0009   O2X1Y1  ADD   INCR2          * d = d + incr2
F0272  05E5 6007          SACL  D              * write right pixel only
F0273  05E6 CAF0          LACK  >00F0
F0274  05E7 4E8B          AND   *,3
F0275  05E8 400A          OR    COLORL
F0276  05E9 6088          SACL  *,0,0
F0277  05EA FB99          BANZ  OCT2HI,*-,1
       05EB 05C4
F0278  05EC FF80          B     DONE
       05ED 0667
F0279
F0280
F0281                      * DRAW A LINE IN OCTANT #3
F0282  05EE D300   OCT3    LRLK  AR3,XY1N0+XDY1
       05EF 0830
F0283  05F0 200D          LAC   BFLAG
F0284  05F1 F380          BLZ   OCT3LO
       05F2 060A

F0286
F0287                      * OCTANT3 (LEFT PIXEL)

              +----+----+
              : O O2: O2O :
              +----+----+
              : O O : X O :
              +----+----+

F0291
F0292  05F3 2007   OCT3HI  LAC   D
F0293  05F4 F480          BGEZ  O3XDY1
       05F5 0600
F0294  05F6 0008   O3X0Y1  ADD   INCR1          * d = d + incr1
F0295  05F7 6007          SACL  D              * write left pixel only
F0296  05F8 CAOF          LACK  >000F
F0297  05F9 4E8A          AND   *,2
```

```
F0298 05FA 4D0B          OR    COLORH
F0299 05FB 6088          SACL  *,0,0
F0300 05FC FB99          BANZ  OCT3HI,*-,1
      05FD 05F3
F0301 05FE FF80          B     DONE
      05FF 0667
F0302 0600 0009  O3XDY1  ADD   INCR2        * d = d + incr2
F0303 0601 6007          SACL  D            * write left pixel only
F0304 0602 CA0F          LACK  >00F0
F0305 0603 4E8B          AND   COLORL
F0306 0604 4D0B          OR    COLORH
F0307 0605 6088          SACL  *,0,0
F0308 0606 FB99          BANZ  OCT3LO,*-,1
      0607 060A
F0309 0608 FF80          B     DONE
      0609 0667
F0310
F0311                    * * * *
F0312                    * OCTANT 3 (RIGHT PIXEL)
F0313                    * * * *
F0314
F0315
F0316
F0317 060A CAF0  OCT3LO  LACK  >00F0
F0318 060B 4E8A          AND   COLORL
F0319 060C 4D0A          OR    COLORH
F0320 060D 6088          SACL  *,0,0
F0321 060E 2007          LAC   D
F0322 060F F480          BGEZ  O31NC2
      0610 0617
F0323 0611 0008  O31NC1  ADD   INCR1        * write right pixel only
F0324 0612 6007          SACL  D            * d = d + incr1
F0325 0613 FB99          BANZ  OCT3LO,*-,1
      0614 060A
F0326 0615 FF80          B     DONE
      0616 0667
F0327 0617 0009  O31NC2  ADD   INCR2        * d = d + incr2
F0328 0618 6007          SACL  D
F0329 0619 FB99          BANZ  OCT3HI,*-,1
      061A 05F3
F0330 061B FF80          B     DONE
      061C 0667
F0331
F0332                    * * * *
F0333                    * OCTANT 4 (LEFT PIXEL)
F0334                    * * * *
F0335
F0336
F0334 061D 0300  OCT4    LRLK  AR3,XYIND+XDYI   * DRAW A LINE IN OCTANT #4
      061E 0830
F0335 061F 0400          LRLK  AR4,XYIND+XDEC
      0620 0810
F0336 0621 200D          LAC   BFLAG
F0337 0622 F380          BLZ   OCT4LO
      0623 063B
F0338
F0339
F0340
F0341
F0342
```

```
+-----+-----+
| 0 02| 0 0 |
+-----+-----+
| 0 02| X 0 |
+-----+-----+
```

```
F0343                    *
F0344                    *
F0345 0624 2007  OCT4HI  LAC   D
F0346 0625 F480          BGEZ  O4XDY1
      0626 0631
F0347 0627 0008  O4XDY0  ADD   INCR1        * d = d + incr1
F0348 0628 6007          SACL  D            * write left pixel only
F0349 0629 CA0F          LACK  >000F
F0350 062A 4E8C          AND   COLORH
F0351 062B 4D0B          OR    *,4
F0352 062C 6088          SACL  *,0,0
F0353 062D FB99          BANZ  OCT4LO,*-,1
      062E 063B
F0354 062F FF80          B     DONE
      0630 0667
F0355 0631 0009  O4XDY1  ADD   INCR2        * d = d + incr2
F0356 0632 6007          SACL  D            * write left pixel only
F0357 0633 CA0F          LACK  >000F
F0358 0634 4E8B          AND   COLORH
F0359 0635 4D0B          OR    *,3
F0360 0636 6088          SACL  *,0,0
F0361 0637 FB99          BANZ  OCT4LO,*-,1
      0638 063B
F0362 0639 FF80          B     DONE
      063A 0667
F0363
F0364                    * * * *
F0365                    * OCTANT 4 (RIGHT PIXEL)
F0366                    * * * *
F0367
F0368
F0369
F0370 063B 2007  OCT4LO  LAC   D
F0371 063C F480          BGEZ  O4INC2
      063D 064A
F0372 063E 0008          ADD   INCR1        * d = d + incr1
F0373 063F 5588          MAR   *,4          * write right pixel only
F0374 0640 FB9C          BANZ  OCT4HI,*-,4
      0641 0654
F0375 0642 5589          MAR   *,1
F0376 0643 6007          SACL  D
F0377 0644 CAF0          LACK  >00F0
F0378 0645 4E80          AND   COLORL
F0379 0646 4D0A          OR    COLORH
F0380 0647 6080          SACL  *
F0381 0648 FF80          B     DONE
      0649 0667
F0382 064A 0009  O4INC2  ADD   INCR2        * d = d + incr2
F0383 064B 6007          SACL  D            * write right pixel only
F0384 064C CAF0          LACK  >00F0
F0385 064D 4E8A          AND   COLORL
F0386 064E 4D0A          OR    COLORH
F0387 064F 6088          SACL  *,0,0
F0388 0650 FB99          BANZ  OCT4HI,*-,1
      0651 0624
F0389 0652 FF80          B     DONE
      0653 0667
```

```
+-----+-----+
| 020 | 0 0 |
+-----+-----+
| 02X | 0 0 |
+-----+-----+
```

23. A Graphics Implementation Using the TMS32020 and TMS34061

```
F0390 0654 F380 OCT4HL BLZ  O41NC1
      0655 065F
F0391 0656 0009        ADD  INCR2      * d = d + incr2
F0392 0657 6007        SACL D          * write both pixels of the pair
F0393 0658 200C        LAC  COLOR2
F0394 0659 558B        MAR  *,3
F0395 065A 6088        SACL *,0,0
F0396 065B FB99        BANZ OCT4LO,*-,1
      065C 063B
F0397 065D FF80        B    DONE
      065E 0667
F0398 065F 0008 O41NC1 ADD  INCR1      * d = d + incr1
F0399 0660 6007        SACL D          * write both pixels of the pair
F0400 0661 200C        LAC  COLOR2
F0401 0662 6088        SACL *,0,0
F0402 0663 FB99        BANZ OCT4LO,*-,1
      0664 063B
F0403 0665 FF80        B    DONE
      0666 0667
*
F0404 0667 C800 DONE   LDPK 0          * restore entry data page pointer
F0405 0668 5060        LST  STS0
F0406 0669 CE26        RET
*
*     DRAW A HORIZONTAL LINE
*
F0407
F0408
F0409
F0410
F0411 066A 2002 HLINE  LAC  X2         * dx = x2 - x1
F0412 066B 1000        SUB  X1
F0413 066C 6005        SACL DX
F0414 066D F480        BGEZ DXPOS
      066E 0673
F0415 066F CE23        NEG             * save |dx|
F0416 0670 6005        SACL DX
F0417 0671 2002        LAC  X2         * swap x1 & x2 : draw left to right
F0418 0672 6000        SACL X1
F0419 0673 CA20 DXPOS  LACK >0020      * load X-Y Offset Register (lo byte
F0420 0674 60E0        SACL *0+
F0421 0675 2F00        LAC  X1,15
F0422 0676 600D        SACL BFLAG      * bit flag for even/odd bit (0/1)
F0423 0677 DF04        ANDK >0006,15   * mask for next 2 lsb's of x1
      0678 0006
F0424 0679 68E0        SACH *0+        * load X-Y Offset Register (hi byte
F0425 067A 2D00        LAC  X1,13      * seperate 7 msb's of x1
F0426 067B 6800        SACH X1
F0427 067C 2701        LAC  Y1,7       * add lsb of y1 to x1
F0428 067D 0000        ADD  X1
F0429 067E 60E0        SACL *0+        * load low byte of X-Y Address Reg
F0430 067F 2F01        LAC  Y1,15      * seperate 8 msb's of y1
F0431 0680 6880        SACH *0+        * load high byte of X-Y Address Reg
F0432 0681 3005        LAR  AR0,DX     * AR0 = # of pixels in line
F0433 0682 D100        LRLK AR1,XY1ND+XYNOP
      0683 0800
F0434 0684 D200        LRLK AR2,XY1ND+XINC
      0685 0808
F0435 0686 2000        LAC  BFLAG
F0436 0687 F380        BLZ  HOZLO
      0688 0699
```

```
F0437 0689 5588        LARP AR0
F0438 068A FB9A HOZHI  BANZ FILL2,*-,2
      068B 0693
F0439 068C 5589        LARP AR1        * write left pixel only
F0440 068D CA0F        LACK >000F
F0441 068E 4E80        AND  COLORH
F0442 068F 400B        OR   *
F0443 0690 6080        SACL *
F0444 0691 FF80        B    DONE
      0692 0667
F0445 0693 200C FILL2  LAC  COLOR2     * write both pixels of the pair
F0446 0694 6088        SACL *,0,0
F0447 0695 FB98        BANZ HOZH1,*-,0
      0696 068A
F0448 0697 FF80        B    DONE
      0698 0667
F0449 0699 5589 HOZLO  LARP AR1        * write right pixel only
F0450 069A CAF0        LACK >00F0
F0451 069B 4E8A        AND  COLORL
F0452 069C 4D0A        OR   *,2
F0453 069D 6088        SACL *,0,0
F0454 069E FB98        BANZ HOZH1,*-,0
      069F 068A
F0455 06A0 FF80        B    DONE
      06A1 0667
*
*     DRAW A VERTICAL LINE
*
F0456
F0457
F0458
F0459 06A2 3006 VLINE  LAR  AR0,DY     * AR0 = # of pixels in line
F0460 06A3 D100        LRLK AR1,XY1ND+XYNOP
      06A4 0800
F0461 06A5 D200        LRLK AR2,XY1ND+YINC
      06A6 0820
F0462 06A7 5589        LARP AR1        * write left pixel
F0463 06A8 200D        LAC  BFLAG
F0464 06A9 F380        BLZ  VRTLO
      06AA 06B3
F0465 06AB CA0F VRTHI  LACK >000F      * write left pixel
F0466 06AC 4E80        AND  COLORH
F0467 06AD 400B        OR   *,0,0
F0468 06AE 6088        SACL *,0,0
F0469 06AF FB99        BANZ VRTH1,*-,1
      06B0 06AB
F0470 06B1 FF80        B    DONE
      06B2 0667
F0471 06B3 CAF0 VRTLO  LACK >00F0      * write right pixel
F0472 06B4 4E8A        AND  COLORL
F0473 06B5 400A        OR   *,2
F0474 06B6 6088        SACL *,0,0
F0475 06B7 FB99        BANZ VRTLO,*-,1
      06B8 06B3
F0476 06B9 FF80        B    DONE
      06BA 0667
*
      0007             COPY FELIPS2.ASM
```

23. A Graphics Implementation Using the TMS32020 and TMS34061

```
G0002        *****************************************
G0003        *
G0004        * ROUTINE: FELIPS
G0005        * This routine is called with the following information
G0006        * located in 5 words labeled as XC, YC, A, B,
G0007        * COLOR. Using this information, a solid filled ellipse of
G0008        * the specified color is drawn centered at (xc,yc) with
G0009        * x-axis intercept a and y-axis intercept b. A four-way
G0010        * stepping algorithm is used to determine the boundaries of
G0011        * the ellipse. The x coordinate value of the first quadrant
G0012        * end point of each horizontal line used to construct the
G0013        * ellipse are stored in consecutive locations on internal
G0014        * data ram pages 4 and 5.
G0015        *
G0016        *****************************************
G0017
G0018  06BB        FELIPS  EQU   $
G0019  06BB 7860           SST   STS0         * save current data page pointer
G0020  06BC C809           LDPK  >0009        * VSC register page
G0021  06BD C264           LARK  AR2,SAVREG   * internal mem on DP=0
G0022  06BE 558A           LARP  AR2
G0023  06BF 2030           LAC   CTLR1L       * save Control Register 1
G0024  06C0 60A0           SACL  *+
G0025  06C1 2038           LAC   CTLR1H
G0026  06C2 60A0           SACL  *+
G0027  06C3 2040           LAC   CTLR2L       * save Control Register 2
G0028  06C4 60A0           SACL  *+
G0029  06C5 2048           LAC   CTLR2H
G0030  06C6 60A0           SACL  *+
G0031  06C7 C806           LDPK  6
G0032  06C8 D100           LRLK  AR1,>0200    * set-up access to data stack
       06C9 0200
G0033  06CA 5589           LARP  AR1
G0034  06CB 2004           LAC   COLOR
G0035  06CC 600A           SACL  COLORL
G0036  06CD 640B           SACL  COLORH,4
G0037  06CE 0404           ADD   COLOR,4
G0038  06CF 600C           SACL  COLOR2
G0039
G0040        * * *
G0041          CALCULATE CURVE DRAWING PARAMETERS
G0042  0600 3212           LAR   AR2,A        * first stack value
G0043  0601 3C12           LT    A
G0044  0602 3812           MPY   A            * square horizontal axis value A
G0045  0603 CE14           PAC
G0046  0604 6814           SACH  T1H          * T1 = A**2
G0047  0605 6015           SACL  T1L
G0048  0606 CE15           APAC
G0049  0607 6816           SACH  T2H          * T2 = 2*T1
G0050  0608 6017           SACL  T2L
G0051  0609 3C13           LT    B
G0052  060A 3813           MPY   B            * square vertical axis value B
G0053  060B CE14           PAC
G0054  060C 6818           SACH  T3H          * T3 = B**2
G0055  060D 6019           SACL  T3L
G0056  060E CE15           APAC
G0057  060F 681A           SACH  T4H          * T4 = 2*T3
```

```
G0058  06E0 601B           SACL  T4L
G0059  06E1 3C12           LT    A
G0060  06E2 3818           MPY   T3H          * T5 = T3*A
G0061  06E3 CE14           PAC
G0062  06E4 CB0F           RPTK  15
G0063  06E5 CE18           SFL
G0064  06E6 3819           MPY   T3L
G0065  06E7 CE15           APAC
G0066  06E8 681C           SACH  T5H
G0067  06E9 601D           SACL  T5L
G0068  06EA 691E           SACH  T6H,1        * T6 = 2*T5
G0069  06EB 611F           SACL  T6L,1
G0070  06EC CA00           ZAC
G0071  06ED 6020           SACL  T7L          * T7 = 0
G0072  06EE 6021           SACL  T7H
G0073  06EF 4014           ZALH  T1H
G0074  06F0 4915           ADDS  T1L
G0075  06F1 4818           ADDH  T3H
G0076  06F2 4919           ADDS  T3L
G0077  06F3 CE19           SFR
G0078  06F4 CE19           SFR
G0079  06F5 441C           SUBH  T5H          * D = (T1 + T3)/4 - T5
G0080  06F6 451D           SUBS  T5L
G0081  06F7 6826           SACH  DH
G0082  06F8 6027           SACL  DL
G0083
G0084          CALCULATE ELLIPSE BOUNDARY FOR ONE QUADRANT
G0085
G0086  06F9 72A0           SAR   AR2,*+
G0087  06FA 2026           LAC   DH
G0088  06FB F480           BGEZ  STEPX
       06FC 071B
G0089  06FD 4020    STEPY  ZALH  T7H
G0090  06FE 4921           ADDS  T7L
G0091  06FF 4816           ADDH  T2H
G0092  0700 4917           ADDS  T2L          * T7 = T7 + T2
G0093  0701 6820           SACH  T7H
G0094  0702 6021           SACL  T7L
G0095  0703 4826           ADDH  DH
G0096  0704 4927           ADDS  DL
G0097  0705 6826           SACH  DH           * D = D + T7
G0098  0706 6027           SACL  DL
G0099  0707 72A0           SAR   AR2,*+       * save x-value on data stack
G0100  0708 2026           LAC   DH
G0101  0709 F380           BLZ   STEPY
       070A 06FD
G0102  070B FF80           B     STEPX
       070C 071B
G0103  070D 401E    LOOPX  ZALH  T6H
G0104  070E 491E           ADDS  T6L
G0105  070F 441A           SUBH  T4H
G0106  0710 451B           SUBS  T4L
G0107  0711 681E           SACH  T6H
G0108  0712 601F           SACL  T6L          * T6 = T6 - T4
G0109  0713 CE23           NEG
G0110  0714 4826           ADDH  DH
G0111  0715 4927           ADDS  DL
```

23. A Graphics Implementation Using the TMS32020 and TMS34061

```
G0112 0716 6826       SACH  DH          * D = D - T6
G0113 0717 6027       SACL  DL
G0114 0718 2026       LAC   DH
G0115 0719 F380       BLZ   STEPY
      071A 06FD
G0116 071B 558A STEPX LARP  2           * continue until X at origin
G0117 071C FB99       BANZ  LOOPX,*-,1
      071D 0700
G0118 071E 712C PLOT1 SAR   AR1,PIXCNT
G0119 071F 2013       LAC   B
G0120 0720 6046       SACL  SAVEB
G0121 0721 092E       ADD   ONE6,9      * plotting stack = b ?
G0122 0722 102C       SUB   PIXCNT
G0123 0723 F380       BLZ   PLOT2
      0724 072B
G0124 0725 602C       SACL  PIXCNT
G0125 0726 322C       LAR   AR2,PIXCNT
G0126 0727 CA00       ZAC               * zero remainder of stack
G0127 0728 60AA NILPT SACL  *+,0,2
G0128 0729 FB99       BANZ  NILPT,*-,1
      072A 0728
*
*                     DRAW "SHORT" HORIZONTAL LINES FOR UPPER HALF
*
G0129
G0130
G0131
G0132 072B D400 PLOT2 LRLK  AR4,VSCREG+>0080+XYOFFL
      072C 04E0
G0133 072D 558C       LARP  AR4
G0134 072E CA20       LACK  >0020
G0135 072F 6089       SACL  *,0,1       * load X-Y Offset Register (lo byte)
G0136 0730 3213       LAR   AR2,B       * number of values on data stack
G0137 0731 0300       LRLK  AR3,XY1ND+XYNOP
      0732 0800
G0138 0733 5590 LOOP1 MAR   *-
G0139 0734 7213       SAR   AR2,B
G0140 0735 2010       LAC   XC
G0141 0736 1080       SUB   *
G0142 0737 F480       BGEZ  OSCRX1
      0738 073A
G0143 0739 CA00 OSCRX1 ZAC              * X1 = XC - X >= 0
G0144 073A 6000       SACL  X1
G0145 073B 2010       LAC   XC
G0146 073C 009C       ADD   *-,0,4      * X2 = XC + X
G0147 073D 6002       SACL  X2
G0148 073E 2002       LAC   X2,13       * seperate 7 MSBs from 3 LSBs (X2)
G0149 073F 682A       SACH  X2H
G0150 0740 D004       ANDK  >FFFF
      0741 FFFF
G0151 0742 6C2B       SACH  X2L,4       * seperate 7 MSBs from 3 LSBs (X1)
G0152 0743 2000       LAC   X1,13
G0153 0744 6828       SACH  X1H
G0154 0745 D004       ANDK  >FFFF
      0746 FFFF
G0155 0747 6C29       SACH  X1L,4
G0156 0748 202A       LAC   X2H
G0157 0749 1028       SUB   X1H
G0158 074A F580       BNZ   WIDE        * endpoints in same 8 pixel group ?
      074B 0797
```

```
                                                              * "short" line
G0159 074C 2011       LAC   YC
G0160 074D 1013       SUB   B
G0161 074E F480       BGEZ  OSCRY1
      074F 0751
G0162 0750 CA00 OSCRY1 ZAC              * Y1 = YC - Y >= 0
G0163 0751 6001       SACL  Y1
G0164 0752 C008       LARK  AR0,8       * set-up access to VSC registers
G0165 0753 D400       LRLK  AR4,VSCREG+>0080+XYOFFH
      0754 04E8
G0166 0755 558C       LARP  AR4         * load XY coordinates for upper lin
G0167 0756 2E29       LAC   X1L,14
G0168 0757 68E0       SACH  *+
G0169 0758 2701       LAC   Y1,7
G0170 0759 0028       ADD   X1H
G0171 075A 60E0       SACL  *+
G0172 075B 2F01       LAC   Y1,15
G0173 075C 6888       SACH  *,0,0
G0174 075D 2F2B       LAC   X2L,15
G0175 075E 1F29       SUB   X1L,15
G0176 075F 682C       SACH  PIXCNT      * # pixels = X2 - X1
G0177 0760 302C       LAR   AR0,PIXCNT
G0178 0761 D400       LRLK  AR4,XY1ND+XINC
      0762 0808
G0179 0763 9129       BIT   X1L,1       * start on an odd pixel ?
G0180 0764 F980       BBNZ  SLHIU
      0765 0776
G0181 0766 558B SLLOU LARP  AR3
G0182 0767 CAF0       LACK  >00F0       * pixel in lsnibble only
G0183 0768 4E8C       AND   *,4
G0184 0769 4D0A       OR    COLORL
G0185 076A 6088       SACL  *,0,0
G0186 076B FB90       BANZ  SLHIU
      076C 0776
G0187 076D FF80       B     DONE1U
      076E 077C
G0188 076F 558C SLHLU LARP  AR4
G0189 0770 200C       LAC   COLOR2      * pixel pair
G0190 0771 6088       SACL  *,0,0
G0191 0772 FB90       BANZ  SLHIU
      0773 0776
G0192 0774 FF80       B     DONE1U
      0775 077C
G0193 0776 FB9B SLHIU BANZ  SLHLU,*-,3
      0777 076F
G0194 0778 CA0F       LACK  >000F       * pixel in msnibble only
G0195 0779 4E80       AND   *
G0196 077A 4D0B       OR    COLORH
G0197 077B 6080       SACL  *
G0198 077C 558A DONE1U LARP AR2
G0199 077D FB99       BANZ  LOOP1,*-,1
      077E 0734
G0200 077F FF80       B     PLOT3
      0780 07CD
*
*                     DRAW "LONG" HORIZONTAL LINES FOR UPPER HALF
*
G0201
G0202
G0203
G0204 0781 7213 LOOP2 SAR   AR2,B
```

23. A Graphics Implementation Using the TMS32020 and TMS34061

```
GRAPHIC    32020 FAMILY MACRO ASSEMBLER   PC 1.0 85.157    15:17:30 12-05-85
TMS32020 - TMS34061 DEMONSTRATION                            PAGE 0045

G0205 0782 2010          LAC   XC
G0206 0783 1080          SUB   *
G0207 0784 F480          BGEZ  OSCRX2
      0785 0787
G0208 0786 CA00          ZAC                  * X1 = XC - X >= 0
G0209 0787 6000  OSCRX2  SACL  X1
G0210 0788 2010          LAC   XC
G0211 0789 009C          ADD   *-,0,4
G0212 078A 6002          SACL  X2             * X2 = XC + X
G0213 078B 2002          LAC   X2,13
G0214 078C 682A          SACH  X2H            * seperate 7 MSBs from 3 LSBs (X2)
G0215 078D 0004          ANDK  >FFFF
      078E FFFF
G0216 078F 6C2B          SACH  X2L,4
G0217 0790 2000          LAC   X1,13
G0218 0791 6828          SACH  X1H
G0219 0792 0004          ANDK  >FFFF
      0793 FFFF
G0220 0794 6C29          SACH  X1L,4          * seperate 7 MSBs from 3 LSBs (X1)
G0221 0795 202A          LAC   X2H
G0222 0796 1028          SUB   X1H
G0223 0797 112E          SUB   ONE6,1         * "long" line
G0224 0798 602D  WIDE    SACL  PAXCNT         * # 8 pixel groups
G0225 0799 2011          LAC   YC
G0226 079A 1013          SUB   B
G0227 079B F480          BGEZ  OSCRY2
      079C 079E
G0228 079D CA00          ZAC                  * Y1 = YC - Y >= 0
G0229 079E 6001  OSCRY2  SACL  Y1
G0230 079F C008          LARK  AR0,8
G0231 07A0 D400          LRLK  AR4,VSCREG+>0080+XYOFFH
      07A1 04E8
G0232 07A2 558C          LARP  AR4
G0233 07A3 2E29          LAC   X1L,14         * load XY coordinates for upper lin
G0234 07A4 68E0          SACH  *+
G0235 07A5 2701          LAC   Y1,7
G0236 07A6 0028          ADD   X1H
G0237 07A7 60E0          SACL  *+
G0238 07A8 2F01          LAC   Y1,15
G0239 07A9 6888          SACH  *+,0,0
G0240 07AA D000          LRLK  AR0,VSCREG+>0080+CTLR2L
      07AB 04C0
G0241 07AC D400          LRLK  AR4,XYIN0+XINC
      07AD 0808
G0242 07AE CA00          ZAC
G0243 07AF 608B          SACL  *,0,3          * load Control Register 2
G0244 07B0 D001          LALK  LSBASE
      07B1 0878
G0245 07B2 0029          ADD   X1L
G0246 07B3 CE24          CALA
G0247 07B4 CABC          LACK  >00BC          * write pixels left of blocks
G0248 07B5 608C          SACL  *,0,4          * X-Y mode RAS override
G0249 07B6 202D          LAC   PAXCNT         * load Control Register 2
G0250 07B7 F380          BLZ   RIGHT1
      07B8 07BC
G0251 07B9 200C          LAC   COLOR2
G0252 07BA 4B2D          RPT   PAXCNT         * fill 8 pixel blocks
```

```
GRAPHIC    32020 FAMILY MACRO ASSEMBLER   PC 1.0 85.157    15:17:30 12-05-85
TMS32020 - TMS34061 DEMONSTRATION                            PAGE 0046

G0253 07BB 6080          SACL  *
G0254 07BC 5588  RIGHT1  LARP  AR0
G0255 07BD CA00          ZAC
G0256 07BE 6080          SACL  *              * load Control Register 2
G0257 07BF D000          LRLK  AR0,VSCREG+>0080+XYOFFH
      07C0 04E8
G0258 07C1 6080          SACL  *
G0259 07C2 D000          LRLK  AR0,VSCREG+>0080+XYADRL
      07C3 04F0
G0260 07C4 2080          LAC   *
G0261 07C5 608C          SACL  *,0,4
G0262 07C6 0001          ADD   RSBASE
      07C7 0888
G0263 07C8 002B          ADD   X2L
G0264 07C9 CE24          CALA                 * write pixels right of blocks
G0265 07CA 558A          LARP  AR2
G0266 07CB F899          BANZ  LOOP2,*-,1
      07CC 0781

* DRAW "LONG" HORIZONTAL LINES FOR LOWER HALF
*
*
G0267
G0268
G0269
G0270 07CD D400  PLOT3   LRLK  AR4,VSCREG+>0080+XYOFFL
      07CE 04E0
G0271 07CF 558C          LARP  AR4
G0272 07D0 CA20          LACK  >0020
G0273 07D1 6089          SACL  *,0,1          * load X-Y Offset Register(lo byte)
G0274 07D2 3246          LAR   AR2,SAVEB      * # of values on data stack
G0275 07D3 D300          LRLK  AR3,XYIN0+XYNOP
      07D4 0800
G0276 07D5 55A0  LOOP3   MAR   *+
G0277 07D6 55A0          MAR   *+
G0278 07D7 2011          LAC   YC
G0279 07D8 0046          ADD   SAVEB
G0280 07D9 6011          SACL  YC
G0281 07DA FF80          B     LOOPE
      07DB 0824
G0282 07DC 7213          SAR   AR2,B
G0283 07DD 2010          LAC   XC
G0284 07DE 1080          SUB   *
G0285 07DF F480          BGEZ  OSCRX3
      07E0 07E2
G0286 07E1 CA00          ZAC
G0287 07E2 6000  OSCRX3  SACL  X1             * X1 = XC - X >= 0
G0288 07E3 2010          LAC   XC
G0289 07E4 00AC          ADD   *+,0,4
G0290 07E5 6002          SACL  X2             * X2 = XC + X
G0291 07E6 2002          LAC   X2,13
G0292 07E7 682A          SACH  X2H
G0293 07E8 0004          ANDK  >FFFF          * seperate 7 MSBs from 3 LSBs (X2)
      07E9 FFFF
G0294 07EA 6C2B          SACH  X2L,4
G0295 07EB 2000          LAC   X1,13
G0296 07EC 6828          SACH  X1H
G0297 07ED 0004          ANDK  >FFFF
      07EE FFFF
G0298 07EF 6C29          SACH  X1L,4          * seperate 7 MSBs from 3 LSBs (X1)
G0299 07F0 202A          LAC   X2H
```

23. A Graphics Implementation Using the TMS32020 and TMS34061

```
G0300 07F1 1028    SUB    X1H
G0301 07F2 F680    BZ     NARROW    * endpoints in same 8 pixel group ?
      07F3 083D
G0302 07F4 112E    SUB    ONE6,1    * "long" line
G0303 07F5 602D    SACL   PAXCNT    * # 8 pixel groups
G0304 07F6 2011    LAC    YC
G0305 07F7 1013    SUB    B         * Y2 = YC + Y
G0306 07F8 6003    SACL   Y2
G0307 07F9 C008    LARK   AR0,8     * set-up access to VSC registers
G0308 07FA 0400    LRLK   AR4,VSCREG+>0080+XYOFFH
      07FB 04E8
G0309 07FC 558C    LARP   AR4       * load XY coordinates for lower lin
G0310 07FD 2E29    LAC    X1L,14
G0311 07FE 68E0    SACH   *0+
G0312 07FF 2703    LAC    Y2,7
G0313 0800 0028    ADD    X1H
G0314 0801 60E0    SACL   *0+
G0315 0802 2F03    LAC    Y2,15
G0316 0803 68B8    SACH   *,0,0
G0317 0804 D000    LRLK   AR0,VSCREG+>0080+CTLR2L
      0805 04C0
G0318 0806 D400    LRLK   AR4,XYIND+XINC
      0807 0808
G0319 0808 CA00    ZAC
G0320 0809 6088    SACL   *,0,3     * load Control Register 2
G0321 080A D001    LALK   LSBASE
      080B 0878
G0322 080C 0029    ADD    X1L
G0323 080C CE24    CALA
G0324 080E CABC    LACK   >00BC     * write pixels left of the blocks
G0325 080F 608C    SACL   *         * X-Y mode RAS override
G0326 0810 202D    LAC    PAXCNT    * load Control Register 2
G0327 0811 F380    BLZ    RIGHT2
      0812 0816
G0328 0813 200C    LAC    COLOR2
G0329 0814 4B2D    RPT    PAXCNT    * fill 8 pixel blocks
G0330 0815 6080    SACL   *
G0331 0816 5588    LARP   AR0
G0332 0817 CA00 RIGHT2 ZAC
G0333 0818 6080    SACL   *,0,3     * load Control Register 2
G0334 0819 D000    LRLK   AR0,VSCREG+>0080+XYOFFH
      081A 04E8
G0335 081B 6080    SACL   *
G0336 081C D000    LRLK   AR0,VSCREG+>0080+XYADRL
      081D 04F0
G0337 081E 2080    LAC    *,0,4
G0338 081F 60BC    SACL   *
G0339 0820 D001    LALK   RSBASE
      0821 08B8
G0340 0822 002B    ADD    X2L       * write pixels right of the blocks
G0341 0823 CE24    CALA
G0342 0824 558A LOOPE LARP   AR2
G0343 0825 FB99    BANZ   LOOP3,*-,1
      0826 07DC
G0344 0827 FF80    B      ALDONE
      0828 086A
G0345           *
```

```
                   * DRAW "SHORT" HORIZONTAL LINES FOR LOWER HALF
                   *
G0346           *
G0347
G0348 0829 7213 LOOP4 SAR    AR2,B
G0349 082A 2010    LAC    XC
G0350 082B 1080    SUB    *
G0351 082C F480    BGEZ   OSCRX4
      082D 082F
G0352 082E CA00    ZAC
G0353 082F 6000 OSCRX4 SACL   X1      * X1 = XC - X >= 0
G0354 0830 2010    LAC    XC
G0355 0831 00AC    ADD    *+,0,4      * X2 = XC + X
G0356 0832 6002    SACL   X2
G0357 0833 2002    LAC    X2,13       * seperate 7 MSBs from 3 LSBs (X2)
G0358 0834 682A    SACH   X2H
G0359 0835 D004    ANDK   >FFFF
      0836 FFFF
G0360 0837 6C2B    SACH   X2L,4
G0361 0838 2000    LAC    X1,13       * seperate 7 MSBs from 3 LSBs (X1)
G0362 0839 6828    SACH   X1H
G0363 083A D004    ANDK   >FFFF
      083B FFFF
G0364 083C 6C29    SACH   X1L,4
G0365 083D 2011    LAC    YC          * "short" line
G0366 083E 1013    SUB    B
G0367 083F 6003 NARROW SACL   Y2      * Y2 = YC + Y
G0368 0840 C008    LARK   AR0,8       * set-up access to VSC registers
G0369 0841 D400    LRLK   AR4,VSCREG+>0080+XYOFFH
      0842 04E8
G0370 0843 558C    LARP   AR4         * load XY coordinates for lower lin
G0371 0844 2E29    LAC    X1L,14
G0372 0845 68E0    SACH   *0+
G0373 0846 2703    LAC    Y2,7
G0374 0847 0028    ADD    X1H
G0375 0848 60E0    SACL   *0+
G0376 0849 2F03    LAC    Y2,15
G0377 084A 68B8    SACH   *,0,0
G0378 084B 302C    LAR    AR0,PIXCNT
G0379 084C D400    LRLK   AR4,XYIND+XINC
      084D 0808
G0380 084E 9129    BIT    X1L,1       * start on an odd pixel ?
G0381 084F F980    BBNZ   SLLOL
      0850 0861
G0382 0851 558B SLLOL LARP   AR3
G0383 0852 CAF0    LACK   >00F0       * pixel in lsnibble only
G0384 0853 4E8C    AND    *,4
G0385 0854 400A    OR     COLORL
G0386 0855 6088    SACL   *,0,0
G0387 0856 FB90    BANZ   SLHIL
      0857 0861
G0388 0858 FF80    B      DONEIL
      0859 0867
G0389 085A 558C SLHLL LARP   AR4
G0390 085B 200C    LAC    COLOR2      * pixel pair
G0391 085C 6088    SACL   *,0,0
G0392 085D FB90    BANZ   SLHIL
      085E 0861
G0393 085F FF80    B      DONEIL
```

23. A Graphics Implementation Using the TMS32020 and TMS34061

```
G0394 0860 0867
G0394 0861 FB9B  SLHIL  SLHLL,*-,3
      0862 085A
G0395 0863 CA0F         LACK   >000F    * pixel in msnibble only
G0396 0864 4E80         AND
G0397 0865 4D0B         OR     COLORH
G0398 0866 6080         SACL
G0399 0867 558A         LARP   AR2
G0400 0868 FB99  DONEIL BANZ   LOOP4,*-,1
      0869 0829

G0401 *
G0402 086A 558A  ALDONE LARP   AR2
G0403 086B C264         LARK   AR2,SAVREG
G0404 086C C809         LDPK   >0009    * VSC register page
G0405 086D 20A0         LAC    *+
G0406 086E 6030         SACL   CTLR1L   * load Control Register 1
G0407 086F 20A0         LAC    *+
G0408 0870 6038         SACL   CTLR1H
G0409 0871 20A0         LAC    *+
G0410 0872 6040         SACL   CTLR2L   * load Control Register 2
G0411 0873 20A0         LAC    *+
G0412 0874 6048         SACL   CTLR2H
G0413 0875 C800         LDPK   0
G0414 0876 5060         LST    STS0     * restore data page pointer
G0415 0877 CE26         RET

G0416 *
G0417 0878 FF80  LSBASE B      LE0
      0879 0898
G0418 087A FF80         B      LE1
      087B 08A0
G0419 087C FF80         B      LE2
      087D 08A9
G0420 087E FF80         B      LE3
      087F 08AF
G0421 0880 FF80         B      LE4
      0881 08B7
G0422 0882 FF80         B      LE5
      0883 08BC
G0423 0884 FF80         B      LE6
      0885 08C3
G0424 0886 FF80         B      LE7
      0887 08C7

G0425 *
G0426 0888 FF80  RSBASE B      RE0
      0889 08CC
G0427 088A FF80         B      RE1
      088B 08D2
G0428 088C FF80         B      RE2
      088D 08D5
G0429 088E FF80         B      RE3
      088F 08DC
G0430 0890 FF80         B      RE4
      0891 08E0
G0431 0892 FF80         B      RE5
      0893 08E8
G0432 0894 FF80         B      RE6
      0895 08ED
```

```
G0433 0896 FF80         B      RE7
      0897 08F6
G0434 *
G0435 *        WRITE 8 PIXELS LEFT OF BLOCK
G0436 *              (0 UNWRITTEN PIXELS)
G0437 *
G0438 *        !* * * * * * * *; 8 PIXEL BLOCKS;
G0439 *
G0440 0898 558C  LE0    LARP   AR4
G0441 0899 200C         LAC    COLOR2
G0442 089A 6080         SACL   *
G0443 089B 6080         SACL   *
G0444 089C 6080         SACL   *
G0445 089D 6088         SACL   *,0,0
G0446 089E CE26         RET
G0447 089F CE26         RET
G0448 *
G0449 *        WRITE 7 PIXELS LEFT OF BLOCK
G0450 *              (1 UNWRITTEN PIXELS)
G0451 *
G0452 *        !? * * * * * * *; 8 PIXEL BLOCKS;
G0453 *
G0454 08A0 CAF0  LE1    LACK   >00F0
G0455 08A1 4E8C         AND    *,4
G0456 08A2 400A         OR     COLORL
G0457 08A3 6080         SACL   *
G0458 08A4 200C         LAC    COLOR2
G0459 08A5 6080         SACL   *
G0460 08A6 6080         SACL   *
G0461 08A7 6088         SACL   *,0,0
G0462 08A8 CE26         RET
G0463 *
G0464 *        WRITE 6 PIXELS LEFT OF BLOCK
G0465 *              (2 UNWRITTEN PIXELS)
G0466 *
G0467 *        !? ? * * * * * *; 8 PIXEL BLOCKS;
G0468 *
G0469 08A9 558C  LE2    LARP   AR4
G0470 08AA 200C         LAC    COLOR2
G0471 08AB 6080         SACL   *
G0472 08AC 6080         SACL   *
G0473 08AD 6088         SACL   *,0,0
G0474 08AE CE26         RET
G0475 *
G0476 *        WRITE 5 PIXELS LEFT OF BLOCK
G0477 *              (3 UNWRITTEN PIXELS)
G0478 *
G0479 *        !? ? ? * * * * *; 8 PIXEL BLOCKS;
G0480 *
G0481 08AF CAF0  LE3    LACK   >00F0
G0482 08B0 4E8C         AND    *,4
G0483 08B1 400A         OR     COLORL
G0484 08B2 6080         SACL   *
G0485 08B3 200C         LAC    COLOR2
G0486 08B4 6080         SACL   *
G0487 08B5 6088         SACL   *,0,0
G0488 08B6 CE26         RET
```

23. A Graphics Implementation Using the TMS32020 and TMS34061

```
G0489
G0490          *
G0491          *      WRITE 4 PIXELS LEFT OF BLOCK
G0492          *          (4 UNWRITTEN PIXELS)
G0493          *
G0494                  !? ? ? ? * * * * !  8 PIXEL BLOCKS!
G0495 08B7 558C  LE4   LARP   AR4
G0496 08B8 200C        LAC    COLOR2
G0497 08B9 6080        SACL   *
G0498 08BA 6088        SACL   *,0,0
G0499 08BB CE26        RET
G0500
G0501          *
G0502          *      WRITE 3 PIXELS LEFT OF BLOCK
G0503          *          (5 UNWRITTEN PIXELS)
G0504          *
G0505                  !? ? ? ? ? * * * !  8 PIXEL BLOCKS!
G0506 08BC CAF0  LE5   LACK   >00F0
G0507 08BD 4E8C        AND    *,4
G0508 08BE 400A        OR     COLORL
G0509 08BF 6080        SACL   *
G0510 08C0 200C        LAC    COLOR2
G0511 08C1 6088        SACL   *,0,0
G0512 08C2 CE26        RET
G0513
G0514          *
G0515          *      WRITE 2 PIXELS LEFT OF BLOCK
G0516          *          (6 UNWRITTEN PIXELS)
G0517          *
G0518                  !? ? ? ? ? ? * * !  8 PIXEL BLOCKS!
G0519 08C3 558C  LE6   LARP   AR4
G0520 08C4 200C        LAC    COLOR2
G0521 08C5 6088        SACL   *,0,0
G0522 08C6 CE26        RET
G0523
G0524          *
G0525          *      WRITE 1 PIXELS LEFT OF BLOCK
G0526          *          (7 UNWRITTEN PIXELS)
G0527          *
G0528                  !? ? ? ? ? ? ? * !  8 PIXEL BLOCKS!
G0529 08C7 CAF0  LE7   LACK   >00F0
G0530 08C8 4E8C        AND    *,4
G0531 08C9 400A        OR     COLORL
G0532 08CA 6088        SACL   *,0,0
G0533 08CB CE26        RET
G0534
G0535          *
G0536          *      WRITE 1 PIXELS RIGHT OF BLOCK
G0537          *          (7 UNWRITTEN PIXELS)
G0538          *
G0539                  !8 PIXEL BLOCKS !* ? ? ? ? ? ? ?!
G0540 08CC 558B  RE0   LARP   AR3
G0541 08CD CAOF        LACK   >000F
G0542 08CE 4E8C        AND    *,4
G0543 08CF 4D0B        OR     COLORH
G0544 08D0 6088        SACL   *,0,0
G0545 08D1 CE26        RET
```

```
G0546          *
G0547          *
G0548          *      WRITE 2 PIXELS RIGHT OF BLOCK
G0549          *          (6 UNWRITTEN PIXELS)
G0550          *
G0551                  !8 PIXEL BLOCKS !* * ? ? ? ? ? ?!
G0552 08D2 200C  RE1   LAC    COLOR2
G0553 08D3 6080        SACL   *
G0554 08D4 CE26        RET
G0555
G0556          *
G0557          *      WRITE 3 PIXELS RIGHT OF BLOCK
G0558          *          (5 UNWRITTEN PIXELS)
G0559          *
G0560                  !8 PIXEL BLOCKS !* * * ? ? ? ? ?!
G0561 08D5 200C  RE2   LAC    COLOR2
G0562 08D6 608B        SACL   *,0,3
G0563 08D7 CAOF        LACK   >000F
G0564 08D8 4E8C        AND    *,4
G0565 08D9 4D0B        OR     COLORH
G0566 08DA 6088        SACL   *,0,0
G0567 08DB CE26        RET
G0568
G0569          *
G0570          *      WRITE 4 PIXELS RIGHT OF BLOCK
G0571          *          (4 UNWRITTEN PIXELS)
G0572          *
G0573                  !8 PIXEL BLOCKS !* * * * ? ? ? ?!
G0574 08DC 200C  RE3   LAC    COLOR2
G0575 08DD 6080        SACL   *
G0576 08DE 6088        SACL   *,0,0
G0577 08DF CE26        RET
G0578
G0579          *
G0580          *      WRITE 5 PIXELS RIGHT OF BLOCK
G0581          *          (3 UNWRITTEN PIXELS)
G0582          *
G0583                  !8 PIXEL BLOCKS !* * * * * ? ? ?!
G0584 08E0 200C  RE4   LAC    COLOR2
G0585 08E1 6080        SACL   *
G0586 08E2 608B        SACL   *,0,3
G0587 08E3 CAOF        LACK   >000F
G0588 08E4 4E8C        AND    *,4
G0589 08E5 4D0B        OR     COLORH
G0590 08E6 6088        SACL   *,0,0
G0591 08E7 CE26        RET
G0592
G0593          *
G0594          *      WRITE 6 PIXELS RIGHT OF BLOCK
G0595          *          (2 UNWRITTEN PIXELS)
G0596          *
G0597                  !8 PIXEL BLOCKS !* * * * * * ? ?!
G0598 08E8 200C  RE5   LAC    COLOR2
G0599 08E9 6080        SACL   *
G0600 08EA 6080        SACL   *
G0601 08EB 6088        SACL   *,0,0
G0602 08EC CE26        RET
```

```
G0603        *
G0604        *
G0605        *        WRITE 7 PIXELS RIGHT OF BLOCK
G0606        *           (1 UNWRITTEN PIXELS)
G0607        *
G0608        *8 PIXEL BLOCKS ;* * * * * * ?;
G0609 08ED 200C   RE6    LAC    COLOR2
G0610 08EE 6080          SACL   *
G0611 08EF 608B          SACL   *,0,3
G0612 08F0 608B          SACL   *
G0613 08F1 CA0F          LACK   >000F
G0614 08F2 4E8C          AND    *,4
G0615 08F3 4D08          OR     COLORH
G0616 08F4 6088          SACL   *,0,0
G0617 08F5 CE26          RET
G0618        *
G0619        *
G0620        *        WRITE 8 PIXELS RIGHT OF BLOCK
G0621        *           (0 UNWRITTEN PIXELS)
G0622        *
G0623        *8 PIXEL BLOCKS ;* * * * * * * *;
G0624 08F6 200C   RE7    LAC    COLOR2
G0625 08F7 6080          SACL   *
G0626 08F8 6080          SACL   *
G0627 08F9 6080          SACL   *
G0628 08FA 6088          SACL   *,0,0
G0629 08FB CE26          RET
0008                     COPY   CUBER.ASM
```

```
H0002  *
H0003  *
H0004  *
H0005  *
H0006  * ROUTINE:  CUBER
H0007  *
H0008  *   This routine controls the rotation and perspective
H0009  *   projection of a normalized cube. The cube is defined by
H0010  *   the vertices (0.5,-0.5,-0.5), (0.5,0.5,-0.5),
H0011  *   (-0.5,0.5,-0.5), (-0.5,-0.5,-0.5), (0.5,-0.5,0.5),
H0012  *   (0.5,0.5,0.5), (-0.5,0.5,0.5), and (-0.5,-0.5,0.5). The
H0013  *   cube is rotated about the z-axis at an angular rate equal
H0014  *   to R/128 degrees. i.e. The angle of rotation per display
H0015  *   time is any one of 128 integral quantities determined by
H0016  *   dividing PI/2 into 128 equal parts.
H0017  *
H0018  *
```

```
        ^ z
        |
   7 --------- 6
  /           /
 4 --------- 5
          3 ------ 2   --> y
        0 ------ 1
   x
   v
```

```
H0034  *   This routine loads (initializes) the matrices used in the
H0035  *   rotation, scaling, and projection of the cube, initializes
H0036  *   the vertices, and then controls the modification and
H0037  *   display of the cube. The modification and display are
H0038  *   actually two sections of a larger loop. In the first
H0039  *   section, each point in normalized coordinates is rotated,
H0040  *   scaled, and projected to determine the display
H0041  *   coordinates. Once this has been completed for all
H0042  *   vertices in the cube, the processing begins in the second
H0043  *   section. The second section processing consists of
H0044  *   selecting vertex pairs constituting edges of the cube and
H0045  *   calling the line drawing algorithm to plot the line
H0046  *   defining the edge.
H0047  *
H0048  *
H0049  *
H0050 0004        RATE    EQU    4
H0051 0400        RPTCNT  EQU    >0400
H0052  *
H0053  *           INITIALIZATION OF MATRICES
H0054  *
H0055 08FC        CUBER   EQU    $
H0056 08FC D200           LRLK   AR2,NRMPTS+DP6
      08FD 0347
H0057 08FE 558A           LARP   AR2
```

23. A Graphics Implementation Using the TMS32020 and TMS34061

```
H0058 08FF CB17          RPTK   23
H0059 0900 FCA0          BLKP   VRTICS,*+       * transfer the normalized
H0060 0901 0B73                                 * points to data memory
           0902   CUBER2 EQU    $
H0061 0902               LARP   AR2
H0062 0902 558A          LRLK   AR2,C11+DP4
H0063 0903 D200
           0904 0216
H0064 0905 CB0B          RPTK   11
H0065 0906 FCA0          BLKP   CMTRX,*+        * load the viewpoint matrix
           0907 0BBB
H0066 0908 C804          LDPK   4
H0067 0909 D001          LALK   COSTBL          * initialize the COS address
           090A 0AAD
H0068 090B 6027          SACL   COS
H0069 090C D001          LALK   SINTBL          * initialize the SINE address
           090D 0A6D
H0070 090E 6026          SACL   SIN
H0071 090F CA04          LACK   RATE            * load the rate of rotation
H0072 0910 6028          SACL   THETA
H0073 0911 D200          LRLK   AR2,VSX+DP4
           0912 0222
H0074 0913 CB03          RPTK   3
H0075 0914 FCA0          BLKP   SCALE,*+        * load the projection matrix
           0915 0B97
H0076 0916 C806          LDPK   6
H0077 0917 2B2E          LAC    ONE6,11         * numerator = >0800
H0078 0918 6071          SACL   NUMR8R
H0079 0919 D000          LRLK   AR0,RPTCNT
           091A 0400

H0080   *                CUBE MODIFICATION AND DISPLAY CONTROL LOOP
H0081   *
H0082   *
H0083 091B 702F   CTLOOP  SAR    AR0,SAVAR0
H0084 091C D200           LRLK   AR2,NRMPTS+DP6   * initialize indirect access
           091D 0347
H0085 091E D300           LRLK   AR3,DSPPTS+DP6
           091F 035F
H0086 0920 D400           LRLK   AR4,NRMPTS+DP6
           0921 0347
H0087 0922 C007           LARK   AR0,7
H0088 0923 C804           LDPK   4
H0089 0924 2027           LAC    COS
H0090 0925 5812           TBLR   R11
H0091 0926 5815           TBLR   R22
H0092 0927 2026           LAC    SIN
H0093 0928 5814           TBLR   R12
H0094 0929 CA00           ZAC
H0095 092A 1014           SUB    R12
H0096 092B 6013           SACL   R21
H0097 092C C806           LDPK   6
H0098 092D 558A           LARP   AR2

H0099   *                MATRIX MODIFICATION OF CUBE VERTICES
H0100   *
H0101   *                                       * fetch the x coordinate
H0102 092E 20A0   CLOOP1  LAC    *+
H0103 092F 6022           SACL   X
```

```
H0104 0930 20A0          LAC    *+              * fetch the y coordinate
H0105 0931 6023          SACL   Y
H0106 0932 20A0          LAC    *+              * fetch the z coordinate
H0107 0933 6024          SACL   Z               * z not affected by rotation
H0108 0934 6027          SACL   ZR
H0109 0935 FE80          CALL   ROTZ            * rotate the vertex
           0936 0962
H0110 0937 FE80          CALL   MTRX4           * apply viewpoint matrix
           0938 0973
H0111 0939 FE80          CALL   PRJCTN          * project the vertex
           093A 0993
H0112 093B 558B          LARP   AR3
H0113 093C 2000          LAC    XP
H0114 093D 60A0          SACL   *+              * store the X screen address
H0115 093E 2001          LAC    YP
H0116 093F 60A8          SACL   *+,0,0          * store the Y screen address
H0117 0940 FB9A          BANZ   CLOOP1,*-,2     * all vertices processed?
           0941 092E

H0118   *                CUBE DISPLAY
H0119   *
H0120   *
H0121 0942 CA00          ZAC
H0122 0943 FE80          CALL   CLEAR           * clear the screen
           0944 04DC
H0123 0945 FE80          CALL   PLOT            * draw the cube
           0946 09B3

H0124   *                ROTATION MATRIX UPDATE
H0125   *
H0126   *
H0127 0947 C804          LDPK   4
H0128 0948 2026          LAC    SIN
H0129 0949 0028          ADD    THETA           * update sine angle address
H0130 094A 6026          SACL   SIN
H0131 094B D003          SBLK   ENDTBL
           094C 0B6C
H0132 094D F280          BLEZ   DOCOS
           094E 0952
H0133 094F D002          ADLK   SINTBL          * reset to beginning of tabl
           0950 0A6D
H0134 0951 6026          SACL   SIN
H0135 0952 2027   DOCOS  LAC    COS
H0136 0953 0028          ADD    THETA           * update cosine angle addres
H0137 0954 6027          SACL   COS
H0138 0955 D003          SBLK   ENDTBL
           0956 0B6C
H0139 0957 F280          BLEZ   NXTROT
           0958 095C
H0140 0959 D002          ADLK   SINTBL          * reset to beginning of tabl
           095A 0A6D
H0141 095B 6027          SACL   COS
H0142 095C C806   NXTROT LDPK   6
H0143 095D 302F          LAR    AR0,SAVAR0
H0144 095E 558B          LARP   AR0
H0145 095F FB90          BANZ   CTLOOP          * rotation sequence complete
           0960 091B
H0146 0961 CE26          RET
           0009          COPY   ROTATEZ.ASM
```

23. A Graphics Implementation Using the TMS32020 and TMS34061

```
* ROUTINE: ROTZ
*
*       This routine is performs the function of rotation about
*       the z-axis. When the routine is called it expects that
*       rotation matrix, 2 x 2, is located on page 4 in
*       consecutive locations (R11, R21, R12, R22) and the data
*       point matrices are on page 6 in locations labeled X, Y,
*       XR, YR.
*
*                             +-        -+
*       [ XR YR ] = [ X Y ] * : R11 R12 :
*                             : R21 R22 :
*                             +-        -+
*
*       This shortcut can be taken since we are defining a simple
*       rotation about the z-axis only and the object is assumed
*       to be located about the axis of rotation. This permits
*       reduction to a two-dimensional rotation and elimination of
*       the homogenous coordinate system for this operation.
*
*       It is assumed that the points to be rotated are
*       normalized on a unit basis, with values in the range of
*       (-1,+1) and supplied in a Q15 format. The same assumption
*       is placed on the rotational values of cos 0 and sin 0. The
*       output points are therby also guaranteed to be in a Q15
*       format.
*
*       It is also assumed that the data page pointer is equal to
*       6 and that the block B0 is configured as data memory
*       before the routine is called.
*
*       This routine modifies the T-reg, P-reg, ACcumulator, and
*       Auxiliary Register 1.
*
I0040                    ROTZ    EQU    $
I0041  0962 D100                 LRLK   AR1,R11+DP4
       0963 0212
I0042  0964 5589                 LARP   AR1
I0043  0965 CA00                 ZAC            * zero the accumulator
I0044  0966 3CA0                 LT     *+
I0045  0967 3822                 MPY    X        * X*R11
I0046  0968 30A0                 LTA    *+
I0047  0969 3823                 MPY    Y        * Y*R21
I0048  096A 30A0                 LTA    *+       * X*R11 + Y*R21
I0049  096B 6925                 SACH   XR,1
I0050  096C CA00                 ZAC            * zero the accumulator
I0051  096D 3822                 MPY    X        * X*R12
I0052  096E 3080                 LTA    Y
I0053  096F 3823                 MPY    Y        * Y*R22
I0054  0970 CE15                 APAC           * X*R12 + Y*R22
I0055  0971 6926                 SACH   YR,1
I0056  0972 CE26                 RET
0010                             COPY   MATRIX4.ASM
```

```
* ROUTINE: MTRX4
*
*       This routine is performs the function of a combination
*       (translations, rotations, scalings, etc.) of matrix
*       operations on a three-dimensional point in a homogenous
*       coordinate system. The computations executed by the
*       routine are the matrix multiplication of a [1x4] x [4x4].
*       In actuality, since the last point is only the homogeneous
*       coordinate, the matrix multiplication is reduced to a
*       [1x4] x [4x3].
*
*       When the routine is called it expects that the combination
*       matrix, 4 x 3, is located on page 4 in consecutive
*       locations (C11, C21, C31, C41, C12, C22, C32, C42, C13,
*       C23, C33, C43) and the data point matrices are on page 6
*       in locations labeled XR, YR, ZR, XC, YC, ZC.
*
*                                        +-             -+
*       [ XC YC ZC 1 ] = [ XR YR ZR 1 ] * : C11 C12 C13 0 :
*                                        : C21 C22 C23 0 :
*                                        : C31 C32 C33 0 :
*                                        : C41 C42 C43 1 :
*                                        +-             -+
*
*       It is assumed that the points to be manipulated are
*       normalized on a unit basis, with values in the range of
*       (-1,+1) and supplied in a Q15 format. The combination
*       matrix is assumed to be in a Q11 format, with values in
*       the range of (-16,+16). The output points are thereby
*       guaranteed to be in a Q11 format.
*
*       It is also assumed that the data page pointer is equal to
*       6 and that the block B0 is configured as data memory
*       before the routine is called.
*
*       This routine modifies the T-reg, P-reg, ACcumulator, and
*       Auxiliary Register 1.
*
J0043                    MTRX4   EQU    $
J0044  0973 D100                 LRLK   AR1,C11+DP4
       0974 0216
J0045  0975 5589                 LARP   AR1
J0046  0976 CA00                 ZAC            * zero the accumulator
J0047  0977 3CA0                 LT     *+
J0048  0978 3825                 MPY    XR       * XR*C11
J0049  0979 30A0                 LTA    *+
J0050  097A 3826                 MPY    YR       * YR*C21
J0051  097B 30A0                 LTA    *+       * XR*C11 + YR*C21
J0052  097C 3827                 MPY    ZR       * ZR*C31
J0053  097D 0FA0                 ADD    *+,15    * XR*C11 + YR*C21 + ZR*C31 + C41
J0054  097E 30A0                 LTA    *+
J0055  097F 6910                 SACH   XC,1     * XR*C11 + YR*C21 + ZR*C31 + C41
J0056  0980 CA00                 ZAC            * zero the accumulator
```

23. A Graphics Implementation Using the TMS32020 and TMS34061

```
J0058 0981 3825   MPY   XR      * XR*C12
J0059 0982 3DA0   LTA   *+
J0060 0983 3826   MPY   YR      * YR*C22
J0061 0984 3DA0   LTA   *+      * XR*C12 + YR*C22
J0062 0985 3827   MPY   ZR      * ZR*C32
J0063 0986 0FA0   ADD   *+,15   * XR*C12 + YR*C22          + C42
J0064 0987 3DA0   LTA   *+      * XR*C12 + YR*C22 + ZR*C32 + C42
J0065 0988 6911   SACH  YC,1
J0066 0989 CA00   ZAC           * zero the accumulator
J0067 098A 3825   MPY   XR      * XR*C13
J0068 098B 3DA0   LTA   *+
J0069 098C 3826   MPY   YR      * YR*C23
J0070 098D 3DA0   LTA   *+      * XR*C13 + YR*C23
J0071 098E 3827   MPY   ZR      * ZR*C33
J0072 098F 0F80   ADD   *,15    * XR*C13 + YR*C23          + C43
J0073 0990 CE15   APAC          * XR*C13 + YR*C23 + ZR*C33 + C43
J0074 0991 6912   SACH  ZC,1
J0075 0992 CE26   RET
0011              COPY  PRJCTN.ASM
```

```
K0002 *****************************************************************
K0003 * ROUTINE: PRJCTN
K0004 *
K0005 *
K0006 * This routine is performs the function of a projecting a
K0007 * three-dimensional image onto a two-dimensional display
K0008 * map.  In general this involves the matrix multiplication
K0009 * of a [1x4] x [4x4].  Since this is the final operation to
K0010 * displaying a point, the homogeneous coordinate column can
K0011 * be deleted from the matrix.  Also this is a scaling
K0012 * (diagonal) and translation matrix, allowing the
K0013 * computations to be further reduced by avoiding multiply
K0014 * accumulates involving the known zeroes.
K0015 *
K0016 * When the routine is called it expects that the matrix
K0017 * parameters to be located on page 4 in consecutive
K0018 * locations (VSX, VCX, VSY, VCY) and the data point matrices
K0019 * to be on page 6 in locations labeled XC, YC, ZC, XP, YP.
K0020 *
K0021 *
K0022 *                                    +-                 -+
K0023 * [ XP YP ZP 1 ] = [ XC YC ZC 1 ] *  : VSX'  0    0    0
K0024 *                                    :  0   VSY'  0    0
K0025 *                                    :  0    0   VSZ'  0
K0026 *                                    : -VCX  VCY  VCZ  1
K0027 *                                    +-                 -+
K0028 * VCX, VCY, VCZ are the location of the center of object on
K0029 * the screen and, of course, VCZ = 0.  VSX', VSY', and VSZ'
K0030 * constitute the scaling of the object to meet the display
K0031 * size or resolution and to include the factor of the
K0032 * projection of the z-coordinate onto the x,y-plane.  Again
K0033 * VSZ' = 0 and VSX' and VSY' are given by the equations:
K0034 *
K0035 *    VSX' = VSX/ZC      and     VSY' = VSY/ZC
K0036 *
K0037 * It is assumed that the points to be scaled and projected
K0038 * have been provided in a Q11 format as a result of a
K0039 * combination matrix operation.  It is furthered assumed
K0040 * that the ZC parameter in particular has some integer
K0041 * portion (either positive or negative).  This is important
K0042 * since we perform a division and expect to return the
K0043 * result for additional computations as a Q15 number.  The
K0044 * scaling and centering values should be provided such that
K0045 * the multiplication and addition generate a Q0 value as
K0046 * output from the original Q11 input value when retrieved
K0047 * from the accumulator with a SACH <dma>,1.  With a distance
K0048 * to screen size ratio of 4 and a view point at (6,8,7.5),
K0049 * the scaling numbers are represented in Q6 and the
K0050 * centering values in a Q0 format.
K0051 *
K0052 * It is also assummed that the data page pointer is equal to
K0053 * 6 and that the block B0 is configured as data memory
K0054 * before the routine is called.
K0055 *
K0056 * This routine modifies the T-reg, P-reg, ACcumulator, and
K0057 * Auxiliary Register 1.
K0058 *
```

23. A Graphics Implementation Using the TMS32020 and TMS34061

```
K0059
K0060           *
K0061     0993      PRJCTN  EQU   $
K0062  0993 2012            LAC   ZC          * save sign of ZC
K0063  0994 6814            SACH  ZCSIGN
K0064  0995 CE1B            ABS
K0065  0996 6012            SACL  ZC          * save absolute value of ZC
K0066  0997 4071            ZALH  NUMR8R      * load a Q11 value equal 1
K0067  0998 CB0E            RPTK  14
K0068  0999 4712            SUBC  ZC          * divide by ZC
K0069  099A 6013            SACL  ZCI
K0070  099B 2013            LAC   ZCI         * save :ZC: inverse (1/:ZC:)
K0071  099C 4C14            XOR   ZCSIGN      * result is in Q15 format
K0072  099D 1014            SUB   ZCSIGN      * sign adjustment
K0073  099E 6013            SACL  ZCI
K0074  099F 3C13            LT    ZCI         * save ZC inverse (1/ZC)
K0075  09A0 3810            MPY   XC          * XC/ZC
K0076  09A1 CE14            PAC
K0077  09A2 6910            SACH  XC,1
K0078  09A3 3811            MPY   YC          * YC/ZC
K0079  09A4 CE14            PAC
K0080  09A5 6911            SACH  YC,1
K0081  09A6 D100            LRLK  AR1,VSX+DP4
       09A7 0222
K0082  09A8 5589            LARP  AR1
K0083  09A9 3CA0            LT    *+
K0084  09AA 2FA0            LAC   *+,15       * XC*VSX       VCX
K0085  09AB 3810            MPY   XC          * XC*VSX + VCX
K0086  09AC 30A0            LTA   XP,1
K0087  09AD 6900            SACH  XP,1
K0088  09AE 2F80            LAC   *,15        * YC*VSY       VCY
K0089  09AF 3811            MPY   YC          * YC*VSY + VCY
K0090  09B0 CE15            APAC
K0091  09B1 6901            SACH  YP,1
K0092  09B2 CE26            RET
0012                        COPY  OBJPLOT.ASM
```

23. A Graphics Implementation Using the TMS32020 and TMS34061

```
L0002  **************************************************************
L0003  *
L0004  * ROUTINE: PLOT
L0005  *
L0006  * This routine controls the line drawing (plotting) required
L0007  * to display a wire-frame cube. The vertices are expected
L0008  * to be stored in a screen location table located on page 6
L0009  * beginning with the address DSPPTS. The line drawing
L0010  * routine is called to draw each of the 12 lines connecting
L0011  * the 8 vertices and defining the edges of cube as shown
L0012  * below.
L0013  *
```

```
        ^ z
        |
   7 +------+ 6
    /|     /|
   / |    / |          y
  4 +------+ 5   +------>
  |  +----|--+ 3
  | /     | /
  |/      |/
  0 +------+ 1
       /
      v  x
```

```
L0014  *
L0015  * The routine proceeds by first drawing lines from points
L0016  * 0-1-2-3-0. Next the lines using point pairs 0-4, 1-5,
L0017  * 2-6, and 3-7 are drawn. Finally lines are drawn
L0018  * connecting the points 4-5-6-7-4 in sequence.
L0019  *
L0020  **************************************************************
L0021
L0022
L0023
L0024
L0025
L0026
L0027
L0028
L0029
L0030
L0031
L0032
L0033
L0034
L0035
L0036
L0037     09B3      PLOT    EQU   $
L0038  09B3 CA01            LACK  1
L0039  09B4 6004            SACL  COLOR          * initialize color to RED
L0040  09B5 D200            LRLK  AR2,DSPPTS+DP6
       09B6 035F
L0041  09B7 20A0            LAC   *+
L0042  09B8 6000            SACL  X1             * fetch X0
L0043  09B9 20A0            LAC   *+
L0044  09BA 6001            SACL  Y1             * fetch Y0
L0045  09BB 20A0            LAC   *+
L0046  09BC 6002            SACL  X2
L0047  09BD 6024            SACH  XB             * fetch X1
L0048  09BE 20A0            LAC   *+
L0049  09BF 6003            SACL  Y2             * save X1 for next line
L0050  09C0 6025            SACH  YB
L0051  09C1 726F            SAR   AR2,SAVAR2     * fetch Y1
L0052  09C2 FE80            CALL  LINE           * save Y1 for next line
       09C3 0524
L0053  09C4 558A            LARP  AR2            * draw line connecting 0 & 1
L0054  09C5 326F            LAR   AR2,SAVAR2
L0055  09C6 2024            LAC   XB             * fetch X1
L0056  09C7 6000            SACL  X1
```

```
L0057 09C8 2025    LAC   YB
L0058 09C9 6001    SACL  Y1          * fetch Y1
L0059 09CA 20A0    LAC   *+
L0060 09CB 6002    SACL  X2          * fetch X2
L0061 09CC 6024    SACL  XB          * save X2 for next line
L0062 09CD 20A0    LAC   *+
L0063 09CE 6003    SACL  Y2          * fetch Y2
L0064 09CF 6025    SACL  YB          * save Y2 for next line
L0065 09D0 726F    SAR   AR2,SAVAR2
L0066 09D1 FE80    CALL  LINE        * draw line connecting 1 & 2
      09D2 0524
L0067 09D3 558A    LARP  AR2
L0068 09D4 326F    LAR   AR2,SAVAR2
L0069 09D5 2024    LAC   XB
L0070 09D6 6000    SACL  X1          * fetch X2
L0071 09D7 2025    LAC   YB
L0072 09D8 6001    SACL  Y1          * fetch Y2
L0073 09D9 20A0    LAC   *+
L0074 09DA 6002    SACL  X2          * fetch X3
L0075 09DB 6024    SACL  XB          * save X3 for next line
L0076 09DC 20A0    LAC   *+
L0077 09DD 6003    SACL  Y2          * fetch Y3
L0078 09DE 6025    SACL  YB          * save Y3 for next line
L0079 09DF 726F    SAR   AR2,SAVAR2
L0080 09E0 FE80    CALL  LINE        * draw line connecting 2 & 3
      09E1 0524
L0081 09E2 558A    LARP  AR2
L0082 09E3 D200    LRLK  AR2,DSPPTS+DP6
      09E4 035F
L0083 09E5 2024    LAC   XB
L0084 09E6 6000    SACL  X1          * fetch X3
L0085 09E7 2025    LAC   YB
L0086 09E8 6001    SACL  Y1          * fetch Y3
L0087 09E9 20A0    LAC   *+
L0088 09EA 6002    SACL  X2          * fetch X0
L0089 09EB 20A0    LAC   *+
L0090 09EC 6003    SACL  Y2          * fetch Y0
L0091 09ED 726F    SAR   AR2,SAVAR2
L0092 09EE FE80    CALL  LINE        * draw line connecting 3 & 0
      09EF 0524
L0093 09F0 D200    LRLK  AR2,DSPPTS+DP6
      09F1 035F
L0094 09F2 D300    LRLK  AR3,DSPPTZ+DP6
      09F3 0367
L0095 09F4 558A    LARP  AR2
L0096 09F5 20A0    LAC   *+
L0097 09F6 6000    SACL  X1          * fetch X0
L0098 09F7 20AB    LAC   *+,0,3
L0099 09F8 6001    SACL  Y1          * fetch Y0
L0100 09F9 20A0    LAC   *+
L0101 09FA 6002    SACL  X2          * fetch X4
L0102 09FB 20A0    LAC   *+
L0103 09FC 6003    SACL  Y2          * fetch Y4
L0104 09FD 726F    SAR   AR2,SAVAR2
L0105 09FE 7370    SAR   AR3,SAVAR3
L0106 09FF FE80    CALL  LINE        * draw line connecting 0 & 4
      0A00 0524
```

```
L0107 0A01 326F    LAR   AR2,SAVAR2
L0108 0A02 3370    LAR   AR3,SAVAR3
L0109 0A03 558A    LARP  AR2
L0110 0A04 20A0    LAC   *+
L0111 0A05 6000    SACL  X1          * fetch X1
L0112 0A06 20AB    LAC   *+,0,3
L0113 0A07 6001    SACL  Y1          * fetch Y1
L0114 0A08 20A0    LAC   *+
L0115 0A09 6002    SACL  X2          * fetch X5
L0116 0A0A 20A0    LAC   *+
L0117 0A0B 6003    SACL  Y2          * fetch Y5
L0118 0A0C 726F    SAR   AR2,SAVAR2
L0119 0A0D 7370    SAR   AR3,SAVAR3
L0120 0A0E FE80    CALL  LINE        * draw line connecting 1 & 5
      0A0F 0524
L0121 0A10 326F    LAR   AR2,SAVAR2
L0122 0A11 3370    LAR   AR3,SAVAR3
L0123 0A12 558A    LARP  AR2
L0124 0A13 20A0    LAC   *+
L0125 0A14 6000    SACL  X1          * fetch X2
L0126 0A15 20AB    LAC   *+,0,3
L0127 0A16 6001    SACL  Y1          * fetch Y2
L0128 0A17 20A0    LAC   *+
L0129 0A18 6002    SACL  X2          * fetch X6
L0130 0A19 20A0    LAC   *+
L0131 0A1A 6003    SACL  Y2          * fetch Y6
L0132 0A1B 726F    SAR   AR2,SAVAR2
L0133 0A1C 7370    SAR   AR3,SAVAR3
L0134 0A1D FE80    CALL  LINE        * draw line connecting 2 & 6
      0A1E 0524
L0135 0A1F 326F    LAR   AR2,SAVAR2
L0136 0A20 3370    LAR   AR3,SAVAR3
L0137 0A21 558A    LARP  AR2
L0138 0A22 20A0    LAC   *+
L0139 0A23 6000    SACL  X1          * fetch X3
L0140 0A24 20AB    LAC   *+,0,3
L0141 0A25 6001    SACL  Y1          * fetch Y3
L0142 0A26 20A0    LAC   *+
L0143 0A27 6002    SACL  X2          * fetch X7
L0144 0A28 20A0    LAC   *+
L0145 0A29 6003    SACL  Y2          * fetch Y7
L0146 0A2A 726F    SAR   AR2,SAVAR2
L0147 0A2B 7370    SAR   AR3,SAVAR3
L0148 0A2C FE80    CALL  LINE        * draw line connecting 3 & 7
      0A2D 0524
L0149 0A2E CA07    LACK  7
L0150 0A2F 6004    SACL  COLOR       * initialize color to WHITE
L0151 0A30 558A    LARP  AR2
L0152 0A31 D200    LRLK  AR2,DSPPTZ+DP6
      0A32 0367
L0153 0A33 20A0    LAC   *+
L0154 0A34 6000    SACL  X1          * fetch X4
L0155 0A35 20A0    LAC   *+
L0156 0A36 6001    SACL  Y1          * fetch Y4
L0157 0A37 20A0    LAC   *+
L0158 0A38 6002    SACL  X2          * fetch X5
L0159 0A39 6024    SACL  XB          * save X5 for next line
```

```
L0160 0A3A 20A0          LAC   *+            * fetch Y5
L0161 0A3B 6003          SACL  Y2            * save Y5 for next line
L0162 0A3C 6025          SACL  YB
L0163 0A3D 726F          SAR   AR2,SAVAR2
L0164 0A3E FE80          CALL  LINE          * draw line connecting 4 & 5
      0A3F 0524
L0165 0A40 558A          LARP  AR2
L0166 0A41 326F          LAR   AR2,SAVAR2
L0167 0A42 2024          LAC   XB
L0168 0A43 6000          SACL  X1            * fetch X5
L0169 0A44 2025          LAC   YB
L0170 0A45 6001          SACL  Y1            * fetch Y5
L0171 0A46 20A0          LAC   *+
L0172 0A47 6002          SACL  X2            * fetch X6
L0173 0A48 6024          SACL  XB            * save X6 for next line
L0174 0A49 20A0          LAC   *+
L0175 0A4A 6003          SACL  Y2            * fetch Y6
L0176 0A4B 6025          SACL  YB            * save Y6 for next line
L0177 0A4C 726F          SAR   AR2,SAVAR2
L0178 0A4D FE80          CALL  LINE          * draw line connecting 5 & 6
      0A4E 0524
L0179 0A4F 558A          LARP  AR2
L0180 0A50 326F          LAR   AR2,SAVAR2
L0181 0A51 2024          LAC   XB
L0182 0A52 6000          SACL  X1            * fetch X6
L0183 0A53 2025          LAC   YB
L0184 0A54 6001          SACL  Y1            * fetch Y6
L0185 0A55 20A0          LAC   *+
L0186 0A56 6002          SACL  X2            * fetch X7
L0187 0A57 6024          SACL  XB            * save X7 for next line
L0188 0A58 20A0          LAC   *+
L0189 0A59 6003          SACL  Y2            * fetch Y7
L0190 0A5A 6025          SACL  YB            * save Y7 for next line
L0191 0A5B 726F          SAR   AR2,SAVAR2
L0192 0A5C FE80          CALL  LINE          * draw line connecting 6 & 7
      0A5D 0524
L0193 0A5E 558A          LARP  AR2
L0194 0A5F D200          LRLK  AR2,DSPPTZ+DP6
      0A60 0367
L0195 0A61 2024          LAC   XB
L0196 0A62 6000          SACL  X1            * fetch X7
L0197 0A63 2025          LAC   YB
L0198 0A64 6001          SACL  Y1            * fetch Y7
L0199 0A65 20A0          LAC   *+
L0200 0A66 6002          SACL  X2            * fetch X4
L0201 0A67 20A0          LAC   *+
L0202 0A68 6003          SACL  Y2            * fetch Y4
L0203 0A69 726F          SAR   AR2,SAVAR2
L0204 0A6A FE80          CALL  LINE          * draw line connecting 7 & 4
      0A6B 0524
L0205 0A6C CE26          RET
0013                     COPY  SINTBL.ASM
```

                                        SINE   COSINE

```
M0002
M0003
      *
      *
      *
M0004 0A6D 0000  SINTBL  DATA  0         * 0/360 degrees   270 degrees
M0005 0A6E 0324          DATA  804
M0006 0A6F 0648          DATA  1608
M0007 0A70 096A          DATA  2410
M0008 0A71 0C8C          DATA  3212
M0009 0A72 0FAB          DATA  4011
M0010 0A73 12C8          DATA  4808
M0011 0A74 15E2          DATA  5602
M0012 0A75 18F9          DATA  6393
M0013 0A76 1C0B          DATA  7179
M0014 0A77 1F1A          DATA  7962
M0015 0A78 2223          DATA  8739
M0016 0A79 2528          DATA  9512
M0017 0A7A 2826          DATA  10278
M0018 0A7B 2B1F          DATA  11039
M0019 0A7C 2E11          DATA  11793
M0020 0A7D 30FB          DATA  12539
M0021 0A7E 33DF          DATA  13279
M0022 0A7F 36BA          DATA  14010
M0023 0A80 398C          DATA  14732
M0024 0A81 3C56          DATA  15446
M0025 0A82 3F17          DATA  16151
M0026 0A83 41CE          DATA  16846
M0027 0A84 447A          DATA  17530
M0028 0A85 471C          DATA  18204
M0029 0A86 49B4          DATA  18868
M0030 0A87 4C3F          DATA  19519
M0031 0A88 4EBF          DATA  20159
M0032 0A89 5133          DATA  20787
M0033 0A8A 539B          DATA  21403
M0034 0A8B 55F5          DATA  22005
M0035 0A8C 5842          DATA  22594
M0036 0A8D 5A82          DATA  23170
M0037 0A8E 5CB3          DATA  23731     * 45 degrees        305 degrees
M0038 0A8F 5ED7          DATA  24279
M0039 0A90 60EB          DATA  24811
M0040 0A91 62F1          DATA  25329
M0041 0A92 64E8          DATA  25832
M0042 0A93 66CF          DATA  26319
M0043 0A94 68A6          DATA  26790
M0044 0A95 6A6D          DATA  27245
M0045 0A96 6C23          DATA  27683
M0046 0A97 6DC9          DATA  28105
M0047 0A98 6F5E          DATA  28510
M0048 0A99 70E2          DATA  28898
M0049 0A9A 7254          DATA  29268
M0050 0A9B 73B5          DATA  29621
M0051 0A9C 7504          DATA  29956
M0052 0A9D 7641          DATA  30273
M0053 0A9E 776B          DATA  30571
M0054 0A9F 7884          DATA  30852
M0055 0AA0 7989          DATA  31113
M0056 0AA1 7A7C          DATA  31356
M0057 0AA2 7B5C          DATA  31580
M0058 0AA3 7C29          DATA  31785
```

```
M0059 0AA4 7CE3    DATA 31971
M0060 0AA5 7D89    DATA 32137
M0061 0AA6 7E1D    DATA 32285
M0062 0AA7 7E9C    DATA 32412
M0063 0AA8 7F09    DATA 32521
M0064 0AA9 7F61    DATA 32609
M0065 0AAA 7FA6    DATA 32678
M0066 0AAB 7FD8    DATA 32728
M0067 0AAC 7FF5    DATA 32757
M0068 0AAD 7FFF COSTBL DATA 32767      * 90 degrees
M0069 0AAE 7FF5    DATA 32757
M0070 0AAF 7FD8    DATA 32728
M0071 0AB0 7FA6    DATA 32678
M0072 0AB1 7F61    DATA 32609
M0073 0AB2 7F09    DATA 32521
M0074 0AB3 7E9C    DATA 32412
M0075 0AB4 7E1D    DATA 32285
M0076 0AB5 7D89    DATA 32137
M0077 0AB6 7CE3    DATA 31971
M0078 0AB7 7C29    DATA 31785
M0079 0AB8 7B5C    DATA 31580
M0080 0AB9 7A7C    DATA 31356
M0081 0ABA 7989    DATA 31113
M0082 0ABB 7884    DATA 30852
M0083 0ABC 776B    DATA 30571
M0084 0ABD 7641    DATA 30273
M0085 0ABE 7504    DATA 29956
M0086 0ABF 73B5    DATA 29621
M0087 0AC0 7254    DATA 29268
M0088 0AC1 70E2    DATA 28898
M0089 0AC2 6F5E    DATA 28510
M0090 0AC3 6DC9    DATA 28105
M0091 0AC4 6C23    DATA 27683
M0092 0AC5 6A6D    DATA 27245
M0093 0AC6 68A6    DATA 26790
M0094 0AC7 66CF    DATA 26319
M0095 0AC8 64E8    DATA 25832
M0096 0AC9 62F1    DATA 25329
M0097 0ACA 60EB    DATA 24811
M0098 0ACB 5ED7    DATA 24279
M0099 0ACC 5CB3    DATA 23731
M0100 0ACD 5A82    DATA 23170      * 135 degrees
M0101 0ACE 5842    DATA 22594
M0102 0ACF 55F5    DATA 22005
M0103 0AD0 539B    DATA 21403
M0104 0AD1 5133    DATA 20787
M0105 0AD2 4EBF    DATA 20159
M0106 0AD3 4C3F    DATA 19519
M0107 0AD4 49B4    DATA 18868
M0108 0AD5 471C    DATA 18204
M0109 0AD6 447A    DATA 17530
M0110 0AD7 41CE    DATA 16846
M0111 0AD8 3F17    DATA 16151
M0112 0AD9 3C56    DATA 15446
M0113 0ADA 398C    DATA 14732
M0114 0ADB 36BA    DATA 14010
M0115 0ADC 33DF    DATA 13279
```

0/360 degrees   45 degrees

```
M0116 0ADD 30FB    DATA 12539
M0117 0ADE 2E11    DATA 11793
M0118 0ADF 2B1F    DATA 11039
M0119 0AE0 2826    DATA 10278
M0120 0AE1 2528    DATA 9512
M0121 0AE2 2223    DATA 8739
M0122 0AE3 1F1A    DATA 7962
M0123 0AE4 1C0B    DATA 7179
M0124 0AE5 18F9    DATA 6393
M0125 0AE6 15E2    DATA 5602
M0126 0AE7 12C8    DATA 4808      * 180 degrees
M0127 0AE8 0FAB    DATA 4011
M0128 0AE9 0C8C    DATA 3212
M0129 0AEA 096A    DATA 2410
M0130 0AEB 0648    DATA 1608
M0131 0AEC 0324    DATA 804
M0132 0AED 0000    DATA 0
M0133 0AEE FCDC    DATA -804
M0134 0AEF F9B8    DATA -1608
M0135 0AF0 F696    DATA -2410
M0136 0AF1 F374    DATA -3212
M0137 0AF2 F055    DATA -4011
M0138 0AF3 E038    DATA -4808
M0139 0AF4 EA1E    DATA -5602
M0140 0AF5 E707    DATA -6393
M0141 0AF6 E3F5    DATA -7179
M0142 0AF7 E0E6    DATA -7962
M0143 0AF8 DDDD    DATA -8739
M0144 0AF9 DAD8    DATA -9512
M0145 0AFA D7DA    DATA -10278
M0146 0AFB D4E1    DATA -11039
M0147 0AFC D1EF    DATA -11793
M0148 0AFD CF05    DATA -12539
M0149 0AFE CC21    DATA -13279
M0150 0AFF C946    DATA -14010
M0151 0B00 C674    DATA -14732
M0152 0B01 C3AA    DATA -15446
M0153 0B02 C0E9    DATA -16151
M0154 0B03 BE32    DATA -16846
M0155 0B04 BB86    DATA -17530
M0156 0B05 B8E4    DATA -18204
M0157 0B06 B64C    DATA -18868
M0158 0B07 B3C1    DATA -19519
M0159 0B08 B141    DATA -20159
M0160 0B09 AECD    DATA -20787
M0161 0B0A AC65    DATA -21403
M0162 0B0B AA0B    DATA -22005
M0163 0B0C A7BE    DATA -22594
M0164 0B0D A57E    DATA -23170      * 215 degrees
M0165 0B0E A34D    DATA -23731
M0166 0B0F A129    DATA -24279
M0167 0B10 9F15    DATA -24811
M0168 0B11 900F    DATA -25329
M0169 0B12 9B18    DATA -25832
M0170 0B13 9931    DATA -26319
M0171 0B14 975A    DATA -26790
M0172 0B15 9593    DATA -27245
```

90 degrees   135 degrees

23. A Graphics Implementation Using the TMS32020 and TMS34061

```
M0173  0B16  93DD    DATA    -27683
M0174  0B17  9237    DATA    -28105
M0175  0B18  90A2    DATA    -28510
M0176  0B19  8F1E    DATA    -28898
M0177  0B1A  80AC    DATA    -29268
M0178  0B1B  8C4B    DATA    -29621
M0179  0B1C  8AFC    DATA    -29956
M0180  0B1D  89BF    DATA    -30273
M0181  0B1E  8895    DATA    -30571
M0182  0B1F  877C    DATA    -30852
M0183  0B20  8677    DATA    -31113
M0184  0B21  8584    DATA    -31356
M0185  0B22  84A4    DATA    -31580
M0186  0B23  83D7    DATA    -31785
M0187  0B24  831D    DATA    -31971
M0188  0B25  8277    DATA    -32137
M0189  0B26  81E3    DATA    -32285
M0190  0B27  8164    DATA    -32412
M0191  0B28  80F7    DATA    -32521
M0192  0B29  809F    DATA    -32609
M0193  0B2A  805A    DATA    -32678
M0194  0B2B  8028    DATA    -32728
M0195  0B2C  800B    DATA    -32757
M0196  0B2D  8001    DATA    -32767      * 270 degrees
M0197  0B2E  800B    DATA    -32757
M0198  0B2F  8028    DATA    -32728
M0199  0B30  805A    DATA    -32678
M0200  0B31  809F    DATA    -32609
M0201  0B32  80F7    DATA    -32521
M0202  0B33  8164    DATA    -32412
M0203  0B34  81E3    DATA    -32285
M0204  0B35  8277    DATA    -32137
M0205  0B36  831D    DATA    -31971
M0206  0B37  83D7    DATA    -31785
M0207  0B38  84A4    DATA    -31580
M0208  0B39  8584    DATA    -31356
M0209  0B3A  8677    DATA    -31113
M0210  0B3B  877C    DATA    -30852
M0211  0B3C  8895    DATA    -30571
M0212  0B3D  89BF    DATA    -30273
M0213  0B3E  8AFC    DATA    -29956
M0214  0B3F  8C4B    DATA    -29621
M0215  0B40  80AC    DATA    -29268
M0216  0B41  8F1E    DATA    -28898
M0217  0B42  90A2    DATA    -28510
M0218  0B43  9237    DATA    -28105
M0219  0B44  93DD    DATA    -27683
M0220  0B45  9593    DATA    -27245
M0221  0B46  975A    DATA    -26790      * 305 degrees
M0222  0B47  9931    DATA    -26319
M0223  0B48  9B18    DATA    -25832
M0224  0B49  9D0F    DATA    -25329
M0225  0B4A  9F15    DATA    -24811
M0226  0B4B  A129    DATA    -24279
M0227  0B4C  A34D    DATA    -23731
M0228  0B4D  A57E    DATA    -23170
M0229  0B4E  A7BE    DATA    -22594
```

180 degrees

215 degrees

```
M0230  0B4F  AA0B    DATA    -22005
M0231  0B50  AC66    DATA    -21402
M0232  0B51  AECD    DATA    -20787
M0233  0B52  B141    DATA    -20159
M0234  0B53  B3C1    DATA    -19519
M0235  0B54  B64D    DATA    -18867
M0236  0B55  B8E4    DATA    -18204
M0237  0B56  BB86    DATA    -17530
M0238  0B57  BE32    DATA    -16846
M0239  0B58  C0E9    DATA    -16151
M0240  0B59  C3AA    DATA    -15446
M0241  0B5A  C674    DATA    -14732
M0242  0B5B  C946    DATA    -14010
M0243  0B5C  CC21    DATA    -13279
M0244  0B5D  CF05    DATA    -12539
M0245  0B5E  D1EF    DATA    -11793
M0246  0B5F  D4E1    DATA    -11039
M0247  0B60  D7DA    DATA    -10278
M0248  0B61  DAD8    DATA    -9512
M0249  0B62  DDDD    DATA    -8739
M0250  0B63  E0E6    DATA    -7962
M0251  0B64  E3F5    DATA    -7179
M0252  0B65  E707    DATA    -6393
M0253  0B66  EA1E    DATA    -5602
M0254  0B67  ED38    DATA    -4808
M0255  0B68  F055    DATA    -4011
M0256  0B69  F374    DATA    -3212
M0257  0B6A  F696    DATA    -2410
M0258  0B6B  F9B8    DATA    -1608
M0259  0B6C  FCDC    DATA    -804
0014                 ENDTBL  COPY    ROM.ASM
```

23. A Graphics Implementation Using the TMS32020 and TMS34061

```
N0002      ***********************************************************
N0003      *
N0004      *          ROM - RAM DATA INITIALIZATION
N0005      *
N0006      ***********************************************************
N0007  0B6D 02CF  ROMDAT  DATA  719        XMAX
N0008  0B6E 012B          DATA  299        YMAX
N0009  0B6F 0095          DATA  149        AMAX
N0010  0B70 002A          DATA  42         AMIN
N0011  0B71 0061          DATA  97         BMAX
N0012  0B72 0017          DATA  23         BMIN
N0013
N0014  0B73 4000  VRTICS  DATA  >4000      X0,Y0,Z0
N0015  0B74 C000          DATA  >C000
N0016  0B75 C000          DATA  >C000
N0017  0B76 4000          DATA  >4000      X1,Y1,Z1
N0018  0B77 4000          DATA  >4000
N0019  0B78 C000          DATA  >C000
N0020  0B79 C000          DATA  >C000      X2,Y2,Z2
N0021  0B7A 4000          DATA  >4000
N0022  0B7B C000          DATA  >C000
N0023  0B7C C000          DATA  >C000
N0024  0B7D C000          DATA  >C000      X3,Y3,Z3
N0025  0B7E C000          DATA  >C000
N0026  0B7F 4000          DATA  >4000
N0027  0B80 C000          DATA  >C000      X4,Y4,Z4
N0028  0B81 4000          DATA  >4000
N0029  0B82 4000          DATA  >4000
N0030  0B83 4000          DATA  >4000      X5,Y5,Z5
N0031  0B84 4000          DATA  >4000
N0032  0B85 C000          DATA  >C000
N0033  0B86 4000          DATA  >4000      X6,Y6,Z6
N0034  0B87 4000          DATA  >4000
N0035  0B88 C000          DATA  >C000
N0036  0B89 C000          DATA  >C000      X7,Y7,Z7
N0037  0B8A 4000          DATA  >4000
N0038      *
N0039  0B8B E666  CMTRX   DATA  >E666      C11,C21,C31,C41
N0040  0B8C 1333          DATA  >1333
N0041  0B8D 0000          DATA  0
N0042  0B8E 0000          DATA  0
N0043  0B8F 0B85          DATA  >0B85      C12,C22,C32,C42
N0044  0B90 0F5C          DATA  >0F5C
N0045  0B91 E666          DATA  >E666
N0046  0B92 0000          DATA  0
N0047  0B93 FC29          DATA  >FC29      C13,C23,C33,C43
N0048  0B94 FAE1          DATA  >FAE1
N0049  0B95 FB33          DATA  >FB33
N0050  0B96 6400          DATA  >6400
N0051      *
N0052  0B97 2880  SCALE   DATA  162*64     VSX
N0053  0B98 0167          DATA  359        VCX
N0054  0B99 1680          DATA  90*64      VSY
N0055  0B9A 009D          DATA  157        VCY
0015                      COPY  CONTROL.ASM
```

```
N0002      ***********************************************************
N0003      *
N0004      *             DEMONSTRATION COMMAND STRING
N0005      *
N0006      ***********************************************************
N0007      *    This file provides a series of data statements which are
N0008      *    used to control the graphics demonstration for the
N0009      *    TMS32020-VSC system. The data statements are used to
N0010      *    arbitrarily select the various fundamental algorithms
N0011      *    (pixel, lines, ellipses, fills, etc.) along with the
N0012      *    appropriate data to generate a video graphics
N0013      *    demonstration. The data word 32020 in decimal is used as
N0014      *    a synchronization value at the start and between commands
N0015      *    to allow recovery from errors in entering the data
N0016      *    sequence. The synchronization word is followed by a
N0017      *    single command word selecting the appropriate graphics or
N0018      *    control function and as many words as necessary to provide
N0019      *    all of the required information to the selected function.
N0020      *
N0021      ***********************************************************
N0022
N0023  0B9B 7D14  BEGIN   DATA  32020
N0024  0B9C 0002          DATA  >0002      CLEAR SCREEN
N0025  0B9D 0000          DATA  >0000
N0026      *
N0027  0B9E 7D14          DATA  32020
N0028  0B9F 0016          DATA  >0016      ROTATING CUBE
N0029      *
N0030  0BA0 7D14          DATA  32020
N0031  0BA1 0002          DATA  >0002      CLEAR SCREEN
N0032  0BA2 0000          DATA  >0000
N0033      *
N0034  0BA3 7D14          DATA  32020
N0035  0BA4 000A          DATA  >000A      SPIN FILLED ELLIPSE
N0036  0BA5 0168          DATA  360
N0037  0BA6 00A0          DATA  160
N0038  0BA7 006C          DATA  108
N0039  0BA8 003C          DATA  60
N0040  0BA9 0201          DATA  >0201
N0041  0BAA 0010          DATA  >0010
N0042      *
N0043  0BAB 7D14          DATA  32020
N0044  0BAC 0009          DATA  >0009      FILLED ELLIPSE
N0045  0BAD 01F3          DATA  499
N0046  0BAE 0090          DATA  144
N0047  0BAF 00DC          DATA  220
N0048  0BB0 0073          DATA  115
N0049  0BB1 0005          DATA  >0005
N0050      *
N0051  0BB2 7D14          DATA  32020
N0052  0BB3 0009          DATA  >0009      FILLED ELLIPSE
N0053  0BB4 01FD          DATA  509
N0054  0BB5 0103          DATA  259
N0055  0BB6 00C8          DATA  200
N0056  0BB7 001E          DATA  30
N0057  0BB8 0006          DATA  >0006
N0058      *
```

```
00059 0BB9 7D14    DATA  32020
00060 0BBA 0009    DATA  >0009    FILLED ELLIPSE
00061 0BBB 0118    DATA  280
00062 0BBC 00C7    DATA  199
00063 0BBD 0064    DATA  100
00064 0BBE 001E    DATA  30
00065 0BBF 0002    DATA  >0002
00066
00067 0BC0 7D14    DATA  32020
00068 0BC1 0009    DATA  >0009    FILLED ELLIPSE
00069 0BC2 0064    DATA  100
00070 0BC3 0064    DATA  100
00071 0BC4 005A    DATA  90
00072 0BC5 0055    DATA  85
00073 0BC6 0001    DATA  >0001
00074
00075 0BC7 7D14    DATA  32020
00076 0BC8 0009    DATA  >0009    FILLED ELLIPSE
00077 0BC9 00C8    DATA  200
00078 0BCA 0028    DATA  40
00079 0BCB 003C    DATA  60
00080 0BCC 001E    DATA  30
00081 0BCD 0007    DATA  >0007
00082
00083 0BCE 7D14    DATA  32020
00084 0BCF 0009    DATA  >0009    FILLED ELLIPSE
00085 0BD0 0154    DATA  340
00086 0BD1 0258    DATA  600
00087 0BD2 0032    DATA  50
00088 0BD3 001E    DATA  30
00089 0BD4 0003    DATA  >0003
00090
00091 0BD5 7D14    DATA  32020
00092 0BD6 0009    DATA  >0009    FILLED ELLIPSE
00093 0BD7 01CC    DATA  460
00094 0BD8 0055    DATA  85
00095 0BD9 0028    DATA  40
00096 0BDA 0016    DATA  22
00097 0BDB 0003    DATA  >0003
00098
00099 0BDC 7D14    DATA  32020
00100 0BDD 0009    DATA  >0009    FILLED ELLIPSE
00101 0BDE 0226    DATA  550
00102 0BDF 0073    DATA  115
00103 0BE0 001E    DATA  30
00104 0BE1 0012    DATA  18
00105 0BE2 0003    DATA  >0003
00106
00107 0BE3 7D14    DATA  32020
00108 0BE4 0009    DATA  >0009    FILLED ELLIPSE
00109 0BE5 026C    DATA  620
00110 0BE6 008C    DATA  140
00111 0BE7 0018    DATA  24
00112 0BE8 000E    DATA  14
00113 0BE9 0003    DATA  >0003
00114
00115 0BEA 7D14    DATA  32020
```

```
00116 0BEB 0009    DATA  >0009    FILLED ELLIPSE
00117 0BEC 029E    DATA  670
00118 0BED 00A5    DATA  165
00119 0BEE 0014    DATA  20
00120 0BEF 000C    DATA  12
00121 0BF0 0003    DATA  >0003
00122
00123 0BF1 7D14    DATA  32020
00124 0BF2 0009    DATA  >0009    FILLED ELLIPSE
00125 0BF3 02C6    DATA  710
00126 0BF4 00B9    DATA  185
00127 0BF5 000A    DATA  10
00128 0BF6 0006    DATA  6
00129 0BF7 0003    DATA  >0003
00130
00131 0BF8 7D14    DATA  32020
00132 0BF9 0009    DATA  >0009    FILLED ELLIPSE
00133 0BFA 006E    DATA  110
00134 0BFB 00FA    DATA  250
00135 0BFC 0050    DATA  80
00136 0BFD 002D    DATA  45
00137 0BFE 0004    DATA  >0004
00138
00139 0BFF 7D14    DATA  32020    DELAY
00140 0C00 0001    DATA  >0001
00141 0C01 012C    DATA  60*5
00142
00143 0C02 7D14    DATA  32020    CLEAR SCREEN
00144 0C03 0002    DATA  >0002
00145 0C04 0000    DATA  >0000
00146
00147 0C05 7D14    DATA  32020    FILLED ELLIPSE SETS
00148 0C06 000B    DATA  >000B
00149 0C07 0167    DATA  359
00150 0C08 0095    DATA  149
00151 0C09 0018    DATA  24
00152 0C0A 0096    DATA  150
00153 0C0B 0006    DATA  >0006
00154 0C0C 0018    DATA  24
00155 0C0D FFF6    DATA  -10
00156 0C0E 000F    DATA  15
00157
00158 0C0F 7D14    DATA  32020    DELAY
00159 0C10 0001    DATA  >0001
00160 0C11 012C    DATA  60*5
00161
00162 0C12 7D14    DATA  32020    CLEAR SCREEN
00163 0C13 0002    DATA  >0002
00164 0C14 0000    DATA  >0000
00165
00166 0C15 7D14    DATA  32020    FILLED ELLIPSE DEMO
00167 0C16 000C    DATA  >000C
00168 0C17 016F    DATA  367
00169 0C18 0061    DATA  97
00170 0C19 0055    DATA  2*42+1
00171 0C1A 004A    DATA  97-23
00172 0C1B 0001    DATA  >0001
```

23. A Graphics Implementation Using the TMS32020 and TMS34061

```
00173 0C1C 0013      DATA    19
00174 0C1D FFF5      DATA    -11
00175 0C1E 0028      DATA    40
00176 0C1F FFE9      DATA    -23
00177 0C20 0960      DATA    300*8
00178              *
00179 0C21 7D14      DATA    32020      DELAY
00180 0C22 0001      DATA    >0001
00181 0C23 012C      DATA    60*5
00182              *
00183 0C24 7D14      DATA    32020      CLEAR SCREEN
00184 0C25 0002      DATA    >0002
00185 0C26 0000      DATA    >0000
00186              *
00187 0C27 7D14      DATA    32020      LINE DEMO 1
00188 0C28 0011      DATA    >0011
00189
00190 0C29 7D14      DATA    32020      DELAY
00191 0C2A 0001      DATA    >0001
00192 0C2B 012C      DATA    60*5
00193
00194 0C2C 7D14      DATA    32020      CLEAR SCREEN
00195 0C2D 0002      DATA    >0002
00196 0C2E 0000      DATA    >0000
00197
00198 0C2F 7D14      DATA    32020      LINE DEMO 2
00199 0C30 0012      DATA    >0012
00200              *
00201 0C31 7D14      DATA    32020      DELAY
00202 0C32 0001      DATA    >0001
00203 0C33 012C      DATA    60*5
00204              *
00205 0C34 7D14      DATA    32020      CLEAR SCREEN
00206 0C35 0002      DATA    >0002
00207 0C36 0000      DATA    >0000
00208              *
00209 0C37 7D14      DATA    32020      LINE DEMO 3
00210 0C38 0013      DATA    >0013
00211
00212 0C39 7D14      DATA    32020      DELAY
00213 0C3A 0001      DATA    >0001
00214 0C3B 012C      DATA    60*5
00215              *
00216 0C3C 7D14      DATA    32020      CLEAR SCREEN
00217 0C3D 0002      DATA    >0002
00218 0C3E 0000      DATA    >0000
00219              *
00220 0C3F 7D14      DATA    32020      LINE DEMO 4
00221 0C40 0014      DATA    >0014
00222              *
00223 0C41 7D14      DATA    32020      DELAY
00224 0C42 0001      DATA    >0001
00225 0C43 012C      DATA    60*5
00226              *
00227 0C44 7D14      DATA    32020      CLEAR SCREEN
00228 0C45 0002      DATA    >0002
00229 0C46 0000      DATA    >0000
```

```
00230
00231 0C47 7D14      DATA    32020      LINE DEMO 5
00232 0C48 0015      DATA    >0015
00233              *
00234 0C49 7D14      DATA    32020      DELAY
00235 0C4A 0001      DATA    >0001
00236 0C4B 012C      DATA    60*5
00237              *
00238 0C4C 7D14      DATA    32020      RESET
00239 0C4D 0000      DATA    >0000
00240              *
00241 0C4E 7D14      DATA    32020      RESET
00242 0C4F 0000      DATA    >0000
00243              *
00244 0C50 7D14      DATA    32020      RESET
00245 0C51 0000      DATA    >0000
00246              *
00247 0C52 7D14      DATA    32020      RESET
00248 0C53 0000      DATA    >0000
00249              *
00250 0C54 7D14      DATA    32020      RESET
00251 0C55 0000      DATA    >0000
00252              *
00253 0C56 7D14      DATA    32020      RESET
00254 0C57 0000      DATA    >0000
00255              *
00256 0C58 7D14      DATA    32020      RESET
00257 0C59 0000      DATA    >0000
00258              *
00259 0C5A 7D14      DATA    32020      RESET
00260 0C5B 0000      DATA    >0000
0016                 COPY    RAMB2.ASM
```

23. A Graphics Implementation Using the TMS32020 and TMS34061

```
GRAPHIC    32020 FAMILY MACRO ASSEMBLER   PC 1.0 85.157    15:17:30  12-05-85
TMS32020 - TMS34061 DEMONSTRATION                                    PAGE 0077

P0002       *
P0003       ********************************************************
P0004       *       RAM - MEMORY-MAPPED REGISTERS AND BLOCK B2
P0005       ********************************************************
P0006       *
P0007                  LEGEND FOR INTERNAL UTILIZATION
P0008
P0009             I = Initialization
P0010             P = Parser and demo setup routines
P0011             C = Clear screen routines
P0012             L = Line drawing routines
P0013             E = Ellipse drawing routines
P0014             R = Rotating cube routines
P0015
P0016  0C5C  DP0      EQU     $          I   P   C   L   E   R
P0017
P0018  0000           DORG    >0000
P0019
P0020       * Memory-Mapped Register #00
P0021  0000 0000  DRR      DATA    0   * Serial port data receive register
P0022
P0023       * Memory-Mapped Register #01
P0024  0001 0000  DXR      DATA    0   * Serial port data transmit registe
P0025
P0026       * Memory-Mapped Register #02
P0027  0002 0000  TIM      DATA    0   * Timer register
P0028
P0029       * Memory-Mapped Register #03
P0030  0003 0000  PRD      DATA    0   * Period register
P0031
P0032       * Memory-Mapped Register #04
P0033  0004 0000  IMR      DATA    0   * Interrupt mask register
P0034
P0035       * Memory-Mapped Register #05
P0036  0005 0000  GREG     DATA    0   * Global memory allocation register
P0037
P0038
P0039  0060           DORG    >0060
P0040
P0041       * RAM Location #0060
P0042  0060 0000  STS0     DATA    0           C   L   E
P0043
P0044       * RAM Location #0061
P0045  0061 0000  STS1     DATA    0
P0046
P0047       * RAM Location #0062
P0048  0062 0000  ONE      DATA    0       P   C
P0049
P0050       * RAM Location #0063
P0051  0063 0000  FILLC    DATA    0           C
P0052
P0053       * RAM Location #0064
P0054  0064 0000  SAVREG   DATA    0           C       E
P0055
P0056       * RAM Location #0065
P0057  0065 0000           DATA    0           C       E
P0058       *


GRAPHIC    32020 FAMILY MACRO ASSEMBLER   PC 1.0 85.157    15:17:30  12-05-85
TMS32020 - TMS34061 DEMONSTRATION                                    PAGE 0078

P0059       * RAM Location #0066
P0060  0066 0000           DATA    0               E
P0061       *
P0062       * RAM Location #0067
P0063  0067 0000           DATA    0               E
P0064       *
P0065       * RAM Location #0068
P0066  0068 0000  TMP1     DATA    0       P
P0067       *
P0068       * RAM Location #0069
P0069  0069 0000  SYNCH    DATA    0       P
P0070       *
P0071       * RAM Location #006A
P0072  006A 0000  TIMDLY   DATA    0       P
P0073       *
P0074       * RAM Location #006B
P0075  006B 0000  CMND     DATA    0       P
P0076       *
P0077       * RAM Location #006C
P0078  006C 0000  CMDADR   DATA    0
0017                       COPY    RAMB0.ASM
```

23. A Graphics Implementation Using the TMS32020 and TMS34061

```
Q0002  *
Q0003  *  *****************************************
Q0004  *  *        RAM – BLOCK B0 (PAGES 4 AND 5)     *
Q0005  *  *            I   P   C   L   E   R          *
Q0006  *  *****************************************
Q0007  0200            DP4    DORG   >80*4
Q0008  0200                   EQU    $                    E
Q0009
Q0010  0000                   DORG   >0000
Q0067  *
Q0068  *       RAM Location #0212
Q0069  0012 0000      R11     DATA   0                    R
Q0070  *
Q0071  *       RAM Location #0213
Q0072  0013 0000      R21     DATA   0
Q0073  *
Q0074  *       RAM Location #0214
Q0075  0014 0000      R12     DATA   0                    R
Q0076  *
Q0077  *       RAM Location #0215
Q0078  0015 0000      R22     DATA   0                    R
Q0079  *
Q0080  *       RAM Location #0216
Q0081  0016 0000      C11     DATA   0                    R
Q0082  *
Q0083  *       RAM Location #0217
Q0084  0017 0000      C21     DATA   0                    R
Q0085  *
Q0086  *       RAM Location #0218
Q0087  0018 0000      C31     DATA   0                    R
Q0088  *
Q0089  *       RAM Location #0219
Q0090  0019 0000      C41     DATA   0                    R
Q0091  *
Q0092  *       RAM Location #021A
Q0093  001A 0000      C12     DATA   0                    R
Q0094  *
Q0095  *       RAM Location #021B
Q0096  001B 0000      C22     DATA   0                    R
Q0097  *
Q0098  *       RAM Location #021C
Q0099  001C 0000      C32     DATA   0                    R
Q0100  *
Q0101  *       RAM Location #021D
Q0102  001D 0000      C42     DATA   0                    R
Q0103  *
Q0104  *       RAM Location #021E
Q0105  001E 0000      C13     DATA   0                    R
Q0106  *
Q0107  *       RAM Location #021F
Q0108  001F 0000      C23     DATA   0                    R
Q0109  *
Q0110  *       RAM Location #0220
Q0111  0020 0000      C33     DATA   0                    R
Q0112  *
Q0113  *       RAM Location #0221
Q0114  0021 0000      C43     DATA   0                    R
```

```
Q0115  *
Q0116  *       RAM Location #0222
Q0117  0022 0000      VSX     DATA   0                    R
Q0118  *
Q0119  *       RAM Location #0223
Q0120  0023 0000      VCX     DATA   0
Q0121  *
Q0122  *       RAM Location #0224
Q0123  0024 0000      VSY     DATA   0                    R
Q0124  *
Q0125  *       RAM Location #0225
Q0126  0025 0000      VCY     DATA   0
Q0127  *
Q0128  *       RAM Location #0226
Q0129  0026 0000      SIN     DATA   0                    R
Q0130  *
Q0131  *       RAM Location #0227
Q0132  0027 0000      COS     DATA   0
Q0133  *
Q0134  *       RAM Location #0228
Q0135  0028 0000      THETA   DATA   0                    R
0018                          COPY   RAMB1.ASM
```

```
R0002                 *
R0003                 *
R0004                 *        RAM - BLOCK B1 (PAGES 6 AND 7)
R0005                 *
R0006
R0007   0300          DP6      DORG   >80*6
R0008   0300                    EQU    $
R0009
R0010   0000                    DORG   >0000
R0011
R0012                 * RAM Location #0300
R0013   0000          XP       BSS    0
R0014   0000 0000     X1       DATA   0
R0015
R0016                 * RAM Location #0301
R0017   0001          YP       BSS    0
R0018   0001 0000     Y1       DATA   0
R0019
R0020                 * RAM Location #0302
R0021   0002 0000     X2       DATA   0
R0022
R0023                 * RAM Location #0303
R0024   0003 0000     Y2       DATA   0
R0025
R0026                 * RAM Location #0304
R0027   0004 0000     COLOR    DATA   0
R0028
R0029                 * RAM Location #0305
R0030   0005 0000     DX       DATA   0
R0031
R0032                 * RAM Location #0306
R0033   0006 0000     DY       DATA   0
R0034
R0035                 * RAM Location #0307
R0036   0007 0000     D        DATA   0
R0037
R0038                 * RAM Location #0308
R0039   0008 0000     INCR1    DATA   0
R0040
R0041                 * RAM Location #0309
R0042   0009 0000     INCR2    DATA   0
R0043
R0044                 * RAM Location #030A
R0045   000A 0000     COLORL   DATA   0
R0046
R0047                 * RAM Location #030B
R0048   000B 0000     COLORH   DATA   0
R0049
R0050                 * RAM Location #030C
R0051   000C 0000     COLOR2   DATA   0
R0052
R0053                 * RAM Location #030D
R0054   000D 0000     BFLAG    DATA   0
R0055
R0056                 * RAM Location #030E
R0057   000E 0000     ASIGN    DATA   0
R0058
```

```
R0059                 * RAM Location #030F
R0060   000F 0000     ABSIGN   DATA   0
R0061
R0062                 * RAM Location #0310
R0063   0010 0000     XC       DATA   0
R0064
R0065                 * RAM Location #0311
R0066   0011 0000     YC       DATA   0
R0067
R0068                 * RAM Location #0312
R0069   0012          ZC       BSS    0
R0070   0012 0000     A        DATA   0
R0071
R0072                 * RAM Location #0313
R0073   0013          ZCI      BSS    0
R0074   0013 0000     B        DATA   0
R0075
R0076                 * RAM Location #0314
R0077   0014          ZCSIGN   BSS    0
R0078   0014 0000     T1H      DATA   0
R0079
R0080                 * RAM Location #0315
R0081   0015 0000     T1L      DATA   0
R0082
R0083                 * RAM Location #0316
R0084   0016 0000     T2H      DATA   0
R0085
R0086                 * RAM Location #0317
R0087   0017 0000     T2L      DATA   0
R0088
R0089                 * RAM Location #0318
R0090   0018 0000     T3H      DATA   0
R0091
R0092                 * RAM Location #0319
R0093   0019 0000     T3L      DATA   0
R0094
R0095                 * RAM Location #031A
R0096   001A 0000     T4H      DATA   0
R0097
R0098                 * RAM Location #031B
R0099   001B 0000     T4L      DATA   0
R0100
R0101                 * RAM Location #031C
R0102   001C 0000     T5H      DATA   0
R0103
R0104                 * RAM Location #031D
R0105   001D 0000     T5L      DATA   0
R0106
R0107                 * RAM Location #031E
R0108   001E 0000     T6H      DATA   0
R0109
R0110                 * RAM Location #031F
R0111   001F 0000     T6L      DATA   0
R0112
R0113                 * RAM Location #0320
R0114   0020 0000     T7H      DATA   0
R0115
```

23. A Graphics Implementation Using the TMS32020 and TMS34061

```
GRAPHIC      32020 FAMILY MACRO ASSEMBLER   PC 1.0 85.157    15:17:30 12-05-85
TMS32020 - TMS34061 DEMONSTRATION                                    PAGE 0083

R0116                    * RAM Location #0321
R0117  0021 0000  T7L    DATA   0                              E
R0118                    *
R0119                    * RAM Location #0322                P     R
R0120  0022       X      BSS    0
R0121  0022 0000  XA     DATA   0
R0122                    *
R0123                    * RAM Location #0323                P     R
R0124  0023       Y      BSS    0
R0125  0023 0000  YA     DATA   0
R0126                    *
R0127                    * RAM Location #0324                P     R
R0128  0024       Z      BSS    0
R0129  0024 0000  XB     DATA   0
R0130                    *
R0131                    * RAM Location #0325                P     R
R0132  0025       XR     BSS    0
R0133  0025       XM     BSS    0
R0134  0025 0000  YB     DATA   0                              E
R0135                    *
R0136                    * RAM Location #0326                      R
R0137  0026       YR     BSS    0
R0138  0026       YM     BSS    0
R0139  0026 0000  DH     DATA   0                              E
R0140                    *
R0141                    * RAM Location #0327                      R
R0142  0027       ZR     BSS    0
R0143  0027       ZM     BSS    0
R0144  0027 0000  DL     DATA   0                              E
R0145                    *
R0146                    * RAM Location #0328                      R
R0147  0028 0000  X1H    DATA   0                              E
R0148                    *
R0149                    * RAM Location #0329                      R
R0150  0029 0000  X1L    DATA   0                              E
R0151                    *
R0152                    * RAM Location #032A                      R
R0153  002A 0000  X2H    DATA   0                              E
R0154                    *
R0155                    * RAM Location #032B                      R
R0156  002B 0000  X2L    DATA   0                              E
R0157                    *
R0158                    * RAM Location #032C                P
R0159  002C 0000  PIXCNT DATA   0
R0160                    *
R0161                    * RAM Location #032D
R0162  002D 0000  PAXCNT DATA   0
R0163                    *
R0164                    * RAM Location #032E                I   P
R0165  002E 0000  ONE6   DATA   0
R0166                    *
R0167                    * RAM Location #032F                    P
R0168  002F 0000  SAVAR0 DATA   0                              R
R0169                    *
R0170                    * RAM Location #0330                    P
R0171  0030 0000  NPOINT DATA   0                              R
R0172                    *
```

```
GRAPHIC      32020 FAMILY MACRO ASSEMBLER   PC 1.0 85.157    15:17:30 12-05-85
TMS32020 - TMS34061 DEMONSTRATION                                    PAGE 0084

R0173                    * RAM Location #0331                    P
R0174  0031 0000  SAVCLR DATA   0
R0175                    *
R0176                    * RAM Location #0332                    P
R0177  0032 0000  SAVEA  DATA   0
R0178                    *
R0179                    * RAM Location #0333                    P
R0180  0033 0000  CLRSAV DATA   0
R0181                    *
R0182                    * RAM Location #0334                    P
R0183  0034 0000  SPEED  DATA   0
R0184                    *
R0185                    * RAM Location #0335                    P
R0186  0035 0000  DA     DATA   0
R0187                    *
R0188                    * RAM Location #0336                    P
R0189  0036 0000  DB     DATA   0
R0190                    *
R0191                    * RAM Location #0337                  I P
R0192  0037 0000  XMAX   DATA   0
R0193                    *
R0194                    * RAM Location #0338                  I P
R0195  0038 0000  XMIN   DATA   0
R0196                    *
R0197                    * RAM Location #0339                  I P
R0198  0039 0000  YMAX   DATA   0
R0199                    *
R0200                    * RAM Location #033A                  I P
R0201  003A 0000  YMIN   DATA   0
R0202                    *
R0203                    * RAM Location #033B                    P
R0204  003B 0000  XMAXA  DATA   0
R0205                    *
R0206                    * RAM Location #033C                    P
R0207  003C 0000  XMINA  DATA   0
R0208                    *
R0209                    * RAM Location #033D                    P
R0210  003D 0000  YMAXB  DATA   0
R0211                    *
R0212                    * RAM Location #033E                    P
R0213  003E 0000  YMINB  DATA   0
R0214                    *
R0215                    * RAM Location #033F                  I P
R0216  003F 0000  AMAX   DATA   0
R0217                    *
R0218                    * RAM Location #0340                  I P
R0219  0040 0000  AMIN   DATA   0
R0220                    *
R0221                    * RAM Location #0341                  I P
R0222  0041 0000  BMAX   DATA   0
R0223                    *
R0224                    * RAM Location #0342                  I P
R0225  0042 0000  BMIN   DATA   0
R0226                    *
R0227                    * RAM Location #0343                  I P
R0228  0043 0000  THREE6 DATA   0
R0229                    *
```

```
R0230             * RAM Location #0344
R0231 0044 0000  I P  FIVE6   DATA    0
R0232             I P  *
R0233             * RAM Location #0345
R0234 0045 0000       NINE6   DATA    0
R0235             *
R0236             * RAM Location #0346
R0237 0046 0000  E    SAVEB   DATA    0
R0238             *
R0239             * RAM Location #0347
R0240 0047            NRMPTS  BSS
R0241 0047 0000       XN000   DATA    0
R0242             *
R0243             * RAM Location #0348                      R
R0244 0048 0000       YN000   DATA    0
R0245             *
R0246             * RAM Location #0349                      R
R0247 0049 0000       ZN000   DATA    0
R0248             *
R0249             * RAM Location #034A                      R
R0250 004A 0000       XN001   DATA    0
R0251             *
R0252             * RAM Location #034B                      R
R0253 004B 0000       YN001   DATA    0
R0254             *
R0255             * RAM Location #034C                      R
R0256 004C 0000       ZN001   DATA    0
R0257             *
R0258             * RAM Location #034D                      R
R0259 004D 0000       XN002   DATA    0
R0260             *
R0261             * RAM Location #034E                      R
R0262 004E 0000       YN002   DATA    0
R0263             *
R0264             * RAM Location #034F                      R
R0265 004F 0000       ZN002   DATA    0
R0266             *
R0267             * RAM Location #0350                      R
R0268 0050 0000       XN003   DATA    0
R0269             *
R0270             * RAM Location #0351                      R
R0271 0051 0000       YN003   DATA    0
R0272             *
R0273             * RAM Location #0352                      R
R0274 0052 0000       ZN003   DATA    0
R0275             *
R0276             * RAM Location #0353                      R
R0277 0053 0000       XN004   DATA    0
R0278             *
R0279             * RAM Location #0354                      R
R0280 0054 0000       YN004   DATA    0
R0281             *
R0282             * RAM Location #0355                      R
R0283 0055 0000       ZN004   DATA    0
R0284             *
R0285             * RAM Location #0356                      R
R0286 0056 0000       XN005   DATA    0
```

```
R0287             *
R0288             * RAM Location #0357                      R
R0289 0057 0000       YN005   DATA    0
R0290             *
R0291             * RAM Location #0358                      R
R0292 0058 0000       ZN005   DATA    0
R0293             *
R0294             * RAM Location #0359                      R
R0295 0059 0000       XN006   DATA    0
R0296             *
R0297             * RAM Location #035A                      R
R0298 005A 0000       YN006   DATA    0
R0299             *
R0300             * RAM Location #035B                      R
R0301 005B 0000       ZN006   DATA    0
R0302             *
R0303             * RAM Location #035C                      R
R0304 005C 0000       XN007   DATA    0
R0305             *
R0306             * RAM Location #035D                      R
R0307 005D 0000       YN007   DATA    0
R0308             *
R0309             * RAM Location #035E                      R
R0310 005E 0000       ZN007   DATA    0
R0311             *
R0312             * RAM Location #035F                      R
R0313 005F            DSPPTS  BSS     0
R0314 005F 0000       XD000   DATA    0
R0315             *
R0316             * RAM Location #0360                      R
R0317 0060 0000       YD000   DATA    0
R0318             *
R0319             * RAM Location #0361                      R
R0320 0061 0000       XD001   DATA    0
R0321             *
R0322             * RAM Location #0362                      R
R0323 0062 0000       YD001   DATA    0
R0324             *
R0325             * RAM Location #0363                      R
R0326 0063 0000       XD002   DATA    0
R0327             *
R0328             * RAM Location #0364                      R
R0329 0064 0000       YD002   DATA    0
R0330             *
R0331             * RAM Location #0365                      R
R0332 0065 0000       XD003   DATA    0
R0333             *
R0334             * RAM Location #0366                      R
R0335 0066 0000       YD003   DATA    0
R0336             *
R0337             * RAM Location #0367                      R
R0338 0067            DSPPTZ  BSS     0
R0339 0067 0000       XD004   DATA    0
R0340             *
R0341             * RAM Location #0368                      R
R0342 0068 0000       YD004   DATA    0
R0343             *
```

23. A Graphics Implementation Using the TMS32020 and TMS34061

```
R0344                    * RAM Location #0369
R0345 0069 0000  XD005    DATA    0
R0346                    *
R0347                    * RAM Location #036A
R0348 006A 0000  YD005    DATA    0
R0349                    *
R0350                    * RAM Location #036B
R0351 006B 0000  XD006    DATA    0
R0352                    *
R0353                    * RAM Location #036C
R0354 006C 0000  YD006    DATA    0
R0355                    *
R0356                    * RAM Location #036D
R0357 006D 0000  XD007    DATA    0
R0358                    *
R0359                    * RAM Location #036E
R0360 006E 0000  YD007    DATA    0
R0361                    *
R0362                    * RAM Location #036F
R0363 006F 0000  SAVAR2   DATA    0
R0364                    *
R0365                    * RAM Location #0370
R0366 0070 0000  SAVAR3   DATA    0
R0367                    *
R0368                    * RAM Location #0371
R0369 0071 0000  NUMR8R   DATA    0
0019                     COPY    DATAMEM.ASM
```

```
S0002            *****************************************
S0003            * External Data-Memory Allocations
S0004            *****************************************
S0005            *
S0006            *
S0007 0400               DORG    >0400                        R
S0008            *
S0009 0400  VSCREG       BSS     0                            R
S0010            *
S0011 0000               DORG    0         * data page 08
S0012            *
S0013 0000  HESYL        BSS     8                            R
S0014            *
S0015 0008  HESYH        BSS     8
S0016            *
S0017 0010  HEBLL        BSS     8                            R
S0018            *
S0019 0018  HEBLH        BSS     8
S0020            *
S0021 0020  HSBLL        BSS     8                            R
S0022            *
S0023 0028  HSBLH        BSS     8
S0024            *
S0025 0030  HTOTL        BSS     8                            R
S0026            *
S0027 0038  HTOTH        BSS     8
S0028            *
S0029 0040  VESYL        BSS     8                            R
S0030            *
S0031 0048  VESYH        BSS     8
S0032            *
S0033 0050  VEBLL        BSS     8
S0034            *
S0035 0058  VEBLH        BSS     8
S0036            *
S0037 0060  VSBLL        BSS     8
S0038            *
S0039 0068  VSBLH        BSS     8
S0040            *
S0041 0070  VTOTL        BSS     8
S0042            *
S0043 0078  VTOTH        BSS     8
S0044            *
S0045 0000               DORG    0         * data page 09
S0046            *
S0047 0000  DADUPL       BSS     8
S0048            *
S0049 0008  DADUPH       BSS     8
S0050            *
S0051 0010  DADSL        BSS     8
S0052            *
S0053 0018  DADSH        BSS     8
S0054            *
S0055 0020  VINTL        BSS     8
S0056            *
S0057 0028  VINTH        BSS     8
S0058            *
```

23. A Graphics Implementation Using the TMS32020 and TMS34061

```
S0059 0030       CTLRIL  BSS    8
S0060            *
S0061 0038       CTLR1H  BSS    8
S0062            *
S0063 0040       CTLR2L  BSS    8
S0064            *
S0065 0048       CTLR2H  BSS    8
S0066            *
S0067 0050       STATL   BSS    8
S0068            *
S0069 0058       STATH   BSS    8
S0070            *
S0071 0060       XYOFFL  BSS    8
S0072            *
S0073 0068       XYOFFH  BSS    8
S0074            *
S0075 0070       XYADRL  BSS    8
S0076            *
S0077 0078       XYADRH  BSS    8
S0078            *
S0079 0000               DORG   0
S0080            * data page 0A
S0081 0000       DADRL   BSS    8
S0082            *
S0083 0008       DADRH   BSS    8
S0084            *
S0085 0010       VCNTL   BSS    8
S0086            *
S0087 0018       VCNTH   BSS    8
S0088            *
```

```
S0090            *
S0091 0800               DORG   >0800
S0092            *
S0093 0800       XYIND   BSS    0
S0094            *
S0095 0000               DORG   0
S0096            * data page 10
S0097 0000       XYNOP   BSS    8
S0098            *
S0099 0008       XINC    BSS    8
S0100            *
S0101 0010       XDEC    BSS    8
S0102            *
S0103 0018       XCLR    BSS    8
S0104            *
S0105 0020       YINC    BSS    8
S0106            *
S0107 0028       X1Y1    BSS    8
S0108            *
S0109 0030       XDY1    BSS    8
S0110            *
S0111 0038       XCY1    BSS    8
S0112            *
S0113 0040       YDEC    BSS    8
S0114            *
S0115 0048       X1YD    BSS    8
S0116            *
S0117 0050       XDYD    BSS    8
S0118            *
S0119 0058       XCYD    BSS    8
S0120            *
S0121 0060       YCLR    BSS    8
S0122            *
S0123 0068       X1YC    BSS    8
S0124            *
S0125 0070       XDYC    BSS    8
S0126            *
S0127 0078       XCYC    BSS    8
0020                     END
NO ERRORS, NO WARNINGS
```

23. A Graphics Implementation Using the TMS32020 and TMS34061

32020 FAMILY MACRO ASSEMBLER   PC 1.0 85.157   15:17:30 12-05-85   PAGE 0091

GRAPHIC LABEL	VALUE	DEFN	REFERENCES
A	0012	R0070	C0217 C0269 C0312 C0372 C0438 G0042 G0043 G0044 G0059
ABS1GN	000F	R0060	C0132 F0147
ALDONE	086A	G0402	G0344
AMAX	003F	R0216	B0104 C0441 C0443
AMIN	0040	R0219	B0106 C0444
AS1GN	000E	R0057	F0126 F0149 F0225
B	0013	R0074	C0219 C0273 C0316 C0376 C0450 G0051 G0052 G0119 G0136 G0139 G0160 G0204 G0226 G0282 G0305 G0348 G0366
BEGIN	0B9B	R0023	C0023
BFLAG	000D	R0054	F0115 F0156 F0231 F0283 F0336 F0422 F0435 F0463
BMAX	0041	R0222	B0108 C0453 C0455
BMIN	0042	R0225	B0110 C0456
C11	001A	Q0081	H0063 J0045
C12	0016	Q0084	
C13	001E	Q0093	
C21	0017	Q0105	
C22	001B	Q0108	
C23	0018	Q0096	
C31	001C	Q0087	
C32	0020	Q0099	
C33	0019	Q0081	
C41	001D	Q0111	
C42	0021	Q0090	
C43	001E	Q0102	
CLEAR	04DC	E0043	B0083 C0113 C0281 C0301 C0324 H0122 H0117
CLOOP1	092E	H0102	C0251 C0317 C0330 C0331
CLRSAV	0033	R0180	C0051
CLRSCR	00F7	C0107	
CMDADR	006C	P0078	C0024 C0028 C0037 C0081 C0084 C0107 C0110 C0118 C0129 C0135 C0148 C0154 C0173 C0178 C0187 C0206 C0211 C0224 C0230 C0246 C0341 C0360 C0395 C0418
CMDLEN	0017	C0072	C0038
CMDTBL	00AB	C0049	C0042 C0072
CMND	0B6B	P0075	C0035 C0039 C0043
CMTRX	0BBB	N0039	H0065
COLOR	0004	R0027	C0126 C0145 C0158 C0221 C0275 C0295 C0318 C0380 C0486 C0033 D0060 D0085 D0112 D0153 D0180 D0221 D0248 D0273 C0300 D0337 D0371 D0403 D0437 D0480 D0511 D0540 D0571 F0090 F0093 G0034 G0037 L0039 L0150 F0094 F0213 F0220 F0393 F0400 F0445 G0038 G0189 G0251 G0328 G0390 G0441 G0458 G0470 G0485 G0496 G0510 G0520 G0552 G0561 G0574 G0584 G0598 G0609 G0624
COLOR2	000C	R0051	F0092 F0174 F0181 F0242 F0298 F0306 F0351 F0359 F0442 F0467 G0196 G0397 G0543 G0565 G0589 G0615
COLORH	000B	R0048	F0091 F0198 F0206 F0267 F0275 F0319 F0379 F0386 F0452 F0473 G0035 G0184 G0385 G0456 G0483 G0508 G0531
COLORL	000A	R0045	H0068 H0089 H0135 H0137 H0141 H0067
COS	0027	Q0132	H0145
COSTBL	0AA0	M0068	E0051 E0101 G0025 G0408
CTLOOP	091B	H0083	E0081 E0083 E0099 G0023 G0406
CTLR1H	0038	S0061	E0049 E0086 E0105 G0029 G0412
CTLR1L	0030	S0059	E0055 E0086 E0105
CTLR2H	0048	S0065	E0053 E0058 E0103 G0027 G0317 G0410
CTLR2L	0040	S0063	C0071
CUBER	08FC	H0055	
CUBER2	0902	H0061	

32020 FAMILY MACRO ASSEMBLER   PC 1.0 85.157   15:17:30 12-05-85   PAGE 0092

GRAPHIC LABEL	VALUE	DEFN	REFERENCES
D	0007	R0036	F0141 F0165 F0171 F0178 F0192 F0195 F0203 F0212 F0219 F0244 F0247 F0251 F0261 F0264 F0272 F0292 F0295 F0303 F0321 F0324 F0328 F0345 F0348 F0356 F0370 F0376 F0383 F0392 F0399
D1L1	028C	D0019	D0043 D0037
D1L1A	02A2	D0040	
D1L2	02A9	D0046	D0068 D0062
D1L2A	02BC	D0064	
D1L3	02C4	D0071	D0095 D0089
D1L3A	020A	D0092	
D1L4	02E1	D0098	D0120 D0114
D1L4A	02F4	D0116	
D2L1	0309	D0139	D0163 D0157
D2L1A	031F	D0160	
D2L2	0326	D0166	D0190 D0184
D2L2A	033C	D0187	
D3L1	034F	D0207	D0231 D0225
D3L1A	0365	D0228	
D3L2	036C	D0234	D0256 D0250
D3L2A	037F	D0252	
D3L3	0387	D0259	D0283 D0277
D3L3A	0390	D0280	
D3L4	03A4	D0286	D0308 D0302
D3L4A	03B7	D0304	
D4L1	03C9	D0323	D0347 D0341
D4L1A	03DF	D0344	
D4L2	03EE	D0357	D0379 D0373
D4L2A	0401	D0375	
D4L3	0410	D0389	D0413 D0407
D4L3A	0426	D0410	
D4L4	0435	D0423	D0445 D0439
D4L4A	0448	D0441	
D5L1	045C	D0462	D0490 D0484
D5L1A	0476	D0487	
D5L2	047D	D0493	D0519 D0513
D5L2A	0494	D0515	
D5L3	049C	D0522	D0550 D0544
D5L3A	04B6	D0547	
D5L4	04BD	D0553	D0579 D0573
D5L4A	04D4	D0575	
DA	0035	R0189	D0353 C0355 C0373 C0377 C0407 C0409 C0439 C0446 C0448 C0451 C0458 C0460
DADRH	0008	S0083	C0088
DADRL	0000	S0081	C0050
DADSH	0018	S0053	C0278 C0298 C0321
DADSL	0010	S0051	C0065
DADUPL	0008	S0049	C0007
DB	0000	S0047	C0125
DECOLY	0036	C0102	C0195
DELAY	00F4	C0081	C0313
DELAY1	000C	C0085	
DELTA	0102	F0095	F0106
DEMO0	052F	D0003	
DEMO1	0280	C0066	
DEMO2	0281	C0067	
DEMO3	02FC	C0068	
DEMO4	0343	C0069	

DEMO5 — 32020 FAMILY MACRO ASSEMBLER PC 1.0 85.157   15:17:30 12-05-85   PAGE 0093

GRAPHIC LABEL	VALUE	DEFN	REFERENCES
DEMO5	0450	D0450	C0070
DH	0026	R0139	G0081 G0087 G0095 G0097 G0100 G0110 G0112 G0114
DL	0027	R0144	G0082 G0096 G0098 G0111 G0113
DOCOS	0952	H0135	H0132
DONE	0667	F0405	F0176 F0184 F0201 F0209 F0217 F0223 F0249 F0253 F0270 F0278 F0301 F0309 F0326 F0330 F0354 F0362 F0381 F0389 F0397 F0403 F0444 F0448 F0455 F0470 F0476
DONEIL	0867	G0399	G0388 G0393
DONEIU	077C	G0198	G0187 G0192
DPO	0C5C	P0016	H0063 H0073 I0041 J0045 K0081
DP4	0200	Q0008	H0056 H0084 H0085 H0086 L0040 L0082 L0093 L0094 L0152
DP6	0300	R0008	L0194
DRR	0000	P0021	H0085 L0040 L0082 L0093 L0094 L0152 L0194
DSPPTS	005F	R0313	
DSPPTZ	0067	R0338	
DUMMY	0020	A0042	A0026 A0028 A0030 A0034 A0036 A0038 A0040 A0042
DUMONE	00DB	C0077	C0054 C0056 C0057 C0062 C0063
DX	0005	R0030	C0411 C0463 C0470 C0472 F0111 F0125 F0130 F0134 F0137 F0140 F0142 F0144 F0413 F0416 F0432
DXPOS	0673	F0419	F0414
DXR	0001		
DY	0006	R0033	C0413 C0475 C0482 C0484 F0107 F0131 F0135 F0136 F0138 F0459
ELPSET	01DD	C0341	C0060
ENDTBL	0B6C	M0259	H0131 H0138
FELIPS	06BB	C0018	C0225 C0276 C0319 C0387 C0490
FELPSC	0158	C0211	C0058
FILL2	0693	P0445	E0046 E0062
FILLC	0063	P0051	B0114 D0173 D0177 D0216 D0243 D0268 D0295 D0332 D0366 D0398 D0432 D0473 D0477 D0504 D0508 D0533 D0537 D0564 D0568
FIVE6	0044	R0231	B0025
GREG	0005	P0036	B0024
HEBLH	0018	S0019	F0139 F0167 F0202 F0211 F0246 F0263 F0294 F0323 F0347 F0372 F0398
HEBLL	0010	S0017	F0143 F0177 F0194 F0218 F0250 F0271 F0302 F0327 F0355 F0382 F0391
HESYH	0008	S0015	A0024 C0049
HESYL	0000	S0013	
HLINE	066A	F0411	F0108
HOZHI	068A	F0438	F0447 F0454
HOZLO	0699	F0449	F0436
HSBLH	0028	S0023	
HSBLL	0020	S0021	
HTOTH	0038	S0027	
HTOTL	0030	S0025	
IMR	0004	P0033	
INCR1	0008	R0039	
INCR2	0009	R0042	
INIT	0022	B0011	
INT0	0002	A0026	
INT1	0004	A0028	
INT2	0006	A0030	
LEO	0898	G0440	G0417
LE1	08A0	G0454	G0418
LE2	08A9	G0469	G0419

32020 FAMILY MACRO ASSEMBLER PC 1.0 85.157   15:17:30 12-05-85   PAGE 0094

GRAPHIC LABEL	VALUE	DEFN	REFERENCES
LE3	0BAF	G0481	G0420
LE4	0BB7	G0495	G0421
LE5	0BBC	G0506	G0422
LE6	0BC3	G0519	G0423
LE7	0BC7	G0529	G0424
LINE	0524	F0084	C0130 C0149 C0188 C0201 C0296 D0040 D0065 D0092 D0117 D0160 D0187 D0228 D0253 D0280 D0305 D0344 D0376 D0410 D0442 D0487 D0516 D0547 D0576 L0052 L0066 L0080 L0092 L0106 L0120 L0134 L0148 L0164 L0178 L0192 L0204
LINEC	010F	C0135	C0053
LOOP1	0734	G0139	G0199
LOOP2	0781	G0204	G0266
LOOP3	07DC	G0282	G0343
LOOPA	0829	G0348	G0400
LOOPE	0824	G0342	G0281
LOOPX	0700	G0103	G0117
LSBASE	0878	G0417	G0244 G0321
MTRX4	0973	J0044	H0110
NARROW	0830	G0365	G0301
NILPT	0728	G0127	G0128
NINE6	0045	G0234	B0116 D0142 D0146
NPOINT	0030	R0171	C0156 C0160 C0165 C0176 C0195
NRMPTS	0047	R0240	H0056 H0084 H0086
NUMR8R	0071	R0369	H0078 K0066
NXTROT	095C	H0142	H0139
O1INC1	059F	F0202	F0193
O1INC2	0588	F0177	F0166
O1OR4	056F	F0149	
O1X1Y1	05B4	F0218	F0210
O2INC1	05CB	F0246	
O2INC2	05D1	F0250	F0245
O2OR3	05BC	F0225	F0148
O2X0Y1	05DA	F0263	
O2X1Y1	05E4	F0271	F0262
O3INC1	0611	F0323	
O3INC2	0617	F0327	F0322
O3X0Y1	05F6	F0294	
O4INC1	0600	F0302	F0293
O4INC2	065F	F0398	F0390
O4X0Y0	064A	F0347	F0371
O4X0Y1	0627	F0355	
OCT1	0631	F0154	F0346
OCT1HI	0572	F0165	F0200 F0208 F0216 F0222
OCT1HL	0579	F0210	F0169
OCT1LO	05A9	F0192	F0157 F0183
OCT2	0592	F0230	
OCT2HI	05BF	F0261	F0248 F0277
OCT2LO	05C4	F0282	F0232 F0252 F0269
OCT3	0507	F0292	F0226
OCT3HI	05EE	F0317	F0284 F0308 F0325
OCT3LO	05F3	F0334	F0300 F0329
OCT4	060A	F0345	F0150
OCT4HI	061D	F0390	F0388
OCT4HL	0624	F0370	F0374
OCT4LO	0654		F0337 F0353 F0361 F0396 F0402

PAGE 0095

GRAPHIC LABEL	VALUE	DEFN	REFERENCES
ONE	0062	P0048	B0027 C0030 C0036 C0083 C0109 C0128 C0147 C0172 C0186 C0205 C0223 C0245 C0247 C0359 C0417
ONE6	002E	R0165	B0098 B0101 B0103 B0105 B0107 B0109 C0122 C0125 C0138 C0140 C0142 C0144 C0157 C0159 C0168 C0182 C0214 C0216 C0218 C0220 C0233 C0235 C0238 C0240 C0242 C0334 C0344 C0346 C0348 C0350 C0352 C0354 C0356 C0383 C0385 C0398 C0400 C0402 C0404 C0406 C0408 C0410 C0412 C0414 C0487 D0022 D0034 D0036 D0038 D0053 D0061 D0074 D0086 D0088 D0090 D0105 D0113 D0130 D0133 D0135 D0154 D0156 D0158 D0181 D0183 D0185 D0222 D0224 D0226 D0249 D0274 D0276 D0278 D0301 D0338 D0340 D0342 D0372 D0404 D0406 D0408 D0438 D0481 D0483 D0485 D0512 D0543 D0545 D0572 G0121 G0223 G0302 H0077
OSCRX1	073A	G0144	G0142
OSCRX2	0787	G0209	G0207
OSCRX3	07E2	G0287	G0285
OSCRX4	082F	G0353	G0351
OSCRY1	0751	G0163	G0161
OSCRY2	079E	G0229	G0227
OTEST	0561	F0138	F0133
PARSER	008F	F0021	B0117
PAXCNT	002D	R0162	G0224 G0249 G0252 G0303 G0326 G0329
PIXCNT	002C	R0159	G0118 G0122 G0124 G0125 G0176 G0177 G0378
PIXELC	0100	C0118	C0052
PLOOP	0093	C0025	C0040 C0045
PLOT	09B3	L0037	H0123
PLOT1	071E	G0118	G0123
PLOT2	072B	G0132	G0200
PLOT3	07C0	G0165	C0162
POLY1	012C	C0154	C0055
POLYC	0120	C0204	C0163
POLYL	0154	C0176	C0196
POLYN	0137	C0190	C0174
POSDX	0145	F0130	F0128
POSY	0558	F0107	F0097
PRD	053D	F0097	B0022
PRJCTN	0003	B0022	H0090 H0093 H0096 H0091 H0071
R11	0993	H0111	I0041 H0095
R12	0012	Q0069	G0426
R21	0014	Q0075	G0427
R22	0013	Q0072	G0428
RATE	0015	Q0078	G0429
RE0	0004	H0050	G0430
RE1	08CC	G0540	G0431
RE2	0802	G0552	G0432
RE3	0805	G0561	G0433
RE4	08DC	G0574	K0061
RE5	08E0	G0584	
RE6	08E8	G0598	
RE7	08ED	G0609	
RESET	08F6	G0624	G0250
RIGHT1	0000	A0024	G0327
RIGHT2	07BC	G0331	
RINT	0816	A0036	
ROMDAT	001A	N0007	B0099
ROTZ	086D	I0040	H0109
	0962		

PAGE 0096

GRAPHIC LABEL	VALUE	DEFN	REFERENCES
RPTCNT	0400	H0051	H0079
RSBASE	0888	G0426	G0262 G0339
SAVAR0	002F	R0168	C0263 C0282 C0306 C0325 C0366 C0388 C0436 C0491 D0019 D0041 D0046 D0066 D0071 D0093 D0098 D0118 D0139 D0161 D0166 D0188 D0207 D0224 D0234 D0254 D0259 D0281 D0286 D0306 D0323 D0345 D0357 D0377 D0389 D0411 D0423 D0443 D0462 D0488 D0493 D0517 D0522 D0548 D0553 D0577 H0083 H0143 L0051 L0054 L0065 L0068 L0079 L0091 L0104 L0107 L0118 L0121 L0132 L0135 L0146 L0163 L0166 L0177 L0180 L0191 L0203
SAVAR2	006F	R0363	L0105 L0108 L0119 L0122 L0133 L0136 L0147
SAVAR3	0070	R0366	C0241 C0250 C0253 C0254 C0274 C0294 C0329 C0332 C0351 C0379 C0382 C0386 C0405 C0485 C0489 D0016 D0032 D0035 D0039 D0052 D0064 D0084 D0087 D0091 D0111 D0116 D0136 D0152 D0155 D0159 D0179 D0182 D0186 D0204 D0220 D0223 D0227 D0247 D0252 D0272 D0275 D0279 D0299 D0304 D0321 D0336 D0339 D0343 D0370 D0375 D0402 D0405 D0409 D0436 D0441 D0459 D0479 D0482 D0486 D0510 D0515 D0539 D0542 D0546 D0570 D0575
SAVCLR	0031	R0174	C0237 C0255 C0258 C0260 C0303 C0357 C0363 C0415 C0433
SAVEA	0032	R0177	G0120 G0274 G0279
SAVEB	0046	R0237	G0021 G0403
SAVREG	0064	P0054	H0075
SCALE	0897	N0052	C0390
SETE0	01F2	C0363	C0364
SETE1	01F5	C0366	C0384
SETE1E	020E	C0389	
SETE2	020B	C0387	
SIN	0026	Q0129	H0070 H0092 H0128 H0130 H0134
SINTBL	046D	M0004	H0069 H0133 H0140
SLHIL	0861	G0394	G0381 G0387 G0392
SLHIU	0776	G0193	G0180 G0186 G0191
SLHLL	085A	G0389	G0394
SLHLU	076F	G0188	G0193
SLLOL	0852	G0383	
SLLOU	0767	G0182	
SPEED	0034	R0183	
SPIN	0169	C0230	C0243 C0257 C0270 C0313
SPIN0	0186	C0260	C0059
SPIN1	0189	C0263	C0336
SPIN1E	01A0	C0283	C0284
SPIN2	01B9	C0306	C0261
SPIN2E	0100	C0326	C0327
STATH	0058	S0069	C0304
STATL	0050	S0067	C0098 C0099 E0077 E0078
STEPX	071B	S0116	G0088 G0102
STEPY	06FD	G0089	G0101 G0115
STOP	0009	C0076	C0064 C0076
STS0	0060	P0042	E0044 E0107 F0085 F0406 G0019 G0414
STS1	0061	P0045	C0033
SYNC	0098	S0029	C0029 C0031
SYNCH	0069	P0069	G0046 G0073
T1H	0014	R0078	G0047 G0074
T1L	0015	R0081	G0049 G0091
T2H	0016	R0084	G0050 G0092
T2L	0017	R0087	G0054 G0060 G0075
T3H	0018	R0090	

GRAPHIC LABEL	VALUE	DEFN	REFERENCES
T3L	0019	R0093	G0055 G0064 G0076
T4H	001A	R0096	G0057 G0105
T4L	001B	R0099	G0058 G0106
T5H	001C	R0102	G0066 G0079
T5L	001D	R0105	G0067 G0080
T6H	001E	R0108	G0068 G0103 G0107
T6L	001F	R0111	G0069 G0104 G0108
T7H	0020	R0114	G0071 G0089 G0093
T7L	0021	R0117	G0072 G0090 G0094
THETA	0028	Q0135	H0072 H0129 H0136
THREE6	0043	R0228	B0112 D0026 D0057 D0078 D0109 D0212 D0239 D0264 D0291 D0328 D0362 D0394 D0428 D0445 D0469 D0496 D0500 D0525 D0529 D0556 D0560
TIM	0002	P0027	C0082 C0085 C0248
TIMDLY	006A	P0072	
TINT	0018	A0034	
TMP1	0068	P0066	C0086 C0108 C0112
TRAP	001E	A0040	
VCNTL	0018	S0087	
VCX	0010	S0085	
VCY	0023	Q0120	
VEBLH	0025	Q0126	
VEBLL	0058	S0035	
VESYH	0050	S0033	
VESYL	0048	S0031	
VESYL	0040	S0029	
VINTH	0028	S0057	C0097 E0076
VINTL	0020	S0055	C0095 E0074
VLINE	0025	S0059	F0127
VRTHI	06A2	F0459	F0469
VRTICS	06AB	F0465	H0059
VRTLO	0B73	N0014	F0464 F0475
VSBLH	06B3	F0471	
VSBLL	0068	S0039	C0092 E0070
VSCIN	0060	S0037	C0091 E0068
VSCREG	0400	S0009	B0032 F0088 G0132 G0165 G0231 G0240 G0257 G0259 G0270 G0308 G0317 G0334 G0336 G0369 H0073 K0081
VSX	0022	Q0117	
VSY	0024	Q0123	
VTOTH	0078	S0043	
VTOTL	0070	S0041	
WAIT1	00E5	C0089	C0102
WAITD	00EF	C0099	C0101
WAITI	04FF	E0078	E0080
WELP0	0237	E0080	C0493
WELP1	023A	C0433	C0434
WELP2	027C	C0436	C0442
WELP3	0246	C0442	C0445
WELP4	0249	C0446	C0454
WELP5	0254	C0449	C0457
WELP6	0257	C0458	C0461
WELP7	0262	C0461	C0466
WELP8	0265	C0470	C0469
WELP9	0270	C0473	C0478
WELPSC	0273	C0482	C0481
WIDE	0212	C0485	C0061
	0797	C0395 G0223	G0158

GRAPHIC LABEL	VALUE	DEFN	REFERENCES
X	0022	R0120	H0103 I0045 I0051
X1	0000	R0014	C0120 C0137 C0192 C0287 D0021 D0048 D0073 D0100 D0141 D0168 D0209 D0236 D0261 D0288 D0325 D0359 D0391 D0425 D0464 D0495 D0524 D0555 F0102 F0105 F0110 F0114 F0118 F0119 F0121 F0412 F0418 F0421 F0425 F0426 F0428 G0144 G0152 G0209 G0217 G0287 G0295 G0353 G0361 L0042 L0056 L0070 L0084 L0097 L0111 L0125 L0139 L0154 L0168 L0182 L0196
X1H	0028	R0147	G0153 G0157 G0170 G0218 G0222 G0236 G0296 G0300 G0313 G0362 G0374
X1L	0029	R0150	G0155 G0167 G0175 G0179 G0220 G0233 G0245 G0298 G0310 G0322 G0364 G0371 G0380
X2	0002	R0021	C0121 C0141 C0180 C0198 C0288 D0025 D0050 D0077 D0102 D0145 D0170 D0211 D0238 D0263 D0290 D0327 D0361 D0393 D0427 D0468 D0499 D0528 D0559 F0103 F0104 F0109 F0411 F0417 G0147 G0148 G0212 G0213 G0290 G0291 G0356 G0357 L0046 L0060 L0074 L0088 L0101 L0115 L0129 L0143 L0158 L0172 L0186 L0200
X2H	002A	R0153	G0149 G0156 G0214 G0221 G0292 G0299 G0358
X2L	002B	R0156	G0151 G0174 G0216 G0263 G0294 G0340 G0360
XA	0022	R0121	C0166 C0197 C0232 C0264 C0286 C0307 C0343 C0367 C0397 C0461 C0464 C0009 D0020 D0023 D0047 D0072 D0075 D0099 C0127 D0140 D0143 D0167 D0197 D0208 D0235 D0260 D0287 D0315 D0324 D0350 D0358 D0386 D0390 D0416 D0424 D0452 D0463 D0466 D0494 D0497 D0523 D0526 D0554 D0557
XB	0024	R0129	C0167 C0181 C0191 C0236 C0268 C0271 C0311 C0314 C0347 C0371 C0374 C0401 C0421 C0424 C0437 C0440 D0013 D0024 C0027 D0049 D0076 D0079 D0101 D0131 D0144 D0147 D0169 D0198 D0210 D0213 D0237 D0240 D0262 D0265 D0289 D0395 D0418 D0426 D0429 D0455 D0467 D0470 D0498 D0501 D0527 D0530 D0558 D0561 L0047 L0055 L0061 L0069 L0075 L0083 L0159 L0167 L0173 L0181 L0187 L0195
XC	0010	R0063	C0213 C0265 C0308 C0368 C0462 C0140 G0145 G0205 G0210 G0283 G0288 G0349 G0354 J0056 K0075 K0077 K0085
XCLR	0018	S0103	
XCYC	0078	S0127	
XCYD	0058	S0119	
XCY1	0038	S0111	
XD000	005F	R0314	
XD001	0061	R0320	
XD002	0063	R0326	
XD003	0065	R0332	
XD004	0067	R0339	
XD005	0069	R0345	
XD006	006B	R0351	
XD007	006D	R0357	
XDEC	0010	S0101	
XDYC	0070	S0125	
XDYD	0050	S0117	F0335
XDY1	0030	S0109	
XIY1	0008	S0099	
XINC	001C	A0038	F0282 F0334
XINT	0068	S0123	E0066 F0155 F0434 G0178 G0241 G0318 G0379
XIYC	0068	S0115	
XIYD	0048	S0107	
XIY1	0028	S0107	F0154 F0230
XM	0025	R0133	

GRAPHIC LABEL	VALUE	DEFN	REFERENCES
XMAX	0037	R0192	B0100 C0420 D0129 D0349 D0385
XMAXA	003B	R0204	C0422 C0465 C0467
XMIN	0038	R0195	C0423
XMINA	003C	R0207	C0425 C0468
XN000	0047	R0241	
XN001	004A	R0250	
XN002	004D	R0259	
XN003	0050	R0268	
XN004	0053	R0277	
XN005	0056	R0286	
XN006	0059	R0295	
XN007	005C	R0304	
XP	0000	R0013	H0113 K0087
XR	0025	R0132	I0049 J0049 J0058 J0067
XYADRH	0078	S0077	E0061
XYADRL	0070	S0075	E0060 C0259 G0336
XY1ND	0800	S0093	F0145 F0146 F0154 F0155 F0230 F0282 F0334 F0335 F0433 F0434 F0460 F0461 G0137 G0178 G0241 G0275 G0318 G0379
XYNOP	0000	S0097	F0145 F0433 F0460 G0137 G0257 G0275 G0308 G0334 G0369
XYOFFH	0068	S0073	G0165 G0231 G0257 G0270
XYOFFL	0060	S0071	F0088 F0132 G0132 G0270
Y	0023	R0124	H0105 I0047 I0053
Y1	0001	R0018	C0123 C0139 C0194 C0291 D0029 D0052 D0081 D0104 D0149 D0172 D0215 D0242 D0267 D0294 D0331 D0365 D0397 D0431 D0472 D0503 D0532 D0563 F0096 F0098 F0101 F0120 F0123 F0427 F0430 G0163 G0169 G0172 G0229 G0235 G0238 L0044 L0058 L0072 L0086 L0099 L0113 L0127 L0141 L0156 L0170 L0184 L0198
Y2	0003	R0024	C0124 C0143 C0183 C0200 C0291 D0029 D0031 D0056 D0083 D0108 D0151 D0176 D0219 D0246 D0271 D0294 D0298 D0335 D0369 D0401 D0435 D0476 D0507 D0536 D0567 F0095 F0099 F0100 G0306 G0312 G0315 G0367 G0373 G0376 L0049 L0063 L0077 L0090 L0103 L0117 L0131 L0145 L0161 L0175 L0189 L0202
YA	0023	R0125	C0169 C0199 C0234 C0266 C0289 C0309 C0345 C0369 C0399 C0473 C0476 D0011 D0028 D0051 D0054 D0080 D0103 D0106 D0128 D0148 D0171 D0174 D0200 D0214 D0217 D0241 D0244 D0266 D0269 D0293 D0296 D0319 D0330 D0333 D0354 D0364 D0367 D0384 D0396 D0399 D0420 D0430 D0433 D0453 D0471 D0474 D0502 D0505 D0531 D0534 D0562 D0565
YB	0025	R0134	C0170 C0184 C0193 C0239 C0272 C0290 C0292 C0315 C0333 C0335 C0349 C0375 C0378 C0403 C0427 C0430 C0449 C0452 D0014 D0030 D0055 D0058 D0082 D0107 D0110 D0134 D0150 D0175 D0178 D0202 D0218 D0245 D0270 D0297 D0317 D0334 D0355 D0368 D0382 D0400 D0434 D0457 D0475 D0478 D0506 D0509 D0535 D0538 D0566 D0569 L0050 L0057 L0064 L0071 L0078 L0085 L0162 L0169 L0176 L0183 L0190 L0197 G0304 G0365 J0065 K0078 K0080 K0089
YC	0011	R0066	C0215 C0267 C0310 C0370 C0474 G0159 G0225 G0278 G0280
YCLR	0060	S0121	
YD000	0060	R0317	
YD001	0062	R0323	
YD002	0064	R0329	
YD003	0066	R0335	
YD004	0068	R0342	
YD005	006A	R0348	
YD006	006C	R0354	
YD007	006E	R0360	

GRAPHIC LABEL	VALUE	DEFN	REFERENCES
YDEC	0040	S0113	
YINC	0020	S0105	
YM	0026	R0138	
YMAX	0039	R0198	F0146 F0461
YMAXB	003D	R0207	
YMIN	003A	R0201	B0102 C0426 D0132 D0381 D0419
YMINB	003E	R0213	C0428 C0477 C0479
YN000	0048	R0244	C0429
YN001	004B	R0253	C0431 C0480
YN002	004E	R0262	
YN003	0051	R0271	
YN004	0054	R0280	
YN005	0057	R0289	
YN006	005A	R0298	
YN007	005D	R0307	
YP	0000	R0017	H0115 K0091
YR	0025	R0137	I0055 J0051 J0060 J0069
Z	0001	R0128	H0107
ZC	0024	R0069	J0074 K0062 K0065 K0068
ZC1	0012	R0073	K0069 K0070 K0073 K0074
ZCSIGN	0013	R0077	K0063 K0071 K0072
ZM	0014	R0143	
ZN000	0049	R0247	
ZN001	004C	R0256	
ZN002	004F	R0265	
ZN003	0052	R0274	
ZN004	0055	R0283	
ZN005	0058	R0292	
ZN006	005B	R0301	
ZN007	005E	R0310	
ZR	0027	R0142	H0108 J0053 J0062 J0071

24. Control System Compensation and Implementation with the TMS32010

Charles Slivinsky
Department of Electrical Engineering
University of Missouri - Columbia

Jack Borninski
Digital Signal Processing - Semiconductor Group
Texas Instruments

INTRODUCTION

Algorithms, software, and hardware for designing and implementing a digital control system using the Texas Instruments TMS32010 signal-processing microprocessor are presented in this application report. Microprocessors, such as the TMS32010, are increasingly being used to implement algorithms for the control of feedback systems. The major factors contributing to this trend are increased availability and lower cost of suitable digital hardware.

Current and potential applications include servo motor control, process control, robot arm and disk head controllers, and temperature and pressure controllers. Military and aerospace applications include stabilized platforms, flight control and autopilot systems, inertial reference systems, and general servomechanisms.

In meeting control system requirements, designers face many alternatives. Cost, size, weight, power, and reliability decisions are typically application dependent. This report highlights the tradeoffs in developing algorithms, software, and hardware for a digital control system.

DIGITAL CONTROL SYSTEMS

General Considerations

A digital controller is a signal processing system that executes algebraic algorithms inherent to the control of feedback systems (i.e., compensator and filter algorithms). Together with the plant (system to be controlled) and signal-acquisition circuitry, the digital controller makes up a digital control system such as the one shown in Figure 1.

Note that the system requires analog-to-digital (A/D) converters for the external (command) inputs and for the state-variable feedback inputs to the digital controller. The system also requires a digital-to-analog (D/A) converter for the control outputs to the plant.

The advantages of the digital control approach over the analog approach are:

1. Ability to implement advanced control algorithms with software rather than special-purpose hardware
2. Ability to change the design without changing the hardware
3. Reduced size, weight, and power, along with low cost
4. Greater reliability, maintainability, and testability
5. Increased noise immunity.

Microprocessor Selection and System Development Cycle

Choosing an appropriate microprocessor is an important factor in efficiently implementing a digital control design. A class of special-purpose (as opposed to general-purpose) digital signal-processing microprocessors has been developed to enable fast execution of digital control algorithms. The Texas Instruments TMS32010 provides several beneficial features for implementing digital control system elements through its architecture, speed, and instruction set.

A prominent feature of the TMS32010 is the on-chip, 16×16-bit multiplier that performs two's-complement multiplication and produces a 32-bit product in a single 200-ns instruction cycle. The TMS32010 instruction set includes special instructions necessary for fast implementation of sum-of-products computations encountered in digital filtering/compensation and Fourier transform calculations. Most of the instructions critical to signal processing execute in one instruction cycle. References [1,2] give full details of the TMS32010 hardware and software considerations.

Many system development tools are available and may be used for digital control system design.[3] Figure 2 outlines

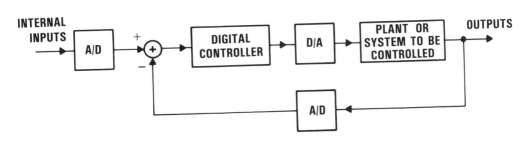

Figure 1. Digital Control System

24. Control System Compensation and Implementation with the TMS32010

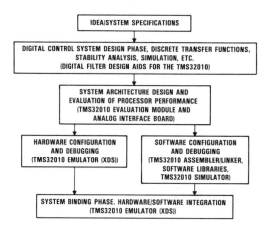

Figure 2. Digital Control System Development Phases

the system development cycle and ties the development tools to different project phases.

Among the non-TI development aids, an interactive program called the Digital Filter Design Package developed by Atlanta Signal Processors Incorporated, may be useful in digital control system design. This program designs various types of filters, compensators, and other structures. It can be used when, for example, there is a need for several notch filters to filter out unwanted frequencies in a digital feedback loop.

DIGITAL COMPENSATOR DESIGN

Alternative methods exist for designing digital compensators. This section outlines several approaches to digital controller design and points out the analytical tools useful in the design process.

Design Based on Analog Prototype

A commonly used method of designing a digital control system is to first design an equivalent analog control system using one of the well-known design procedures. The resulting analog controller (analog prototype) is then transformed to a digital controller by the use of one the transformations described below.

The design of the analog controller may be carried out in the s-plane using design methods such as root-locus techniques, Bode plots, the Routh-Hurwitz criterion, state-variable techniques, and other graphic or algebraic methods. The purpose is to devise a suitable analog compensator transfer function which is transformed to a digital transfer function. This digital transfer function is then inverse z-transformed to produce a difference equation that can be implemented as an algorithm to be executed on a digital computer. Two of the analog-to-digital transformation

methods, the matched pole-zero and the bilinear transformation, are described as follows:

1. The matched pole-zero (matched Z-transform) method maps all poles and zeroes of the compensator transfer function from the s-plane to the z-plane according to the relation:

$$z = e^{sT}$$

where T is the sampling period.

If more poles than zeroes exist, additional zeroes are added at $z = -1$, and the gain of the digital filter is adjusted to match the gain of the analog filter at some critical frequency (e.g., at DC for a lowpass filter). This method is somewhat heuristic and may or may not produce a suitable compensator.

2. The bilinear (Tustin) transformation method approximates the s-domain transfer function with a z-domain transfer function by use of the substitution:

$$s = \frac{2}{T} \frac{z-1}{z+1}$$

As in the matched pole-zero method, the bilinear transformation method requires substitution for s. Compensators in parallel or in cascade maintain their respective structures when transformed to their digital counterparts. This substitution maps low analog frequencies into approximately the same digital frequencies, but produces a highly nonlinear mapping for the high frequencies.

To correct this distortion, a frequency prewarping scheme is used before the bilinear transformation. The frequency prewarping operation results in matching the single critical frequency between the analog domain and the digital domain. To achieve this result, the prewarping operation replaces each s in the analog transfer function with $(\omega_0/\omega_p)s$ where ω_0 is the frequency to be matched in the digital transfer function and

$$\omega_p = \frac{2}{T} \tan \frac{\omega_0 T}{2}$$

Bilinear transformation with frequency prewarping provides a close approximation to the analog compensator[5] and is the most commonly used technique. Other methods for converting a transfer function from the analog to the digital domain are: the method of mapping differentials[18], the impulse-invariance method[6], the step-invariance method[18], and the zero-order hold technique.[6]

The basic disadvantage of design based on an analog prototype is that the discrete compensator is only an approximation to the analog prototype. This analog prototype is an upper bound on the effectiveness of the closed-loop response of the digital compensator.

Direct Digital Design

The direct digital design technique is a design method where a digital control system design is carried out from the very beginning in the z-domain to produce a digital control algorithm. Various pole-placement and relocation techniques are used to position the poles and zeroes of the system's z-domain transfer function to yield the required system performance.

The root-locus method in the z-plane is similar to the root-locus design in the s-plane in that both are based on observing the position of the closed-loop poles as a function of the system gain. However, the effect of the locations of the poles and zeroes on system performance is not the same as in the s-plane. Knowledge of such correspondences in the z-plane allows those familiar with continuous system design to design digital compensators. For example, in stabilizing an unstable system, the adjustable gain is used to move the poles inside the unit circle instead of inside the left-half plane. A phase-lead controller may be employed to shift the root-locus to the left, which results in a system that responds faster. A phase-lag controller may be designed to allow for higher loop gain to produce smaller steady-state errors and improved disturbance rejection.[10]

The z-domain designer may also use pole-zero cancellation. In this technique, some of the poles and zeroes of the digital transfer function of the plant may be cancelled by zeroes and poles of the digital compensator. The compensator then introduces additional poles and zeroes at locations that enable the designer to achieve the desired performance characteristics. This method may affect system stability due to "inexact cancellation". If the poles of the plant that are to be cancelled lie close to the unit-circle, inexact cancellation may cause the system root-locus to go outside the unit circle at some point. This makes the system conditionally stable or unstable.

In a system where fast response to the control input is required, a "deadbeat" approach may be taken. The deadbeat controller cancels all the zeroes and poles of the plant and introduces a pole at z = 1. The result is that the system output reaches its steady-state value in one sampling period with no overshoot. In practice, an ideal deadbeat controller is difficult to implement because of inexact cancellation. Although the output does not experience overshoot, it may oscillate between sampling instants. The design is "tuned" in the sense that the system response may be acceptable for a step input but not acceptable for other inputs.

With z-plane design, conventional design techniques can be used to place the closed-loop system poles exactly where desired. The approximations associated with digitizing

an analog prototype are thereby eliminated. The disadvantage of this technique is the relative difficulty of visualizing the effect of pole-zero locations in the z-plane on system performance. To overcome the disadvantage of the z-plane technique, the designer can use the w-plane or, better yet, the w'-plane design technique.[5] Both techniques transform the design to a plane similar to the s-plane by means of the same kind of substitution as in the bilinear transformation described earlier. This procedure thereby allows the use of the familiar s-plane and frequency-domain methods of continuous-system design. The designer proceeds by first transforming the continuous plant to the z-domain and thence to the w-plane or w'-plane. The appropriate compensator is then devised and transformed back to the z-plane, where it is used to specify the corresponding computational algorithm.

State-Variable Design Methods

State-variable design methods can also be used, including state-variable feedback and optimal control based on quadratic synthesis.

In the state-variable feedback technique, all of the states are measured and fed back through constant gains. This allows all of the closed-loop poles to be positioned at any desired locations in the z-plane, but does not affect the positions of the system zeroes.

In a design based on quadratic synthesis, a performance index or cost function is minimized by proper choices of the control law or feedback compensator. In the most practical design, the resulting compensator has the same form as that resulting from the application of direct digital design techniques.[7]

DIGITAL COMPENSATOR IMPLEMENTATION

Digital compensator algorithms execute on processors that use finite-precision arithmetic. The signal-quantization errors associated with finite-precision computations and the methods for the handling of these errors are presented in this section.

Fixed-Point Arithmetic and Scaling

Computation with the TMS32010 is based on the fixed-point two's-complement representation of numbers. Each 16-bit number has a sign bit, i integer bits, and 15-i fractional bits. For example, the decimal fraction $+0.5$ may be represented in binary as

$$0.100\ 0000\ 0000\ 0000$$

This is Q15 format since it has 15 fractional bits, one sign bit, and no integer bits. The decimal fraction $+0.5$ may equivalently be represented in Q12 format as:

$$0000.1000\ 0000\ 0000$$

This number is in the Q12 format because it has 12 fractional bits, one sign bit, and three integer bits. Note that the Q15 notation allows higher precision while the Q12 notation allows direct representation of larger numbers.

For implementing signal-processing algorithms, the Q15 representation is advantageous because the basic operation is multiply-accumulate and the product of two fractions remains a fraction with no possible overflows during multiplication. When the Q12 format is used, a software check for overflow is necessary. The subsection, OVERFLOW AND UNDERFLOW HANDLING, provides a detailed analysis of overflow handling.

In the case where two numbers in Q15 are multiplied, the resulting product has 30 fractional bits, two sign bits, and (as expected) no integer bits. To store this product as a 16-bit result in Q15, the product must be shifted left by one bit and the most-significant 16 bits stored. The TMS32010 instruction SACH allows for this one-bit shift.

In the case where a Q15 number is to be multiplied by a 13-bit fractional signed constant represented as a Q12 number, the result (to correspond with Q15) must be left-shifted four bits to maintain full precision. The TMS32010 instruction SACH allows for the appropriate shift. The Q15 and Q12 representations are used in the example in the section, DESIGN EXAMPLE: RATE-INTEGRATING GYRO STABILIZATION LOOP.

When fixed-point representations are used, the control system designer must determine the largest magnitudes that can occur for all variables involved in the computations required by the digital compensator. (Floating-point representations allow larger magnitudes, but take more time for the microprocessor to perform the required computations.) Once these largest magnitudes are known, scaling constants can be used to attenuate the compensator input as much as necessary to ensure that all variables stay within the range that can be expressed in the given representation.

Several methods are used to determine bounds on the magnitudes of the variables. One method, called upper-bound scaling, provides a useful, although sometimes too conservative, bound on the magnitude, yet it is straightforward to calculate. Consider a variable y(n) that is obtained as the output of a digital compensator H(z) when the input is the sequence x(n). The bound on y(n) is given by

$$y_{max} = |x_{max}| \sum_{n=1}^{\infty} |h(n)|$$

where x_{max} is the maximum value in x(n) and the sequence h(n) is the unit-sample response sequence for the digital compensator H(z).

Other methods for estimating the upper bound are L_p-norm scaling, unit-step scaling, and the averaging method.[9,10]

After y_{max} is determined, the scale factor can be chosen as the multiplier that is applied to x(n) prior to the compensator computations to ensure that y(n) remains within the required bounds. In addition, the control system designer may have knowledge concerning the bounds on the compensator variables based on prior experience, the characteristics of the corresponding variables of analog prototypes, and simulation results.

Finite-Wordlength Effects

All variables involved in the digital compensator — the input, the compensator coefficients, the intermediate variables, and the output — are represented as finite-wordlength numbers. This restriction gives rise to errors. Another source of errors is the truncation or rounding that takes place when the 32-bit product of two 16-bit numbers is stored as a 16-bit number. Both of these errors give rise to the finite-wordlength effects discussed in this section.

The representation of the compensator input as a finite-precision (quantized) number produces an input-quantization error. The size of this error for a rounding scheme can be anywhere from $-(2^{-B})/2$ to $(2^{-B})/2$ where B is the number of bits in a word. The input-quantization error is usefully modeled as a zero-mean random variable uniformly distributed between its positive and negative bounds. A technique[5,10] is available to calculate the variance of the corresponding error at the compensator output (its mean is zero). In this manner, the designer can determine the effect of input quantization on the compensator output.

Similar quantization errors are associated with the multiplication process. Each multiplication is assumed to produce the "true" product with an error that is a zero-mean, uniformly distributed random variable. The variance of the corresponding error at the compensator output can be calculated in the same manner as for the error due to input quantization. These individual variances are then added to measure the total effect at the compensator output for each truncation or rounding.

Another way to describe the effects of truncation or rounding is in terms of "limit cycles" which are sustained oscillations in the closed-loop system. These oscillations are caused by nonlinearities within the loop. In this case, the nonlinear quantizations are associated with the multiplications. Limit cycles persist even when the system input goes to zero, and their amplitude can be sizeable. No general theory is available to treat this nonlinear phenomenon. Bit-level simulations which model the compensator and the complete closed-loop system are used to ascertain their presence and effect on the closed-loop performance.

When a digital compensator is implemented as an algorithm to be executed on finite-precision hardware, a problem arises with implementing the coefficients present in the corresponding transfer function (see section, TMS32010 IMPLEMENTATION OF COMPENSATORS AND FILTERS). The infinite-precision compensator coefficients must be rounded and stored using a finite-length, fixed-point binary representation. Due to this coefficient-quantization effect, the performance of the implemented filter will deviate from the performance of the designed digital filter.

The deviation in performance can be estimated by computing the filter's pole and zero locations and the corresponding frequency response magnitude and phase for the compensator with the quantized coefficients. Coefficient quantization forces the filter's poles and zeroes into a finite number of possible locations in the z-plane and is of most concern for filters with stringent specifications, such as narrow transition regions.

The designer must choose the filter structure least sensitive to inaccurate coefficient representation. The choice should be of a modular rather than a direct filter structure. For example, a higher-order filter should be implemented as a cascade or parallel combination of first-order and second-order blocks. The reason for this choice is the lesser sensitivity to coefficient variations of the roots of low-degree polynomials in comparison with high-degree polynomials. Several methods for selecting the filter structures least affected by coefficient quantization are available.[9]

To quantitatively evaluate the effect of coefficient quantization on the position of the poles or zeroes of a digital transfer function, a "root sensitivity function" can be computed.[6]

Overflow and Underflow Handling

Digital control system algorithms are usually implemented using two's-complement, fixed-point arithmetic. This convention designates a certain number of integer and fractional bits. The fixed-point arithmetic computations may, at some point, produce a result that is too large to be represented in a chosen form of fixed-point notation (e.g., Q12). The resulting overflow, if untreated, may cause degraded performance such as limit cycles and large noise spikes at the filter's output which may contribute to the system's instability. The system must be able to recover from the overflow condition, i.e., return to its normal, nonoverflow state.

Consider an example of the Q12 representation. The number 7.5 multiplied by itself gives the result of 56.25, an overflow in Q12. However, no hardware overflow occurs in the accumulator; i.e.,

$$
\begin{array}{ll}
7.5 & 0111.1000\ 0000\ 0000 \\
\times\ \ 7.5 & 0111.1000\ 0000\ 0000 \\
\hline
56.25 & 0011\ 1000.0100\ 0000\ 0000\ 0000\ 0000\ 0000
\end{array}
$$

For the Q12 representation, the above 32-bit product is shifted left four bits and the left-most 16 bits are retained:

$$1000.0100\ 0000\ 0000$$

The correct answer is 56.25, but the number stored in the Q12 representation is -7.75.

The TMS32010 has a built-in overflow mode of operation that, if enabled, causes the accumulator to saturate upon detection of an overflow during addition when the accumulator register overflows. During multiplication, an overflow of the fixed-point notation may also occur even though the hardware overflow of the accumulator register does not occur. This is because the 32-bit result of a multiplication of two 16-bit numbers must be stored in a 16-bit memory word in the form consistent with the chosen fixed-point notation (see above example). To adjust the location of the binary point, the storing operation requires that the number in the accumulator be shifted left and truncated on the right before storing. If the most significant bits shifted out contain magnitude information in addition to sign information, an overflow in the chosen fixed-point notation results.

To track overflows associated with the number representation, the control system software should contain an appropriate overflow-checking routine in those places where multiplications and additions occur. This routine should not rely exclusively on the TMS32010's overflow mode to intercept and correct the overflow occurrences.

Two approaches may be used to handle overflows. The first is to prevent the overflow from occurring by choosing conservative scaling factors for the numbers used in computations, as described in the subsection, FIXED-POINT ARITHMETIC AND SCALING. These scaling factors are used to limit the range of inputs to each of the basic building blocks of the compensator, namely, the first- and second-order filter sections. The scaling factor chosen reduces the input magnitude and consequently all other signal levels, thereby enabling the compensator coefficients, the expected inputs, and their products and sums, all to be represented without overflow. The scaling must also maintain the signal levels well above the quantization noise.

The second approach for handling overflow is to adjust the sum or product each time an overflow occurs. To accomplish this, an overflow checking routine must be written and executed at certain points along the computational path. The routine must check whether the number just computed and residing in the 32-bit accumulator can be stored without overflow in a 16-bit memory location in accord with the chosen fixed-point notation. Once the routine detects an overflow condition, it should replace the computed number with the maximum or minimum representable two's-complement number. This scheme simulates a saturation condition present in analog control systems. To prevent overflow limit cycles, the saturation overflow characteristic is preferred to the two's-complement, "wrap-around" characteristic.[9]

An example of the overflow checking and correcting technique for a first- and second-order filter subroutine is provided in the section, TMS32010 IMPLEMENTATION OF COMPENSATORS AND FILTERS. This Direct-Form II implementation subroutine checks for overflow occurrences upon computation of the filter's intermediate state variable and again upon computation of the filter's output.

In a digital control system, the first- and second-order building blocks are either cascaded or connected in parallel to compute a series of control algorithms. The first- and second-order filter subroutine, called to compute each of the control system's elements, uses 16-bit memory locations as storage media for its intermediate values, in which case it is appropriate to check for overflow in each block.

At the end of a computational chain — before the final, computed digital output is ready for transfer to the analog domain — it is necessary to check that the number being sent to the digital-to-analog converter is within the range based on the manner in which the converter is interfaced to the processor data bus. For example, if a 12-bit converter is wired to the 12 least significant bits (LSBs) of the 16-bit processor data bus, then the 12 LSBs must contain both magnitude and sign information, which may require that the original 16-bit number be adjusted or limited before being sent to the converter.

Underflow conditions, which can also appear during digital control algorithms, are conceptually similar to overflows in that the computed value contained in the 32-bit accumulator is too small to be accurately represented in a 16-bit memory word in the chosen fixed-point notation. One possible solution to this problem is to multiply the small result by a gain constant to raise its value to a representable level. The appropriately chosen gain constant may come as a result of gain distribution throughout the digital control system, whereby large gains from some of the building blocks get uniformly distributed over a range of the system's sections.

DIGITAL CONTROL SYSTEM SOFTWARE DESIGN PHILOSOPHY

To maximize the manageability and portability of the system software, a modular or top-down design technique should be used. This section shows how the modular software structure and the proper layout of system memory contribute to the efficient implementation of a digital control design.

Modular Software Structure

The concept of modular software design is a technique developed to make system software more manageable and portable. Top-down design is used to break up a large task into a series of smaller tasks or building blocks, which in turn are used for structuring a total system in a level-by-level form. At the end of a top-down design process, a number of modules are linked together which, under the control of a main program, perform as a complete system.

In addition to making the software-development and software-modification processes more manageable, modular design also enhances software portability. Digital control systems use a number of standard functional blocks such as compensators, notch filters, and demodulators. It is therefore likely that a designer who already has access to one digital control system will want to "borrow" some of its functional building blocks to quickly implement a new, different control unit or reconfigure the existing one. The designer who has access to these functional blocks or modules needs only modify the main program by providing a different sequence of subroutine calls. An initialization routine, a first- and second-order filter routine, a roundoff routine, and an overflow checking routine are examples of functional building blocks.

Each software module is written as a subroutine with a clear and efficient interface (for parameter passing, stack use, etc.) with the main program. In order to maintain the general-purpose function of the module, the data used in computations within a module (i.e., filter coefficients, state-variable values, etc.) should be accessed using indirect addressing rather than direct addressing. Only those variables whose values remain unchanged should be addressed directly.

Layout of TMS32010 Data Memory

The layout of the TMS32010 data memory in a digital control system implementation should be defined in accordance with the requirements of the software modules used in the implementation of the system. The procedure is illustrated by the first- and second-order filter subroutine of the section, TMS32010 IMPLEMENTATION OF COMPENSATORS AND FILTERS. This subroutine manipulates its pointer registers so that upon completion of the computations in one filter section, the registers automatically point to the set of coefficients and state variables of the next filter section.

If the software designer arranges his filter coefficient and state-variable sections in the order of execution of the control-system algorithms, a sequence of compensators and filters may be executed with a single subroutine call for each element. This scheme enables faster execution of the control algorithms since there is no need to explicitly reload the pointers in order to match the requirements of the current software module being called.

A designer must define all of the data memory locations (set up a system memory map) at the beginning of the program. An efficient way to accomplish this is to use the TMS32010 assembler's DORG (dummy origin) directive. This directive does not cause code generation. DORG defines a data structure to be used by the system; i.e., it generates values corresponding to the labels of consecutive data memory locations. Using the DORG directive, as opposed to equating labels with data memory locations through the EQU directive, provides flexibility when the data structure needs to be modified. For example, when defining a number of new data memory locations, the labels are inserted in the middle of the "dummy" block and the assembler assigns the values automatically. This function would have to be performed manually if the EQU directive were used.

The software designer must also build a table in the TMS32010 program memory that corresponds to the previously defined data memory map. The table is then loaded into the data memory by the initialization routine during system startup.

These techniques are illustrated in the next section, TMS32010 IMPLEMENTATION OF COMPENSATORS AND FILTERS. Note that in the example program in Appendix B, location ONE has to be the last location in the table. Note also that the states and the coefficients of the filters are defined in reverse order to the order in which the filters execute. This is due to the way the initialization and filter routines are written.

TMS32010 IMPLEMENTATION OF COMPENSATORS AND FILTERS

Design procedure and error handling for the standard first- and second-order compensator and filter subroutine are described in this section. Methods for implementing higher-order structures, implementation tradeoffs, and examples of typical compensators and filters are also given.

Standard First- and Second-Order Block as a Subroutine

A standard first- and second-order compensator section is a prime example of the building block philosophy discussed earlier. The routine presented here computes first- and second-order IIR filter sections using the Direct-Form II network structure[12] and performs roundoffs and overflow checking. The Direct-Form II, although it somewhat obscures the definition of the variables, is chosen over the Direct-Form I because it requires fewer "delays", i.e., data storage locations, in its computational algorithm.

Consider the second-order transfer function:

$$D(z) = \frac{N0 + N1\ z^{-1} + N2\ z^{-2}}{1 + D1\ z^{-1} + D2\ z^{-2}} = \frac{U(z)}{E(z)}$$

For Direct-Form II, the corresponding difference equations are:

$$x(n) = e(n) - D1\ x(n-1) - D2\ x(n-2)$$

$$u(n) = N0\ x(n) + N1\ x(n-1) + N2\ x(n-2)$$

The signal flowgraph for this transfer function is shown in Figure 3.

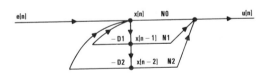

Figure 3. Direct-Form II Compensator/Filter

The filter routine accommodates a scaling scheme as defined on the main program level; i.e., the values can be scaled by 2^{15} or 2^{12}. The routine is written so that chain implementation of a number of compensators and filters is possible if the data structure, namely, the coefficient and state-variable tables, are properly arranged (see the previous section, CONTROL SYSTEM SOFTWARE DESIGN PHILOSOPHY). The routine takes its input from the accumulator and outputs the result to the accumulator so that the main program can efficiently call for the successive execution of the filter routine with a different set of parameters each time.

Two checks for overflow are made within the filter routine. One is made upon computing the value of the intermediate (state) variable and the other upon computing the filter's output. Each overflow check determines whether the 32-bit computed result can be stored in a 16-bit memory location under the adopted scaling scheme. If an overflow condition occurs, the routine "saturates" the output; i.e., it returns the maximum or minimum representable value.

A drawback of overflow checking upon computing the output is the loss of precision of the least-significant bits of the accumulator, which are truncated during the accumulator storing operation. This loss of precision is insignificant, however, in comparison with the loss of precision due to an overflow condition.

The first-order filter section is computed exactly like the second-order section. The two coefficients N2 and $-D2$ that multiply the "oldest" value of the intermediate (state) variable, i.e., the bottom branch of the filter in Figure 3, are equal to zero. This scheme reduces the second-order digital filter to the first-order filter. An example program that uses the first- and second-order filter routine to compute several elements of a digital control system is given in Appendix B.

Higher-Order Filters: Cascade Versus Parallel Tradeoffs

A higher-order filter or compensator in a digital control system can be implemented either as a single section or as a combination of first- and second-order sections. The single section or direct implementation form is easier to implement and executes faster, but it generates a larger numerical error. The larger error occurs because the long filter computation process involves a substantial accumulation of errors resulting from multiplications by quantized coefficients and because the roots of high-order polynomials are increasingly sensitive to changes in their (quantized) coefficients. For this reason, the direct realization form is not recommended except for a very low-order controller.

The suggested method of implementing a high-order transfer function is to decompose it into first-order blocks (to accommodate single real poles) and second-order blocks (to accommodate complex conjugate poles or pairs of real poles), and connect these blocks either in a cascade or a parallel configuration.

For the cascade realization (see Figure 4), the transfer function must be decomposed into a product of first-order and second-order functions of the form:

$$D(z) = K\ D_1(z)\ D_2(z) \ldots D_n(z)$$

Each second-order block has the form:

$$D_i(z) = \frac{N0 + N1\ z^{-1} + N2\ z^{-2}}{1 + D1\ z^{-1} + D2\ z^{-2}}$$

Each first-order block is obtained by equating the coefficients of (z^{-2}), i.e., N2 and D2, to zero.

Figure 4. Cascade Implementation of a High-Order Transfer Function

The designer must decide how to pair the poles and zeroes in forming the $D_i(z)$. A pole-zero pairing algorithm that minimizes the output noise is available.[10] The ordering of the $D_i(z)$ also affects output noise due to quantization and whether or not limit cycles are present.[10]

For the parallel realization (see Figure 5), the transfer function is expanded as a sum of first-order and second-order expressions of the form:

$$D(z) = K + D_1(z) + D_2(z) + \ldots + D_n(z)$$

where the first-order blocks have the form:

$$D_i(z) = \frac{N0}{1 + D1\,z^{-1}}$$

and the second-order blocks have the form:

$$D_i(z) = \frac{N0 + N1\,z^{-1}}{1 + D1\,z^{-1} + D2\,z^{-2}}$$

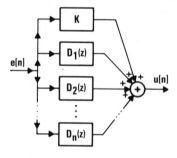

Figure 5. Parallel Implementation of a High-Order Transfer Function

The concept of ordering of $D_i(z)$ does not apply to the parallel configuration.

Pole-zero pairing is fixed by the constraints imposed by the partial-fraction expansion.

The parallel realization has an obvious advantage over the cascade form when an algorithm executes on a multiprocessor system where the filter algorithm can be split up among the processors and run concurrently.

No significant difference is apparent between the parallel and cascade realizations in performance factors such as execution speed and program/data memory use when the algorithms execute on a single processor.[13] The parallel algorithm could possibly provide more precision in computing the filter's output if the designer decided to save the double-precision (32-bit) results from each of the first- and second-order sections and perform a double-precision addition to calculate the final filter output.

Examples of Typical Compensators and Filters

Examples of structures used in digital control systems: compensators (lowpass filters), and notch filters are shown in Table 1. The analog and digital versions of the transfer functions, along with the scaled form of the transfer functions and their coefficient values, are given. A complete example of a transformation from the analog to the digital domain using bilinear transformation with frequency prewarping is presented in Appendix A.

The gain constant that appears in some transfer functions can be implemented either by integrating it into its transfer function or by distributing it over a number of filter sections in a cascade implementation scheme.

PROCESSOR INTERFACE CONSIDERATIONS

Alternatives should be considered when designing the data-acquisition portion of the digital controller hardware. This section addresses the A/D and D/A converter selection, different analog sensor interface methods, and communication with the host processor.

Table 1. Examples of Analog and Digital Versions of Common Transfer Functions

Analog Prototype Transfer Function		Digital Transfer Function (Computed by Bilinear Transformation with Frequency Prewarping; $f_s = 4020$ Hz)	Scaled Digital Transfer Function	Scaled Coefficients and Order of Storage in Data Memory (as required by First- and Second-Order Filter Routine)
First-Order Compensator: $G(s) = 100$	$\dfrac{1}{S+1}$	$D(z) = 0.012436 \dfrac{1.0+1.0\,z^{-1}}{1.0-0.99975\,z^{-1}}$	Scaling Factor $= 2^{15}$ $D(z) = \dfrac{408+408\,z^{-1}}{32768-32760\,z^{-1}}$	$N0 = 408$ $N1 = 408$ $D1 = 32760$
Second-Order Compensator: $G(s) = 1000$	$\dfrac{S^2+68.25+3943}{S^2+2512S+6.31\times10^6}$	$D(z) = 870.77 \dfrac{1.0-1.9824\,z^{-1}+0.9826\,z^{-2}}{1.0-1.2548\,z^{-1}+0.5474\,z^{-2}}$	Scaling Factor $= 2^{12}$ $D(z) = 870.77 \dfrac{4096-8120\,z^{-1}+4025\,z^{-2}}{4096-5140\,z^{-1}+2242\,z^{-2}}$	$N0 = 4096$ $N1 = -8120$ $N2 = 4025$ $D1 = 5140$ $D2 = -2242$
100-Hz Notch Filter: $G(s) = $	$\dfrac{S^2+3.9478\times10^5}{S^2+125.664S+3.9478\times10^5}$	$D(z) = \dfrac{0.98467-1.94534\,z^{-1}+0.98467\,z^{-2}}{1.0-1.94534\,z^{-1}+0.96935\,z^{-2}}$	Scaling Factor $= 2^{12}$ $D(z) = \dfrac{4033-7968\,z^{-1}+4033\,z^{-2}}{4096-7968\,z^{-1}+3970\,z^{-2}}$	$N0 = 4033$ $N1 = -7968$ $N2 = 4033$ $D1 = 7968$ $D2 = -3970$

24. Control System Compensation and Implementation with the TMS32010

A/D and D/A Conversions and Integrated Circuits

The A/D and D/A converter selection for a control system design may be based on several factors. Among the most crucial factors are the maximum conversion speed of the converter and the wordlength of the device.

The A/D conversion speed relates directly to the required sampling rate of the specific application. This rate is determined by the need to sample fast enough to prevent aliasing and excessive phase lag and to sample slow enough to avoid the unnecessary expense and accuracy of high data rates.

The A/D wordlength should be chosen based on a worst-case analysis using the following two criteria:

1. The dynamic range of the continuous input signal, and
2. The quantization noise of the A/D converter.

For dynamic range, the designer should determine the minimum and maximum values of the continuous input that need to be accurately represented and select the A/D wordlength in bits based on the resolution required within this range.

Quantization noise is due to the quantization effect of the A/D. The value of this noise during a single conversion can be represented by the difference between the exact analog value and the value allowable with the finite resolution of the A/D. This quantization noise may assume any value in the range $-q/2$ to $+q/2$ for a rounding converter or 0 to q for a truncating A/D converter where q is the quantization level. The quantization level q is equal to the full-scale voltage range divided by 2^B where B is the number of bits in the converter. The quantization noise may be modeled as uniformly distributed noise. The designer should make his choice based on the maximum acceptable quantization level.

The D/A converter wordlength should be chosen in a similar manner to choosing the A/D wordlength by considering the dynamic range of the output signal.

The effects of A/D and D/A converter wordlength on the performance of a high-speed control system are detailed in the University of Arkansas study (see the section on the design example of the TMS32010-based rate-integrating gyro positioning system). The study analyzed the time-domain performance of the system (unit impulse, step, ramp, and torque-disturbance response) as a function of A/D and D/A wordlengths. Twelve-, fourteen-, and sixteen-bit converters were used. The only significant difference found between them was the steady-state error. Twelve-bit converters were found to be adequate.

In a multi-input digital control system, the signal acquisition portion of the digital controller must provide for the multiplexing of several analog inputs into a single A/D converter. Consequently, some external devices are needed to pre-filter (antialiasing filters), sample and hold the analog signals from each channel (S/H circuits), and multiplex the signals onto the A/D converter (analog multiplexer). Multiplexing and filtering may also be necessary at the output in cases where the digital control system computes multiple outputs for the control of the plant.

Two configurations of a cost-effective, multichannel data acquisition system for a digital controller are shown in Figure 6. The first accommodates up to eight inputs; the second can accommodate up to 32 inputs. Note that in these two systems, only one S/H per eight inputs exists, and the variables are sampled in sequence with the same sampling interval between successive samples of a given signal. There will be a "skew" in time between the samples of the various

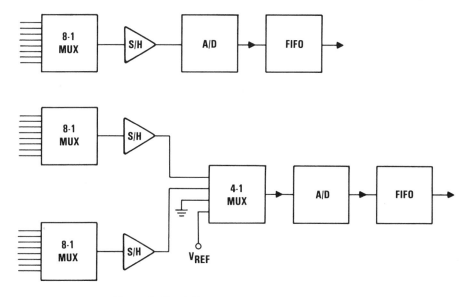

Figure 6. Cost-Effective Data Acquisition Systems

inputs with a possible unwanted effect on the system performance. In this case, the use of a fast A/D converter may be justified to minimize this effect.

If truly simultaneous sampling is required, an array of S/H circuits may be used to capture the values of all the inputs concurrently. This solution, shown in Figure 7, is more expensive due to the cost of S/Hs. In such simultaneous sampling systems, a fast conversion must be performed before the signal values present on the S/Hs start to droop. Therefore, the maximum conversion rate must be fast enough to accommodate this constraint.

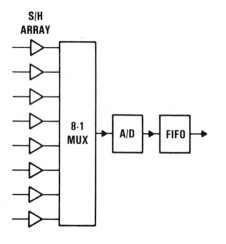

Figure 7. Data Acquisition System with Simultaneous Sampling

In some cases, high-speed A/D converters (100- to 500-kHz conversion rates) are required. Two 12-bit A/D devices that will accommodate these speed requirements are the ADC85 and the AD5240 from Analog Devices. The ADC85 allows a conversion rate of up to 100 kHz and the AD5240 up to 200 kHz. There are other converters available from several manufacturers.

D/A converters, on the other hand, are inherently faster, the selection is much broader, and the cost is less.

When high-speed converters are not needed (when only a few input channels exist or a single converter per channel is justified), devices such as the TCM2913 and TCM2914 codecs may be useful in digital control applications. Although telecommunications-oriented, these devices are low in cost and provide on-chip antialiasing and smoothing filters. The TCM2913 or TCM2914 both contain A/D and D/A converters. The 8-bit digital output of the A/D and the 8-bit digital input to the D/A are both arranged in a companded (compressed/expanded) form using μ-law or A-law companding techniques. The μ-law and A-law companding techniques allow small numbers to be represented with maximum accuracy, but require a conversion routine before the companded samples can be used in two's-complement computations. Such conversion routines are based on lookup tables and need only a few TMS32010 instruction cycles to execute.[14] The devices interface to the processor in a serial form and convert the data at a maximum rate of 8 kHz.

All of these data acquisition systems can accommodate differential inputs from analog transducers, such as pressure sensors, strain gauges, and others. To maintain accuracy in the case of a low-level input signal and to minimize noise effects, twisted-pair leads can be used to connect the transducer output to an instrumentation amplifier (differential-to-single-ended conversion circuit) that in turn is connected to the analog multiplexer. Alternatively, balanced twisted-pair leads can be connected to a differential analog multiplexer which drives an instrumentation amplifier of the same kind. The amplifier rejects the common-mode noise and presents the single-ended output to the S/H circuit and the A/D converter.[15,16] These two configurations are shown in Figure 8.

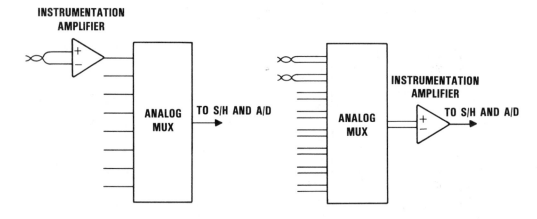

Figure 8. Differential Input Configurations

Synchronization of the Processor and External Devices

In a multichannel data acquisition system with one A/D converter, a designer must generate a sequence of timing signals to synchronize the S/H circuits, the analog multiplexer, the A/D converter, and the input/output latches with the operation of the TMS32010. The designer may decide to generate the timing signals from the TMS32010 clock by subdividing its frequency or using a timing and control circuit based on its own clock.

An alternative way to build a multi-channel data acquisition system is to designate a separate A/D for each channel. In this case, the timing-signal generation is simpler and the A/D converters used may be slower and less costly, although more of them are necessary. The designer should perform a tradeoff analysis based on board space, overall system cost, and power consumption.

Communication with Host Computer

In addition to having a fast signal-processing microprocessor, a need may exist for an executive processor to monitor the system's operation. Such an executive processor would be used for system startup/initialization (coefficient and initial-condition loading), responding to emergency conditions such as overflow and underflow, system reprogramming/reconfiguration (loading a new program or a new set of coefficients), and a thorough system test and calibration. The system should be constructed so that the executive processor can interrupt, halt, or alter the execution of the signal processor at any time in response to contingency situations.

DESIGN EXAMPLE: RATE-INTEGRATING GYRO STABILIZATION LOOP

An example of using the TMS32010 processor to implement a digital control system is presented in this section. Sampling rate selection, the system's hardware and software, and system performance are discussed.

System Description

The system used as an example of the application of the TMS32010 is a servo-control system for stabilizing a large, two-axis gimbaled platform with a DC-motor drive. Inertial rate-integrating gyroscopes mounted directly on the platform serve as angular motion sensors. Such systems are required for the precise control of line-of-sight (LOS) and line-of-sight rate for use in pointing and tracking applications for laser, video, inertial navigation, and radar systems.

At present, digital control is not normally used in systems of this type because of the fast throughput rates and computational accuracy required to perform the control computations and notch filtering. Current line-of-sight stabilization systems continue to use analog electronics to implement servo-compensation functions and error-signal conditioning. Thus, the system is representative in complexity and performance of typical systems currently in use by the aerospace industry and are candidates for microprocessor-based digital control.

The digital control system was designed as part of a research contract carried out by the University of Arkansas under the sponsorship of Texas Instruments from February 1982 to February 1984.[18,19]

System Model and Control Compensation

A single axis of the stabilization system has two primary control loops: the rate loop and the position loop. In addition, a tachometer loop exists within the position loop. The rate and position loops are identified in Figure 9, a diagram of the elevation axis of the system. In its analog version, the system employs analog electronics to implement all control compensation and signal conditioning functions.

Figure 10 identifies those filters and compensators in the rate loop that are to be incorporated into the digital control system.

This study's approach provides a digital implementation of the designated analog elements of the rate loop without sacrificing closed-loop performance. In keeping with the recommendations of the DIGITAL COMPENSATOR DESIGN section, the technique for the conversions of the analog compensators and notch filters to their digital counterparts is the bilinear transformation with frequency prewarping. Within the rate loop, the transfer functions to be implemented digitally consist of a first-order and a second-order compensator, along with six notch filters. Within the position loop, there is one first-order compensator and one notch filter. The transfer functions, shown in Table 2, list both the analog prototypes and their digital equivalents.

The sampling rate chosen is 4020 samples per second (sampling period is 249 μs). This rate is more than twice the highest frequency of consequence (1800 Hz, the highest rate-loop notch frequency) to prevent aliasing. The rate is fast enough to prevent excessive phase lag in the rate loop and is more than ten times the closed rate-loop bandwidth (approximately 80 Hz).[5] The rate was also chosen to be an integer multiple of 30 Hz, which is a commonly used update rate of the video and infrared imaging/tracker devices that provide the line-of-sight rate command to the stabilization system's rate loop. The update rate of the imaging device and the sampling rate within the rate loop are thus synchronized.

After simulating the closed rate loop, the phase margin was found to be five degrees less than it was for the all-analog system, due to the computational and other delays associated with sampling. To overcome this deterioration in phase margin, the second-order rate-loop compensator was redesigned to provide additional phase lead. The compensator was modified to provide enough additional phase lead so that the phase margin of the digital system matched that of the analog system. The modified compensator is listed in the table.

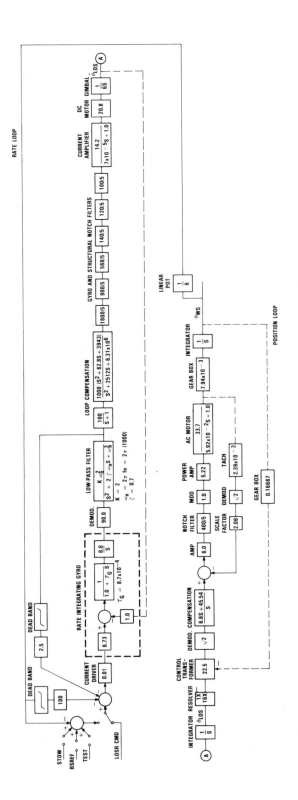

Figure 9. Line-of-Sight Stabilization/Pointing System Elevation Axis

24. Control System Compensation and Implementation with the TMS32010

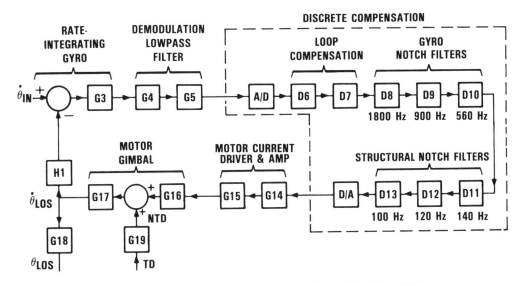

Figure 10. Line-of-Sight Stabilization/Pointing System Rate-Loop [19]

Table 2. Analog and Digital Compensators and Notch Filters

Compensator/ Filter Element	Analog Transfer Function		Digital Transfer Function (f_s = 4020 Hz)	
Rate-Loop 1st-Order Compensator	G6 =	$\dfrac{100}{S+1}$	D6 =	$\dfrac{0.1244\,(1.0+1.0\,Z^{-1})}{1.0-0.99975\,Z^{-1}}$
Rate-Loop 2nd-Order Compensator	G7 =	$\dfrac{1000\,(S^2+62.8S+3943)}{S^2+2512S+6.31\times10^6}$	D7 =	$\dfrac{754.7101\,(1.0-1.98426\,Z^{-1}+0.9845\,Z^{-2})}{1.0-1.255\,Z^{-1}+0.5474\,Z^{-2}}$
Rate-Loop 1800-Hz Notch Filter	G8 =	$\dfrac{S^2+(2\pi\,1800)^2}{S^2+\dfrac{2\pi\,1800}{5}\,S+(2\pi\,1800)^2}$	D8 =	$\dfrac{0.96877+1.83411\,Z^{-1}+0.96877\,Z^{-2}}{1.0+1.83411\,Z^{-1}+0.93754\,Z^{-2}}$
Rate-Loop 900-Hz Notch Filter	G9 =	$\dfrac{S^2+(2\pi\,900)^2}{S^2+\dfrac{2\pi\,900}{2.5}\,S+(2\pi\,900)^2}$	D9 =	$\dfrac{0.8352-0.27291\,Z^{-1}+0.8352\,Z^{-2}}{1.0-0.27291\,Z^{-1}+0.67041\,Z^{-2}}$
Rate-Loop 560-Hz Notch Filter	G10 =	$\dfrac{S^2+(2\pi\,560)^2}{S^2+\dfrac{2\pi\,560}{5}\,S+(2\pi\,560)^2}$	D10 =	$\dfrac{0.9287-1.19021\,Z^{-1}+0.9287\,Z^{-2}}{1.0-1.19021\,Z^{-1}+0.8574\,Z^{-2}}$
Rate-Loop 140-Hz Notch Filter	G11 =	$\dfrac{S^2+(2\pi\,140)^2}{S^2+\dfrac{2\pi\,140}{5}\,S+(2\pi\,140)^2}$	D11 =	$\dfrac{0.97875-1.91083\,Z^{-1}+0.97875\,Z^{-2}}{1.0-1.91083\,Z^{-1}+0.95751\,Z^{-2}}$
Rate-Loop 120-Hz Notch Filter	G12 =	$\dfrac{S^2+(2\pi\,120)^2}{S^2+\dfrac{2\pi\,120}{5}\,S+(2\pi\,120)^2}$	D12 =	$\dfrac{0.9817-1.92896\,Z^{-1}+0.9817\,Z^{-2}}{1.0-1.92896\,Z^{-1}+0.96339\,Z^{-2}}$
Rate-Loop 100-Hz Notch Filter	G13 =	$\dfrac{S^2+(2\pi\,100)^2}{S^2+\dfrac{2\pi\,100}{5}\,S+(2\pi\,100)^2}$	D13 =	$\dfrac{0.98467-1.94534\,Z^{-1}+0.98467\,Z^{-2}}{1.0-1.94534\,Z^{-1}+0.96935\,Z^{-2}}$
Position-Loop 1st-Order Compensator	G22 =	$\dfrac{6.6\,S+45.54}{S}$	D22 =	$\dfrac{6.60566-6.59434\,Z^{-1}}{1.0-Z^{-1}}$
Position-Loop 400-Hz Notch Filter	G24 =	$\dfrac{S^2+(2\pi\,400)^2}{S^2+\dfrac{2\pi\,400}{5}\,S+(2\pi\,400)^2}$	D24 =	$\dfrac{0.94471-1.53204\,Z^{-1}+0.94471\,Z^{-2}}{1.0-1.53204\,Z^{-1}+0.88942\,Z^{-2}}$

Hardware

In the digital control system, the analog compensators and the notch filters are replaced by a digital signal processor, the TMS32010, along with the additional interface hardware needed to provide the digital input signals to the controller and the analog signals to the plant. Figure 11 shows the system hardware block diagram.

The hardware was packaged onto five wirewrap boards. It was fabricated as a prototype test bed, and was constructed from commercially available components that have military-specification counterparts. Twelve-bit A/D and D/A converters were used, based on the studies of time-domain performance characteristics of the system.

Software

The TMS32010 software is composed of four modules: Initialization Routine, Main Program, Rate-Loop Subprogram, and Subroutines. Figure 12 shows the system software block diagram.

The Initialization software disables and enables interrupts, loads data memory with filter coefficients, program constants, and gain terms, and initializes the TMS32010 registers.

The Main Program software calls the Delay Subroutine at the beginning of each sample period to wait for the A/D to complete conversion of all input variables. It then does on-line compensation for the error signal sensor variation by executing the A/D Drift Subroutine. The Main Program then reads the value of the input variable, calls the Rate-Loop Subprogram to compute the control output, and, when that subprogram returns the output variable, loads it into the appropriate output register.

The Rate Loop Subprogram calls subroutines that perform each compensator and notch filter computation and checks the computed output for overflow.

The Subroutines consist of a single routine for performing any of the compensator or notch-filter computations (first- and second-order filter routine), along with routines for checking overflow, providing delay, and performing multiplication of low-precision numbers by a constant.

The A/D Drift Subroutine compensates on-line for the variations in the rate-loop error signal sensor (as a function of time and temperature). The subroutine uses an external calibration input and follows the model of the sensor variations to estimate the true value of the A/D input.

The digital control system is interrupt-driven. An interrupt occurs every 1/4020 seconds (approximately 250 μs). This starts the A/D conversion of a new set of sample inputs and restarts the TMS32010 on a new pass through its software.

A TMS32010 Evaluation Module (EVM) and Emulator (XDS) were used in the software development to permit single-step execution of the software for comparison with the corresponding computations produced by simulations written with the aid of the Continuous System modeling Program (CSMP). These simulations take into account the input/output signal quantization levels, microprocessor architecture, memory and internal register lengths.

Other software functions associated with a complete, self-contained control module include:

1. System calibration, testing, and startup
2. Error checking and contingency responses
3. Setting of gains, time constants, and other programmable or adjustable parameters
4. System shutdown.

These functions are implemented by a general-purpose executive processor (SBP9989), thus allowing the TMS32010 to handle computation-intensive tasks.

System Performance

The system performance was evaluated in the following two-step procedure:

1. A hybrid computer system was constructed

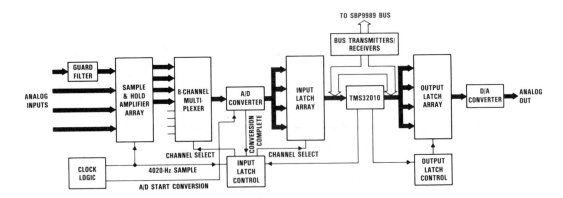

Figure 11. Digital Controller Hardware Block Diagram [19]

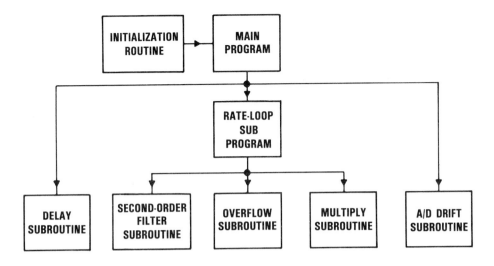

Figure 12. Digital Controller Software Block Diagram [19]

consisting of an analog-computer implementation of part of the nondigital portion of the rate loop coupled with the TMS32010-based digital controller.

2. A full-scale CSMP simulation of the entire rate loop was conducted.

The closed-loop performance of the rate loop was characterized by the following responses: rate-command step response, torque-disturbance step response, and torque-disturbance frequency response. The results are shown in Figures 13 through 15.

In the rate-command step response, the percent overshoot, peak time, and settling time are similar for both the discrete and continuous systems, but the continuous system is slightly smoother. The torque-disturbance step response shows that the discrete system is slightly slower in correcting for a torque disturbance input. In addition, the discrete system has a low-level oscillation (limit cycle). The frequency responses for a torque-disturbance input are also similar, with the continuous system having slightly better torque disturbance rejection in the low-frequency region. These results show that the analog and the digital systems are comparable even though no special efforts were made to take advantage of the capability that digital control offers.

The flexibility of the digital control system was demonstrated by programming the digital system with the capability to correct for a variation in the sensor input to the A/D converter. The system was able to correct on-line (by using a known standard, calculating the gain, and dividing it out) for a 50 percent sinusoidal variation in the sensor gain.

The conversion between two different stabilization systems serves as another flexibility example. The software

of a small, two-axis stabilization system was converted to the software of the higher-precision, large, two-axis gimbaled-platform stabilization loop described earlier. The only modification required was in the Main Program and the Rate-Loop Subprogram for the latter system. The modular software design procedure made possible the use of most of the building blocks (subroutines) in the implementation of the new controller.

In general, the study demonstrated the technical feasibility of digital control for a wide-bandwidth, high-precision type of system. Due to the limited scope of the study, the full power of digital control was not utilized, in that the control algorithms were constrained by the design to emulate their analog prototypes. It is likely that significant performance improvements could be achieved by advanced control techniques.

Additional capacity in the TMS32010 remains to accommodate improved, more sophisticated compensators. Table 3 shows the TMS32010 utilization.

Table 3. TMS32010 Utilization
(LOS Stabilization System Rate-Loop)

	Used	Available	% Use
Program Memory	275 words	4096 words	7%
Data Memory	76 words	144 words	53%
Execution Time	73 µs	250 µs	29% *

*Based on a 16-MHz (i.e., less than maximum) clock rate.

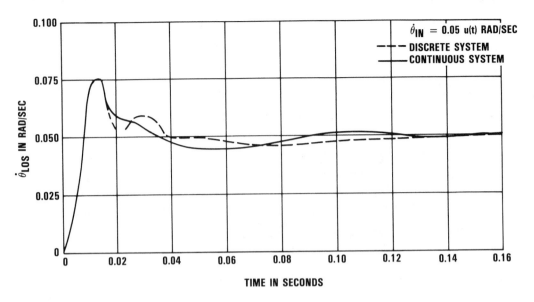

Figure 13. Rate-Loop Rate Command Step Response [19]

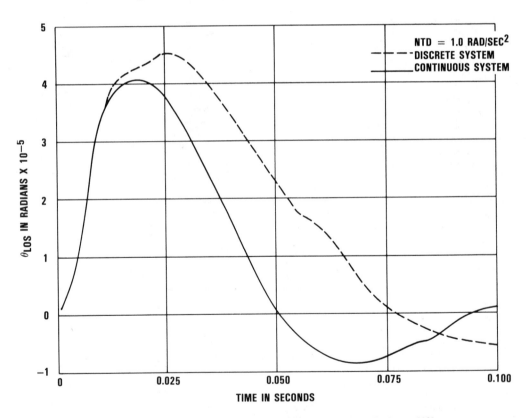

Figure 14. Rate-Loop Normalized Torque-Disturbance Step Response [19]

24. Control System Compensation and Implementation with the TMS32010

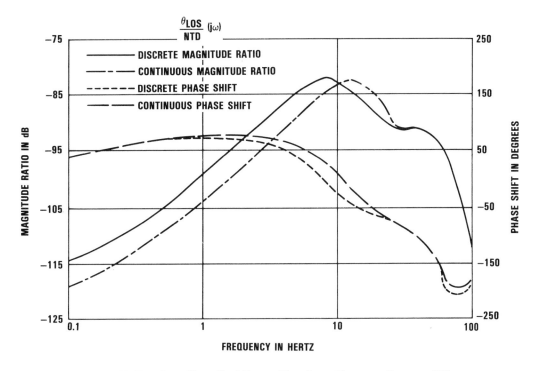

$$\frac{\theta_{LOS}}{NTD}(j\omega)$$

——— DISCRETE MAGNITUDE RATIO
— - — CONTINUOUS MAGNITUDE RATIO
— - - - - DISCRETE PHASE SHIFT
—— —— CONTINUOUS PHASE SHIFT

MAGNITUDE RATIO IN dB

PHASE SHIFT IN DEGREES

FREQUENCY IN HERTZ

Figure 15. Rate-Loop Normalized Torque-Disturbance Frequency Response [19]

Other microprocessors that were considered for implementing this digital controller system were: Intel 8086, Zilog Z-8000, Motorola 68000, and the Fairchild 9445. These microprocessors were unable to meet the criterion that the maximum allowable time between samples for processing be 250 μs. Among the signal-processing microprocessors, the AMI 2811, while apparently fast enough, has only a 12 × 12 multiplier; and the Intel 2920 has only four inputs and no branching instructions.

The principal limitation of the TMS32010 was that of having eight inputs and eight outputs. Except for this restriction, the processor would have been able to carry out the processing for both axes of the two-axis gimbaled platform. This limitation could be removed by the addition of logic circuitry.

ACKNOWLEDGMENTS

The authors would like to thank Larry Chatelain of Texas Instruments Incorporated, Dallas, Texas, and Stanley Stephenson of the University of Arkansas, Fayetteville, Arkansas, for their help in the preparation of this document.

REFERENCES

TI TMS32010 Documentation

1. *TMS32010 Assembly Language Programmer's Guide*, Texas Instruments Incorporated (1983).
2. *TMS32010 User's Guide*, Texas Instruments Incorporated (1983).
3. *TMS32010 Development Support Reference Guide*, Texas Instruments Incorporated (1984).
4. *Digital Filter Design Package: Interactive Software for Digital Filter Design and Automatic Code Generation for the Texas Instruments TMS32010*, Atlanta Signal Processors Incorporated, 770 Spring Street, Suite 208, Atalanta, Georgia 30332 (1984).

References on Digital Control

5. P. Katz, *Digital Control Using Microprocessors*, Prentice-Hall (1981).
6. R. Jacquot, *Modern Digital Control Systems*, Marcel Dekker, Inc. (1981).
7. P. Moroney, *Issues in the Implementation of Digital Feedback Compensators*, MIT Press, pp. 16-17 (1983).

8. B.C. Kuo, *Digital Control Systems*, Holt, Rinehart, & Winston (1980).

9. P. Moroney, *Issues in the Implementation of Digital Feedback Compensators*, MIT Press (1983).

10. C. Phillips, H. Nagle, *Digital Control System Analysis and Design*, Prentice-Hall, Inc. (1984).

11. M. Masten, "Rate-Aiding for Line-of-Sight Stabilization/Tracking Systems," *Texas Instruments Engineering Journal*, **Vol 1,** No. 2 (September-October 1984).

References on Digital Signal Processing

12. A. Oppenheim and R. Schafer, *Digital Signal Processing*, Prentice-Hall, Inc. (1975).

13. *The Implementation of FIR/FIR Filters with the TMS32010*, Texas Instruments Incorporated (1983).

14. *Companding Routines for the TMS32010*, Texas Instruments Incorporated (1984).

15. S. Moore, S. Pietkiewicz, "Monolithic A/D Converter Interfaces Directly With Most Microprocessors", *Electronic Design* (September 6, 1984).

16. R. Jaeger, "Tutorial: Analog Data Acquisition Technology Part III," *IEEE Micro* (November 1982).

17. A. Oppenheim, A. Willsky, *Signals and Systems*, Prentice-Hall, Inc. (1983).

References for the Design Example

18. *Digital Servo Techniques, Phase One*, Department of Electrical Engineering, University of Arkansas, Fayetteville, Arkansas 72701 (1982).

19. *Digital Control Techniques, Phase Two, Feasibility Demonstration*, **Vol I and II,** Department of Electrical Engineering, University of Arkansas, Fayetteville, Arkansas 72701 (1984).

APPENDIX A
Development of a Digital Compensator Transfer Function

The development of a digital equivalent of an analog compensator transfer function using the bilinear transformation with frequency prewarping is shown in this appendix. The technique is described in the section, DIGITAL COMPENSATOR DESIGN.

Beginning with an analog prototype transfer function,

$$G(s) = 1000 \frac{S^2 + 68.2\, S + 3943}{S^2 + 2512\, S + 6.31 \times 10^6}$$

The sampling frequency to be used in converting to a digital equivalent is $f = 4020$ Hz (i.e., the sampling period $T_s = 1/4020$ s $= 248.76 \times 10^{-6}$ s).

The characteristic equation of this analog transfer function is:

$$S^2 + 2512\, S + 6.31 \times 10^6 = 0$$

which fits the standard, second-order form:

$$S^2 + 2\, \zeta\, \omega_n\, S + \omega_n^2 = 0$$

The natural frequency $\omega_n = \sqrt{6.31 \times 10^6} = 2511.9713$ rad/s

To compensate for nonlinear mapping of analog-to-digital frequencies by the bilinear transformation method, the natural frequency is prewarped according to the formula:

$$\omega_p = \frac{2}{T}\tan\frac{\omega_0 T}{2} = \frac{2}{248.76 \times 10^{-6}}\tan\frac{2511.9713 \times 248.76 \times 10^{-6}}{2} = 2597.03 \text{ rad/s}$$

This prewarping scheme matches exactly the natural frequency in the analog and digital domains for the compensator.

To obtain the prewarped version of the analog transfer function, the complex variable s in the original transfer function is replaced with $(\omega_0/\omega_p)s$. It is therefore convenient to compute the ratio:

$$\frac{\omega_0}{\omega_p} = \frac{2511.9713}{2597.03} = 0.9672$$

The prewarped G(s), i.e., $G_p(s)$ is then computed as:

$$G_p(s) = 1000 \ \frac{(0.9672\, s)^2 + 68.2\,(0.9672\, s) + 3943}{(0.9672\, s)^2 + 2512\,(0.9672\, s) + 6.31 \times 10^6}$$

$$= 1000 \ \frac{s^2 + 70.51\, s + 4214.87}{s^2 + 2597.16\, s + 6.75 \times 10^6}$$

Bilinear transformation is next applied to $G_p(s)$ whereby the continuous variable s is replaced by the expression that involves the discrete variable z:

$$S = \frac{2}{T} \frac{z-1}{z+1}$$

This produces the discrete transfer function D(z).

For the compensator,

$$D(z) = G_p(s) \Big|_{s = \frac{2}{T} \frac{z-1}{z+1}} =$$

$$= 1000 \; \frac{s^2 + 70.51\,s + 4214.87}{s^2 + 2597.16\,s + 6.75 \times 10^6} \Bigg|_{s = \frac{2}{248.76 \times 10^{-6}} \frac{z-1}{z+1}}$$

After further computations,

$$D(z) = 706.76 \; \frac{1.0 - 1.9824\,z^{-1} + 0.9826\,z^{-2}}{1.0 - 1.2548\,z^{-1} + 0.5474\,z^{-2}}$$

The final step is the gain adjustment in the digital transfer function. This can be accomplished by matching the analog and digital gains at some predetermined frequency, for example, DC.

For the DC case, $s = j\omega = 0$ and from the bilinear transformation:

$$z = \frac{2 + sT}{2 - sT} = 1$$

Therefore, at DC, $G(0) = D(1)$.

For this transfer function, $G(0) = 0.6249$, $D(1) = 0.5072$. If $G(0) = K \times D(1)$, then the constant K becomes

$$\frac{0.6249}{0.5072} = 1.2321$$

The final form of the digital equivalent transfer function is:

$$D(z) = 870.77 \; \frac{1.0 - 1.9824\,z^{-1} + 0.9826\,z^{-2}}{1.0 - 1.2548\,z^{-1} + 0.5474\,z^{-2}}$$

where the gain of 870.77 is the product of $K \times$ (the unadjusted digital gain), i.e., $870.77 = 1.2321 \times 706.76$.

APPENDIX B
TMS32010 Example Program

An example TMS32010 program that uses the first- and second-order filter routine to compute several elements of a digital control system is provided in this appendix. The program illustrates the concepts of modular software design, data memory layout, and cascade implementation of high-order transfer functions. The program was executed on a combination of the TMS32010 Emulator (XDS) and Analog Interface Board (AIB) using random noise as input. The input was sampled at a 4000-Hz rate.

The following transfer functions are implemented with Q12 scaling:

900-Hz Notch Filter $\qquad D(z) = \dfrac{0.8352 - 0.2729\ z^{-1} + 0.8352\ z^{-2}}{1.0 - 0.2729\ z^{-1} + 0.6704\ z^{-2}}$

1800-Hz Notch Filter $\qquad D(z) = \dfrac{0.9688 + 1.8341\ z^{-1} + 0.9688\ z^{-2}}{1.0 + 1.8341\ z^{-1} + 0.9375\ z^{-2}}$

Other transfer functions (compensators, notch filters) can be implemented in identical fashion by expanding the data structure (filter coefficients and states) and making additional filter routine calls to compute these elements.

The output, as observed on a spectrum analyzer, is shown in Figure B-1.

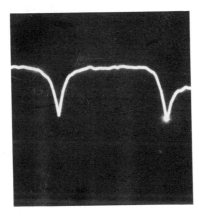

Figure B-1. Spectrum Analyzer output (900-Hz and 1800-Hz Notch Filters)

The first notch from the left is at 900 Hz, the second is at 1800 Hz. The attenuation of the notch frequencies is about 23 dB in reference to the passband region.

The program that produced this output is as follows:

```
0001                         IDT      'RIG SYS'
0002                         OPTION   DUNLST,TUNLST
0003               *
0004               *    The program computes a few sections of the LOS
0005               *    Stabilization System Rate Loop.
0006               *
0007               *    Constants
0008               *    The constant SCALE relates to the scaling factor
0009               *    through a relation:
0010               *    scaling factor = 2 ** SCALE (ex. 4096 = 2 ** 12).
0011               *    Scale constants 12 and 15 are available.
0012      000C  SCALE EQU     12              SCALING FACTOR (Q12)
0013               *
0014               *    Data memory map
0015 0000                    DORG     0              DATA MEM START ADRS
0016 0000 0000  MODE  DATA   $               AIB MODE
0017 0001 0001  RATE  DATA   $               AIB RATE
0018               *    Coefficient table
0019 0002 0002  N03   DATA   $               FILTER #3 COEFF'S
0020 0003 0003  N13   DATA   $
0021 0004 0004  N23   DATA   $
0022 0005 0005  D13   DATA   $
0023 0006 0006  D23   DATA   $
0024 0007 0007  N02   DATA   $               FILTER #2 COEFF'S
0025 0008 0008  N12   DATA   $
0026 0009 0009  N22   DATA   $
0027 000A 000A  D12   DATA   $
0028 000B 000B  D22   DATA   $
0029 000C 000C  N01   DATA   $               FILTER #1 COEFF'S
0030 000D 000D  N11   DATA   $
0031 000E 000E  N21   DATA   $
0032 000F 000F  D11   DATA   $
0033 0010 0010  D21   DATA   $
0034      0010  COEFFS EQU    $-1
0035               *    State var table
0036 0011 0011  X03   DATA   $               FILTER #3 STATES
0037 0012 0012  X13   DATA   $
0038 0013 0013  X23   DATA   $
0039 0014 0014  X02   DATA   $               FILTER #2 STATES
0040 0015 0015  X12   DATA   $
0041 0016 0016  X22   DATA   $
0042 0017 0017  X01   DATA   $               FILTER #1 STATES
0043 0018 0018  X11   DATA   $
0044 0019 0019  X21   DATA   $
0045      0019  STATES EQU    $-1
0046               *
0047 001A 001A  COMAND DATA   $              COMMAND INPUT
0048 001B 001B  OUTPUT DATA   $              SYSTEM OUTPUT
0049               *
0050 001C 001C  MAX16 DATA   $               MAX 2-COMPLEMENT NUM IN 16 BITS
0051 001D 001D  MIN16 DATA   $               MIN 2-COMPLEMENT NUM IN 16 BITS
0052      001C  MASK1 EQU     MAX16           MASK
0053      001D  MASK2 EQU     MIN16           MASK
0054 001E 001E  ONE   DATA   $               ONE
0055               *
0056               *
0057 0000                    AORG     0
```

24. Control System Compensation and Implementation with the TMS32010

```
0058 0000 F900          B      START           RESTART VECTOR
     0001 0021
0059               *
0060               *      Table in prog memory
0061      0002 TABLE EQU      $
0062 0002 000A          DATA   >A,>4DB          AIB MODE AND RATE
0063 0004 0000          DATA   0,0,0,0          FILTER #3 COEFF'S
0064 0009 0D5D          DATA   3421,-1118,3421,1118,-2746  900 HZ NOTCH FILTER
0065 000E 0F80          DATA   3968,7513,3968,-7513,-3840  1800 HZ NOTCH FILTER
0066 0013 0000          DATA   0,0,0            FILTER #3 INITIAL STATES
0067 0016 0000          DATA   0,0,0            900 HZ NOTCH INITIAL STATES
0068 0019 0000          DATA   0,0,0            1800 HZ NOTCH INITIAL STATES
0069 001C 0000          DATA   0,0              COMMAND INPUT, SYSTEM OUTPUT
0070 001E 7FFF          DATA   32767,-32768     MAX AND MIN 16 BIT NUMBERS
0071 0020 0001          DATA   1                ONE
0072      0020 TBLEND EQU     $-1
0073               *
0074               *
0075               *      Initialize the system
0076      0021 START EQU      $
0077 0021 F800          CALL   INIT             INITIALIZATION ROUTINE
     0022 0039
0078               *
0079               *      Wait on sample
0080 0023 F600 WAIT     BIOZ   GET
     0024 0027
0081 0025 F900          B      WAIT
     0026 0023
0082               *
0083               *      Input sample
0084      0027 GET      EQU    $
0085 0027 421A          IN     COMAND,PA2       INPUT COMMAND
0086 0028 661A          ZALS   COMAND           GET COMMAND
0087 0029 781C          XOR    MASK1            CORRECT A/D FORMAT
0088 002A 501A          SACL   COMAND           UPDATE COMMAND
0089               *
0090               *      Process sample
0091 002B 2C1A          LAC    COMAND,SCALE     LOAD SCALED COMMAND
0092 002C 7010          LARK   AR0,COEFFS       AR0 = PTR TO COEFF TABLE
0093 002D 7119          LARK   AR1,STATES       AR1 = PTR TP STATE VAR TABLE
0094 002E F800          CALL   FILTR2           1800 HZ NOTCH FILTER
     002F 0046
0095 0030 F800          CALL   FILTR2           900 HZ NOTCH FILTER
     0031 0046
0096               *
0097               *      Output sample
0098 0032 5C1B          SACH   OUTPUT,16-SCALE STORE OUTPUT
0099 0033 661B          ZALS   OUTPUT           GET OUTPUT
0100 0034 781D          XOR    MASK2            CORRECT FOR D/A FORMAT
0101 0035 501B          SACL   OUTPUT           UPDATE OUTPUT
0102 0036 4A1B          OUT    OUTPUT,PA2       AND SEND IT OUT
0103               *
0104               *      Repeat the sequence
0105 0037 F900          B      WAIT             GO GET NEXT SAMPLE
     0038 0023
0106               *
0107               *
```

24. Control System Compensation and Implementation with the TMS32010

```
0108              *        System initialization routine. The routine initializes
0109              *        the TMS32010 and other system components.
0110              *        The calling sequence is:
0111              *        CALL    INIT
0112              *
0113              *        Initialize the 32010
0114      0039  INIT  EQU     $
0115 0039 7F81          DINT                      DISABLE INTERRUPTS
0116 003A 7F8B          SOVM                      SET OVERFLOW MODE
0117              *
0118              *        Initialize Data Memory
0119 003B 6E00          LDPK    0                 USE PAGE 0
0120 003C 6880          LARP    0                 USE AR0
0121 003D 701E          LARK    AR0,TBLEND-TABLE       INIT PTR TO END OF DATA
0122 003E 7E20          LACK    TBLEND            INIT PTR TO END OF TABLE
0123      003F  XFER  EQU     $
0124 003F 6788          TBLR    *                 XFER FROM PROG TO DATA MEM
0125 0040 101E          SUB     ONE               BUMP PTR DOWN
0126 0041 F400          BANZ    XFER              GO XFER MORE
     0042 003F
0127              *
0128              *        Inititialize AIB
0129 0043 4800          OUT     MODE,PA0          AIB mode
0130 0044 4B01          OUT     RATE,PA3          AIB RATE
0131              *
0132 0045 7F8D          RET                       RETURN
0133              *
0134              *
0135              *        First and second order filter routine. Computes an IIR
0136              *        filter  using Direct Form II algorithm and adapts to a
0137              *        scaling scheme defined in the calling program.
0138              *
0139              *        The routine incorporates overflow handling code upon
0140              *        computing the intermediate value and the output.
0141              *
0142              *        The calling sequence is:
0143              *        ACC = scaled filter input
0144              *        AR0 = ptr to coeff table
0145              *        AR1 = ptr to state var table
0146              *        CALL    FILTR2
0147              *        ACC = scaled filter output
0148              *        AR0 = ptr to next set of coeff's
0149              *        AR1 = ptr to next set of state var's
0150              *
0151      0046  FILTR2 EQU    $
0152 0046 6881          LARP    AR1               USE AR1
0153              *
0154              *        Compute intermediate value
0155 0047 6A90          LT      *-,AR0            T=X2
0156 0048 6D91          MPY     *-,AR1            MPY X2*D2
0157 0049 6CA0          LTA     *+,AR0            T=X1, ACC=KU+X2*D2
0158 004A 6D91          MPY     *-,AR1            MPY X1*D1
0159 004B 6C98          LTA     *-                T=X2, ACC=KU+X2*D2+X1*D1
0160 004C 6898          MAR     *-                AR1=PTR TO X0
0161              *
0162              *        Round, store and check for intermediate overflow
0163 004D FA00          BLZ     LBL10             CHECK FOR +/- RESULT
```

24. Control System Compensation and Implementation with the TMS32010

```
        004E 0058
0164 004F 0B1E       ADD    ONE,SCALE-1     ROUND
0165 0050 5C88       SACH   *,16-SCALE      UPDATE INTERMEDIATE VAL
0166 0051 1C1C       SUB    MAX16,SCALE     SUBTRACT SCALED MAX POS NUMBER
0167 0052 FB00       BLEZ   LBL20           IF ACC<=0 THEN NO OVERFLOW
        0053 005F
0168 0054 661C       ZALS   MAX16           OVERFLOW, LOAD MAX POS NUMBER
0169 0055 5088       SACL   *               UPDATE INTERMEDIATE VALUE
0170 0056 F900       B      LBL20           GO, COMPUTE OUTPUT
        0057 005F
0171     0058 LBL10  EQU    $
0172 0058 1B1E       SUB    ONE,SCALE-1     ROUND
0173 0059 5C88       SACH   *,16-SCALE      UPDATE INTERMEDIATE VAL
0174 005A 1C1D       SUB    MIN16,SCALE     SUBTRACT SCALED MIN NEG NUMBER
0175 005B FD00       BGEZ   LBL20           IF ACC>=0 THEN NO OVERFLOW
        005C 005F
0176 005D 661D       ZALS   MIN16           OVERFLOW, LOAD MIN NEG NUMBER
0177 005E 5088       SACL   *               UPDATE INTERMEDIATE VALUE
0178          *
0179          *      Compute filter output
0180     005F LBL20  EQU    $
0181 005F 68A0       MAR    *+,AR0          USE AR0
0182 0060 6D91       MPY    *-,AR1          MPY X2*N2
0183 0061 7F89       ZAC                    CLR ACC
0184 0062 6B90       LTD    *-,AR0          T=X1, ACC=X2*N2, UPDATE X2
0185 0063 6D91       MPY    *-,AR1          MPY X1*N1
0186 0064 6B90       LTD    *-,AR0          T=X0, ACC=X2*N2+X1*N1, UPDATE X1
0187 0065 6D91       MPY    *-,AR1          MPY X0*N0
0188 0066 7F8F       APAC                   ACC=X2*N2+X1*N1+X0*N0
0189          *
0190          *      Check for output overflow
0191 0067 FA00       BLZ    LBL30           CHECK FOR +/- RESULT
        0068 0071
0192 0069 0B1E       ADD    ONE,SCALE-1     ROUND
0193 006A 5C1B       SACH   OUTPUT,16-SCALE UPDATE OUTPUT
0194 006B 1C1C       SUB    MAX16,SCALE     SUBTRACT SCALED MAX POS NUMBER
0195 006C FB00       BLEZ   LBL40           IF ACC<=0 THEN NO OVERFLOW
        006D 0079
0196 006E 2C1C       LAC    MAX16,SCALE     OVERFLOW, LOAD MAX POS NUMBER
0197 006F F900       B      LBL50           GO, RETURN
        0070 007A
0198     0071 LBL30  EQU    $
0199 0071 1B1E       SUB    ONE,SCALE-1     ROUND
0200 0072 5C1B       SACH   OUTPUT,16-SCALE UPDATE OUTPUT
0201 0073 1C1D       SUB    MIN16,SCALE     SUBTRACT SCALED MIN NEG NUMBER
0202 0074 FD00       BGEZ   LBL40           IF ACC>=0 THEN NO OVERFLOW
        0075 0079
0203 0076 2C1D       LAC    MIN16,SCALE     OVERFLOW, LOAD MIN NEG NUMBER
0204 0077 F900       B      LBL50           GO, RETURN
        0078 007A
0205          *
0206     0079 LBL40  EQU    $
0207 0079 2C1B       LAC    OUTPUT,SCALE    RESTORE ACC
0208     007A LBL50  EQU    $
0209 007A 7F8D       RET                    RETURN
0210          *
0211          *
0212                 END
NO ERRORS, NO WARNINGS
```

24. Control System Compensation and Implementation with the TMS32010

715

TMS320 BIBLIOGRAPHY

The following articles and papers have been published since 1982 regarding the Texas Instruments TMS320 Digital Signal Processors. They are listed here for readers who are interested in getting further information about these processors and their applications.

1982

1. K. McDonough, E. Caudel, S. Magar, and A. Leigh, "Microcomputer with 32-Bit Arithmetic Does High-Precision Number Crunching," *Electronics,* **Vol 55**, No. 4, 105-10 (February 1982).

2. T. Schalk and M. McMahan, "Firmware-Programmable μc Aids Speech Recognition," *Electronic Design,* **Vol 30**, No. 15, 143-7 (July 1982).

3. S. Magar, R. Hester, and R. Simpson, "Signal-Processing μc Builds FFT-Based Spectrum Analyzer," *Electronic Design,* **Vol 30**, No. 17, 149-54 (August 1982).

4. G. Farber, "Microelectronics-Developmental Trends and Effects on Automation Techniques," *Regelungstechnik Praxis* (Germany), **Vol 24**, No. 10, 326-36 (October 1982).

5. K. McDonough and S. Magar, "A Single Chip Microcomputer Architecture Optimized for Signal Processing," *Electro/82 Conference Record* (1982).

6. S. Magar, "Trends in Digital Signal Processing Architectures," *Wescon/82 Conference Record* (1982).

7. L. Kaplan, "Signal Processing with the TMS320 Family," *Midcon/82 Conference Record (1982).*

1983

1. L. Dusek, T. Schalk, and M. McMahan, "Voice Recognition Joins Speech on Programmable Board," *Electronics,* **Vol 56**, No. 8, 128-32 (April 1983).

2. R. Wyckoff, "A Forth Simulator for the TMS320 IC," *Rochester Forth Applications Conference,* 141-50 (June 1983).

3. R. Cushman, "Sophisticated Development Tool Simplifies DSP-Chip Programming," *Electronic Design,* **Vol 28**, No. 20, 165-78 (September 1983)

4. W. Loges, "Digital Controls Using Signal Processors," *Elektronik* (Germany), **Vol 32**, No. 19, 51-4 (September 1983).

5. J. Elder and S. Magar, "Single-Chip Approach to Digital Signal Processing," *Wescon/83 Electronic Show and Convention* (November 1983).

6. M. Malcangi, "VLSI Technology for Signal Processing. III," *Elettronica Oggi* (Italy), No. 11, 129-38 (November 1983).

7. P. Strzelcki, "Digital Filtering," *Systems International* (Great Britain), **Vol 11**, No. 11, 116-17 (November 1983).

8. J. So, "TMS320 - A Step Forward in Digital Signal Processing," *Microprocessors and Microsystems* (Great Britain), **Vol 7,** No. 10, 451-60 (December 1983).

9. W. Loges, "Higher-Order Control Systems with Signal Processor TMS320," *Elektronik* (Germany), **Vol 32**, No. 25, 53-5 (December 1983).

10. A. Kumarkanchan, "Microprocessors Provide Speech to Instruments," *Journal of Institute of Electronic and Telecommunication Engineers* (India), **Vol 29**, No. 12 (December 1983).

11. D. Daly and L. Bergeron, "A Programmable Voice Digitizer Using the TI TMS320 Microcomputer," *Proceedings of IEEE International Conference on Acoustics, Speech and Signal Processing* (1983).

12. L. Morris, "A Tale of Two Architectures: TI TMS320 SPC vs. DEC Micro/J-11," *Proceedings of IEEE International Conference on Acoustics, Speech and Signal Processing* (1983).

13. L. Pagnucco and D. Garcia, "A 16/32 Bit Architecture for Signal Processing," *Mini/Micro West 1983 Computer Conference and Exhibition* (1983).

14. L. Adams, "TMS320 Family 16/32-Bit Digital Signal Processor, an Architecture for Breaking Performance Barriers," *Mini/Micro West 1983 Computer Conference and Exhibition* (1983).

15. C. Erskine, "New VLSI Co-Processors Increase System Throughput," *Mini/Micro Midwest Conference Record* (1983).

16. R. Simar, "Performance of Harvard Architecture in TMS320," *Mini/Micro West 1983 Computer Conference and Exhibition* (1983).

17. W. Gass, "The TMS32010 Provides Speech I/O for the Personal Computer," *Mini/Micro Northeast Electronics Show and Convention* (1983).

18. J. Potts, "A Versatile High Performance Digital Signal Processor," *Ohmcon/83 Conference Record* (1983).

19. A. Holck, "Low-Cost Speech Processing with TMS32010," *Midcon/83 Conference Record* (1983).

20. S. Mehrgardt, "Signal Processing with a Fast Microcomputer System," *Proceedings of Eusipco-83 Second European Signal Processing Conference,* Netherlands (1983).

21. H. Strube, R. Wilhelms, and P. Meyer, "Towards Quasiarticulatory Speech Synthesis in Real Time," *Proceedings of Eusipco-83 Second European Signal Processing Conference,* Netherlands (1983).

22. R. Blasco, "Floating-Point Digital Signal Processing Using a Fixed-Point Processor," *Southcon/83 Electronics Show and Convention* (1983).

23. W. Gass and M. McMahan, "Software Development Techniques for the TMS320," *Southcon/83 Electronics Show and Convention* (1983).

24. L. Kaplan, "Flexible Single Chip Solution Paves Way for Low Cost DSP," *Northcon/83 Electronics Show and Convention* (1983).

25. J. Potts, "New 16/32-Bit Microcomputer Offers 200-ns Performance," *Northcon/83 Electronics Show and Convention* (1983).

26. R. Dratch, "A Practical Approach to Digital Signal Processing Using an Innovative Digital Microcomputer in Advanced Applications," *Electro '83 Electronics Show and Convention* (1983).

27. L. Kaplan, "The TMS32010: A New Approach to Digital Signal Processing," *Electro '83 Electronics Show and Convention* (1983).

APPENDIX: TMS320 BIBLIOGRAPHY

1984

1. O. Ericsson, "Special Processor Did Not Meet Requirements - Built Own Synthesizer," *Elteknik Aktuell Elektronik* (Sweden), No. 3, 32-6 (February 1984).

2. S. Mehrgardt, "General-Purpose Processor System for Digital Signal Processing," *Elektronik* (Germany), **Vol 33**, No. 3, 49-53 (February 1984).

3. H. Strube, "Synthesis Part of a 'Log Area Ratio' Vocoder Implemented on a Signal-Processing Microcomputer," *IEEE Transaction on Acoustics, Speech and Signal Processing*, **Vol ASSP-32**, No. 1, 183-5 (February 1984).

4. E. Catier, "Listening Cards or Speech Recognition," *Electronique Industrielle* (France), No. 67, 72-6 (March 1984).

5. J. Bucy, W. Anderson, M. McMahan, R. Tarrant, and H. Tennant, "Ease-of-Use Features in the Texas Instruments Professional Computer," *Proceedings of IEEE*, **Vol 72**, No. 3, 269-82 (March 1984).

6. S. Magar, "Signal Processing Chips Invite Design Comparisons," *Computer Design*, **Vol 23**, No. 4, 179-86 (April 1984).

7. S. Mehrgardt, "32-Bit Processor Produces Analog Signals," *Elektronik* (Germany), **Vol 33**, No. 7, 77-82 (April 1984).

8. M. Hutchins and L. Dusek, "Advanced ICs Spawn Practical Speech Recognition," *Computer Design*, **Vol 23**, No. 5, 133-9, (May 1984).

9. R. Cushman, "Easy-to-Use DSP Converter ICs Simplify Industrial-Control Tasks," *Electronic Design*, **Vol 29**, No. 17, 218-28 (August 1984).

10. P. Rojek and W. Wetzel, "Multiprocessor Concept for Industrial Robots: Multivariable Control with Signal Processors," *Elektronik* (Germany), **Vol 33**, No. 16, 109-13 (August 1984).

11. D. Lee, T. Moran, and R. Crane, "Practical Considerations for Estimating Flaw Sizes from Ultrasonic Data," *Materials Evaluation*, **Vol 42**, No. 9, 1150-8 (August 1984).

12. P.K. Rajasekaran and G.R. Doddington, "Real-Time Factoring of the Linear Prediction Polynomial of Speech Signals," *Digital Signal Processing - 1984: Proceedings of the International Conference*, 405-10 (September 1984).

13. A. Casini, G. Castellini, P.L. Emiliani, and S. Rocchi, "An Auxiliary Processor for Biomedical Signals Based on a Signal Processing Chip," *Digital Signal Processing - 1984: Proceedings of the International Conference*, 228-32 (September 1984).

14. Keun-Ho Ryoo, "On the Recent Digital Signal Processors," *Journal of Korean Institute of Electrical Engineering* (Korea), **Vol 33**, No. 9, 540-9 (September 1984).

15. G. Pawle and T. Faherty, "DSP/Development Board Offers Host Independence," *Computer Design*, **Vol 23**, No. 12, 109-16 (October 15, 1984).

16. V. Kroneck, "Conversing with the Computer," *Elektrotechnik* (Germany), **Vol 66**, No. 20, 16-18 (October 1984).

APPENDIX: TMS320 BIBLIOGRAPHY

17. M. Malcangi, "Programmable VLSI's for Vocal Signals," *Electronica Oggi* (Italy), No. 10, 103-13 (October 1984).

18. G. Gaillat, " The CAPITAN Parallel Processor: 600 MIPS for Use in Real Time Imagery," *Traitement de Signal* (France), **Vol 1**, No. 1, 19-30 (October-December 1984).

19. W. Loges, "A Code Generator Sets up the Automatic Controller Program for the TMS320," *Elektronik* (Germany), **Vol 33**, No. 22, 154-8 (November 1984).

20. H. Volkers, "Fast Fourier Transforms with the TMS320 as Coprocessor," *Elektronik* (Germany), **Vol 33**, No. 23, 109-12 (November 1984).

21. R. Schafer, R. Mersereau, and T. Barnwell, "Software Package Brings Filter Design to PCs," *Computer Design*, **Vol 23**, No. 13, 119-25 (November 1984).

22. W. Gass and M. Arjmand, " Real-Time 9600 Bits/sec Speech Coding on the TI Professional Computer," *Proceedings of IEEE International Conference on Acoustics, Speech and Signal Processing* (1984).

23. M. Dankberg, R. Iltis, D. Saxton, and P. Wilson, "Implementation of the RELP Vocoder Using the TMS320," *Proceedings of IEEE International Conference on Acoustics, Speech and Signal Processing* (1984).

24. A. Holck and W. Anderson, "A Single-Processor LPC Vocoder," *Proceedings of IEEE International Conference on Acoustics, Speech and Signal Processing* (1984).

25. B. Bryden and H. Hassanein, "Implementation of Full Duplex 2.4 kbps LPC Vocoder on a Single TMS320 Microprocessor Chip," *Proceedings of IEEE International Conference on Acoustics, Speech and Signal Processing* (1984).

26. S. Magar, "Architecture and Applications of a Programmable Monolithic Digital Signal Processor - A Tutorial Review," *Proceedings of IEEE International Symposium on Circuits and Systems* (1984).

27. E. Fernandez, "Comparison and Evaluation of 32-Bit Microprocessors," *Mini/Micro Southeast Computer Conference and Exhibition* (1984).

28. J. Perl, "Channel Coding in a Self-Optimizing HF Modem," *International Zurich Seminar on Digital Communications; Applications of Source Coding, Channel Coding and Secrecy Coding: Proceedings*, 101-6 (1984).

29. D. Quarmby (Editor), *Signal Processor Chips*, Granada, England (1984).

30. W. Loges, "Signal Processor as High-Speed Digital Controller," *Elektronik Industrie* (Germany), **Vol 15**, No. 5, 30-2 (1984).

31. R. Mersereau, R. Schafer, T. Barnwell, and D. Smith, "A Digital Filter Design Package for PCs and TMS320," *Midcon/84 Electronic Show and Convention* (1984).

32. J. Bradley and P. Ehlig, "Applications of the TMS32010 Digital Signal Processor and Their Tradeoffs," *Midcon/84 Electronic Show and Convention* (1984).

33. R. Steves, "A Signal Processor with Distributed Control and Multidimensional Scalability," *Proceedings of IEEE National Aerospace and Electronics Conference* (1984).

APPENDIX: TMS320 BIBLIOGRAPHY

34. N. Morgan, *Talking Chips*, McGraw-Hill (1984).

35. V. Vagarshakyan and L. Gustin, "On a Single Class of Continuous Systems - A Solution to the Problem on the Diagnosis of Output Signal Characteristics Recognition Procedures," *IZV. AKAD. NAUK ARM. SSR, SER. TEKH. NAUK* (USSR) **Vol 37**, No. 3, 22-7 (1984).

36. T.R. Myers, "A Portable Digital Speech Processor for an Auditory Prosthesis," *Wescon/84 Conference Record* (1984).

37. J. Bradley and P. Ehlig, "Tradeoffs in the Use of the TMS32010 as a Digital Signal Processing Element," *Wescon/84 Conference Record* (1984).

38. D. Garcia, "Multiprocessing with the TMS32010," *Wescon/84 Conference Record* (1984).

1985

1. C. Erskine, S. Magar, E. Caudel, D. Essig, and A. Levinspuhl, "A Second-Generation Digital Signal Processor TMS32020: Architecture and Applications," *Traitement de Signal* (France), **Vol 2**, No. 1, 79-83 (January-March 1985).

2. V. Milutinovic, "4800 Bit/s Microprocessor-Based CCITT Compatible Data Modem," *Microprocessing and Microprogramming*, **Vol 15**, No. 2, 57-74 (February 1985).

3. S. Magar, D. Essig, E. Caudel, S. Marshall and R. Peters, "An NMOS Digital Signal Processor with Multiprocessing Capability," *Digest of IEEE International Solid-State Circuits Conference* (February 1985).

4. S. Magar, E. Caudel, D. Essig, and C. Erskine, "Digital Signal Processor Borrows from μP to Step up Performance," *Electronic Design*, **Vol 33**, No. 4, 175-84 (February 21, 1985).

5. S. Magar, S.J. Robertson, and W. Gass, "Interface Arrangement Suits Digital Processor to Multiprocessing," *Electronic Design*, **Vol 33**, No. 5, 189-98 (March 7, 1985).

6. G. Kropp, "Signal Processor Offers Multiprocessor Capability," *Elektronik* (Germany), **Vol 34**, No. 6, 53-8 (March 1985).

7. P. Mock, "Add DTMF Generation and Decoding to DSP-μP Designs," *Electronic Design*, **Vol 30**, No. 6, 205-20 (March 1985).

8. C. D. Crowell and R. Simar, "Digital Signal Processor Boosts Speed of Graphics Display Systems," *Electronic Design*, **Vol 33**, No. 7, 205-9 (March 1985).

9. K. Lin and G. Frantz, "Speech Applications with a General Purpose Digital Signal Processor," *IEEE Region 5 Conference Record* (March 1985).

10. G. Frantz and K. Lin, "The TMS320 Family Design Tools," *Proceedings of Speech Tech 85*, 238-40 (April 1985).

11. M. McMahan, "A Complete Speech Application Development Environment," *Proceedings of Speech Tech 85*, 293-95 (April 1985).

12. J. Reimer, M. McMahan and M. Arjmand, "ADPCM on a TMS320 DSP Chip," *Proceedings of Speech Tech 85*, 246-49 (April 1985).

APPENDIX: TMS320 BIBLIOGRAPHY

13. W.J. Christmas, "A Microprocessor-Based Digital Audio Coder and Decoder," *International Conference on Digital Processing of Signals in Communications*, No. 62, 22-26 (April 1985).

14. P. Ehlig, "DSP Chip Adds Multitasking Telecomm Capability to Engineering Workstation," *Electronic Design*, **Vol 33**, No. 10, 173-84 (May 2, 1985).

15. W.W. Smith, Jr., "Agile Development System, Running on PCs, Builds TMS320-Based FIR Filter," *Electronic Design*, **Vol 33**, No. 13, 129-38 (June 6, 1985).

16. R.H. Cushman, "Third-Generation DSPs Put Advanced Functions On-Chip," *EDN*, **Vol 30**, No. 16, 58-68 (July 11, 1985).

17. K. Lin and G. Frantz, "Speech Applications Created by a Microcomputer," *IEEE Potentials* (December 1985).

18. C. Erskine and S. Magar, "Architecture and Applications of a Second-Generation Digital Signal Processor," *Proceedings of IEEE International Conference on Acoustics, Speech and Signal Processing* (1985).

19. J. Reimer and A. Lovrich, "Graphics with the TMS32020," *Wescon/85 Conference Record* (1985).

20. H. Hassanein and B. Bryden, "Implementation of the Gold-Rabiner Pitch Detector in a Real Time Environment Using an Improved Voicing Detector," *Proceedings of IEEE International Conference on Acoustics, Speech and Signal Processing*, **Vol ASSP-33**, No. 1, 319-20 (1985).

21. T. Fjallbrant, "A TMS320 Implementation of a Short Primary Block ATC-System with Pitch Analysis," *International Conference on Digital Processing of Signals in Communications*, No. 62, 93-96 (1985).

APPENDIX: TMS320 BIBLIOGRAPHY

INDEX

The following topics are included in the application reports or articles and are referenced by the section numbers used in this book.

TEAR OUT THIS PAGE TO ORDER ADDITIONAL COPIES OF THIS TITLE AS WELL AS COPIES OF THE OTHER TITLES IN THE <u>PRENTICE-HALL AND TEXAS INSTRUMENTS DIGITAL SIGNAL PROCESSING SERIES</u>.

<u>Prentice-Hall and Texas Instruments Digital Signal Processing Series</u>

<u>Quant.</u>	<u>Title/Author</u>	<u>ISBN</u>	<u>Price</u>	<u>Total</u>
____	**A Digital Signal Processing Laboratory Using the TMS32010,** Jones/Parks	013–212391–6	$20.33	_____
____	**Digital Signal Processing Applications with the TMS320 Family, Vol. 1,** Kun-Shan Lin, Editor	013–212466–1	$30.67	_____
____	**Digital Signal Processing Applications with the TMS320 Family, Vol. 2,** Kun-Shan Lin, Editor	013–214321–6	$30.67	_____
____	**First-Generation TMS320 User's Guide,** Texas Instruments DSP Engineering Staff	013–922188–3	$18.95	_____
____	**Second-Generation TMS320 User's Guide,** Texas Instruments DSP Engineering Staff	013–922196–4	$18.95	_____
____	**Practical Approaches to Speech Coding,** Papamichalis	013–689019–9	$39.95	_____

TOTAL $ _____

−Discount (if appropriate) _____

NEW TOTAL $ _____

AND TAKE ADVANTAGE OF THESE SPECIAL OFFERS!

When ordering 3 or 4 copies (of the same or different titles) take 10% off the total list price.

When ordering 5 to 20 copies (of the same or different titles) take 15% off the total list price.

To receive a greater discount when ordering more than 20 copies, call or write: Special Sales Department, College Marketing, Prentice Hall, Englewood Cliffs, N.J. 07632 (201–592–2406).

SAVE!

If payment accompanies order, plus your state's sales tax where applicable, Prentice Hall pays postage and handling charges. Same return privilege refund guaranteed. Please do not mail in cash.

☐ **PAYMENT ENCLOSED**—shipping and handling to be paid by publisher (please include your state's tax where applicable).

☐ **SEND BOOKS ON 15-DAY TRIAL BASIS** & bill me (with small charge for shipping and handling).

Name _____

Address _____

City _____ State _____ Zip _____

I prefer to charge my ☐ Visa ☐ MasterCard

Card Number _____ Expiration Date _____

Signature _____

All prices listed are subject to change without notice. Offer not valid outside U.S.

Mail your order to: Prentice Hall, Book Distribution Center, Route 59 at Brook Hill Drive, West Nyack, NY 10994

D-LFPC-FJ (3)